QUANTO-GEOMETRIC THEORY

UNIFIED PHYSICAL LAW THROUGH DERIVATION OF THE FUNDAMENTAL PHYSICAL CONSTANTS

Joseph J. Jean-Claude

Quantometrix, Inc.

Worcester - USA

Publisher's Cataloging-In-Publication Data

Names: Jean-Claude, Joseph J.
Title: Quanto-Geometric Theory : unified physical law through derivation of the fundamental physical constants / Joseph J. Jean-Claude.
Description: Edition I. | Worcester, MA : Quantometrix, Inc., [2017] | Originally published in 2023. | Includes bibliographical references and index.
Identifiers: ISBN 978-0-9647466-2-6
Subjects: LCSH: Quantum theory. | General relativity (Physics) | Consciousness. | Cosmology. | Matter. | Life (Biology)
Classification: LCC QC174.12 .J43 2017 | DDC 530.12--dc23

Worcester, MA 01605

Authorship and content of this book have not been aided or crafted by any Artificial Intelligence tool.

International Standard Book Number: 978-0-9647466-2-6

Acknowledgements

To my son, Joel Jean-Claude,
who was kind enough to break his busy schedule for a pre-press reading of this text, and to provide comments and a Preface which no doubt contributed to its betterment and completeness.

To my brother, Frantzy N. Jean-Claude,
for the endless late-night discussions over matter, philosophy and life, a continuing chapter of our lives and an inspiration.

In memory of my cousin Wilbert Forestal, innately gifted to be a mechanical engineering genius as he was.

RECOMMENDATIONS

The thesis is this Treatise enjoys a formidable level of acceptation among graduates, post-graduates, professorial academics and industry experts worldwide who have been initially exposed to either part of the content in the form of pre-published Chapters or individual books of the series. The academics listed below count among the many more who have expressly recommended Chapters or Books of this monographic series. The same Chapters have received an average research interest score of 20 and a reading figure of 90%, for a total of nearly 6000 readers as of this edition, among the few and far between, on the ResearchGate academic network alone. The reading stats below faithfully testify to the same. While the readership is spread over the entire spectrum of academic STEM disciplines, the bulk of these readers is spotted to belong to the disciplines of Theoretical, Atomic, Molecular and Optical Physics, Quantum Chemistry, Physical Chemistry, Molecular Biology, Mathematics and Information Science. The recommendations all spontaneously come from a list of readers too long to reproduce, leaving us to report only a few as transcribed below.

Richard H Alsenz, MSc Theoretical Physics, Department of Physics, University of Houston, Texas, USA

Mustafa Ali Khan-Mussaini, BSc Theoretical Physics, MD Neurology and Neuroimmunology, Stony Brook University, USA

Boris M Menin, PhD Engineering Physics, Expert Consultant at Mechanical & Refrigeration Engineering, Israel

Amir W. Al-Khafaji, BSCE, MSCE, MSc, PhD Applied Mathematics, Emeritus Professor at Bradley University, USA

Abdul Kuddus, PhD Science and Engineering, Senior Researcher at Ritsumeikan University BKC - Kusatsu, Japan

Muhammad Aledeh, Clinical Psychiatric Nurse, College of Health, Psychology and Social Care, University of Derby, UK

Mirjana Vukovic, PhD Mathematics, Academy of Science and Arts of Bosnia and Herzegovina

Fernando M. S. P. Neves, PhD Aeronautical Engineering, University of Beira Interior, Portugal

Konstantin M Golubev, MSc Mathematical Statistics, Moscow State University, Russia

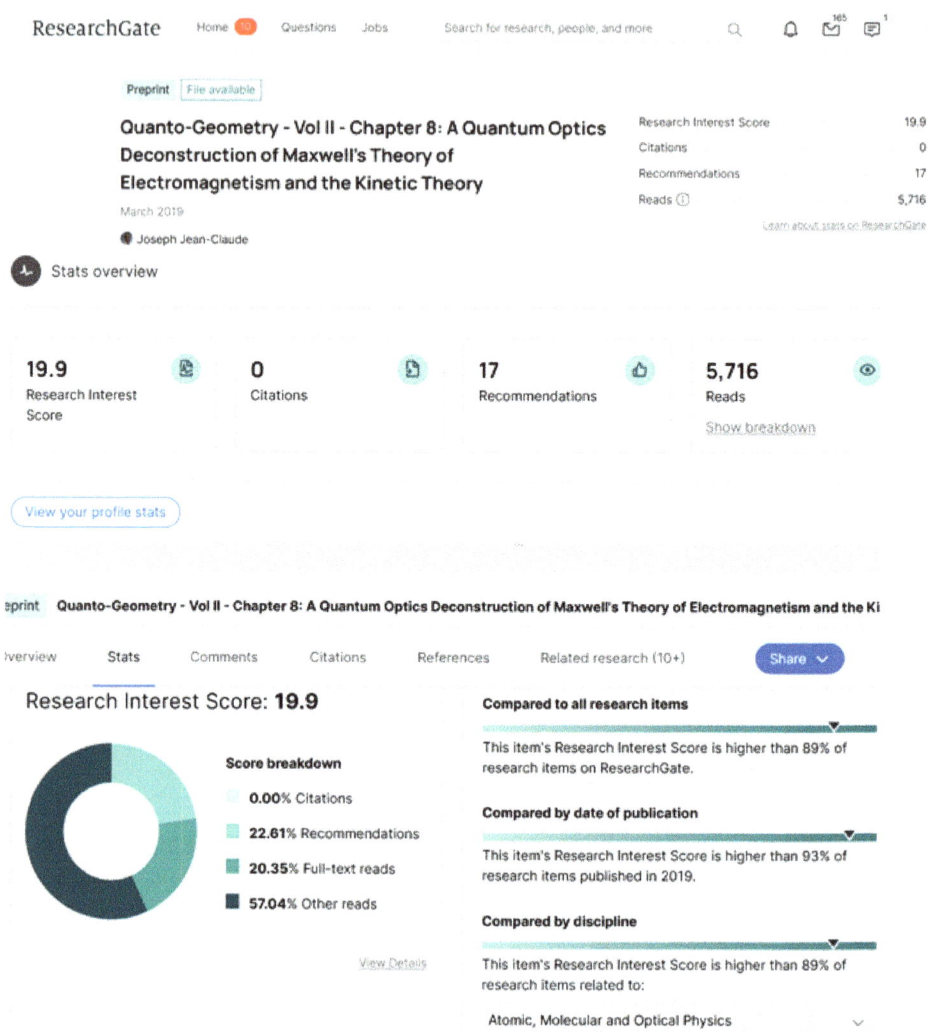

The author has garnered a total readership of nearly 40,000 around the Quanto-Geometric Theory of his inception on the ResearchGate network alone since he started publishing on that platform in 2017, as evidenced by the account screenshot hereby reported. An outstanding research interest score of nearly 240 and 1700 recommendations:

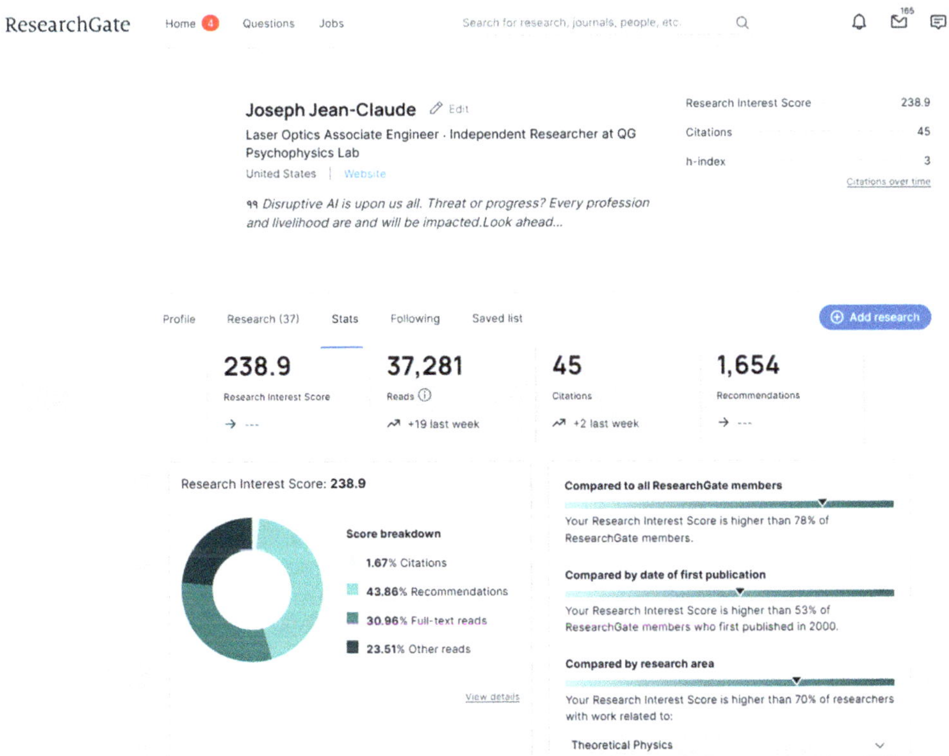
ResearchGate
Home 4
Questions
Jobs
Search for research, journals, people, etc.
165
1
Joseph Jean-Claude
Edit
Laser Optics Associate Engineer · Independent Researcher at QG Psychophysics Lab
United States
Website
Disruptive AI is upon us all. Threat or progress? Every profession and livelihood are and will be impacted.Look ahead...
Research Interest Score 238.9
Citations 45
h-index 3
Citations over time
Profile
Research (37)
Stats
Following
Saved list
Add research
238.9
Research Interest Score

37,281
Reads
+19 last week
45
Citations
+2 last week
1,654
Recommendations

Research Interest Score: 238.9
Score breakdown
1.67% Citations
43.86% Recommendations
30.96% Full-text reads
23.51% Other reads
View details
Compared to all ResearchGate members
Your Research Interest Score is higher than 78% of ResearchGate members.
Compared by date of first publication
Your Research Interest Score is higher than 53% of ResearchGate members who first published in 2000.
Compared by research area
Your Research Interest Score is higher than 70% of researchers with work related to:
Theoretical Physics

PREFACE

This Treatise is primarily aimed at researchers, scientists, physicists, chemists and mathematicians who are intimately familiar with many of the long-standing theories in modern physics, i.e., the Theory of Relativity, Big Bang Theory, etc. However, anyone who has covered courses in at least Calculus 2, Physics 102 and Chemistry 101 will intellectually benefit as well from the discussions carried out in this study. The publication is structured in three Books, representing each of the separate **Quanto-Geometry** volumes independently published. To a certain extent, this presentation may be read the way the reader chooses to approach it. In particular, the reader may choose to start with any Chapter of greatest interest and comprehensively walk back his way or her way into the preliminaries.

For Book I, background in the referred level of Calculus and Physics courses is important since the analysis stems from these very root notions in the physical sciences. While knowledge of Quantum Mechanics (QM) and Quantum Field Theory (QFT) is not a requirement (the author does a fairly good job at succinctly explaining the basics when necessary), it will help take stock of the contrasts between Quanto-Geometric constructs and their parallel QM and QFT constructs whenever applicable.

The first three Chapters of Book I represent a necessary introduction to the main theme, the derivation of fundamental physical constants from first theoretical principles. Primarily aimed at those with an undergraduate level understanding of physics, the discussion in those Chapters lays the foundation for and spells out those principles in a clear and granular fashion. The graduate audience may skip the first three chapters and directly start with Chapter 4, if they so desire.

Chapter 4 is the beginning of what some of the more novice readers might consider "the good stuff", where the author presents a new perspective on things learned in math in earlier cycle of education (formally known as Number Theory), and other curious things that one may have noticed along the way, but could never explain why. There is an intertwining of sorts between these math abstractions and the Quanto-Geometric principles.

Chapter 5 represents an alternative introduction to the main subject, especially written for graduates. The undergraduate audience may skip Chapter 4 and proceed to Chapter 5 from Chapter 3, if they so desire.

Chapter 6 and 7 are a showcase of theoretical discoveries and where the bulk of the contributions germane to this proposal in Book I are delivered. Speed of light constant: derived! Planck constant: derived! Gravitational constant: derived! Rydberg atomic orbital constant: derived! And so on and so forth. It is no less interesting to witness the unraveling of the notion of time, under assault by the author throughout the entire content, and its ultimate consequences as spelled in Chapter 6.

Chapter 8 proposes a new vision of statistical physics, fills in the gaps in the Quantum Theory which caused generations of theoreticians to tax it as analytically incomplete, while demonstrating that Statistical Theory always contained the promising seeds of the Grand Unified Theory.

In Book II, the author undertakes an all-out derivation of not only the mass values of the plethora of known elementary particles but their associated wavefunctions as well. He equally uncovers the long-awaited mathematical origin of the most important symmetries underlying particle physics as well as the symmetries underpinning quantum chemistry.

In Book III, the author presents a large-scale biophysical analysis of phylogenesis and ontogenesis of living species and karyotic individuals respectively, culminating with the theoretical derivation of the physical life constant. This crucial quantity in life sciences, supported by genomic data gathered in basic stem cell research, emerges as the first orchestrator of all life processes. In the first three Chapters of Book III, the author carries out a deep and insightful revision of integral calculus, differential geometry and foundations of symmetry in geometry, in preparation for the study of the foundations of life.

All in all, this Book compendium should be informative, enlightening, engaging and, most of all, exciting: a complete breakthrough of crucial physics conundrums that have challenged generations of physicists and mathematical physicists since early 20th century. Let's begin.

Table of Contents

BOOK I

Chapter 1

Mathematics knows no race
or geographic boundaries; for mathematics
the cultural world is one country.
David Hilbert

RESOLVING THE CHALLENGE OF DAVID HILBERT SIXTH PROBLEM

In Fundamental Physics, whether mainstream or alternative, what we have been pressingly interested in since early 20th century, is to achieve Unified Physical Law, a goal that comes under a variety of designations in physical science, such as the Theory of Everything or Unification of the Quantum Theory and General Relativity. The primary mission of Quantum Field Theory (QFT) was in fact to achieve the coveted theoretical Unification of physics. However, from Maxwell to Renormalization, our physical field theories have failed to meet the challenge. All things considered in theoretical physics, the problem of reaching Unity in Physical Law is the same as the problem of a theoretical derivation of the fundamental constants of Nature. What's more, the last is in itself no other than the problem of formulating all physics in a mathematical axiomatic program, as raised by David Hilbert since 1900 in the Congress of Mathematicians held that yester-year, in what had been famously known since then as David Hilbert Sixth Problem. We are going to discuss in detail in this introductory Chapter the essence and contour of this paramount proposition in modern physics before we undertake thereafter to forcefully take on the challenge of delivering this very mathematical physics program.

1.1. Alliance Between Physics and Mathematics

The essence of mathematics is two-fold: numbers and geometry. Its sole method is philosophical logic or Aristotelian logic. Thus, mathematical logic is the application of philosophical logic to the realm of numbers and geometry. That is the reason why I have repeatedly argued that mathematics, despite its modern designation of mathematical science, is matter-of-factly not a science, but a form of philosophy. Because it bears no concern at all for empirical observation or experimental procedures, and is not subject to the scientific method, it cannot be considered a science or a scientific discipline per se. It is instead a form of tightly constrained philosophy that stands in opposition to speculative and contemplative philosophy. The measure of this rigor is the measure of its usefulness, in particular for physics. It is interesting to correlatively note that physics, from natural philosophy as it was 2000 years ago out of Aristotle's impulse, has evolved into a science in the modern era thanks to the adoption of mathematical logic. There is no modern physics without its peg to mathematical logic and its abstract procedures. All Physics is in essence, and irrevocably so, mathematical physics.

David Hilbert, widely acknowledged to be the greatest 20th Century mathematician, nurtured a deep sense of stature of the discipline and its preponderance over Physics. Having participated in the movement of axiomatization of geometry, which used to pertain to physical science, and having notably worked in the area of mathematical formulation of physical theories as he did with A. Einstein, he had envisioned a program of axiomatization of physical science which he exposed in the International Congress of Mathematicians in 1900. He wrote:

"6. MATHEMATICAL TREATMENT OF THE AXIOMS OF PHYSICS

The investigations on the foundations of geometry suggest the problem : To treat in the same manner, by means of axioms, those physical sciences in which mathematics plays an important part ; in the first rank are the theory of probabilities and mechanics. As to the axioms of the theory of probabilities, it seems to me desirable that their logical investigation should be accompanied by a rigorous and satisfactory development of the method of mean values in mathematical physics, and in particular in the kinetic theory of gases. Important investigations by physicists on the foundations of mechanics are at hand ; I refer to the writings of Mach,f Hertz, J Boltzmann § and Volkmann.|| I t is therefore very desirable that the discussion of the foundations of mechanics be taken up by mathematicians also. Thus Boltzmann's work on the principles of mechanics suggests the problem of developing mathematically the limiting processes, there merely indicated, which lead from the atomistic view to the laws of motion of continua. Conversely one might try to derive the laws of the motion of rigid bodies by a limiting process from a system of axioms depending upon the idea of continuously varying conditions of a material filling all space continuously, these conditions being defined by parameters. For the question as to the equivalence of different systems of axioms is always of great theoretical interest. If geometry is to serve as a model for the treatment of physical axioms, we shall try first by* *a small number of axioms to include as large a class as possible of physical phenomena, and then by adjoining new axioms to arrive gradually at the more special theories. At the same time Lie's a principle of subdivision can perhaps be derived from profound theory of infinite transformation groups. The mathematician will have also to take account not only of those theories coming near to reality, but also, as in geometry, of all logically possible theories. He must be always alert to obtain a complete survey of all conclusions derivable from the system of axioms as assumed."*

This vision is perhaps the most important insight of any mathematician or physicist or so-called mathematical physicist in the last 150 years. The mathematical axioms that D. Hilbert was referring to are the Laws of Physics and his ambition was to come up with an entire mathematical canon made up of first mathematical principles to derive the Laws of physics in their numeric formulation. No mathematician, even less no physicist whatsoever, had taken on this challenge ever since its pronouncement at the International Congress of Mathematicians in 1900. No doubt physicists would barely even take interest in such an undertaking, let alone pursue it, given their concern for the independence of their own academic discipline. These concerns of sorts are pervasive throughout the spectrum of academic disciplines making academia a known departmental topology of silos.

So, a long-sought goal of theoretical physics is to find first principles in Theory from which all of the fundamental dimensionless constants can be calculated and compared to their measured experimental values for corroboration. Up until the inception of Quanto-Geometric Theory, we did not know how to calculate the values of the many dimensionless physical constants, which seem to be the most suitable ones for the exercise, from first mathematical physical principles. They could only be determined by physical measurement. The question which thereby usually follows is: what is the minimum number of dimensionless physical constants that gives rise to all other dimensionless physical constants? Which in turn begs the question: are dimensional physical constants necessary at all? Anyone outside of physics or in alternative physics may review the accuracy of the above from Wikipedia, per the published *List of Unsolved Problems in Physics*. It is interesting from a sociological viewpoint that, as of this writing, this List had not been updated since 2015 to reflect the significant contributions by Quanto-Geometric Theory in this respect, despite the fact that *New Trends in Physical Science Research - Vol. 6 : D. Yu. Tsipenyuk W. B. Belayev (2022)* had reported several outcomes of this approach.

1.2. Sanitizing the Research Procedures in Physics

D. Hilbert's Sixth Problem goes on to stipulate this:

Further, the mathematician has the duty to test exactly in each instance whether the new axioms are compatible with the previous ones. The physicist, as his theories develop, often finds himself forced by the results of his experiments to make new hypotheses, while he depends, with respect to the compatibility of the new hypotheses with the old axioms, solely upon these experiments or upon a certain physical intuition, a practice which in the rigorously logical building up of a theory is not admissible. The desired proof of the compatibility of all assumptions seems to me also of importance, because the effort to obtain such proof always forces us most effectually to an exact formulation of the axioms.

The pattern being subtly decried here by D. Hilbert has been at the heart of physical science for the last 150 years. In modern-day physics, it is flatly and clearly explained by physicist Sabine Hossenfelder in a criticism directed at particle physicists in particular in the following terms:

"The pattern is this: particle physicists invent particles, make predictions for those invented particles, and when these particles are falsified, they change the model and make new predictions. They say it's good science because these hypotheses are falsifiable. I am afraid most of them believe this. But just because a hypothesis is falsifiable doesn't mean it's good science. And no, Popper did not say that a hypothesis which is falsifiable is also scientific. He said that a hypothesis which is scientific is also falsifiable... The best way to see that what particle physicists are doing isn't good science is by noting that it is not working. Good scientists should learn from their failures, but particles physicists have been making the same mistakes for over 50 years."

This practice is the trial-and-error approach which is the only option you are left with when you lack the determination to tackle mother nature's conundrums at the towering axiomatic level due to the enormous difficulties implied. Therefore, after the Solvay generation, theoretical physics increasingly and resolutely took the route of theoretical inferences driven by experimentation or empiricism. Along that line, physics unfortunately became extremely speculative and devolved into a realm of uncorroborated hypotheses dictated as Sacro saint truths for the sake of imposing a theory or the set of ideas of an influential group. A blatant example of this practice is the dark matter conjecture. The physicists at the US Office of Science[1] website so very clearly exemplified this problem when they publish the following about the Higgs particle:

"HIGGS BOSON FACTS

1.- The Higgs boson gets its mass just like other particles—from its own interactions with the Higgs field.
2.- There may be more than one Higgs boson. One theoretical model of new physics predicts five Higgs bosons.

---

[1] https://www.energy.gov/science/doe-explainsthe-higgs-boson

3.- Fundamental particles in our universe acquire mass through their interactions with the Higgs field.
4.- The Higgs boson can be a unique portal to finding signs of dark matter due to its own distinctive characteristics and properties."

As a matter of course, none of these things are facts or scientific facts as we understand them in science. There are no experimental observations to date that support item 1 as an experimental fact. An uncorroborated prediction cannot be a fact as stated in item 2. Just like item 1, the supposed fact in item 3 has not been observed either and is consequently not an experimental or scientific fact. Item 4 as stated is a conjecture and obviously will not qualify as a fact.

It is hard to understand how physicists in an official office of science can use language about facts this ambiguous and misleading. One is compelled to say this is as bad as it gets when the discipline in the last 50 years has been merrily driven by conjectures, from one on to the next, and to the next. These conjectures barely qualify as scientific hypotheses due to the lack of formulative rigor as sought by the scientific method which calls for a direct and coherent response to polemic observables. They certainly do not comply with the stringency dictated by mathematical logic, particularly when a speculative projection is stated as corroborated fact, or imposed as such. There is a failed mathematical logic, in our judgement, in calling for experiments to verify a hypothesis borne out of other hypotheses, which appears to be at the opposite of straight observation of a polemic fact in the natural world, and subsequently formulating a hypothesis to elucidate the conundrum with new logic. And, to finalize, utilize the calculational methods of mathematics to seal the deal, all in the good old tradition of the scientific method. Perhaps, instead of blaming physicists for this deviance in physical research practice, one can arguably see it as an aberration of the scientific method in physics, when driven to its logical extreme.

Therefore, the virtue of the enterprise of Mathematical Axiomatization of Physics is to sanitize this entire bid in the field, while opening the way to a deeper level of knowledge acquisition due to true synchrony with mother nature and superseding of the hypotheses.

1.3. Why Physics Chose Goëdel Over Hilbert

The debate between D. Hilbert and Kurt Goëdel is an indelible part of the history of mathematics. When it comes to physics, despite the overwhelming influence of Hilbert over the discipline, Goëdel's principle of Incompleteness of Mathematics won the day over Hilbert's vision of mathematical axiomatization of physics.

In essence, Kurt Goëdel invented a certain scheme known as Goëdel Numbering by which he used Real numbers to map verbal statements of Arithmetic formulae. These formulae stood for the rules in his scheme [Goëdel, Escher, Bach]. Since real numbers require an endless life to arrive at real counting convergence, he concluded that Arithmetic has so been "proved" to be "incomplete". Goëdel however lived in the ignorance of the profound ontological order inherent to both scalar abstractions and the geometric regime in Mathematics, which we extensively discuss in Chapter 4 of this treatise. From that perspective, his statements and demonstrations are pure err and unfortunately completely worthless, since they lack the elevation and breadth required to become valid axioms. Therefore, they cannot make any reliable or credible assessment about the overall completeness of Mathematics. Our successful developments in Quanto-Geometric Theory in terms of the foundations of Mathematics and the tight contact they make with physical reality prove Goëdel's conclusion misguided and inadequate a posteriori.

We are coming to this debate between Goëdel and Hilbert from a totally different perspective than usual. Specifically, the Incompleteness Theorem only reveals an incomplete knowledge about the foundation of number abstractions and number systems, as well as the full extent of mathematical operators. Goëdel did not know the fundamental quanto-geometric logic in mathematics that is the province of arithmetic operators and that there exist more calculational operators than those we have been familiar with. We shall further anticipatedly state here that numbers are more than cardinal and ordinal entities, as examined in Chapter 4. They contain ontological types which make them in and of themselves true operators on top of the quantum or scalar abstractions they represent. Further, the primal order of operators includes a number of other self-adjoint operators whose remote origin are the typologies giving birth to i.e. a transcendental number, or an algebraic number or an imaginary number. Hence the logical statements that K. Goëdel presented are a restrictive view of the association between numbers, entities or categories of the numeric realm. It shall be clear that we are not coming to this assessment from conjectures, but empowered by our results and findings, all profusely exposed starting from Chapter 4 of this publication.

Hilbert did not know any of the above concepts either. Nevertheless, his inspiration came from better heaven than Goëdel's, who was not inspired at all! And if you are a physicist who struggle to answer the theoretical challenge raised from the inner workings of your own science, yet also wish your science not to be overtaken by another powerful discipline, you opportunistically side with Goëdel's incompleteness view despite your love for Hilbert's constructs such as abstract spaces which are widespread in physics. It is therefore quite understandable that the physics establishment has professed its faith in **Goëdel's Incompleteness Theorem**, given those motivations, and despite a clear understanding of the transcendent role of theoretical derivation of the constants of nature in physics, which is obviously a mathematical development. Mathematics cannot be incomplete and yet absolutely suitable for the universal transcription of physical phenomena and manipulation of the natural world. Incompleteness of Mathematics is an inconsistent truism in physics as much as the common Platonism charge, I might add. Effectively Plato's view of the transcendence of numbers in the formation of Reality and Hilbert's vision of mathematical axiomatization of physics are in line with one another.

1.4. The Imperative of Correct Numbers Toward Unification

Should one come up with yet another formula to compute a constant involving other known physical experimental quantities, that has not the stature of a theoretical derivation in the mathematical sense advocated by D. Hilbert. Such a formula may be another interesting way to compute the quantity or constant, but remains another empirical figure among the many possible known others. The reason why we want these experimental constants to emerge from an axiomatic system is because the latter is poised to reveal the origin of the fundamental physical laws, in particular the origin of the fundamental physical symmetries, long coveted stake in physics, out of the principles encapsulated by the axiomatic set. D. Hilbert's program of axiomatization of physics was aimed at this realization. There have been a few squint efforts in physics to cast some of those quantities computationally, such as the well-known Koide formula, or computations of particle masses by Sergey Efimov. Such efforts have not been considered adequate response to the challenge of mathematical axiomatization of physics, however. Famed physicist Lee Smolin makes the stakes very clear as he writes: *"We don't understand the values the fundamental constants take in our universe... we don't understand the reason for the value of any of the roughly 30 or so parameters [aka the fundamental physical constants] we need to write the laws of physics."* Those are the legitimate motivations behind the theoretical derivation of the fundamental physical constants and they are tightly connected with the ambition in D. Hilbert's 6th Problem aiming at axiomatizing physics.

Over the long course of its development, String Theory has not been able to resolve this problem and attempted to camouflage its inability to do so behind the Anthropic principle and the Multiverse proposition. Lee Smolin to *Scientific American's* John Hogan (Jan 4, 2015):

"This is a sleigh of hand by which they hope to convert an explanatory failure into an explanatory success. If we don't understand the values the fundamental constants take in our universe, just presume our universe is a member of an infinite and unobservable ensemble of universes each with randomly chosen parameters. Our universe has the values it does because those make it hospitable to life.

There is so much wrong with this as a scientific hypothesis. As I have explained in detail in three of my books and several papers, it is hard to see how it could make any falsifiable predictions for doable experiments. Claims to the contrary are fallacious, as I and others have explained in detail. I won't impose the details on your readers but just mention that these criticisms have not been answered."

One of the top String Theory proponents, physicist Brian Greene, insistently argues against that charge with a claim that the theory has fully unified Quantum Mechanics with General Relativity:

"You do quantum mechanics of a string without any knowledge of General Relativity, and just through the process of studying that quantum mechanical description of the string General Relativity emerges and is forced upon you, it comes right out of the mathematics. If Einstein had not discovered General Relativity, String Theorists studying the quantum mechanical motion of a string would have found General Relativity within the formulation itself."

There are several problems with this argument. First, a string has not yet been positively discovered, despite having reached experimentally the level of energies at which it had been hypothesized, and thus does not represent a true quantum mechanical object. Second, and most significantly, you have not fully unified Quantum Mechanics with General Relativity unless and until the putative Unification Theory has successively and credibly derived:

1) The Planck constant h, cornerstone of Quantum Mechanics.
2) The Newton Gravitational constant G or the cosmological constant Λ, which are two pillars of the larger-scale arena as province of General Relativity.
3) The speed of light constant c which we know to be a universal constant all throughout the 3D World, at both microscopic and macroscopic scales.

This set of requirements altogether stands for a sine qua non condition of Unification.

1.5. Can Fundamental Constants Gain the Status of Mathematical Scalars

On the matter of theoretical derivation of fundamental constants there are no standards, no established evaluation procedures, because this subject is quasi unknown territory in physics except for one known official attempt in QED with the derivation of the Electron gyromagnetic moment. A mitigated success for several reasons, as we shall see, principally because of the convoluted manner in which the computation had been achieved. Save the fact that first theoretical principles were not involved, there is also of course the attempt to compute the cosmological constant, mainly from experimental quantities and data, which has been qualified as the greatest predictive failure in physics and the cause of a long-last cataclysm in mainstream physics yet to be subsumed.

When this exercise of theoretical derivation of constants is examined in the light of mathematical axiomatization of physics, one quickly realizes that if the latter should preside over the derivation exercise, the experimental constants must then be in nature mathematical scalars, akin to π and e. In other words, the most fundamental physical constants of nature must be, from that vantage point, mathematical scalars which can be computed from first axiomatic principles if D. Hilbert is right. Except for the fine structure constant α, all of these quantities are dimensional or expressed in unit-dimensions. One may thus legitimately wonder whether it will ever be possible to see a portrayal of these figures in a close figuration to their known numeric cardinality.

The idea of deriving these quantities in term of ratios is a criterion that has been floated around, trusting that ratios between two different quantities, which by being dimensionless, would yield a better chance to cast scalar quantities from the theoretical derivation exercise. This idea rejoins the expectation that a mathematical axiomatic system must cast scalar quantities as well.

What's more, some have attempted to establish a list of the quantities that ought to be derived. Our view is that in the absence of the knowledge of what the sought-for axiomatic might actually look like and poised to yield, it is difficult to credibly rule or pre-determine authoritatively what should come out of it other than one's best wishes. There is the expectation in the field that development of a more fundamental theory might allow the derivation of several constants from a more fundamental one. This reasoning stands for an inference that the axiomatic system might have to uncover more than one quantity and show a hierarchy by which one quantity gives rise to a suite of others, introducing thereby a criteria of consistency between the quantities. In addition, should the derived quantities have completely different numeric portrayals to their experimental pars, an agency of sorts to compare the calculated values to the known experimental values is another expectation.

What follows in this monograph systematically resolves all of these unknowns.

1.6. World Standard Bodies and the Fundamental Physical Constants

Is the late adoption of exact arbitrary figures for the fundamental physical constants justified? Practical consideration relating to maintenance of etalons have justifiably guided the trend by the Standard Bodies to move in 2017 in the direction of establishing pure reference quantities for the magnitudes of the fundamental physical constants. However, if these quantities have to be arbitrarily chosen, then the Standard Bodies are at great risk of engaging in a circular course whereby a revisional regime of these quantities will again become necessary as the need for adjustments and outright changes arise over time. The arbitrary origin of these quantities assuredly leads to those requisites.

Systematic study of the natural fundamental constants has led the World's Standard Bodies, within their recent 2017 redefinition initiative, to a classification of these constants that relates them to our adopted unit-dimensional measurement system (Fig. 1.1). They have found 7 constants to be "fundamental" or foundational from a physical classification viewpoint, namely:

- h, the Planck constant – Defines the *kilogram* and the constants c and $\Delta\nu_{CS}$.

  - $\Delta\nu_{CS}$, the hyperfine transition frequency of cesium-133 – Defines the *second* as unit of time.
  - c, the speed of light in vacuum – Defines the *meter* and $\Delta\nu_{CS}$.
- e, elementary charge – Defines the unit of *electric current A* and $\Delta\nu_{CS}$.
- k, the Boltzmann constant – Defines the *Kelvin* as unit of temperature.
- N_A, the Avogadro constant – Defines the *mole*.
- K_{cd}: the *luminous efficacy* of monochromatic radiation of frequency 540×10^{12} *Hertz* – Defines the candela as unit of radiance.

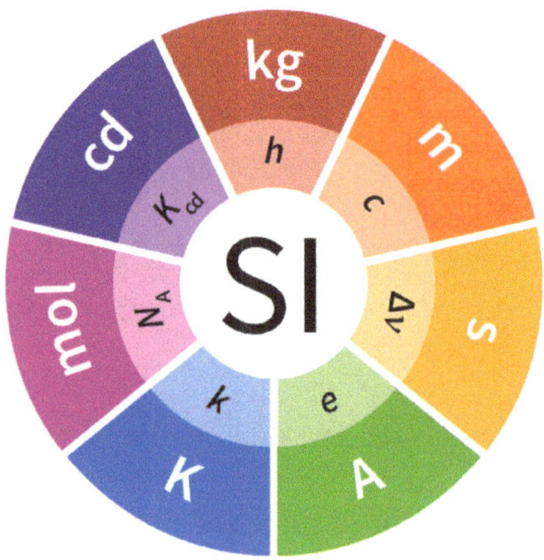

Fig.1.1 A wheel by the Standard Bodies portraying the Système International of Units

It is not difficult to see that some of these definitions are as circular as the taxonomic diagram itself as well as the ultimate attempt to break up the Pepitio Principii flavor of the classification with the imposition of arbitrary exact numbers for some of these constants, as formerly noted. Not to mention the definition of a false constant and its associate unit, which are the constant and unit of time. The Standard does acknowledge that K_{cd} is *"arguably one of the most artificial and human-centered constants"* of the classification. Nevertheless, this redefinition attempt represents best-effort rationalization of what we know and have of natural physical units and their measures (Fig 1.2).

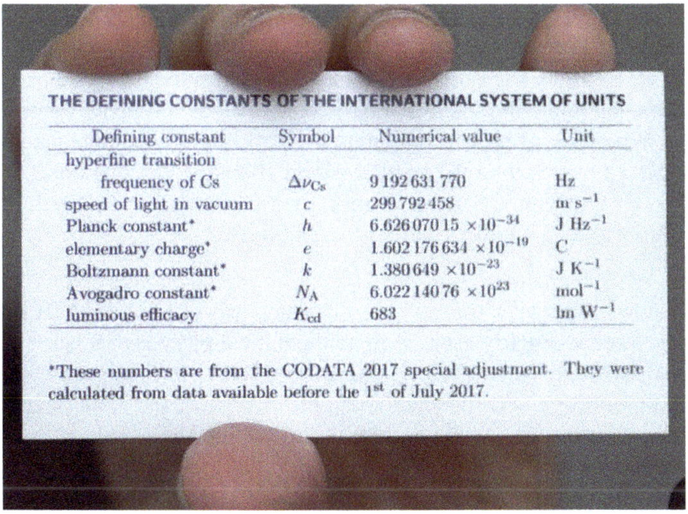

Fig. 1.2 The SI units by the NIST

All in all, the only condition that can guarantee the inalterability of these pure referential numeric quantities and exemption of all arbitrariness in the SI system, as an overarching goal, is for these quantities to be legitimate mathematical constants, the likes of π and e, if they can be. Mathematical constants have not, do not and will never change. In order for the magnitude of the fundamental physical constants to legitimately remain inalterable, they must satisfy the *sine qua non* condition of being able to endorse the same identity as mathematical constants. Therefore, only a theoretical derivation of the fundamental physical constants on exclusive axiomatic grounds per David Hilbert's call of old for Mathematical Axiomatic Physics can furnish the standard required for the transmutation of physical constants to mathematical constants, pursuant to the numeric standardization objectives set by the Standard Bodies. The Quanto-Geometric Framework exactly and abundantly accomplishes just that predicament. No further comment.

1.7. What the New Physics Truly Is

Theoretical physicists frequently invoke new physics or project discovery of new physics. More often than not, what they mean is discovery of new particles, or energy evanescing from a process which must be recovered in the form of a particle to be newly discovered. The extraordinary observation in 2017 of an object later dubbed Oumuamua by astrophysicists, whose zenithal dynamics totally defied most of what we know in cosmological physics, was a matter of lots of talk around water coolers in all physics departments around the world, without rising however to the formal identification of new physics in reach to our physical science. Instead, many attempted to dial back to explanations rooted in current cosmological knowledge which arguably amounted to no more than far-fetched conjectures.

1.7.1 New Physics From Internal Proceedings of the Academy

As an example of new physics prospects arising from inside the Academy's activities, supersymmetric particles were the new physics expected to be observed at the LHC by the vast community of String Theoreticians. The novelty in the discovery of perturbations in the fabric of spacetime, also known as gravitational waves, was not quite claimed as new physics despite the potential, perhaps because it had been predicted so long ago in General Relativity. The ability to allow for a different form of observation and screening of the heavens could of course facilitate the uncovering of "new physics." When it comes to the subject of theoretical derivation of physical quantities, we must acknowledge the existing strong covet in particle physics for a theory that would allow the calculation of particle masses. Admittedly, it represents a core motivation for the search for so-called "Physics beyond the Standard Model", another nomer for new physics.

In regards to the entire history of physics, the new physics has always truly been Mathematical Axiomatic Physics, as an explicitly set goal in mathematical physics that has never been met and not even seriously researched. The Quanto-Geometric Theory is a response to this goal of old and represents the most authentic incarnation of the New Physics in contemporary fundamental physics. Whether physics survives as an independent discipline or becomes a Chapter of Mathematics, akin to the chronicle of the former science of geometry, is anyone's guess. From the viewpoint of acquisition of knowledge, it is not a relevant or significant question, having more to do with form and packaging than substantive content. What we have already seen with String Theory is an overwhelming inoculation of mathematics to Physics anyhow, a point of insistent and entrenched criticism from those outside of the String community. It could be that this very effort paradoxically represents String Theory's best legacy, notwithstanding the unsuitability in the development of a mountain of applied mathematics unable to make contact with reality in an appropriate fashion if at all. In all, the inception of Quanto-Geometric Theory stands for a deep dive into the existing science of mathematical physics and spells out as the true essence of physics of the natural world while putting before the world a novel form of 'New Physics'.

1.7.2 New Physics from Drivers External to the Academy

In 2020 the US Department of Defense has put before the American and the World public an extraordinary video record captured in 2017 by US Navy pilots in exercise of common military duty. Flight dynamics of these UAP objects are quite flabbergasting and outclass anything ever achieved in human technology resulting from fundamental physics. Government officials and pilots qualify these observations as *"physics we do not understand."* While Oumuamua may seem a hard problem to tackle yet one we may have the tools for from our astrophysical knowledge toolset, UAP observations pose an enormous problem to Knowledge and Culture that seem to completely supersede the current capabilities of the Academy.

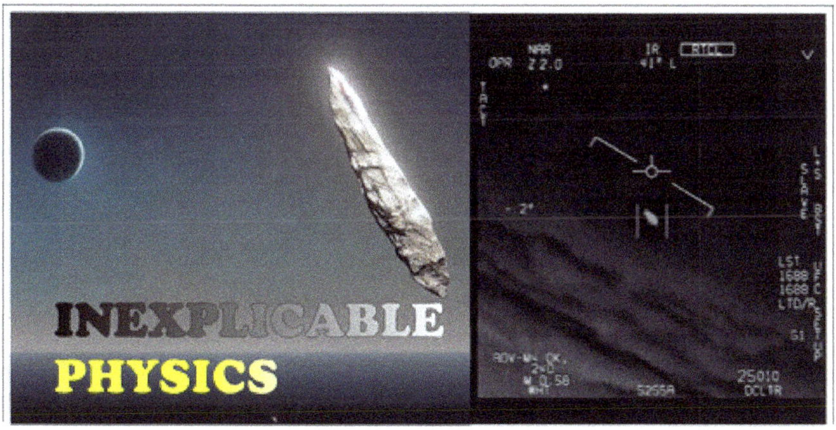

Fig. 1.2 Oumuamua object cruising through solar system at left. At right a UAP on radar of a US Navy pilot.

One thing that must remain clear is that all qualified "Problems" in fundamental physics must be solved before we can ascertain a modicum of elucidation of the conundrums before us. The launchpad for this new phase in physics of the Academy and Culture in general is the singular superseding of that entire set of formally identified problems at once. Those problems are far from disconnected from one another. That is why solving the most profound ones also momentously helps solve all the rest of them. This all-encompassing solution further yields the path to solid apprehension of the inexplicable physics related to UAP-UFO's. There is indeed a continuum between the expected new physics from the evolution of proceedings inside the Academy of physical science as a challenge and the inexplicable physics put before us by the US Department of Defense in UAP-UFO sightings. The ambition of the Quanto-Geometric Theory is to respond to this challenge in fullness, as we believe we have, along the line of the most important mission of all physics, which is Unified Physical Law thru the theoretical derivation of the fundamental physical constants.

Chapter 2

To discover the higher Truth requires
the courage to put everything on the table,
at least for once, to whatever extent possible.
René Descartes

GENERIC PROPERTIES OF SPACE VACUUM

This Treatise strives to significantly contribute to an intractable overture of cosmic consciousness towards a Space era in the evolution of mankind as a top priority. It sets forth a novel vision of the universe through a deeper understanding of Space, or what is denoted in physics coordinate space, aka the observable void, namely its properties and phenotypes. The understanding of the ontology and behavior of space vacuum is a paradigmatic enterprise in human knowledge development and acquisition. Indeed, the imminent elucidation of our accumulated philosophical, religious and scientific paradigms to date stands for the gateway to the next era. It will require a profound revolution in the way we formally and informally visualize the many different scales of the natural world as regimented by the continuum we call the vacuum, which will confer a higher degree in the capacity to manipulate and interact with the world around us along with the universe beyond us.

In initiating this study, our immediate aim is to set forth a novel physical model capable of undertaking a successful interpretation of space deployments in the universe, from microcosm to macrocosm, and to subsequently put it to task. We expect said model to specifically and fruitfully assist us in undertaking the long-sought theoretical derivation of the fundamental physical constants, a paramount goal of modern physics as discussed in Chapter 1. To any mind, it is quite forthcoming to appreciate the significance of that objective, given that the physical constants represent the structural pillars of the universe that surrounds us and gives us haven. As we arrive at the last Chapters of this publication, the developed model will help shed light on an astonishing array of matters germane to Biology and the Human Sciences, beyond the strict physical order, attesting thereby to the true signature of the long-awaited *Theory of Everything*. This rendez-vous with the space entity is one for the layman, better accustomed to common sense interpretations of events in everyday life, as well as for the educated and well-rounded mind, better trained to elaborate interpretations of life on an individual and collective level. So much far-reaching is this model anchored on the physical vacuum because as an ontological entity the latter traverses all levels of expression of matter and life. Let us further take a more detailed look at this model.

2.1. Mass v. Space

Gravitation, the force of attraction between two macroscopic objects, has tricked our common sense into believing that the only way things come to existence is through the property of weight. We find ourselves naturally tied to the surface of planet Earth, along with all other objects around us subject to the same condition. The interaction between these many earth-bound entities produces the events and phenomena that are meaningful to us. Thus, to common sense the property of weight defines the existence of any one object: if some *thing* can hit or be hit, we take it as an existing object and vice versa. However, what qualifies all objects around us to cause an impact or be impacted upon, is rated about 10 times less than what common sense tends to make us believe. In effect, according to Newtonian physics, we must divide the weight of an object by 10 approximately in order to circumscribe the real quantity that makes it subject to impact or collision.

Sir Isaac Newton posed that:

$$W_{eight} = m_{ass} \times g_{ravity}$$

In other words, when we see an object fall from the sky, the quantity inscribed within the object that will make it **hit** the ground is in reality approximately 10 times less than what our common sense tends to appreciate, or perhaps to fear if the object is about to hit us. This quantity, which is embedded in all objects and qualifies them as such according to Newtonian mechanics, is called their *mass m*. Thus, from a more careful interpretation than common sense appreciation, from a Newtonian viewpoint that is, an object exists inasmuch as it exhibits mass. This definition is mathematically described by manipulating the above equation and converting it into the following:

$$m_{ass} = \frac{W_{eight}}{g_{ravity}}$$

The value of $g_{\text{ravitation}}$ is $9.8\ m/s^2$, that is, almost *10*, if m_{ass} is measured in kilograms. But what is the definition of mass? It is interesting to note that mass as a primary property of objects has never been defined intrinsically but in relation to other quantities. The truth is, despite all the elaborate developments brought about by Newtonian physics and later by relativistic mechanics, no one to date has ever conceived or proposed an *exact definition* of mass, because no one knows what mass is *per se*.

Thus far, no physics school, neither Newtonian, nor quantum or relativistic has ever proposed an intrinsic definition of mass (the stuff that it consists of), notwithstanding its stature as a fundamental physical quantity. Even though our inner senses tell us that it relates to the property of compactness, entirety or indivisibility of an object or the components of thereof, the difficulty to explain explicitly *what* it is still remains wholesome. As we attempt harder and harder to conceptualize the notion of mass, it becomes interchangeable with the concept of quantity, a truly mathematical concept. There lies indeed the reason why in physics mass is categorized as a *scalar quantity*, in other words, a pure quantity. Paradoxically, the only reason why we know of mass is because of another existent called space or *vacuum*, which we usually disregard. This is to say that mass veritably can only be defined by opposition to space or vacuum: mass is *some thing* while space is *no thing*.

Hence, by relating mass to space it is possible to set forth a formal basis of interpretation of its physical meaning. It is only because of the existence of space that we can have a witful representation of mass. We know of *some thing* because we know it is not *nothing*. The inverse is as much true. Space or vacuum is truly the absolute absence of *some thing*. Physicists contend that the problem of definition of mass is a philosophical matter which physics as a scientific discipline should not be concerned with. However, there is much to learn from such a philosophical exercise, since it may indeed yield a new framework of interpretation of the natural world. Clearly, from our perspective, said space or vacuum entity acquires paramount significance. Space not only assists us with a more formal interpretation of mass, but it also stands for a crucial component in the makeup of all objects in the universe.

Space or vacuum deployments are omnipresent in the universe, from quarks to solar systems to galaxies... Space is meaningful, however absolute its absence of contents for which we call it the void. Our common form of appreciation of objects gives prominence to their property of mass. Nevertheless, their vacuum or space content does play an important role in their physical behavior. This result is one of the most important consequences of relativistic physics. Albert Einstein proposed that in the vicinity of large masses the structure of *spacetime* is curved. From our standpoint, however, there is no such thing as *spacetime* but simply space, and the properties of space are not limited to cosmological space but generalized throughout all of its forms of deployments both in the macrocosm and the microcosm. An understanding of the *nothingness* in coordinate space or that the state of vacuum is endowed with specific physical properties, no matter the type of space deployment under consideration cannot be overstated.

2.2. Space and Time

The concept of time has surreptitiously misled us into misconstruing coordinate space. From a Newtonian standpoint, time stands for an absolute and universal reference which can be used to measure all events occurring in the universe. This intuitive concept of time proves to simply be incorrect. If you are moving and observing a moving object, while someone else, being at rest or not, is equally observing the same object, one of you will see the object moving faster than the other observer can appreciate. This outcomes means that there is no absolute value for the quantity that we call speed of an object as observed. Because no one is at the center of the universe and every one thing in it is in motion, no one can admittedly claim his or her observed speed to be the proper or correct one. Since speed is mathematically dependent upon one constant and only one variable, which is time, no absolute speed directly means no absolute time. Therefore, in our common way of interpreting nature, which involves the idea of speed, something will be at fault. Our intuitive idea of absolute and universal time is what is at fault. If there is no absolute time, is there any time at all?

What's more, no matter how fast you and any other observer are moving, assuming that both of you could measure the speed of a travel pulse of light, you would both see it move at exactly the same speed! This second result even obliterates the idea of time itself in our sense. Were the observers still to appreciate different speeds for the light pulse, the idea of time might then still linger with some level of validity. The invariance of the speed of the object under observation, independently of the observers' travel speeds, represents a demise of the idea of absolute speed and thus of the implied idea of time. Albert Einstein imagined and proposed an interpretation for defaulting time: the concept of *dilation* of time. He was merely not ready to ring the bell for it yet. Let us look further into the notion of time and how Albert Einstein dealt with it in constructing Special Relativity.

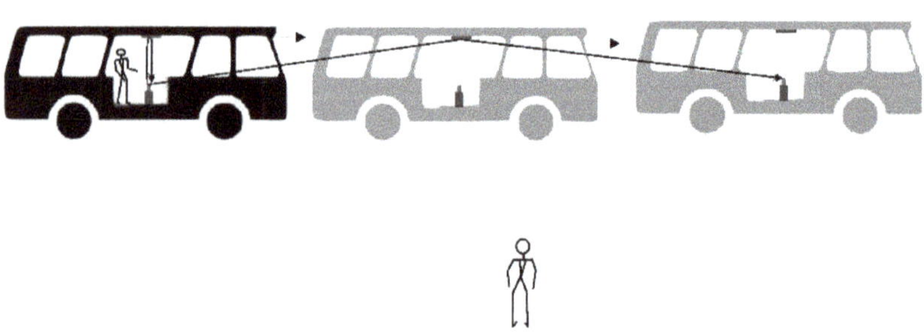

Fig. 1.1 Travel path of a laser beam in a fast-moving vehicle as seen by two observers

Time basically means for us all *duration*. Suppose that you are in a moving train (Fig. 1.1), and hold in your hand a small laser pen whose beam of light you direct upward to a mirror attached to the ceiling so that the light reflects back into the aperture of the laser device. An amount of time T_1 will have elapsed between the instant the light left the laser and the instant it reaches back the aperture of the laser pen. Now, to anyone standing outside the train, a friend maybe, whether he or she is at rest or not, the beam will appear split in two beams. Due to the movement of the entire train, the external observer will also see two different positions for the pen, one at the time the light left the laser pen and another one at the time it returns to it. The total path of the light will appear as two sides of a triangle to the external observer, not anymore as a straight line like you, the first observer located in the train, duly perceive it. Obviously, this second path will be much longer than the one you observe in the train yourself. Since the speed *c* of the beam of light remains the same, this also means that the light will take longer to cover that distance. Consequently, the total time duration T_2 measured by your friend, the second observer, is larger than the time duration T_1 that you observed in the train for the round trip. This is a contradictory situation where the same event is measured to last different amounts of time or different time lapses. Albert Einstein proposed to conceptualize this outcome as a dilation of time and showed how to relate T_2 to T_1 with the following relation:

$$T_2 = \frac{T_1}{\sqrt{1 - \left(\frac{v}{c}\right)^2}}$$

where v is the relative velocity between the reference frames, and c the speed of light.

Therefore, time is also relative. The rate at which time elapses for you depends on the speed at which you are moving. Clearly, from that perspective, every observer may have his own scale of time. While the concept of relative speeds is acceptable because they are made subject to a common reference which is time, the concept of relative time is not quite acceptable unless we can provide an explanation of what these relative time durations are referenced to. We express speed as a rate of change of distance in time in terms of *dx/dt*. If time is relative, how shall we express its rate of change? The idea of rate at which time passes is formally unclear and to further make it depend on the speed of the observers themselves is to create a vicious circle whereby speed depends on time and time depends on speed. Deemed to resolve this serious problem, A. Einstein was led to conceive that space and time are not independent from one another, that they are *entangled*. That is to say, the ultimate implication of Special Relativity is to make the rate *dx/dt* equal to *dt/dx*. He thus agreed with Minkowski to reference time to space. This is implicitly the basis for the relativistic entanglement of space and time, the structure coined *spacetime* by Minkowski.

From either a Newtonian or a relativistic vantage, if time is not a primary, exclusive and independent state of matter (or a dimension as it is traditionally visualized), the only other state of which it may become a variable is space, not mass. Mass is naturally and definitely best conceived as a pure unredeemable scalar quantity in all forms of mechanics. As previously stated, the mathematical formulation of construing speed as dependent on time and time as dependent on speed is:

$$\frac{dx}{dt} = \frac{dt}{dx}$$

This relation can only be true if the differentials *dx* and *dt* can be equated to one another: $dx = dt$. This would translate into equating time and space, or simply ringing the bell for time. A. Einstein was not ready for a mathematical description of the macroscopic universe in complete dismissal of the time variable. Despite the great success of Special and General Theory of Relativity, essentially attributed to their ability to correctly predict various observations in the physical order, the Theory remained of difficult application at the quantum level, in our sense because time or time-driven dynamics do not occur in that realm, much to the dislike of many. In the context of quantum physics, A. Einstein proposed an experiment that could not be carried out at the time. To our good fortune, the developments in laser-optics technology in the 1980's made it possible to realize in the laboratory a good approximation to the original concept (the EPR Paradox). Take two identical units of the smallest particle of light, two photons, and allow them to travel in opposite directions to each other in deep vacuum, can anyone influence by any means the other? Let us first remember that the epitome of Relativity is the invariance of the speed of light and that this speed *c* is the maximum possibly attainable by any physical object. Since each particle is moving away from each other at speed *c*, the maximum attainable, any interaction or information from one particle toward the other must travel at a higher speed than *c* in order to reach its mate and bring any information to it. A hypothesis held impossible by one of the fundamental postulates of Special Relativity. However, it has been observed over and over for now four decades that both particles always respond in the same way to individual random events affecting first only one of them, as if they could indeed inform one another in some unknown fashion.

Other variants of this experiment have been realized, notably the one by physicist Raymond Chiao, all of which always end up casting some faster-than-light result. Physicists have been somewhat uncomfortable with these results because they suggest a violation of the cornerstone rule of Special Relativity. Instead of retrocausal activity in quantum physics as increasingly hypothesized of late, could there be non-observable *locality* beyond 3-dimensional space where faster-than-light speed is possible and can in fact produce *effects* that are observable in 3-dimensional space? Could the Theory of Relativity as a whole be no more than a particular case of a more general physical theory from which it can be deduced when the speeds under treatment are neighboring the speed of light? A somewhat similar case to Newtonian physics which can be deduced from a relativistic treatment when the speeds involved are only a small fraction of the speed of light.

Those hypotheses would suggest the death of time as we know it. The dismissal of the time variable is well established in pure Quantum Mechanics, the study of events in the atomic and, to some extent, the sub-atomic world. Because of a lack of understanding of the larger implications of Quantum Mechanics, quantum physicists have not been able to settle those debates around the EPR paradox satisfactorily, since they have to evoke un-physical effects for an explanation. Furthermore, Quantum Mechanics does not include a treatment of gravitation, a major form of interaction generalized at cosmic scales, in what has been considered a significant physics problem. Despite its atemporal approach, which particularly suits the effect described above, Quantum Mechanics is unable to predict the gravitational behavior of celestial bodies. It is interesting to note that the quest for a unified theory, with the capability to predict the universe as a whole and in its details, has suggested the eventual fusion of time and space. Effectively, in Richard Feynman's incipient proposal to formulate Quantum Mechanics in terms of the integral of all possible trajectories, because we would need to express time with imaginary numbers instead of real numbers (earliest strain toward the retrocausal interpretation to which we do not subscribe however), one may arguably conclude that the historical distinction between time and space totally vanishes and ultimately time is reduced to space for all practical purposes.

2.3. Properties and Faculties of Space and Mass

Let us convince ourselves: *there is no time anywhere but space everywhere.* We have stipulated earlier that space can only be defined in opposition to mass and vice versa. There is no prior principle from which an ontological definition of mass and space can be derived. There is no absolute physical or conceptual separation between them. From our standpoint, they are prime values non-deducible from any previous form of reality and they relate to one another in exclusion of any other possible form of reality. Nothing else can affect space and mass, and from nowhere do they come. Mass, call it a *thing*, and space, call it *nothingness*, are the basis of all forms of objects as well as events in our reality.

In studying the microcosm, physicists had pinpointed the existence of small particles of matter, neutrons, protons, electrons, ... which we originally believed to be small quantities of mass. Soon we discovered that these particles could exhibit properties not attributable to their mass. This discovery contributed to the further development of quantum physics through the work of Louis Victor de Broglie, Erwin Schrödinger and many other brilliant minds. In particular, sometime after Max Planck's discovery of quantization of energy, particles were experimentally found endowed with wave properties. Since the particle was visualized as pure mass, or a pure scalar, its capability to harbor a wave function came as a surprise and somewhat of a conundrum. In effect, something different from the mass must exist within the particle to yield the wave behavior. To date quantum physicists do no hold an all-clear physical representation of the wave function (Ψ) embodied in particles, notwithstanding the adequacy of the related mathematical developments. The interpretation of the physical meaning of involved functions and operators did not come as an easy task. Hence, physicists have resorted to many different interpretations of these results which are often at odds with each other. We contend that the wave function contained in particles is an expression of their space component, simply put the *vacuum* or *void* that is a part of their ontology.

Going beyond the laws of gravitation exposed by Isaac Newton, A. Einstein developed a mathematical basis, in the so-called Einstein Field Equations (EFE) in General Relativity to explain gravitational effects occurring in the macrocosm. He concluded that gravitation is a geometric effect released by the inner structure of *spacetime*, say just space. According to this view, when a planet for instance orbits the sun, it does so not because of a force of interaction between the sun and the specific planet, but because the latter follows a particular spatial path that is curved. From Newtonian mechanics, the closer two bodies are to each other, the stronger the force of gravitation (or attraction) between them. Consequently, when a planet orbiting the sun arrives at its closest point to the star it should never be allowed to move away from it. Nevertheless, this is the case with Mercury which, instead of following its elliptical path as it gets closer to the sun, surprisingly diverges from the trajectory, describing a divergent arc of a circle around the sun. A postulate of relativistic treatment is that around large masses space is curved. Since the planet follows the inner structure of the space tapestry, it can only diverge in close vicinity of the sun. Therefore, space in the macrocosm must be endowed with geometric properties.

Let us now generalize by establishing that geometry and motility are two primal properties of coordinate space at all scales of the universe. Things do not move because of their mass. They move because of their space component. Movement is ultimately transmitted from one body to another not through mass but through space. When something is moving, it is the space distributed within the object that produces the movement. How can that be possible however since space is *nothingness* and infinitely *dilatable*? Because space is nothingness, in a sense it does not make much of a difference in terms of its fabric whether it is stretched or shrunk, it remains nothing. Conversely, if something can be stretched or shrunk and remain same, this *thing* must be nothingness, because nothing we know of in the universe, from our usual quantum or mass point of view, can demonstrate these qualities. This thought exercise is far from trivial and is key to an understanding of the structure of space. The following is a law: *space is infinitely elastic*. It is so infinitely elastic that, no matter its degree of propagation, it always appears to us as nothing and has never deserved our slightest attention. The dilation of space means that it is in propagation. From our standpoint, those two terms are interchangeable. Space is always in propagation, it is indeed a *propagation field*, everywhere it appears, from microcosm to macrocosm. One way to look at any physical object is to consider it a population of discrete masses or scalar elements (dense quanta) articulated to a propagation space field which they follow because they are subject to it. **In its purest state** (an unreachable asymptotic limit)**, the degree of propagation of space would be absolute and would give rise to instantaneousness**. In physical reality however, where it is coupled to scalar elements, it loses innate dynamism and develops distinct degrees of propagation, as we shall see later. We will give mathematical support to these concepts further down in this exposé.

From a quantum-mass standpoint, the universe is populated at all scales with individual objects, distinct and segregated from one another. From a space vantage, all objects are directly related to one another, not by juxtaposition, but through immersion or superposition. Recall that space is infinite and so the particular propagation field of an object expands to infinity. Consequently, no matter the distance between two objects, they are directly related because they are embedded in one another through their ontological space component. Fourier analysis has instructed us that there is no limit to the superposition of waves. It tells us that what appears to be one particular wave of a specific frequency may be treated as an infinite number of specific harmonics. Fourier treatment is not just a theoretical model to perform electrical analysis, it reflects the exact behavior of coordinate space, and only because space is the essence of the wave phenomenon does the Fourier treatment become possible. In other words, when we see one "form of space" or one thing with wave characteristics, it is in fact the ultimate sum of an infinite number of space forms which bear a definite symmetric relationship between one another. We will further elaborate on this particular topic later in this text.

This naturally brings us to consider the second property of coordinate space: *symmetry*. Space is not only a propagation field, it also holds latent states of symmetry that become apparent as geometric forms, or just forms, if you will. There is no objecting to space as the origin of the dimensioning of physical objects. Space veritably provides the form or shape, simple or complex, extrinsic to the dimensioning of objects. It is an inherent property of the space field which organizes the geometry of the object. This property of space which becomes apparent in the geometry of objects is what we call state of symmetry. A more practical approach to this very abstract concept is probably to start from the end toward the beginning. Consider this: all forms are forms of symmetry. Whether an object exhibits a very sophisticated sinuous shape or simply a linear or planar shape, it is eliciting symmetry. It is easy to understand that a circular or spheroid shape can be broken down into parts that are all similar to each other. These fractional parts can be superimposed on each other in perfect match, they are interchangeable. The most complex sinuous shapes can be broken down into circular, elliptic, parabolic... fractional parts that are interchangeable with each other, whereas all of the parts from a linear or planar shape are almost absolutely interchangeable. Therefore, all forms are forms of symmetry. Now, if by some magical means, one could take out of an object and its components all of their discrete masses, what would be left behind? You would hypothetically leave behind a space field, which, no matter how deeply intangible, did articulate the masses into the particular geometry or form of the object. Hence, it would hypothetically remain with this specific potentiality of geometrization. This potentiality, simple or complex, is a state of symmetry. Much of what is called a wave function in Quantum Mechanics is attributable to it. By simple observation, you would never know of the geometric potentiality of this space field, if conserved, until it gets involved with a re-distribution of discrete masses. It will then re-inform them according to its specific inner geometric orientation or qualities.

2.4. The Spatial Nature of Harmonic Waves

Our notion of waves stems from the observation of oscillatory motion. We have come to use trigonometry to give an exact account of this type of motion. Consider a point on a circle and let it travel eternally on the perimeter of the circle (Fig. 2.1). The projection of this point on the x-axis of the trigonometric referential describes in time an oscillatory motion, travelling endlessly from one end of the axis to the other. Let us now presume that you lay an endless stripe of paper through the x-axis perpendicularly to the plane of the circle, and let the paper move at constant speed. If the projections of the circling point could be set to cause a manner of dots on the paper, we would quickly start to notice a continuous sinuous form on it, which our mathematics designates the eternal sinusoid. All that we know of oscillatory motions is captured in this sinusoid and the sinusoid describes the sum of our knowledge of wave motion. We visualize the eternal sinusoid in time by laying it down on a coordinate system where the horizontal axis stands for the time variable t and the vertical axis the function of time $f(t)$. We will now briefly show how to arrive at the usual time expression of the eternal sinusoid.

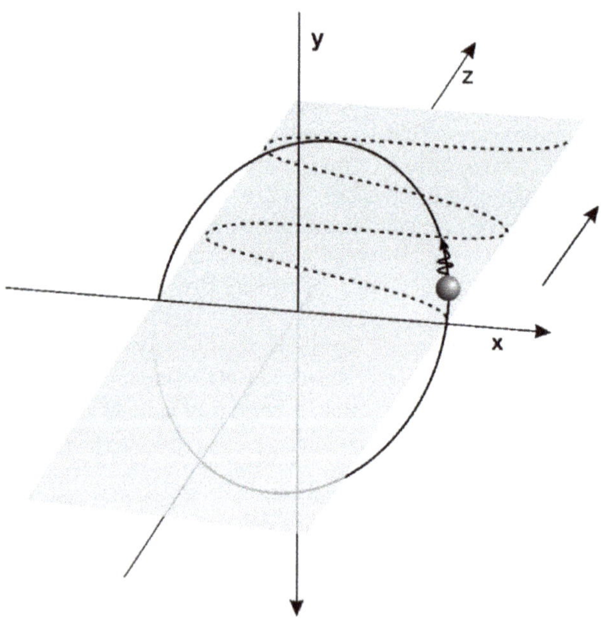

Fig. 2.1 Eternal sinusoid resulting from projection of a circling point

The equation describing the position of an object traveling in uniform rectilinear motion is:

$$x(t) = v_0 t + x_0$$

where *x(t)* is the travel distance at any time, v_0 the linear velocity of the object at time zero, the initial time, and x_0 the distance already traveled at time zero, if any.

Looking back at our circling point, we can express in the same fashion the distance it travels around the circle as:

$$\theta(t) = \omega_0 t + \theta_0$$

where $\theta(t)$ is the travel circular distance at any time expressed in radian, ω_0 the angular velocity of the point at time zero and θ_0 the distance already travelled at time zero, if any, and expressed in radian.

Both equations are seemingly identical. The only difference concerns linear quantities in the first one as opposed to angular quantities in the second one. We know, according to the trigonometric law of sine, how to express an angle:

$$\sin\theta = \frac{opposite\_side}{hypothenus} = \frac{y}{x}$$

When θ is expressed in degrees it means the angle formed by the point on the circle, the origin of the referential and the right-side portion of the x-axis. When θ is expressed in radian it simply means the arc subtended by the angle, which is the angular distance whose time expression we are looking for. Whether the angle is expressed in degree or radian, the law of sine holds always true. We can rewrite the last equation in the form:

$$y = x\sin\theta(t)$$

and by replacing $\theta(t)$ by the value we have found earlier, we obtain:

$$y = x\sin(\omega_0 t + \theta_0)$$

We replace y by *f(t)* because the expression on the right side is a function of time. In addition, we can easily see that *x* is the maximum value that the function can reach: since the maximum value of a sine is 1, the maximum value that the right side of the equation can reach is x itself. For that matter x is usually called the amplitude of the function, it is preferentially designated by A and takes the unit associated with the nature of the phenomenon, distance, current, voltage, flux, etc. The final general expression of the oscillatory phenomenon is then:

$$f(t) = A_m \sin(\omega_0 t + \theta_0)$$

This expression describes and gathers all we know about the physical nature of wave phenomena in general. The details of its derivation have been reported here in order to set a baseline for the discussion. There is no denying that all wave phenomena involve motion and repetition. Repetition means symmetry or interchangeability. The period of a wave is defined as the time it takes to complete itself before the cycle starts over. The half-period itself is a second element of symmetry. Any set of two consecutive periods of the eternal sinusoid are absolutely interchangeable, and within the period there is odd symmetry (symmetry with respect to the axis-crossing point) between the two half-cycles. Our usual mathematical expression for wave phenomena encloses indeed an account for symmetry. The odd symmetry between the half cycles springs from the even symmetry between the half-circles of the trigonometric circle itself, which stands for the roots of the model; the symmetry between the periods for themselves derives directly from the trigonometric circle as a whole. Consequently, the forms of wave phenomena we know how to model mathematically are those enclosing circular/spherical symmetry.

The ordinary implicit assumption that the nature of all wave functions is always the same is totally uncertain. We used to believe that the circular model set forth in trigonometry is general and applicable to all forms of oscillatory phenomena. Until we discovered that there exist certain objects in the universe that demonstrate oscillatory properties but do not demonstrate oscillatory motion like the pendulum or electrical current in a wire, for instance. The discovery that elementary particles moving in straight line could also exhibit oscillatory properties, i.e., interference and diffraction, as do conventional wave phenomena, posed a barrier to our understanding of the atomic and sub-atomic structure of the universe. Despite a fierce intellectual battle conducted through the use of differential calculus and operator mathematics, the wave function associated with particles is still not clearly understood. In addition, Quantum Mechanics can only explain the organization of the hydrogen atom, the simplest of all. It is important to understand that a wave phenomenon is not only a form of motion but also a form of symmetry. The trigonometric model principally takes into account the mobility aspect of the phenomenon. The symmetry facet does need to be accounted for as well. To do so, we may need to replace the trigonometric circle by other forms of symmetry and derive an expression for each form of symmetry. But we would first need to study the different orders of symmetry prevailing in the universe, the different inner states of symmetry that space may be endowed with, that is. This will become the subject of later treatment.

2.5. Wavefunctions in Quantum Mechanics and Relativity

The uttermost contribution of Quantum Mechanics to physical science is arguably not to have modeled the atom, a microscopic object, basically as a wave function, however important this form of representation proved to be. The success of Erwin Schrödinger's treatment is mostly due to the set of conditions that he posed for the wave function to give an adequate account of what the atom appeared to be: a particle limited to move freely in a box. He started with the travel electromagnetic wave equation, boldly defined a wavelength λ for the particle which he incorporated to the equation turning it into:

$$\theta(x,t) = A_m \sin\left(\frac{x}{\lambda} - vt\right)$$

This equation depends on two different variables: x, the distance travelled by the wave (or the particle) and t, the time variable. Since we introduce a second dimension in the wave phenomena, we must ensure that the variations of the wave in this dimension are also regular or "symmetrical", otherwise the motion would no longer be a wave motion. We express this condition mathematically by stipulating a specific condition between the rate of change of the wave $\theta(x, t)$ with respect to the time variable t and the rate of change of the wave with respect to the distance x. The ratio of these second-order partial derivatives must be equal to $(\lambda f)^2$:

$$\frac{d^2\theta}{dt^2} \cdot \frac{d^2\theta}{dx^2} = (\lambda f)^2$$

Or, in proper partial derivative notation:

$$\frac{\partial^2\theta}{\partial t^2} \cdot \frac{\partial^2\theta}{\partial x^2} = (\lambda f)^2$$

This condition should come as no surprise should we remember that, in differential calculus, the process of differentiation affords the study of concavity, in fact the curvature or curvilinear evolution of functions. If t were a tangible dimension, like z in a three-dimensional frame of reference, this condition ensures that the motion would then describe the eternal sinusoid in three dimensions. Imagine a plane folding in sinusoids.

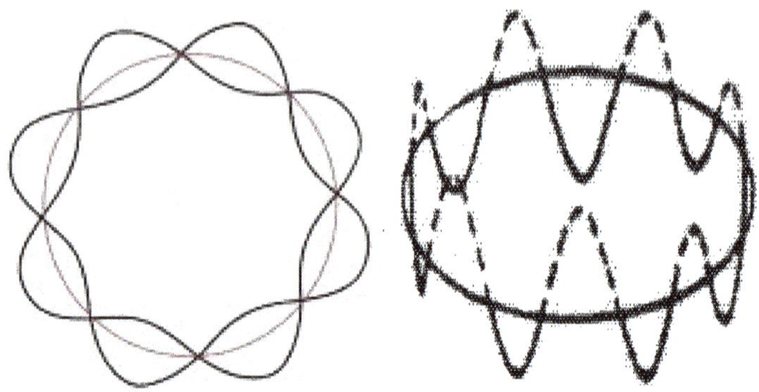

Fig. 2.2 Virtual depiction of a stationary wave from two different angles

E. Schrödinger further established that this wave associated to a particle limited to move within a box must be stationary. Otherwise, this associated wave would end up fading away. If the walls of this imaginary box were tangible, the supposed sinusoids described by the particle would need to hit the wall only in the node phases of the sinusoids. Otherwise, the reflected waves would be out of phase with the incident ones, and the wave phenomena would be quickly absorbed by the intangible walls. However, these walls bear in fact no physical ability to be impacted by the particle, because there is no real box, so that the wave must yield the stationary effect all by itself (Fig. 2.3). Consequently, the original wave must be a composite one: two waves of the same frequency (same sinusoid) but with different phase. The expression of this composite wave is:

$$\theta(x,t) = A_m \sin 2\pi\left(\frac{x}{\lambda} - vt\right) + A_m \sin 2\pi\left(\frac{x}{\lambda} + vt\right)$$

$$= 2A_m \sin 2\pi\left(\frac{x}{\lambda}\right)\cos 2\pi ft$$

Because of the ad hoc introduction of the terms of this specific composite function, we ought to make sure that it obeys the conditions for wave behavior. When we apply the above differential condition, that is to say taking the ratio of the second-order derivative and making it equal to $(\lambda f)^2$, surprisingly the time component of the composite wave ($\cos 2\pi ft$) cancels out and the condition becomes:

$$\frac{d^2(2A^m \sin(2\pi x/\lambda))}{dt^2} = \left(\frac{2\pi}{\lambda}\right)^2 2A_m \sin\left(\frac{2\pi x}{\lambda}\right)$$

Let us make $2A_m sin(2\pi x/\lambda) = \Psi(x)$ giving the function a denomination. The above condition then becomes:

$$\frac{d^2\Psi(x)}{dt^2} = \left(\frac{2\pi}{\lambda}\right)^2 \bullet \Psi(x)$$

Recapitulating, in order for our wave-particle to act as limited within an intangible contour, it must behave like a stationary wave such that the second derivative of this wave function is equal to the wave itself, except for the multiplication by a constant term. If we remember that the first derivative of a sine function is a cosine function and the derivative of the latter a sine function again, we see that the sine function satisfies the posed differential condition for the composite wave. As to the argument of this function, it must be such that when we differentiate it we obtain the coefficient $(2\pi/\lambda)^2$ present in the right-hand side of the equation. As a consequence, the wave function must be:

$$\Psi(x) = A\sin\left(\frac{2\pi x}{\lambda}\right)$$

We do not really know what this wave expression physically means. We know that the particle does not certainly have a classical wave motion, akin to a sinusoidal perturbation moving along a straight rope. The latter is the classical case of a motion as a function of time, whereas Ψ is only a function of x. The argument of the function $(2\pi x/\lambda)$ does not lend itself to a clear interpretation either. What is x/λ? This wave is what is called a De Broglie wave. Louis Victor De Broglie established that to every movement is associated a wave which pilots the movement, no matter how big or small the travel object is. In fact, the argument $2\pi x/\lambda$ is an equivalent of the argument $\omega_0 t$ in the general expression of the sinusoid. We know that:

$$\omega_0 t = 2\pi f t$$

where f is the frequency of the wave or the number of complete circular cycles per second. Since the frequency is the inverse of the period T: $f = 1/T$,

$$\omega_0 t = \left(\frac{2\pi}{T}\right)t$$

One way to make sense out of the De Broglie wave function is to compare it to the general equation of wave motion. That is to equate the arguments:

$$\left(\frac{2\pi}{T}\right)t = \left(\frac{2\pi}{T}\right)x$$

The role played by time is now played by distance or *space.* The variable t may be assimilated to x and the period in time T assimilated to the period in distance or the wavelength λ. That is to say, the movement of the object is not rated to change in time anymore but in space. In other words, the continuous dilation of space produces the movement. Moreover, the cancellation of the time variable is a confirmation of this interpretation. This conclusion is similar to the one drawn from our critical study of the relativistic treatment. Furthermore, the sine expression tells us that the De Broglie wave of the constrained particle possesses circular/spherical symmetry. The translation is that the structure of the space confines where the particle is constrained to remain possesses spherical symmetry. This form of symmetry will become apparent in the coupling of the particle with this extension of the space layout. From our perspective, the particle is not merely somewhere, the space location where the particle is moving is indeed an intrinsic part of the particle. What's more, it is this space location or field which is "piloting" the movement of the particle through its inner process of dilation. Its inner property of circular symmetry will lend to the distribution of the particle the apparent geometry of a sphere. The object is no longer a particle, it is a space sphere with a "distributive" particle.

From a quantum-mass perspective, the reality of the particle restricted to a space location is a kinematic event, from a space perspective it is rather an object because the particle is *coupled* to the immediate space deployment surrounding it. You no longer observe the particle as an object localized somewhere, you observe a particle describing a spherical geometry, a new object. This object is truly an *observable* no matter how close to the speed of light the speed of the inner moving particle is. We call this object a distributive object. It is endowed not only with proper space but also with proper mass. The mass is not localized, it is distributed. If the particle were merely somewhere, no matter how limited the extent of its motion, its mass would be preserved because it is not radiating energy: its pilot wave is stationary. But because the mass of the particle is redistributed throughout its immediate space deployment in what we call its coupling to space proper, the new object is born with less mass than that of the discrete particle itself, its exclusive mass contributor. In other words, the new mass loses localization and as such scalar magnitude. This effect is what is known as a mass defect in Quantum Mechanics.

2.6. Physical Space as Active Void

While the notion that pure vacuum is not an absolute void might still remain quite foreign to the larger public, it has been floating in Quantum Electrodynamics since the early

1930's. This line from Joseph Silk is often cited to explain the representation that the field holds of pure vacuum:

The quantum theory asserts that a vacuum, even the most perfect of vacuum devoid of any matter, is not really empty. Rather the quantum vacuum can be depicted as a sea of continuously appearing and disappearing particles that manifest themselves in the apparent jostling of particles that is quite distinct from their thermal motions... At any given instant, the vacuum is full of such virtual pairs, which leave their signatures behind, by affecting the energy levels of atoms.

Others might equally add fleeting electromagnetic waves to this constant annihilation of particle pairs or pairs of artifacts inside of pure vacuum. While in Quanto-Geometric Theory we may subscribe to this view, our primary intent however is not to describe what there is beyond physical space vacuum and how that affects the latter, but to describe its own seminal structure as it appears in front of us. Quanto-Geometry asserts a whole order of things in pure vacuum or nothingness that amounts to a structural order in its fullness such as:

- Symmetric latencies
- Geometric (Morphological) latencies
- De-ploy (Un-Fold) and splitting capabilities
- Active propagation
- Parameterized entropy and Shell configuration

We will further elaborate along those lines as we develop this proposal. We expressly caution that Quanto-Geometric Theory should not be assimilated to a Quantum Field Theory, as this exposition is substantially developed in contrast to the latter. The distinction between these two views of matter essentially rests on the nature of abstractions and analytical framework at work in both theories, despite their addressing the same objects, as we shall see in the next chapters. We believe that comparatively Quanto-Geometry presents a much wider analytical scope and better targeting, as much as superior results in the task of responding to the string of paramount yet unsolved questions in the physical sciences.

2.7. Conclusion

In this Chapter, we have examined the general properties of space vacuum. We have revisited Special Relativity and Quantum Mechanics to show their limits and their implication in relation with this new vision of space in the universe. We do not wish to overwhelm the uninitiated reader with more mathematical developments than necessary at this point and will tend to emphasize the interpretative aspect of the exposition. All too often, the desire to develop supporting mathematical formalism in order to secure accuracy, all loadable though it is, supersedes the abilities for assertive and sensible interpretation of results cast by the applied formalism. If the late paradigms are to be overcome, physics and physicists ought to reconcile and engage with the grander view furnished in philosophy, and creatively tackle the difficult issue of interpretation of reality from the fullness of the formalism being employed, which includes its philosophical connotations. The biggest problem in the field in our view has been and remains the inability to overcome the notion of time as a consequence of the inability to properly envision the inherent dynamics of space vacuum, despite unrelenting attempts. The final assault toward Grander Unification requires many bridges to be cast and many doors to be knocked open.

Chapter 3

God used beautiful mathematics
in creating the world.
Paul M. Dirac

UNIVERSAL MASS-SPACE COVARIANCE IN QUANTO-GEOMETRIC THEORY

This journey through space brings us now to the specifics of mass-space coupling. Having established the primacy of mass and space in the makeup of all objects of the universe, along with the general properties of coordinate space, our next step is to study the different modalities of articulation between mass and space within physical objects making up the universe. At the end of the previous Chapter, we have shown how both tenets arrive at a mutual articulation. We will now look into the principles that govern this interaction. To that end, we first examine the use of Operators in modern physics.

3.1. Mathematical Operators

Everybody has long been familiar with simple arithmetic. Since our early schooling days, we learned how to perform simple arithmetic operations involving addition, subtraction, multiplication and division as applied to any set of numbers. So, for instance, when we see the + sign between two numbers, we know that refers to the operation of adding one of them to the other. Hence the + sign, as all the other arithmetic signs, means a specific type of *operation* to be applied to two or more quantities. We may generalize this notion by stating that an *operator* sign is a transformation index. It instructs us about a specific transformation to apply to any quantity or set of quantities. An operator is thus in essence a transform. Take the expression:

$$9 = 6 + 3 \quad (3\text{-}1)$$

We may transform expression *(3-1)* by multiplying both sides by 2, for instance. We then obtain:

$$2 \text{ x } 9 = 2 \text{ x } (6 + 3) \quad (3\text{-}2)$$

$$18 = 12 + 6$$

Essentially, the equality remains true but we have transformed it by applying on both terms the exact same operation, which is literally 2 x. Therefore, 2 x as a whole is an operator. It means: do the operation of multiplying the terms of the equality by two. Thus, here we are applying our operator to not only other numbers but numbers that are already involved in a specific relationship, in this case a whole equation. We may avail ourselves of the convenience to rewrite equality *(3-2)* in a more concise format as:

$$2 \text{ x } [9 = (6 + 3)] \quad (3\text{-}3)$$

Here, the expression inside the brackets is a simple arithmetic expression, it might have been a function instead however. Then it would become a specific case within a higher level of abstraction. It might be, among many other alternatives, an instance of the function:

$$f(x) = 2x + 3$$

where x might stand for the number of lies commonly detected in the testimony of a certain category of criminals, in a court trial! The function means to multiply by 2 the number of lies found and add 3 to that result, in order to obtain the total amount of lies f(x) in a testimony. In the instance that you detect 3 lies, the expression becomes $f(x) = 6 + 3$. In that case, it turns out to be literally equivalent to expression *(3-3)* and reads:

$$2 \text{ x } [f(x)] \quad (3\text{-}4)$$

Giving to 2 x the standard operator notation, expression (4) reads:

$$< 2 \text{ x} > [f(x)]$$

Our operator $< \mathbf{2\ x} >$ is the simplest of operators, merely meaning to multiply the function by 2. We might want to perform that operation in order to obtain information about the age of the criminal. Due to its simplicity, this operation is probably best accounted for inside the function itself, or otherwise by simple arithmetic. However, we might have an operator prescribing to take the first-order, second-order or fourth-order derivative of the function rather and then to multiply it by 2 so that it reads $< 2 \cdot d^4/dx^4 >$. We might want to do so to extract pertinent information about the age of the criminal.

Operators are useful in modeling complex systems where the functional representation is unable by itself to cast proper solutions to the unknown variables under scrutiny. The use of a specific operator allows us to test for a specific pre-conceived behavior of the system. One very widely used operator in Quantum Mechanics is the Hamiltonian Operator. It is an operator of kinetic and potential energy and stands for the only path toward resolution of quantum systems' variables. Hence, operators are specific types of mathematical operations that we generally apply upon complex functions as a first step toward their resolution. If the function reveals itself proper to the operator, then we may resolve the variables through the *eigenvalues* of the operator.

Above all, it is important to understand that an operator is a pre-conceived abstraction about the behavior of a system; it is as such a model. Our contention in Quanto-Geometric Theory is that operators are typed, on the one hand, and that, on the other hand, nine fundamental types of operators exist in the physical universe. Therefore, we may acquire ample knowledge about all properties of a physical system because we know a set of fundamental patterns of *operation* - which we shall discuss shortly - according to which the system must behave.

3.2. Quanto-Geometric Coupling

The death of time ultimately signifies that all events in the universe may be observed as objects and all objects as an articulation between mass and space, regardless of the phase of matter under observation. The analysis conducted in the previous chapter asserts that the implications of the Special Theory of Relativity are equivalent to those induced from Quantum Mechanics. This strongly suggests that there is not a crucial difference between the laws governing the microcosm and those ruling the macrocosm. This general law exists and its name is *Quanto-Geometry.*

Since all objects enclose space, their mass always qualifies as *distributive* scalar, even though in some conditions the mass appears to be localized. We will then gradually abandon the term mass formally used to describe the tangible *thing* that makes up all of universal objects in favor of the term quantum. The quantum notion provides indeed a more general framing of the reality of mass in an object. On another hand, we have seen that the coupling of mass to space gives birth to *geometry*, that is to say the observable form of objects. Hence the term Quanto-Geometry which signifies articulation of mass and space in an object. Quanto-Geometry has a formal mathematical expression that we shall analyze later on in this Chapter. Visualized in its purest form, coordinate space displays an infinite degree of inner propagation. Visualizing mass in its purest form reduces it to an absolute level of concentration or localization likening it to a pure scalar. Nevertheless, none of these absolute forms of mass and space exist in the universe because all masses are coupled to space and all apparent particular extension of space is coupled to at least one discrete mass or quantum packet. The articulation of mass to space means a constraint exerted on space by mass and, concurrently, a constraint exerted on mass by space. The constraint undergone by space translates into a restriction in its degree of propagation which would be absolute without it. This constraint diminishes the extensiveness of a *space spread*[2] and brings it to some level of *tension*, expressible down to the form of standing behavior. In other words, articulated space is compressed. The constraint undergone by mass for itself means that it becomes *distributive*, albeit not discretely distributed, for it must follow the integral of the trajectories made available by its associated space extension. This constraint renders mass less localized, turning it into a quantum or dense scalar. This stands for a mandatory implication of the Quanto-Geometric articulation. The principle of Quanto-Geometry represents the most primal of all laws of conservation. It is a more general form of the law of conservation as visualized in Newtonian physics and as subsequently utilized in all other forms of physical theory. To render a prime account of an object is to specify its quantum-scalar degree and the degree of spread and propagation of its space partner: the specification must thus render a Quanto-Geometric account of the object. This Quanto-Geometric portrait is the primary modality of expression of the existence of an object. We call this foremost Quanto-Geometric principle the *generic Momentum* of an object and express it as follows:

$$q \times s = 1$$

where q is the quantum component of the object and s is its space component. The above formula postulates that the product of the scalar component q of any element, abstract or real, tangible or intangible, by the space (or nothingness) component s is equal to unity, *1* standing for the whole of the element.

The above expression derives into a series of forms of articulation that are best captured in a functional representation of the Quanto-Geometric primal principle. Before we delve into that functional analysis, let us first revisit the use of operators in Quantum Mechanics.

[2] In Chapter 8 we give a formal treatment to the notion of *space spread*, which is not exactly equivalent to the common notion of *field* in physics.

3.3. The Hamiltonian Operator in Quantum Mechanics

In the preceding chapter, we have seen that the last form of the condition imposed on the wave function in order to consider it emulating limitation in space is:

$$\frac{d^2\psi(x)}{dx^2} = \left(\frac{2\pi}{\lambda}\right)^2 \times \psi(x)$$

E. Schrödinger incorporated to this condition, the proposal by Louis Victor de Broglie that all forms of movements are piloted by a wave whose wavelength *is* $\lambda = h/p$, h being Planck's constant and p the momentum of the particle. He did as much with Dirac's constant $\hbar$ which is equal to $h/2\pi$. By substituting h and λ in the differential condition, it becomes:

$$\frac{d^2\psi(x)}{dx^2} = \frac{2m(E-V)^2}{\hbar^2} \times \psi(x)$$

Further manipulation yields:

$$\frac{\hbar^2}{2m} \cdot \frac{d^2\psi(x)}{dx^2} + V\psi(x) = E\psi(x)$$

where E is total mechanical energy of the particle (kinetic + potential), and V the potential energy.

The question posed by this differential equation is: what is the wave function which, when differentiated twice and multiplied by $(-\hbar^2/2m)$ and added to the potential energy of the particle yields the same wave function back but now multiplied by its total mechanical energy? When you find this function, you have found the wave function that the constrained particle actually exhibits. Another way of formulating this question mathematically is:

$$\left[\frac{\hbar^2}{2m} \cdot \frac{d^2}{dx^2} + V\right]\psi(x) - E\psi(x) \qquad (3\text{-}5)$$

Each of the terms within the bracket means a specific operation to perform on the wave function *Ψ(x)* which is not limited, however, to be an arithmetic form of operation. The first term instructs us to differentiate and multiply by a constant, the second one to only multiply by V. Each one of these terms stands for an operator because they do a particular operation on the wave function. The first one is related to the kinetic energy of the particle (the constant term is derived in fact from the kinetic expression: $E = mv^2/2$) and the second one expresses clearly and exclusively the potential energy of the particle. Therefore, the first one is designated the kinetic energy Operator E_k^ and the second one the potential energy Operator V^. Their sum E_k^ + V^, a non-arithmetic operation, is designated the Hamiltonian Operator H^. That is:

$$\left[\frac{\hbar^2}{2m} \cdot \frac{d^2}{dx^2} + V\right] = H$$

So that:

$$\mathbf{H}\, \Psi(x) = \mathbf{E}\, \Psi(x)$$

It is remarkable that the original question regarding the nature of the wave function which satisfies the confining behavior of the particle is hitherto still not answered and, in the aggregate, will not be answered directly. Concisely stated, the answer is: *"well, I don't know what this wave is like, but since the wave originates from a particle with mass and momentum, I know something that will provide further knowledge of some of the dynamic variables of this wave-like particle. I know an operator, the Hamiltonian operator, which is capable of interrogating it successfully."* Despite the systematic method developed by Erwin Schrödinger to obtain the operator corresponding to a specific dynamic variable in any system of particles, it remains universally accepted that Quantum Mechanics cannot bring a direct and precise answer to all of the unknowns of a dynamic system. Quantum mechanics even postulates that all of the variables of a given system cannot be determined with absolute precision but only statistically. It is fair to say that this science is more interpretative than predictive, notwithstanding its success in modeling adequately a number of known physical and chemical effects at the atomic and molecular scale.

3.4. The Quanto-Geometric Principle

In interpreting the quantum Hamiltonian operator, we may appropriately liken it to a lens that allows the observer to screen the kinematics of atomic elements. In order to render their observation possible, this screening lens must be transparent to their inner structure. Put differently, this model must *mold* the atom close enough in order to yield an adequate account of its structure and its subsequent kinematics. In summary, the

Hamiltonian operator proposes a form of bi-structure of atomic objects and correlated phenomena. Since it is a conjunction of two operators, it so proposes to screen every atomic event by considering both their kinetic and potential energy implications on the basis of a specific mathematical scheme. The order of application of each operator to the unknown is inconsequential, since the expression of the operator dramatically excludes the time variable. In that regard, it makes no difference whether you use on the unknown the kinetic operator first and then the potential operator or conversely. The result will remain invariantly the same in either case. This makes of the Hamiltonian treatment at the outset a structural analysis rather than a historic or kinematic analysis.

The principle embedded in the Hamiltonian is valid not only for the study of the atomic world but for all structures present in the universe. It is a particular form of the more general law we call **Quanto-Geometry**. In previous discussion, we have stipulated and hereby reiterate that all objects are made up of mass and space. We have stipulated as well that they both influence each other to give birth to Quanto-Geometry. That is to say, in the process of mutual interaction, mass becomes intrinsically distributive and space becomes extrinsically geometric. We have equally stated that there exists no pure mass and no pure space in the universe because all apparent discrete masses and all apparent pure space are always coupled to one opposite gender partner. Quanto-Geometry is as such generalized in the universe and represents the fundamental modality of all forms of existence in the universe.

In giving a formal Quanto-Geometric account of an object beyond the generic momentum, we may elect to express independently what happens to "pure" mass due to its coupling to space and likewise what happens to "pure" space due to its association with mass. (The term pure here is relative, it means the pre-birth state of the coupled mass and coupled space as well). The exercise of the articulation is such that mass then becomes *spatialized* or distributive and space becomes *compressed* or *tensorized*. If we establish q as the quantum (scalar) value and s as the compressed space value, per previous notation, the Quantum part of the object is:

$$Q_2 = q \times s \quad (3\text{-}6)$$

We call this quantum part Q_2 because it is the product of an initial state Q_1 where the mass and the space deployment were "purer". We may then conceive that the quantum part Q_1 of the object involves an integrative sum, and the space deployment as well. We have earlier posed that space, albeit homogeneous, is variable. Therefore, the variable of integration will be space. The quantum quantity will be a variable of space, and the space quantity will be a variable of *itself*. So that the initial or "purer" state of the quantum will be:

$$Q_1 = q\int s \cdot ds$$

$$= q \cdot \frac{s^2}{2} + C_q$$

It is important to comprehend the physical meaning of the integration of *s* with respect to itself. Space, being nothingness, is a continuum. The continuum does not disappear while space is in dilation or in propagation. This basically means that space can always be referenced to itself. That implies *s* to be a function in or of itself such as *s(s*)*, where *s** is space as the independent variable and *s* the dependent variable or the function. In this development however, we are going to keep *s* as the independent variable of the *Q(s)* function.

As to the quantum part *q* of the object, some might be uneasy with the representation of a macroscopic object holding a distributive mass, so deserving to be treated as a quantum. We do know however that the mass of any macroscopic object engaged in a non-linear motion, for instance, loses its point-like character and acquires inertia. The mass must then be related to the rotational characteristic of the object and become as such a *moment of inertia I*. We so take into account how the mass is distributed around an axis of symmetry of the motion and no longer speak of mass as such but of the quantity of inertia of the body.

Let us now take a look at the situation of the space spread part of the object in the coupling process. Since space experiences a compression within the articulation due to the constraint exerted by the mass partner, a formal expression for this compressed state must include a quantum factor or a coefficient of compression we call k. So that:

$$S_2 = k \times s$$

and the primary state of the space deployment of the object comes to be:

$$S_1 = k \int s \cdot ds$$

$$= k \cdot \frac{s^2}{2} + C_s$$

The expressions for Q_1 and S_1 are in fact operators. In analyzing the constitution of an object, these expressions formulate the request to square the variable part, divide this squared quantity by 2 and then multiply it by the involved constant, which may even be a coefficient. This operation is very common in physics and it becomes possible and meaningful only because of the generalization of Quanto-Geometry. This form of expression directly applies to the Quanto-Geometric composition of all universal objects. Let us now revisit the utilization of this form of expression throughout the entire discipline of physics.

a.- The linear and rotational Kinetic Energy

$$K_1 = m \cdot \frac{v^2}{2}, \quad K_r = I \cdot \frac{\omega^2}{2}$$

In these two equations, the variables are v, the linear velocity in a linear motion, and ω the angular velocity in non-linear motion. The coefficient m stands for the mass or, if you will, the translational inertia of the body in motion and I for its rotational inertia. They play both the role of the quantum part of the object. K_l and K_r are the respective kinetic linear and rotational energy of the object in motion.

b.- The energy stored in the magnetic field of an Inductor

$$U = L \cdot \frac{i^2}{2}$$

The variable here is the current i, which is squared, divided by 2 and multiplied by the inductance L to produce the energy stored in the magnetic field of the inductor.

c.- The energy density in an Inductor

$$U = \mu_0 n^2 \cdot \frac{i^2}{2}$$

The variable is still i in this equation, the constant or quantum part is $\mu_0 n^2$. The quantity $\mu_0 n^2$ being a product of the permeability μ_0, a constant and the number of turns n of the inductor, equally a constant, remains a constant as a whole and plays thoroughly the quantum role.

d.- The energy stored in a Capacitor

$$U_1 = C \cdot \frac{V^2}{2}$$

In this situation, the variable is the voltage V across the plate of the capacitor and the quantum part is played by C, the cumulative capacity of the Capacitor called the capacitance.

e.- The energy density in the electric field of a charged Surface

$$U = \varepsilon_0 \cdot \frac{E^2}{2}$$

E, the electric field between the plates of the capacitor, is actually the variable in this situation, and ε_0, the permittivity constant, exhibits the quantum part.

f.- The kinetic energy of a fluid in motion

$$K = \sigma \cdot \frac{v^2}{2}$$

σ stands for the fluid density constant and v the velocity of the fluid, a variable. This equation is used in fluid mechanics and is part of Bernoulli's condition.

g.- The elastic potential energy of a spring submitted to a force

$$U = k \cdot \frac{x^2}{2}$$

U is the potential energy at any point of the spring. It depends on the distance x of the point from its initial position. This distance is obviously a variable. The spring constant k exhibits the quantum part here.

There may be more of these relationships than the above reported ones. It is all the more surprising that this form of expression applies to the kinetic energy and the potential energy alike. From our usual kinematic point of view, kinetic energy and potential energy are two radically different physical realities. The similarity of their expressions strongly suggests an equivalence between the variables they hold as well as the constants. That is to say velocity, current, voltage, electric field and distance would be, if not equivalent, yet of the same origin. Since we know velocity and current as definite representatives of movement, the above implies that the variables of voltage, electric field and distance (or space) must hold movement alike. Similarly, inertia, capacitance, energy density, permittivity and inductance are implicated to be of the same nature, if not equal. The best way to account for all of these symmetries is through the Quanto-Geometric premises set forth in the study in course. By the same token, we are pressed to thoroughly review our common notion of energy as well.

First of all, we ought to understand that the static form of the energy is not the potential but instead what we have commonly designated the kinetic energy, the term that includes the scalar quantity. The kinetic energy is very much indeed the potential or static form of the energy. Secondly, considering space as an integral part of the object, the energy is not an intangible quantity or entity hosted by the physical objects but instead an ontological agency of the object. Thirdly, the energy is not a temporal quantity such that at one moment the object may hold it in a potential form and one instant later release it as kinetic energy. The physical object is always a sum of potential and kinetic energy concomitantly. The state of energy is the primary level of existence of the object. This is formally expressed in Quanto-Geometry as:

$$\varphi = \langle S_1 + Q_1 \rangle = q\int s \cdot ds + k\int s \cdot ds$$

$$= (q+k)\int s \cdot ds$$

$$= (q+k)\frac{s^2}{2} \quad (3\text{-}7)$$

We now differentiate equation *(3-7)*:

$$\frac{d\varphi}{ds} = (q+k)\cdot s \quad (3\text{-}8)$$

Equation *(3-8)*, with q being a scalar and k a scalar coefficient, expresses a momentum Q that we rewrite in the form:

$$Q = (q+k)\cdot s \quad (3\text{-}9)$$

So that by replacing *(3-9)* in *(3-8)*:

$$\varphi = (q + k) \cdot \frac{Q^2}{2(q + k)^2}$$

$$\varphi = \frac{Q^2}{2(q + k)}$$

The scalar sum of *(q+k)* may be rewritten as:

$$M = q + k \quad (3\text{-}10)$$

To finally yield by replacement:

$$\varphi = \frac{Q^2}{2M} \quad (3\text{-}11)$$

We have elected to do away with the constant of integrations C_q and C_s in equation *(2-6)*, because we are considering the object in native or ground state. In the excited state of the object, equation *(3-11)* for the expression of the object becomes:

$$\varphi = \frac{Q^2}{2M} + C_{sq} \quad (3\text{-}12)$$

The above development characterizes a set of definitions that frame the endogenous organization of an object. In some ways expression *(3-12)* stands for an approximation to the Hamiltonian operator. Equation *(3-12)* expresses a preconceived Quanto-Geometric abstraction that we may apply to objects in order to elucidate, not only their structure, but their behavior as well, because the quantum-space organization of an object reflects on its spatial behavior or kinematics.

Effectively the form of the term $Q^2/2M$ in the equation is a very common expression in the rendering of physical phenomena whereby Q corresponds to a momentum or other similar two-term product quantity and M a physical scalar. Reconsidering the above energy identities, we find that they may be transformed to yield the form of the rational term $Q^2/2M$. The following will make it even more apparent.

3.4.1 Linear Kinetic Energy

$K_l = m\ v^2/2,\ p = m \times v,\ K_l = p^2/2m,$

where p is the linear momentum of the object in motion.

3.4.2 Rotational Kinetic Energy

$K_r = I\ \omega^2/2, \quad H = I \times \omega, K_r = H^2/2I,$

where H is namely the rotational momentum of the object in motion.

3.4.1 Energy in Magnetic Flux of Inductor

$U = L\, i^2/2, \quad Li = N\Phi, \quad U = (N\Phi)^2/2L,$

where the quantity $N\Phi$ is defined as the flux linkage of the inductor.

3.4.2 Energy Density in an Inductor

$U = \mu_0 n^2\ i^2/2, \quad \beta = \mu_0 n\ i, U = \beta^2/\ 2\mu_0$

where β is the strength of the magnetic field set up by a current-carrying inductor.

3.4.3 Energy Stored in a Capacitor

$U = C\ V^2/2,\ Q = CV,\ U = Q^2/2C,$ Q is the total amount of charge stored on the plates of a capacitor.

3.4.4 Energy Density in Field of a Charged Surface

$U = \varepsilon_0\ E^2/2,\ \sigma = \varepsilon_0 E,\ U = \sigma^2/2\varepsilon_0,$

where σ is defined as the charge density on a surface.

3.4.5 Kinetic Energy in Liquid Flow

$K = \sigma v^2 /2, F = \sigma v, K = F^2 /2\sigma,$

where F is the flow rate of a liquid.

3.4.6 Elastic Energy Stored in a Spring

$U = k x^2 /2, \; F = k x, U = F^2 /2k$

The second equation is known as Hooke's law, and F the restoring force of a spring.

The difference between the incipient Quanto-Geometric operator above and the Hamiltonian lies in that the latter instructs about how to look at an oscillatory event in order to acquire knowledge about the characteristics of its variable, while the former screens objects (or events), whether oscillatory or not, from a scalar-space prism and directly expresses the structural order of the incipient observed object, which stands as a preview to its kinematics. In light of the unmitigated scope of the Quanto-Geometric bi-focal formalism, one shall give primacy to the Schrödinger's Hamiltonian of older Quantum Mechanics over Dirac's Lagrangian in the treatment of dynamics within the world of atoms. The latter shall be considered not as a substitute but an extension to the former, despite the two key merits of the Dirac Lagrangian. For purpose of reference, the extremely complex Dirac Lagrangian equation is reported in Appendix II.

3.5. Quanto-Geometric Normalization

Catering to the property of absolute homogeneity of space, the first law of Newton stating conservation of linear momentum *($p = mv$)* has been viewed by and large as the most fundamental physical law. One might confer to the incipient Guanto-Geometric operator, which is based on the energetic distribution in an object, the standing of most fundamental law of conservation from the Quanto-Geometric perspective. Notwithstanding its primacy, it only represents one among many other forms of Quanto-Geometric articulation.

We will now study the functional expression dictating the specific forms of Quanto-Geometric articulation. At the outset, the expression of this function is as follows:

$$Q(s) = \pm \left| \frac{s^0}{\sigma\sqrt{2\pi}} \bullet e^{\left(\frac{-s^2}{2\sigma^2}\right)} \right|$$

Despite its relative similarity with the Gaussian, this function bears significant specifics solely proper to Quanto-Geometry. Its dependent variable is the quantum and is denoted *Q(s)*. Its independent variable is space and is denoted s. The *Q(s)* function takes positive and negative values (Fig. 3.1). It is defined for all values of s except $s = 0$. When $s = 0$, a form of indetermination (0^0) appears in the numerator leaving the function undefined. The s-axis represents an asymptote. When s takes infinitely large values, *Q(s)* approaches 0 without ever reaching it, converting the s-axis into an asymptote. This behavior of the function as conceived ensures that the Quanto-Geometric premises are fully respected. A primary postulate of Quanto-Geometry stated that no absolute mass and no absolute space can be present in the universe. We stipulated that pure mass was absolutely concentrated and pure space infinite or in infinite propagation. These unreachable absolutes, of which we can think nonetheless, can thus only stand for a reference, they may not be part of reality itself. The function perfectly models these conditions as well as the referential[3] or coordinate system does. We will later show further implications of the nature of the referential on the plane of number theory.

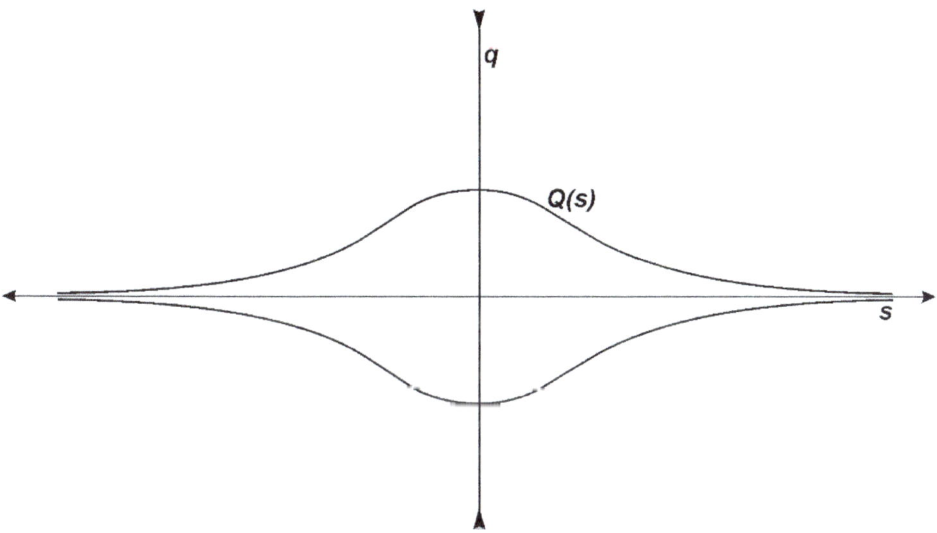

Fig. 3.1 Graph of the Quanto-Geometric function of one variable

The parameter σ stands for the accustomed standard deviation. It defines the evasiveness of the curve. Its real physical meaning is a coefficient of Quanto-Geometric articulation. Since *Q(s)* has a discontinuity, thus formally speaking it is not a function and cannot

[3] In this exposé we use the terms referential and coordinate system interchangeably. We introduce the term referential to signify that we assign generic physical properties to the axes of the coordinate system beyond simple mathematical ordinalization.

be studied as such. To acquire the properties of a function, it must be expressed as shown below:

$$Q(s) = \pm\left|\frac{s^0}{\sigma\sqrt{2\pi}} \bullet e^{\left(\frac{-s^2}{2\sigma^2}\right)}\right|, s \neq 0$$

Notwithstanding the first part of the expression which looks even more similar to the Gaussian, the function is not the Gaussian. Most notably, because of the discontinuity at $s = 0$, this function calls for a different form of normalization than the conventional Gaussian. Its eventual mean value would fall within the area of discontinuity. For that matter, the modality of interpretation by the mean must be discarded. Consequently, the variance also becomes meaningless. The function must be graphed on an orthogonal referential or coordinate system, with both axes equally normed, so that its inner geometric properties remain preserved. By no means can it be reduced to the ordinary "bell-shaped" curve. In a word, it is not a tool for the study of probability as we know it. Instead, it introduces a new form of normal distribution which goes beyond the Gaussian distribution, the Quanto-Geometric distribution.

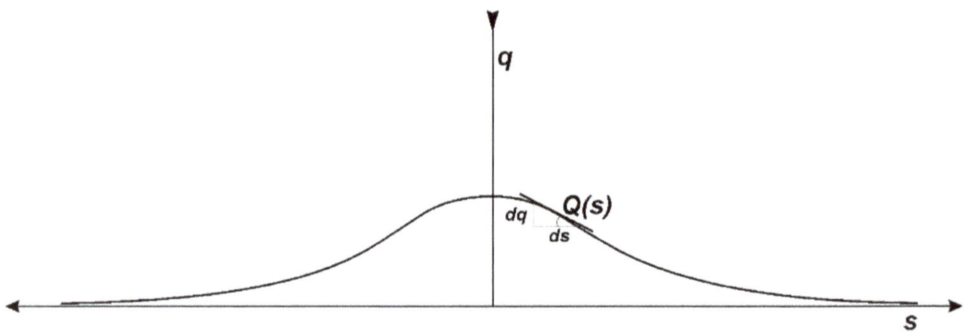

Fig. 3.2 Derivation by tangent line method

The discontinuity constrains us to study the function either on the range]0, ∞] or]0, -∞] of the variable. We know that *Q(s)* can never be equal to 1, the maximum attainable if there were no discontinuity. Starting in the neighborhood of 1, the function approaches 0 as *s* tends to infinity. In analyzing graphically the process of differentiation of this function, we notice that the tangential line to the curve, whose slope is nothing else than the rate of change ***dq/ds*** of the function, is external to the curve from the summit to the standard deviation point. As the function progresses downward to 0, the tangential line becomes internal to the curve. The rate of change ***dq/ds*** thus divides the function into two distinct sectors. The slope of the tangential line implies a relationship between *s* and *Q(s)* that we will shortly investigate.

By constructing two lines, one parallel to the positive s-axis and the other parallel to the positive q-axis respectively, we form together with the tangential line a right triangle (Fig. 3.2). The slope of the tangential line is equal to the tangent of the angle θ.

$$\tan\theta = \frac{opposite\_side}{adjacent\_side} = \frac{dq}{ds}$$

We also note that the rate of change dq/ds is always larger than 1 from the summit of the curve to the standard deviation point:

$$\frac{dq}{ds} > 1$$

In this region the quantum differential is larger than the space differential. In other words, for a small change in space there is a larger change in the quantum. This means that the quantum imprisons space, it compresses it. Stated differently, within the articulation, the quantum is the dominant tenet. Sliding the tangential line down, we reach the *standard deviation* (inflection) point or σ where:

$$\tan\theta = \frac{dq}{ds} = 1$$

$$\text{or } ds = dq$$

Here the differential dq is equal to ds, signifying that a change in space purports an equal change in the quantum. Within the Quanto-Geometric articulation at the standard deviation point, the constraint exerted by space on the quantum is equivalent to the one exerted by the quantum on space: mutual influence is equated at that point. In other words, there is no resulting *imposition* on the part of any of the two factors. This is gathered in the null value of the second-order derivative at that point (Fig. 3.3):

$$\frac{d^2Q(s)}{ds^2} = 0 \rightarrow s = \sigma$$

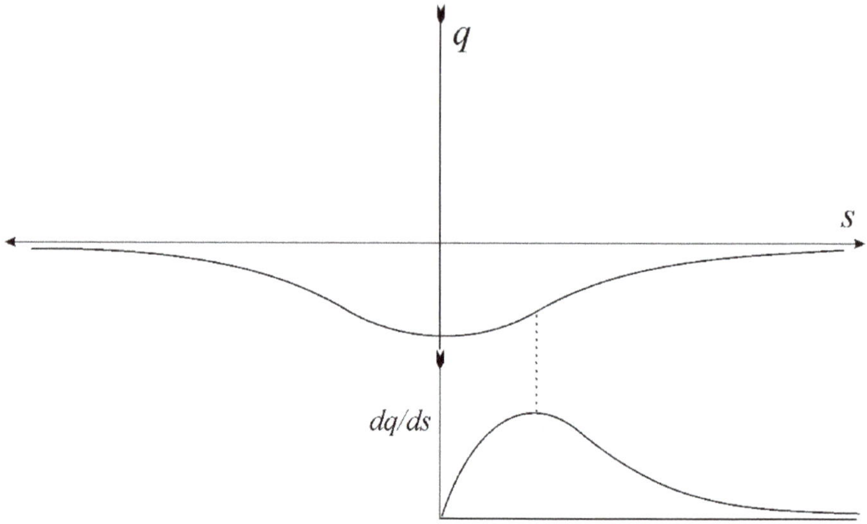

Fig. 3.3 Derivative of the Quanto-Geometric function

From the standard deviation point to infinity, the differential relation is:

$$\tan\theta = \frac{dq}{ds} < 1$$

$$\text{or } dq < ds$$

We can clearly read on the graph that the differential *ds* is larger than *dq* in this region. It is space now that jugulates the quantum. This physically means that the quantum becomes unapparent and submits itself totally to the degree of mobility of its space partner.

We understand from the above analysis that, within the framework of our *Q(s)* function, there exist three *quintessential* states of Quanto-Geometric articulation (Fig. 3.4). Further scrutiny of the curvature of the function tells us however that these states include 9 different granular forms of articulation. Each responds to a particular form of curvature contained within the geometry of the function. The following sequence, a paramount entity, will help us delineate these fractional forms of curvature:

$$\{q_n\} = \left\{ \frac{e^{(1-n)/(10-n)}}{\sigma\sqrt{2\pi}} \right\}, \quad \textit{where } 1 \le n \le 9 \quad \textit{(3-12)}$$

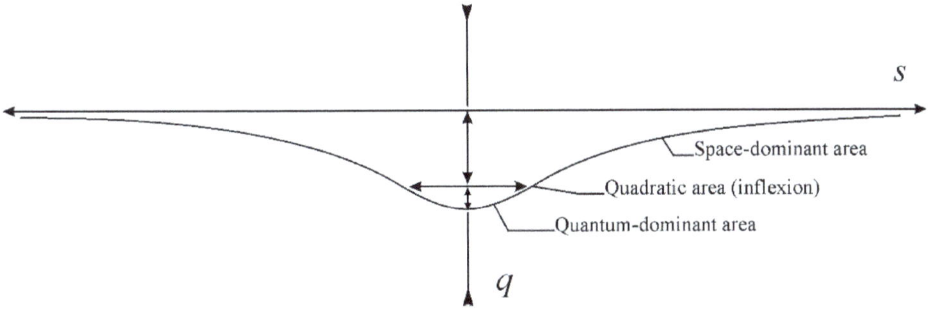

Fig. 3.4 Dissection of the Quanto-Geometric function in 3 main areas

This sequence is of capital importance, surpassed only in significance by the following pilot geometric sequence which it stems from:

$$\{s_n\} = \left\{\frac{1-n}{10-n}\right\}, \quad \textit{where } 1 \le n \le 9 \quad \textit{(3-13)}$$

Those sequences determine the higher norm of ***symmetry-breaking*** that ultimately presides over the development of all objects modeled by the Quanto-Geometric Function. The $\{q_n\}$ sequence sets the end point of each specific form of symmetry embedded in the function throughout its graphical course, so defining the intervals of Quanto-Geometric *norm-alization*. This sequence defines the set of real numbers which will dissect the curve of the $Q(s)$ function into its 9 *natural* discrete forms of curvature, whereas the $\{s_n\}$ sequence determines the pivotal s points on the space line that give rise to the specific s domains of normalization. So therefore, the first sequence yields the values taken by the function at the points of separation of curvature sectionals, the defining points of symmetry-breaking. By replacing n with the values in its defined range, the resulting terms of the sequence follow (see section 4.6 for relationship between n and s):

$$\text{for } n = 1, \quad \frac{e^0}{\sigma\sqrt{2\pi}} = \frac{1}{\sigma\sqrt{2\pi}} = Q_1(s)$$

$$\text{for } n = 2, \quad \frac{e^{-1/8}}{\sigma\sqrt{2\pi}}, Q_2(s)$$

for $n = 3$, $\dfrac{e^{-2/7}}{\sigma\sqrt{2\pi}}, Q_3(s)$

for $n = 4$, $\dfrac{e^{-1/2}}{\sigma\sqrt{2\pi}}, Q_4(s)$

for $n = 5$, $\dfrac{e^{-4/5}}{\sigma\sqrt{2\pi}}, Q_5(s)$

for $n = 6$, $\dfrac{e^{-5/4}}{\sigma\sqrt{2\pi}}, Q_6(s)$

for $n = 7$, $\dfrac{e^{-2}}{\sigma\sqrt{2\pi}}, Q_7(s)$

for $n = 8$, $\dfrac{e^{-7/2}}{\sigma\sqrt{2\pi}}, Q_8(s)$

for $n = 9$, $\dfrac{e^{-8}}{\sigma\sqrt{2\pi}}, Q_9(s)$

We obtain these values of the function by substituting the corresponding values of s for n in the expression of the function (See Section 6.5 of Chapter 6 for relation between n and s.):

$n = 1, \quad s_1 = 0$

$n = 2, \quad s_2 = \pm\,\sigma/2$

$n = 3, \quad s_3 = \pm\,2\sigma/7^{1/2}$

$n = 4, \quad s_4 = \pm\sigma$

$n = 5, \quad s_5 = \pm\sigma\,(8/5)^{1/2}$

$n = 6, \quad s_6 = \pm\sigma(10)^{1/2}/2$

$n = 7, \quad s_7 = \pm 2\sigma$

$n = 8, \quad s_8 = \pm\sigma 7^{1/2}$

$n = 9, \quad s_9 = \pm 4\sigma$

The value of the sequence for $n = 1$ determines a value of the Quanto-Geometric function which is in fact a limit:

$$\lim_{s \to 0} Q(s) = 1$$

At $s = 0$, the corresponding value of the variable for n = 1, the Quanto-Geometric function approaches a limit. It is important to realize that the function is not defined at $s = 0$, even though our graphism of representation thus far does not portray that discontinuity, simply for practical graphing purposes. In the interval $]0, \sigma/2[$, the curvature of the graph is circular or **monocentric**. In the next interval $[\sigma/2, 2\sigma/7^{1/2}[$, the geometry of the curvature is elliptical or **bipolar**. The section of curvature laying in the interval $[2\sigma/7^{1/2}, \sigma[$ is parabolic or **associative**. In the interval $[\sigma, \sigma 8/5^{1/2}[$ the function is of hyperbolic or **ambivalent** curvature. The forms of curvatures we know as conic sections end up with this interval. In the fifth interval $[\sigma 8/5^{1/2}, \sigma 10^{1/2}/2[$ the function displays a form of curvature we call the **entropic** form of curvature. The sixth interval $[\sigma 10^{1/2}/2, 2\sigma[$ delineates a form of curvature we denote the **enthalpic** curvature. In the seventh interval $[2\sigma, \sigma 7^{1/2}[$ the function describes a form of curvature that we call **paradoxical** curvature. The eighth interval $[\sigma 7^{1/2}, 4\sigma[$ encompasses a fraction of curvature we coin the **perfect** curvature. And finally in the ninth interval $[4\sigma, \infty[$ the function displays a section of curvature we denote linear or **infinite** curvature. The choice of these attributes will be completely justified later in this exposition.

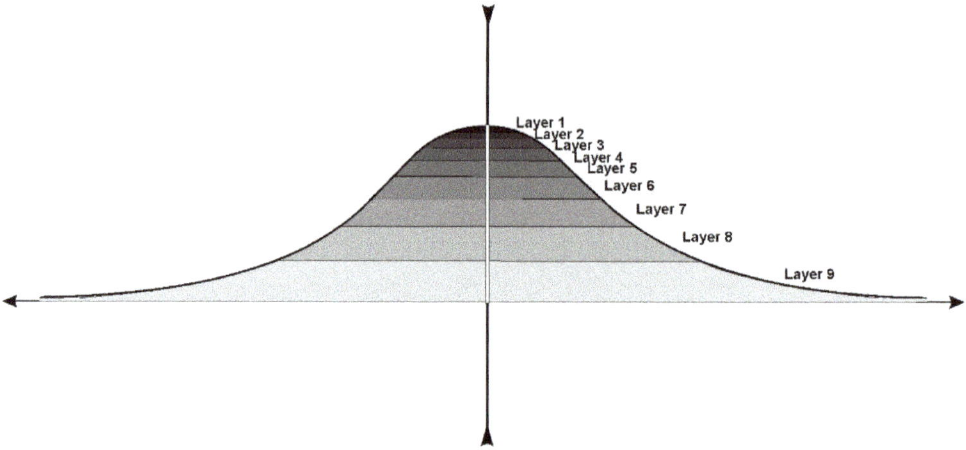

Fig. 3.5 The 9 essential layers of the Quanto-Geometric function

These intervals define the *normalization* modes of the Quanto-Geometric function. Taken horizontally, they define the different levels of Quanto-Geometric articulation. The areas that make sense under the curve are not vertically oriented as the normal Gaussian distribution has it. The meaningful and useful areas, from the Quanto-Geometric perspective, are the ones horizontally delineated by the natural quantum values of the function (Fig. 3.5). These areas, termed layers in the above illustration, are Quanto-Geometric *normals*. One way to visualize these levels, from a conventional modern physics standpoint, is in terms of energy levels. We have seen earlier that what we call energy is the structural expression of an object. We may now understand that our usual concept of mechanical energy (kinetic + potential) corresponds in fact to the first level of the Quanto-Geometric function or the first *normal*. In Quantum Mechanics, as the energy level increases the particle is able to remain preferably farther in space than its regular "ground" level location. Our model readily explains this effect with the visualization that the higher the Quanto-Geometric level displayed in the graph of the function, the larger the related space extent *s* or the *normal*. We also know, in Quantum Mechanics, that the energy associated with a level is always given as a difference between two quantities. This scheme is part of the Quanto-Geometric model on account of the area included in one horizontal stripe. In comparison with the energy levels, the size of that area is equal to the difference between a larger area bounded by the upper limit of the interval and a smaller area bounded by the lower limit of the interval.

On another hand, our model readily explains why there is a specific geometry associated with each energy level. We will not, however, formally utilize the term energy in our exposition, since each level represents indeed a preeminent physical law in the light of Quanto-Geometry.

3.6. The Nine Quanto-Geometric Primitives

If it indeed holds true that the Quanto-Geometric function contains the conservation operator, in the form that we have expressed it in section 2.4, we must be able to retrieve the operator from the expression of the function. Let us start by writing the function in an implicit form by replacing the notation *Q(s)* by *q*, in the same way that we traditionally replace the notation *f(x)* by *y*. The Gaussian-like part of the Quanto-Geometric function may then be written as:

$$q = \frac{1}{\sigma\sqrt{2\pi}} e^{\frac{-s^2}{2\sigma^2}}$$

$$q\sigma\sqrt{2\pi} = e^{\frac{-s^2}{2\sigma^2}}$$

$$\frac{-s^2}{2\sigma^2} = \ln\left[q\sigma\sqrt{2\pi}\right]$$

$$\frac{-s^2}{2\sigma^2} = \ln q + \ln\sigma + \ln\sqrt{2\pi}$$

$$\frac{-s^2}{2\sigma^2} = \ln q + \ln\sigma + \frac{1}{2}\ln(2\pi)$$

$$\frac{-s^2}{2\sigma^2} = \frac{1}{2}\left[2\ln q + 2\ln\sigma + \ln(2\pi)\right]$$

$$\frac{-s^2}{\sigma^2} = 2\ln q + 2\ln\sigma + \ln(2\pi)$$

By multiplying numerator and denominator of the first term by $2q^2$:

$$\frac{-s^2}{\sigma^2}\cdot\frac{2q^2}{2q^2} = 2\ln q + 2\ln\sigma + \ln(2\pi)$$

$$\frac{-s^2q^2}{2q}\cdot\frac{2}{q\sigma^2} = 2\ln q + 2\ln\sigma + \ln(2\pi)$$

$$\frac{-s^2q^2}{2q}\cdot\frac{2}{q\sigma^2[2\ln q + 2\ln\sigma + \ln(2\pi)]} = 1$$

$$\frac{-(sq)^2}{2q}\cdot\frac{2}{q\sigma^2[2\ln q + 2\ln\sigma + \ln(2\pi)]} = 1$$

Replacing sq by Q ($Q = s \times q$ from equation *2-1*):

$$\frac{-Q^2}{2q}\cdot\frac{2}{q\sigma^2[2\ln q + 2\ln\sigma + \ln(2\pi)]} = 1$$

Let: $2\ln q = a$,

then $\frac{a}{2} = \ln q$

and $q = e^{\frac{a}{2}}$.

Let: $2\ln\sigma = b$,

then $\frac{b}{2} = \ln\sigma$

and $\sigma = e^{\frac{b}{2}}$.

Let: $\ln(2\pi) = c$

The equation then becomes:

$$\frac{-Q^2}{2q} \bullet \frac{2}{e^{\left(\frac{a}{2}\right)} \cdot e^{\left(\frac{b^2}{4}\right)}[a+b+c]} = 1$$

$$\frac{-Q^2}{2q} \bullet \frac{2}{e^{a/2} \cdot e^{b^2/4}[a+b+c]} = 1$$

$$\frac{-Q^2}{2q} \bullet \frac{2}{e^{\left(\frac{2a+b^2}{4}\right)}[a+b+c]} = 1$$

$$\frac{-Q^2}{2q} \bullet \frac{2}{ae^{\left(\frac{2a+b^2}{4}\right)} + be^{\left(\frac{2a+b^2}{4}\right)} + ce^{\left(\frac{2a+b^2}{4}\right)}} = 1$$

First, notice that the equation holds true only if $s \neq 0$, which is an explicit condition of the Quanto-Geometric function. If s equals *0*, then Q equals *0* and we end up with the absurd result of $0 = 1$. So therefore, *s* cannot be equal to *0*. In its usual form, the Gaussian function does not incorporate this condition, for it is defined for all values of *s*. It is all the more significant that when the Gaussian expression, as a part of the Quanto-Geometric function, is rewritten in an implicit form the condition pops out. Even though unapparent, the condition is therefore implicit in the Gaussian function as we conventionally know it, and further poses significant implications regarding the physical meaning of the variable *s*. These implications are but the Quanto-Geometric premises set forth at the beginning of this discussion.

Also notice that the equation is equal to 1, a condition commonly known as the condition of normalization. It means that the equation expresses the totality of reality, or the entirety of the particular physical phenomenon which it is modeling. Totality means 100%, that is the ratio 100/100, which is equal to1. We can perfectly replace 1 by ψ because we Quanto-Geometrically think of this totality as an object, in the same manner as we did for the former Quanto-Geometric operator equation. We then express the above equation in the following fashion:

$$\frac{-Q^2}{2q} \bullet \frac{2}{ae^{\left(\frac{2a+b^2}{4}\right)} + be^{\left(\frac{2a+b^2}{4}\right)} + ce^{\left(\frac{2a+b^2}{4}\right)}} = \|\Psi\| \quad (3\text{-}14)$$

The characteristic term $Q^2/2q$ is again present as is and needs no further consideration. The term $2/(\, a\cdot e^{(2a+b^\wedge 2)/4} + b\cdot e^{(2a+b^\wedge 2)/4} + c\cdot e^{(2a+b^\wedge 2)/4})$, expresses the specificities of a relationship between the quantum q of circulation and the standard deviation value σ of the space under tension. Exclusively made up of q and σ, the exponential term represents in fact the footprint of a transcendental function of the hyperbolic type. As we know, the hyperbolic functions are generally based on a ratio between the number 2 and a sum of exponentials, commonly two of them and each raised to an x and an $-x$ power (x being the independent variable). The transcendental part of the above *(3-14)* expression is a general formulation for the precursor of the family of hyperbolic functions and contains all of the possible forms of symmetry. In support of this claim, one should remain wholly mindful that there is knowingly complete correspondence between the family of trigonometric functions and a subset of the family of hyperbolic functions.

In addition, there is a set of values for parameters a, b and c in the *(2-8)* equation that will give us 9 different normalized values of $\|\Psi\|$, such as $\|\Psi_{g1}\|$ ... to $\|\Psi_{g9}\|$ all representing the 9 prime Quanto-Geometric Operators. However, we will not pursue operator-based analysis in this text, remaining in favor of functional analysis in a bid to avoid the *uncertainties* generally associated with Operator-based analysis. We must keep in mind that an Operator is essentially a pre-conceived notion about a system, the solutions cast by that treatment being as good as the abstractions implicated by the Operator. We will make use of Operator treatment only when absolutely necessary. Furthermore, in that event we will use explanatory language instead of the nomenclature reserved to Operator-based analysis in an effort to keep this text accessible to the largest audience possible. Alternatively, we will equally refer to the 9 Quanto-Geometric normals in terms of **Primitives**.

In sum, we have established thus far a certain number of Quanto-Geometric principles, proposed a Gaussian-like functional expression of these principles and demonstrated the close equivalence between the implications of the Gaussian and the Quanto-Geometric platform.

3.7. Quanto-Geometric Tri-Valent Operator

The examination of the geometric progression of the curve of the Quanto-Geometric function has revealed 3 grand moments or phases of the function. One where the value of the derivative was larger than 1 *($dq/ds > 1$)*, another where the value of the derivative was equal to 1 *($dq/ds = 1$)* and lastly one where the derivative was less than 1 *($dq/ds < 1$)*.

This is a result of the *quintessential* Quanto-Geometric articulation incarnated in the function. This tri-partite phasing of the Function goes beyond functional mathematical representation and is best expressed in terms of mathematical operators. We gather this concept therefore under the term Quanto-Geometric Tri-Valent Operator, in the understanding that it is an Operator that encompasses 3 sub-operators. The first of these sub-operators we designate the Quintessential Quantum Operator or Hyper Quantum Operator (Fig. 3.6).

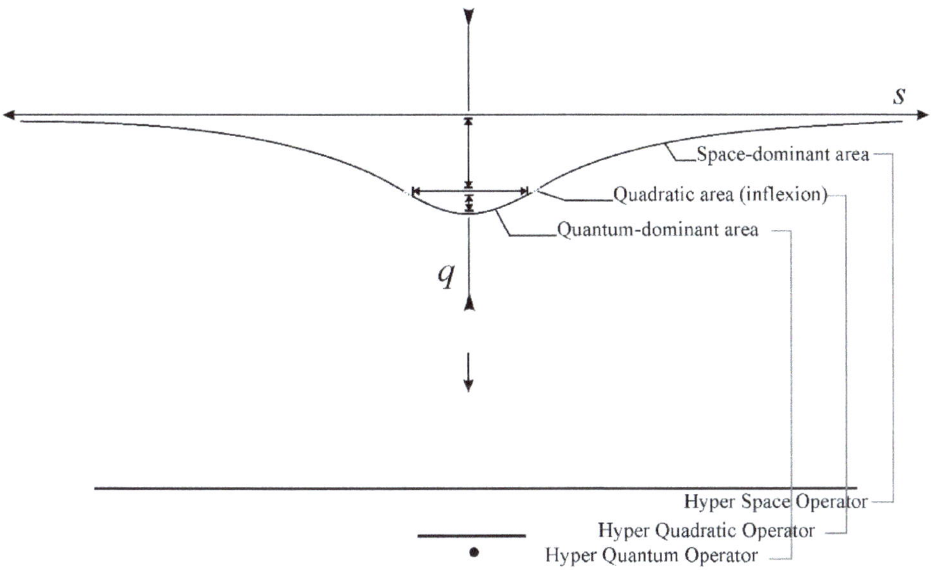

Fig. 3.6 Quanto-Geometric Quintessential Trivalent Operator

The second of the sub-operators we coin the Quintessential Quadratic Operator or Hyper Quadratic Operator. The third of these sub-operators we dub the Quintessential Space or Hyper Space Operator. Whenever they participate in the configuration of matter, these operators do not each act alone or individually, they all three intervene concomitantly. We have given them the full designation of Operators, because the transformations they operate are first-order Operator's transforms. Yet it must remain completely clear that they each belong to the larger entity we designate the Tri-Valent Operator.

The *norm* of the Hyper Quantum Operator is the perpetration of utmost *quantification* or *densification*, and the overloading of everything pertaining to the signature of Quanto-Geometric type 1. The *norm* of the Hyper Quadratic Operator is the perpetration of utmost *ambi-valence* and the overloading of everything pertaining to the signature of

Quanto-Geometric type 4. And finally, the *norm* of the Hyper Space Operator is the perpetration of utmost *spatiation* and *spatialization* and the overloading of everything pertaining to the signature of Quanto-Geometric type 9. The details of these signatures we shall study later in the exposé addressing Quanto-Geometric Number Theory. As a reminder, the *norm* of a mathematical operator is the typical feature or behavior that the Operator propitiates.

3.8. An Alternative Notional View of Probability Theory

Before we proceed with the analysis of the other Quanto-Geometric primitives, let us briefly revisit the statistical notion of probability. We now know for certain that the Gaussian function is much more than what it purports to be at face value. Through Quanto-Geometry we have raised it to the category of a model capable of describing not only probability distribution but the physical universe and all of its contents. The paramount reason why we have created the concept of probability in parlance and mathematics is our ignorance of the essential nature of space and its phenomenological behavior. We look at the contents of the universe as discrete components and whenever we find ourselves unable to identify a causal relationship accountable for an observable physical effect, we interpret the latter as a statistical phenomenon permeable to intelligence only through the law of probability.

Let us now consider an ordinary statistical phenomenon. Gather any number of individuals in a room, give them a ruler and assign them the task of measuring the length of any of the walls of the room. You will find that the number of individuals versus of the measured length values describe a Gaussian distribution. In other words, most of them will hold values that are close to a specific one, namely the mean. As you consider a lower and lower number of individuals with same measurements, you will find that their error, or the dispersion around the mean, becomes larger and larger. That is to say, few individuals will have extremely large margins of error. In short, the correlation between a number of these individuals and the length they measured for the wall is modeled by the Gaussian function. Take a value of measured length, make it equal to the variable and you will find a value of the function yielding directly the number of individuals who measured that same length. Notwithstanding the beautiful mathematical operation allowed by the function, it does not explain why, no matter the number of individuals, no matter their temper, their race or social characteristics, no matter the size of the room, independently of everything else, the correlation between the measured values and the number of individuals with same measured values always holds the same form, or the same form of dispersion.

If we look at the phenomenon "individuals measuring the length of a wall" in the light of Quanto-Geometry, we may obtain a better understanding of the phenomenon than a mere statistical account. Quanto-Geometry poses that these individuals are not really discrete, absolutely separated from each other. They are made up of quanta, which unfortunately are the only parts we can perceive of them, and of space, which we unfortunately always disregard. They interact with each other even though they don't hit each other or they don't speak to each other. They intrinsically interact through their space component, and for that matter they can be together visualized as a whole or one individual. They make up a Quanto-Geometric object, that is to say a space field inhabited by the quanta, their observable bodily parts. Whatever they do "as a whole or in common", whatever they do "together", they will do it according to the fundamental Quanto-Geometric articulation for any object. There will be a point of quantum-space articulation where the value of the inertia is extremely high and jugulates space.

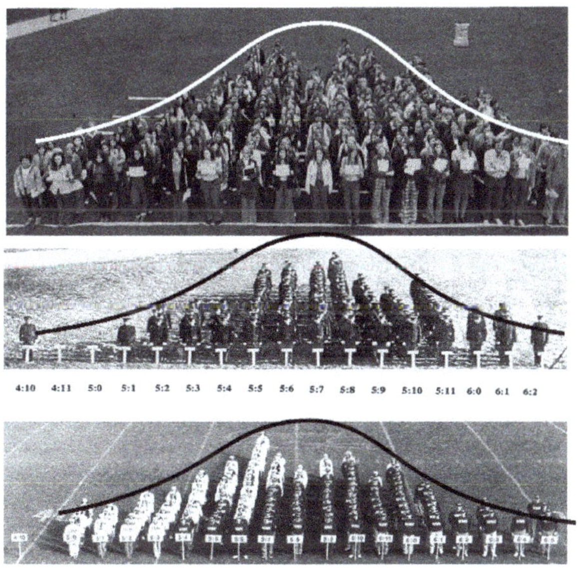

Fig. 3.7 Examples of height distributions in limited random samples of population

These individuals are very much "localized" and in an event of measurement their deviation will be minimal producing the most accurate measurement. Notice that the prediction of the model is that the inertia will be the highest at that point. It certainly does not mean that the heaviest individual will occupy this level of the distribution but that there will be many individuals in this particular group (Fig. 3.7). Remember that inertia means to us distributive mass akin to the quantum and we have made these two terms interchangeable. At the other extreme, we have a point of articulation where the space component jugulates the inertia. These individuals are very dispersive and, in an event of measurement, the deviation from the mean will be so significantly high that the measurement will be extremely inaccurate. The quantum-individuals making up this level of the "object", which is the highest one, will be scarce: the quantum is overwhelmed, it is unapparent, there is "almost" no quantum in this state. In the intermediate position, there is a point of articulation where the quanta are in equalization with the articulated space, these individuals are in the quadratic number and they measure the wall with the "standard" deviation σ. All others follow the Quanto-Geometric pattern alike in terms of their grouping and the dispersion in their measurements.

Quanto-Geometry further tells us that there will never be one group of individuals to measure the wall with 100% accuracy. We can fully take stock of the adequacy of this assumption because no measurement can admittedly ever be done with 100% accuracy. The real length of the wall is a limit that can only be approached. Quanto-Geometry also tells us that no matter how much deviated a measurement turns out to be, it is never wrong. There will always be at least one group of two individuals with an extremely deviated value, but they are never totally wrong. This is an irrevocable Quanto-Geometric stipulation. So it is, because the Quanto-Geometric vision assesses that every object possesses a space component which is limitless, notwithstanding its level of compression or tension. As such the object is in essence not measurable or confinable in the absolute.

This is indeed an example of the presence of quantization right in the midst of the macroscopic world, among a myriad of others such as the height distribution in human population (Fig. 2.7). Many may wonder how a theory describing the quantum world of particles may yield interpretative elements for the macrocosm, knowingly a continuum domain. By allowing us to divide the individuals of our example into specific groups of inertia, which undoubtedly correlate with the fundamental quantum numbers, this treatment has demonstrated how quantization works in the macrocosm in spite of its apparent continuum. Each of these individuals, no matter the kind of actions taken while performing the measurements, (one might have stopped for some coffee, while another might have stepped aside for a cigarette or some vapor), was part of a quantum group, or a group of inertia if you will, which belongs in a specific Quanto-Geometric level. While they were moving up and down busy at their measurements, you could have used Newtonian physics to measure their weight, the friction of their shoes against the floor, the pitch of the sound of their voices, etc., nevertheless they were all involved in a higher frame of event where they were in fact quantized. If you do not invoke space and its articulation with the quantum-bodies, there will be no possible explanation for such a peculiar and perhaps magical form of reality, which we call the Gaussian distribution, right within the macroscopic world, and we will be left at the end only with a statistical account of the event as is usually done. In the last Chapter of this text, we provide a more formal framework to this view.

The hidden variables invoked by A. Einstein to sustain an objection to the philosophy implied by Quantum Mechanics are no less than the Quanto-Geometric principles. The basic problem of Quantum Mechanics, a science heavily based on probability density distribution, dwell not in its mathematical results, which have always been universally accepted, but in the interpretation of these results. The uncertainty in the philosophical implications of quantum science had sparked a debate which divided physicists in two schools, the school of Copenhagen including among others Neils Bohr, Max Born, Paul Dirac, Wolfgang Pauli and Werner Heisenberg, and the school of determinism including Albert Einstein, Nathan Rosen, Boris Podolsky and John Wheeler among others. They have all strongly contributed to the onset of Quantum Mechanics however. While the school of Copenhagen developed a philosophy of uncertainty as a prevailing implication of Quantum Mechanics and as an intrinsic component of reality, the other stream fiercely resisted this view advocating the existence of unknown or hidden variables accountable for physical effects with yet no recognizable causes. They termed Quantum Mechanics an incomplete science insofar as an explanation of what makes physical objects or events behave or occur statistically remained to be found. A. Einstein was even quoted saying that he cannot "imagine the electron hopping like a bug", like the individuals of our previous example, that is!

We hope to have presented thus far enough novel elements of interpretation for the reader to at least perceive where the truth lies. The panorama will become much clearer after the study of the remaining eight Quanto-Geometric operators, and the kinetic-symmetric properties of space within each operator as well.

3.9. The Quanto-Geometric Coordinate System

The Quanto-Geometric coordinate system encompasses the fundamental concepts of space and mass at the most generic level that we may possibly visualize them (Fig. 2.8). Its vertical q-axis holds the regency of the scalar domain, the purview of the real number line, whereas the horizontal s-axis embodies the scale of void, both abstract and physical. The referential is orthonormed in relation to all axes. This referential or coordinate system is not a Cartesian coordinate system, although it may appear of great similarity with the latter. Further the s-axis may be represented either as a single line for a two-dimensional referential overall or as a double axis which together with the q-axis makes for a 3-dimensional referential (Fig. 3.8). In its incarnation as a double axis, the s linear run makes explicit the two inner dimensions of void, which are symmetry and mobility as discussed later, thereby eliciting the Quanto-Geometric principle that void in its quality of a tenet of matter contains two degrees of freedom.

What's more, while the q-axis is finite, the s-axis is infinite. That is the reason why the symbology of the q-axis portrays an inward-pointing arrow, whereas the s-axis portrays an outward-pointing arrow. The q-axis is engaged in the business of *discretization* and the *infinitesimal*, in contrast to the regency of *continuum* and *infinity* endorsed by the s-axis (although restricted by symmetry-breaking intervention as we shall see later). The graphism we use in Quanto-Geometry may facultatively omit the depiction of the arrows as a matter of practicality. When it does however, one must keep in mind the nature of the principles embodied in the axes of the coordinate system. In its role of representative of the real number line the q-axis grandfathers the entire suite of real numbers from the largest integers to complex numbers to transcendental numbers.

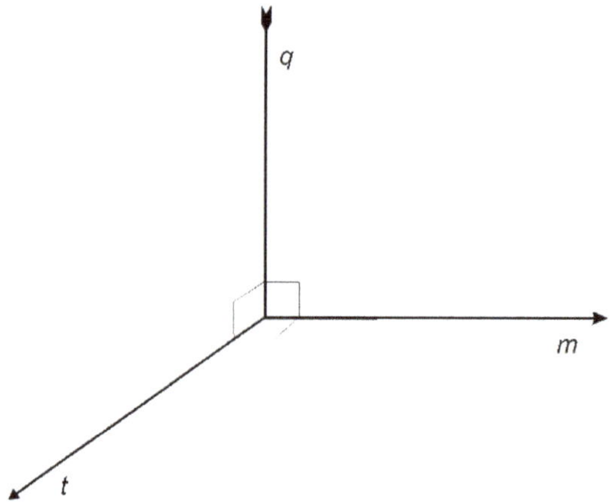

Fig. 3.8 Three-dimensional Quanto-Geometric coordinate system

While one may easily comprehend how integers, rational numbers as well as irrational numbers are modeled by a scaling entity such as a line, it may be less intuitive to visualize how that line may equally model or scale the set of complex numbers. The imaginary part of those numbers would presumably pose an obstacle to the scaling by the real axis of representation. One ought to remember the towering Quanto-Geometric principle that posits all scalar to be necessarily accompanied by a degree of void, no matter how minimal, as an existential partner. Consequently, the densest scalar cannot be devoid of a space component altogether. That is the reason why the real number line in Quanto-Geometry (in traditional mathematics as well I might add) is indeed a density referential. When it comes to the imaginary part of a complex number, it is no more than an expression of the void component of the number, which takes the embodiment of a geometric entity both in the rectangular form $(a + bj)$ of the complex number and in its exponential form $(ce^{j\theta})$.

Effectively in differential calculus, incidentally also called *infinitesimal* calculus, the instantaneous rate of change of a curvature is equal to the tangent of the angle subtended by the curvature, expressed as below:

$$\frac{dy}{dx} = \tan\theta\,,$$

so that: $$\theta = \tan^{-1}\left(\frac{dy}{dx}\right).$$

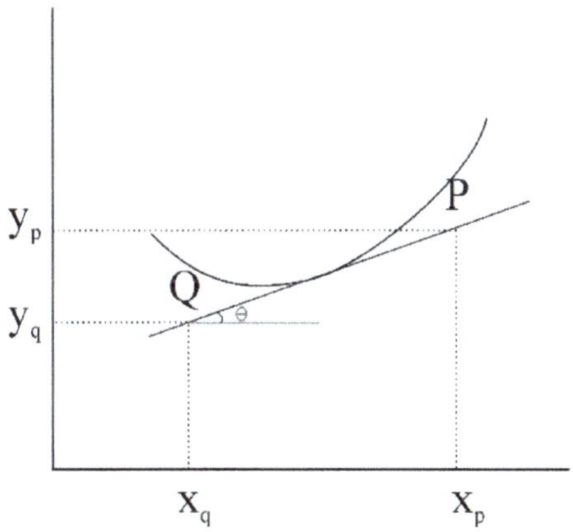

Fig. 3.9 Definition of a derivative at one point of a function in differential calculus

This θ angle is exactly the same as the one pertaining to the rectangular form of a complex number as well as the exponential form (Fig. 3.9):

$$\theta = \tan^{-1}\left(\frac{b}{a}\right)$$

and $ce^{j\theta}$.

As a reminder c relates to a and b by:

$$c = \sqrt{a^2 + b^2}$$

Consequently, it is important to keep in mind that the imaginary part of a complex number is equivalent to a form of the *infinitesimal*, and as such fully scalable on the traditional real number line and on the Quanto-Geometric q-axis. The above entails that the imaginary part of a complex number, if it must remain complex, cannot become 0 and understandably so. So does the affine increment ds in the ratio dq/ds, a condition that ultimately springs from the nature of the scalar itself. Although point 0 exists in the coordinate system on all axes, no function is allowed to take that value, both on the independent variable side and on the dependent variable side. It is a long-standing debate in traditional mathematics as well whether or not 0 belongs to the real number line.

Undoubtedly, the Quanto-Geometric coordinate system is significantly different from the Cartesian coordinate system. We have made it a total referential that is capable of scaling all scalar quantities in a very explicit way. All axes must be orthonormed and normal to each other. Should that condition not be raised on the referential, the Quanto-Geometric functions would lose all of their quantum and geometric properties. They would become meaningless and unable to reflect graphically the inner quantum and geometric qualities innate to their respective expressions.

3.10. Quanto-Geometric Symmetry Groups

We are now going to introduce one last but not least significant descriptor to the Quanto-Geometric model. Because the physical variable of space purports dimensionality, any space descriptor must assume this essential property of *multiplicity*. One other way of looking at multiplicity is *variability*.

For that matter, we have primarily set space as the independent variable of the function itself. However, we must further expand the expression of multiplicity embedded in the mathematical construct of *variable* by multiplying or surrogating the variable. Consequently, we must establish the Quanto-Geometric function as a function of two variables. The space variable *s* becomes as such a composite variable in order to incarnate the two basic sub-components of space: *form* and *mobility*. As we pointed out earlier, all forms are forms of symmetry. The end result, on the mathematical plane, is that the composite variable *s* may be cast into two separate variables, denoted *m* and *t*, the first standing for symmetry and the second for mobility. Our Quanto-Geometric function of one variable *Q(s)* then becomes *Q(t,m)*, a function of two variables. We may appropriately rewrite the function as follows:

$$Q(m,t) = \pm\left|\frac{(m+t)^0}{\sigma\cdot\sqrt{2\pi}} \bullet e^{-(m+t)^2 / 2\sigma^2}\right|, \ (m,t) \neq (0,0), (m = \pm|t|) and \ (-m = \pm|t|) \quad (3\text{-}15)$$

Although the same expression might be alternatively written in polar coordinates, we do not intend to perform any further development to the function in that direction. The variables *t* and *m* introduce two new dimensions to the Quanto-Geometric function, making it apt to be displayed in a three-dimensional referential with two independent variables substituting the one *s* variable. Albeit modeling each a different construct, these variables behave exactly alike. Shall we take a look at the graph of *Q(t,m)*?

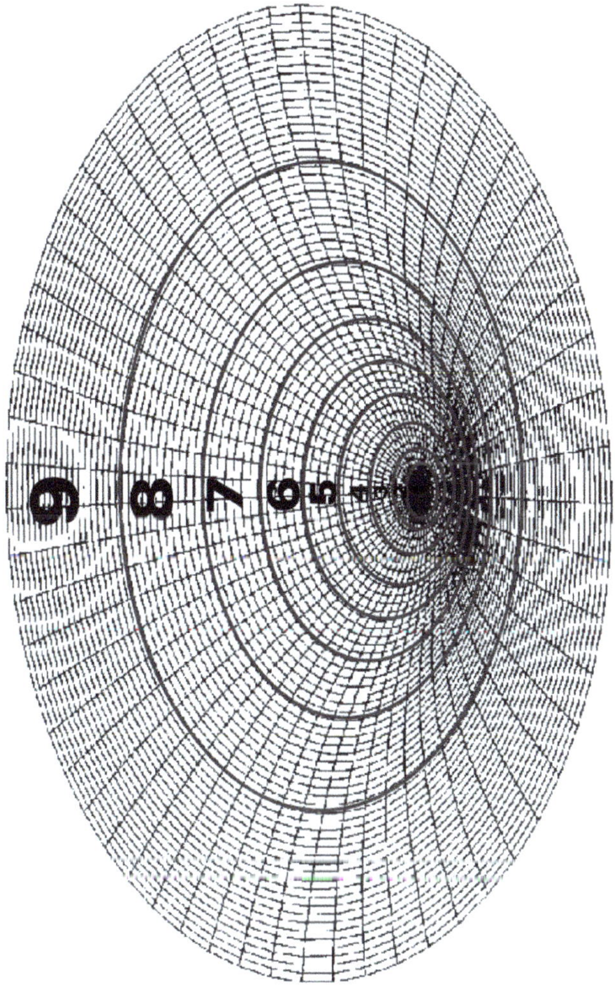

Fig. 3.10 The nine base shells of the Quanto-Geometric function of two variables.

We build the graph by holding constant one variable, t for example, assigning a value m_i to m, the other variable, and computing the value of $Q(t,m_i)$ (Fig. 3.10). When we do this for one constant value of t and successive values m_i of m, we obtain a Gaussian-like curve centered on that value of t, which is perpendicular to the t-axis and symmetric about it. By repeating the procedure with a span of t values, we obtain the three-dimensional graphical layout of the function.

The function is thus as symmetric about the *t*-axis as it is about the *m*-axis. This form of symmetry we generically call *transverse symmetry*. The elements in one quadrant will have common behavior among themselves, but each will have an opposite equivalent in the opposite quadrant, exactly like a 10° angle is transversally symmetric to a 170° angle. As one can see, these are descriptors derived from the trigonometry of the function. It could be the case that expressing the function in polar coordinates makes even more apparent the property of transverse symmetry. So instead of translating s as $s = (m,t)$ we might want to express it as $s = (r, \theta)$, where r is the magnitude of a desired radius and θ its angle with the s axis. So therefore, the point (r_1, θ) is transversally symmetric to the point $(r_1, \pi\text{-}\theta)$, about an s axis construed as identical to the trigonometric cosine axis. This is likely to be the technique used by the computer program that built the graph in Fig. 3.11. We see therein a series of concentric circles (different radii) around the q-axis and a series of top-down and bottom-up stripes (different angles) all across the circles.

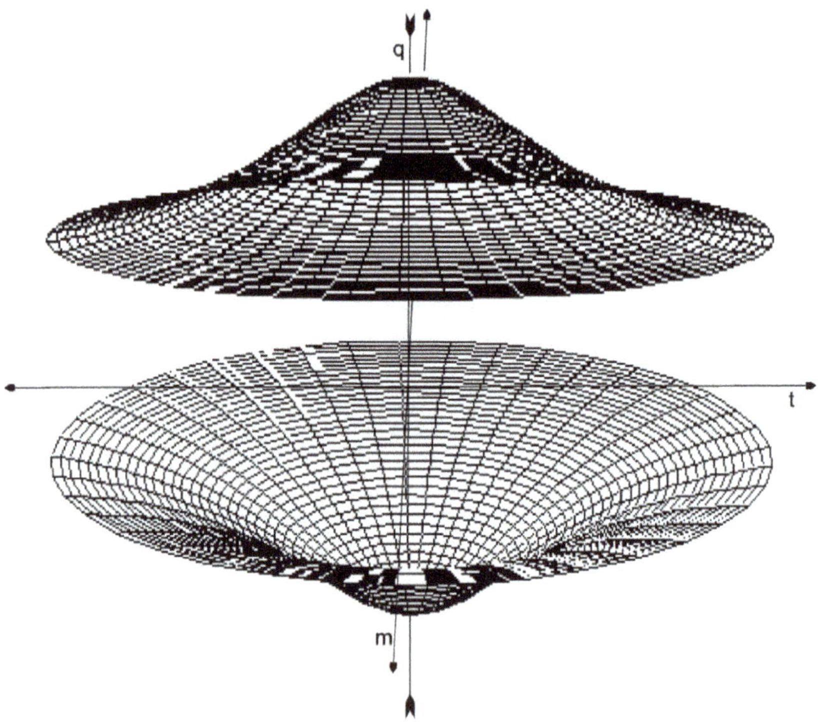

Fig. 3.11 A complete view of the graph of the Quanto-Geometric function of two variables.

If we look at the graph atop the vertical q-axis, we would see a series of 9 concentric stripes (washers) on an *m-t* plane (Fig. 3.10), every other stripe being larger than the previous as determined by the specific extent of the intervals of normalization. Hence, for every quarter of stripe the function exhibits a transversally symmetric quarter of stripe about one or the other horizontal axis. In Fig. 2.13 we see that symmetry of sorts displayed about the *m* axis.

The rendered graph in Fig. 2.13 is similar to an age-old hand fan. Transverse symmetry is the one-to-one similarity of the slices displayed about the *q*-axis and opposite to one another. This modeling artifact means to express that for one descriptor of a kind, the elements pertaining to opposite slices will behave in some kind of anti-par fashion.

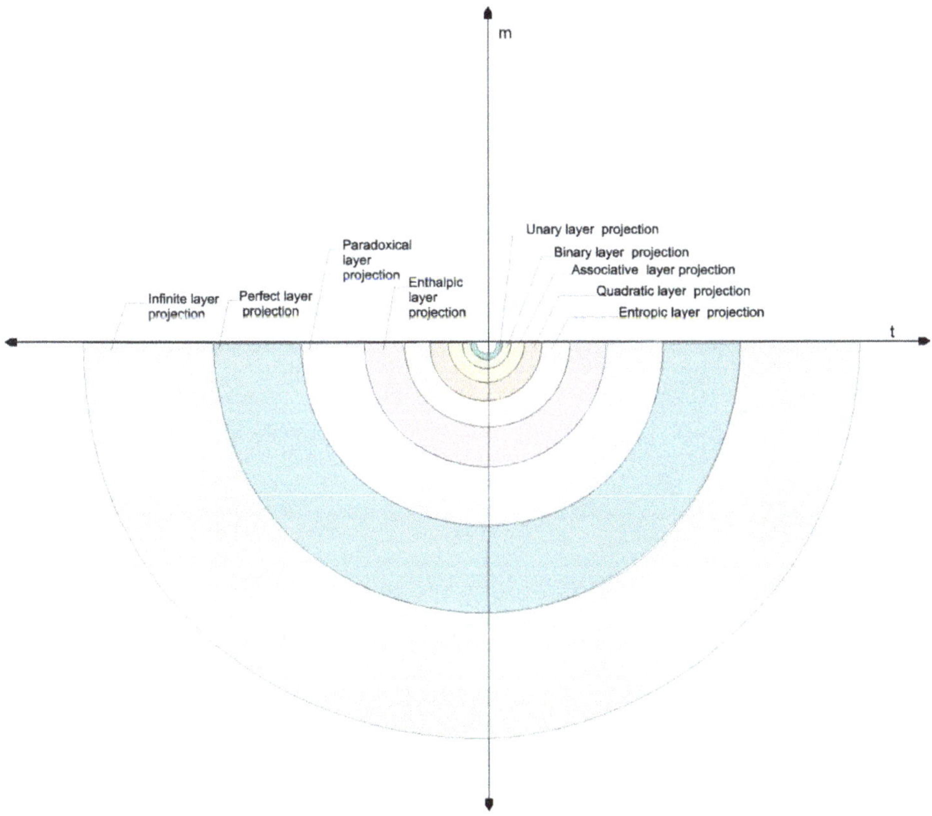

Fig. 3.12 Projection of the three-dimensional Quanto-Geometric graph on the m-t plane. Each washer represents the projection of one Quanto-Geometric layer.

The complete model showcases 4 domains of transverse symmetry as delineated in Fig. 3.14. The complete span of positive values of the function *(+q)* casts four equal quarters, as many as the span of negative values. There is *group symmetry* between the paired quarters such that all the elements of one quarter behave in the same way as opposed to the group of elements in the other matching quarter which behave in a different or opposite way, you might say in an *anti-symmetric* way. Inside of each matching quarter, we have elements in slices or threads that behave in a *transversally symmetric* fashion to one another. These are important descriptors that translate into clear and specific physical properties of matter as we shall see later.

By the same token, elements belonging to the same slice of a thread present among themselves a trait of commonality which we will call *thread symmetry* or *family symmetry*. Those elements are scattered vertically across the 9 normal Quanto-Geometric layers within the thread.

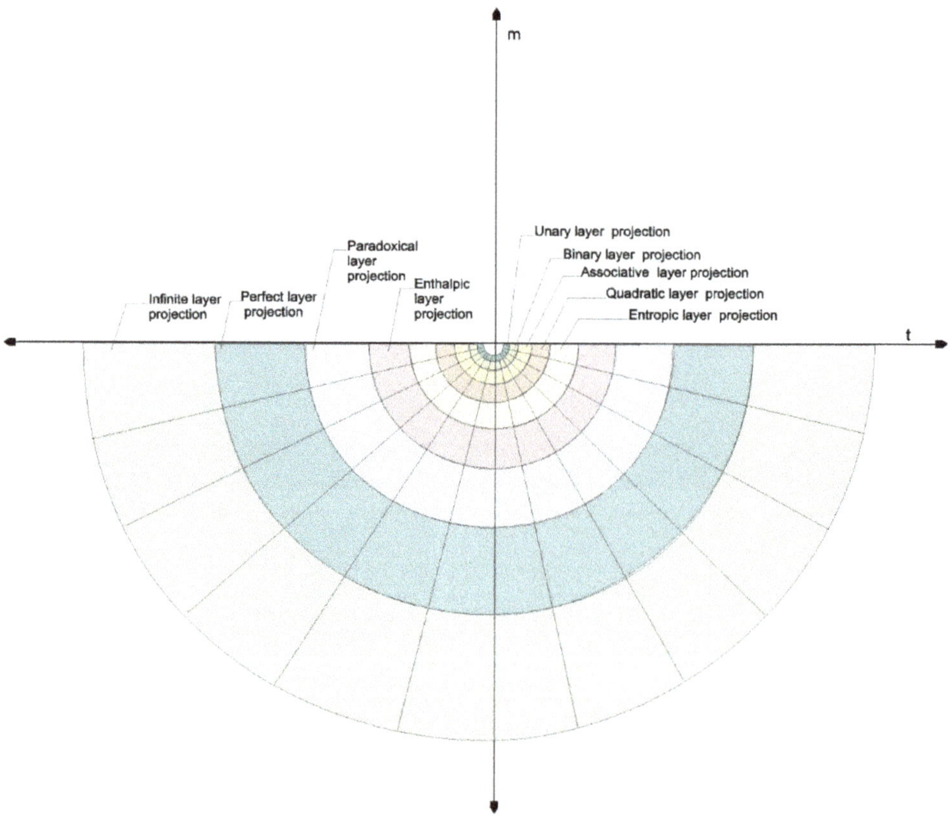

Fig. 3.13 Projection of the three-dimensional Quanto-Geometric graph on the m-t plane with the radii of transverse symmetry about the m axis. Each washer represents the projection of one Quanto-Geometric layer.

To recap, the symmetry groups inherent to the Quanto-Geometric model are:

1. anti-symmetry between pairing quarters (Fig. 3.12)

2. transverse symmetry between threads belonging to pairing quarters (Fig. 3.13)

3. family symmetry between elements belonging to the same thread

The meaning of the term *symmetry group* is therefore for us very different from that conventionally attributed to it thus far in mathematical physics, in terms of groups of objects experiencing a form of invariance in their physical properties under a particular spatial transformation, rotational or otherwise. In later Chapter we use a specific naming convention to refer to that which otherwise qualifies a usual transformation *Symmetry Group (SU)*.

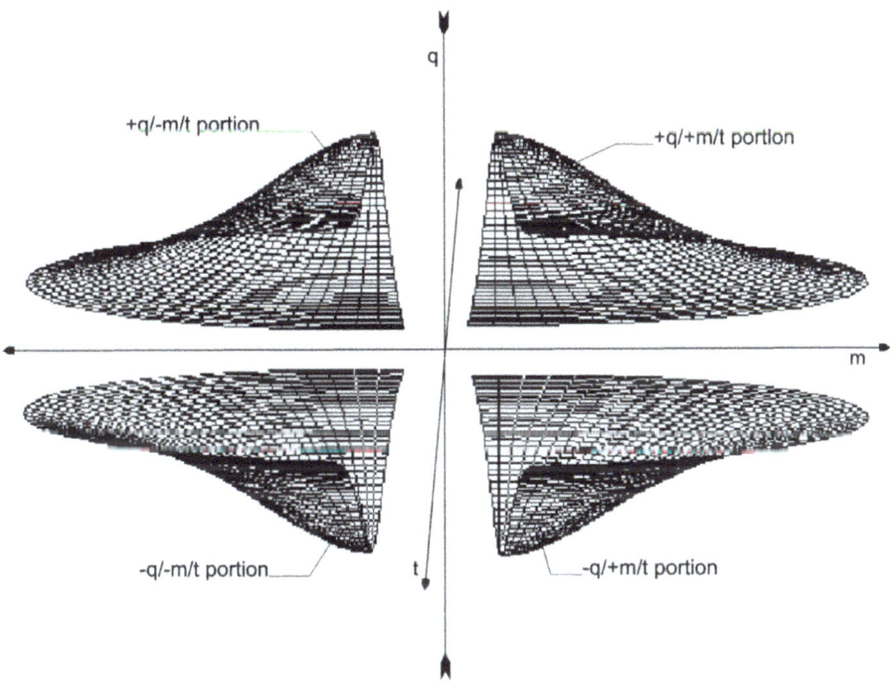

Fig. 3.14 A separate view of all 4 half portions of the graph of *Q(s)*. Each half portion contains two anti-symmetric quarters (symmetry with respect to the *m-axis*).

3.11. Conclusion

In this Chapter we have gone beyond the preliminary principles of Quanto-Geometry and established a formal mathematical formulation of the theory. We have conducted a detailed graphical analysis of the concavity of the Quanto-Geometric function and laid down its properties of symmetry as well as its primal degrees of quantization. We have also looked into one of the 9 Quanto-geometric primitives, the conservation operator, and the correlation with the Gaussian function studied throughout. Finally, we have re-examined the Quanto-Geometric referential and established the conditions it ought to meet in order to respond to the specific requisites of the Quanto-Geometric function.

Chapter 4

Numbers represent the highest degree
of the intelligible.
Plato

QUANTO-GEOMETRIC NUMBER THEORY

Our primary notion of numbers concerns their cardinality, a quality by which they are applicable to objects, real or virtual, as a way to garner their inventory. And that is what we ask of numbers for the most part. When we state that there are 5 chairs in a room, we are saying that there are chair A, chair B, chair C, chair D and chair E. However, numbers are not just numbers, that is to say quantum descriptors of quantity or amounts.

Every number represents a form of quantity that is not only exogenous to the number as an expression of count applicable to objects in our world, but this form of quantity is equally intrinsic to the number. That intrinsic state of quantity qualifies as a quantum abstraction or an inner state to the number which may be described as a *scalar state*. As such in the numeric domain, there is no continuum, contrarily to what traditional number theory dogma posits. This is true for the numeric domain as a whole (the cardinal domain), including the real, the rational, the irrational, the algebraic and the transcendental ensembles.

It follows from that view that numbers are typed as well. The reason why it is so is ultimately because numbers participate in a Quanto-Geometric spectrum. In all, numbers ought to be visualized as quantum abstractions with specific types in structure and endowed with behavior pattern. This is all to mean that numbers are Operators.

4.1. Number Bases

There are all kinds of number systems in use, from the binary system to the decimal system, and beyond, each established on a base containing a set number of units as building blocks. A number base is whatever you want to make of it in terms of the number of base units. We are going to utilize the decimal number base (0 to 9), which is familiar to most, in order to undertake our Quanto-Geometric analysis of numeric quantum abstractions.

From the Quanto-Geometric vantage, base numbers, decimal or not, do not exist one after another on a timeline. There are all cast at the same time in a Quanto-Geometric spectrum. By relating the base numbers to one another, we are able to study their identity, their attributes and their behavior, all of which make up their specific Quanto-Geometric signature.

Rules of divisibility are paramount in helping us shed light on the signature of these base numbers. If a number, pertaining to the numeric system but localized beyond the base units, is divisible by a base unit, this is an indication that the base unit or base number represents the building block of that larger number, then designated its multiple. In a sense, if we know the properties of the multiple, we will acquire information about the base number itself, because the properties of the multiple stand in part for a reflection of the intrinsic properties of the base number. That clearly stands to reason since the base number in that case represents the building block of the multiple. Another fruitful exercise is to analyze the results of divisibility of the base numbers among themselves, particularly from 1, in order to establish the primitives for these signatures.

4.2. Decimal Base Number 1

All numbers are divisible by 1. They divide by 1 to remain **integrally** themselves. Thus, any number divided by 1 returns whole out of the operation: 1 cannot fraction any number. Number 1 is a monolith whose realm is the Whole. Additionally, all numbers may be built from 1 by **integration** or aggregation. Those characteristics make number 1 a ubiquitous entity and a master key.

4.2.1 Quanto-Geometric Signature:

Identity: Totality – Integrality

Attributes: omnipresence - monolithic

Behavior: As a master key, creates overtures. Centralizes. Aggregates. Individualizes. Conserves. Purports the operation of addition. Begets the ensemble of prime numbers.

4.3. Decimal Base Number 2

The conventional rule states that a number is divisible by base number 2 if it is an even number. However, since the definition of an even number is that it is divisible by 2, the rule is not very much explanatory. An alternative rule is this one: any number ending in 2, 4, 6 and 8 (and 0 as well) is divisible by base number 2. If we restrict the analysis strictly to the base numbers, the statement quickly becomes completely moot as a definition. The following has more of a defining value than the previous: any number that is the sum of two identical numbers is divisible by base number 2. This is expressed as:

$$X = Y + Y \text{ or } X - Y = Y, \text{ in order for X to be even (where X is even)}$$

Hence Base element 2 is a primitive for parity.

4.3.1 Quanto-Geometric Signature:

Identity: Duality

Attributes: Equality – Equivalence

Behavior: polarizes, dichotomizes. Purports the operation of subtraction. Begets the ensemble of even composite numbers.

4.4. Decimal Base Number 3

A number is stated to be divisible by 3 if the sum of its digits is divisible by 3. Once again while this statement can be an expedient rule for dealing with rather small numbers, it bears no merits for a definition since it is calling for a vicious circle of re-divisibility by 3. A better definition is this: a number is divisible by 3 if the ultimate sum or cross sum of its member digits is 3, twice 3 or thrice 3. The ultimate sum means the most non-reducible sum which is obtained after a cascade of summation of the containing digits by resulting numbers if necessary. As a consequence, divisibility by 3 requires a relationship between the constituent digits of the number. This relationship may be any of 3 different ones and they all involve a direct association with 3. Another way of expressing this condition is that they must all have *one thing in common* as per expression below:

$$ABC = (A+B+C)/3, \quad \text{where ABC is a multiple of 3.}$$

Base element 3 is a keen proponent of association.

4.4.1 Quanto-Geometric Signature:

Identity: Association

Attributes: Relational

Behavior: Purports the operation of multiplication. Begets the realm of odd composite numbers.

4.5. Decimal Base Number 4

The conventional definition states that a number is divisible by 4 if the last two digits are divisible by 4. Again, a statement with little merits for a definition. A better definition yet: a number is divisible by 4 if it can be written as 2^n where $n>1$. This is a statement of how many equal times the abstract quantum can be fractioned, with the requirement that this number of times be paired. In other words, element 4 **doubles** down on divisibility by requiring a multiplication of the pairing. That is how it introduces the operation of taking the *root* or its inverse, which is raising to a *power*. The number must be of the form:

$$A \times A \times A \times A \times A$$

Base element 4 thus is a proponent of concatenation.

4.5.1 Quanto-Geometric Signature:

Identity: Ambivalence

Attributes: Transitory – Cross-Linking

Behavior: Concatenates. Transitions. Purports the operation of raising to *power* and taking the *root*.

4.6. Decimal Base Number 5

Number 5 poses a rather strange condition of divisibility. Classic number theory has it that a number is divisible by 5 if the last digit of the number is either 0 or 5 itself. That rule has inarguably the perfect value of a definition. The requirement for 0 to be part of the expression of the number is a requirement for the abstract quantum of type 5 to be accompanied by or made up of *nothingness*. It is an invocation of nothingness in the midst of scalar quantity. Furthermore, the requirement that 5 itself be part of the number as an image of itself is an invocation of virtualization. By all accounts the abstract quantum of type 5 is a proponent of the imaginary.

4.6.1 Quanto-Geometric Signature:

Identity: Newness

Attributes: Virtual, metaphoric, metamorphic

Behavior: Transforms. Purports the operation of *complexification* as construed in complex numbers, that is to say adding the imaginary to.

4.7. Decimal Base Number 6

As the rule has it, for a number to be divisible by 6 it has to meet both tests of divisibility by 2 and 3. In other words the abstract quantum of type 6 is calling for both duality and association. Translate *association* by *pair*. This is a requirement for coupling or matching. So therefore, the abstract quantum of type 6 is the precursor of all coupling arithmetic, namely matrixial mathematics.

4.7.1 Quanto-Geometric Signature:

Identity: Couple

Attributes: Matchmaker. Orderly.

Behavior: Harmonizes. Purports the operation of *matrification.*

4.8. Decimal Base Number 7

Decimal number 7 is a case of oddities all around. The test for divisibility by 7 rules that all component digits of the testing number be added together, to then test that sum for divisibility by 7, and repeat the whole process until the smallest 2-digit multiple of 7 is possibly reached. Once again, this procedure has the value of a test and stands not for a definition. Examining divisibility of base number 1 by 7 gives us a better chance to understand the characteristics of the inherent abstract quantum that makes for its building block. One divided by 7 yields 0.142857. Taking the decimal part of this fractional number two digits at a time, we see 14, 28 and 57. Fourteen is 2 x 7, 28 is 4 x 7. So, we see 7 as part of the inner constitution of the number, a paradox that expresses the number's refusal to be fractioned. From that trend, the 3rd set would expectedly be 56, a multiple of 7. However, it is 57 instead, in an attempt to showcase very explicitly this time number 7 itself, so portraying the number's determined refusal to be fractioned. In all, all base numbers divided by 7 return the same rotating decimal 142857 as reported below.

1/7 = 0.142857142857142857142857142857714
2/7 = 0.285714285714285714285714285714 29
3/7 = 0.428571428571428571428571428571 43
4/7 = 0.571428571428571428571428571428 57
5/7 = 0.714285714285714285714285714285 71
6/7 = 0.857142857142857142857142857142 86
7/7 = 1
8/7 = 1.1428571428571428571428571428571
9/7 = 1.2857142857142857142857142857143

Note that ends of above expressions are rounded off. Many more odd adventures are well known to this number set (142857) in traditional number theory and I could add to the record a few more so far unknown or unpublished. However, I will spare this exercise to the reader at this time in the interest of keeping this exposition amenable to all. I will however relate this one, which from the standpoint of our analysis is testimony to the refusal by the number to *construct* or produce straight multiples. If you multiply 142857, which personifies the abstract quantum of type 7, by the units of the number base, the operation returns the same 142857 set in a revolving fashion, except for one case.

142857 x 1 = 142857
142857 x 2 = 285714
142857 x 3 = 428571
142857 x 4 = 571428
142857 x 5 = 714285
142857 x 6 = 857142
142857 x 7 = 999999
142857 x 8 = 1142856
142857 x 9 = 1285713

That result means that quantum abstraction of type 7 cannot *construct* since it fully shows itself back as a result of the construction operation. Thus, it is only apt at *de-construction*. In the case of multiplication by 9, you can distinguish among the digital group the sets 28 – 57 – 13 (1), 13 being located at the trailing end and 1 at the beginning. 13 (1) or (13+1) is definitely an attempt to produce the usual 14 which the arithmetic here forbids. In the case of multiplication by 7, because you are already making the intent of multiplying by 7, which is to *de-construct*, the operation is already satisfied. This is arguably the reason why it produces the strange result of 999999 which nevertheless lacks multiplicity as a construct.

4.8.1 Quanto-Geometric Signature:

Identity: Paradox, Chaos

Attributes: Odd – Chaotic

Behavior: Operator of Destruction, Antagonism, Conflicts.

4.9. Decimal Base Number 8

Let us directly start by analyzing the results of divisibility of all the base numbers by 8, the conventional rule of divisibility by 8 not helping much our analytical purposes (rule has it that last 3 digits of number must be divisible by 8).

1/8 = 0.125 *(4-1)*
2/8 = 0.25 *(4-2)*
3/8 = 0.375 *(4-3)*
4/8 = 0.5 *(4-4)*
5/8 = 0.625 *(4-5)*
6/8 = 0.75 *(4-6)*
7/8 = 0.875 *(4-7)*
8/8 = 1 *(4-8)*
9/8 = 1.125 *(4-9)*

We may notice at first glance in those results that the digital set 125 is either visibly present in those numbers or portrays itself in a hidden fashion. It is clearly present in *(4-1)* and *(4-9)*. We can see it in *(4-2)* although the number 1 is absent there. We can still see it in *(4-4)* although 1 and 2 are absent. In *(4-5)*, with the number 0.625, it is indirectly present if we realize that 6 x 2 = 12. It is also present in 0.75 *(4-6)* if we realize that 7 + 5 = 12. It is equally present in *(4-7)* with the number 0.875, although deeply hidden here: let's realize that 8 + 7 is 15, which would turn it into 0.155 whereby it could be read as **1** and **2** times **5**. Shall we also say that the figurate 12 is also there under the disguise of 75, which is 7 + 5. We may at least conclude from the above that base number 8 shows a keen awareness for the digital set 125 as if they could be its building block in some fashion. The minimal conclusion at this stage is that 8 is perfectly divisible by the other 8 base numbers in terminating decimals involving the digital set 125.

It is noteworthy that these 3 integers, 1, 2 and 5 are the ones involved in the golden ratio. We remind the reader that the golden ratio is given by:

$$\frac{1}{x} = \frac{x}{1-x}$$

which when developed casts the polynomial: $x^2 + x - 1 = 0$ *(4-10)*

The golden ratio is a perfectly normalized ratio. It has come to embody perfect proportioning or apportionment in applied sciences. The solution to this 2nd degree polynomial is given by the quadratic as:

$$x = \frac{-1 \pm \sqrt{5}}{2}$$

which yields the numbers Φ (phi) and Ψ (psi) where we note once more the digital set 1, 2 and 5 . There is equivalence between the standard golden ratio expression and the known condition of finding 2 numbers whose product and difference are both equal to 1. That condition stands for an *irrationalization* in the real realm. It is expressed as follows:

$$m - n = 1 \quad (4\text{-}11)$$

$$m \times n = 1 \quad (4\text{-}12)$$

From *(4-11)* $m = 1 + n$. If we replace *m* by this value in *(4-12)*, we obtain:

$$n^2 + n = 1$$

and $n^2 + n - 1 = 0$ *(4-13)*

This polynomial is equivalent to *(3-10)*. Consequently, both polynomials pose the necessity of two integers engaged in an algebraic relationship. We are going to show that that algebraic relationship is intimately linked to number 8. The condition for resolution of this relationship (the polynomial) is given by the quadratic root. Let's do the following manipulations to the quadratic.

$$x = \frac{-b \pm \sqrt{b^2 - 4ac}}{2a}$$

$$x = \frac{-b \pm \sqrt{b^2 - 4ac}}{2a} \times \frac{4}{4}$$

$$x = \frac{-4b \pm 4\sqrt{b^2 - 4ac}}{8a}$$

$$x = \frac{-4b \pm \sqrt{16(b^2 - 4ac)}}{8a}$$

$$x = \frac{-4b \pm \sqrt{16b^2 - 64ac}}{8a}$$

$$x = \frac{-4b \pm \sqrt{8 \times 2b^2 - 8a \times 8c}}{8a}$$

$$x = \frac{-4b \pm \sqrt{8 \times 2b^2 - 8a \times 8c}}{8a}$$

$$x = \frac{-4b \pm \sqrt{8[2b^2] - [8a][8c]}}{8a}$$

If b is even, that is $b = 2d$, then

$$x = \frac{-[8d] \pm \sqrt{[8][8d^2] - [8a][8c]}}{8a}$$

Therefore, we went from the quest of the nature of the relationship between integers 1, 2 and 5, a number set derived from divisibility by 8, to the quadratic root which incarnates a stunning multiplicity of integer 8. In case you did not notice: 1 + 2 + 5 = 8 ! In conclusion, abstract quantum of type 8 is a promoter of perfectibility and begets the ensemble of algebraic numbers.

4.9.1 Quanto-Geometric Signature:

Identity: Per-fection

Attribute: Irrational

Behavior: Operator of or endows: Completion, Fulfillment.

4.10. Decimal Base Number 9

The rule of divisibility states that a number is divisible by 9 if the sum of its digits is divisible by 9. Since this rule will not help us delve into the ultra-structure of this number, let us start by examining the results of divisibility of the base numbers by 9.

1/9 = 0.1111111111111111111111111111111111111111111111111111111
2/9 = 0.2222222222222222222222222222222222222222222222222222222
3/9 = 0.3333333333333333333333333333333333333333333333333333333
4/9 = 0.4444444444444444444444444444444444444444444444444444444
5/9 = 0.5555555555555555555555555555555555555555555555555555555
6/9 = 0.6666666666666666666666666666666666666666666666666666666
7/9 = 0.7777777777777777777777777777777777777777777777777777777
8/9 = 0.8888888888888888888888888888888888888888888888888888888
9/9 = 000000000000000000000000000000000000000000000000000000001

All base numbers divided by 9 return an endless suite of digits of the same dividing base number, demonstrating an affinity for serialization. By the same token, it so demonstrates the greatest level of linearity among all base numbers, since the series are made up of the same numbers. One may think of linearity as the inverse of multiplicity, here at its minimal expression.

The serial sum of all base numbers returns an image of 9 as shown below.

$$1 + 2 + 3 + 4 + 5 + 6 + 7 + 8 + 9 = 45$$

which yields a multiple of 9 and the cross sum 4 + 5 = 9. Likewise, if we multiply all base numbers together in a string, we obtain an image of 9.

$$1 \times 2 \times 3 \times 4 \times 5 \times 6 \times 7 \times 8 \times 9 = 362880.$$

The image resides in the fact that the cross sum:

$$3 + 6 + 2 + 8 + 8 + 0 = 27$$

which yields the other cross sum 2 + 7 = 9. In addition, the number 362880 is a multiple of 9. Now of course the digital series 1 x 2 x 3 x 4 x 5 x 6 x 7 x 8 x 9 is **9!** (9 factorial). Let us remind the reader that a factorial is defined as a downward serial product starting with the designated number.

Hence, we shall define divisibility by 9 as the ability for any number to meet the following test: may be written as: **n!** for all n ≥ 6. Hence factorialization, a form of serialization, becomes an important expression of abstract quantum of type 9.

Let us now consider the following series, known as a series of the power type, that is to say a power series:

$$\sum_{n=0}^{\infty} \frac{x^n}{n!}$$

This is the expression of elevating a number x to a power n, divide that power number x^n by the power factorial $n!$, and finally taking the serial sum of all the terms starting with power n equals 0. Clearly, even though the number that we are acting on is a power, we are primarily serializing that number, because firstly we are adding a suite of terms in sequence with one another, and, second, we are additionally introducing another layer of serialization as well with the division of the number by a sequential factorial. The developed expression of the series commonly reads like the following:

$$\sum_{n=0}^{\infty} \frac{x^n}{n!} = 1 + x + \frac{x^2}{2!} + \frac{x^3}{3!} + \ldots$$

We are going to rewrite the right side of the expression to make sure it is understood as a thorough suite of progressing terms.

$$\sum_{n=0}^{\infty} \frac{x^n}{n!} = \frac{x^0}{0!} + \frac{x^1}{1!} + \frac{x^2}{2!} + \frac{x^3}{3!} + \ldots$$

Any number raised to 0 power is equal to 1 ($x^0 = 1$) and 0! is defined equal to 1 by convention. We end up with 1/1 which is equal to 1 for the first term of the series. As to the second term, since 1! is straightforwardly equal to 1, the second term is clearly equal to x. This series is an infinite sum, we can push it to the largest n integer that we are possibly able to conceive. The numerators of our fractions are serializing as much as the denominators are. The denominators are furthermore factorialized. It is not so much about the powers at the numerators but the fact that they are serializing. Now, this infinite series, sum of an infinite suite of serialized terms, converges to one number, which is *transcendental e*. Effectively, this series is equal to e^x as stated below, no matter how far we push n.

$$\sum_{n=0}^{\infty} \frac{x^n}{n!} = e^x$$

Hence, the operation of serialization bears the essential potential of creating transcendental numbers. Those are the inner workings of abstract quantum of type 9.

4.10.1 Quanto-Geometric Signature:

Identity: Endlessness

Attribute: Serial. Transcendental.

Behavior: Operator of Infinition or Asymptotization.

4.11. Quintessential Geometric Order

Akin to the scalar order that we have just studied, there exists a space or vacuum state order as well. Although the nature of physical space is distinct from mind, absolute pure space, which does not physically exist, can be perfectly assimilated to the abstraction of nothingness. One may wonder how can an order of things in state of nothingness even be formulated? Truth is that the ultra-structure of state of nothingness can indeed become tangible thru quintessential geometry or the order of forms, just like scalar order may be elicited by abstract quanta representing its building blocks per previous discussion.

A simple thought exercise may help us initiate that analysis. We choose the Euclidean plane to develop our abstraction. What is the simplest thing we can do on that plane to start out? As a minimum we can draw a line segment. Not a rectangle (defined by two line segments) as contented in traditional formalism. Not even a dot, because a dot is a concentric element, hence a scalar entity. If you insistently contend that it must be a dot, then conceded, but it must be two dots, the reason being that our Euclidean plane only recognizes linear elements, and two dots are very much a condition for a line. Consequently, we initialize our abstraction with an A line as our element 1. This line is Unique, Total and Autonomous.

As long as we remain on that line, no further geometrization is possible since the line is limitless on both ends. We must step aside from the line in order to strike the next geometric degree. The least subsequent operation is to draw a second line. It can either be parallel to the first or cross it thru. If a parallel line, we may liken this operation to a symmetric movement which amounts to a repetition. In that event there is no new operation. Therefore, our second line must necessarily cross the first. No matter the orientation we give the second line, it will always be interchangeable with the first one. Consequently, our B line materializes the abstraction of Similitude.

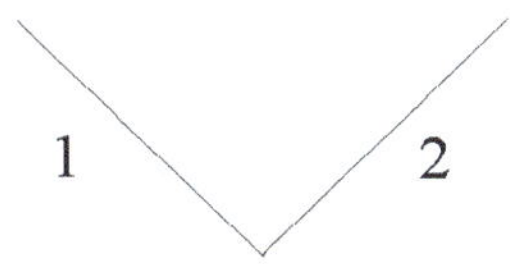

B

Next step in our movement leads us to leave the second line in order to draw a third one. It will mandatorily cross the first two, either in an irregular fashion, or in a regular fashion by forming an isosceles or even equilateral triangle. Whatever the case, the 3rd line establishes a reunion of the first two, therefore positing the axiom of Association or Reunion.

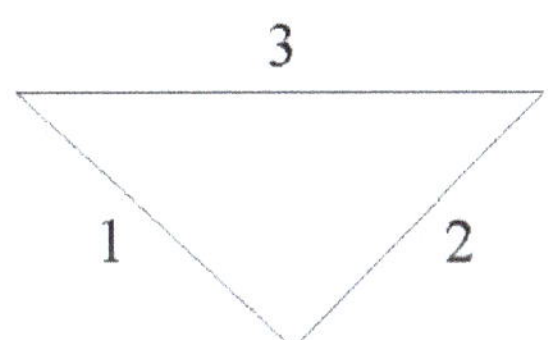

C

In continuing this abstract exercise, let us suppose that the triangle resulting from the above movement is regular, isosceles for instance. The triangle figure, whether regular or not, represents a closed space eliciting completion. Hence, the next geometric step must occur externally to the triangle.

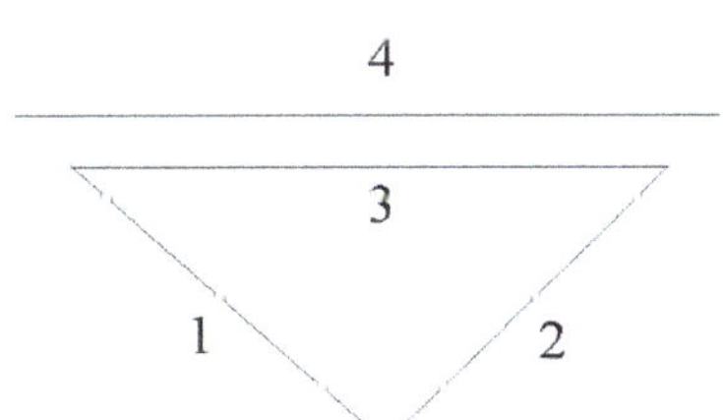

D

Our next line must be drawn with no interaction with the triangle. This obligated rupture and absence of common measure with the fore places the fourth line in the isolation of an end. Thus, it must be drawn in parallel with the third line. However, this end can only be Transitory since we may continue with the movement.

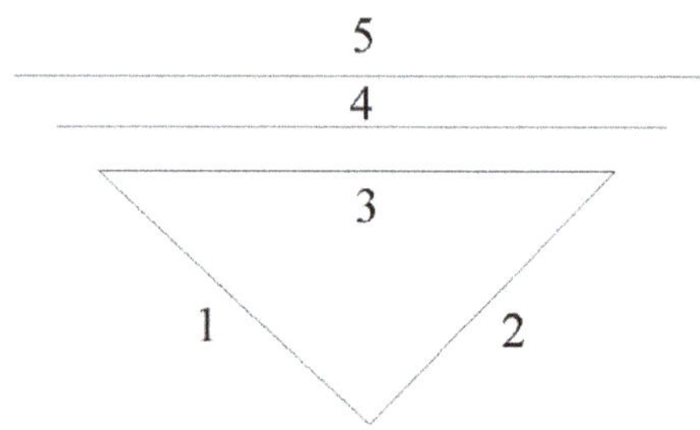

E

The rupture and isolation posited by the fourth geometric unit lends to the next one an origin and differentiation value from which we may reinitiate our geometric movement. Therefore, the fifth must be drawn beyond the fourth and must remain parallel to it in order to avoid a violation of the Rupture value incarnated by the fourth. Being a representative of origin and differentiation, the fifth ought to embody Novelty, Change or Transformation.

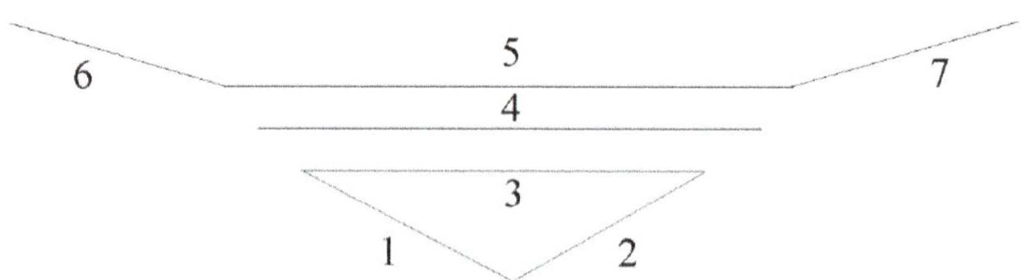

F

What are the criteria that may allow us to continue with our abstract exercise? We know two lines were necessary in order to make up the prime angle (angle formed by A and B). We may now emulate the primordial geometry by drawing two lines in an angular fashion from the fifth, as though they originated from the virtual continuation of side 1 and 2 of the prime triangle. In doing so, we remain in linkage with the primordial geometry while striking a new (post-geometric) operation. Yet we must obligatorily introduce a distinction of sorts to this new geometric step: we will have for such the sixth and the seventh drawn per a divergent angle from the primordial angle as shown in Fig F.

The sixth and the seventh lines gather these notions and become the expression of Concordance and Divergence, the first characterization being applicable to the sixth geometric unit and the second to the seventh unit.

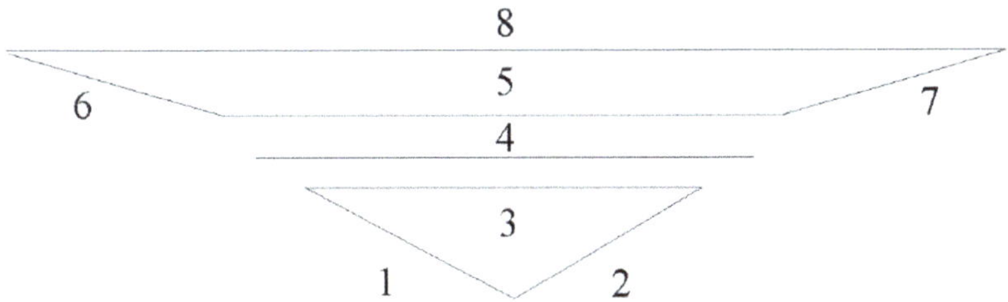

G

We may continue the exercise by uniting the sixth and the seventh with an eighth line. The eighth so confers to the post-geometric movement its utmost completion. We may say that the eighth affords its highest degree of wholeness to the post-geometric movement by configuring a closed trapezoid. It grants the post-geometric movement with its highest level of concretion and becomes the representative of Completion and Per-fection (see Fig G).

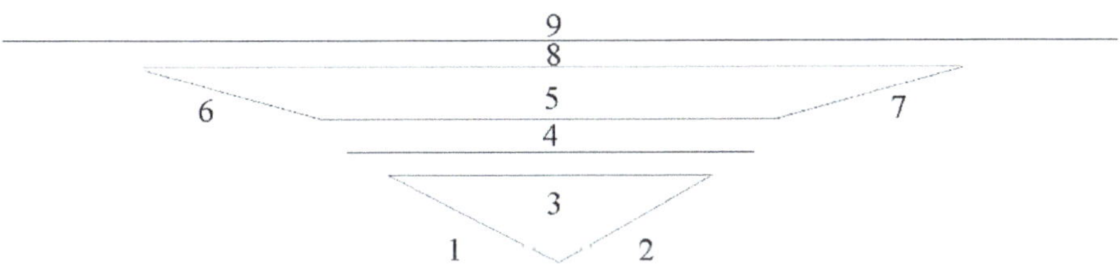

H

Being complete in and of itself, the eighth permits no more crosses. Its wholeness forces the ninth to remain independent by parallelism, in a movement that is akin to the fourth. However, after the most Perfect Completion, only an absolute ending may occur, that is to say an end from which nothing else can originate, lest we repeat all previous operations. Consequently, the ninth rejects all crosses as well as stops. It extends in one or the other direction indefinitely and unconditionally. That is how the paradox of endless end comes about: the Asymptotic End.

This abstraction which we conducted on the basis of a Euclidian plane may be realized as well on a tri-dimensional reference system with the exact results and conclusions. There undoubtedly exists an ultra-structure within the state of nothingness based on 9 geometric primitives in the terms that we have exposed above. The ensuing geometry elicits in forms these inner states of the order of nothingness. One can further see that these inner states are tightly matched with the abstract quanta of the scalar order. More on that correspondence later.

4.12. States of Symmetry in Nothingness

The inner states of the order of nothingness are states of symmetry. It is important for us to recognize that forms of one kind or another always amount to degrees of symmetry. A geometric form is a *form* of *symmetry*. The reason why things get to hold a form is because of the existence or primacy of these abstract states of symmetry which make up the order of nothingness. Just like every object automatically has a count which springs from the scalar order, every object automatically has a form that is rooted on the inner states of symmetry pertaining to the order of nothingness germane to the object.

It is by articulation with quanta of the scalar order that the states of symmetry pertaining to the order of nothingness come to tangibility under the wrapper of forms (Fig. 4.1). Geometry is a materialization of these forms in the physical world, it is a *formal* representation. Later on we will further discuss the terms of this Quanto-Geometric articulation.

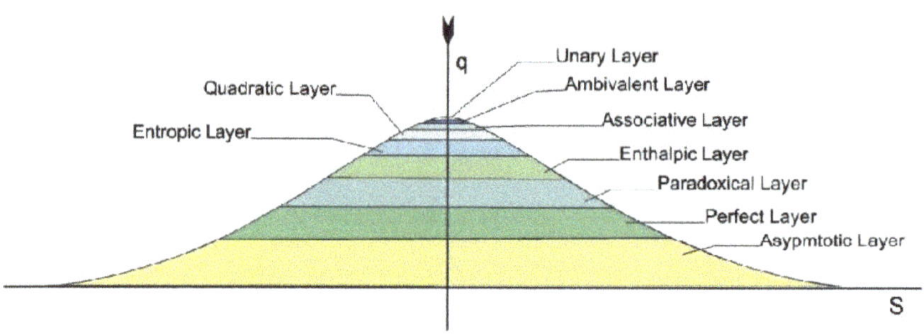

Fig. 4.1 The nine primal Quanto-Geometric types represented by the intrinsic layers of the *Q(s)* function

Notwithstanding their abstract nature, it is not possible to completely separate from one another the two primal orders, the scalar order and the order of nothingness. Therefore, geometric formation appears within the ultra-structure of numbers, as much as scalar behavior is perceptible within the order of nothingness per our former exercise of geometrization. We can *count*, I repeat *count*, the geometric states and this is not a trivial statement. This means that they have a *cardinal* reality which is mapped to the scalar units of the scalar order in a one-to-one relationship. You may as well *ordinalize* the geometric states, that is to say establish a progressive rank among them. Traditional Number Theory professes that cardinalization and ordinalization are the two foremost qualities of the number domain.

The most important consequence from the above proposition to bear in mind is that the qualities that we have derived for each state of symmetry match the qualities that we have derived for scalar units from the scalar order. This is because there is a tight Quanto-Geometric articulation between the two orders, whereby they both influence each other. It is the influence exerted on the order of symmetry by the scalar order which produces its cardinalization and ordinalization. Conversely the number domain presents geometric properties as a result of influence exerted by the order of nothingness or symmetry. Shall we take a look at these geometric properties elicited by the scalar order.

4.13. Geometric Properties of the Scalar Order

It is well-known that real numbers show a handful of geometric properties. For instance, real numbers characterized as *palindromic* are numbers that read the same forward or backward. The number 12721 i.e. is palindromic. We can easily see that a palindromic number with axial symmetry must have an odd number of digits so that the mid integer may stand as its center of symmetry. If the number is turned around about that mid digit as its center of symmetry, it reads exactly the same. We can also invoke *figurate* numbers, which are numbers that are structured according to a geometric pattern. For instance, pentagonal numbers, sexagonal numbers, heptagonal numbers and octogonal numbers are all numeric entities well described in the literature (Fig. 4.2).

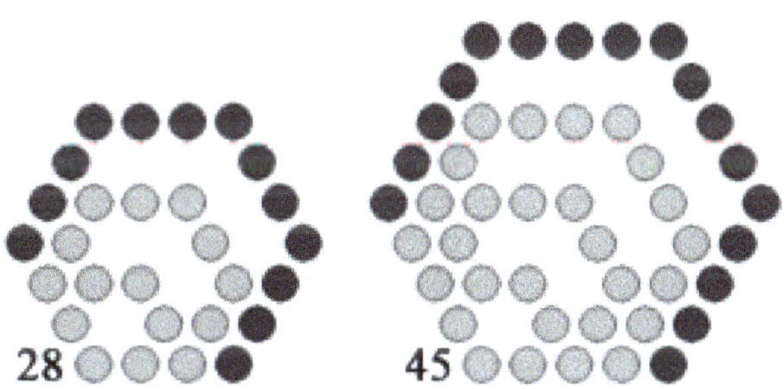

Fig. 4.2 Polygonal Numbers

Our point of interest herein however is to show the geometric properties evidenced by the intrinsic scalar order itself. We are going to study the nature and distribution of numbers in the real domain in order to develop this thesis. The dogmatic approach followed in the classroom and in text in presenting mathematics in general tends to make us lose sight that this science is all about the nature and distribution of numbers.

0	1	2	3	4	5	6	7	8	9	10
11	12	13	14	15	16	17	18	19	20	21
22	23	24	25	26	27	28	29	30	31	32
33	34	35	36	37	38	39	40	41	42	43
44	45	46	47	48	49	50	51	52	53	54
55	56	57	58	59	60	61	62	63	64	65
66	67	68	69	70	71	72	73	74	75	76
77	78	79	80	81	82	83	84	85	86	87
88	89	90	91	92	93	94	95	96	97	98
99	100	101	102	103	104	105	106	107	108	109
110	111	112	113	114	115	116	117	118	119	120
121	122	123	124	125	126	127	128	129	130	131
132	133	134	135	136	137	138	139	140	141	142
143	144	145	146	147	148	149	150	151	152	153
154	155	156	157	158	159	160	161	162	163	164
165	166	167	168	169	170	171	172	173	174	175
176	177	178	179	180	181	182	183	184	185	186
187	188	189	190	191	192	193	194	195	196	197
198	199	200	201	202	203	204	205	206	207	208
209	210	211	212	213	214	215	216	217	218	219
220	221	222	223	224	225	226	227	228	229	230
231	232	233	234	235	236	237	238	239	240	241
242	243	244	245	246	247	248	249	250	251	252
253	254	255	256	257	258	259	260	261	262	263
264	265	266	267	268	269	270	271	272	273	274
275	276	277	278	279	280	281	282	283	284	285
286	287	288	289	290	291	292	293	294	295	296
297	298	299	300	301	302	303	304	305	306	307
308	309	310	311	312	313	314	315	316	317	318
319	320	321	322	323	324	325	326	327	328	329
330	331	332	333	334	335	336	337	338	339	340
341	342	343	344	345	346	347	348	349	350	351
352	353	354	355	356	357	358	359	360	361	362
363	364	365	366	367	368	369	370	371	372	373
374	375	376	377	378	379	380	381	382	383	384
385	386	387	388	389	390	391	392	393	394	395
396	397	398	399	400	401	402	403	404	405	406
407	408	409	410	411	412	413	414	415	416	417
418	419	420	421	422	423	424	425	426	427	428
429	430	431	432	433	434	435	436	437	438	439
440	441	442	443	444	445	446	447	448	449	450
451	452	453	454	455	456	457	458	459	460	461
462	463	464	465	466	467	468	469	470	471	472
473	474	475	476	477	478	479	480	481	482	483
484	485	486	487	488	489	490	491	492	493	494
495	496	497	498	499	500	501	502	503	504	505

Fig. 4.3 Status of the first 505 positive integers on the real number line: 96 primes, 155 odd composites and 254 even composites

It is almost always about finding a number or set of numbers which satisfy a certain condition, that condition being that which is expressed in a function, a matrix, a series or other manners of mathematical expressions presented upfront.

Two large classes of numbers are known to make up the real number domain: integers and decimals. These two classes are distributed in layers which we shall review shortly.

1.- At the top of this hierarchy are the *Prime* numbers. Primes are numbers that can only be divided by 1 and themselves. They are indivisible monoliths. Although there are an infinite number of primes, there are less primes than any other types of numbers in the whole number realm. The table in Fig. 4.3 above shows the primes among the first 505 real numbers (those in yellow or light gray cells) of the number line. There are only 96 primes among the first 505 real numbers.

2.- The next layer is that of composite numbers, specifically odd composite numbers. Hence at layer 2, we are first of all dealing with *composition*, that is to say elements made up by two or more elements. *Composite* numbers are multiples of other numbers. Secondly, one may wonder why we are restricting this class to odd numbers. So we do because we are executing the inverse of a densification function, we are breaking down the monoliths from layer 1. All composite numbers, including odd composite numbers, are the result of multiplying primes together.

Because composite numbers result from combination of primes, there will be more composite numbers than the monolithic primes. There will even be more odd composite numbers than primes.

3.- Layer 3 represents the class of even composite numbers, which are numbers divisible by 2. This class will have more numbers than the previous, the set of odd composite, for 3 main reasons:

1. Element 2 of the base despite being a prime is an even number (the only even prime); as a multiplicand (factor) it will make every number it generates an even number.

2. Element 0 or 10 of the base is an even number and as a multiplicand it will make every number it creates an even number.

3. While every even number multiplied by an even number creates another even number and every odd number multiplied by an even number creates an even number, every odd number multiplied by an odd number creates another odd number however.

Therefore, the class of even composite will be a lot more prolific than the class of odd composite.

4.- Layer 4 showcases root or power numbers. These are a sort of special composite numbers which are the result of identical multiplicands. For instance, 5 x 5 = 25 = 5^2, so 25 is a power number of which 5 is the root. Another example: 6 x 6 x 6 = 216 = 6^3, so therefore 216 is a power number of which 6 is the root.

Is this class more prolific than the previous? It sure is and the reason is simple: since any integer, not only the base elements (2 to 9), may be multiplied by itself 2, 3, 4, 5... a limitless number of times, to produce power numbers, then from only one strain of power (i.e. 8 x 8 x 8 ...) you are already generating a divergent series of numbers toward infinite. If the previous classes were open sets toward infinite from one, two or three streams, the class of power numbers opens up to infinite from a larger and larger number of streams that are themselves limitless. This characterization brings the cardinality of this class to experience an explosion of sorts which places it much beyond the cardinality of the previous class. By all accounts this behavior represents an inflexion point in the cardinality of the vertical hierarchy of classes herein under scrutiny.

5.- Layer 5 is the province of complex numbers. These are numbers made up by a *real* part added to an imaginary part, that imaginary part being made up of a coefficient of *i* (*i* is equal to $\sqrt{-1}$) as in *6i* or *25i.* The real part can be an integer belonging to any one of the previous classes. On that count alone, this class would surpass the previous one in cardinality. But when you add the imaginary part, the coefficient of which may belong to any of the previous classes as well, then we can see that the number of possible combinations between the real part and the imaginary part puts us well beyond the count of the previous class at layer 4.

6.- Layer 6 showcases the class of rational numbers or decimals resulting from ratios, specifically those numbers with terminating decimals (i.e., 6.5) and repeating decimals (0.16666666666666̶). One can easily appreciate that the variety of decimals that may be achieved in these two manners can give way to a very high level of cardinality that surpasses the cardinality of layer 5. Let us first realize that the domain of integers (elements from layer 1 to 4) can take the place of the whole part in these decimal numbers. Secondly, to each of these whole integers, we may add an infinite number of terminating decimals of different lengths and an infinite number of repeating decimals of different lengths as well. And obviously if we wanted to turn these numbers into complex numbers, we could add these same sets to the imaginary part of these numbers as coefficients of *i*.

7.- Layer 7 is the province of irrational numbers, which are decimals with nonterminating, nonrepeating decimal portions endowed with no particular patterns. They are not the result of integer ratios. The most surprising account about irrational numbers is that there are more of them than rational numbers on the real number line, as proven by Danish-German mathematician Georg Cantor in the 19th century. Common sense then and still today would tend to make us believe that there would have to be more rational numbers than irrational numbers in existence. It is shockingly and strictly not so however as proven by Cantor.

8.- Layer 8 represents the class of algebraic numbers. These numbers are defined by clear geometrical relationships as we have shown in the study of the order of nothingness. It is well established that their cardinality is higher than that of irrational numbers on the real number line.

9.- Finally Layer 9 portrays the class of transcendental numbers. These numbers are defined as well by geometric relationships. The most famous among them, π and e, are respectively defined by relationships between elements of a circle and elements of a function ($1/x$). It is equally well established that among irrational numbers (those starting from layer 7) the set of transcendental numbers have the highest cardinality.

Despite the fact that all 9 classes of numbers on the number line form infinite sets, the cardinality of each set augments from class to class down the vertical hierarchy, with the important fact that class 4 shows an explosion in cardinality over the previous class (class 3). That explosion we have identified as an inflexion in the numerical analysis above. Therefore, there is a clear geometric pattern overall to the distribution of numbers in the realm of real numbers. You may even liken it, if you will, to the distribution figure shown below in Fig. 4.4.

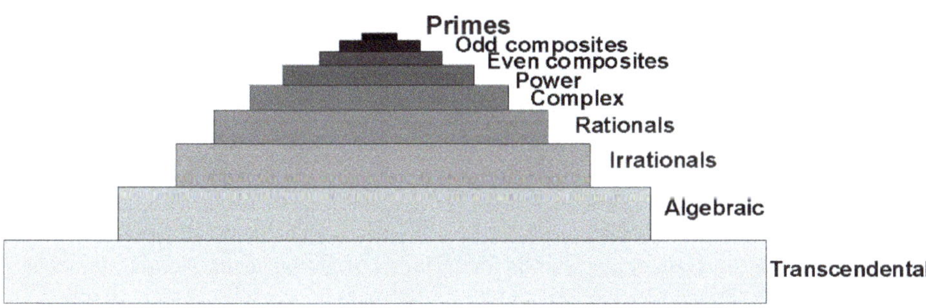

Fig. 4.4 Real Number Stack

At any rate, this pattern is nothing else than the Quanto-Geometric pattern or the bell-shaped pattern (see Fig.3.5). The inflexion point at layer 4 is indeed a standard deviation point in nature. What causes the various sets of numbers to behave in that manner is the intrinsic state of symmetry embedded in the constitution of these numbers within their sets. In other words, it is the reflection or influence of the order of symmetry over the scalar order within the Quanto-Geometric articulation which causes the scalar order to behave geometrically. It is not quite possible to plot a graph of the scalar distribution of real numbers that would capture the fullness of the number stack, simply because each layer takes origin from - ∞ and tends toward + ∞. Nonetheless it remains utterly clear that the concept of the number line is far insufficient and probably inadequate to capture the ultra-structure of real number elements.

By the same token, the geometric distribution of real numbers per above analysis has an overwhelmingly important consequence: it reveals that *the number line is not a continuum, the number line is discontinuous*. Effectively the distribution of real numbers studied above represents the highest level of abstraction possible if one is to appropriately model the domain of real numbers. This abstraction represents a meta-model of the overarching structure and behavior of real numbers. The ensuing distribution cannot be plotted on any of known coordinate systems, perhaps only the Quanto-Geometric one set forth in this study, because it is utterly difficult if inappropriate to use numbers to model numbers. Our best attempt at a pictorial representation of the distribution spectrum of the realm of real numbers is the graph shown in Fig. 4.5.

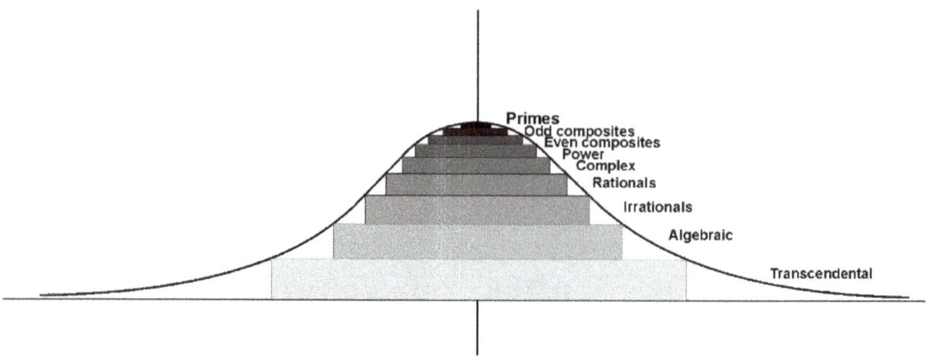

Fig. 4.5 Quanto-Geometric envelope of the real number stack

The horizontal axis stands for an index of cardinality which speaks of generic orders of magnitude while the vertical axis stands for an index of global scalar weight (or scalar modulus) so as to image the relative amount of numbers in sets with respect to unity. Unity in this context is represented by the set of monoliths (Primes). That approach makes the vertical axis a modeler for density figures. On the basis of that coordinate system the distribution graph naturally follows as depicted in our representation. Thus, the monoliths represented by Primes therein have the highest index of scalar weight but the lowest count or index of magnitude, whereas transcendental numbers hold the lowest index of scalar weight but highest count or index of cardinality (the magnitude). The graph shows an inflection point and reveals itself symmetric to the Quanto-Geometric function. Again, more than a functional representation, this characterization stands rather for the depiction of a graphical envelope to the real number stack (Fig. 3.5).

4.14. States of Symmetry in the Quanto-Geometric Function

In closing this Chapter, it is worth revisiting the Quanto-Geometric function with the aims of further eliciting the inner articulation between scalar states and states of symmetry. That function is based on transcendental numbers such as e, π, and $\sqrt{2}$, located at the highest level of the real number distribution, thereby making it apt to describe developments at all the other layers of the distribution.

I cannot overstate at the outset that objects as well as phenomena in physical reality are **numbers' applications**. That is to say that all things physical do not only exist thru number cardinality (their intrinsic numeric count) but also thru the scalar state and state of symmetry that make up their cardinality. On that account, every physical object is typed by the scalar and symmetric qualities pertaining to their cardinality, which endow them with specific properties emanating from the inner numeric ultra-structure. Those qualities and properties are the *Quanto-Geometric* descriptors set forth in previous chapters.

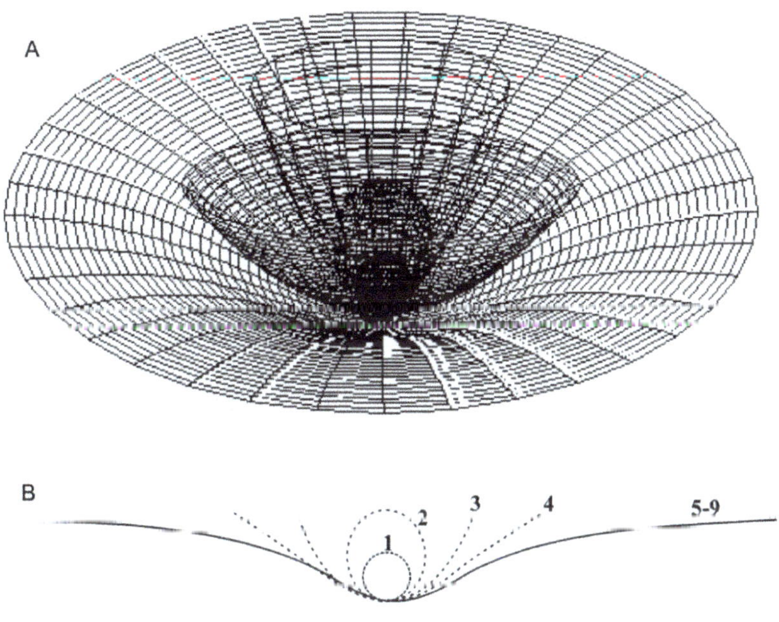

Fig. 4.6 Delimitation and typification of normalized intervals of the Quanto-Geometric function.

We are going to delve somewhat deeper into the correlation between scalar states and symmetric states from our abstract framework as reflected in the Quanto-Geometric function. We are going to revisit the *s* intervals of the Quanto-Geometric function and the curvature of the function across those intervals (Fig. 4.6).

One may easily observe a distinct form of curvature of the function for each interval (Fig. 3.7). At s = 0, the Quanto-Geometric function approaches a limit. In the interval $]0, \sigma/2[$, the curvature of the graph is circular. A circle, having only one focus which is its point of origin, translates into all points of the circle having only one way to compare to or resemble with one another. That one center of symmetry is a **singular** state of symmetry which matches the cardinality of that layer: numeral 1. In other words, scalar state 1 correlates with state of symmetry 1 at layer 1.

In the next interval $[\sigma/2, 2\sigma/7^{1/2}[$, the geometry of the curvature is elliptical. Since an ellipse is based on two foci, one may assert that there are two ways that the points of an ellipse may compare to or resemble with each other. This also means that there is two times more symmetry in the ellipse figure than there is in the circle. Thus, the ellipse contains a **dual** state of symmetry which matches the cardinality of that layer: numeral 2. And consequently, scalar state 2 correlates with state of symmetry 2.

The section of curvature laying within the interval $[2\sigma/7^{1/2}, \sigma[$ is parabolic. Since a parabola in defined by one line, the directrix, and one focal point, the both of which are reducible to 3 points in total, one may conclude that the points of a parabola may compare to or resemble with each other in three different ways. This also means that there is three times more symmetry in the parabola than there is in the circle. The parabola is therefore endowed with a **triple** state of symmetry which matches the cardinality of that layer: numeral 3. Thus, scalar state 3 correlates with state of symmetry 3.

In the interval $[\sigma, \sigma(8/5)^{1/2}[$ the function is of hyperbolic curvature. Since the geometry of the hyperbola is based on 4 centers of symmetry, the rectangle that defines the two axes of representation thru its diagonals, one may affirm that the hyperbola possesses a **quaternary** state of symmetry which matches the cardinality of that layer: numeral 4. Consequently, scalar state 4 correlates with state of symmetry 4. The forms of curvatures we know as conic sections end up with this interval.

The next five *s* intervals delineate forms of curvature known as hyperbolic from the hyperbolic functions, specifically the *sech* function. We have shown in Chapter 1 how the expression in *e* of the Quanto-Geometric function may be rewritten in implicit form to show the expression of the hyperbolic *sech* function. Assigning different values to the variable *a*, *b* and *c* in the implicit expression will yield graphical curves falling into interval 5, interval 6, interval 7, interval 8 and interval 9 of the Function per our form of normalization.

In the fifth *s* interval $[\sigma(8/5)^{1/2}, \sigma 10^{1/2}/2[$ the function displays a form of curvature we call the entropic form of curvature, which stands for a state of symmetry based on 5 centers of symmetry. Consequently, state of symmetry 5 correlates and articulates with numeral 5 or scalar state 5.

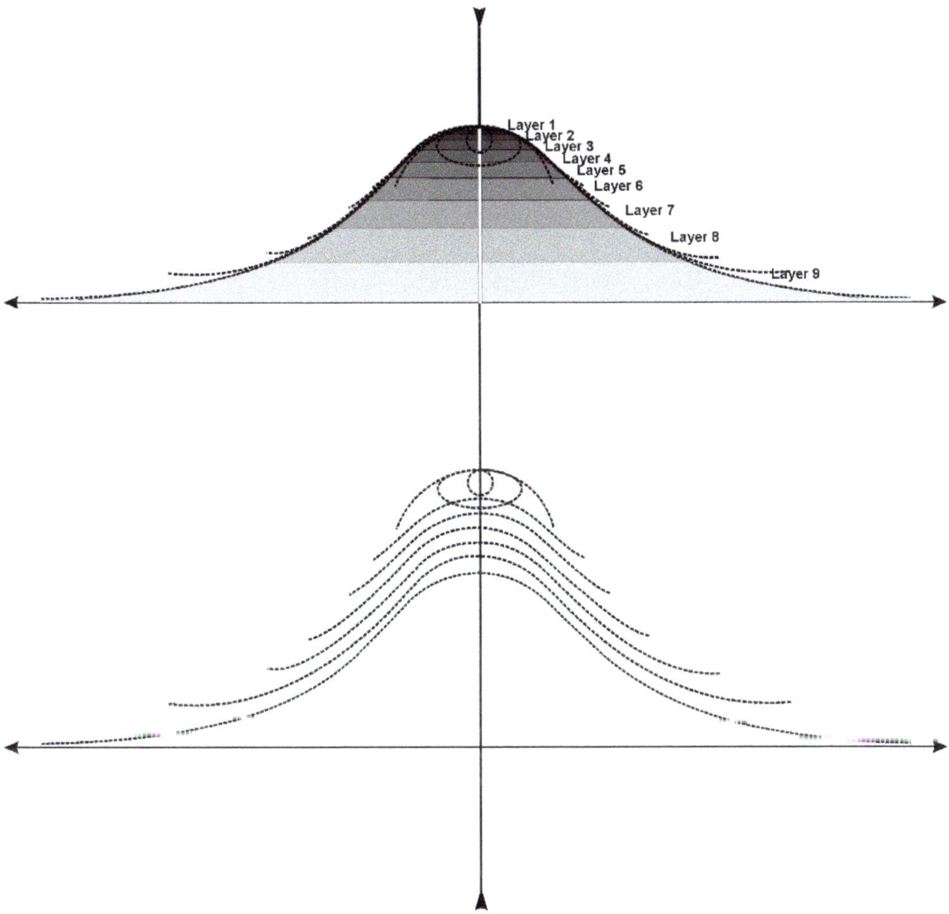

Fig. 3.7 Exploded view of embedded symmetries within the Quanto-Geometric function

The sixth interval $[\sigma 10^{1/2}/2, 2\sigma[$ delineates a form of curvature we denote the enthalpic curvature, which stands for a state of symmetry based on 6 centers of symmetry. Thus, state of symmetry 6 correlates and articulates with scalar state 6.

In the seventh interval [2σ, $\sigma 7^{1/2}$[the function describes a form of curvature that we call paradoxical curvature, which stands for a state of symmetry based on 7 centers of symmetry. Thus, state of symmetry 7 is in articulation with scalar state 7.

The eighth interval [$\sigma 7^{1/2}$, 4σ[encompasses a fraction of curvature which we call the perfect curvature. This state of symmetry is based on 8 centers of symmetry. Hence, state of symmetry 8 is in articulation with scalar state 8.

And finally in the ninth interval [4σ, ∞[the function displays a section of curvature we denote linear or infinite curvature, which stands for a state of symmetry based on 9 centers of symmetry. Hence, state of symmetry 9 is in articulation with scalar state 9.

What justifies our assigning an attribute to those states of symmetry is the correspondence that we have observed between these attributes when determined by either the geometric order or the scalar order. It is now therefore firmly established that the Quanto-Geometric articulation, between the two prime elements of the ultra-structure and across the abstract numeric distribution, is *typed.* This fact or artifact indeed raises each layer of the distribution to the value of an Operator when the number system is applied to physical reality, or, do we dare say, when the number system is put to work to create physical reality. Is this a reflection of the Mind of God at Creation? I let you make that call.

4.15. Conclusion

The numerical analysis that we have conducted above is based per choice on the decimal number system with 10 base units. The identity of base unit 0 has been in dispute for quite a while now, some mathematicians having removed it altogether from the number line as origin. The prevalent supporting argument has always been that *no-thing* cannot belong to a set of *some-things*.

The best way to formulate an abstraction about 0 is thru the philosophical principles that we have proposed in Quanto-Geometry, as expressed in Chapter 1. There is no absolute 0 in physical reality. There is only absolute 0 in complete and total abstraction, and that is why we have subsumed that 0 into the coordinate system's *m-t* plane or the *s* axis which acts as an unreachable asymptote. When the abstraction comes down to the number realm, 0 starts having some life or becomes part of that abstract life as an order of nothingness.

It is because of that 0 that numbers of the scalar order behaves the way they do in terms of their geometric properties and their distribution. Even if we were to remove 0 from the set of base units, making it a base of 9 units, the properties and the distribution would remain the same, because there cannot be any scalar order without an order of nothingness. From the Quanto-Geometric perspective, the base of 9 units (1 to 9) is the most natural of base number systems. However, there is no prejudice whatsoever in keeping 0 within a number base, as does the decimal base, because the meta-level abstraction that presides on all number systems does not formally forbid its inclusion. I shall equally add that the conclusions and results that this numerical analysis has reached would not have changed had we used a different number system than the decimal.

By and large, all the facts known in Number Theory, particularly those known in *recreational* number theory, are relevant facts within the context of the Quanto-Geometric vision. Thus far the many facts known about a long string of specific numbers which have received rather exotic attributes and have been very loosely categorized, reason for their denomination as *recreational*, will find fully coherent and systemic interpretation within the Quanto-Geometric framework. We are refraining ourselves from further exposition in that direction in the interest of not overwhelming readers with possibly little interest in this topic, no matter how recreational mathematicians think this section of Number Theory is. Nevertheless, it must remain absolutely clear that numeric abstraction is the foundation of all physical reality as they appear under the umbrella of both objects and phenomena. The most towering vision that should remain stamped in the reader's mind is the artifact that the ultra-structural signatures of base number units of the real number realm are the source from which spring the behavior, attributes and identity of all real physical objects and phenomena. This fact or artifact has been altogether ignored in the West despite the 2000-year-old warning from the brilliant Greek philosopher Aristocles, nicknamed **Plato**, that *numbers represent the highest level of the intelligible.*

Chapter 5

Everything should be as simple as possible,
but not simpler.
A. Einstein

QUANTO-GEOMETRIC TENSORS AND OPERATORS

Physicists have long realized that the idea of holding one theory for the treatment of dynamic systems in the microcosm and one other for the dynamics taking place at larger scale physical structures is simply untenable, moreover if both families of theories are completely opaque to one another. Despite the proliferation of gauge theories, we are still unable to explain the continuum of Nature as revealed before us and it may be time to realize that the bridging notion of quantum gravity is just ill-founded and non-falsifiable as framed. As argued early on in this text, Grand Unification requires an elucidation of the origin of the most important traits of the continuum, which is the plethora of fundamental physical constants that make up the continuum of Nature, those which we have arrived at by pure empiricism for the most part. In this Chapter, I lay down an augmentation of conceptual framework we have dubbed **Quanto-Geometry or Quanto-Geometric Theory** while granting an exacting numeric basis to the ontology of coordinate space and the quantum-scalar along with their correlation throughout the continuum, under the umbrella of pivotal Tensors and Operators acting on the continuum.

5.1. Duality and Codependency in Quanto-Geometry

To achieve the afore-mentioned goals, we recognize that the Theory must be built on an abstraction that lies even higher than Representation Theory and traditional Differential Geometry for historical reasons, on the one hand, and on the other, this formalism must be closely related to the most ubiquitous form of distribution in the physical sciences, the normal Gaussian density function.

This theoretical rendez-vous is located at the intersection between a certain form of Number Theory and the functional Gaussian formalism from which they emerge mainly as a system of tensors and operators. The most towering umbrella raised by this system of constructs is the principle of duality for all physical ontologies, which considers every object a reunion between an irreducible tangible tenet, or a *scalar* element, and an irreducible intangible tenet, or the *spatial void* element, both seized in a covariant relationship. This duality is captured in the following formula:

$$Q \times S = 1 \text{ or } Q = S^{-1} \quad (5\text{-}1)$$

where Q is the scalar and S the space tenet. The Quanto-Geometric function that incarnates the fullness of the principle of duality and codependency between the dual tenets is the following:

$$Q(s) = \pm\left|\frac{s^0}{\sigma\sqrt{2\pi}} \bullet e^{\left(\frac{-s^2}{2\sigma^2}\right)}\right|, s \neq 0 \quad (5\text{-}2)$$

where $Q(s)$ is the dependent variable portraying the scalar tenet and s the independent variable portraying the space tenet (Fig. 5.1).

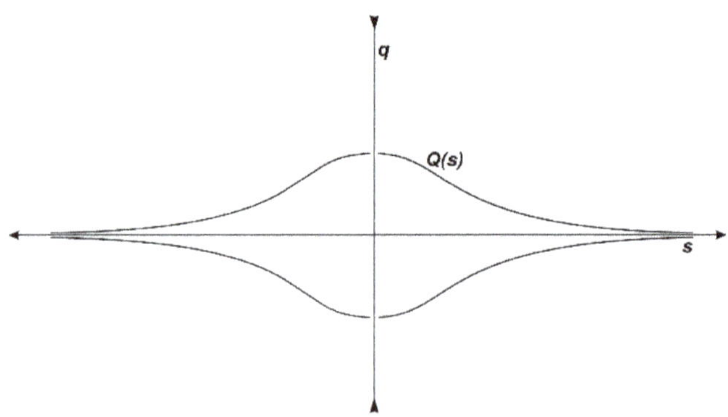

Fig. 5.1 Graph of the Quanto-Geometric function of one variable

Because in philosophy there cannot exist any absolute tangible scalar, nor any absolute intangible void, the function is expressed with a discontinuity at the q-axis introduced with the unconventional s^0 term. The q-axis is further mandated to surrogate pure scalar state. For the very same reason, the function displays an asymptotic behavior as s approaches 0, with the s-axis mandated to surrogate absolute pure void. Therefore, our system of coordinates is not merely cardinal but it additionally and inherently reflects the background axiomatic proposition upheld in the Theory.

This set of directives and constraints result in the demotion or de-emphasis of the orthodox norms of distribution based on the standard deviation traditionally known to the Gaussian Function, as much as they call for a new form of normalization of the Function based on the conic or curvilinear symmetries delineated in its graphical pattern.

5.2. Natural Symmetry Groups

The association between the conic or curvilinear symmetries and their enclosing layered distribution Groups make for the most natural of symmetry groups, lying far above traditional transformation groups because they provide the normative basis for all forms of variance or transformations.

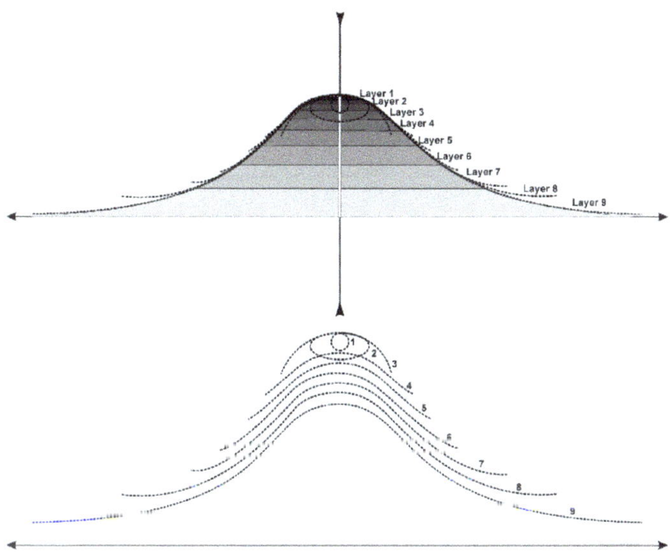

Fig. 5.2 The 9 essential layers of the Quanto-Geometric function with their associated manifolds or forms of symmetry

In this capacity, the Quanto-Geometric Function establishes a hierarchy of symmetry that ranks all known forms of symmetry in the following order (Fig.5.2):

1. circular form of symmetry
2. elliptic form of symmetry
3. parabolic form of symmetry
4. hyperbolic form of symmetry
5. hyperboloid forms of symmetry

The hyperboloid forms of symmetry regroup 5 distinct forms of symmetries that stand each in its very own right. The hierarchy so posits the existence of a total of 9 different fundamental forms of symmetry. This order of forms amounts indeed to an order of manifolds in the sense that the 4 conic forms and the 5 hyperboloid forms are the building blocks of all conceivable manifolds. In other words, the geometric composition of manifolds is strictly based on these unit-forms of symmetry, in the understanding that each one of them is fully embedded in the following one from unit-form 1 to unit-form 9.

5.3. Toward a Natural Hierarchy of Operators

Unit-norms are imprinted in the curvilinear metric or arc-metric of the Quanto-Geometric Function, giving rise to a set of 9 natural typified Quanto-Geometric tensors. These norms permeate the correlating sectionals or layers of the Function while giving rise to a set of 9 corresponding operators. To be sure, the referred norms constitute the binding agency that makes up the entire sets of numbers in all known forms of mathematical representation. The normal layers, as sets or groups, do no more than reflect the nature of the primal Numeric operators.

The Quanto-Geometric Operators represent ontological Norms applicable to physical observables in order to query their states and dynamics. This set of operators constitutes one of the higher orders of certainties that subsume all statistical order of uncertainties encompassing physical observables, that which is traditionally implemented from an integrative view of the function. We contrastingly give precedence to a layered view of the function thru its curvilinear sectionals.

5.4. Quanto-Geometric Coordinate System and its Metrics

In general, the metric tensor and its manifolds represent a broad program in differential geometry essentially aimed at describing the geometric properties of surface sheets. The seldom raised question of why curvature symmetries exist in 3 broad classes is one that finds natural fit within The Quanto-Geometric framework, these classes being:

1. Differentiable or Smooth continuous Manifolds (Riemannian)
2. Finsler Manifolds
3. Topologies (or Discrete Manifolds)

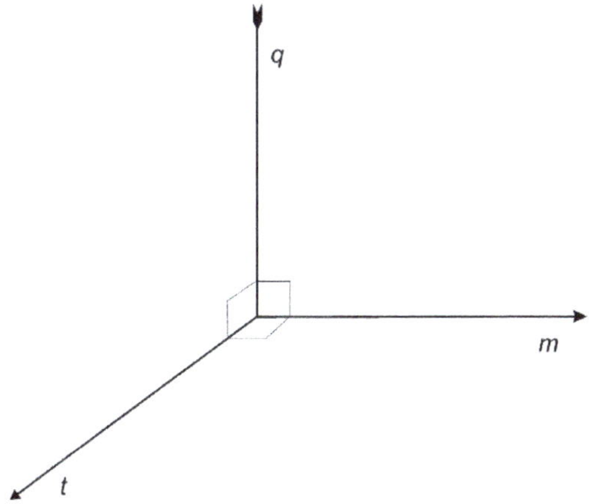

Fig. 5.3 Quanto-Geometric 3-space coordinate system

In Quanto-Geometry, the 3 dimensions of 3-space, despite the fact that they are isotropic to one another, portray 3 different metrics (Fig.5.3). In the 3-space background, two of the spatial degrees of freedom are associates or Abelian associates that are symmetric to one another. One represents the very archetype of the metric of symmetry and the other a kinematic metric. These two degrees of freedom, spatial depth *(t)* and spatial width *(m)*, together with their metrics are engenders of 3-space itself. The third degree of freedom, spatial height *(q)*, that is, is not a direct product of 3-space and its inner workings, but stems from interaction between innate 3-space and the Q_3 scalar (the value of $Q(s)$ at s_3 or 3-space). In representation, the third degree of freedom in 3-space becomes the scalar axis.

In 2-space background, space is two-dimensional, having a degree of freedom of height *(q)* and a degree of freedom of width *(s)*. The metric of one of the 2-space dimensions *(s)* represents a conflation of the distance (or symmetry) metric and the kinematic metric. This conflated metric is the innate expression of 2-space in and of itself. The second degree of freedom in 2-space, spatial height, that is, is not a direct product of 2-space and its inner workings, but stems from interaction between 2-space and the Q_3 scalar (the value of $Q(s)$ at s_2 in the graph). In representation, the second degree of freedom in 2-space becomes the scalar axis.

The Quanto-Geometric Theory advances both a 2-space Model or irreducible analytic image in 2 dimensions and a 3-space Model or irreducible analytic image in 3 dimensions, both of them being instrumental in the description of physical observables leading to the theoretical derivation of several different classes of fundamental physical constants. The important point here to bear in mind is that the nature of the coordinate system that one chooses to uphold in analysis is of utmost significance and that the familiar Cartesian system with the appended 4th dimension of time, for a 4-dimensional reference system, constitutes the source of many of the serious drawbacks and dead ends we have experienced in theoretical physics thus far. From the classical flat intuitive Cartesian system, the reference frame had acquired behavior in Einstein's theories of Relativity in very important ways. In the Quanto-Geometric framework, the dimensional background reference frame becomes intrinsically endowed with both ontological and phenomenological properties, which enhance its imaging capabilities as a system of representation and allow for very high towering observation and scrutiny of physical variables.

It is a fact of note that when importing mathematical constructs to physical observables, physicists have arguably not placed sufficient emphasis on the role of the very graphism of functions as expressions of the correlation between the independent and dependent variables at play. In our view the manifolds have become all too often abstract spaces obscuring interpretation instead of acting as direct elucidating tools. The geometric design of a function gives a direct measure of first-order correlation between its variables. Indeed, much before the study of area under a curve by dividing it into infinitesimal boxes comes the study of curvilinear behavior of a curve by breaking it into infinitesimal sloped segments. The pertinence of this approach has led to the foundation of the entire mathematical science of differential geometry with important consequences in its physical application.

5.5. Specificities of the Quanto-Geometric Metrics

For all the importance and attention given to 3-space metric, 2-space Euclidean metric remains the most important of all metrics, if for no other reason because it constitutes the ultimate reference for all tensor categories arising from other metrics, such as the Ricci and scalar curvature tensors. Euclidean metric is in fact the most complex of all metrics being indeed the symmetry birthplace of all metrics.

All metrics, including Euclidean metric, are based on the distance metric which had been defined as stated below:

$d(x,y) = 0 \leftrightarrow x = y$, positing the identity of indiscernibles

$d(x,y) = d(y,x) \leftrightarrow x = y$, positing the property of symmetry

$d(x,z) \leq d(x,y) + d(y,z)$, positing sub-additivity or triangle inequality.

From the abstract axiomatic at its foundation, the Quanto-Geometric metrics are based on slightly different grounds in terms of their distance metric and on substantially different grounds when it comes to their differential metric. In Quanto-Geometry, the distance metric is based on the space between two points as stated below:

$$S : X \times Y \to \left]0,+\infty\right[$$

The axiomatic so forbids the inception of 0 as a representation of pure void in physical reality on the claim that all ontologies are irreducibly dual in nature as formed by a scalar and a space element[4], all of which makes absolute void or 0 a natively vanishing entity. Furthermore, the nature of the distance element between two points is not only *geometrical*, as given in a straight line or a geodesic curve, but *dynamical* as well. The kinematic element is embedded in a degree of propagation coupled to the spread element incarnated in the distance figure. Therefore, the *s*-metric predicates a *wavefunction metric* whose units are quantified in *wavenumbers.* Hence in Quanto-Geometry the distance between two points in space is a wavenumber. Going even further than conventional differential geometry, we pose that the Quanto-Geometric metric bears an etalon that is quantified in the standard deviation figure of σ, a distribution parameter, which makes it a seemingly running etalon.

5.6. Degrees of Freedom in Normalized Manifolds

The wavefunction dimension of the *s*-metric produces a scale of wavenumbers based on the σ etalon whose intervals are normalized and further represent unit-norms of distance and harmonic propagation. The primeval sequence s_n that produces this scale is:

$$\{s_n\} = \left\{\frac{1-n}{10-n}\right\}, \quad with\ 1 \le n \le 9 \qquad (5\text{-}3)$$

where n is a modulus of symmetry and/or degree of propagation. In its generalized application over the distance dimension of 2-space metric, its formulation becomes:

[4] We choose to apply the straightforward term *space* to the physical observable that is formally denoted *coordinate space* in the physical science literature by contrast to abstract spaces such as *momentum space* or *phase space*. We do not peruse abstract fictitious spaces in Quanto-Geometry.

$$s = \pm\sigma\sqrt{\frac{(2n-2)}{10-n}}, \quad (5\text{-}4)$$

where s is the 2-space variable (See Section 5.5 for derivation of relation 5.*4*).

In that manner it yields the following super-imposed scale over the metric:

for $n = 1$, $s_1 = 0$

for $n = 2$, $s_2 = \pm\sigma/2$

for $n = 3$, $s_3 = \pm\sigma 2/7^{1/2}$

for $n = 4$, $s_4 = \pm\sigma$

for $n = 5$, $s_5 = \pm\sigma\cdot(8/5)^{1/2}$

for $n = 6$, $s_6 = \pm\sigma(10)^{1/2}/2$

for $n = 7$, $s_7 = \pm 2\sigma$

for $n = 8$, $s_8 = \pm\sigma 7^{1/2}$

for $n = 9$, $s_9 = \pm 4\sigma$

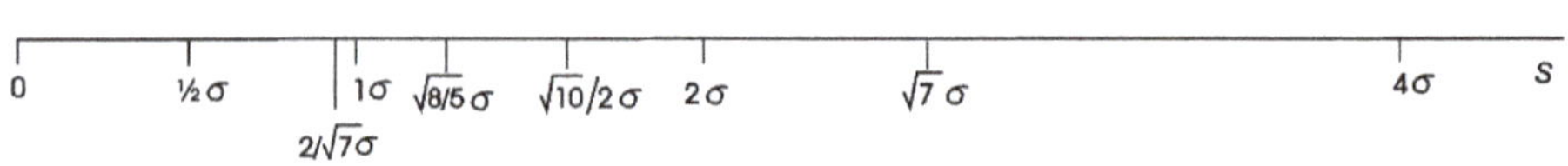

Fig. 5.4 Primeval sequence overscale of degrees of freedom laid on the s-axis

The s_i sequence of points represents the special wavenumber points that make up the scale (Fig.5.4). At the same time these points represent archetypical symmetry-breaking points delimiting the normal sectional curvatures of the curvilinear pattern of the Quanto-Geometric Function. They mark phase transitions from one form of the symmetry to another. If one cares to understand that all minimal sectional forms are typified forms of symmetry, then the sectional breakdown of the curvilinear design of the Quanto-Geometric Function ostensibly shows the natural hierarchy of symmetries ultimately dictated by 2-space metric. Those archetypical forms of symmetry constitute the basis of the sectional curvatures which grade the manifolds associated with all of the common differential geometry tensors. To make this point even clearer, we shall consider that degree of freedom Mod1 *(n = 1)* ultimately maps to the circular sectional of the curve thru interval]0, 1/2σ[of the *s*-metric. Below is reported the full suite of normal intervals on the metric with their correlation with the *n* degrees of freedom:

Interval 1 → $]0, \sigma/2[$, for $n = 1$

Interval 2 → $[\sigma/2, 2\sigma/7^{1/2}[$, for $n = 2$

Interval 3 → $[2\sigma/7^{1/2}, \sigma[$, for $n = 3$

Interval 4 → $[\sigma, \sigma\cdot(8/5)^{1/2}[$, for $n = 4$

Interval 5 → $[\sigma\cdot(8/5)^{1/2}, \sigma(10)^{1/2}/2[$, for $n = 5$

Interval 6 → $[\sigma(10)^{1/2}/2, 2\sigma[$, for $n = 6$

Interval 7 → $[2\sigma, \sigma 7^{1/2}[$, for $n = 7$

Interval 8 → $[\sigma 7^{1/2}, 4\sigma[$, for $n = 8$

Interval 9 → $[4\sigma, \infty[$, for $n = 9$

We have therefore established a direct relationship between the number of degrees of freedom and the number of focal points of the conic and hyperboloid forms making up the curvilinear design of the Function. This correlation gives us:

- One degree of freedom for one focal point of symmetry in the *circular* curvature
- Two degrees of freedom for two focal points of symmetry or foci in the *elliptic* curvature
- Three degrees of freedom for three focal points of symmetry or foci in the *parabolic* curvature

- Four degrees of freedom for four focal points of symmetry or foci in the *hyperbolic* curvature
- Five degrees of freedom for five focal points of symmetry or foci in the *entropic* curvature
- Six degrees of freedom for six focal points of symmetry or foci in the *enthalpic* curvature
- Seven degrees of freedom for seven focal points of symmetry or foci in the *paradoxical* curvature
- Eight degrees of freedom for eight focal points of symmetry or foci in the *perfect* curvature
- Nine degrees of freedom for nine focal points of symmetry or foci in the *infinite* curvature

There are good reasons to the epithets chosen for the hyperboloid forms of symmetry (5 to 9) which we address in the next Chapter. Suffice it to say for now that they stem from numeric signatures arising from the Number Stack that presides over the development of the framework. While the first 4 results may be of little surprise due to our familiarity with the conic sections, the degrees of freedom dictating the five hyperboloid forms of symmetry may likely come as a surprise due to their complete counter-intuitiveness. Very particularly, the last sectional, based on 9 degrees of freedom, practically portrays a linear element since this sectional infinitely lives in the neighborhood of the asymptotic *s*-axis. As such it represents, and counter-intuitively so, the most complex archetypical form of symmetry, whereas the circular sectional of curvature stands for the simplest archetypical form of symmetry, contrary to common wisdom in differential geometry. In that manner, this symmetry characterization makes explicit the reason that assists us in making the linear element with its stature of highest Modulus of freedom the most complex form of symmetry, as much as it explains why the linear element is indeed the irreducible in all metrics and why Euclidean-type frames are sufficient for the representation of the background of the Universal Arena.

5.7. Quanto-Geometric Primal Tensors

From the rectilinear distance metric, we now move to the curvilinear metric that substantiates the curvilinear pattern of the Quanto-Geometric $Q(s)$ Function.

5.7.1 2-Space Tensors

The most irreducible representation of two points on the graph of the Quanto-Geometric Function implies a distance and a slope, which a vector entity perfectly incarnates. We may define the unit curvilinear metric by examining the behavior of a secant line thru any chosen point A on the curve (Fig.4.5).

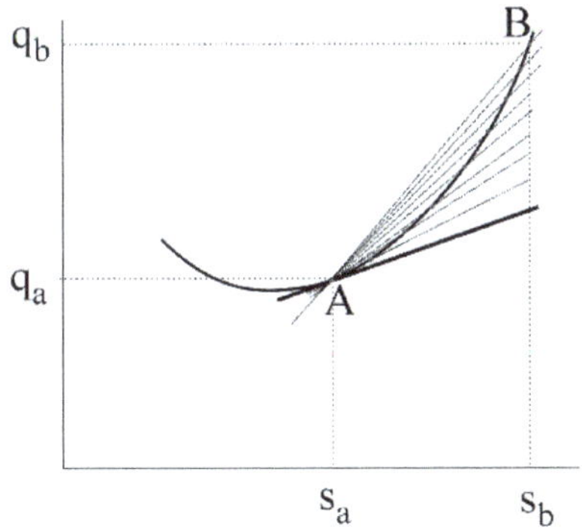

Fig. 4.5 Secant line approaching the tangential location at one end point

The unit curvilinear metric is equivalent to the length or magnitude of the secant segment just prior to the secant line becoming a tangent to the curve at point A thru affine approximations of the secant to the tangential direction. We further clarify that in that scenario point B is the next dot in the curvilinear movement of the curve from point A and that segment [AB] is equivalent to the circular arc $A\hat{B}$ (or $\hat{a}$) of *infinite* radius, an arc section which may be assimilated to a linear element for all intents and purposes. Therefore,

$$[AB]^2 = ds^2 + dq^2$$

$$[AB] = \sqrt{ds^2 + dq^2}$$

$$[AB] = \frac{1}{ds}\sqrt{ds^2 + dq^2} \cdot ds$$, which becomes the arc-length:

$$\hat{a} = \int_B^A \sqrt{\frac{ds^2}{ds^2} + \frac{dq^2}{ds^2}} \cdot ds$$

$$\hat{a} = \int_B^A \sqrt{1 + \frac{dq^2}{ds^2}} \cdot ds$$

$$\hat{a} = \int_B^A \sqrt{1 + \left(\frac{dq}{ds}\right)^2} \cdot ds$$

We express the curvilinear metric or 2-space metric as:

$$g_i^2 = \int_B^A \sqrt{1 + \left(\frac{dQ(s)}{ds}\right)^2} \, ds \quad (5\text{-}5)$$

On the basis of this metric, we may determine the length of any desired curvilinear sectional of the *Q(s)* Function, which emerges as the measure of tensors or the tensile strength of the scalar-space correlation (or covariance) represented by the Function. The singlet tangent bundles that span each *q-s* normal interval, or the individual arc-lengths of these normals, give us the fundamental one-dimensional tensors that inhabit the curvilinear rundown of the Quanto-Geometric Function. We express the Quanto-Geometric 2-space tensors as integrals of the unit bundles, per expression *(5-5)*, as follows:

Monolithic Tensor:
$$g_1^2 = \int_0^{\sigma/2} \sqrt{1 + \left(\frac{dQ(s)}{ds}\right)^2} \, ds$$

Parity Tensor:
$$g_2^2 = \int_{\sigma/2}^{2\sigma/\sqrt{7}} \sqrt{1 + \left(\frac{dQ(s)}{ds}\right)^2} \, ds$$

Associative Tensor:
$$g_3^2 = \int_{2\sigma/\sqrt{7}}^{\sigma} \sqrt{1 + \left(\frac{dQ(s)}{ds}\right)^2} \, ds$$

Power Tensor:
$$g_4^2 = \int_{\sigma}^{\sigma\sqrt{8/5}} \sqrt{1 + \left(\frac{dQ(s)}{ds}\right)^2} \, ds$$

Imaginary Tensor:

$$g_5^2 = \int_{\sigma\sqrt{8/5}}^{\sigma\sqrt{10}/2} \sqrt{1+\left(\frac{dQ(s)}{ds}\right)^2}\, ds$$

Matrixial Tensor:

$$g_6^2 = \int_{\sigma\sqrt{10}/2}^{2\sigma} \sqrt{1+\left(\frac{dQ(s)}{ds}\right)^2}\, ds$$

Rationalization Tensor:

$$g_7^2 = \int_{2\sigma b}^{\sigma\sqrt{7}} \sqrt{1+\left(\frac{dQ(s)}{ds}\right)^2}\, ds$$

Perfect Tensor:

$$g_8^2 = \int_{\sigma\sqrt{7}}^{4\sigma} \sqrt{1+\left(\frac{dQ(s)}{ds}\right)^2}\, ds$$

Transcendental Tensor:

$$g_9^2 = \int_{4\sigma}^{\infty} \sqrt{1+\left(\frac{dQ(s)}{ds}\right)^2}\, ds$$

Although the curve is a product of a 2-space reference frame or two-axis coordinate system, the curve itself is one-dimensional in the sense that the variations that make up the curvilinear pattern are realized only in the dimension of height of the coordinate system. Therefore, the arc-lengths of the sectional curvature normals corresponding to the overscale of the *s*-metric implanted in the *s*-axis constitute the preeminent quanto-geometric 2-space Tensors at the heart of the Theory.

5.7.2 3-Space Metric

When it comes to 3-space, per the two-variable expression of the Function *(expression 2-14)*, with its coordinate system bearing the properties that we have already discussed, the metric becomes bilinear, primarily informed by the two specialized spatial axes (Fig.5.6).

We can perfectly use the definition of the metric in 2-space as a root for the 3-space formulation of the metric. From 2-space to 3-space an axial transformation takes place, however, that splits the *s*-axis into two symmetric *m*- and *t*-axes, *m* and *t* being the new independent variables correlating to the one dependent variable *Q(s)*. So therefore, *ds* becomes *(dm,dt)*.

We now have the minimal segment of the curvilinear metric split into two equal segments which then form the triangular surface constituting the unit-surface for the metric. We may assimilate this unit-surface to the sectional curved surface of infinite radius (Fig.5.7). The area A enclosed in an equilateral triangle bears the following expression:

$$A = \frac{\sqrt{3}}{4} l^2, \; l \text{ being a side of the triangle.}$$

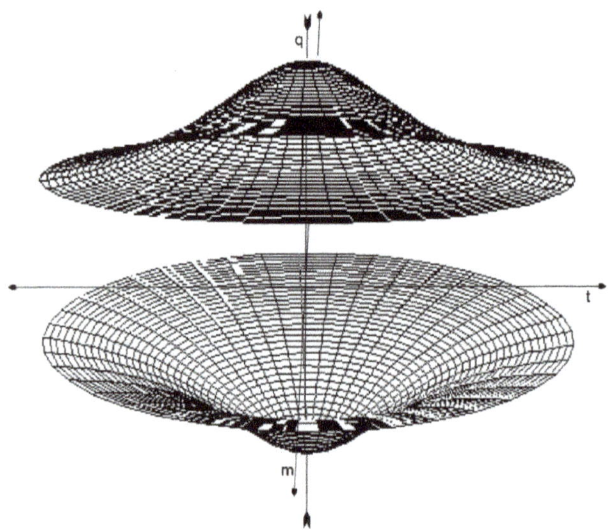

Fig. 5.6 A complete view of the graph of the Quanto-Geometric function of two variables *(m,t)*

Therefore, the unit-surface of 3-space metric may be expressed as follows:

$$A = g_{im}^{3} = \frac{\sqrt{3}}{4} \left(\int_a^b \sqrt{1 + \left(\frac{\partial q}{\partial m} \right)^2} \right)^2,$$

point b being the next dot to point a in the curvilinear movement of the generating curve on the partial q-m reference frame. Since m and t are symmetric, the 3-space metric may perfectly have the following alternate expression:

$$g_{it}^{3} = \frac{\sqrt{3}}{4}\left(\int_{a}^{b}\sqrt{1+\left(\frac{\partial q}{\partial t}\right)^{2}}\right)^{2}$$

point b being the next dot to point a in the curvilinear movement of the generating curve on the partial q-t reference frame. The generating segments are duplet tangents that are the defining elements for the tangent bundles characterizing every point of the planar surface of 3-space manifolds, in accordance with the definition of 3-space manifolds in traditional differential geometry.

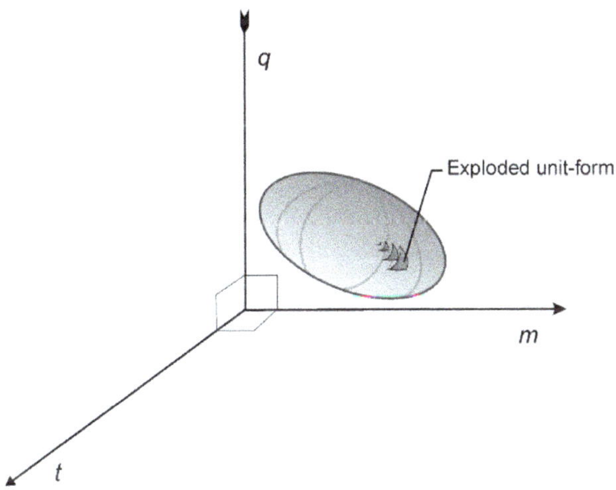

Fig. 5.7 Exploded view of a 3-space unit-form in an object

It is a notable fact however that the vast majority of parameters and the so-called fundamental constants of nature, which are at the foundation of 3-space or the 3-D universal domain, are artifacts of 2-space metric and its tensor derivates, 3-space metric contributing in a very limited manner in fashioning those quantities. One may intuitively deduce the expression of the 3-space tensors from that of the 2-space tensors, the list of which we will spare this exposé.

It is important to remember that despite the fact that we are codifying them as a length in following the method long in use in differential geometry, the basis metric units represent, beyond the mere length or arc-length element, a wave-length of a *wavefunction* (a wavenumber to be precise), as well as a *degree of propagation*. This follows from the properties initially incarnated within the Quanto-Geometric coordinate system. In that sense differential quanto-geometry represents a higher framework of generalization of which Riemannian geometry and non-Riemannian differential geometry are only particular cases.

At the height of its utility, the application of the transition map method in traditional differential geometry allows one to charter a course of interpretation of physical observables that is as comprehensive and complete as afforded by the atlas of representation. However, we do not apply this technique in Quanto-Geometry because the mathematical-physics nature of our coordinate systems for every *n*-space, with the properties expressly subsumed in their axes, affords us to directly compute the quantities borne out by the states and dynamics of the physical observables, or alternatively by application of the Quanto-Geometric Operators. All of which traditional Differential Geometry is unable to materialize. One can already see that Quanto-Geometry solidly encapsulates both Euclidean Geometry and Differential Geometry.

5.8. Quintessential Tri-Valent Tensor

In continuing to scrutinize it along the line of differential geometry formalism, one more important characterization springs from the geometric design of the Function. First order derivation, which stands for the second order of the quanto-geometric correlation between the scalar and space tenets (the function itself being the first correlational order), gives us three broad sectors dividing the graphical pattern:

In one phase → $\frac{dq}{ds}>1$ or in 3-space $\left.\frac{\partial q}{\partial t}\right|_m >1$ and $\left.\frac{\partial q}{\partial m}\right|_t >1$, which translates into the scalar tenet dominating or eclipsing the space tenet in the correlation.

In the second phase → $\frac{dq}{ds}<1$ or in 3-space $\left.\frac{\partial q}{\partial t}\right|_m <1$ and $\left.\frac{\partial q}{\partial m}\right|_t <1$, which translates into the space tenet dominating or eclipsing the scalar tenet in the correlation.

In the third or mid-phase → $\frac{dq}{ds}=1$ or in 3-space $\left.\frac{\partial q}{\partial t}\right|_m =1$ and $\left.\frac{\partial q}{\partial m}\right|_t =1$, which means equated covariant influence between the scalar tenet and the space tenet. The first-order derivative monotonically increases in phase 1, monotonically decreases in phase 2 and remains invariant in the mid-phase.

In the first sector, the scalar overwhelms its space partner in the covariant relationship (Fig. 5.8). In the 3rd sector, the space partner overwhelms the scalar. And in the 2nd or mid-sector the covariant relationship between the two partners is equalized. The nature of these differential sectors is equivalent to *singlet* tangent bundles. We denote them as a whole the *Quintessential Trivalent Tensor* in the understanding that the latter represents 3 different tangent bundles or 3 component trivalent tensors, namely the *Quantum-dominant* or *Scalar-Dominant* Tensor, the *Space-dominant* Tensor and the *Quadratic* Tensor. This characterization is one of the most pivotal concepts of the Quanto-Geometric framework with far reaching consequences.

Hyper Quantum Quintessence is a generator portending to the formation of closed spaces. Hyper Space Quintessence is a generator of surface sheets or open spaces. Hyper Quadratic Quintessence is a generator of semi-closed/semi-open spaces. If the first generator has its way, it will make everything circular in creating volumes, whether 2-dimensional (circle), or 3-dimensional (sphere) thru the nth dimensional order. If the secondly mentioned generator has its way, it will make everything straightly linear by creating flatness, whether 2-dimensional (planes) or plane-like in all dimensional orders. And if the thirdly mentioned generator has its way, it will turn everything into thickness in creating layers by giving equalized weight to vertical and horizontal deploys.

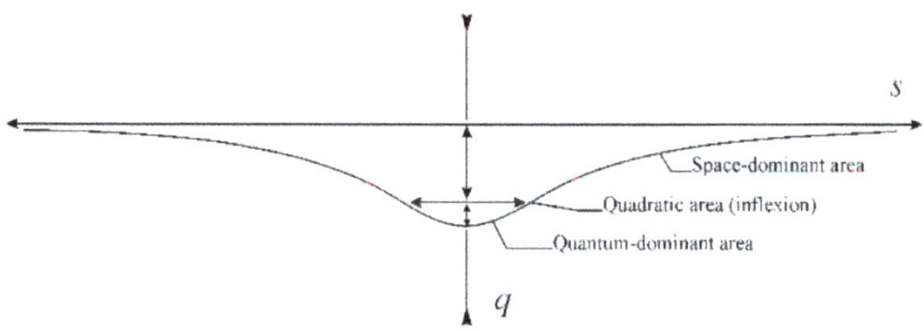

Fig. 5.8 The three sectors of the Quanto-Geometric Function that grandfathers the Tri-Valent Operator

The determinations of this symmetric order are of paramount significance. They are not quite in line with common wisdom in traditional differential geometry. For instance, one may argue that a circle is a plane figure. However, if we define a volume a closed space or a closed-like space, per our familiarity with conformal 3-space, in a two-dimensional world the circle must be homeomorphic to the sphere as the simplest form of closed space or volume. While the charts and transition maps are probably a useful tool to study the parallels between 2-space (or R^2) manifolds and other higher n-space (or R^n) manifolds, they do not establish the higher order of symmetry at the foundation of the distribution of manifolds. Tri-Valent Quintessence is an ultra-structural categorization that reveals the higher class of symmetry to which manifolds and tensors belong, in a manner that is similar to group covers in Group Representation Theory. In that sense Tri-Valent Quintessence reveals the three super-classes that regroup manifolds and tensors in differential geometry. It is by virtue of Tri-Valent Quintessence that three and only three broad classes or super classes of these elements are in existence.

In establishing a parallel between our framework and Category Theory abstractions, we may point out that there is a homeomorphic relationship between the order of surface sheets in 2-space and the class of Riemannian (differentiable) manifolds in 3-space. There is a homeomorphic relationship between the order of closed spaces in 2-space and the order of topological manifolds, generator of volumes, in 3-space. There is a homeomorphic relationship between the order of semi-closed/semi-open spaces or layers and the class of Finsler manifolds in 3-space. Tri-Valent Quintessence starts in 2-space and propagates thru n-space by homeomorphism, keeping in mind that a specific axis transformation is required to transition from one space form to another. This homeomorphism is a result of the fact that at each subsequent n^{th}-space the spatial entity develops a richer multiplet state than the previous.

5.9. Quanto-Geometric Primal Operators

In a nutshell, consider this: arithmetic calculus is based on 4 simple operators of primarily passive agency:

- The Monolithic Operator that promotes the method of addition [+].
- The Parity Operator that promotes the method of Sci-Parity or Se-paration, which we call subtraction [-]. The method of division [/] for its own is a poster child to the method of subtraction.
- The Associative Operator which promotes the relational method of multiplication [$\times$].
- The Radical Operator which promotes the methods of square root (or roots) [$\sqrt{}$] and squaring $[x]^2$.

There exist 5 additional Operators, more complex and primarily of active agency, that are at the root of the calculational methods put forth in functional calculus otherwise called functional analysis. They are:

- The Complex Operator which promotes calculus of the imaginary.
- The Harmonic Operator which promotes matrix calculus.
- The Rationalization Operator which promotes irrational calculus.
- The Golden Operator which promotes algebraic calculus.
- The Transcendental Operator which promotes serial calculus.

The norms that are rooted in the Quanto-Geometric Number Stack (Chapter 3) create not only the class of Normalized Tensors that we have previously studied but the parallel class of Normalized Operators above enumerated, this thru the Quanto-Geometric Function. One can see that the natural bounds of the span of *operation* are the punctual Monolith at one end of the spectrum and the Infinite series at the other end. The Quanto-Geometric formal expression for these irreducible put under the same umbrella not only the functionally-woven Operators (the last 5) but the arithmetic Operators as well.

From mathematical study undertaken over the Quanto-Geometric Function (Chapter 3), the ensuing generic expression of the Quanto-Geometric Operators is expressed below:

$$\|\Psi\| = \left[\frac{-Q^2}{2q}\right] \bullet \left[\frac{2}{ae^{\left(\frac{2a+b^2}{4}\right)} + be^{\left(\frac{2a+b^2}{4}\right)} + ce^{\left(\frac{2a+b^2}{4}\right)}}\right] \quad (5\text{-}6)$$

where:

q is the scalar and equates $e^{a/2}$ or: $q = e^{\left(\frac{a}{2}\right)}$

Q is the primal quanto-geometric momentum $q \times s$ or: $Q = q \times s$

a is a coefficient related to q and equates $2 \ln q$ or: $a = 2\ln q$

b is a coefficient related to the standard deviation parameter and equates $\ln \sigma$ or: $b = \ln\sigma$

c is the constant $\ln 2\pi$ or: $c = \ln 2\pi$

The idea that originally motivated the inception of functional Operators in mathematical science was the question of finding the shape of a vibrating string in a musical string instrument for a certain family of notes. The string is fixed at its boundaries and has a certain length. Those are the invariants that condition the variance that will be given in the ensuing standing waves the chord will be subject to. It is by posing the problem in terms of a differential operator denoting the displacements of the infinitesimal portions of the string that we obtain a set of eigenfunctions and eigenvalues to this operator as the solution to the problem. The operator is of the form of a second-order derivative:

$\frac{\partial^2}{\partial x^2}$ in the length dimension

and

$\frac{\partial^2}{\partial t^2}$ in the time dimension

acting on a function h of the displacement and time variables: $h(x,t)$.

Likewise, we may pose the question of finding the set of wavefunctions or harmonic oscillations experienced by a space spread (field) in codependent relationship with a scalar partner as the most generic question surrounding the ontology of any physical object. To that aim we may bring the Quanto-Geometric Operator to the fore, recognizing that the family of functions embedded in the Quanto-Geometric Function are the eigenfunctions along with their eigenvalues that are proper to the generic Quanto-Geometric Operator.

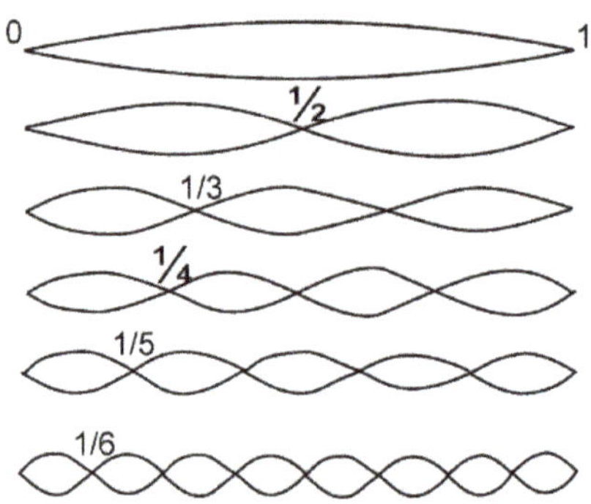

Fig. 5.9 Overtone waveforms of a vibrating string or drum as eigenfunctions and eigenvalues

The harmonic pattern in Fig. 5-10 is very similar to the spectral pattern (Fig. 5-9) imputing the analytic notion of normed function space with intervening Operator, except for a few important distinctions. The distribution in Fig.5.10 portrays a sequence of harmonic waveforms involving different families of functions. Every eigenvalue is thus associated with a waveform of a distinct nature from the next. The eigenvalues are not just a coefficient to the period of the waveform or even to the function but come as cosets implicating both the wavenumber (or wavelength) and the scalar density associated with the waveform. What this characterization reflects is that the Quanto-Geometric eigenspectrum represents a level of generalization that lies far above abstract Hilbertian function space, that which plays a central role in traditional Operator analysis.

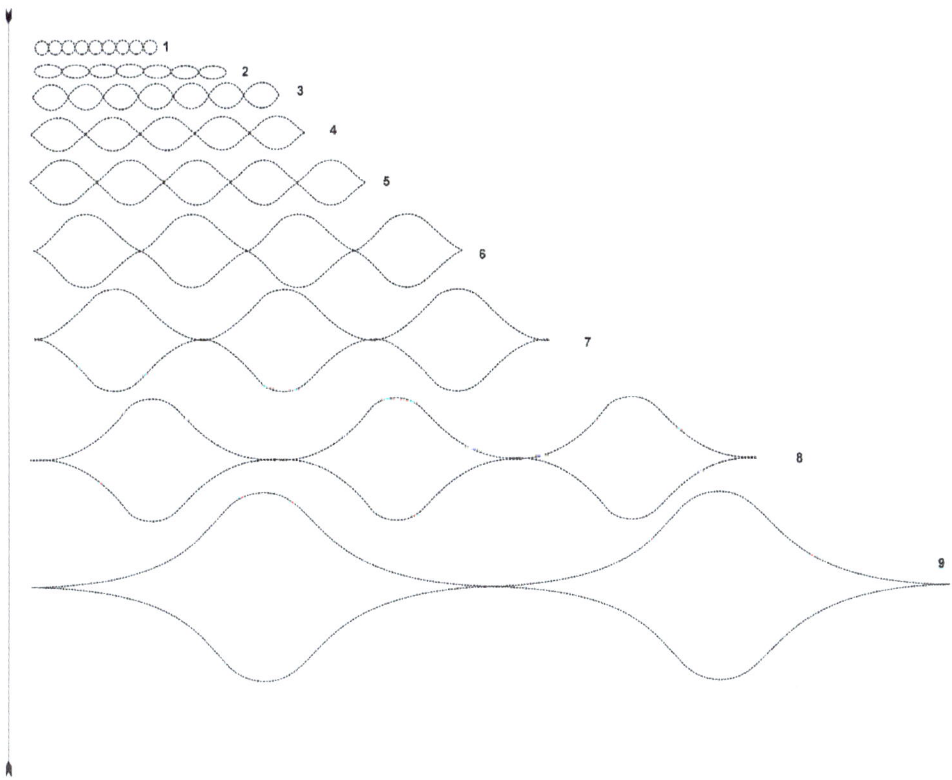

Fig. 5.10 Waveform spectral of families of functions embedded in the Quanto-Geometric Function

Every layer of waveforms represents in fact a whole functional domain, which then might host different harmonic spectra of Hilbert spaces. Every functional layer of the Quanto-Geometric spectrum constitutes the Primitive or Generator of a normed function space, or, if you will, an Operator primitive. One of the properties that significantly distinguish the Quanto-Geometric Operator from Hermitian Operators is the ontological character subsumed in the variables of their function basis. The main quality of Hermitian Operators is the geometric orthogonality of their L^2 eigenfunction basis. Comparatively, the orthogonality of the Quanto-Geometric Operators rests on the dual ontological set inscribed in the unit function of their function basis, while their norm is given by the typology of the set. Expression *(4-1)*, which mandates the scalar tenet of the function and its space tenet to be the reciprocal of one another, is the most iconic representation of this orthogonality. In that sense there is homomorphism between the orthonormal criteria advocated in both frameworks.

Hilbert orthogonality is between an element f and its conjugate g or $\bar{f}$ forming an inner product. The duality is between two conjugates and their relationship is measured in terms of a geometric angular element, specifically the cosine of the angle formed by their representative vectors. The angle must be *90°* or the cosine of *null* value for orthogonality to hold in the vector space. In Quanto-Geometry, the function space is an entire functional domain and every eigenfunction primitive of the different domains inherits the qualities set forth by the towering Quanto-Geometric relation expressed in *(5-1),* which in and of itself is seminally orthonormal.

In summary, the Quanto-Geometric functional system is based on a 9-dimensional real-coordinate space, with each of the 9 eigenfunctions in the role of generator of a family of functions with typified *eigenforms* (circle, ellipse, etc.). The generic Quanto-Geometric Operator is the master key to the 9-dimensional eigenspectrum and grandfathers 9 typified Operators, each acting on or adjoined to the specific functional family in their domain. All known or theorized abstract spaces are lower-level entities of partial or limited scope which pertains to the different functional domains germane to the Quanto-Geometric eigenspectrum.

5.10. Expressions of the Nine Specific Quanto-Geometric Operators

Consonant with the above, we may now express the generic Quanto-Geometric Operator as follows:

$$\|\Psi\|_{a_i b_i} = \left[\frac{-Q^2}{2q}\right] \bullet \left[\frac{2}{a_i e^{\left(\frac{2a_i+b_i^2}{4}\right)} + b_i e^{\left(\frac{2a_i+b_i^2}{4}\right)} + c e^{\left(\frac{2a_i+b_i^2}{4}\right)}}\right]$$

This Operator applies to the whole eigenspectrum portrayed in Fig.4.10. Every sub-index i in $a_i b_i$ represents a *q-s* set, yielding the following suite of generators or primitive Operators:

Monolithic Operator:

$$\|\Psi\|_{a_1 b_1} = \left[\frac{-Q^2}{2q}\right] \bullet \left[\frac{2}{a_i e^{\left(\frac{2a_1+b_1^2}{4}\right)} + b_i e^{\left(\frac{2a_1+b_1^2}{4}\right)} + c e^{\left(\frac{2a_1+b_1^2}{4}\right)}}\right]$$

Parity Operator:

$$\|\Psi\|_{a_2b_2} = \left[\frac{-Q^2}{2q}\right] \bullet \left[\frac{2}{a_2e^{\left(\frac{2a_2+b_2{}^2}{4}\right)} + b_2e^{\left(\frac{2a_2+b_2{}^2}{4}\right)} + ce^{\left(\frac{2a_2+b_2{}^2}{4}\right)}}\right]$$

Associative Operator:

$$\|\Psi\|_{a_3b_3} = \left[\frac{-Q^2}{2q}\right] \bullet \left[\frac{2}{a_3e^{\left(\frac{2a_3+b_3{}^2}{4}\right)} + b_3e^{\left(\frac{2a_3+b_3{}^2}{4}\right)} + ce^{\left(\frac{2a_3+b_3{}^2}{4}\right)}}\right]$$

Radical Operator:

$$\|\Psi\|_{a_4b_4} = \left[\frac{-Q^2}{2q}\right] \bullet \left[\frac{2}{a_4e^{\left(\frac{2a_4+b_4{}^2}{4}\right)} + b_4e^{\left(\frac{2a_4+b_4{}^2}{4}\right)} + ce^{\left(\frac{2a_4+b_4{}^2}{4}\right)}}\right]$$

Imaginary Operator:

$$\|\Psi\|_{a_5b_5} = \left[\frac{-Q^2}{2q}\right] \bullet \left[\frac{2}{a_5e^{\left(\frac{2a_5+b_5{}^2}{4}\right)} + b_5e^{\left(\frac{2a_5+b_5{}^2}{4}\right)} + ce^{\left(\frac{2a_5+b_5{}^2}{4}\right)}}\right]$$

Matrixial Operator:

$$\|\Psi\|_{a_6b_6} = \left[\frac{-Q^2}{2q}\right] \bullet \left[\frac{2}{a_6e^{\left(\frac{2a_6+b_6{}^2}{4}\right)} + b_6e^{\left(\frac{2a_6+b_6{}^2}{4}\right)} + ce^{\left(\frac{2a_6+b_6{}^2}{4}\right)}}\right]$$

Rationalization Operator:

$$\|\Psi\|_{a_7 b_7} = \left[\frac{-Q^2}{2q}\right] \bullet \left[\frac{2}{a_7 e^{\left(\frac{2a_7 + b_7{}^2}{4}\right)} + b_7 e^{\left(\frac{2a_7 + b_7{}^2}{4}\right)} + c e^{\left(\frac{2a_7 + b_7{}^2}{4}\right)}}\right]$$

Golden Operator:

$$\|\Psi\|_{a_8 b_8} = \left[\frac{-Q^2}{2q}\right] \bullet \left[\frac{2}{a_8 e^{\left(\frac{2a_8 + b_8{}^2}{4}\right)} + b_8 e^{\left(\frac{2a_8 + b_8{}^2}{4}\right)} + c e^{\left(\frac{2a_8 + b_8{}^2}{4}\right)}}\right]$$

Transcendental Operator:

$$\|\Psi\|_{a_9 b_9} = \left[\frac{-Q^2}{2q}\right] \bullet \left[\frac{2}{a_9 e^{\left(\frac{2a_9 + b_9{}^2}{4}\right)} + b_9 e^{\left(\frac{2a_9 + b_9{}^2}{4}\right)} + c e^{\left(\frac{2a_9 + b_9{}^2}{4}\right)}}\right]$$

Let me note in passing that the term Golden for type-8 Operator is not a phenomenological attribute but a tribute to the key role played by the Golden Ratio in fomenting algebraic calculus, per numeric analysis of inner structures of the Higher Number Stack exposed in Chapter 4.

The functional domain associated with the Monolithic Operator is the family of circles or circular functions.

The functional domain associated with the Parity Operator is the family of elliptic functions.

The functional domain associated with the Associative Operator is the family of parabolic functions.

The functional domain associated with the Radical Operator is the family of hyperbolae.

The functional domain associated with the Hyperboloid Operators consists in the five distinct families of hyperbolic functions.

In support of this view, a direct exponential expression of each of the first 4 families of functions, the so-called conic functions, is well known in different forms. The root Jacobi theta function that gives rise to the family of so-called *elliptic functions* has the following exponential expression:

$$\theta_{ab}(x\,|\,\tau)=\sum_{n\in Z}\exp\left[\pi i\tau\left(n+\frac{a}{2}\right)^2+2\pi i\left(n+\frac{a}{2}\right)\left(x+\frac{b}{2}\right)\right]$$

where τ is the modulus or eccentricity and $a,b = 0,1$.

The general class of Jacobi elliptic functions may be viewed as the root for the exponential expression of all 4 known conic. Effectively those functions are closely related to the geometry of *elliptic curves*, in particular the familiar conic ellipse which historically motivated their development amidst the quest for methods to compute the arc-length of the latter. Further their rectangular arguments can be parameterized to produce 4th-, 3rd- and particularly 2nd-degree polynomials, which are conic parabolic, while they may drop down to circles as well. Needless to point out their relationship with the hyperbola, since they hold a parallel trigonometric expression.

It is noteworthy that these exponential expressions almost always include a complex element, while they may alternatively include matrix terms as well. In that regard, the quanto-geometric spectrum duly shows that there is no direct path from exponential calculus to quadratic calculus on down. Thus, the expression of a specific level of the stack from elements of a higher level stratum must include any intermediate calculational figure.

5.11. Self-Adjoint Quanto-Geometric Operation

We must stress that the functional representation of these Operators does not supersede or do away with the arithmetic operators but simply re-formulate them. The first 4 expressions of these Operators, namely the Monolithic Operator, the Parity Operator, the Associative Operator and the Radical Operator are respectively completely equivalent to the familiar arithmetic operators. Consider this: when we see for instance a set of 2 chairs in a corner of a room and another set of 3 chairs in the next corner of the room, and say there are 5 chairs in the room, which we express as 2 [+] 3 – 5, what we are doing as observers or analysts is applying the Monolithic Norm to the scalar state of these observables by which we aim to conceptualize them as **one (mono)** group of objects. Note that there is operation as well on the spatial state of these objects from the spatial mono-centric symmetry element encapsulated by the Operator, which we normally do not consider part of the summation operator in elementary Number Theory, and perhaps wrongly so. If we further physically join these chairs, then we have realized or made explicit the *transformation* implicit in the Monolithic Norm. So therefore, it is perfectly correct to write:

$$[+] \equiv \|\Psi\|_{a_1 b_1}$$

to mean equivalency between the arithmetic operator of summation and the Monolithic Quanto-Geometric Operator. Likewise, we may apply the same rationale to the phenomenology of these objects in the evoked context insofar as the two subsequent arithmetic operators, which would give us:

$$[-] \equiv \|\Psi\|_{a_2 b_2}$$

$$[\times] \equiv \|\Psi\|_{a_3 b_3}$$

If there were only 3 chairs in one and only one corner of the room, under the condition that there are 9 people needing accommodation in the room, we might consider as agents a replicate of these 3 chairs in two other desired corners of the room. In thinking of the original group of chairs, we might conceptualize a squaring of the scalar state of those chairs by applying one of the conjugate components of the Radical Operator to their scalar state, specifically the Square, for the desired transformation of obtaining more of these chairs. Thus, we may justifiably write:

$$\left[\sqrt[n]{} : \langle m^n \rangle\right] \equiv \|\Psi\|_{a_4 b_4}$$

to mean equivalency between the arithmetic operator of squaring and the Radical Quanto-Geometric Operator. The reason why the 5 hyperboloid Operators do not have simple arithmetic replicates is because of the preeminent role played by the exponential wavefunction component in the transformations realized by these operators, which is expressed in the wealth of methods associated with their realization. The great diversity of these methods constitutes the province of Functional Analysis. However, when a molecular entity of the biosphere such as a gene is able to infinitely and exactly replicate itself across an endless number of generations, one can be sure that a Transcendental Operator is at work within the seminal structure of this object, whereby the wavefunction plays an absolutely preeminent role vis-à-vis that one played by the scalar tenet of the Operator.

In mathematical science, we have not thus far considered the many different functional methods in Functional Analysis as calculational operators deserving nomenclature based on operator symbols similar to the ones adopted for the four arithmetic operators. The sum of these methods pertains to the hierarchy of the hyperboloid Operators. And therefore, all nine Quanto-Geometric Operators, in their condition as direct expression of arithmetic and calculational operators, are all primal *self-adjoint* operators.

5.12. Quanto-Geometric Linear and Non-Linear Operators

The Transcendental Operator presents a special attribute that uniquely distinguishes it from all others, in particular from the other functional Operators of the spectrum. With its inherent methods of derivative and integrative operation, this entity presents the unique characteristic of linearity primarily due to the exponential nature of its quanto-geometric eigenfunction. We attribute the character of linearity to an operator A acting on a function f whenever the operator returns from the transformation the same exact function except for multiplication by a scalar constant, as expressed below:

$$Df = af$$

where D is the Operator, f the function and a the scalar constant. Because the exponential function is the only known one that is at the same time its own derivative and its own integral, it becomes the eigenfunction of the derivative operator and the integral operator.

Therefore, the function space originated in the transcendental layer of the Quanto-Geometric Function, being the province of exponential calculus, becomes the haven of Linear Operation. Analysis of the structure of this layer of the Quanto-Geometric Function is quick to reveal that because of the eternal vicinity of the curve to the asymptotic s-axis in that sector, the Function behaves linearly or with effective linear symmetry therein, due to straight parallelism with the asymptote. This essential character of Linear symmetry reflects on the function space and thus onto the Operator, making it not an abstract but a real character.

All other Quanto-Geometric Operators (types 8 to 1) are non-linear because the symmetry *mani-folded* in their function spaces is curvilinear, since those spaces are strictly made up of conic and more or less curved hyperboloid sectionals.

5.13. Inner Ontological Properties of the Quanto-Geometric Operators

The very specific values taken by the set *[a, b]* in the expression of these Operators determines the Primitive identity or Norm of the operator, in other words one among the 9 Quanto-Geometric *normals*. It cannot be overstressed that none of these function spaces is an abstraction or an abstract Hilbert space, but represents each a property of real coordinate space. By the same token, and in catering to the smooth span of these 9 basis function spaces, one can inarguably concord with the alternative designation of Grand Eigenfunction that we have attributed to the Quanto-Geometric Function itself.

Delving somewhat further into the structure of the generic Quanto-Geometric Operator, let us note that the leftmost term in its expression may be assimilated to an energy density term that establishes the order of potential energy in an object as its highest scalar state, a static state:

$$\left[-\frac{Q^2}{2q}\right].$$

The next term at the left side is the wavefunction term that grandfathers the kinetic energy of the object in its most elevated form while defining its form of symmetry:

$$\left[\frac{2}{a_i e^{\left(\frac{2a_i+b_i^2}{4}\right)}+b_i e^{\left(\frac{2a_i+b_i^2}{4}\right)}+c e^{\left(\frac{2a_i+b_i^2}{4}\right)}}\right].$$

One can readily notice the hyperbolic form of expression in that term as the ratio between 2 and a sum of exponentials typical of the hyperbolic functions, very particularly the *sech* function. Each of the two terms gives us a composition between *q* and *s*, but with one of the two variables with a preeminent weight in each term, thereby giving the specific term its final ontological identity. In that regard, we must remember that *e* is the numeric expression of space in transcendentalization and that σ is a value of *s*.

5.14. A Parallel Between the Generic Quanto-Geometric Operator and the Hamiltonian Operator

When scrutinizing the nature of the generic Quanto-Geometric Operator, it is interesting to compare it to the Quantum Mechanics Hamiltonian:

$$\left[-\frac{\hbar^2}{2m}\cdot\frac{d^2}{dx^2}+V\right]=H \quad (5\text{-}7)$$

The core term that conforms the *operation* in the Hamiltonian is:

$$\left[-\frac{\hbar^2}{2m}\cdot\frac{d^2}{dx^2}\right],$$

the kinetic energy Operator, which equation *(5-7)* above drops down to when the potential energy V equals *0*. The Hamiltonian Operator is indeed a reflection of the Quanto-Geometric operator canon in that it is a composition between an energy density term, the first one at left of the expression, identical in form to the Quanto-Geometric scalar state term, and a *wavefunction* term, the second one, here expressed as the second-order derivative of a *function*. Rewriting the core term as expressed below makes even more apparent the parenthood relationship between the Hamiltonian and the generic Quanto-Geometric Operator:

$$\left[-\frac{\hbar^2}{2m}\right]\bullet\left[\frac{d^2}{dx^2}\right].$$

It is by virtue of narrow compliance to the higher canon that the eigenfunction of the Hamiltonian Operator turns out to be:

$$\phi_m(\varphi)=\frac{1}{\sqrt{2\pi}}\cdot e^{im\phi},$$

a functional expression in polar coordinate that is a faithful image of the Quanto-Geometric Function. So it is, not just because the Hamiltonian is a linear operator. In particular, the wavefunction term of the generic Quanto-Geometric Operator contributes to a significant extent to clarify the nature of the Quantum Mechanics wavefunction, a subject of historically unrelenting controversy. The wavefunction term of the Quanto-Geometric Operator initially instantiates the construct that the Quantum wavefunction is not an abstract vector space, but a real-valued *eigenform*, specifically one of 9 possible distinct universal *eigenforms*. Therefore, at the very least, it proposes that:

1. A wave of any nature is not necessarily, if at all, an expression in *time* of the eternal sinusoid but a specific spatial *eigenform* of which the sinusoid as a circular manifold is only a class.

2. A wave is dynamically a complex function of two quantities: one that is both coupled to and expressed by the *eigenform* with vibrational or oscillatory manifestation, and a quite separate entity which is its degree of propagation, *0* being a possible value of the latter.

3. The very fabric of the void is the originator of all harmonic dynamics thru the modality of *de-ploys* of its intangible continuum. In that regard, consider that a differential operator of any n^{th} order, progeny of the spatial term of the Quanto-Geometric Operator, is an entity of active agency engendered thru auto-dynamism, as it were, of a metaphoric secant line to a curve moving to its single-point tangential location.

These notions are all implicit in the Quanto-Geometric *wavenumbers* in their s-space form or (m,t)-space form, the independent variable to the Quanto-Geometric coordinate system respectively in 2-space and 3-space. There is for certain far more to the notion of wavefunction than we have expounded here. Much more in Chapter 7.

With the analytical order of Operators concluding with the notion of *eigenforms*, we have circled back to the analytic order of Tensors and Manifolds, in a tight demonstration of the consistency of the whole Quanto-Geometric system of constructs. Effectively there is within the framework a direct Operator expression of the normalized Tensors. However, we will spare it to this exposé.

5.15. Quintessential Tri-Valent Operators

Similar to the Tensor order, the triplet partition of the Quanto-Geometric Function equally has an echo in the order of *operation* subjacent to the Function. Despite the fact that we have repeatedly used the functional figure of first-order derivative to extol the quintessential triplet characterization, it is important to understand that the latter is not subject to nor is it intrinsically defined by functional derivatives.

It remains obvious that the first 3 conic families of functions that are ostensibly manifest in the graphical rundown of the Function, namely the circle family, the elliptic family and the parabolic family, have in common a "closed" form of symmetry, explicit in the orientation of the tangential angle to that curvilinear sector of the Function, while the last 5 families share an "open" form of symmetry manifest in the orientation of the sweeping tangential angle as well. We say that each normal sectional is a family because they are all parameterized on the basis of a, b and σ, and indeed beyond those parameters, as previously explained, to produce their own spectrum. The quadratic sector is a Hilbert space as well for the same reason. The partial function space represented by each sector is smooth and orthonormal, and therefore represents each the partial spectrum of eigenfunctions of an adjoint operator.

We denote these Operators:

1. The Hyper Quantum Operator for the first sector

2. The Hyper Space Operator for the second sector

3. The Hyper Quadratic Operator for the mid-sector

The intervention of each of these three operators on physical observables respectively causes an **accrued** scalar densification, an **accrued** linearization or *superficialization* of the object and an **accrued** form of ambivalence or mirror imaging within the same.

While we normally and principally use functional linear operators, of Hermitian nature or else, to query state and dynamics of observables, the entire suite of Quanto-Geometric Operators is equally applicable either to *query* the unknowns about observables or to *inform* an observable with a quality emanating from the Norm of the Operator, thereby *creating* a *transformation.* In the latter condition, these functional Operators exert their *transformative norm* in a way that is relatively similar to the arithmetic operators.

5.16. Universality of the Quanto-Geometric Operators

It is important to realize that the Quanto-Geometric Operators apply not only at the level of the quantum realm but at all subsequent levels of the development of physical matter.

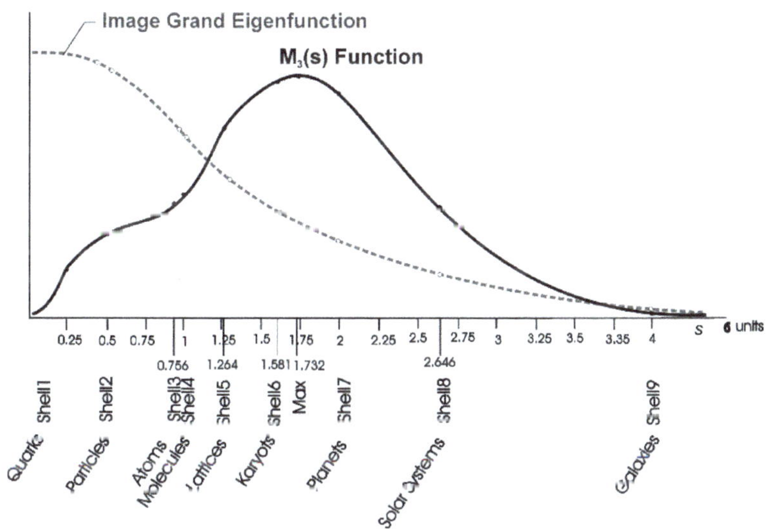

Fig. 5.11 An incipient view of the suite of Shells in the transverse order

In their genericity, they represent an essential scheme that informs all ontologies thru the course of the buildout of the several different Shells of matter as they appear before us, namely the Quark Shell, the Particle Shell, the Atom Shell, the Molecular Shell, the Material or Lattice Shell, all the way up to the Galactic Shells (Fig. 5.11).

Therefore, on that account, they duly inform the ontology of objects in the Astral Shell (individual planets, moons or satellites, comets and stars), habitat of all Karyots. Classical physics has ascertained the following notable quantities from dynamical observables in our immediate environment:

Linear Kinetic Energy

From $K_l = m\ v^2/2,\ p = m \times v \rightarrow \quad \mathbf{K_l = p^2/2m},$

where p is the linear momentum of the object in motion.

Rotational Kinetic Energy

From $K_r = I\ \omega^2/2, \quad H = I \times \omega \rightarrow \mathbf{K_r = H^2/2I},$

where H is namely the rotational momentum of the object in motion.

Energy in Magnetic Flux of Inductor

$U = L\ i^2/2, \qquad Li = N\Phi \quad \rightarrow \qquad \mathbf{U = (N\Phi)^2/2L},$

where the quantity $N\Phi$ is defined as the flux linkage of the inductor.

Energy Density in an Inductor

$U = \mu_0 n^2\ i^2/2, \quad \beta = \mu_0 n\ i \quad \rightarrow \qquad \mathbf{U = \beta^2/2\mu_0}$

where β is the strength of the magnetic field set up by a current-carrying inductor.

Energy Stored in a Capacitor

$U = C\ V^2/2,\ Q = CV \qquad \rightarrow \qquad \mathbf{U = Q^2/2C},$

where Q is the total amount of charge stored on the plates of a capacitor.

Energy Density in Field of a Charged Surface

$$U = \varepsilon_0 \, E^2/2, \quad \sigma = \varepsilon_0 E \quad \rightarrow \quad \boldsymbol{U = \sigma^2/2\varepsilon_0},$$

where σ is defined as the charge density on a surface.

Kinetic Energy in Liquid Flow

$$K = \sigma v^2/2, \; F = \sigma v \quad \rightarrow \quad \boldsymbol{K = F^2/2\sigma},$$

where F is the flow rate of a liquid.

Elastic Energy Stored in a Spring

$$U = k\,x^2/2, \quad F = k\,x, \quad \rightarrow \quad \boldsymbol{U = F^2/2k}$$

The second expression is known as Hooke's law, and F the restoring force of a spring.

Save the negative sign, these quantities are completely conformal to the scalar component of the generic Quanto-Geometric Operator. It follows from this treatment that the Newton's First Law approach, which essentially creates the momentum-like figure as a product between the scalar and its motion, is only phenomenological in appearance (or descriptive of exogenous dynamics otherwise called interactions) in some cases. From the Quanto-Geometric vision the momentum figure emerges instead as an ontological quantity about the internal composition of an object as part of its scalar component. The Quanto-Geometric Operator predicates that what appears to be the compactified component in an observable residing within any Shell of matter, including our own Shell of habitation, that which we call mass or quantum or energy packet is never a pure monolith but always a density quantity in the form of an integral sum or packet modeled by $\left[-\frac{Q^2}{2q}\right]$. It represents the first ontological level of the scalarity of the observable. In that regard, the energy nomenclature for the above reported quantities would be quite misplaced if we do not consider the notion of energy as the very primal nature of the object or observable in question. It is quite in accordance with the *Hamilton principle* which ultimately produces a description of all classical physics from an *Energy* view of those physical observables. Furthermore, the Quanto-Geometric Operator prescribes that there is always a dynamic spatial component to the observable, directly reflected in its eigenform, whether pure or composite, which is the ontological wavefunction component of the observable as modeled by the term:

$$\left[\frac{2}{a_i e^{\left(\frac{2a_i+b_i^2}{4}\right)}+b_i e^{\left(\frac{2a_i+b_i^2}{4}\right)}+ce^{\left(\frac{2a_i+b_i^2}{4}\right)}}\right].$$

Thus far in modern physics, we have been completely oblivious to this representation having maintained a predominant view of pure and quasi exclusive compactification about the ontology of physical observables. For that matter, the above reported quantities from our physics have no accompaniment in the space realm.

Contrastingly, this treatment allows us to capture the universality of the background throughout the span of Shells of matter in our 3-dimensional universe, while preparing us to capture the eigenvalue elements that make up the specific invariants upon which each Shell is built, those which modern physics have denoted the fundamental physical constants of Nature. The visualization of the transverse Shells of matter (Fig.5.11) beyond the important but insufficient Einsteinian concepts of Lorentzian transformations on Minkowski spacetime contemplated in the macroscopic arena, along with the ensuing computation of the pivotal invariants that sustain them, constitutes the royal way that takes us to Grand Unification.

To elicit the application modality of the Quanto-Geometric operators across all Shells of matter is to:

1. Apply to the observable any necessary mathematical transformation that results in turning up its implicit functional expression, if said expression is clearly unapparent at the outset.

2. Scrutinize the implicit functional expression for conformity to any one of the two ontological tenets of the Quanto-Geometric principle in the form depicted above for the Operator for a direct representation of these forms or thru integration, according to the prescription below:

$$\|QG\|=\left[\left(\int_B^A f\cdot ds\right)\bullet g(s)\right]\equiv\left[-\frac{Q^2}{2q}\right]\bullet\left[\frac{2}{a_i e^{\left(\frac{2a_i+b_i^2}{4}\right)}+b_i e^{\left(\frac{2a_i+b_i^2}{4}\right)}+ce^{\left(\frac{2a_i+b_i^2}{4}\right)}}\right],$$

f being the functional expression for the scalar component and g the wavefunction component.

3. Analyze the coefficients of these terms in comparison with the Quanto-Geometric eigenvalue cosets in order to determine the exact Operator among the set of 9 the observable conforms to.

When it comes to the much-celebrated relativistic Energy formula $E = mc^2$, in the light of the above constructs, one might wonder why it did not turn out to be instead $E = mc^2/2$ in straight conformance to the scalar component of the generic Quanto-Geometric Operator.

In attempting a response to this hypothetical, it is important to recognize that the relativistic Energy formula does not refer to the wavefunction component of objects but strictly to the scalar component and appears as a degeneracy of the latter. Effectively the relativistic energy formula is admittedly a statement of the energy harbored by an object as a result of its mass. Since this expression is also valid for an object at rest $(E = m_0 c^2)$, it is therefore a statement that c must be an attribute of a spatial property participating in the scalar ontology of the object. The best way to interpret this notion is to consider mass or all physical scalar indeed as a density quantity, as posited by the scalar component of the Quanto-Geometric Operator. On that account the expression of the relativistic energy is largely equivalent to the former.

Furthermore, the Relativistic Kinetic Energy expression *(KE)* is going to put us closer to the Quanto-Geometric scalar expression:

$$KE = mc^2 - m_0 c^2 \quad (5\text{-}8)$$

which so defines the relativistic kinetic energy of a particle as energy in excess of its ontological or rest-mass energy. Because there cannot be a conflict between the internal wavefunction of the scalar tenet and the actual explicit wavefunction component of the moving object, there must be a proportionality relationship between the propagation rate associated with the ontological wavefunction in all facets *(c)* and the exogenous speed of the moving object *(v)*. In other words, the moving object can never move at a rate that is larger than the propagation rate of its basis wavefunction (both as part of its scalar packet and its space spread tenet). That is where the Lorentz transformation factor comes into play as well as the specifics relating to the internal quanto-geometric covariance at play. We will not address here the rather involved study of quanto-geometric covariance in this context which is undertaken in Book II of this publication. We will proceed only by calling the direct ratio between the two speeds as well as the Lorentz factor into consideration. When v is much smaller than c, we have:

for $v/c << 1 \rightarrow \quad KE = mc^2 - m_0c^2 \approx \dfrac{m_0 v^2}{2}$ (or $\dfrac{p^2}{2m_0}$)

For $v/c \rightarrow 1 \rightarrow KE = \dfrac{1}{2} m_0 v^2 + \dfrac{3}{8}\dfrac{m_0 v^4}{c^2} + \dfrac{5}{16}\dfrac{m_0 v^6}{c^4} + \ldots$ *(5-9)*

We arrive at expression *(5-9)* for v approaching c thru the Lorentz conversion factor:

$\gamma = \sqrt{1 - \left(\dfrac{v}{c}\right)^2}$ with its factoring into the Relativistic Kinetic Energy expression, and additionally thru the binomial expansion of the latter. The preeminent role played by the standard scalar term in both expressions is inarguable and need no further argument.

The above analysis requires us to firmly bear in mind two key points:

- The quanto-geometric scalar is always a dimensional scalar. That is to say that all mass is in the form of density in and of itself.
- When all tensors vanish, including within the dimensional scalar component of an object, the parabolic tensor inalterably persists as a property of the background basis space (Universal 3-space) that participates in the mass component buildout of the object. The single most important dynamic property associated with the parabolic tensor is c.

I will conclude this section by stating that while Lorentzian transformation based on the invariance of c in all reference frames is the foundation of relativistic mechanics, we are going to supersede that vision by directly deriving c from the very axiomatic of the Quanto-Geometric framework, in a demonstration undertaken in the next Chapter.

5.17. Phenomenological Primitives

In this brief survey of analytic structures and methods inherent to the Quanto-Geometric Theory, we shall mention that the class of 9 ontologies in the Functional order, deriving into the orders of Operators and Tensors, has a precise echo in phenomenology. If we assume the ontologies to be Primitive Identities, their strict numeric signatures give rise to behavioral or phenomenological attributes.

We have chosen the nomenclature summarized in Table 5.1 for the Quanto-Geometric Primitives in phenomenology. Barring some variance in the denominations, they are exactly equivalent one to one to the Primitives studied in the previous Chapter. We are only making here a narrower connection in the nomenclature with self-adjoint or calculational operators. One should read the first column of the Table as Scalar State-1, Scalar State-2, etc. or Primitive-1, Primitive-2 and so on and so forth. The integer symbol is not to mean an ordinal for the purpose of ranking but stands for a scalar state or a Primitive.

Scalar State/Primitive	Numeric Identity	Phenomenology
1	Monolithic	Egocentrism
2	Binary	Cooperative
3	Associative	Relational
4	Radical	Ambivalence
5	Imaginary	Metamorphism
6	Matrixial	Harmony
7	Irrational	Paradox
8	Algebraic	Perfection
9	Transcendental	Infinity

Table 5.1 Nomenclature of Phenomenological Primitives

This characterization directly applies to all physical observables but is very particularly useful when approaching unknown phenomena initially unyielding to Functional characterization. In the life sciences, or the karyotic span in the 3-D universal arena, where strict and full mathematical formalism may be of difficult application at first, the phenomenological approach is well granted in the attempt to elucidate their dynamics, well beyond the usual statistical accounts. Nevertheless, one must remain fully aware that there is not one single aspect of physical reality that escapes full Quanto-Geometric characterization, as much as there exists no physical observable without a direct numeric expression. Every numeric or scalar state is a direct path to phenomenology and from phenomenology the observer may directly funnel into the deep realms of Functional analysis. All in all, the analyst must be convinced that every physical object is quanto-geometrically instantiated in terms of identity, attribute and behavior (in accordance with the Tri-Valent canon), all three of which make up their ontological signature.

5.18. Conclusion

In this presentation of the Quanto-Geometric Theory, we have exposed an all-encompassing framework that addresses the nature of space and the quantum-scalar along with their correlation throughout the continuum of the Universal Arena. We have elicited an order of Operators and Tensors directly applicable to the entire continuum of the 3-dimensional Arena. This demonstration represents a decisive step leading to the achievement of theoretical Unification long sought thru the momentous derivation of the entire suite of fundamental physical constants of Nature. In the aggregate, in assessing the value of any system of constructs toward Grand Unification, one must remain keenly aware that the gauging barometer is not at all the proliferation of "beautiful" mathematical tours de force for their own sake but the ostensible ability to derive consistently, effectively and unequivocally from that theoretical framework the fundamental constants of Nature that modern physics has over time arrived at only empirically.

Chapter 6

All things of your ignorance together make up
a realm that is much grander than you.
*Proverb by the **Fon-Yoruba People***
of Haiti

DERIVATION OF THE FUNDAMENTAL PHYSICAL CONSTANTS OF NATURE

In this Chapter we will attempt to put to rest the long debate between the meta-physical view of the Universe contended by occultists and religion in general and the strictly physical view upheld by rationalists such as scientists, mathematicians and others. This discussion posits a framework of formalism from which a whole suite of the most fundamental physical constants is directly derived. It further sets rational grounds for an interpretation of the meta-physical realms from where paranormal experience develops. I must warn the reader however that this discussion requires sound knowledge of intermediate-level mathematics to be amenable, lest the reader is willing to simply lend credence to results and conclusions and overlook the mathematical developments. A sound grasp of the concepts discussed in any of the two previous Chapters is key to an understanding of what is to follow.

6.1. Mappa Mundi of the Universe

The primary Quanto-Geometric function, in its role of universal modeler of all things in existence, represents the grand map of the Universe in its fullness.

$$Q(s) = \pm\left|\frac{s^0}{\sigma \cdot \sqrt{2\pi}} \bullet e^{-s^2/2\sigma^2}\right|, \ s \neq 0$$

The reader recalls the primeval q_n symmetry-breaking sequence introduced in Chapter 2 which helps delineate the suite of natural curvatures embedded in the function:

$$\{q_n\} = \left\{\frac{e^{(1-n)/(10-n)}}{\sigma\sqrt{2\pi}}\right\}, \text{ where } 1 \leq n \leq 9$$

The quantity n relates to s and determines the *degrees of freedom of coordinate space.* By that measure there exist 9 different dimensions of matter, each based on its own fundamental Quantum type. That generic quantum is a scalar object that may be a *quantum* proper as visualized in Quantum Theory or somewhat of a Mass as visualized in classical and relativistic physics. The value of that distributive *quantum* entity defines each Universal Domain as displayed below.

$$Q_1(s) = q_1 = \frac{e^0}{\sigma\sqrt{2\pi}} = \frac{1}{\sigma\sqrt{2\pi}}$$

→for $n = 1$

$$Q_2(s) = q_2 = \frac{e^{-1/8}}{\sigma\sqrt{2\pi}}$$

→for $n = 2$

$$Q_3(s) = q_3 = \frac{e^{-2/7}}{\sigma\sqrt{2\pi}}$$

→for $n = 3$

$$Q_4(s) = q_4 = \frac{e^{-1/2}}{\sigma\sqrt{2\pi}}$$

→for $n = 4$

$$Q_5(s) = q_5 = \frac{e^{-4/5}}{\sigma\sqrt{2\pi}}$$

→for $n = 5$

$$Q_6(s) = q_6 = \frac{e^{-5/4}}{\sigma\sqrt{2\pi}}$$

→for $n = 6$

$$Q_7(s) = q_7 = \frac{e^{-2}}{\sigma\sqrt{2\pi}}$$

→for $n = 7$

$$Q_8(s) = q_8 = \frac{e^{-7/2}}{\sigma\sqrt{2\pi}}$$

→for $n = 8$

$$Q_9(s) = q_9 = \frac{e^{-8}}{\sigma\sqrt{2\pi}}$$

→for $n = 9$

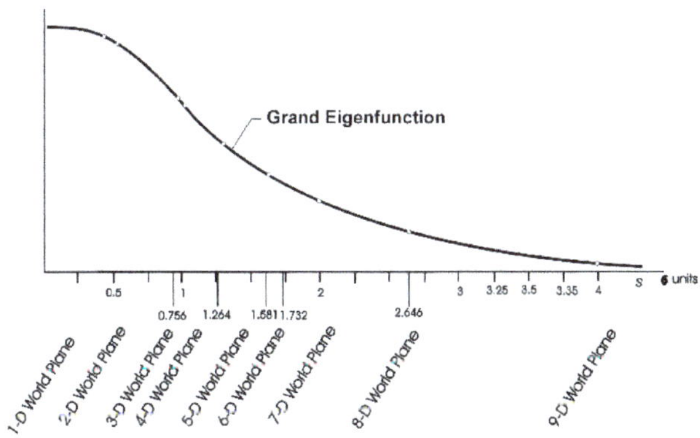

Fig. 6.1 World Planes composing the Omniverse

In other words, the Universe as a whole is made up of 9 distinct domains (Fig. 6.1), each based on a specific *degree of freedom* or *dimensional set.* This characterization sets the basis for the visualization of the universe as an Omniverse with 9 horizontally layered domains or World Planes. The above set of relationships may be rewritten to cast a figure of momentum deriving from the Quanto-Geometric articulation inherent to the function. We denote this momentum the *standard* Quanto-Geometric Momentum of a Universal Domain or World Plane. An important point to keep in mind is that the independent variable s here is not a simple distance but a *degree of space propagation or dilation,* which ultimately gives rise to distances or lengths. The best physical representation for s is what a descriptor for a wave refers to, a *wavelength (or a wavenumber).* It is a figure that characterizes both motion or propagation and the length associated with a wave phenomenon as a *moving periodic* entity. **This moving periodic entity is physical space itself.** Rewriting the equations above in order to make explicit the momentum figures, we obtain:

$$q_1 \times \sigma = \frac{1}{\sqrt{2\pi}} \text{ , for } n = 1$$

$$q_2 \times \sigma = \frac{e^{-1/8}}{\sqrt{2\pi}}, \text{ for } n = 2$$

$$q_3 \times \sigma = \frac{e^{-2/7}}{\sqrt{2\pi}}, \text{ for } n = 3$$

$$q_4 \times \sigma = \frac{e^{-1/2}}{\sqrt{2\pi}}, \text{ for } n = 4$$

$$q_5 \times \sigma = \frac{e^{-4/5}}{\sqrt{2\pi}}, \text{ for } n = 5$$

$$q_6 \times \sigma = \frac{e^{-5/4}}{\sqrt{2\pi}}, \text{ for } n = 6$$

$$q_7 \times \sigma = \frac{e^{-2}}{\sqrt{2\pi}}, \text{ for } n = 7$$

$$q_8 \times \sigma = \frac{e^{-7/2}}{\sqrt{2\pi}}, \text{ for } n = 8$$

$$q_9 \times \sigma = \frac{e^{-8}}{\sqrt{2\pi}}, \text{ for } n = 9$$

The *standard* momentum is represented by the expression $q_i \times \sigma$, where σ, a value of *s*, is the standard deviation for the Global Universe. The latter lends its name to the chosen denomination for this figure. In all, q_i is the figure that specifically defines every Universal Domain.

The one-dimensional Universe is given by $Q_1(s)$ or q_1 and is represented in the numerous existing Black Holes and Neutron Stars throughout the cosmos. These entities are knowingly one-dimensional.

The 2-dimensional Universe is given by $Q_2(s)$ or q_2 and is represented in the deployments of Plasmic Matter in Stars and at the center of planets. Plasma, of which fire is a type, is knowingly 2-dimensional.

The 3-dimensional Universe is given by $Q_3(s)$ or q_3 and is represented in what we commonly call *our* physical universe, the visible Cosmos.

The 4-dimensional Universe is given by $Q_4(s)$ or q_4 and represents the first of metaphysical universes. It is a trans-physical universe that cannot fully fall under our observation for reasons we will explain later.

The 5-dimensional Universe to the 9-dimensional Universe are respectively given by $Q_5(s)$ or q_5, $Q_6(s)$ or q_6, $Q_7(s)$ or q_7, $Q_8(s)$ or q_8 and $Q_9(s)$ or q_9, each representing a meta-physical domain that cannot fall under our observation at all because they pertain to the larger post Quanto-Geometric hierarchy. Each of these Universal Domains or World Planes are contained within one another in accordance with the Quanto-Geometric mapping function. While we can observe but not penetrate into the domains that are below us in the order of embedding, it is understandably not possible to observe the ones above which our domain is embedded in.

6.2. Theoretical Derivation of the Speed of Light

How do we know that the above map is correct and not just a speculation? For an answer, let us start by scrutinizing further the value of expression $q_3 \times \sigma$, the standard Quanto-Geometric Momentum for our 3-dimensional universe, by developing the expression in completely rational form:

$$q_3 \times \sigma = \frac{e^{-2/7}}{\sqrt{2\pi}}$$

$$q_3 \times \sigma = 299796065 \times 10^{-9}$$

It is not difficult to recognize in this number the figure or digital set that we know as the speed of light *c: 299,792,458 m/s*. From relation *(5-4)* we have picked the value of the dimensional set for the 3-D World Plane and replaced it in *(5-5)* to obtain the s_3 wave-number which characterizes space in 3-D, that is:

$$\text{For } n = 3\,,\ s_3 = \pm\frac{2\sigma}{\sqrt{7}}$$

Once we obtain s_3, we replace it in expression *(5-2)* of the Grand Eigenfunction to obtain the value of the Function at that point. So:

$$\text{For } s_3 = \pm\frac{2\sigma}{\sqrt{7}},\ Q_3(s) = \frac{e^{-2/7}}{\sigma\sqrt{2\pi}} = \frac{0.299796065}{\sigma}$$

$$\text{or } q_3 \times \sigma = 0.299796065$$

$$\text{or } q_3 = 0.299796065 \cdot \sigma^{-1}$$

In summarized form:

$$n_3 = 3 = D_3$$
$$s_3^{-1} = 1.322875656 \cdot \sigma^{-1}$$
$$q_3 = 0.299796065 \cdot \sigma^{-1}$$

For the purpose of the demonstration, once we arrive at the 3rd expression, we could squarely conclude the analysis. It is possible, however, that a professional physicist might not have been swayed by this result due to sheer unfamiliarity with the expressed quantity for q_3. In proceeding with dimensional-unit analysis for further clarification, our first observation of these results is that, if σ^{-1} is to be taken for a unit of measure, the value of the parabolic scalar and that of reciprocal space faithfully and consistently reflect the principle encapsulated in expression *(4-1)*, namely that nothing exists beyond the two irreducible tenets of matter and that as such they can only be formulated in terms of one another. The system is wonderfully self-contained because we see that the figure of the standard deviation σ arises as a natural endogenous unit of measure within the system, here in its reciprocal dimension.

In our quest for *c*, let's note that the quantity appearing in the 3rd expression is a clear precursor of the quantity experimentally known to represent *c*:

$$\mathbf{0.299796}065\cdot\sigma^{-1} \rightarrow \mathbf{0.299792}458 \cdot 10^{9}$$

on account of a large figurate match. This expression gives us the value of the parabolic scalar in reciprocal standard deviation σ^{-1} unit. The reciprocal standard deviation of the Omniverse is a quanto-geometric wavenumber in nature, representing both a length and a propagating wave, that is. We set out to look for a quantity whose known nature gives us a match with the identity expectation that we hold for σ^{-1} or σ, no matter the cardinality or magnitude of the quantity.

This number is indeed an already known quantity in physics in the personification of the *nano-meter*, which is a sub-unit element of the meter scale. The meter being no longer just a unit of measure based on a physical etalon but a fundamental constant, the nanometer sub-element inherits that condition and thus equally stands for a constant in its own right. Most importantly, the nanometer is the metric for radiation wavelengths in spectroscopy, although the BIPM (Bureau International des Poids et Mesures) has given it the general definition of a *unit of length in the metric system equal to 1 billionth of a meter*:

$$1nm = 1.0000\overline{0} \times 10^{-9}\,m$$

The metrology of this unit is typically utilized in condensed matter physics and below, very particularly in the field of spectroscopy, for the measurement of **light** radiation (not a coincidence) resulting from energy transitions of particles in atoms and molecules. Therefore, the match in the "dimensional" nature of the *nanometer* constant and the nature of the standard deviation figure σ is a solid given and is completely natural. We shall add as a note that the Angstrom unit (Ä) with the value of *0.1 nanometer* has also been in use in spectroscopy but has been fairly superseded by the *nanometer* due to better suitability to task.

Our motivation being the derivation of the exact value known for *c*, we are going to allow ourselves to fine-tune ever so slightly the value of the *nanometer* to the following magnitude:

$$1nm = 1.00001203 \times 10^{-9}\,m$$

We may well do so because the official value of $1.0 \times 10^{-9} m$ of the nanometer, as a dependency on the meter, is not known to be absolutely exact and a change in the order of the 5th decimal place is, from that standpoint, absolutely inconsequential to the applicability of the sub-unit. Furthermore, and most importantly, we primarily port this adjustment in magnitude to σ, an element in our own system, so that:

$$\sigma = 1.00001203 \times 10^{-9} m$$

We thus register the value of the standard deviation of the Omniverse in its reciprocal form to be exactly the above reported quantity. By substitution of σ^{-1} in the parabolic scalar relation, we obtain:

$$q_3 = \frac{0.299796065}{1.00001203} \times 10^9 \quad \text{or}$$

$$q_3 = 299{,}792{,}458$$

exact magnitude of the speed of light! This quantity, however, is an atemporal figure! It is important to recognize that we have not arrived at this quantity by introduction of speeds or ratios of *m/s* in the algebra, but strictly by a legitimate manipulation of the metric, and in complete compliance to the time-free condition of the framework. This quantity is the measure of the *Moment of Propagation* of the parabolic *eigenwave* that makes up dimensional background space in the 3-D World Plane. The wavelength of the propagating *eigenwave* is given in the value of reciprocal space $1/s_3$. In other words, c is a *quantity of motion*, the motion *inherent* to background 3-space in the Omniverse. All the transversal 3-D Shells of matter, from the quark scale to the large-scale cosmic structures, have developed over and from this spatial background, and all physical motion occurs by degeneracy from this dynamical fabric. All manifolds develop from this dynamical fabric as well and that is why at any point in the fabric when all tensors vanish the parabolic *eigenform*, otherwise called the Einstein tensor, indefectibly subsists. The latter is however matter for another analysis. By the same token, this analysis patently explains why c as the moment of propagation of the basis spatial eigenwave of the 3-D World Plane cannot be surpassed as long as we remain within the boundaries of the 3-D realm.

If the value of the physical constant c is indeed *299,792,458*, then the value of the standard deviation of the Omniverse *(σ)* of $1.00001203 \times 10^{-9}$ must consequently be an outright true physical quantity in its own right (not just a mathematical scalar) and an effectual one as well. The best way to interpret this result is to visualize the parabolic scalar quantity *(q_3)* of $\mathbf{0.299796065}\sigma^{-1}$ as a pure mathematical quantity, while the same quantity expressed as the scalar $\mathbf{0.299792458} \cdot 10^9$ is a mathematical-physics quantity that comes into existence when we give physical meaning to the standard deviation scalar. This all means that the meter etalon, originally devised by BIPM as a certain distance between two marks on an iridium bar held at a certain temperature, holds an incredible level of accuracy and is true to nature in a highly accurate manner.

The theoretical derivation of the quantity representing the speed of light (or its moment of propagation) stands for the sole and exclusive measure of theoretic support a-posteriori to the decision by BIPM to set that quantity as the standard for the definition of the meter, because it removes arbitrariness in a very large measure from the metric standard. In addition, if the quantity of $1.00001203 \times 10^{-9} m$ is found to be effectual in experimentation or observation in the predicted condition, then we would have realized complete synchrony between human theory and the physics of nature under the auspices of a mathematical-physics system, the Quanto-Geometric system! In that regard, we will come back to this number in Chapter 6 in order to ground this contention as well as further elucidate one other important dimension of this quantity in relation to the cosmological expansion of the universal arena.

6.3. *Misadventures of c in Time*

Table 6.1 below shows the history of measurements of the speed of light since the 17th century to date. Ever since Albert Einstein's Theory of Relativity became officially accepted in the first half of the 20th century, physicists recognized the towering importance of the speed of light in our understanding of physical reality, and endeavored to measure it with as much accuracy as could be possible. The latest measurements employed the best technological means available such as coherent laser beams.

Nevertheless, the task of accurately measuring the speed of light proved to be somewhat elusive to the standard bodies as they recognized that such accuracy depended on the accuracy with which we measure both distance and time, notably the latter. The measurement of time has had its own shortcomings, references for the time unit, the second, having had to be abandoned one after another due their inconstancy, to the point that currently two different scales of time are in use in the United States of America. The standard bodies had even adopted in the past a time unit pegged to a set date in time (year 1900) in order to remain clear of the variations that were naturally occurring in etalons and measurement conditions. It soon became clear that measure was equally unsustainable.

As a consequence of these shortcomings, the standard bodies, CGPM (Conférence Générale des Poids et Mesures) and NIST (National Institute of Science and Technology) finally decided in 1975 to establish the best measurement taken then of light's travel speed as a reference in itself, and make the unit distance, the meter, and the unit of time, the second, secondary constants which were defined by the very speed of light. Consequently, the speed of light constant has been conventionally set to be *exact*. Now one may wonder what happened to the universality of time, which was established in Newtonian mechanics as "universal, uniform and true", and which has survived to this day thanks to Einstein's constructs of Relativity? It is obvious that the notion of time, just as the notion of distance, has preponderance over the notion of speed of any physical object or artifact. Making time, acknowledged thus far to be a dimension in itself, dependent on *c* seems to be a counter-intuitive proposition that amounts to putting the cart before the horse.

DATE	INVESTIGATOR	METHOD	MEASURED SPEED (m/s)
1676	Ole Roemer	Jupiter's Moons	220,000
1726	James Bradley	Stellar Aberration	301,000
1834	Charles Wheatstone	Rotating Mirror	402,336
1849	Armand Fizeau	Rotating Wheel	315,000
1862	Leon Foucault	Rotating Mirror	298,000
1868	James Clerk Maxwell	Theoretical Calculations	284,000
1875	Marie-Alfred Cornu	Rotating Mirror	299,990
1879	Albert Michelson	Rotating Mirror	299,910
1888	Heinrich Rudolf Hertz	Electromagnetic Radiation	300,000
1889	Edward Bennett Rosa	Electrical Measurements	300,000
1890's	Henry Rowland	Spectroscopy	301,800
1907	Edward Bennett Rosa and Noah Dorsey	Electrical Measurements	299,788
1923	Andre Mercier	Electrical Measurements	299,795
1926	Albert Michelson	Rotating Mirror (Interferometer)	299,798
1928	August Karolus and Otto Mittelstaedt	Kerr Cell Shutter	299,778
1932 to 1935	Michelson and Pease	Rotating Mirror (Interferometer)	299,774
1947	Louis Essen	Cavity Resonator	299,792
1949	Carl I. Aslakson	Shoran Radar	299,792.40
1951	Keith Davy Froome	Radio Interferometer	299,792.75
1973	Kenneth M. Evenson	Laser	299,792.4570
1978	Peter Woods and Colleagues	Laser	299,792.4588

Table 6.1 Historical measurements of the speed of light c

Well aware of this inconsistency, physicists have justifiably sought to make *c* as much an independent constant as possible by turning it into an abstraction, thereby giving it a life of its own. From Maxwell theory of electromagnetism, it ensued that:

$$c = \frac{1}{\sqrt{\mu_0 \varepsilon_0}}$$

where μ_0 is the magnetic permeability of space and ε_0 the electric permittivity of space. The values assigned to μ_0, and subsequently that of ε_0, were conventionally chosen in order to satisfy the above equation with the *exact* value that was established for *c*. Therefore, non-physicist readers should not be disabused when they see these two constants in Table of Physical Constants published in the literature as *exact* constants. They are rather *conventionally exact*. What's more, these two constants do not describe, and are not intended to describe, any physical property of space whatsoever. Shall we say that the Quanto-Geometric architecture provides the most natural derivation of *c* from pure abstraction there is.

In accordance with its preeminence within the hierarchy of physical constants, the speed of light constant has become a ubiquitous entity in modern physics, appearing as a parameter in many contexts unrelated to light phenomenon per se. The Lorentz transformation equations are the first of such examples where speeds of objects are formulated in terms of *c*. General Relativity predicts the propagation speed of gravitation to be *c*, although gravitational waves have yet to be observed (as of first publication date of this text). The constant *c* is not only the speed of light in vacuum but generically the speed of all "mass-less" particles in vacuum. All of this clearly means to tell us that it is not about *c*, but about space. We remain blind to those important facts because we are unable to formulate a model of space as the mother of all motion and we are determined to give fit and life to *c* at all costs in a world of time of our creation which it does not belong to. Again, the Quanto-Geometric framework has been able to properly authenticate and quantify *c* by derivation because it has established the necessary foundation for that lacking space model as implied by the behavior of *c*. We must once and for all get rid of the false notion of time to visualize the inherent properties of space. And yes, *spacetime* is the wrong notion.

Individually we are all innately cognizant of the fact that the universe around us is a 3-dimensional domain, that is to say it is endowed with 3 isotropic dimensions of space: length, width and height. A dimension of time is nowhere to be found, except in our human minds as a counter-productive abstraction. These 3 dimensions of space are what ultimately explains the specific value of *c* and its insurmountable stature in our universe.

6.4. World Planes and Space Propagation

Just as we have seen that the Quanto-Geometric momentum $q_3 \times \sigma$ or $Q_3(s)$ remits to a constant for space propagation in the 3-dimensional Universe, the other momenta of the Mappa Mundi Function remit to specific levels of space propagation in each of the other 8 Universal Domains (World Planes) as well. As we go up the hierarchy of Universes from U_1 to U_9, the value of the standard Momentum should increase and we would expect the suite of equations to reflect an increasing *degree of space propagation*, in their comparable figure to *c*, from U_1 to U_9. That may not be quite apparent however, unless we call for the *generic* Quanto-Geometric momentum and its specific values from U_1 to U_9.

Below is the derivation of all q_i from U_1 to U_9. We use the physical nanometer quantity for σ in the computations.

For $n = 1$, $q_1 \times \sigma = \frac{1}{\sqrt{2\pi}} = 0.39894228$

and $q_1 = \frac{1}{\sigma\sqrt{2\pi}} = 398937481\!2$

For $n = 2$, $q_2 \times \sigma = \frac{e^{-1/8}}{\sqrt{2\pi}} = 0.352065327$

and $q_2 = \frac{e^{-1/8}}{\sigma\sqrt{2\pi}} = 352061091.5$

For $\boldsymbol{n = 3}$**,** $q_3 \times \sigma = \frac{e^{-2/7}}{\sqrt{2\pi}} = 0.299796065$

and $q_3 = \frac{e^{-2/7}}{\sigma\sqrt{2\pi}} = 299792458$

For $n = 4$, $q_4 \times \sigma = \frac{e^{-1/2}}{\sqrt{2\pi}} = 0.241970725$

and $q_4 = \frac{e^{-1/2}}{\sigma\sqrt{2\pi}} = 2419677813.6$

For $n = 5$, $q_5 \times \sigma = \frac{e^{-4/5}}{\sqrt{2\pi}} = 0.179256322$

and $q_5 = \dfrac{e^{-4/5}}{\sigma\sqrt{2\pi}} = 179254165.2$

For $n = 6$, $q_6 \times \sigma = \dfrac{e^{-5/4}}{\sqrt{2\pi}} = 0.114298877$

and $q_6 = \dfrac{e^{-5/4}}{\sigma\sqrt{2\pi}} = 114297502$

For $n = 7$, $q_7 \times \sigma = \dfrac{e^{-2}}{\sqrt{2\pi}} = 0.053990967$

and $q_7 = \dfrac{e^{-2}}{\sigma\sqrt{2\pi}} = 53990317.01$

For $n = 8$, $q_8 \times \sigma = \dfrac{e^{-7/2}}{\sqrt{2\pi}} = 0.012047013$

and $q_8 = \dfrac{e^{-7/2}}{\sigma\sqrt{2\pi}} = 12046868.08$

For $n = 9$, $q_9 \times \sigma = \dfrac{e^{-8}}{\sqrt{2\pi}} = 0.00013383$

and $q_9 = \dfrac{e^{-8}}{\sigma\sqrt{2\pi}} = 133828.61$

This list gives us a decreasing suite of numbers from q_1 to q_9. If we recall the expression for the *generic* Quanto-Geometric Momentum:

$$q \times s = 1,$$

$$\text{then } s = \frac{1}{q}$$

$$\text{or } s_i = \frac{1}{q_i}$$

Thus, the value for each of these degrees *(s_i)* of space propagation is the inverse of the momentum *(q_i)*, which sets them to be on an increasing slope from U_1 to U_9. In other words, the maximum attainable speed in each Universe of the Omniverse is higher from one to the next without ever reaching either infinite (instantaneity) at U_9 or 0 (quasi-absolute immobility) at U_1.

Furthermore, there is a distinct quality of the meta-physical universes (from U_4 to U_9) as compared to the physical universes (U_1 to U_3). The fact that we, as observers from the 3-dimensional realm, cannot perceive them at all is not a trivial fact and testifies to their distinct quality in terms of their Quanto-Geometric momenta. We can only become observers to these other World Planes by using technological transforms such as based on the superheterodyne principle but would never be able to physically penetrate them. We will later see what that level of detection consists in. All of the Quanto-Geometric principles fully apply here, including the signatures attributable to each layer as qualifiers and descriptors for each of these Universes individually. It is worth adding that although we may have an idea of what space may be like in U_4 to U_9 thru the cardinality of their different dimensions, we are and will remain completely unable to conceptualize what the 4-th, 5-th ... 8-th and 9-th dimension really mean from the boundaries of 3-dimensional abstractions. Theosophy gives a path to that knowledge which, however, implies a completely different vision of ourselves as living entities, a matter that we shall address later in this work. As elusive as these meta-physical universes might be to us, they remain nevertheless completely real domains, as real as our 3-dimensional Universe, in accordance to their signature within the Quanto-Geometric framework.

6.5. Horizontal Hierarchy of Domain Shells

Let us start by positing the following relationship, which equates the exponent of *e* in the Grand Mappa Mundi function of one variable to the primeval sequence:

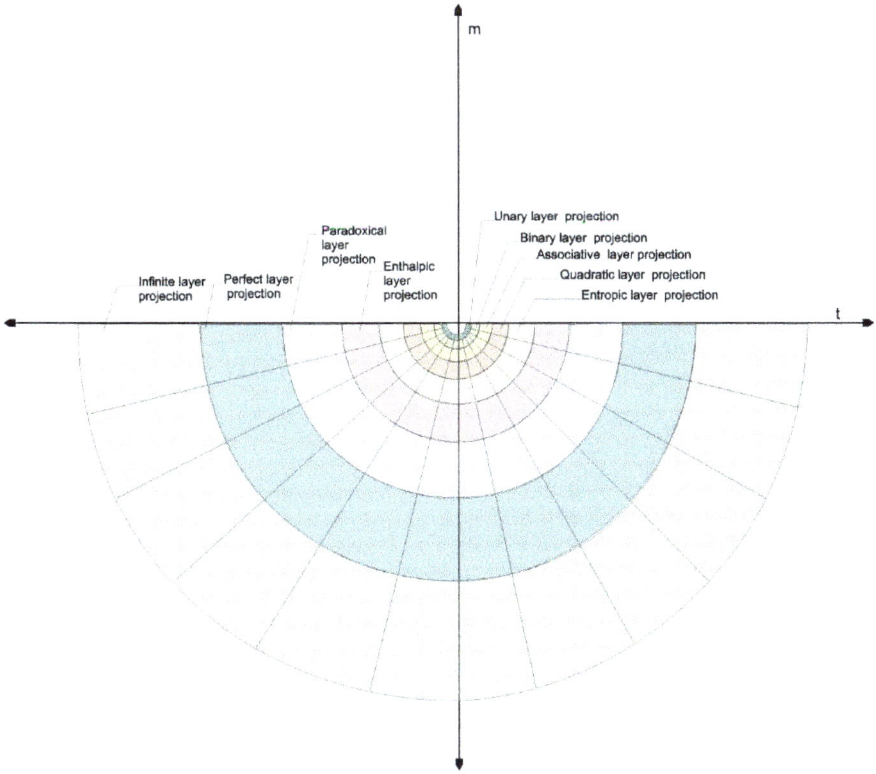

Fig. 6.2 Projection of the graph of Quanto-Geometric function on the *m-t* plane displaying the normalized groups

$$-\frac{s^2}{2\sigma^2} = \frac{1-n}{10-n}, \text{ where } 1 \le n \le 9$$

$$s^2 = -2\sigma^2 \frac{1-n}{10-n}, \quad s^2 = \sigma^2 \frac{-2(1-n)}{10-n}$$

$$s^2 = -\sigma^2 \frac{2(1-n)}{10-n}, \quad s^2 = \sigma^2 \frac{2n-2}{10-n}$$

$$s^2 = \pm\sqrt{\frac{\sigma^2(2n-2)}{10-n}}$$

$$s = \pm\sigma\sqrt{\frac{(2n-2)}{10-n}}$$

In other words, to every individual Universe in the vertical Hierarchy, going from U_1 to U_9 and in accordance with their respective generic Momentum, corresponds an interval of matching *s* values. We know, from formerly discussed Quanto-Geometric principles, that there are 9 critical *s* values, by which the *Q(s)* function can be horizontally normalized. To find those 9 critical values of *s*, we may replace integer *n* by an allowed value within its allocated range, in the last of above equations. After performing those replacements, the list of corresponding *s* values are as follows:

for $n = 1$, $s_1 = 0$

for $n = 2$, $s_2 = \pm\sigma/2$

for $n = 3$, $s_3 = \pm\sigma 2/7^{1/2}$

for $n = 4$, $s_4 = \pm\sigma$

for $n = 5$, $s_5 = \pm\sigma(8/5)^{1/2}$

for $n = 6$, $s_6 = \pm\sigma(10)^{1/2}/2$

for $n = 7$, $s_7 = \pm 2\sigma$

for $n = 8$, $s_8 = \pm\sigma 7^{1/2}$

for $n = 9$, $s_9 = \pm 4\sigma$

Again, it cannot be overstated that these values of *s* mean more than just a distance in space but also a *form* of space. They stand for *wavenumbers*, derivates of wave-lengths which characterize both a distance and a *degree of propagation* of space. From one interval to the next, space experiences more freedom or more extensiveness (longer wavelength) resulting in a larger capsule or a larger domain as we go down the line. We recall from Chapter 3 that the list of space points reported above emanate from a primeval symmetry-breaking sequence. So therefore, the different n degrees of freedom of physical space yield the following space Intervals for the whole Omniverse:

Interval 1 $\rightarrow$ $]0, \sigma/2[$, for $n = 1$

Interval 2 $\rightarrow$ $[\sigma/2, 2\sigma/7^{1/2}[$, for $n = 2$

Interval 3 $\rightarrow$ $[2\sigma/7^{1/2}, \sigma[$, for $n = 3$

Interval 4 $\rightarrow$ $[\sigma, \sigma(8)^{1/2}/5[$, for $n = 4$

Interval 5 $\rightarrow$ $[\sigma(8)^{1/2}/5, \sigma(10)^{1/2}/2[$, for $n = 5$

Interval 6 $\rightarrow$ $[\sigma(10)^{1/2}/2, 2\sigma[$, for $n = 6$

Interval 7 $\rightarrow$ $[2\sigma, \sigma 7^{1/2}[$, for $n = 7$

Interval 8 $\rightarrow$ $[\sigma 7^{1/2}, 4\sigma[$, for $n = 8$

Interval 9 $\rightarrow$ $[4\sigma, \infty[$, for $n = 9$

The rotation of the Quanto-Geometric function of one variable around the q-axis in order to recover its full 3-dimensional layout is going to transversally spread these intervals over circular spans around the q-axis as depicted in Fig. 6.2. By that account alone, each universe would have a single uniform type of space and a single class of scalar unit (quantum/mass), as it pertains to the postulated generic Quanto-Geometric momentum. However, our experience in the 3-dimensional universe tells us that it is not quite so: the universe around us is a deployment of space capsules. This is because in order to generate an individual physical Universe, the *work of creation* puts forth a transform and an operator, the nature of which we shall shortly discuss.

6.6. Operators and Transforms of Creation

The work of creation instantiates 3 quintessential Moments under the regency of the Tri-Valent Quanto-Geometric Operator. The first instantiation originates from the Hyper Quantum Operator and puts forth Uni-Form Operation per its defined norm. The second instantiation is an inheritance of the Hyper Quadratic Operator and puts forth Distributive Operation according to its norm of Ambi-valence. The third instantiation takes origin from the Hyper Space Operator and puts forth the Operation of Space Overture in accordance with its norm of *spatiation* and *spatialization.* The Operator treatment that follows requires us to keep in mind that all Quanto-Geometric operators are self-adjoint operators.

6.7. Hyper Quantum Tri-Valent Operator and the Laws of Conservation

Space in U_3 is characterized as a three-dimensional form of space owing to its origin of parabolic symmetry. The embedded parabolic conic section within the geometry of the Quanto-Geometric function is its native realm. Within the work of creation of U_3, the Uni-Form Operator acts on the dimensional configuration of U_3 space in order to further modulate it, notwithstanding the constancy of its dimensional modulus, which remains inalterably 3.

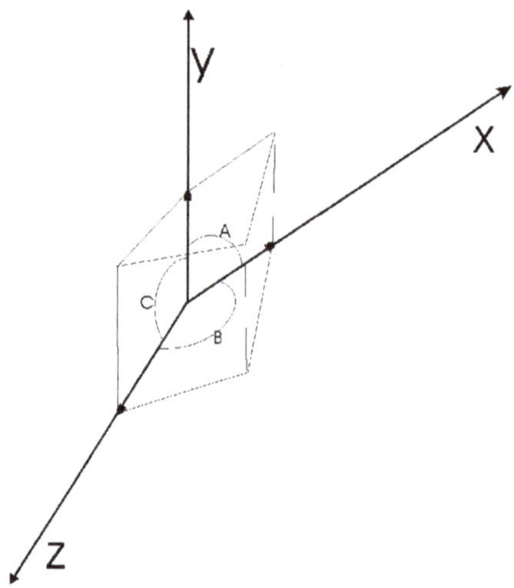

Fig. 6.3 Anisotropic U_3 space in the absence of the Hyper Quantum Operator

Were it not for the intervention of this operator however, the 3 dimensions of U_3 space would be irregular or *an-isotropic* as depicted in Fig. 6.3. (*Tropism* means direction, *iso* means same, *an-* means non). While two of the dimensions would necessarily fall in the same plane, the third dimension would be cast in any irregular direction and the angle they all make with one another could be any angle of any value. Furthermore, the unit-length on each dimensional axis would be different, as displayed in Fig. 6.3.

Uni-Form operation by the Hyper Quantum Operator, whose signature is type 1 symmetry or mono-centric symmetry, as studied in Quanto-Geometric Number Theory, compels the three dimensions of U_3 space to exhibit mono-centric symmetry, a constraint whereby all directions must exhibit same unit-length, and all angles between them must lay equal to one another. The only possible configuration that responds to this constraint is U_3 space cast from an ortho-normed Euclidean coordinate system, as shown in Fig. 6.4.

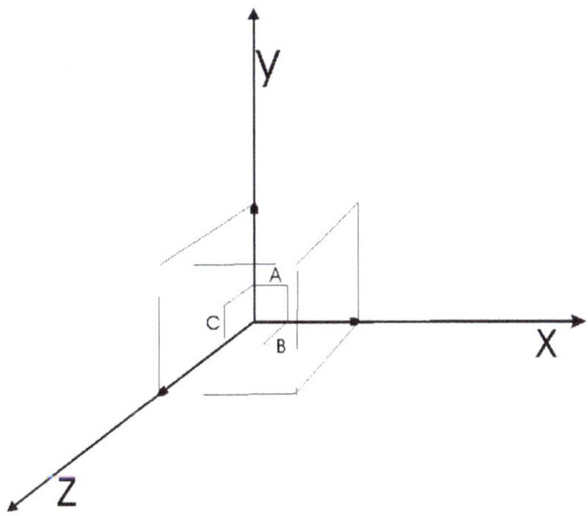

Fig. 6.4 Spatial cube of action in U_3

The Hyper Quantum Operator thereby constrains U_3 space to demonstrate complete *isotropic* properties and to develop from a perfectly regular cube of action. In that manner, the intervention of the Hyper Quantum Operator in the work of creation is responsible for all the known **laws of conservation** in our 3-dimensional universe thru the type 1 symmetry constraints imposed on the configuration of space in U_3.

1. It is responsible for the *conservation of linear momentum*, a law whereby any object thrown in pure vacuum at a constant velocity is going to keep moving at that same velocity and along the same direction forever.

2. It is responsible for the *conservation of angular momentum*, a physical law whereby any object thrown in pure vacuum at a constant angular velocity is going to keep moving maintaining that same angle of progression forever.

3. It is responsible for the *conservation of energy*, a law whereby any process of a certain type containing a certain amount of energy can change to another process of a different type with the total amount of energy present in the initial state still fully present in the final state after the conversion.

4. It is responsible for the *instantiation and allocation of all masses* to all elemental objects in U_3, in particular the sub-components of particles as well as particles themselves.

Were it not for the intervention of the Hyper Quantum Operator, an object thrown at a certain velocity in the purest of vacuum could end up coming to a stop or increasing its travel velocity in and of itself. Absent the Hyper Quantum Operator, an object launched in purest vacuum from a circular movement could end up moving instead according to an ellipse and no longer a circle, or not even in the same plane of rotation, this absent any other intervening factor. Were it not for the Hyper Quantum Operator, no energy transformation process in U_3 could take place without always ending in either a loss or a gain of energy according to the direction along which the process takes place. Absent the isotropic constraints imposed by the Hyper Quantum Operator, and because of the eventual resulting imbalance between dimensional directions in U_3 in that condition, no adiabatic process would most probably be feasible in our universe either.

The instantiation and allocation of masses making up the foundation of U_3 by the Hyper Quantum Operation we shall address in the Book II of this publication. We will only anticipate here that it is based on a Transform T_1 such that:

$$T_1 \rightarrow \frac{d[Q(s)]}{ds}, \text{ where } s \rightarrow 0$$

the parameter s setting the spin (rotational wavefunction) of the scalar element.

6.8. Derivation of the Planck Constant and the Stephan-Boltzmann Constant

Hyper Quadratic Operator takes two distinct actions in the generation of U_3. Firstly, it instantiates a transform, which is the derivative of the Grand Mappa Mundi function, taken as an image of the entire 9-plane Omniverse, and utilizes that image as a sub-map to drive the generation of sub-universe U_3. Secondly it selects two specific values of that function from two particular s points in order to create the first two physical pillars upon which the edifice of U_3 is called to rest: the physical constant of potential energy density and the physical constant of kinetic energy density, namely the *Plank constant h* and the *Stefan-Boltzmann constant σ* respectively. This σ symbol is not the same as the standard deviation s value of the Quanto-Geometric function. We shall designate in this exposition the Stephan-Boltzmann constant σ^* rather in order to distinguish it from the Quanto-Geometric standard deviation symbol. The Hyper Quadratic Operator sets the scale and dynamics of quantization, the basis for all forms of segregation and integration in the composition of U_3. By that norm, two elements are to preside over this form of distribution, *Thermodynamic Temperature* and *Energy Quanta.*

Therefore, the Transform T_4 such that:

$$T_4 \rightarrow \frac{d[Q(s)]}{ds}, \text{ where } s = (\tfrac{1}{2}\sigma, \sigma)$$

represents the Hyper Quadratic Operator's product put forth for the configuration of the 3-dimensional universe. The choice of the two mentioned values of s for the transform is far and away from being arbitrary. We recall that the standard deviation point σ is a critical s value where the derivative of the Quanto-Geometric function or its differential experience shows unique behavior: at that point the first-order derivative function reaches its highest value and the second-order derivative function $d^2Q(s)/ds^2$ is null.

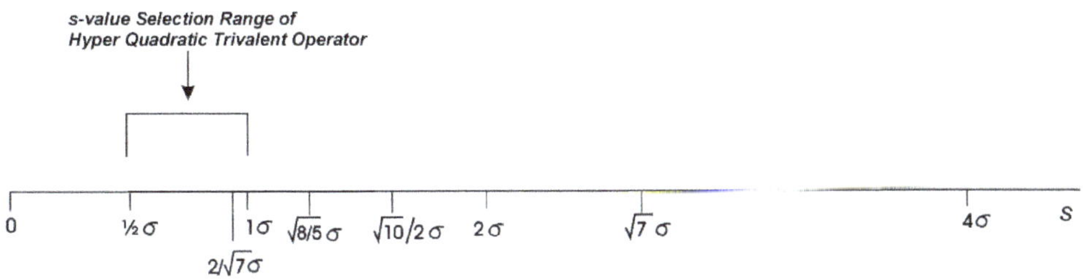

Fig. 6.5 Selection range of s values for the T_4 Transform instantiated by Hyper Quadratic Trivalent Operator

It is all the more natural that this point is selected for the Transform by a Hyper *Quadratic* Operator, which already has a defining *quintessential* life set at that point (Fig. 6.5). Secondly, it stands all the more natural that an Operator whose norm is *Ambi*-Valence instantiates *two* points of action to develop its operation. And finally, it stands no less to reason and conformity that, given a choice, this second point be ½ as the one half of the two ambi-valent tenets of the articulation. Therefore, even though the point $2/7^{1/2}$ belongs within the range of selectable values, it is summarily discarded by the Operator.

In further detailing the analysis of the transformation, the Hyper Quadratic Operator notably forces the original trinary object into equalized behavior by its two components Q_3 and S_3 according to the very norm of Ambi-Valent action by the Operator. By that norm, Q_3 is forced to induce more space into its own makeup, becoming thereby more distributive and S_3 is forced to borrow from or acquire Q_3's scalarity, inducing tensors into its own texture. This co-valent interaction develops within the boundaries of the primary or generic Quanto-Geometric Momentum of U_3 objects whereby $Q_3 \, x \, S_3 = 1$, a principle that cannot be violated. The ultimate consequence of these constraints is that the new Q_3 and S_3 components of the U_3 unit object acquire comparable equity with one another, although not becoming identical in nature. This is well captured in the popular Eastern ying-yang symbol (Fig. 6.6), a cryptogram that had understandably long fascinated Niels Bohr, one of the lead proponents of Quantum Mechanics, world of conjugates.

Fig. 6.6 Covalent interaction between 3-D scalar and space

They each come to exhibit a Momentum of sorts because they both now have significant space/scalar constitution which is further equalized. This Momentum figure for Q_3 may be expressed as Potential Energy, or to be more exact, Potential Energy Density (because of the derivative backdrop) and its mathematical expression cast as:

$$E_p = h \cdot v \quad (6\text{-}1)$$

where h is the distributive scalar and v the acquired space wave function. Likewise, the Momentum figure for S_3 may be expressed as Kinetic Energy density and its numeric definition cast as:

$$E_k = \sigma^* \cdot T \quad (6\text{-}2)$$

where σ^* is the acquired scalar and T what the principal wave function has become (*tensored* space). The above transforms set the stage for many known transformations in classical energy physics and quantum electrodynamics, notably the reversibility between mass and energy as well as the reciprocal conversion of potential energy and kinetic energy. Let us now take a look at the numbers involved in these relations by starting with the first-order derivative expression of *Q(s)*. When developed the first-order derivative of *Q(s)* shows:

$$\frac{dQ(s)}{ds} = \frac{1}{\sigma\sqrt{2\pi}} \cdot \frac{-2s}{2\sigma^2} \cdot e^{\frac{-s^2}{2\sigma^2}}$$

$$\frac{dQ(s)}{ds} = \frac{-s}{\sigma^3\sqrt{2\pi}} \cdot e^{\frac{-s^2}{2\sigma^2}}$$

At $s = \sigma/2$, $\frac{dQ(s)}{ds} = \frac{0.17603266}{\sigma^2}$ and $\left(\frac{dQ(s)}{ds}\right)^{-1} = 5.680763 \cdot \sigma^2$

At $s = \sigma$, $\frac{dQ(s)}{ds} = \frac{0.2419707245}{\sigma^2}$ and $\left(\frac{dQ(s)}{ds}\right)^{-1} = 0.413273135 \cdot \sigma^2$

In the above relations, σ is the Quanto-Geometric standard deviation parameter. The official value of the Stephan-Boltzmann constant is: **5.6705·10^1 W m^{-2} K^{-4}**. And the official value for the Plank Constant is: **0.4135666751 10^{-16} eV·s**. Both of these values are in a satisfactory match with our values for the inverse derivatives above, except of course for the units used, as per determination of the transforms previously discussed. The first relations *(5-1)* is well known in Quantum Electrodynamics and indeed there is equivalence between h in that equation, standing as the distributive scalar of the unit object in U_3, and the Planck's quantum of action, the smallest energy packet in U_3. In passing, one should know that the Planck Constant is admittedly the fundamental physical constant measured to the worst level of precision. What may be new territory are the implications of the second relation *(5-2)*, only a variation of which is known. Let us compare that relation to the Stephan-Boltzmann law:

$$j^* = \sigma^* \cdot T^4 \quad (6\text{-}3)$$

where j^* is the emissive power or energy flux emanating from an irradiated black body and T the thermodynamic temperature. We can see that the underlying notion subtended by j^* is a fair match to the notion of kinetic energy density, the only apparent discrepancy relating to the 4th power to which T is elevated. Is it any surprise however that a *Quadratic* operator, with numeric definition emanating from the set of power numbers on the real number line, establishes a relationship based on a quantity not only elevated to a power but additionally a power of 4? In that relation T is expressed in Kelvin. If we make $T^4 = N$, the relation becomes:

$$j^* = \sigma^* N \quad (6\text{-}4),$$

where N is any real number elevated to the 4th power comprised within the Kelvin scale of temperature. This equivalency gives us a match, under the condition that N must be a number representing a power of 4 within the boundaries of the Kelvin scale ($N = T^{1/4}$). We are consequently driven to three conclusions, namely that:

1. The 4th power of the Kelvin scale of temperature is the *natural* scale of temperature (K^4).
2. Temperature is a physical variable that represents native kinetic energy. This is in fair agreement with classical definition of Temperature, which is cast as the average kinetic energy of all unit objects in a population of such objects.
3. The physical constant σ^* (the Stephan-Boltzmann constant) represents the smallest possible unit or "degree" of heat or *quantum* of heat in U_3 as it were.

Please refer to the Table of high-level constants above (Table 6.2) as well as Table 6.6 of Physical Constants further down in this Chapter, offered in support of this analysis of the theoretical derivation of known physical constants. Also offered in Appendix I for reference is the complete repertory of known physical constants with their current official values released by CODATA 2010 (Committee for Data on Science and Technology by the United States' National Institute of Science and Technology).

QG Function	Spinor Space → $s_i = 1/2 \cdot \sigma$		Spinor Space → $s_i = 1 \cdot \sigma$	
dq/ds	$0.17603266/\sigma^2$	Electron Gyromagnetic Moment	$0.2419707245/\sigma^2$	Electron Compton Wavelength
$(dq/ds)^{-1}$	$5.680763 \cdot \sigma^2$	Stefan-Boltzmann constant	$0.413273135 \cdot \sigma^2$	Planck Constant (in eVs)

Table 6.2 High-level black-body-related physical constants

We remind the reader that a black body is by definition an idealized object able to absorb all the radiation it receives. The interesting fact about such an object is three-fold:

- At a certain point the distribution of energy throughout the different wavelengths of the radiation takes the form of normalized distribution (incidentally and principally the form of the derivative of our image function) independently of the shape of this body.
- At that point the temperature of the body has stabilized and is the same everywhere throughout the emissive object or emissive part of the object as well as the radiation spectrum.
- The total amount of energy distributed over the radiation wavelengths is dependent only on the uniform equilibrium temperature.

The salient fact to take into account from the above is probably not the radiation energetic spectrum or the shape or constitution of the body that is home to the interaction, but the *uniformization* of temperature when the interaction between the radiation and the black body stabilizes and that the temperature level alone is the single most significant dependency at the state of equilibrium (Fig. 6.7). This is captured in the known thermodynamic relation *(5-3)* above.

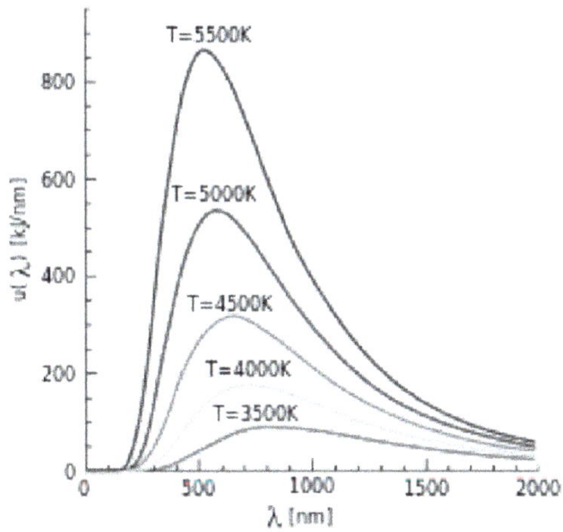

Fig. 6.7 Black body radiation spectrum at different temperatures

An interpretation of this formula within our framework sets σ^*, the Stephan-Boltzmann constant, to be a constant of proportionality for *Temperature* and that the smallest quantity or unit temperature that will ever make sense is σ^*. We shall remember that the Temperature of a macroscopic object is defined as the average kinetic energy of the molecules making up that body. In fact, Temperature is the average kinetic energy of all unit-scalar elements making up an object, whether the object is a population of particles, a population of atoms, molecules, etc. So therefore, Quanto-Geometrically, the Stephan-Boltzmann constant in the portrait of:

$$\left(\frac{dq}{ds}\right)^{-1} = 5.680763 \cdot \sigma^2,$$

is what we might call in its own right the *Temperature constant*. The Stephan-Boltzmann constant, officially measured to be $\mathbf{5.6705 \cdot 10^1\ W\ m^{-2}\ K^{-4}}$, duly shows the temperature unit in the dimensions of the measured value of the constant in K^{-4}. It is so because it expresses the recognition that the constant constitutes the very etalon for temperature or *thermo-dynamic* temperature. By the same token we now also know why the radiation spectrum of hot objects has the peculiar form that it takes as shown in Fig 5.6.

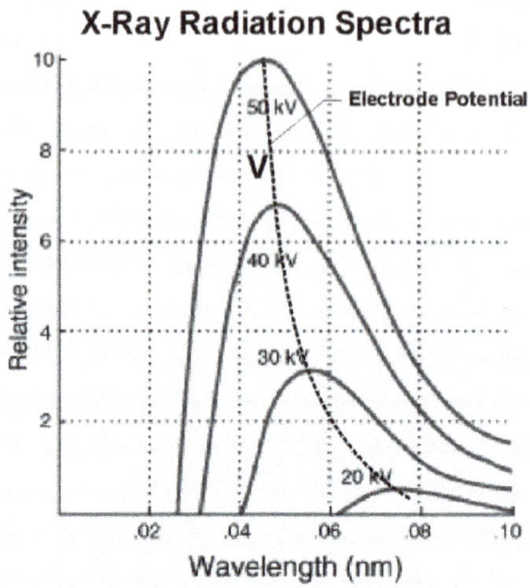

Fig. 6.8 X-ray radiation patterns at different anode-cathode voltages

Effectively one thing is to hold a formula that describes the distribution of energy across a radiation spectrum, and a totally different other of no less significance is to elucidate why the distribution has the specific form or geometry that it portrays and not any other. We know that the geometry of this distribution is principally an inheritance of the geometry of the derivative germane to the generating image function of the Mappa Mundi which is provisioning the entire physical architecture of U_3. To that extent, another radiation distribution pattern, that of X-rays, may become of interest in that it shows a similar pattern (Fig. 6.8).

The uniformizing element of thermodynamic temperature is hereby replaced by the steady voltage between electrodes within the evacuated X-ray chamber. Nevertheless, the distribution follows the same pattern as that of black body radiation of visible light both in the geometry of every individual distribution spectrum and in their displacement as a function of voltage.

6.9. Derivation of the Electron Gyromagnetic Ratio and the Electron Compton Wavelength

One may inadvertently overlook the other two quantities mentioned above, the two straight values of $dQ(s)/ds$, namely:

$$\frac{0.17603266}{\sigma^2} \text{ and } \frac{0.2419707245}{\sigma^2}.$$

The first is indeed what we know as the *Electron Gyromagnetic Ratio* with the official value of **0.176085974** $\cdot 10^{10}$ ***Rad*** s^{-1} T^{-1}, and the second is the *Electron Compton Wavelength*, with the current official value of **0.242631058** $\cdot$ 10^{-11} ***m***.

The *Electron Compton Wavelength* is the equivalent to the wavelength of a photon whose energy is the same as the rest-mass of the electron. The interaction that yields this quantity is the collision between an incident photon and the free electron, which scatters away or recoils after the collision. The Compton wavelength for the electron is given by the formula:

$$\lambda_e = \frac{h}{m_e c}$$

where h is the Plank Constant, m_e the rest-mass of the electron and c the speed of light.

The *Electron Gyromagnetic Ratio* for its own is a measure of the rotational velocity of the free electron, indeed a spinning object, or in proper designation its spin angular momentum. The defining formula for the Gyromagnetic Ratio of a classical rotating body is:

$$\gamma = \frac{q}{2m}$$

where q is its charge and m its mass. Note that the constant known as *Electron Charge-to-Mass Ratio* is another ratio figure quite close in value to the *Electron Gyromagnetic Ratio*, with its official value set at 0.1758812012 $\cdot$ 10^{10} *C/Kg* Although the units are different, the two figures are very similar to one another given that the Electron Gyromagnetic Ratio is equal to *e/2m* and the Electron Charge-to-Mass Ratio is equal to *e/m* . Since the mass is an extremely small value with respect to the charge, the divisions by *m* or *2m* end up casting almost the same result. From the Quanto-Geometric analytical perspective however, the relevant figure is the Electron Gyromagnetic Ratio and that is the one we will shortly devote our attention to.

After taking stock of the above facts, one is left to wonder why these very specific quantities have something to do in particular with the electron to start with. Why do these quantities not qualify as universal physical constants for U_3 just as the *Planck constant* and the *Stephan-Boltzmann constant*? The truth is that this is exactly the case, because the electron is the smallest fermionic particle (mass dominant particle) and because the natural physical quantities associated with it potentially become base units for the fermionic world. So therefore, the physical quantities associated with the *h quantum* represents the natural units for the mass-less (space dominant) world of U_3 objects whereas the quantities associated with the electron represent the natural units for point-like or mass dominant objects. One may justifiably view the latter equally as a form of quantization, as it in fact is.

By giving primacy to the units of our usual measurements of physical quantities, instead of the cardinality of the quantities themselves, most physicists may have never noticed that the Plank constant is the inverse of the Electron Compton Wavelength and that the Stephan-Boltzmann constant is the inverse of the Electron Gyromagnetic Ratio. The premise that we have set forth corroborates this result in many ways. Suffice it to re-examine the relations *(5-1)* and *(5-2)* for that determination. Effectively if we normalize those two relations by setting the Potential Energy Density E_p equal to 1 and the Kinetic Energy Density equal to 1 (their being equal is what quadratic equalization means), then:

$$h = 1/v \quad \text{and} \quad \sigma^* = 1/T$$

where we assimilate v to the Electron Compton wave function and T to the Electron Gyromagnetic Ratio. While the first assignment may be readily understood, the second may seem at first objectionable. Until we realize that Temperature is defined as the average kinetic energy of a population of objects and that the spin component of the motion of these objects accounts for the most part for that average quantity. While both translational or linear momentum and angular momentum of individual elements in a population of free particles may hypothetically be reduced to a minimum or null, their gyromagnetic moment (spin) is a non-reducible, establishing thereby the basis for kinetic energy density or T as a necessary physical quantity or a characteristic of matter of great consideration (even when we are considering a single object). We shall note, by the same token, that while great emphasis is usually placed on the energy distribution of a radiation spectrum because of its historically revolutionary *h quantization* implications, the role of thermodynamic temperature and the shift of the whole spectrum as the thermodynamic temperature increases in accordance with the Stephan-Boltzmann law are usually understated. All four of these constants that we come to witness thru pure radiation spectra denoted black body radiation are equally important and remit to two distinct bases of quantization of matter as normed by the Quanto-Geometric Hyper Quadratic Operator.

Unlike what has become customary in academia, it is important when representing these spectra to not only present the various energy distribution curves but also the thermodynamic temperature variation curve going thru the peaks of these individual spectra as the second significant tenet of these experimental artifacts (Fig. 6.9).

What's more, these two orders of phenomena come again to be manifest in the value of the spin of both the electron and the quantum of action (the minimal photon). It turns out that the expected value for the spin angular momentum from a "classical" approach is quite different from the experimental value, which was found to necessitate a g correction factor and most importantly was found to be a function of $\hbar$. Interestingly enough, the reduced Planck constant figure is in itself a gyric measure *(h/2π),* and the quantum mechanical measure for the electron's spin angular momentum is expressed as:

$$p_s = \sqrt{s(s+1)} \cdot \hbar$$

where *s* is the spin quantum number. The formula makes $\hbar$ a natural scale for that spin. The only value that comes to satisfy experimental results is $s = \frac{1}{2}\,\hbar$. Now we recall that the norm of the Hyper Quadratic Operator in developing the work of creation at this stage was the selection of two *s* values of the Quanto-Geometric Function, one of which was ½ σ. We hold right there the reason for the determination of the spin of the electron at that particular value in Quantum Mechanics. If this is true, the counterpart object, the quantum of action or photon particle, should also have a spin and the value of that spin should be 1 (or *1 σ* or simply *σ).* Well, the photon particle is known to have a full spin value of $1\hbar$, in an unmistakable demonstration that those two points of the real image map of our generating work function are the origin of spinors or the spinor space as known in Quantum Mechanics.

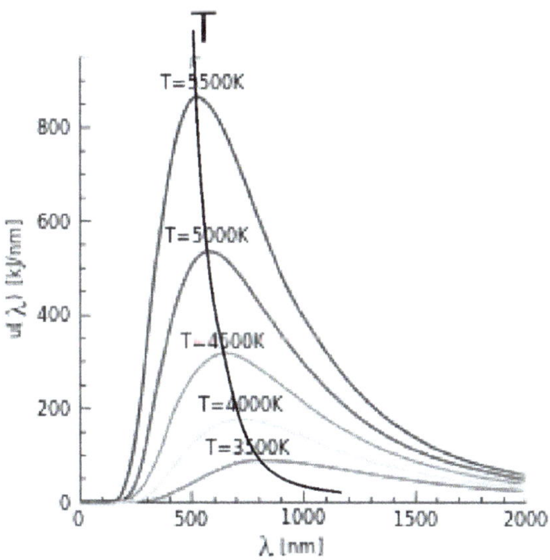

Fig. 6.9 Black body radiation spectrum showing the Temperature function

The above parallel does not necessarily mean that $\hbar$ is equivalent to the standard deviation parameter σ of the Quanto-Geometric Mappa Mundi Function, just as the classical Gyromagnetic Ratio is not exactly interchangeable with the same figure in Quantum Mechanics.

6.10. The Hyper Space Operator and the Interaction Coupling Constants

Within the work of creation, the Tri-Valent Hyper Space Operator is entrusted with the development of space shelling or domain dimensioning in U_3. In its capacity of a space-dominant entity it is best positioned to undertake the task of provisioning spatial encapsulation to the Universe in development. As a reminder the norm of this Operator is *space overture* or *space de-ployment (un-folding)*. This Operator correspondingly takes the following three actions, all consistent with its norm and the nature of the object under operation:

1. It instantiates an image of the Mappa Mundi function, which is appealed to play the role of a template.

2. It combines the Mappa Mundi image's derivative function with a parabolic function in order to generate the final map of construction of U_3. The choice of a parabolic conic function over the other conic functions is geared by the 3-dimensional nature of the Universe in construction *($n = 3$)*, which a figure of symmetry base 3 such as the parabola will provide the best generating matrix for.

3. By prescribing the selection of the parabolic function of construction from within the conic sub-functions embedded in the Mappa Mundi function, it poses thereby a requirement for that function to be a Quanto-Geometrically normalized parabola.

Ultimately the end operation that accomplishes the generation of U_3 is given in a Transform T_9 such that:

$$T_9 \rightarrow \frac{d[Q(s)]}{ds} \bullet y_p$$

We are going to handle that transform as an $M_3(s)$ function because the parabolic function is ultimately reduced to a simple coefficient in the course of the operation, as expressed below:

$$M_3(s) = \frac{dQ(s)}{ds} \bullet k_p$$

We will shortly see the reason for k_p, a parabolic coefficient, instead of the full parabolic function y_p. In general, the resulting individual transformations create 8 *shells* of matter across U_3, namely the Particle shell, the Atom Shell, the Molecular Shell, The Crystal Shell, the Karyot Shell, the Astral Shell, the Star-System or Solar Shell and the Galactic Shell. We might add the Quark Proto-Shell to make justice to it, notwithstanding its limited shell development. What these shells specifically stand for are whole domains of *de-ployed* space or capsules of matter which are embedded within one another. Below is the hierarchical organization they present in our physical universe:

(Shell 0 → Quarks)

Shell 1 → Particles

Shell 2 → Atoms

Shell 3 → Molecules

Shell 4 → Crystals (Materials)

Shell 5 → Karyots

Shell 6 → Astral Bodies

Shell 7 → Star Systems

Shell 8 → Galaxies

The assignment map for the Transform T_9 is:

($s_0 = 0 \rightarrow$ Shell 0 – Quarks)

$s_1 = \pm\sigma/2 \rightarrow$ Shell 1 - Particles

$s_2 = \pm 2\sigma/7^{1/2} \rightarrow$ Shell 2 - Atoms

$s_3 = \pm\sigma \rightarrow$ Shell 3 - Molecules

$s_4 = \pm\sigma(8/5)^{1/2} \rightarrow$ Shell 4 - Crystals (materials) or Lattices

$s_5 = \pm\sigma(10)^{1/2}/2 \rightarrow$ Shell 5 - Karyots

$s_6 = \pm 2\sigma \rightarrow$ Shell 6 - Astral Elements

$s_7 = \pm\sigma 7^{1/2} \rightarrow$ Shell 7 - Star Systems

$s_8 = \pm 4\sigma \rightarrow$ Shell 8 Galaxies

The match per our interval definition for the Quanto-Geometric Function is:

(Shell 0 $\rightarrow]0, \sigma/2[$)

Shell 1 $\rightarrow [\sigma/2, 2\sigma/7^{1/2}[$

Shell 2 $\rightarrow [2\sigma/7^{1/2}, \sigma[$

Shell 3 $\rightarrow [\sigma, \sigma(8)^{1/2}/5[$

Shell 4 $\rightarrow [\sigma(8/5)^{1/2}, \sigma(10)^{1/2}/2[$

Shell 5 $\rightarrow [\sigma(10)^{1/2}/2, 2\sigma[$

Shell 6 $\rightarrow [2\sigma, \sigma 7^{1/2}[$

Shell 7 $\rightarrow [\sigma 7^{1/2}, 4\sigma[$

Shell 8 $\rightarrow [4\sigma, \infty[$

We may notice in the above list of shells that *quarks* are annotated pertaining to *Shell 0.* Yet they do not matter-of-factly form a shell. Shell 0 is not a shell proper per determination of the assigned interval which is open on both ends: it is a *proto-shell.* There is no finite starting point for an *s-interval* for this would-be domain, s equal to 0 being undefined or forbidden by the model. Hence no possible finite *dQ(s)/ds* value at that point which could serve as a constant. Effectively the first value of *s* within each interval is what is going to yield a physical constant, a quantity incarnated by the corresponding value of the $M_3(s)$ function. Because quarks do not or cannot have a defining physical constant, hence a space domain of their own, they come to behave as *glued* to one another and *confined* within the internal or endogenous boundaries of the next space *de-ploy*, the Particle Shell. The so-called phenomenon of *confinement* means that they can never be manifest in free space of their own but only as sub-components of Shell 1 elementary particles. They may only have a life as indivisible parts of particles in what construes as the closest representation of pure scalar elements in U_3, if ever possible. Their inner space value is close to 0, albeit not 0. This circumstance does not preclude them from having behavior however, specifically Quanto-Geometric behavior. They remain thoroughly sub-ject to Quanto-Geometric spectral organization based on the 9 typified primitives and other determinants of the model as we shall see in Book II. This is by all accounts the origin of the phenomenon of *quark confinement*, empirically observed and subsequently posited, but not theoretically derived, in the Standard Model of particle physics.

As to *Quasars*, which are not explicitly mentioned in the above presentation, they are part of the Galaxy class since they are known to have galactic identity as proto-galaxies.

Does this entire construction all amount to just the author's view, a speculation or a mere vanity? Let us examine the creation work function $M_3(s)$ for an answer to that ques-tion.

6.11. The Parabolic Coefficient as a Generator

From previous developments, we know that the Quanto-Geometric $Q(s)$ function en-closes a parabola among its embedded conic sections. We have stated right above that the Tri-Valent Hyper Space Operator has elected to incorporate this parabolic function within the transformation at work to give birth to U_3. On a Cartesian coordinate system, the ex-pression of this embedded parabola would be:

$$y = \frac{1}{\sigma\sqrt{2\pi}} + \frac{s^2}{4p}, \text{ with } \frac{1}{\sigma\sqrt{2\pi}} \text{ being the y intercept.}$$

We know however that the Quanto-Geometric $Q(s)$ function is not defined for s = 0, so that no y intercept exists for the embedded parabola. On the Quanto-Geometric coordi-nate system the expression of this embedded parabola centered on the q axis reads:

$$y = \frac{s^2}{4p}$$

The *latus rectum* point of the parabola also belongs to $Q(s)$ as s_3. We remind the reader that the latus rectum is the cord that goes thru the focal point of the parabola and stands orthogonal to the y axis. Its distance to the vertex of the parabola (a point of discontinuity in our function) is p. It can be demonstrated that the length of the latus rectum at each side of the y-axis is $2p$, for a total length of 4p for that special cord. Since the point $s = 2p$ of the parabola also belongs to the $Q(s)$ as s_3, it follows that:

$s_3 = 2p$ and from $Q(s)$ $s_3 = \frac{2\sigma}{\sqrt{7}}$.

So that: $2p = \frac{2\sigma}{\sqrt{7}}$ or $p = \frac{\sigma}{\sqrt{7}}$.

Replacing p by its value in the equation of normalized parabola yields:

$$y = \frac{s^2}{4p} = \frac{s^2}{4} \cdot \frac{\sqrt{7}}{\sigma}$$

At s_3, $y_p = \frac{\sqrt{7}}{4\sigma} s^2 = \frac{0.6611437827 8}{\sigma} \cdot s^2$ or:

$$y_p = \frac{0.6611437827\ 8}{\sigma} \cdot s^2$$

The eventual graph of y_p follows in Fig. 6.10.

Now, it is important to note that σ does not have the same dimension as s in the above expression of the parabola. The σ parameter is the value of s at inflexion point in the $Q(s)$ function, which lays much beyond any shared point with the embedded conic sections. Therefore, the point $s = \sigma$ does not and cannot belong to the normalized parabola. And thus, although σ is an s value, it cannot stand as such in the expression of the parabolic function. As a result, within that expression, the entire term $0.66114378278/\sigma$ which appears as a coefficient of s^2 can only have the value or weight of a coefficient as a "dimension-less" scalar. We denote that term as follows:

$$k_p = \frac{0.66114378278}{\sigma}$$

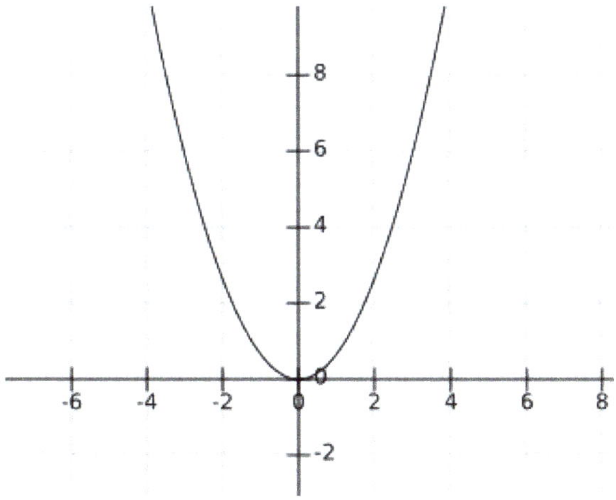

Fig. 6.10 Graph of y_p

The parabolic coefficient k_p is instead going to play that eminent role in the generation of U_3 as we shall see shortly, in substitution of y_p to which the role is *naturally* reserved.

6.12. Derivative of the Grand Mappa Mundi Function

As previously noted, the work of Shell de-ployments in U_3 is performed according to the end function $M_3(s)$:

$$M_3(s) = \frac{dQ(s)}{ds} \bullet k_p$$

The first-order derivative expression of $Q(s)$ reads as follows:

$$\frac{dQ(s)}{ds} = \frac{1}{\sigma\sqrt{2\pi}} \frac{-2s}{2\sigma^2} \cdot e^{\frac{-s^2}{2\sigma^2}}$$

$$\frac{dQ(s)}{ds} = \frac{-s}{\sigma^3\sqrt{2\pi}} \cdot e^{\frac{-s^2}{2\sigma^2}}$$

$M_3(s)$ becomes after due replacements:

$$M_3(s) = \frac{-s}{\sigma^3\sqrt{2\pi}} \cdot e^{\frac{-s^2}{2\sigma^2}} \bullet \frac{0.6611437827\ 8}{\sigma}$$

$$M_3(s) = 0.6611437827\ 8 \bullet \frac{-s}{\sigma^4\sqrt{2\pi}} \cdot e^{\frac{-s^2}{2\sigma^2}}$$

See the graph in Fig. 6.12 for a pictorial representation of the *M₃(s)* function. We may now replace *s* successively by the critical values defining the intervals in order to obtain the defining constants for each physical Shell of U_3 as their building blocks. We shall point out that what we call the defining values of these intervals are the first or initial value of each. The Hyper Space Operator assigns each of these initial values to the generation of a Shell, thereby creating the defining physical constant for each Shell thru the Norm of the Operator, which amounts here to simply the value of the $M_3(s)$ function. Under these circumstances *s* creates the space *de-ploy* or composition and $M_3(s)$ establishes the pillar of the Shell edifice, the physical constant.

6.13. Shells of U_3 and Their Physical Constants

We are going to compute the value of $M_3(s)$ for each interval s_i corresponding to the different physical Shells making up U_3.

At s_2, Shell of elementary Particles or S_2:

$$s = \frac{\sigma}{2},\ M_1(s_1) = 0.6611437827\ 8 \bullet \frac{-s}{\sigma^4\sqrt{2\pi}} \cdot e^{\frac{-s^2}{2\sigma^2}}$$

$$= -\frac{0.6611437827\ 8 \cdot 0.1760326634}{\sigma^3}$$

$$= -\frac{0.1164346625}{\sigma^3}$$

At s_3, Shell of Atoms or S_3:

$$s = \frac{2\sigma}{\sqrt{7}},\ M_2(s_2) = 0.6611437827\,8 \bullet \frac{-s}{\sigma^4\sqrt{2\pi}} \cdot e^{\frac{-s^2}{2\sigma^2}}$$

$$= -\frac{0.6611437827\ 8 \cdot 0.2266245234}{\sigma^3}$$

$$= -\frac{0.1498980325}{\sigma^3}$$

At s_4: Shell of Molecules or S_4:

$$s = \sigma,\ M_3(s_3) = 0.6611437827\,8 \bullet \frac{-s}{\sigma^4\sqrt{2\pi}} \cdot e^{\frac{-s^2}{2\sigma^2}}$$

$$= -\frac{0.6611437827\ 8 \cdot 0.2419707245}{\sigma^3}$$

$$= \frac{0.1600485904}{\sigma^3}$$

A symmetry explosion at s_5 overtakes the parabolic restriction of y_P and transmutes the generation function from $k_P \cdot dq/ds$ to the full $y_P \cdot dq/ds$ or M_3 in order to generate the Shell of Materials (nanomaterials) or S_5:

$$s = \sigma\sqrt{8/5},\ y_P(dq/ds) = M_3(s) = -\frac{\sqrt{7}}{\sigma^4 4\sqrt{2\pi}} \cdot e^{-\frac{s^2}{2\sigma^2}} \cdot s^3$$

$$= -\frac{0.23996255805}{\sigma}$$

See Appendix for the full step-by-step computation of this quantity. The next Shell up is Shell 6, the Shell of Life. Since for the Shells beyond S_6, no physical constant is known, although they very much exist, we will stop the computational report at Shell 6. We offer below the further computation of the constant for Shell 2, the Shell of Atoms.

$$M_2(s_2) = M_2(s_2) = 0.6611437827\,8 \bullet \frac{-s}{\sigma^4\sqrt{2\pi}} \cdot e^{\frac{-s^2}{2\sigma^2}}$$

$$= 0.66114378278 \bullet \frac{-(2\sigma/\sqrt{7}) \cdot e^{\frac{-(2\sigma/\sqrt{7})^2}{2\sigma^2}}}{\sigma^4\sqrt{2\pi}}$$

$$= 0.6611437827\,8 \bullet \frac{-(2\sigma/\sqrt{7}) \cdot e^{\frac{-(4\sigma^2/7)}{2\sigma^2}}}{\sigma^4\sqrt{2\pi}}$$

$$= 0.66114378278 \bullet \frac{-(2\sigma/\sqrt{7}) \cdot e^{-(2/7)}}{\sigma^4\sqrt{2\pi}}$$

$$= 0.66114378278 \bullet \frac{-(2\sigma/\sqrt{7}) \cdot e^{-(2/7)}}{\sigma^4\sqrt{2\pi}}$$

$$= 0.66114378278 \bullet \frac{-2\sigma \cdot e^{-(2/7)}}{\sigma^4\sqrt{7}\sqrt{2\pi}}$$

$$= 0.6611437827\;\;8 \bullet \frac{-0.2266245234}{\sigma^3}$$

$$= -\frac{0.1498980325}{\sigma^3}$$

Reported in Table 6.3 are all these values per Shell 1 up to Shell 6; it includes as well other related figures which contribute to the makeup of all physical constants intervening in the composition of the U_3 universe. Effectively the derivative function itself produces physical constants so much as its inverse and the inverse of the generating function $M_3(s)$ do. This table allows us to appreciate the very important fact that the known physical constants are not random values only computationally related to one another but obey to an order of hierarchies which are the vertical and horizontal Quanto-Geometric hierarchies set forth in this study. Some of these figures may be readily recognizable by the reader but others may not. We will next address each in more details.

QG Function	3-D Universe → $s_3 = 2/\sqrt{7} \cdot \sigma$	
$Q(s) \to s_3$	$0.299796065/\sigma$	Speed of Light

QG Function	Spinor Space → $s_i = 1/2 \cdot \sigma$		Spinor Space → $s_i = 1 \cdot \sigma$	
dq/ds	$0.17603266/\sigma^2$	Electron Gyromagnetic Moment	$0.2419707245/\sigma^2$	Electron Compton Wavelength
$(dq/ds)^{-1}$	$5.680763 \cdot \sigma^2$	Stefan-Boltzmann constant	$0.413273135 \cdot \sigma^2$	Planck Constant (in eVs)

QG Function	Quark Particles – Shell 1	Elementary Particles – Shell 2	Atoms – Shell 3	Molecules – Shell 4
$k_p \cdot (dq/ds)$	No invariant (s = 0 undetermined)	$0.11643466/\sigma^3$ Fine Structure Constant	$0.1498980325/\sigma^3$ Inverse Gravitational constant	$1.600485904 \cdot 10^{-1}/\sigma^3$ Elementary Charge

QG Function	Nano Materials – Shell 5	Karyotes – Shell 6	Celestial Singlets – Shell 7	Star Systems – Shell 8	Galaxies – Shell 9
$y_P \cdot (dq/ds)$	$0.239962558/\sigma$ Van der Vaals constant	$0.298841566/\sigma$ Physical Life constant	$0.2856933408/\sigma$ unknown	$0.14757590/\sigma$ unknown	$0.00566530/\sigma$ unknown

QG Function	Topological Constants	
$(k_P \cdot dq/ds)^2$	$0.1355703/\sigma^6$ Rydberg Energy unit	$0.0224969/\sigma^6$ Molar Volume of Ideal Gas

Table 6.3 Table of physical constants deriving from Quanto-Geometric functional constructs. We have muted the negative sign of the quantities, whose principal meaning concerns the direction of correlational displacements.

We shall also note that these figures are harmonized due to their common origin, being all expressed in parametric form as a composition between a scalar and σ the standard deviation of the Grand Mappa Mundi Function. Consequently, they will not be in an absolutely exact match with the known figures but they will match by an acceptable order of "pure magnitude", or to be more exact as **figurate** numbers. The most striking fact remains how from a single and relatively simple mathematical-physics system all these constants are naturally derived, raising the question of which side needs a correction if need be as to the orders of magnitude.

The reader will find in Appendix I a list of all known physical constants published by the U.S. National Institute of Science and Technology. It is a rather furnished list, however one shall not lose sight of the fact that there is only a handful of foundational physical constants, the most preeminent of which is the speed of light.

6.14. Derivation of the Newton Gravitational Constant

The official value of the gravitational constant G is 6.6720 10^{-11} N-m^2/kg^2, which makes G^{-1}, the inverse of G, equal to: **0.1498800959 10^{-12}** kg^2/N-m^2.

Our table reports G^{-1} as a Shell 2 constant with the value of: **0.1498980325/σ^3**. Now, one can see that the two values are appreciably close to one another except for the dimensions they are expressed in. The dimension N-m^2/kg^2, with the inverse kg^2/N-m^2, is a view of *G*. That is how we historically came to know of it by physical measurements based on our understanding of physical nature.

The most significant flaw of that view consists in that it makes no distinction between the qualities of space in each space shell, nor does it understand the nature of space as dictated by the vertical hierarchy. It is a rather significant fact that *G* represents a physical constant in operation in Shell 2, the shell of Atoms. It means that contrarily to what is expected by both the Standard Model and the Quantum Gravity models, gravity does not exist at the particle level. No such thing there is as gravity being so weak at the particle level that it becomes lost in the presence of stronger forces as the current belief holds. *There is simply NO gravity at the particle level.* If we expect gravity to be detectable as gravitational waves similar to the wave function of particles, and that gravity should be in effect at the level of interaction between free particles, then it is a lost quest. On the other hand, if we restrict the statement strictly to the Atom shell as the lower limit of exercise of gravity, or if we look for gravity within the interaction taking place between particles *in situ* inside the atomic edifice (not free particles), then we are off to a good start.

Finally, the fact that we commonly know of and find use for *G* and not G^{-1} is equally telling. It reveals that the view which leads us to *G* is at the very least a backward view from a rear vantage point, not to say that it is a skewed view. Since we are looking at everything occurring in sub-shells of space from another higher shell without making the necessary preventions, exemptions and distinctions, it is no wonder that the resulting statements or conclusions are somewhat falsified, distorted or otherwise inadequate.

6.15. Derivation of the Electric Charge of Electrons

The official value of the electron charge constant *e* currently is **1.60217656 10^{-19}** Coulomb. As per our table *e* is a Shell 3 constant with the value figure of: **1.600485904/σ^3**.

Remarkably, it is not a constant emanating from Shell 1, the Particle Shell, as one might have expected. It is a building block for the Molecular shell. Be that as it may, as soon as we remember that the electrostatic systematic between atoms' electronic shells is at the heart of molecular elements then the surprise element might fade away. This all means that what is happening in the Molecular Shell at its formation is a condition whereby the interaction between atoms is pre-set by an electrical systematic, the smallest unit or form of which is the unit-Coulomb. In other words, the e constant is a building block for the Molecular Shell, and not a quality inherent to the *electron.* That is the very reason why the proton also demonstrates electrical charge at exactly the same value.

The property of charge is a property of the molecular medium not of the particles residing therein. We recognize that this conclusion will be a difficult fact for physicists to accept, given that so much has been built on the concept of charge being inherent to elementary particles themselves. However, few have historically asked themselves the question: how come the value of the electrical charge of an electron is exactly the same as that of a proton which is a particle with much larger mass? What in the particle makes a charge unit? Is the charge an artifact of mass, is it an artifact of the particle's wave function? Where does it come from?

The nature of space within the Molecular medium is the origin of this constant. It is the driver that orchestrates the systematic between atoms at formation of Molecules and their subsequent behavior during their life-cycle. For those who will still resist the idea of charge not being inherent to particles, we invite them to find a reason or an element capable of explaining the property or quality of their alleged charge, the origin of that charge, not just rely on experimental measurement of quantities, which are designs from our own constructs. Some other polemic facts with no answer are well known about particles in relation to their electric charges or lack thereof. We will defer this analysis for treatment in Book II.

6.16. Derivation of the Fine-Structure Constant and the Fermi Coupling Constant

Interestingly enough, these two constants are very closely related to one another when viewed thru the prism of Quanto-Geometric treatment. For one, they are both coupling constants, and secondly their value is almost identical when estimated or computed suitably. The Fermi coupling constant is a mediator for the electro-weak force (radioactive decay) whereas the fine structure-constant, also known as the electro-magnetic coupling constant, is said to be a mediator for electro-magnetic interaction between elementary charged particles.

6.16.1 The Fine-Structure Constant

The official value of the fine-structure constant α currently is **7.297352568 10^{-3}**. This may be the only wholly computed constant, after the permittivity and permeability of space, both of which are incidentally part of the formula for α, as reported below.

The formula is:

$$\alpha = \frac{e^2/\hbar c}{4\pi\varepsilon_0},$$

where e is the elementary charge, $\hbar$ is the reduced Plank constant. Since c the speed of light and ε_0 the permittivity of space are formulated as:

$$c = \frac{1}{\sqrt{\mu_0 \varepsilon_0}}$$

and the reduced Plank constant is valued at $\hbar = h/2\pi$, the formula for α becomes after these replacements:

$$\alpha = \frac{\mu_0 c e^2}{2h}$$

By replacing these constants with their known values in the α formula above, α comes out to be:

$$\alpha = \frac{(1.602176565 \cdot 10^{-19})^2 \times 299792458 \times 4\pi \cdot 10^{-7}}{2 \times 4.135667516 \cdot 10^{-15}}$$

$$= \frac{2.566969745 \cdot 10^{-38} \times 376.7303135}{8.271335035 \cdot 10^{-15}}$$

$$= \frac{9.670553169 \cdot 10^{-36}}{8.271335035 \cdot 10^{-15}}$$

$$\alpha = 0.116916654 \; 10^{-20}$$

The Plank constant *h* is generally chosen to be **6.6256 10^{-34} J-s** when computing the value of α. We have chosen instead in computing the formula its value expressed in eV-s (electron-volt–second), a more standardized or primary dimension than the J-s (Joule-second) dimension, because it is directly related to *e*, the electron or its charge. As shown on Table 5.3, the Quanto-Geometrically derived value for the constant is: **0.1164346625/σ^3**.

While the value of the fine structure constant can be fully computed from any of the defining formulae, quantum electrodynamics has provided ways to measure this constant, principally via the quantum Hall effect. As usual with measurements of physical constants, they have historically gone from appreciably inaccurate to appreciably accurate as the technological means undergo refinement. It is fair to say that this constant is one that has most fascinated physicists, even to the point of bewilderment, different visions of what it exactly means having been debated. If physicists are loath to numerology, regarded by them as a suspicious pseudoscience, the fine structure constant had them engage openly in that exercise. Here is what Richard Feynman, one of the pioneers of Quantum Electrodynamics, wrote on this subject in 1985:

"There is a most profound and beautiful question associated with the observed coupling constant, e – the amplitude for a real electron to emit or absorb a real photon. It is a simple number that has been experimentally determined to be close to 0.08542455. (My physicist friends won't recognize this number, because they like to remember it as the inverse of its square: about 137.03597 with an uncertainty of about 2 in the last decimal place. It has been a mystery ever since it was discovered more than fifty years ago, and all good theoretical physicists put this number up on their wall and worry about it.) Immediately you would like to know where this number for a coupling comes from: is it related to pi or perhaps to the base of natural logarithms? Nobody knows. It's one of the greatest damn mysteries of physics: a magic number that comes to us with no understanding by man. You might say the "hand of God" wrote that number, and "we don't know how He pushed his pencil." We know what kind of a dance to do experimentally to measure this number very accurately, but we don't know what kind of dance to do on the computer to make this number come out, without putting it in secretly!"

One may note with a smile that the accuracy of the number today puts it at 0.07297352568 as previously mentioned, not his "very accurate" 0.08542455 commonly accepted at the time! But his remarks are admittedly otherwise quite telling. This quantity so intrigued physicists, not just theoretical physicists, that Wolfgang Pauli, known by and large for his contribution to Quantum Mechanics, took it to the point of collaborating with Carl Gustav Jung, a towering contemporaneous figure in the field of Psychology, in order to unravel the transcendental meaning of this quantity

Because this constant lends itself to so many different deriving formulae implicating thereby different genericity, its physical nature has come to be viewed in many different ways by physicists. In some interpretations, it is considered to typify interaction between free particles such as free electrons and photons. When derived from the application of classical methods, it is viewed as a characterization of a *massless* charge on a circular track embedded in its own radiation field and expressing the strength of the electro-dynamic interaction. Quantum Mechanics has it that it characterizes the interaction between a charge and its field, a form of kinematic that replaces the punctual charge by a charge distribution rather which leads to oscillatory behavior. It is a rather remarkable fact that this constant can be valid without any correction both in Classical energy physics and in Quantum Mechanics, knowing the conceptual steep jump laying between the two fields. More on this later. Its application in physics is extremely widespread, perhaps second only to *c,* the speed of light, or *h,* the Planck constant. The truth to this matter is that the fine-structure constant is a characterization of the quality of space intervening in the buildout of not only electrons and photons, but all 3-dimensional particles as pertaining to Shell1 of U_3, the Quanto-Geometric Particle Shell. If we were to be more precise, we would make the statement that it corresponds to the specific value of the so-called $M_3(s)$ composite Quanto-Geometric function at a specific value (s_3) of s.

We do not have to engage in an exercise of numerology about it, having established the Quanto-Geometric number theory that systematically leads us to its province. So the late Richard Feynman, Wolfgang Pauli and all other theoretical physicists with this interest to date are hereby well served! We shall conclude these remarks by stating that the *pencil of God* that R. Feynman referred to and *the way that He pushed it* are indeed the generating Transform materializing the work of Shell generation in U_3:

$$M_3(s) = \frac{dQ(s)}{ds} \bullet k_p$$

6.16.2 The Fermi Coupling Constant

In the Standard Model of particle physics, interaction between sub-particle components that make up particles, which becomes observable in decay or radioactive events where they occur, is developed thru the mediation of a coupling constant. The primary coupling constant for such interaction is the Fermi coupling constant G_F, after the name of its proponent, Enrico Fermi, and proposed since the early days of particle physics. The Electroweak Theory of the Standard Model so restricts the domain of this constant to unstable particles while it constitutes the province of interpretation for all particle decay phenomena, whether they occur in free particles such as the naturally unstable neutron or within the context of a higher edifice such as atomic nuclei.

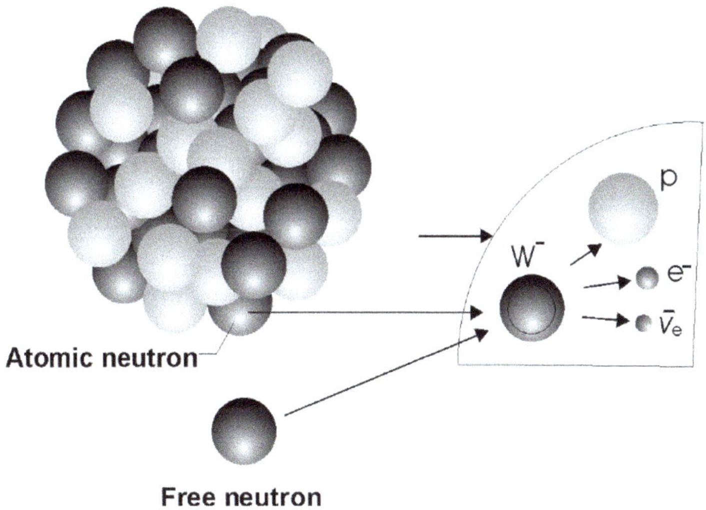

Fig. 6.11 Beta decay of a neutron particle

The formula and value of G_F are as follows:

$$\frac{G_F}{(\hbar c)^3} = \frac{\sqrt{2}}{8}\frac{g^2}{m_W^2} = 0.116637\times 10^{-4}\,GeV^{-2}$$

where g is the coupling constant *(g-factor)* of the so-called electroweak interaction and m_W the mass of the W boson that mediates the decay interaction. The whole term

$$\frac{G_F}{(\hbar c)^3}$$

is the quantity in use by preference in modern physics, not just the original Fermi's G_F.

Is it any surprise that this value is in a match with the fine-structure constant provided the latter is suitably computed, and in a match as well with the Quanto-Geometrically derived value? Beta decay in an atomic nucleus, resulting from the breakdown of a neutron into several other sub-particles, was the original focus of Enrico Fermi (Fig. 6.11). His proposal, originally a theoretical treatment, was initially rejected, but later gained credit when it became experimentally ratified and other forms of particle decay, aside from neutron decay, could be accounted for by the constant. This and other Fermi's findings were instrumental in paving the way for the development of particle physics. Finally, this constant registers as a paramount entity in the composition of all Shell1 elementary particles, whether they are stable or unstable.

6.17. Unifying the Electromagnetic Interaction with the Electroweak Interaction

The many distinct views that physicists have historically held about the identity of the electromagnetic coupling constant have not prevented them from sharing a sense of its possible relationship with the electroweak force. Reason why a so-called *unifying electroweak theory* has been developed in order to establish a single scale for both types of interaction, the difference between one and the other being a simple question of level of energy harbored by each. Effectively in the unified electroweak theory the electromagnetic interaction is part of a mixture of interactions associated with electroweak fields so that the strength of the electromagnetic interaction becomes a function of the strength of the energy field. The observed value of α relates therein to the energy level of the mass of the electron. Since the electron is the lightest charged object, then α is ranked at ground level or energy zero in the scale, leaving the Fermi coupling/energy constant higher up in the scale.

This construction has been plagued with shortcomings because the expected projections do not realize in the actual behavior of matter, forcing physicists to engage in a series of ill-fated extrapolations called *Renormalization*, *Regularization* and other misfit engenders. These developments drew sharp criticism from some of the early and renown formulators of Quantum Electrodynamics. Irate Paul Dirac explains as late as 1975:

Most physicists are very satisfied with the situation. They say: 'Quantum electrodynamics is a good theory and we do not have to worry about it anymore'. I must say that I am very dissatisfied with the situation, because this so-called 'good theory' does involve neglecting infinities which appear in its equations, neglecting them in an ***arbitrary*** *way. This is just* ***not sensible*** *mathematics. Sensible mathematics involves neglecting a quantity when it is small – not neglecting it just because it is infinitely great* ***and you do not want it****!*

Richard Feynman, another lead physicist in the formulation of the theory, further added:

The shell game that we play ... is technically called 'renormalization'. But no matter how clever the word, it is still what I would call a dippy process! Having to resort to such hocus-pocus has prevented us from proving that the theory of quantum electrodynamics is mathematically self-consistent. It's surprising that the theory still hasn't been proved self-consistent one way or the other by now; I suspect that renormalization is ***not mathematically legitimate****.*

Finally, Lewis Ryder had this to say about the state of affairs that lasted to this day:

In the Quantum Theory, these [classical] divergences do not disappear; on the contrary, they appear to get worse. And despite the comparative success of renormalization theory the feeling remains that there ought to be a ***more satisfactory way*** *of doing things.*

There are many reasons to the failure of Quantum Electrodynamics to unify the "electroweak forces", let alone the adjunction of the other two fundamental forces in the act. The primary reason is that its scale of interaction failed to recognize the extent to which space vacuum behaves as a physical variable, and secondly that the energies are distributed across distinct real Shells of space that are completely integral to the composition of matter. Lastly it failed to understand that the electric field does not exist across all Shells except for one. I might add that the recognition that the authentic value of α is not 1/137 but **0.116916** which puts it not only close to but in superposition with the figurate **0.116637** of the Fermi coupling constant would have been a great start! Finally, let it be resoundingly proclaimed, there is a high price to pay for abandoning the key signature that has made Quantum Mechanics otherwise so successful: the uncompromising and unmitigated ban of *time* as a physical variable. To that account one shall add the deep underestimation that Quantum Mechanics for its own ought to imperatively and permanently subscribe to that school of thoughts and adopt its implications as a matter of necessity.

6.18. The Fine-Structure Constant and the g-factor Cluster

In particle physics, one important defining quantity for an elementary particle (leptons and bosons) is its *magnetic moment*, also called *magnetic dipole moment*. This notion is tightly related to that of fine-structure constant (α), otherwise called electromagnetic coupling force, which we have just studied above. The clearest view of what the fine-structure constant represents is that it stands for the originator of the internal binding force operating inside of particles, which the Standard Model of particle physics believes to be of magnetic nature. As such it is the originator of all forms of observable magnetic moments known to particles. The spin of the particle is a contributor to its magnetic moment; for an atomically bound particle, the angular momentum is another contributor, and the composition between spin and angular momentum represents a third contributor in that case. The spin or gyromagnetic moment is the single contributor to the magnetic moment for a free particle and remains the chief contributor to that of an atomically bound particle.

In particle physics, the magnetic moment is generally expressed in terms of a *g*-factor, which measures the difference between the expected or predicted value and the one actually observed or measured experimentally. That differential quantity is termed *anomalous* magnetic moment. The disparity may very well be due to the wrong quantity being used for alpha as we pointed out earlier. It has been fashionable as a matter of practice to use the quantity **7.297 352 5698 10^{-3}** or its inverse in physical calculations. We have shown however that the authentic value of alpha is **0.1164346625/ σ^3**. That stands for the very reason why, after applying the respective *g*-factor correction, all the experimentally measured magnetic momenta of elementary particles narrowly approach the Quanto-Geometrically derived value as reported in Table 6.4 below.

CONSTANT	VALUE
Fermi coupling constant	0.1166 364 × 10^{-4} GeV^2
Electron magnetic moment anomaly	0.1159 65218076 ×10^{-2}
Muon magnetic moment anomaly	0.1165 92091×10^{-2}
Helion magnetic mom. to Bohr magneton	0.1158 740958 ×10^{-2}
Shielded Helion magnetic mom. to Bohr magneton	0.1158 671471 × 10^{-2}
Sakur-Tetrode constant (1K, 101.325 Kpa)	0.1164 8700 ×10^{1}

Table 6.4 The g-factor cluster for the fine-structure constant

Central to those quantities is the standard value of alpha with which they present figurate matches to different extents. We shall point out as well that even the non-standard value of alpha relates to the Quanto-Geometric figure via division by 2π as computed below:

$$\frac{\alpha}{2\pi} = \frac{7.2973525698 \cdot 10^{-3}}{2\pi} = 0.11614097 \cdot 10^{-2} = a,$$

where a is the *anomalous* moment. Within our framework, the division by 2π relates to the 3-dimensional trigonometric symmetry of the Quanto-Geometric model. The above figure is theoretically attributed to the electron's anomalous magnetic moment in the Standard Model. The experimentally measured value is: 0.115965218076 10^{-2}. For the Muon particle, a heavy electron, the anomalous magnetic moment is: 0.116592091 10^{-2}, which puts it quite close in figurate match to the value of the Fermi coupling constant: 0.1166364 10^{-4}. Incidentally, the measurement of the muon lifetime in neutron degeneracy (radioactive decay) provides the most precise determination of the Fermi coupling constant, the vehicle of the Electroweak force. The W boson, the first intermediate degeneracy product, plays a crucial role in neutron decay just as the W and Z bosons play a significant role in the magnetic moment of the muon particle. We may thereby conclude that the fine-structure constant is a first-order regulator of bosonic behavior (photons and interaction bosons). This suggests that the Unifying Electroweak theory, if it is still to be completed to flawlessness, is better orchestrated along the line of bosonic behavior, with alpha as a prime orchestrator, than along the line of magnetic behavior, which is keenly challenged by the neutron electric paradox.

Should we consider here the g-factor for the neutron and the proton as well? This analysis has already been completed in the review of the Fermi constant. For composite particles such as neutrons and protons, the order of the sub-components of these particles is the stage where the fine-structure constant develops its intervention by scaling the energy of interaction of photons and interaction bosons. One may choose to model the systematics of this interaction in the terms undertaken by the Electroweak Theory. However, it is clear that the fine-structure constant is a paramount entity in the internal configuration of all particles, both stable and unstable, since all other quantities that one may construe as a constant in the realm of sub-particles are strictly dependent on alpha as the chief scaling purveyor.

Lastly there is the curious appearance of the Sakur-Tetrode constant in this landscape. The Sakur-Tetrode view is an attempt to frame the state of an ideal classical gas (monoatomic gas) in the extreme condition of temperature as low as 1K, extremely close to the absolute 0 mark. In this regime the classical realization of a gas (its molecular-Shell spread) is suppressed because the quantum mechanical effects take over. The molecular-level entropy entirely collapses into particle-Shell level entropy and the scalarity of the atom singlets is forced to behave similarly to elementary particle scalarity. This is of course a Quanto-Geometric translation of what is reflected in the Sakur-Tetrode equation, below reported, in terms of the mass of gas particles expressed in atomic units and the presence of the Planck constant in the equation. Those inclusions are equivalent to breaking up the atoms into a sea of small particle-like elements (Shell1 elements) as it were. The very figure of the constant, *S/R* or entropy over the gas constant, which attempts to capture entropy per unit volume per particle unit, is indeed an expression of particle-level entropy (distribution of entropy per quantum particle). It is worth noting that this constant may be derived as well from pure Quantum Mechanical treatment. It takes its name however after Hugo Martin Tetrode and Otto Sackur who independently derived it in very early 1900's principally from the then prevalent thermodynamical view of physical states. The equation follows:

$$S = kN\, ln\left[\left(\frac{V}{N}\right)\left(\frac{U}{N}\right)^{\frac{3}{2}}\right] + \frac{3}{2}kN\left(\frac{5}{3} + ln\,\frac{4\pi m}{3h^2}\right)$$

where V is the volume of the gas, U is its internal energy, N the number of gas particles, k the Boltzmann's constant, m is the mass of a gas particle in atomic unit and S the entropy of the gas. At standard condition of pressure *(p= 1 at or 101.325 kPa),*

$$\frac{S_0}{R} = 0.11648708 \cdot 10^1$$

This figure curiously shows an even broader figurate match with the Quanto-Geometrically derived quantity for the fine-structure constant.

It cannot be overstressed while ending this section that the fine-structure constant α plays a capital role in the internal architecture of elementary particles. In the ignorance of the origin of this number, physicists have understandably arranged to develop the physics of elementary particles on a side track to this capital quantity. By doing so, however, they have run into several impasses requiring them to drop this quantity "secretly" and extemporaneously, in the very word of Richard Feynman, into their equations for those to become sensible and effective. Further, the incorrect valuation of alpha overall has also driven physicists to develop different frameworks for the treatment of radioactivity and the physics of the photon, requiring them to find at the end a difficult unifying umbrella in response to the unmistakable perception that they are forces of the same nature. This clearly transpires in the terminology of *anomalous* magnetic moment attributed to particle's magnetic behavior, whereby the expected value from interpretation does not match the real experimental behavior. In all, one has to be persuaded that when nature does not match the view that one upholds in theory, it is not nature that is *anomalous* but the theory itself.

6.19. Fabric of Space as the Mother of All Motion

Now that we are studying the physical manifestation of the derivative of the Grand Mappa Mundi function, we shall take that opportunity to reinforce the notion of space as the mother of all motion, especially for those who might have had to skip the previous Chapter. We had firmly established the principle that the *s* variable of the function represents its **independent** variable. In that capacity it is the originator of change. The function or value of *Q(s)* varies in response to change initiated by the independent variable *s*. So therefore, space itself is auto-dynamic and modeled as such by the independent variable *s* of the Quanto-Geometric function (the Mappa Mundi).

The derivative of the function as applied to the interpretation of the physical Universe provides a corroboration of that principle. What we have established so far is that the resulting values of the $M_3(s)$ function, which is essentially the derivative of *Q(s)*, for specific values of *s*, come to be in agreement with the physical constants that govern the development of the physical Universe. Since the derivative of a function is the study of how the function changes or responds to the changes in values experienced by the independent variable, the conclusion then is that indeed *s* itself is the initiator of change or physical motion. Philosophically the notion of change is at its origin equivalent to or interchangeable with the notion of motion because there *can be no change with no motion,* whether this motion is endogenous or exogenous.

It is easy to see how an exogenous change implies motion since it imperatively implies the passage from one position to another. Less obvious may be the implication of motion in endogenous change since it rather evokes the passage from one state to another. All states, albeit appearing as motionless artifacts, are in fact stationary motion, as all standing waves are. It is because we have not been able to perceive that *state* motion that we have historically assimilated static to absolutely motion-less. Motionless is relative in many ways. Both Fourier analysis and the De Broglie treatment of objects, which associates a wave function to all objects, straightforwardly lead us to that conclusion. Ultimately there is not a single object in absolute rest in the entire Universe, and that alone is a probing fact to our argument.

The problem with the visualization of *nothingness* being able to *move* is a paradigm that defies our capability for high level abstraction, both for the learned and the common man. That is why most prominent physicists such as Nobelist G. t'Hooft or Lee Smolin have acknowledged in their own words that to this day *"we do not understand motion in physics."* The motion originating in nothingness is not the motion of particle pairs beyond the span of space, as visualized in Quantum Field Theory. It is a capability for innate self-replication by coordinate space which spells out as auto-generation of motion. Our inability to assert that representation is what explains why we are all still using the baffling abstraction of *time* in our minds. The motion of space is endogenous and not scaled in time. The scale of the motion emanates from the matrix of space itself, so to speak. The inability to sustain that representation is one major handicap preventing Quantum Field Theory as a whole from reaching the goal of unification with General Relativity. By visualizing "fluctuations" in pure vacuum, Quantum Field Theory makes an incipient step forward, but attempting to describe the kinematic of a point element in that field with the time variable is making a horrible step backward. We can help ourselves with the assimilation of the paradigm of inner dynamics of space by porting everything we visualize about time to the nothingness of space. It is not going to resolve the issue but it sure is a start. One might otherwise refute our argument by making the following observation: well, isn't that what the Laplace transforms do, in shifting the visualization of phenomena in *time* to their visualization in the *s* domain, the frequency domain? That is a start as well, but just a start, because we are still helping ourselves with the time variable in expressing frequency at the end.

If space can *dilate* as stipulated in Relativity, it certainly can move, because something intrinsic to it has to move in order for it to *dilate*. But then the question is: what is the essence of space motion and what caused that intrinsic motion to stop at end of the dilation process? The truth is that the entire universal space is moving, that is why the speed of light is the same everywhere in U_3: the speed of the perturbation that we call light is indeed the speed of space moving isotropically in all directions. This is not to be confused with expansion or inflationary expansion of universal space in the manner contemplated in Standard Model cosmology. Beyond the question of motion or kinematics, there are equally questions to be raised in the order of structure of the space matrix: i.e. how is the matrix graded and what are the processes that take place inside of the matrix? We believe to have answered an important aspect of these questions with the preeminent symmetry-breaking *n* sequence and its subsequent segregation of space in intervals of entropy. Further elucidation we will offer as we go. As *unphysical* as this model might appear to common sense, we ought to remain persuaded that it is the one guiding this unraveling of the physical constants now well underway in this study.

6.20. Derivation of the Van Der Vaals a Constant

To our consternation, we shall state that, setting aside the very extensively developed Kinetic Theory of Gases, the study of inter-molecular forces at the root of the formation of molecular populations in materials is a confusing field of physics with a host of misrepresentation and disagreements. From our vantage point, we are able to discern that the problem arises from a blatant misunderstanding of the space Shelling that presides over the development of matter in our 3-dimensional Universe.

Akin to the particle coupling constant (fine structure constant), the gravitational constant and the electro-static constant, we would expect the field to have established a molecular coupling constant as a pure and simple physical constant setting the scale for the inter-molecular force. The closest thing to such a figure is an *a* constant long established by Danish physicist Diderik Van Der Vaals, which was named after him the *Van Der Vaals constant*. This constant however is not a defining building unit for the crystalline state or more generally the state of Materials as we termed it. Every *structural* material (flat populations of molecules) has been established with its own *a* constant. Furthermore, under the umbrella of "Van Der Vaals force" every academic text puts different constructs, some describing it as the total force between two molecules of the material including the repulsive component, others describing it as the total attractive force. Yet others go so far as to apply the defining Van Der Vaals equation to larger macroscopic objects of any kind. What is worse, the officially published list of values of the *a* constant per material, tallies the gas Helium, Krypton and the other noble gases, which are monoatomic elements, among other multi-molecular gases, demonstrating a complete confusion between state behavior of materials and inter-molecular dynamics in the composition of multi-molecular materials. Although it is fair to say that the original defining equation developed by Van Der Vaals was already a state equation which reads as follows:

$$(p+\frac{n^2a}{V^2})(V-nb)=nRT,$$

where *a* is the inter-molecular force constant. This equation needs an empirical correction, called the Maxwell equal area rule, established by J. C. Maxwell, in order to become thoroughly applicable with sufficient accuracy. The virtue of the Van Der Vaals equation is to provide a relative improvement over the ideal gas law for a generalization. Constant ***a*** provides a correction for the intermolecular forces. Constant ***b*** adjusts for the volume occupied by the gas particles as a correction for finite molecular size. The value of the *b* constant is the volume of one mole of the atoms or molecules in the substance. If a gas behaves ideally, both *a* and *b* are zero, and Van Der Waals equations approaches the ideal gas law:

$$PV = nRT$$

Constants *a* and *b* do not characterize molecules in a population of such entities but moles of molecules or the topology of molecules itself. That is why the unit for constant *a*, i.e., spells out a pressure-volume dimension per **mole square**, which is a specification for the unit intermolar force between the minimum of two moles. The larger the volume holding one mole of substance, the fewer the molecules in a given region and the smaller the attraction between the two molar populations. Should one inquest about the attraction in a small region of the topology, one finds that it inversely varies with the molar volume.

Because the Van Der Vaals treatment remains unable to identify an **invariant** intermolar constant but a running constant which is specific to each substance, it leaves us with no direct correspondent for our theoretically derived figure of $\boldsymbol{0.239962558 / \sigma^3}$ to compare to. The running constant also bears a number of other limitations conspiring against its predictive capacity, as for any gas topology the values of both ***a*** and ***b*** do not always accurately describe the behavior of the substance. It is particularly so in the extreme conditions of high pressure (near phase transitions) and low temperature (quantum behavior), as well as so-called effects yielded by the geometry of constituting molecules.

The list of *a* and *b* values reported in the textbook Tables of Van Der Vaals constants for substances are computed from *P*, *V* and *T* at chosen conditions or from what is called their critical forms. For all this lack of standardization, one should not expect a significant or broad match between the theoretically derived figure and the magnitude of any representative empirical figure, should there be any. In fact, there is none.

In the absence of an official universal intermolar constant, and for the benefit of an approximation, it stands to reason granting the intermolar constant for the smallest topology, Hydrogen molecular moles in gaseous state [H_2 mole – H_2 mole], the role of model unit of the Materials Shell with the initially reported value of $\boldsymbol{0.2444\ atm\ L^2/mol^2}$. This empirical value had then changed in 2012 to $\boldsymbol{0.2476\ atm\ L^2/mol^2}$ with some now reporting it at around $\boldsymbol{0.2453\ atm\ L^2/mol^2}$. Between the theoretical value and the reported empirical ones, we have an understandable rather low match. However, further rounding up our theoretical figure to ***0.24*** puts is exact match with a rounded-up average of the empirical figures of ***0.24***. Keep in mind that the background in the measurement of the empirical figure has become so wild that nowadays some even put under the umbrella of the Van der Vaals constant attractive coupling between large macroscopic objects in complete ignorance of scale physics!

In concluding this section, we wish to emphasize that the Kinetic Theory based on physical states has to remain separate from the study of inter-atomic attractions or forces, which give rise to the Molecular Shell per our visualization in Quanto-Geometric Therory. The state of materials and their associated kinematics represent a higher level of phenomena incarnated in the inter-molecular dynamics cementing the foundation of the Materials Shell in U_3. Furthermore, the Kinetic Theory, despite all its successes, remains in our view incomplete because there exist many more physical states than just the customary three considered (Solid, Liquid and Gases) in accordance with the Quanto-Geometric vertical hierarchy below:

1. → *Amorphous Solids*
2. → *Crystals*
3. → *Powders*
4. → *Waxes*
5. → *Colloids*
6. → *Oils*
7. → *Aqueous Liquids*
8. → *Condensed Vapors/Clouds*
9. → *Volatile Gases*

These definitions are important for such fields as nanotechnology, structural biology, supra-molecular chemistry, polymer science, surface science and perhaps condensed matter physics as a whole. We will further state that these same fields, particularly nanotechnology and structural biology, have much to gain with the experimental application of the physical constants pertaining to the next two Shells, figures thus far unbeknownst to the physical sciences in general. However, for these benefits to become tangible, physicists must exercise sufficient intellectual discipline in keeping separate the notion of interaction forces as the building block of a physical Shell from that of state of the population of building blocks making up Shell objects. More in Chapter 16 of Book II.

6.21. Concordance and Geometry of the Interaction Constants

We shall call the reader's attention to the fact that all 4 of the principal interaction constants belong to the same Quanto-Geometric Hierarchy, a rather remarkable fact attesting to the correctness of the Quanto-Geometric view. What physicists generally hold for the most preeminent constants are those that have to do with the so-called 4 forces: the electromagnetic force, the gravitational force, the strong force and the electroweak force (radioactive). The statute of the strong force, which pertains to both quark interaction within hadron particles and nucleon interaction within atomic nuclides, is ambivalent to say the least, and to that extent may not be classified as foundational as the other three forces at this point. The Quanto-Geometric view offered thru the generating Tri-Valent Hyper Quadratic and Hyper Space Operators offers a harmonized order of these constants because it relates them to their mathematical origin. Hence, the veritable four primary interaction constants are: the fine-structure constant, the gravitational constant, the electrostatic constant and the Van Der Vaals constant as they relate to natural successive Shells of matter in the prime role of forming their building blocks.

This concordance of interaction constants remits to a different view of matter in U_3 by helping us better understand the Shelling that takes place in the organization of 3-dimensional matter. We have been impervious to that shelling thus far because of our lack of understanding of the nature of space in general despite the advances by Albert Einstein's Relativity theories. We hope the above study has made it sufficiently clear that there is a significant distinction, not only between the nature of the scalar element (quantum or mass) in each Shell, given in the specific values of the $M_3(s)$ function itself (under the direction of the Operator), but equally in the multiplicity quality of s_3 space from Shell to Shell, which gives rise to the different forms of intra-shell interactions.

Furthermore, it is no less interesting to realize that the equations for all the interaction forces are almost of the same form and all contain a square term in their expression.

Namely:

1. For the Gravitational Force, $F = G\dfrac{M_1 \times M_2}{r^2}$, where G is the gravitational constant, M_1 and M_2 are the masses of the objects in interaction and r the distance between them.

2. For the Electrostatic Force, $F = k\dfrac{e_1 \times e_2}{r^2}$, where e is electrostatic constant, e_1 and e_2 the charges in interaction and r the distance between them.

3. For the Van Der Vaals Force, $F = a\dfrac{n \times n}{V^2}$, where a is the Van Der Vaals constant, n the total number of molecules and V the volume of the container shared between each molecule.

4. For the so-called Electromagnetic Force, the Force, viewed from a classical vantage point and expressed rather in Power of the so-named Synchrotron Radiation, is captured in this equation:

$$P = \frac{c}{4\pi\varepsilon_0} \cdot \frac{e \times e}{r^2},$$

which we have rearranged to make explicit the product and square terms.

The reason why the interaction forces take that very specific form is because they inherit the geometry of the parabolic component of the generating function, which we recall to be:

$$y = \frac{s^2}{4p}$$

If we choose to elect a scalar *n* to be equal to *2* for the sake of the comparison, the function becomes:

$$y = \frac{1}{p} \cdot \frac{s^2}{n \times n},$$

where *n* is a scalar of value *n = 2*

$$\text{or } y^{-1} = p \cdot \frac{n \times n}{s^2}$$

Whether we make the comparison directly with *y* or the inverse function y^{-1} so that the form of the equation becomes exactly identical to that of the Forces is a simple matter of *point of view* from one side or the other. The match stands overly clear in any chosen case.

Accordingly General Relativity makes a fair statement when it typifies the Gravitational force to be a force of *geometric* origin. Just that it is not only the Gravitational force that this qualifier applies to but all the other interaction coupling forces generated by the $M_3(s)$ function and that, in addition, the geometry of this force is preeminently parabolic.

6.22. Quantum Wave Function: Truth or Myth?

In spite of its central role in Quantum Mechanics, no notion has probably generated more controversy in physics than the notion of wave function. Its philosophical implication and consequences insofar as human representation of nature have divided physicists since early-day Quantum Mechanics thru modern-day Quantum Electrodynamics. The representation of the wave function itself, however intuitively factual and consequential it was granted to be, posed enormous interpretative problems. The principle of uncertainty, which springs from an analysis of the modulus squared of the wave function $(\psi)^2$, was the spark that ignited the controversy, with scientists such as Albert Einstein, who himself made contributions to the theory, remaining unswayed by it until their last breath. If you can tell the position of an electron in an atomic orbital, then you cannot concomitantly tell its momentum. When Werner Heisenberg proposed that $(\psi)^2$ ought to say something about the localization of the associated particle in space, he was solidifying the mathematical quantum framework that gave spectacular account of the orbital shelling of electrons in the mysterious atom element, as undertaken by the Copenhagen school of quantum physicists and those who subscribed to it. Yet this aspect of the theory which matured into today's Quantum Field Theory, remains hampered by serious limitations, not the least of which is the inability, on several accounts, to define the wave function of a single photon, as it already was when Einstein, Podolsky and Rosen challenged it with the so-called EPR paradox.

Further, Quantum Field Theory, the latest extension of Quantum Mechanics, has arrived at an impasse when it comes to formulate the granular interaction taking place in the gravitational field on the basis of the quantum wave function, in any attempt to formulate the so avidly sought-after quantum theory of gravitation. In essence, at all orders of the formulation, gravity essentially *gravitates*, and *re-gravitates*, displaying a nightmarish behavior which at the end makes it impossible to segregate gravitational interaction from that of the other forces. Albert Einstein staunchly opposed the notion of the wave function on the basis of being, in his view, a fuzzy abstract mathematical space, and contended that all description of reality should be based on unequivocal and causal description of space and time or "spacetime". He saw no space as it were for the ghost of uncertainty as implied in the principle of uncertainty inherent to conjugate variables of matter. To Einstein's credit, Louis Victor de Broglie, father of the material wave function, did develop in his later years a *real*-valued wave function formalism, in contrast to the common *complex*-valued one. That treatment, however, did not quite reverberate among quantum physicists, both because the complex-based treatment was no doubt more seductive and because it offered a broader analytical framework.

Despite relentless efforts, Albert Einstein could not come up until his last day with the formalism to support his view of the primacy and higher hierarchy of causal relationships over partial, uncertain and probabilistic effects in the description of physical reality.

6.23. Quantum Mechanics' Wave Function Through Quanto-Geometric Lens

The closest Quanto-Geometric equivalent to the modulus squared of Quantum Mechanics' wave function is the figure:

$$\left[\frac{dQ(s)}{ds} \bullet k_p\right]^2 \text{ or } [M_3(s)]^2$$

as per Table 5.3. In accordance with the premises that we have set forth in this study, this figure stands for a real quantity. The *s* variable in this figure unequivocally refers to *real* space. The most primal description of the Quanto-Geometric wave function is *s(s*)* as previously stated, *s or s** not being an abstract or complex imaginary but a real physical quantity or an observable despite its nothingness and its infinite "de-ployability". The far-down derivative figure of $[M_3(s)]^2$ expresses that *de-ploy* ability starting with the index value of s = ½σ or the particle Shell. Effectively a square integer represents the expression of the splitting or de-ploy of a smaller twin quantity: for instance, *25* is what *5 x 5* looks like as a square. The squaring operation is a transformation. So therefore, the square transform implies the split or conjugation of a quantity that was originally a singlet. Space as a physical variable is the mother of that manner of transform. What's more, the *normal* order inherent to the Quanto-Geometric abstract number stack dictates that the radix quantity (i.e. 5) is smaller in size or dimensioning than its square (25 in this case), but affirmatively larger in scalarity than the square. The Quanto-Geometric function inherits all these properties from the higher number abstraction. In any derivative form where it appears as a composite function, both elements of the composition become individually squared when the whole function is squared since $(ab)^2$ equals $a^2 \times b^2$, which further multiplies the conjugates. This is the prism through which one must understand the modulus squared of wave function in Quantum Mechanics, never to be equated to an abstract mathematical space, once one is able to identify it thru the agency of Quanto-Geometric lens. We further consolidate this interpretation of the material wavefunction in Book II by deriving the known values of many elementary particles' Compton wavelengths.

6.23.1 Derivation of the Rydberg Energy Constant

The comparable Quanto-Geometric wave function squared is such a real quantity that it takes the majestic incarnation of one of the most *fundamental* physical constants: the *Rydberg energy constant* R_y, with which it presents an otherwise acceptable match. The Rydberg energy constant R_y is the energy of the lowest-energy photon capable of breaking the orbital bond between the electron and the proton that make up the simplest of all atoms, the hydrogen atom. In other words, it is the unit of energy over which the whole atom edifice is generically built across the spectrum of atomic elements, a natural *orbitalization unit*. It is equal to the Rydberg constant R_∞ times *hc* per the following formula:

$$R_y = R_\infty \cdot hc = \frac{m_e e^4}{8\varepsilon_0^2 h^3 c} \cdot hc = 0.136056253 \cdot 10^2 eV$$

where *e* is the elementary charge, m_e is the rest mass of the electron, ε_0 is the permittivity of free space, *h* is the Planck constant, and *c* is the speed of light in a vacuum. This is the value published by the National Institute of Science and Technology of the United States in CODATA 2010.

The Quanto-Geometric corresponding value is: **0.1355703/ σ⁶**, portraying an admittedly low figurate match with the currently known value. However, if we instead compute the constant from the Fermi coupling constant as an alpha (α) surrogate, and apply the Quanto-Geometric rule of the square of that quantity to produce the orbital constant, the figurate match significantly improves.

From $\dfrac{G_F}{(\hbar c)^3} = 0.116637 \times 10^{-4}$,

we obtain $\left[\dfrac{G_F}{(\hbar c)^3}\right]^2 = 0.136041898 \times 10^{-10}$.

Per our model, the physical constant alpha (α) and the Rydberg energy constant work hand in hand, in that the second is the square of the first: if you don't get alpha desirably accurate, nor will your orbital constant be satisfactorily accurate. Furthermore, the above formula for the Rydberg energy constant implicitly shows that alpha (α) does belongs or commonly participate in the computation for the orbital energy constant as part of h^2. We fully uphold and stand by the full figure of **0.1355703/ σ⁶** reported in Table 5.3. Modern physics taunts the Rydberg constant R_∞, hence the Rydberg unit of energy R_y, to be the most accurately measured constant ever. The Rydberg constant historically arose as a fitting coefficient in the Rydberg formula for the spectral lines of the hydrogen atom, but Neils Bohr later showed that it could be derived from other more *fundamental* physical constants, thru the relationship reported above, toward supporting his proposed atomic model. From the Quanto-Geometric perspective the Rydberg energy unit R_y qualifies even more as a physical constant than the Rydberg constant R_∞ itself. The assessment of the significance of the square of the M_3 function (for s = ½σ) conducted previously implicates the following:

- Orbitals are a result of two elements operating in the form of conjugates of one another within the atomic edifice.
- Every element is appealed to be paired with a conjugate as a norm.
- Each element develops its own orbitalization or shelling (in fact a sub-shelling) in its own right.
- Each orbital may further become or de-ploy into a twin spatial "habitat"
- Every orbital is leveled as a discrete energy pocket scaled on the Rydberg energy unit.

This sets the stage not for a two-tier atom (nucleus-electrons) as traditionally established, but a three-tier atom with two sets of conjugates: electron-proton and proton neutron. The proton (or protons) is positioned at the mid-tier level and plays an *Ambivalent* role. This architecture not only gathers all the known facts about the atom in terms of electronic distribution and nuclide behavior but has additionally proved able to fully account for all the puzzling *magic numbers* known to the distribution of atomic nuclei of elements, as we have demonstrated elsewhere since 1995[5] and in Chapter 15 of this study.

6.24. Derivation of the Molar Gas Constant

An ideal gas is defined as a gas whose punctual elements:

- do not interact with each other via any force of any nature.
- are of negligible size with respect to the distance separating them.
- are free to move in any direction and collide only randomly with one another in an elastic fashion.

In the classical approach to this question, the above definition applies to ordinary gases based either on single atoms, specifically the noble gases, or molecules of any composition that are volatile under **s**tandard conditions of **p**ressure and **t**emperature (or **STP**).

In order to determine the volume that an ideal gas will occupy in space, three parameters must be taken into account. One is density of the gas or simply the total amount of punctual elements making up the gas. The second is pressure exerted on the sample population and the last one is the temperature of the whole sample. It can be shown that for the same pressure, the same temperature and the same amount of punctual elements, all gases of the ideal type occupy the same spatial volume. Therefore, these three factors, pressure p, temperature T and amount of punctual elements n, may be considered, when taken as a product, as a gas expansion coefficient k. A comparison involving these three factors between two gases will help us gather this fact in a mathematical expression:

$$\frac{p_1V_1}{n_1T_1} = \frac{p_2V_2}{n_2T_2} = k$$

where p_1 and p_2 are pressures of the two gases, V_1 and V_2 are the volumes of the two gases and n_1 and n_2 the amount of elements in each of the two gases. We may then pursue with a generalization for an ideal gas with the following expression:

[5] La Quanto-Géométrie, a volume representing the protogenesis of Quanto-Geometric Theory.

$$\frac{p_i V_i}{n_i T_i} = k$$

So that:

$$\frac{V_i}{n_i} = k \cdot \frac{T_i}{p_i} \quad (5\text{-}2)$$

The above equation gives us the inverse density figure of the ideal gas. As a reminder the density is the amount of elements per unit volume. Consequently, the inverse of that quantity is the Volume per unit element or the total volume per total amount of elements.

The proportionality constant k is an experimentally measurable quantity and is known as the molar gas constant with the value of 8.3143. If we choose a standard value of pressure (1 at or 101305 Pa) and a standard value of temperature (0°C or 273.15K), the above equation yields:

$$\frac{V_i}{n_A} = 8.1314 \cdot \frac{273.15}{101325}$$

$$\frac{V_i}{n_A} = 0.022413968 \Rightarrow m^3 / mol$$

The change of n_i to n_A is due to the value of the molar gas constant which is based on a specific amount or quantity, the Avogadro number. When we introduce the value of 8.1314 for k then n automatically changes to n_A. The quantity 0.022413968 m^3/mol, called the molar volume of an ideal gas, characterizes the space occupancy or localization behavior of an ideal gas. Put differently, every molar amount of an ideal gas will spatialize or exert entropy within an occupancy volume of 0.02241 m^3. As counter-intuitive as this may appear, this act is played by atoms within the molecular Shell, just as the previous act in this transverse kinematic order was played by particles within the atomic Shell. For that very reason do we see the noble gases, which are mono-atomic elements, equally play this act of spatialization configuration to the same extent as molecular elements. This stands for an epi-phenomenon superimposed on the systematics of inter-atomic bonding that make up molecular elements.

The Quanto-Geometrically derived value for V_i/n_A is **0.0224694201/ σ^6**. The two leading zeros in this case were not introduced by using the device of power of 10 adjustment, and must be considered part of the match, making it a 5-digit figurate match. This tight match leads to a remarkable natural validation of the scale of the Pascal and Kelvin units. It also tells us that STP conditions are not an accident on Earth but represent critical points in the development of matter itself.

The high match portrayed by the Quanto-Geometrically derived value of the molar gas constant additionally corroborates the validity of neglecting the possible quantum mechanical effects at STP. Effectively the conditions that would impose quantum restrictions on the modality of occupation of the spatial volume do not exist at STP but only at high pressure and high temperature. Nonetheless even in those extreme conditions, the basic gas law relationship *(5-2)* would not totally collapse, it would only change by an understandable proportionality constant, as in effect it does, since the relationship simply goes:

$$\text{from: } p_i V_i = nkT$$

$$\text{to: } pV = \frac{2}{5} nkT \; .$$

The latter is the relationship that describes the spatial occupancy of the Fermi Gas, a so-called gas of electrons. In those circumstances the departure of the gas behavior away from the classical gas constant anchor is completely understandable, since it constitutes no more than a regression to the atom and particle Shells catalyzed by the shorter scales and high energy levels common to these lower habitats.

6.25. The Cosmological Constant and the Anthropic Principle

A discussion of the Cosmological constant seems to be in order in this text given the endless talk over the last decade regarding that figure in the field. The terms of the discussion are however completely ill-framed from our point of view, since the predicates for this figure have been all but unknown to this point.

What the field currently refers to as the cosmological constant is what is called the "vacuum energy density", which it assumes to be the same wherever you look at the vacuum in the universe, whether it addresses the vacuum between galaxies or the vacuum between atomic elements for instance. Further there is a whole series of considerations and hypotheses that arise from the value of that constant in terms of the configuration of U_3, particularly the question of whether or not the universe is expanding, contracting or static. This constant seems to have been ill-fated since its inception by Albert Einstein in the first half of the 20th century, because he devised it as a figure to justify the stationary universe known at the time when shortly thereafter observational evidence pointed in the direction of an expanding or inflationary universe instead, the "greatest blunder" of his life in his own words.

Discussing a cosmological constant as the generic "energy of pure vacuum" without understanding the Shelling hierarchy based on the states of entropy inherent to each Shell, which give birth to the host of physical constants, is a complete act of blindness. The only kinematic characterization in U_3 that applies as a generalization to all vacua, throughout the integral suite of Shell deployments, is the propagation speed of coordinate space, which the field calls the speed of light. Beyond that figure, every Shell of space vacuum must be treated individually. There is no "cosmological constant" that universally applies to the larger-scale galactic groups, galaxies, solar systems and so on and so forth. There is only a specific constant for each of these Shells per chart in Table 4.3.

The proponents of the notion of cosmological constant realize that there must be a connection between the physics of an expanding universe thru cosmic vacuum and the physical constants that relate to the vacuum at lower scale. In their view, the vacuum at larger scale is the same as the vacuum at lower scale. Consequently, an expanding vacuum at higher scale must imply an expanding vacuum at lower scale as well. This poses the nagging question of the consequences of an expanding lower-scale vacuum over the physical constants, since the vacuum state governs those lower scale physical constants. Another implication is that the physical constants may not only vary over time but most significantly may not retain their exact same value everywhere in the universe.

There are a number of negative consequences to these views. That is why cosmologists and astrophysicists somewhat arbitrarily pose that the vacuum at lower scale starting with the galactic scale is disconnected from the expansion of the most global universe even though we are in total ignorance of the higher super scales there may be beyond galaxy groups. In fact, they seem to believe there is a limiting point to the superstructures which authorizes them to speak of a universe with a superficial contour. Consider that if the global "Universe" may be expanding thru its edges, nothing precludes an expansion at possibly at a much lower rate in the sub-galactic scales. To say that from quarks to galaxies, nothing is expanding while believing that the top end of the universal scale is expanding is playing foul to good logic because consistency with unobserved dimensional alterations in the sub-galactic scales oblige. Humans and planets are continually perceived with the same size correlationally. The energy scale in the subscale gauge groups is not linear, therefore a uniform expansion is naturally disallowed, leaving us to observe disproportionate dimensioning or non-linear expansion in the subscales driven by the global expansion if it existed. The fact that we do not observe that dimensional disproportion in the subscales strongly argue against the idea of an expansion of the global universe.

Furthermore, the visualization that the edge of the Universe does not expand into anything is completely nonsensical and shows that this hatch of ideas does not properly understand the intrinsic qualities of coordinate space. The problem here is that relativists are using differential geometry notions from Gauss to evaluate the ontology and behavior of the Cosmos. We discuss in Chapter 17 foundational problems with many of these notions, in particular the notion of embedding and embedded spaces. The application of these notions to the geometry and dynamics of the Cosmos has led astrophysicists to believe that "the universe is expanding and needs not anything to expand into," because you can supposedly study the properties of an embedded space without knowing anything about the embedding space. Refer to Chapter 17 for a critical analysis of this notion. We will only contend here, as argued in Chapter 4, that geometry appears as a reflection of an inner quality of a patch of coordinate space, which is its state of symmetry. It is thus natural that we can study the properties of an embedded space without knowing anything at all about a possible embedding space. An embedding space must necessarily have spread, if so then it must be receding, shrinking or diminishing in itself in order to create agency for the expanding embedded space. That does not solve the problem of an adequate description of the expansion because it creates a bigger one, which is now to find the laws that explain the behavior of the embedding space.

What one ought to understand regarding the superstructures beyond the galactic scale is that:

1. Coordinate space is still 3-dimensional in those realms.
2. There must be no end to the superstructures as nothing authorizes the dictate of a limit.
3. The porting of the geometric properties of 3D shapes of an embedded 3D space to the dimensional quality of a global or embedding Universal realm is a profound mistake in approaching the ontology of material realms in the Universe.

Therefore, the idea of a Multiverse in the deep heavens beyond the galactic and galactic-group scales with any speculated number of dimensions is totally misplaced and erroneous. So are the projections from the Multiverse idea such as the Anthropic principle, according to which constants of nature are not universal throughout the Cosmos and take different values in different regions of the arena. So that only certain regions of the universe are presumably fit for human life in suitable material habitats, ours being one of them as observed! The arguments that we have put forth in order to refute the possibility of alteration of the fundamental physical constants in the next Chapter of this study also hold against variation of these constants as a function of locality in U_3, and consequently they equally hold against the Anthropic Principle. After taking stock of that development the reader may conveniently revisit this section.

A cosmological constant at larger-scale structure means **not** an expansion of space such that each and every galaxy will ultimately find itself absolutely lonely in a rarefied and deserted Universe. The cosmological constant should be understood instead within the framework of the $M_3(s)$ function and interpreted within the transverse hierarchy of localization. The Hubble law (based on the Doppler effect) does not mean that because galaxies are receding from one another, the vacuum they populate is in inflationary expansion, just as two electrons in an orbital constantly receding from one another do not describe an inflationary orbital vacuum. The Hubble law must be interpreted along the line of localization instead of inflationary expansion, while the cosmological constant will have better success within the interpretative framework of punctual population statistics, otherwise known as Fermi-Dirac statistics, Bose-Einstein statistics, and even the classical Kinetic Theory, all of which have proven very successful and adequate theoretical frameworks.

The current computed value of the cosmological constant is given by a variation of the formula proposed long ago by A. Einstein:

$$\Omega = \frac{8\pi G}{3H^2c^2} \cdot \rho$$

where G is the Newton Gravitation Constant, H is the Hubble constant, c is the speed of light and ρ is the vacuum energy density. After careful and repeated examination of the numbers deriving from this formula in several different dimensions, we could not find any match with any of the constants marked *unknown* in the astrophysical Shells of our model chart. Consequently, we summarily dismiss these Ω numbers as fundamental as we further discuss them in greater detail below. We believe that all the historic cosmological data accumulated to date must be put back on the table and carefully re-examined in the light of population statistics, an exercise that might yield measurable quantities with appreciable matches with the Quanto-Geometrically derived astrophysical constants for their corroboration.

6.26. Quanto-Geometric Cosmogony

The cosmogony that directly emerges from the Quanto-Geometric framework, in particular the Grand Eigenfunction at the heart of the Theory, is a Universe that appears as an *Omniverse*, with 9 basis adiabatic layers or strata in the vertical order, which we have designated World Planes, and in the horizontal order a transverse span of Shells, in principle of boundless extension despite their “nanoscale” or 9-dimensional bundling. We acknowledge the hypothesis that in the vertical order the Omniverse could be bundled as well with a global boundless span.

The World Planes are meta-physical to one another while characterized by their specific degrees of freedom dictated by the primeval $\{s_n\}$ scaling sequence. This sequence defines the numeric value or Modulus of the dimensional set that makes up the background of each of these World Planes. Despite their tight mutual estrangement, their adiabatic partition is not absolute, since the mere virtue that a World Plane agent or observer may envision or formulate vertical trans-ontology is in itself a testimony to inter-penetrability of some sorts, at least at the consciousness level, which is a fact of fully physical nature. One other significant reason is that the σ standard deviation metric is common to all World Planes of the Omniverse.

We intuitively recognize that the Physical Universe that we inhabit as agents and observers is endowed with 3 degrees of freedom. The question of the derivation of dimensional set $D = 3$ has been posed by physicist Martin Rees as a non-trivial question deserving serious attention. In other words, the question as to why our universe is 3-dimensional is a rightful question to pose to human intuition or common sense. We have already explained how the background in this Universal Arena of ours is constructed, that $D = 3$ is actually $D_3 = [2+1]$, we now have to add that the reason why there is a 3-D World Plane is because $n = 3$ belongs in the range of the $\{s_n\}$ sequence which defines the several dimensional sets of the Omniverse. We, as intelligent beings able to pose questions such as these, just happen to occupy the 3-D World Plane. The idea that the other World Planes of different dimensional moduli do not host intelligent agents is unfounded from the Quanto-Geometric viewpoint, because the transverse span of Shells that see matter develop from the infinitesimal quantum to the karyotic forms of the living to the cosmological heavens is common to all World Planes. The adiabatic division that guarantees the independence of each World Plane and sustains the laws of symmetry on which each is built understandably prevents at the same time all forms of direct inter-element interaction.

6.27. Vacuum Energy Density and Standard Deviation of the Omniverse

Gathered Supernova data in the field of astrophysics suggest that the vacuum energy density or "zero-point mass density" of the 3-D universe is greater than zero. From computation focusing on electromagnetic force, one obtains the not so surprising result that empty space in the 3-D universal arena "weighs" 10^{93} *grams per cc*. This result would be quite significant if true. However, the actual average mass density of the 3-D universe is computed to be 10^{-28} *grams per cc*. Cosmic background radiation (CMB) data combined with the measured Hubble constant do confirm the supernova data that suggests a value of 10^{-28} *grams per cc* for the average mass density of the universe. Astrophysicists believe that mass density is equivalent to the vacuum energy density per current understanding in the field.

This discrepancy between expected value and measured value, in the order of 10^{-120}, has created a deep decade-long crisis in the field of cosmology. Astrophysicists find themselves faced to the conundrum of explaining the astonishing fact that the positive and negative contributions to the "cosmological constant" cancel to *120-digit* accuracy, and yet fail to cancel beginning at the *121st digit*. This paradox only comes to corroborate the problem with the visualized nature of the "cosmological constant" as argued previously.

We shall duly notice the coincidence in pure magnitude between the average mass density computed from cosmological data and the primal standard deviation parameter cubed for a quantum-space distribution applying to the 3-dimensional universe, σ^3, that is. If the adjusted nanometer constant indeed represents the physical materialization of σ for the Omniverse, which would equally apply to the 3-D World Plane, then:

$$\sigma^3 = (1.00001203 \cdot 10^{-9})^3$$

$$\sigma^3 = 1.00003609 \cdot 10^{-27}$$

The magnitude of σ^3 is in very close agreement with the 10^{-29} quantity for the vacuum energy density in the 3-D World Plane, as subsequently reported, showing a difference in only 2 orders of magnitude. Furthermore, the intrinsic dimension of the number (m_E^3, wavelength of harmonic fluctuation in cubic meter) is in close agreement as well with the λ_C^3intrinsic unit-dimension (Compton wavelength in cubic meter) of the vacuum energy density figure.

What's more, if the transversal hierarchy of Shells in the 3-D World Plane surrogates in any way the 9-stratrum vertical hierarchy of the Omniverse, as suggested earlier in this study, it should not be any surprise that there three 3 different possible phases for the large-scale geometry of the 3-D World Plane as currently hypothesized in astrophysics. This triplet latency is in complete accordance with the Quanto-Geometric Tri-Valent characterization. Effectively, if we use Ω_Λ to denote the critical limit density and Ω_M to denote the ratio of ordinary matter density to critical density, then:

1. The 3-D World Plane is open if $\Omega_M + \Omega_\Lambda$ is less than one (space-dominant).
2. The 3-D World Plane is closed if $\Omega_M + \Omega_\Lambda$ is greater than one (scalar/mass-dominant).
3. The 3-D World Plane is flat if $\Omega_M + \Omega_\Lambda$ is exactly one (scalar-mass and space covariantly equivalent).

The ratio of ρ_{vacuum} to $\rho_{critical}$ is in offered calculations what is denoted Ω_Λ, expressing the vacuum energy density on the same scale used by the density parameter Ω, leading to the relation:

$$\Omega_\Lambda = \frac{\rho_{vacuum}}{\rho_{critical}}$$

The supernova results have suggested that the vacuum energy density is close to the theorized limit:

$$\rho_{vacuum} = 0.75 \cdot \rho_{critical} = 6 \cdot 10^{-29} gm/cc$$

Thus, the data suggests that $\Omega_\Lambda = 0.75$, and that therefore the 3-D universe is in constant expansion. These views are encapsulated in NASA's depiction of the evolution of the 3-D arena reported in Fig.6.2, which interestingly reflects a quanto-geometric distribution. However, from our standpoint, the sketch should have been drawn as a probabilistic distribution instead of anchored on the usual timeline. That graphical depiction alone should tell astrophysicists that beyond the representation of a putative evolution in time for the 3-D World Plane lies indeed a larger normal probabilistic distribution, which is essentially timeless.

Note further that the current view is that if Ω_Λ is greater than zero, then the Universe will expand forever unless the matter density Ω_M is much larger than current observations have suggested. For the Ω_Λ greater than zero scenario, even a closed Universe would expand forever.

It is important to put in highlight that the above reported numeric details reflect the cumulative inquisitive thinking in cosmology today about the dynamics of the *observable* universe and only enjoy limited support from the sketched Quanto-Geometric perspective. The more fundamental view remains the true cosmological constant figures deriving from the framework exposed in the fundamental physical constant chart and their dynamical implications in terms of the behavior of the entire 3-D World Plane which is not altogether observable.

6.28. Scalar Fields in the Standard Model

The treatment of sup-particle dynamics we defer to the next cast of this study, as previously advised. However, due to the nomenclature being used lately in reference to the latest findings about bosonic scalar fields, such as related to the Higgs boson, we feel compelled to bring a few conceptual adjustments to the discussion. In addition, the notion of "pure scalar fields" in terms of a mechanism to generate mass for elementary particles seems to be in contradiction with the Quanto-Geometric axiom that "pure scalar" cannot exist, per our contention at the beginning of this study, on the basis that a minimum level of vacuum component is a requirement for the physical inception of a scalar. This all begs a clarification.

Let us note, at the outset, that the treatment of "pure physical scalar fields" is entirely based on the mathematical model for scalar fields. A mathematical scalar field is a region U defined by a complex function, distribution or matrix, admitting points with complex coordinates (real and imaginary) where the real part is an invariant independent from the field region and its coordinates, and the imaginary part is fully dependent on the spatial nature of the region or field. The field and the populating scalar form a complex conjugate while the topological point is equally a complex quantity in itself. If the field is a physical tensor or gauge field, it may thus provide scalarity to the point(s) of population by means of interaction with these points thru their imaginary quantity. One can see that this analysis is based on the type **5 Primitive** of the Higher Numeric Abstraction, the realm of complex numbers. Therefore the "pure physical scalar field" that the Standard Model refers to is merely an idealization of a pure scalar, which does not even qualify as a *monolith*. It is at best a form of scalar, a lower-level entity in the scalar hierarchy.

What's more, the mere fact that a scalar packet can be broken by a highly energetic projectile (as it happens in high energy colliders) means that:

1. The packet was a dimensional entity, with a vacuum component that is, which provided the means for the segregation (a pure monolith is conceivably unbreakable).
2. The dichotomized free particle, i.e. a Higgs boson, is still a dimensional element because it can be subsequently observed or accounted for, for it would otherwise vanish as a dimension-less element.

From that standpoint, the idea of mass as a "pure scalar" is untenable and is only sensible as an idealized pure scalar. In general, scalar field theory does not precisely explain how mass is "created", shall we say, but how mass is induced or compounded thru the Lagrangian transform of symmetry breaking that relates the imaginary component of the elemental packet to the field. So it is, since an invariant (tenet with mass dimension) must pre-exist in the quality of the elemental packet for the action or the Lagrangian transformation to take place as a paramount condition for the operation. Philosophically speaking, there is nothing in abstraction or in existence that can be innately motionless other than a very scalar.

It is worth noting that we have also made serious advances in the quest for allocation of mass to elementary particles with the direct derivation of the value of interaction coupling constant α, the Fermi coupling constant and the Rydberg constant. It is important to recognize once again that in neutron decay, where the Rydberg constant intervenes, the first particle emanating from degeneration is the heavy W Boson. The energy of the

Rydberg constant is also the energy or mass of that boson, the value of which was Quanto-Geometrically derived.

In the aggregate, we remain highly skeptical of "pure scalar fields" or "dark matter" in the cosmological heavens in connection with the cosmological constant, as hypothesized in the field, having seen that the phenomenon of mass allocation is essentially orchestrated within the S_0 protoShell. From a cultural point of view, there is more to gain in the manipulation of entropy in the higher Shells than the manipulation of elemental scalars in the S_0 protoShell. Simply because the deconstruction of the foundations leads to infernal chaos whereas the ubiquitous propagation of the activity under the Sun and beyond is glory and majesty.

6.29. Graphical View of the U_3 Creation Function

Toward performing the graphing of the Creation function, we already know of nine critical points of this graph, which are the points that produce the transverse hierarchy of interaction coupling constants.

A.- To further facilitate the graphing (Fig. 6.12), we will compute the *Q(s)* coordinate of the function at s = ¼ σ, as completed below.

$$M_x(s_x) = 0.66114378278 \bullet \frac{-s}{\sigma^4\sqrt{2\pi}} \cdot e^{\frac{-s^2}{2\sigma^2}}$$

$$= 0.66114378278 \bullet \frac{-(\sigma/4) \cdot e^{\frac{-(\sigma)^2}{16 \cdot 2\sigma^2}}}{\sigma^4\sqrt{2\pi}}$$

$$= 0.66114378278 \bullet \frac{-1 \cdot e^{-1/32}}{4\sigma^3\sqrt{2\pi}}$$

$$= 0.66114378278 \bullet \frac{-0.096667029}{\sigma^3}$$

$$= \frac{0.063910805}{\sigma^3}$$

The reader must keep in mind that we have overridden the parabolic truncation of the Function in the shelling representation and harmonized the function for graphical representation. To do so, we have legitimately given the σ parameter the metric value of 1.0.

We are now going to inquire about the minima and maxima of this function with the zeros of its derivative. The function is:

$$M = \frac{dQ(s)}{ds} \bullet y_p = \frac{-s}{\sigma^3\sqrt{2\pi}} \cdot e^{\frac{-s^2}{2\sigma^2}} \bullet \frac{\sqrt{7}}{4\sigma} s^2$$

$$= e^{\frac{-s^2}{2\sigma^2}} \bullet \frac{\sqrt{7}}{4\sigma^4\sqrt{2\pi}} s^3$$

Let's segregate the coefficient into:

$$k = \frac{\sqrt{7}}{4\sigma^4\sqrt{2\pi}},$$

so that $M = ks^3 e^{\frac{-s^2}{2\sigma^2}}$

The derivative of M is:

$$\frac{dM(s)}{ds} = k\left[3s^2 \cdot e^{\frac{-s^2}{2\sigma^2}} + \frac{-2s}{2\sigma^2} \cdot e^{\frac{-s^2}{2\sigma^2}} \cdot s^3\right]$$

$$\frac{dM(s)}{ds} = k\left[3s^2 \cdot e^{\frac{-s^2}{2\sigma^2}} - \frac{s^4}{\sigma^2} \cdot e^{\frac{-s^2}{2\sigma^2}}\right]$$

There are two possible conditions to be met for $\frac{dM(s)}{ds} = 0$.

The first is $k = 0$, which is not a possible or meaningful condition.

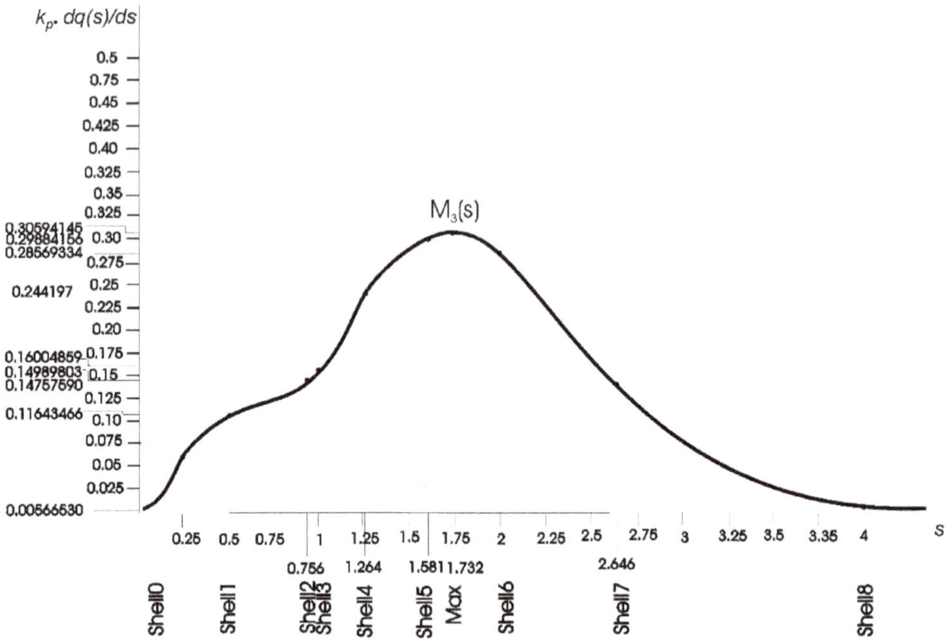

Fig. 6.12 Graph of the $M_3(s)$ creation function

The second is for $\left[3s^2 \cdot e^{\frac{-s^2}{2\sigma^2}} - \frac{s^4}{\sigma^2} \cdot e^{\frac{-s^2}{2\sigma^2}}\right] = 0$

$$\left[3s^2 \cdot e^{\frac{-s^2}{2\sigma^2}} - \frac{s^4}{\sigma^2} \cdot e^{\frac{-s^2}{2\sigma^2}}\right] = 0$$

$$e^{\frac{-s^2}{2\sigma^2}}(3s^2 - \frac{s^4}{\sigma^2}) = 0$$

This implies: $e^{\frac{-s^2}{2\sigma^2}} = 0 \rightarrow s = \infty,$

or $3s^2 - \frac{s^4}{\sigma^2} = 0$.

Which in turn implies: $s^2(3 - \frac{s^2}{\sigma^2}) = 0 \rightarrow$ **s = 0,** or $\frac{s^2}{\sigma^2} = 3$

$$s = \pm\sigma\sqrt{3}$$

Therefore, we have 3 extrema: **s = ∞, s = 0** and $s = \pm\sigma\sqrt{3}$.

B.- Let's compute the *q* coordinate for that last abscissa from *M(s)*:

$$e^{\frac{-s^2}{2\sigma^2}} \bullet \frac{\sqrt{7}}{4\sigma^4\sqrt{2\pi}} s^3$$

By replacing s with $\sigma\sqrt{3}$,

$$M = \frac{0.3059421073}{\sigma}$$

So therefore, that last point is a maximum, and *M* experiences minima at $s = 0$ and $s = \infty$. However, we must remember that the Function is undefined at $s = 0$.

C.- We will also study in this section the indefinite integral of the *M(s)* function. To do so we will employ the so-called method of integration by parts as shown below. Let:

$$T(s) = \int M(s) \cdot ds = \int -ks^3 e^{\frac{-s^2}{2\sigma^2}} ds$$

Let $u = -ks^2$ and $du = -2ks \cdot ds$

Let $dv = s \cdot e^{\frac{-s^2}{2\sigma^2}} \cdot ds$, from which $v = \int dv = -\sigma^2 \cdot e^{\frac{-s^2}{2\sigma^2}} + C$

$$T(s) = uv - \int v \cdot du = (-ks^2)(-\sigma^2 e^{\frac{-s^2}{2\sigma^2}}) - \int (-\sigma^2) \cdot e^{\frac{-s^2}{2\sigma^2}} (-2ks) \cdot ds$$

$$= k\sigma^2 s^2 e^{\frac{-s^2}{2\sigma^2}} + 2k\sigma^4 e^{\frac{-s^2}{2\sigma^2}}$$

$$= k\sigma^2 e^{\frac{-s^2}{2\sigma^2}} (s^2 + 2\sigma^2)$$

Replacing k by its value:

$$T(s) = \frac{\sqrt{7}}{4\sigma^4 \sqrt{2\pi}} \sigma^2 e^{\frac{-s^2}{2\sigma^2}} (s^2 + 2\sigma^2) + C$$

$$T(s) = \frac{\sqrt{7}}{4\sigma^2 \sqrt{2\pi}} e^{\frac{-s^2}{2\sigma^2}} (s^2 + 2\sigma^2) + C$$

This is an important quantification integral to which we will devote more attention in the next cast of this study.

6.30. Role of Mathematics in Physics and State of Physics Today

The purist and trained physicists may not have noticed in the above developments much of the algebrae, the Lagrangians and other complex forms of mathematical structures that they might expect to mandatorily comport the agency of such foundational physical dynamics. This is a demonstration that much of it is useless for the inquest of fundamental constants in foundational physics. I am certainly not about to apologize for Nature preferring to be simple, clear and straightforward, as incidentally noted by A. Einstein. Further, we have expressly chosen to remain clear of complex formalism as much as possible, preferring to adopt simple mathematical structures and language wherever possible in order to keep this text amenable to the most universal audience.

	Quanto-Geometric Theory	Other Frameworks
Philosophy	Quantum-space systematics	Dialectic Idealism
Mathematics	Theory of Hierarchization and Typification of the Number Stack	Traditional Number theory based on cardinality and ordinalization of numbers
Physics	Coupling effect of space over mass	Quantum scalar fields in Quantum Field Theory
	Coupling effect of mass over space	Conversion of mass into energy in Special Relativity
	Ban of the time variable	All abstracts based on time except in Quantum Mechanics
	Horizontal normalization resulting in 9 ontological primitives or typified certainties	Vertical normalization resulting in a probabilistic spread or certainty-incertainty conjugate
	Real-image spinor space for elemental objects	Abstract spinor space in Quantum Mechanics and Standard Model
	Derivation of physical constant figures	None - properties of matter in abstract only
	Representation of Grand Universe staged on 9 Real World Planes	String Theory's representation of 11 universes with only D4 real

Table 6.5 Conceptual correspondence between Quanto-Geometric Theory and other Theoretical Frameworks

In addition, we do wish to bear witness that good physics, otherwise described as a rigorous and coherent description of nature, may be done with no recourse to and seeking no legitimacy in overly complex mathematics, a long-running problem in modern theoretical physics acknowledged to have been acutely sterile in meaningful results since the 1970's. The abundance, elegance and complexity of mathematical developments, all of which goes much to the enjoyment of the sanctimonious cast, can never substitute the formulation of clear and proper abstraction correctly positioned to ultimately expel the final solution to the problem at hand. Mathematical developments with no direct and tangible results in the resolution of a physical problem remain but mathematics.

The preeminence of abstraction that we spoke of in concluding Chapter 4 should not be construed in terms of a free license to ill-fated physics. Analytic models should not seek comfort in intense and sophisticated mathematics in the lack of resolution of physical conundrums they are meant to elucidate. The higher abstraction is and must be the realm that gives rise to the elements of solution to the problem, it can never lay in segregation with the problem in question, and further ought to provide a clear path to its resolution, lest the abstraction becomes useless and almost meaningless aether. What we have argued is that the number stack represents the highest level of abstraction that models physical reality as a whole, but it only makes sense to us strictly in the context of physical reality and the unraveling of its mysteries, never as an end to itself. In that sense mathematics are abstract solutions awaiting to be applied, the exercise being frankly totally dispensable and in the long run valueless if it can never meet physical applications. Of even more contestable value is the physics that does not produce physical solutions but only new provinces within mathematics, which is not at all its mission.

Finally, a brief press review of the current state of the discipline will be quick to reveal a pervasive discomfort at all levels in regards to the long-running squint level of productivity of the field. Lack of new and plural ideas is a matter of concern, overt in some circles and reportedly camouflaged in others, while most of the few ideas enjoying prevalence nowadays stand in many ways for recycled notions that the great generation of physicists of mainly first half 20th century credited with no or little merits. Most importantly, the new and bold ideas remain no less than unconfirmed hypotheses for the most part. This is not an attempt to discredit any one physicist of the day or any work. Indeed, the *trouble with physics* today, as some have put it, goes in our estimation much beyond physicists, despite the above critical remarks to their address, and stems from the prevailing values in the culture, which negatively reflects on institutions and ultimately on physicists and their practices as well. The culture of foremost mercantilism and blind gain, in almost complete detriment of expression of the humanities, is a societal problem that, at its recently attained peak, wreaked havoc of transatlantic and southern economies on the planet, and with them many human lives, notwithstanding the many disingenuous interpretations to the contrary by these actors. In this landscape, the career of a physicist is ultimately faced with increasing institutional pressures that end up suffocating the vibrancy and productivity of the field. This problem is not going to be solved in our physics conference rooms and fora. It is a severe societal problem that requires for its resolution a push for a larger degree of freedom of the humanities vis-à-vis mercantilism, both of which dually and naturally configure the Quanto-Geometric engine that orchestrates societal dynamics.

Lastly, we offer in Table 6.5 a brief parallel between high-level Quanto-Geometric constructs and corresponding constructs in other physics theories and philosophy, which many may find useful to a final reconciliation of ideas.

6.31. Conclusion

In concluding this Chapter, we shall point out that all of the unknown constants listed in Table 6.3, although undiscovered thus far, are completely existent physical quantities in their own rights, to the same extent as the other well-known constants we have just studied. While ideas about the cosmological constant continue to be actively debated and remain in flux, a *life* constant is nowhere under the prospective radar of academia, the assumption in most quarters being that life with its bewildering complexities is probably not submissive to mathematical determinants. The Quanto-Geometric prediction of a physical life or *karyotic* constant is deemed to profoundly alter, if not revolutionize, our idea of the *living* by placing it in the context of the transverse hierarchy of Quanto-Geometric *de-ployments* in the 3-dimensional universe. This is the subject of the last chapters of this Treatise.

Yet the reader may wonder at this end point where is the debate and argumentation in favor of the meta-physical realm promised at the introduction? Well, there is no debate, we have let the exposition speak for itself: the physicality of the meta-physical realm is an inherent property of the entire Quanto-Geometric functional framework; its unraveling of the pillars on which the physical U_3 universe is erected stands for an undismissable validation of the constructs it purports in relation to all predicted U_i meta-physical universes of the Omniverse. The higher transverse Universal hierarchies on the vertical line we have called World Planes are as real as the 3-dimensional transverse hierarchy in the 3-D World Plane. Just as we have seen that there exists a complete program of organizational instantiation of the fundamental constants upon which U_3's Shells are built, there exists a similar program for the foundational organization of every other World Plane of the Omniverse. The repeated attempts that we have seen in theoretical physics in describing multi-dimensional space (or spacetime as ordinarily denoted) may be taken as efforts in that direction but remain unfounded and frankly quite mislaid due to their total lack of connection with the fundamental constants of Nature. It is my personal view that Religion of any revealed strain, well understood, is Cosmic Science. End of discussion.

Finally, it is well worth noting the following remarks by E. Schrödinger for an introduction to his 1950 publication *Structure of Spacetime*:

In Einstein's theory of gravitation, matter and its dynamical interactions are based on the notion of an intrinsic geometric structure of the spacetime continuum. The ideal aspiration, the ultimate goal, of the theory is not more and not less than this: A four-dimensional continuum endowed with a certain intrinsic geometric structure, a structure that is subject to certain inherent purely geometric laws, is to be an adequate model or picture of the "real world around in space and time" with all that it contains and including its total behavior, the display of all events going on in it.

It seems to me that the Quanto-Geometric Theory, in the entire developments that we have hereby undertaken, furnishes for its own a vast response to this goal of old by successive generations of physicists of an encompassing generalized description of Nature from the inner qualities of ubiquitous and universal space. Except that it is cast far and away from *time*, of course.

Chapter 7

Time is what Human measure
on clocks!
(SIC)

THIS IMPOSTOR OF TIME

If time is what human measure on clocks, we then have an absolute, since it has been raised to the status of a physical dimension, that is dependent not only on human but on their clocks as well. This is the most absurd definition ever proposed, it makes a complete disservice to the reputation of human intelligence in this corner of the woods we call the Universe. The notion of time in our minds is a very resilient and tenacious abstraction that has accomplished the feat of resisting the assaults of Quantum Mechanics, which has completely demoted and unseated it insofar as the visualization of one of the physical realms. No one seems to wonder why a whole span of the universe would be so disparate and disjunctive that at one scale it would be entirely bereft of a "universal" variable called time, incidentally its very foundation inasmuch as atoms and sub-atomic particles constitute for the most part its bedrock structures, and at a higher encompassing scale it is completely anchored on this variable. A visualization which further implicates that this variable has started to have a life since the very foundation when the particles themselves were thrown into existence. This stands for the all-brave contention of the Standard Model's Big Bang Theory about the formation of the three-dimensional universe! If time has started to elapse since the initial Big Bang event, why did it suddenly disappear at the atomic and molecular phase of development and suddenly re-appear going forward into the development of the following structures? Is our Universe anchored on a shaky and broken timeline? This picture is very seriously flawed, to say the least. We shall actively continue henceforth our crusade against this false and baffling notion of time. We will also revisit in this Chapter other crucial concepts such as evolution of matter and life as well as the expectations the field has assiduously harbored for a more consistent visualization of matter at all scales, this in the light of the success of the Quanto-Geometric constructs in unraveling the mathematics behind the physical constants.

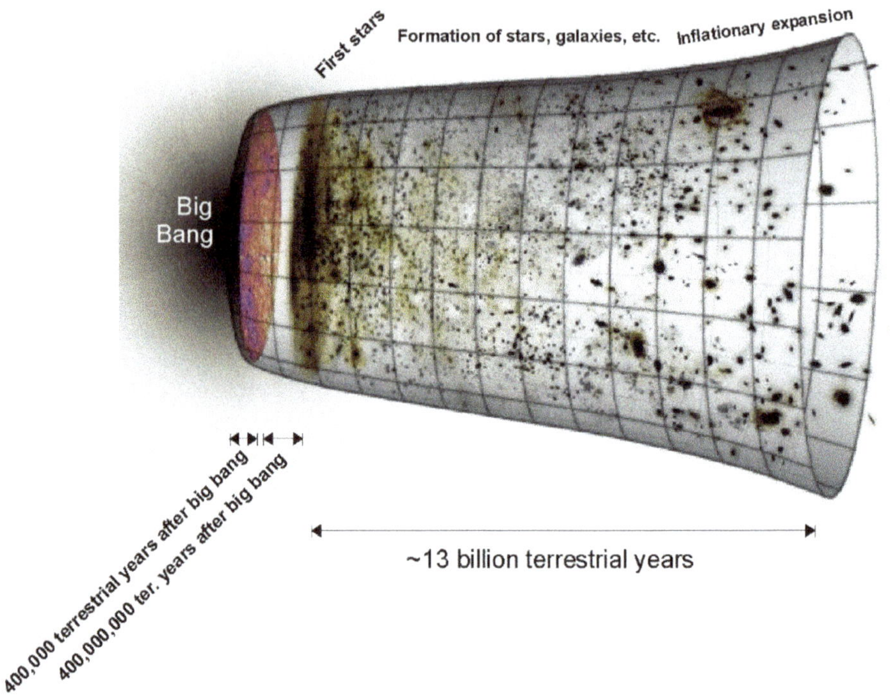

Fig. 7.1 Standard Model's Big Bang Universe

7.1. Apologies But No Big Bang

The observation that the Universe is in expansion in all directions is in direct contradiction with the idea that it has started from a singularity or one single point explosively (Fig. 7.1). If the universe has started from one single point, at a time t_0, then all subsequent expansion must have developed along a timeline starting at t_0 from one point s_0 in space. That point s_0 must have existed even though it could no longer be observable. We would have a sense of where it lies because the "spacetime" arrow would develop from it. In such a scenario all matter would be expanding along virtual cones centered about virtual s_0 as the center of a virtual sphere of expansion. This would give us spacetime with a number of timelines but all equated to one another and all collapsible into one. As a consequence, the scale of time would always be the same to all observers within the universe.

The Theory of Relativity has already broken apart the visualization of one universal time scale to all observers. But since the construct of time in our minds will not die, among some of the ideas that followed suit since then is the notion that, at early times of the universe, travel light has created space, another absurd construct. The late John A. Wheeler, who had been instrumental in the development of these ideas, hypothesizes:

Maybe the universe itself sprang into existence out of nothingness, a gigantic vacuum fluctuation which we know today as the big bang.

Probably a best-effort yet oblivious visualization by an otherwise brilliant and creative mind to frame the bigger picture. Unfortunately, the bigger picture admits no frame and no framing as a condition for successful *in-sight*.

All of these concepts are in truth *monocentric* constructions corresponding to a quantum-dominant form of culture that has been prevalent in the West. We want to find at all costs a *beginning* to the universe, a *singularity* from which it emerged because this is the monocentric *way* we think (Fig. 6.2). The astrophysicist who wants to find a single-point origin to the universe is thinking no different than the religious spirit who wants to find a single God to be the Creator of the universe. Furthermore, the monocentrism goes so far as to stipulate that men must be at the image of this God. The scientist wants the universe to have surged from a single point as much as the religious wants it to have surged from a point-like God as the single primal cause of all things in existence. If God had intelligence, a know-how whereby he could put in existence a world of differentiated things, he must have been himself a very differentiated, complex and organized structure (because in our experience intelligence is complex), something like a self-replicating structure from which the act of creation proceeded. If so, then we are back to the beginning, since the question becomes again: where does He come from, what singularity has given birth to Him? The proponents of the Big Bang singularity are quick to close this endless loop by cleverly stating that t_0 is the farthest instance of time that will ever make sense. Yet I must insistently ask Mr Big Bang, what of You had exploded at time t_0, where does it come from and where is s_0 today?

If we were to look at the Universe from a space point of view instead, we would understand that the hypothetical scalar that exploded at time t_0 could not be an absolutely pure scalar but a composite scalar-space entity. It could only have exploded by virtue of sudden inner space expansion. From the fullness of **type 9 Primitive**, we can appreciate that space or nothingness cannot be a progeny of anything *else*. It is not laid in points, but in infinites. Thus, there is no single point from which it comes. It does not and cannot come from some-*where*, because it is the *where* of itself and all *where* is of its own. Consequently, the only sensible conclusion is that it has *always* been *there* and has always been in expansion if ever. Despite its possible constant expansion, there is nothing beyond it, because every "where" we can think of it is "there", in the infinites. The movements and transformations that we may witness in the observable universe

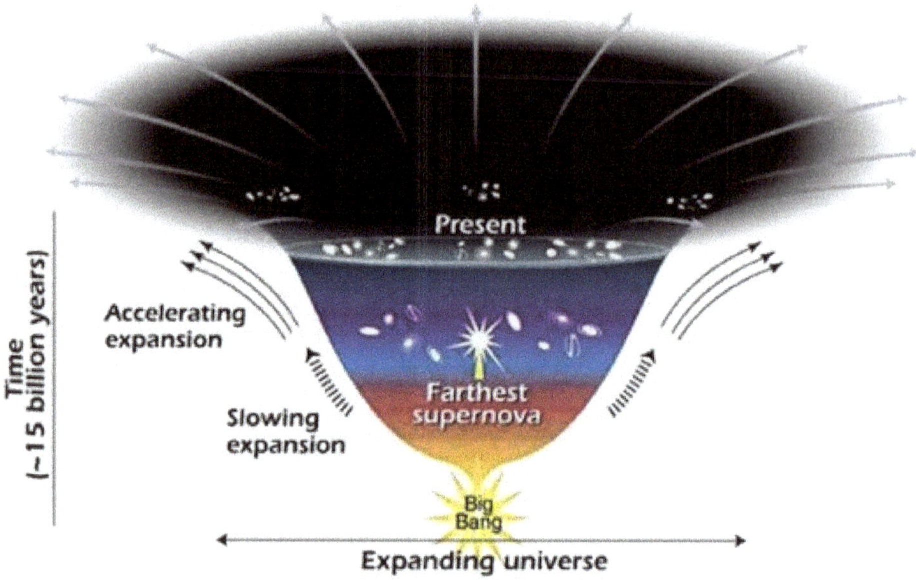

Fig. 7.2 NASA's model of an inflationary universe

are **locally instigated by partitions of space** that materialize the triplet cycle of life-death-rebirth. There will be no end to the larger-scale structures of the 3-dimensional universe because thanks to space the universe has no edge. Emphatic affirmation: there is no edge, no boundary and no exogenous shape to the Universe. We and everything else in it are irremediably thrown into life *forever* and will never absolutely die, as we evolve along a line of eternal transformation. This vision requires a re-dimensioning of our understanding of matter and life, which falls under the purview of the Quanto-Geometric transverse hierarchy in U_3. There is nothing of a referential nature in U_3 that makes one object of the hierarchy more valuable than another in its fullness. A human being is no more valuable than a star or a planet, and vice versa. However, every structure positioned along the transverse hierarchy line is not only greater but phenomenologically more advanced than the previous one in the line or anyone prior. This may be construed as a result of the greater size, which means more expansive dynamics resulting in more phenomenological processes. As intelligent as a human or a *Karyote* may be, he or she is not phenomenologically more advanced than a star or a planet but strictly less.

The current state of the knowledge in the discipline of Astrophysics is at the notion of 2-level-up larger scale cosmic structures above galactic structure in the 3-dimensional universe. Galaxies are currently visualized to be laid in groups, in the form of elements over the surface of vacuum bubbles comparable to foam bubbles. I must caution that these are hypotheses and that, with all due respect to our fellow astrophysicists.

Astrophysics has historically proven to be a tentative science with hypotheses and truths frequently proven to be wrong after the knowledge has settled.

Fig. 7.3 Computer representation of larger-scale galactic groups

Consequently, we are not going to engage into structural analysis of the hypothetical larger-scale galactic structures as conjectured of late. The computer model resulting from these conjectures is shown in Fig. 7.3. Given the regularity, the harmony and frankly the great structural beauty the 3-dimensional universe has accustomed us to, across all Shells up to single galaxies in terms of directly observable elements, I shall confess my refusal to believe that the larger-scale structures could be so messy, so chaotic and so ugly!

We do not at all reject the possible existence of these structures because they respond to a consequence of the Quanto-Geometric model. The reader will remember that we have conferred to the Grand Mappa Mundi Function the value of formal functional representation of the Quanto-Geometric number system or the higher abstraction system. We shall remember as well that in that abstraction every transverse number set is open to infinity. The numeric transverse hierarchy is captured in the functional representation by a circular arc of π length. The symmetry groups delineated over π are infinitely granular in accordance with the higher abstraction so that an infinite line of Shells may be accommodated by the transverse order thru that granularity. As to the s intervals, the whole *normal* interval bundle from *0* to 4σ may be re-distributed as unity over the semi-circle with a particular value assigned to each bundle, thereby making the whole functional representation still hold together. In so doing the model would account for any extension beyond the galactic Shell as their reality is being firmly established. However, we believe there is better and more fruitful adjudication of mind power and material resources in researching the implications and **applications** of the cosmological constants we have hereby unveiled instead of engaging in this breathless race after infinity. We should identify the task at hand as the amplification of the living medium for mankind both into the Solar and Galactic Shells, for the cause of advancement of spirits and means as to ourselves, and as stewards of Creation for higher purpose. Some might prefer to hear me say: the anti-gravity craft should be henceforth our true focus in Physics and Astrophysics, and so I say unreservedly and unapologetically. Perhaps the detection of 4-dimensional and 5-dimensional reality as well in an effort to harness vertical hierarchy.

In summary, by any account rendered from a space-dominant vantage point, an epitomization of the state of the 3-dimensional universe, let alone the Grander Mappa Mundi Universe or the Omniverse as we dubbed it, should irrevocably manifest that it is: *birthless, ageless, timeless, shapeless, edgeless, boundless, infinite and eternal!*

7.2. Apologies But No Billion-Year-Stage Evolutionary Mutations

Despite the continued success in the field of Biology in relation to the decodification of the karyotic genome, the evolutionary model traditionally upheld in that discipline is, in our view, in dire need of correction. So is the visualization of the essentials that make up the living.

First of all, the million- or billion-year timeline along which the model has anchored evolution of species as the propitiator and catalyst of mutational changes is flatly off the mark. In Quanto-Geometric analysis, within the Karyote Shell there is no time variable to effect events or any constitutional element, nor is there any in the Astral Shell, the immediate medium for Karyotes. The time element must summarily be stricken out of the equation.

Secondly, the fact that a karyotic species may tangentially evolve into another does not mean that at the higher levels of the karyotic line intelligence may not re-create lower-level Karyotes in minute quantities or massively.

Third, the express rejection by the field of possible mechanisms by which karyotic genome may sense environmental conditions and orchestrate adequate response is highly questionable. Effectively it is very hard to object to certain phylogenetical artifacts such as mimicry whereby the genome strikingly replicates the exact appearance of the environment for distinct fitting purposes, moreover when the contention is that **billions of contentious years** of fortuitous mutational trials are necessary for such an outcome.

Fourth, the stipulation that mutations are fortuitous events, which amounts to a stipulation of the resulting traits or characters as *untyped*, is unsustainable within the Quanto-Geometric framework. There is nothing in existence, whether it is an abstract or an object, that is not typed along the line of the 9 Primitives and/or subject to intervention of the Tri-Valent Operators. Everything belongs to an order of Quanto-Geometric types. Those types are the deepest level of codification within the karyotic genome, I might add. I shall also point out in passing that here essentially lies the reason why the genome project, despite its large success, surprisingly enough found a lot less gene sequences than expected in the human genome.

Fifth, the field has been developing in the lack of the most significant defining element for an understanding of what the living really is, which is the notion of a *karyotic physical constant*. Human Karyotes are not the best and brightest of creation and do not enjoy something unique to only themselves called life and intelligence as opposed to inanity for the planets, stars, moons, etc. The physical structure of karyotes is anchored on one physical constant to the same extent as all the elements germane to the other Shells (Fig. 7.4). The space wave function operating in karyotic cells has a much lesser level of multiplicity and auto-dynamism than those attributed to elements in the higher Shells. Thrown into ecliptic space, a human as is will shortly die. Thrown into ecliptic space, a planet like mother Earth will live and thrive. Because a planet follows an inalterable orbit around a Sun Star does not mean that it is bereft of higher-level phenotypical processes, part of which may and should be construed as intelligence.

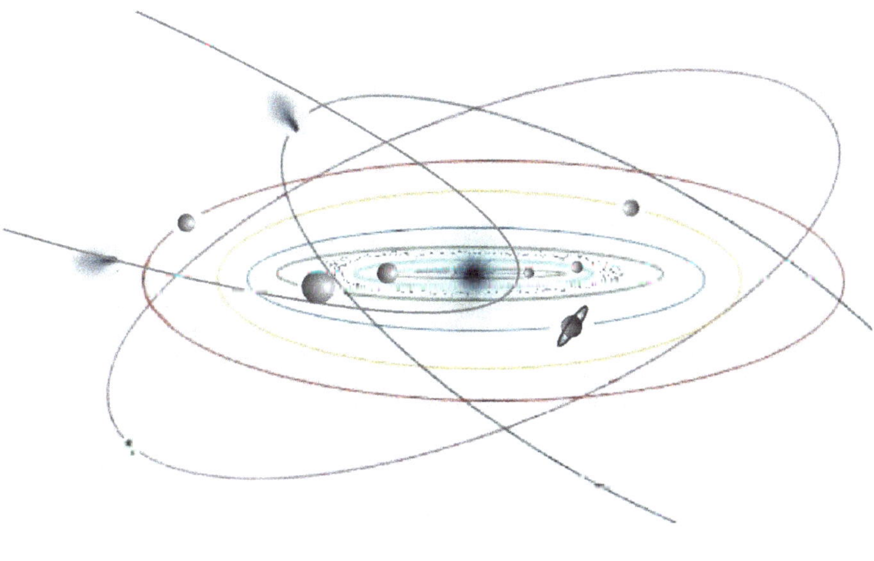

Fig. 7.4 Our solar system with orbital paths of comets and planets

Involuntary motion is inherent to the life of higher Shell beings just as it is to the life of all human Karyotes. The women among us duly know what their uterus cause to their spirits and body on its own in prelude to delivery and during delivery. Furthermore, does the whole delivery event itself turn a woman into a "cyclone", a "hurricane" or a "tornado"? Oh, they never stop being intelligent all the while! Just that they don't stop hollering and drifting like gusty winds and crying like rain showers.

Fig. 7.5 Common Model of a multi-electronic atom

Few if any man have ever imagined their entire life as nothing more than orbiting about their homes or "homing", just as planets or comets do around the Sun Star. We wake up and go away from our homes, in order to create entropy in the world in the form of our daily work routine, doing a lot of intelligent things, and then come back home, and then go away again, perhaps somewhere else, shopping for instance, and come back home, and perhaps sleep, and go away again, and on and on again... If you should now take a look at the orbitals of an electron around an atomic nucleus as illustrated in our physics and chemistry textbooks (Fig. 7.5), or the orbiting of comets around the sun for instance, in light of the above, you now know what you have been doing about your *home* all your life and what your life was mostly about. The great things that you might think about yourself and the great things that you have done for yourself, have all been accomplished over this orbital substrate. The storyline of this orbital substrate constitutes a very large part of the story of our lives. And what conditions those orbitals for all of us is the value of our $[dQ(s)/ds] \times k_p$ function for our s value of $\sigma(8/5)^{1/2}$ as incarnated within our bodies, cells and organs. Now since the *home* is likely to shelter more than one occupant, i.e. parents, children, cousins, nieces and perhaps a friend, the orbital structure becomes more akin to that of a multi-electronic atom (Fig. 7.6) or that of our familiar Solar System. Let me close these remarks by stating once again that there is no flat object anywhere, every Shell object in its fullness is a being, and the beings we call planets and the Sun Star, of which we know so very little, are far more advanced and complex than we are and will ever be as Karyotes.

Last but not at all the least, human life orbitals are thoroughly carried out thru involuntary motion. Our daily circling and wandering in space require the automatism of our legs walking in step with one another, the automatic balancing of our body thru coordination of leg and arm motion when we run, the automatism of tongue and jaws when we speak, and so on and so forth. Even when we employ an artificial craft for our daily space travel a large measure of automatism in steering the craft is a requirement for success in the undertaking. We never have to think about the minute details of coordinating our limbs in the performing of any of those *movements*, while our minds generally wander about other concerns. As soon as we start investing voluntary thoughts into those automatisms, they instantly break down. All higher-level processes are overlaid on these orbitals as epi-phenomena and require the innate automatism of the orbital substrate for their development. So it is, for all Karyotes as it is for all beings in the higher Shells.

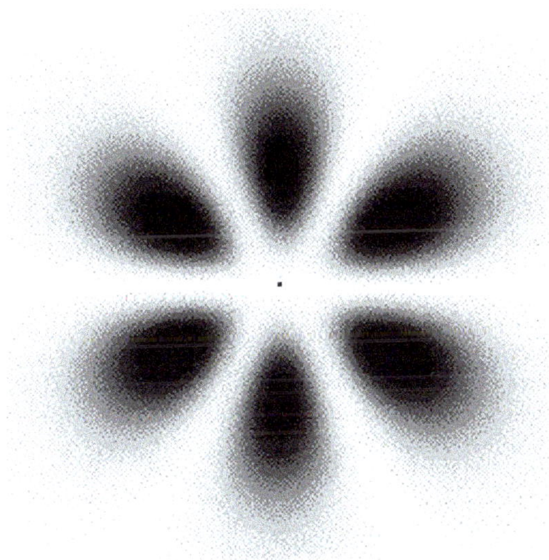

Fig. 7.6 A *4f* atomic orbital with nucleus at center in a planar view

For the readers who may remain unconvinced to this point and view this construction as an extrapolation, I shall invite them to another discussion staged as follows. Consider one of those viruses whose whole body is made up of 3 juxtaposed **crystal** blocks with a **karyotic** nucleus in one of them (Fig. 7.7), an element (the virus) which is able to remain dormant for thousands of our years and become active again under favorable conditions. The question asked is three-fold:

- What of It was life when the state of inane matter was in effect?
- What of He will subsequently make him inane matter when the state of life is in effect, in the understanding that inanity can recur thereafter?
- Can He die? If so, what does that mean?

Bonus question: If He may die, should He be buried?

Fig. 7.7 One of the many human-laden viruses

Finally, the above considerations turn moot the otherwise pungent question between Creation and Evolution that has long created sharp divisions everywhere in culture, whether in academia or the larger society. If indeed the precursor of the genomic molecule was amino-acids formed in Earth's atmosphere at the favor of light rays of thunder (today conjectured of comet origin), we shall conclude that a combination of cosmic intervention between *mother* Earth and the *king* Star as a process created karyotic life. In the context of those larger beings taken appropriately for who and what they are, the terms creation and evolutionary emergence become essentially exactly identical or interchangeable.

7.3. Response to Today's Challenging Action Items in Physics

There is a rather long list of challenging action items in the field pressingly awaiting unequivocal answers. We are going to review some of them in the light of the contributions the Quanto-Geometric Theory has offered thus far for their resolution. It is natural that the call for resolution of these conundrums formulates them from the standpoint of the current perspective the field holds on those questions. However, no one should be surprised that these answers may require, as they indeed historically have, a thoroughly different and wholly new perspective for the unraveling of those puzzles. The continued history of the discipline of physics has proven no less.

7.4. Dimensionless Physical Constants

The maturity attained in the field has made evident to the Standard Bodies the need for moving away from dimensional physical constants, well understood as it is that dimensional units are human constructs and not attributes of matter. The problem, however, with the few dimensionless physical constants that are known is that they cannot be independently calculated or formulated thus far but only estimated from other experimentally measured physical quantities by applying relationship formulae. This has remained a major unresolved problem of physics until Quanto-Geometry has knocked at the door. Effectively we have presented here a theoretical system able enough to predict not one or two physical constants but a whole host of constants in a dimensionless manner. When one rationalizes these known constants by expressing them in a single system of units and making sure they are consistently computed from formulae and relationships in compatible units, one finds complete agreement between the known values of these physical constants and the Quanto-Geometrically derived ones. The non-physicist reader might have indulged in some objection to the fact that the known quantities per their dimensional units are not quite represented in the Quanto-Geometric figures. Matter-of-factly this is exactly what they cannot and, most importantly, **are not expected to be by the field**, and it is a rather remarkable fact that despite the absence of unit dimensionality in the Quanto-Geometric figures, they each bring a figurate match to their par on the table that is frankly no less than astonishing. Not having had anything to do at all with the dimensional units, and given that these units are man-made constructions, one might have rightfully expected an independent theoretical system to cast completely different numbers or quantities. By not doing so and showing up as matching figurates, the Quanto-Geometric figures curtail any argument about identity and correspondence, or lack thereof, with the known dimensional quantities. Otherwise, we would need to engage in a discussion about correspondence and compatibility between the physical constructs and laws behind the known constants and the constructs and abstractions behind the independent system. Therefore, with these results, the perspective and axiomatic behind the new theoretical system ought to bear preponderance over the former, while the most intractable visualization Quanto-Geometry, as the new system, uncompromisingly upholds is the absolute death of time, henceforth passing the weight of all evidence to space. Further, there is no doubt that Quanto-Geometric Theory represents the mathematical physics axiomatic system envisioned by topmost 20th Century mathematician David Hilbert in his famous 6th Problem as discussed in Chapter 1.

7.5. Do Physical Constants Remain Constant Over Time?

Although this interrogation proves to be frankly of no practical significance, despite its hypothesized connection with the alleged cosmological constant for an expanding universe, it must be demoted as a concern because it expresses no more than a moot problem. Emphatic answer: No, physical constants do not change in time because there is no universal timeline for them to change along.

If we move past defunct time however, worthy of attention is the fact that the Quanto-Geometric physical constants are rooted on σ as a parameter: their values exclusively depend on the standard deviation of the Grand Universe which includes U_3. Because the σ parameter may theoretically take a number of different values, the correct answer to the question would seem to be: yes, the values of physical constants may change.

However, the Quanto-Geometric function is not the proper framework for the treatment of this question. This question shall be analyzed within the framework of the Quanto-Geometric numeric abstraction rather because this abstraction is what primarily governs physical reality in accordance with the postulate that physical reality stands for applications of the higher numeric system. Consequently, the question of the invariance or not of physical constants remits to the consideration of whether or not the *normal* stack of number systems may inherently change. From that standpoint, the categorical answer to the question posed is that **physical constants do not vary**. The reasons for the negative are three-fold:

1. Despite their open-ended trending to infinity, the cardinality of each number set in the stack is higher and remain higher than that of the previous.
2. Their individual cardinality must remain proportional to one another, lest the whole system is disjunctive.
3. We do not see any meta-level of abstraction that would pose, cause or justify an intrinsic asymmetric change in the modulus of sets' cardinality across the stack. Any change, if ever, would have to be symmetrically applied across the stack, leaving the standard deviation of the entire stack unchanged.

Consequently, physical reality as a projection of the *normal* numeric system must have a fixed standard deviation figure. Hence the **invariance** of physical constants.

The fact that the actual known constants, attained thru a completely foreign system, present figurate numbers with a high match with the ones derived from functional abstraction seem to be a testament to the above conclusion. We put the visible mismatch beyond the first 4 or 5 significant digits on account of inaccuracies arising from improprieties of our unit system per its theoretical construction as well as the measurements themselves.

Finally, it is worth underscoring that experimental results corroborate our theoretically-driven conclusion of the constancy of physical constants. In effect, in December 2013, physicists studying methanol molecules in a distant galaxy reported that the variation of proton-to-electron mass ratio μ in those molecules measured a $\Delta\mu/\mu$ change ratio equal to 0.0×10^{-7} at red shift $z = 0.89$, consistent with a "null result" in their own terms.

7.6. Fundamentalism in Physical Constants

The reader may have noticed that, in constructing the Quanto-Geometric Theory, we have avoided as much as possible employing the descriptive epithet of "fundamental", an overused term in most literary cultures under the Sun. By that trend, everything and anything of some significance or importance in the global culture emphatically becomes "fundamental" whether in writing or in language. We have shown that in the highest forms of abstraction there is a whole order of things *fundamental*, the term quickly becoming too poor to establish the necessary distinction between one construct and another.

In the spirit of the primacy of dimensionless physical constants, physicists, thru regulatory bodies of the field, have elected to raise some physical constants to the status of *"fundamental physical constants"* and have chosen the dimensionless constants to play that role. The term "fundamental physical constants" however has also been used for such constants as the speed of light, vacuum permittivity, the gravitational constant and Planck constant with regards to their universality and large scope of application, this despite their dimensional nature. That seems to stand as a regrettable but apparently inevitable confusion. The cause of this situation seemingly stems from the field not perceiving admittedly the higher order from which these quantities arise. Quanto-Geometry has now opened the door to that order by showing how a Tri-Valent Operator has formulated the hierarchy of those constants. If any physical constant shall be deemed fundamental, it should be the speed of light constant to deserve that role, in fact the Quanto-Geometric functional figure that directly relates to it rather, which is the value of the Grand Mappa Mundi function for the 3-dimensional universe. The black-body-related constants should play the role of secondary physical constants whereas the interaction coupling constants defining the U_3 Shells, the known ones among them coined fudge factors, should play the role of tertiary physical constants. The next level down of constants should be viewed or ranked as epi-phenomena, as important as they might be in the configuration of matter. As to the mass constants attributable to sub-component particles, they should be regarded as infrastructural constants.

Some physicists have made laudable efforts in framing the questions relating to fundamental constants in terms of what in their judgement ought to be found regarding these constants going forward. Martin Rees's Six Numbers proposal is probably one of the most prominent of these sets of deliberations. I will let readers exercise their own appreciation on these deliberations and re-dimension them within the Quanto-Geometric context by themselves.

In the aggregate, we are in a position to conclude that the long-sought goal in theoretical physics of finding an abstraction that organically yields all the fundamental dimensionless physical constants and successfully weighing them against measured values are hereby significantly met. The remaining constants that address the mass of elementary particles we will address in the next Chapters of this study. By and large, we now may have a wholly stratified *chart* of physical constants because we know the relationship they keep or must keep with one another as much as the hierarchical order of their distribution.

The elusive goal of this Theory of Everything able to perform that far-reaching calculation has been significantly met with Quanto-Geometric Theory at this stage.

7.7. Unification of Physical Units of Measurement

The Quanto-Geometric framework not only adequately orchestrates the unification of physical forces but, interestingly enough, it also unifies the scaling nomenclature for units of measurement, which are otherwise subject to man-made circumstances and representations, hence to a string of inevitable inaccuracies. In effect, the majority of measurement units in use in the physical sciences bear in one way or another the name of their inventors in honor to the scientists who discovered the physical quantities or phenomena they are associated with. While this practice remains perhaps of no harmful consequence, it sure does nothing to alleviate the necessity to keep and secure samples or etalons of matter in special conditions able to guard against the slightest variations. Since those conditions are not easily achieved, the Standard Bodies have decided to move away from the practice of keeping etalons by increasingly electing to peg the value of dimensional physical constants to that of the dimensionless ones, as previously mentioned. The risk in so doing is that the imprecision in one unit of measure may quickly become compounded, adding even more inaccuracies to the unit system as a whole.

By contrast, the Quanto-Geometrically derived constants are all parameterized on σ, the standard deviation figure of the Omniverse, which appears as their natural unit of expression. The many different mathematical dimensions of σ produce the specific units that measure the associated generic physical element, force or physical phenomenon. Table 5.3 reports the following dimensions of σ: σ, σ^{-1}, σ^{2}, σ^{-2}, σ^{3}, σ^{-3} and σ^{6}. The sweeping advantage of this system is that it provides a single minimal measurement nomenclature applicable at all scales of matter from the Sub-Particle Shell to the Galactic Shell, and even across the vertical hierarchy of the Omniverse, the meta-physical World planes. This system would dramatically reduce the necessary number of etalons to one, and only one, simplifying the institutional task of standardizing and safeguarding physical units of measure and their etalons.

To anyone harboring any reservation about the reported physical constant quantities in this study toward the above proposal, I shall further offer the following concession. In the absence of a protocol or standardized estimation method that sets the acceptable level of precision of any quantity or constant theoretically derived, those reservations are perfectly understandable and probably justified. The problem however is that there exists no referential framework against which to contrast those derived quantities. The historically known units of measure do not hold any referential value, unless we expressly choose to establish them as such, with all their known inaccuracies, for the sake of relying on at least one independent system. We recognize this all amounts to a very fine needle to thread but it is an inevitable task for the new Physics.

Furthermore, in assessing the authenticity and accuracy of the Quanto-Geometrically derived figures presented in our chart, one possible method for the estimator is to start out with the best match, opting in doing so to set out with the highest level of certainty. One then analyzes the relationship between this quantity and the next one in its class as well as the top one it is directly derived from. One makes a critical judgement about the level of accuracy of those dependencies and the soundness of these relationships in theory. The estimator walks throughout the chart in that manner until *c*, the primordial constant for U_3, is reached. This exercise, though not sufficiently systematic, may nevertheless provide a satisfactory basis for assessing the validity of these results in the interim.

7.8. The Standard Model and the New Physics

Despite the harsh criticism extended here in regards to several of its concepts, the Standard Model is a Theory that has probably correctly served its original purpose. Essentially of taxonomic nature, it has always been implicitly intended, however, to be a provisional theory, with the mission to adequately shelter and assimilate new experimental findings occurring in particle physics specifically. The intention behind our criticism is solely but uncompromisingly to obliterate the notion of time anywhere it might appear in theory. It is indeed a tall order for the Standard Model to continually stimulate and harmoniously integrate new discoveries, however challenging they might turn out to be, to its core principles. Not to mention the positive support role played by the Standard Model in the effort to standardize and institutionalize modern physics, in particular when it comes to experimentation in high energy physics. In that sense, this Theory has served us well since its inception.

Furthermore, the multiple generations of physicists who have strived under that banner, many of profound intellectual brilliance, but all of whom with an unwavering determination in the quest for unraveling the mysteries of matter, deserve no less than earnest praise for their efforts. So do String, Loop Quantum Theoreticians, and others of other color, who have been engaged in the exceedingly arduous task of building an overarching wrapper of quantum-gravity unification over the Standard Model in order to overcome its lacks and shortcomings. In the community of physicists at large, there has been a long string of physicists who have demonstrated the commendable ability to join personal valor and selflessness to intellectual brilliance at crucial junctures in the history of mankind. The world has the community of physicists to tribute for the Industrial Revolution, the cataclysmic deterrent stamp to the official Nazi-Japanese-Fascist global project of folly following its defeat (regrettable but arguably necessary), human space travel to the Moon, the global Information Technology infrastructure, to name a few of the most salient achievements.

Without a doubt, physics is and will continue to be a paramount force in human culture. Yet, with all the deserved credit granted to the Standard Model, and in light of enormous challenges currently facing the field as we shall see in the following Chapters, it is now the juncture for a new vision to take the stage.

Chapter 8

To see what is in front of one's nose
needs a constant struggle.
George Orwell

QUANTUM PHENOMENA IN THE MACROCOSM

As we mentioned before, the debate of all debates in modern physics has long been the question concerning the completeness of Quantum Mechanics, a controversy owing to the mathematical treatment of statistical nature at its foundation. Because probability or statistical theory is overwhelmingly a numeric science with no contemplation or inference with regards to causation and determinants of action, physicists have always felt that the Quantum theory is at a miss when it comes to intimately intelligible interpretation of dynamic systems. Even the proponents from the Quantum School of Copenhagen felt that a philosophical justification was necessary in order to at least provide a view as to the analytic implications of the Quantum Theory, notwithstanding the pressure brought to bear onto them by compelling critics. Amidst this hurly-burly, little to no attention was ever paid to the dimension of discrete Quantum expression in the macrocosm thru the application of the same science of Probability Theory. In the following, we re-examine Probability Theory to show that it encloses a discrete ontological dimension that is universally applicable to all objects permeable to probabilistic treatment, thereby directly relating macroscopic dynamics to quantum dynamics on the basis of a unique Unification theory.

8.1. Applicability of the Grand Eigenfunction at Large Scale

The question of unification of Quantum Mechanics to General Relativity is most easily resolved by considering how quantum phenomenology is rendered in the macrocosm under the auspices of the Quanto-Geometric Grand Eigenfunction and its derivates. The Quanto-Geometric canon, as a symmetry stack bundled in the Grand Eigenfunction, describes the ontological foundations of the continuum of coordinate space and the discretisation of the quantum-scalar, along with their correlations. These ontologies are implicated in all objects and phenomena permeable to statistical analysis thru the Gaussian distribution, which is a derivate of the Grand Eigenfunction, per incipient analysis in Chapter 2. In that sense everything in our immediate or remote living environment that is permeable to the standard Gaussian distribution constitutes evidence of quantum phenomenology right in the midst of the macrocosm.

8.2. Timelessness in Statistical Analysis

The most notable yet least appreciated result in statistical analysis is the absence of the time variable in standard distributions, for all manner of phenomena, all of which are initially and ordinarily inconceivable to us if removed from the trace of the arrow of flowing time. Effectively, the distribution of heights in a population, to take just one example, is an artifact that is totally independent from the passage of time as we understand it. There is no formal statistical time constraint imposed on sampling in order to test for the standard distribution of this parameter in a population. Populations of humans throughout the ages have developed consistent with this configuration pattern for the inception of individual heights. The correlation of events and all manner of genetic and behavioral associations between individuals that will occur within the sample group, and which one may want to subject to time flow at least descriptively, has no bearing whatsoever on the distribution.

The timelessness artifact implicated in statistical treatment has been underappreciated throughout the many developments in statistical physics, especially in its application to the physics of the macrocosm. One may even further argue that one of the reasons why probability theory fits so well to the resolution of E. Schrodinger's equation independent of time culminating in the quantum Hamiltonian is because of its complete estrangement with the time variable.

8.3. Intricacies of the Gaussian Probability Function

The reader who has previously been exposed to the Quanto-Geometric Eigenfunction may have at first rebuked it charging that it is simply a repackage of the long-known Gaussian Distribution function. However, several characters of that function provide indices for unsuspected features of the Gaussian that are worth analyzing in the context of this re-examination.

From conversion of the Grand Eigenfunction to its implicit form as undertaken in Chapter 2 Section 6, we obtained at one step:

$$\frac{-s^2q^2}{2q}\cdot\frac{2}{q\sigma^2\left[2\ln q+2\ln\sigma+\ln(2\pi)\right]}=1$$

One may quickly notice that if s takes a zero value, then the absurd result of 0 = 1 ensues. The variable q cannot take a zero value, since *ln(0)* is undefined or forbidden leaving the entire left term undefined in that event. Therefore, implicit in the function is the condition that neither s nor q can be equal to zero. These constraints are of paramount significance when it comes to interpretative treatment allowed by the Function. In the explicit expression was already set the condition that the Function may never take a zero value $(q = 0)$ because we could already note that as s becomes infinitely large the function approaches zero value without ever taking that value, making the s-axis equivalent to an asymptote. The implicit expression thus reinforces the constraint on q at the same time that it dictates an equiparable constraint on s $(s \neq 0)$.

The constraint of no q-intercept is not known in Statistical Theory. However, it bears significant phenomenological consequences in analysis inspired by the Function. Should we take q for representing probability density and if s models measurement spread for instance, the constraint of no q-value for s becoming extremely small remits to two significant conclusions:

- No matter how many iterations in measurement in the pursuit of precision, there will always be a probability gradient of error, as small as it might be. In other words, absolute certainty (100%) is not possible.
- There will always be a minimal probability gradient of precision, no matter how removed or inaccurate the measurement. In other words, zero certainty, which is equivalent to absolute uncertainty, is not possible.

Very particularly, because the Function admits no q-intercept by reason of discontinuity at s = 0, the notion of total area under the curve and that of said area equaling unity completely vanish. The notion of the area under the curve equaling unity, however, stands for the most fundamental axiom of traditional Statistical Theory. Since the function is simply not smooth, thus not integrable, over the full domain of $[-\infty, +\infty]$, the total area under the curve of the Quanto-Geometric function cannot be formulated.

8.4. Discretisation in Probability Theory as a Poster Child of Quanto-Geometric Theory

In quanto-geometric analysis, q or *Q(s)* represents the universal scalar which ontologically gives rises to mass, and s represents universal space or the void spread. The above characterization poses the first restraint on both the continuum of space in any envisaged dimension and the discretisation of the scalar wherever manifest.

Therefore, if measurements of any kind will always incorporate a gradient of uncertainty or error, if data of all kinds gathered in macroscopic phenomenology will always miss a certain mark, it is not just a fact of nature without cause. They are discretisation restraints imposed by the foremost form of distribution at the root of all ontologies and phenomenologies.

8.5. The Problem of Meaning and Causation in Statistical Analysis

While Statistical Theory have been widely used historically throughout the sciences, it is fair to say that it has always left scientists with the unsatisfactory after-taste of incomplete grasp of the actual physical dynamics effectively correlating the variables at play. Statistical accounts do not explain the "why" it happened and only gives a countable account of events, devoid of underlying dynamical meaning.

A large number of groups in the physics of matter at the macroscopic level and the life sciences follow the distribution pattern depicted by the Probability Density Function. Its use is indeed very wide spread in science in general. Below we report just a few of the well-known areas where prospected data is known to conform to the Gaussian distribution:

- Individuals' heights in populations of the living
- Blood pressure in human populations
- Measurement errors in science
- Test grades in education
- IQ scores in psychometrics
- Workplace salaries
- Societal Polling Data

While statisticians do not ask more of the capabilities of this Function and the empirical analytic techniques developed around that Function, in the physical sciences and in particular in certain quarters of theoretical physics, statistical analysis of dynamic systems had historically been considered incomplete, leaving analysts in search for deeper meaning. The nature of statistical physics had been at the center of the famous debate over Quantum Mechanics ever since the publication of the EPR paper. Effectively causation and determinism are completely absent from this form of knowledge. Most feel that the sole rendition of a probabilistic percentile of events does not constitute thorough explanation.

8.6. Standard Deviation Metric

The ignorance of the inner meanings or ultra-structure of the variables of distribution has led to empirical techniques put in place in statistical theory, thereby obscuring the higher order of certainties inherent to macroscopic physical dynamics.

The standard deviation value represents the most critical parameter set forth in statistical theory for measurement metrics. The empirical practice has it that the abscissa of the putative coordinate system for the graph is measured in integer multiples of σ, setting the basis for the so-called empirical rule of 99-95-68. There is nothing in nature or inherent to the theory, however, that confers any particular weight to these percentiles. According to this rule:

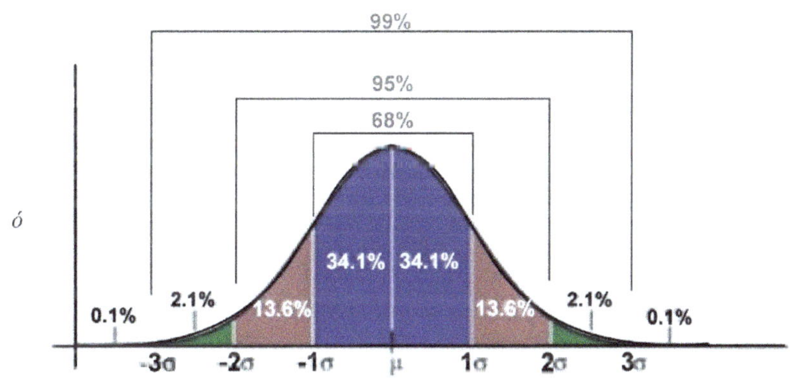

Fig. 8.1 The empirical rule of percentile distribution in Probability Theory

99% of the data corresponds to the area under the curve delimited by -3σ and $+3\sigma$.
95% of the data corresponds to the area under the curve delimited by -2σ and $+2\sigma$.
68% of the data corresponds to the area under the curve delimited by -1σ and $+1\sigma$.

An empirical σ metric had been devised for a seemingly rational treatment of sampled data. The one particularity about the standard deviation point that grants it a manner of referential weight is its singular inflexion gradient relating to the curvature of the curvilinear pattern of the function, reflected in its unit value of the abscissa scale (Fig. 8.1).

8.7. Overwhelming Numeric Trend in Statistical Theory

The emphasis on number play is such in Statistical Theory that it might almost be considered a form of Number Theory in its own right. It is inarguably a science more heavily based on querying probability figures than on studying *behavior* per se or phenomenology if at all. Accordingly, the queries are as follows.

A dice is rolled. What is the probability for an even number among the six to show when it stops?

If the probability for the sought event is $p(E)$, $n(E)$ the total count of the sought event and $n(S)$ the total number of possible events, the probability of the sought event is given by the formula:

$$p(E) = \frac{n(E)}{n(S)}$$

In this case:

$$n(E) = \{2,4,6\} = 3$$

$$n(S) = \{1,2,3,4,5,6\} = 6$$

Therefore:

$$p(E) = \frac{n(E)}{n(S)} = \frac{3}{6} = \frac{1}{2}$$

$$p(E) = 0.5 = 50\%$$

The above allows us to coin the more general formula:

$$\textit{Probability of Specific Events} = \frac{\textit{Number of Specific Events}}{\textit{Number of Possible Events}}$$

A probability is thus a figure that quantifies the specific with respect to the general, at the same time that it principally concerns physical *events*. Furthermore, all probability of specific events must be less than or equal to 1. So it is, simply because the number of specific events constitutes a subset of the number of possible events, making thereby the fraction always less than or equal to 1. Additionally, because the probability is a ratio figure and that division by zero is forbidden, the probability cannot be equal to 0. Which leads us to the following conditional statement:

$0 < p(E) \leq 1$, where *p(E)* is the probability of specific events.

The pioneers of statistical theory (Poisson, De Moivre, Lebesgue, etc.) have found that the Gaussian distribution function perfectly encapsulates all of these principles and have erected this function as the pivotal mathematical scheme for the study of facts or events of chance. Statistical theory views a statistical fact or event as a simple mathematical entity whose only virtue is to contribute to a probability figure. It is not concerned, nor does it consider that it should be concerned, with the underlying physics implicated in the event.

Once the statistical principles have been transposed onto the Gaussian distribution function, statisticians have built an interpretative structure allowing for the computation of probability percentile figures from the data that might be empirically available from a population under scrutiny. Particularly useful to that aim is the z-score formula:

$z = \frac{X - \mu}{\sigma}$, where μ is the mean, σ the standard deviation and X the variable's space separation being studied.

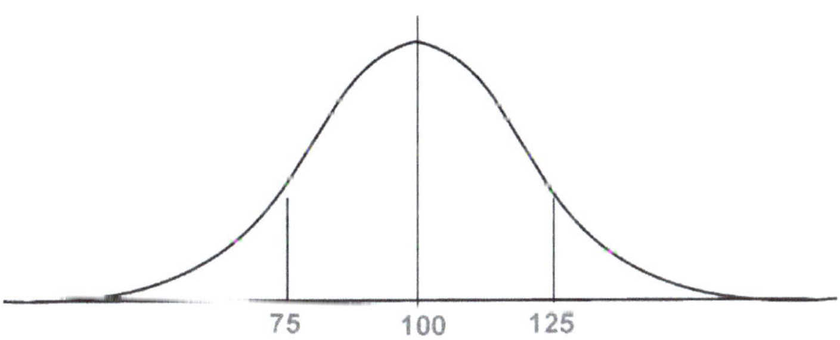

Fig. 8.2 In Probability Theory the graph of the Gaussian distribution is not based on a Cartesian coordinate system

Note that the treatment of the abscissa incarnating the space variable is somewhat unorthodox insofar as the Cartesian coordinate system, since in the Cartesian system the mean is always 0, mid-point between the set of negative real numbers and that of positive real, in accordance with the real number line. In that sense, the abscissa on the probability coordinate system is only a surrogate of the Cartesian abscissa (Fig. 8.2).

From the z-tables, to the standard deviation figure, to the μ figure, statisticians have developed empirical techniques aiding in the computation of probable percentiles. While this is all good and well from a utilitarian purpose, the question concerning the meaning of the underlying physics of statistical events has never been addressed, not even in mathematical-physics. The problems posed tend to become merely numeric in nature, such as illustrated below for the use of the empirical formula:

$$s = \sqrt{\frac{\sum(x-\bar{x})^2}{n-1}}$$ *(1)*, so-called unbiased version or its equivalent:

$$s = \sqrt{\frac{\sum(X-\mu)^2}{n}}$$ *(2)*, so-called biased version.

A typical reductionist problem follows:

Find the standard deviation for the following distribution of numbers: 13, 16, 18, 21, 31, 44, 45 and 55.

Step 1: Using expression *(1)* and adding up the 9 numbers in the sample or data set:

$x = 13 + 16 + 18 + 21 + 31 + 44 + 45 + 55 = 243.$

Step 2: $\frac{x^2}{n} = \frac{243^2}{9} = \frac{59049}{9} = 6561$

Step 3: Squaring and adding up the set of original numbers:

$\bar{x}^2 = 13^2 + 16^2 + 18^2 + 21^2 + 31^2 + 44^2 + 45^2 + 55^2$

$\bar{x}^2 = 9137$

Step 4: Subtracting the amount in Step 2 from the amount in Step 3 yields:

$$\frac{x^2}{n} - \bar{x}^2 = 9137 - 6561 = 2576$$

Step 5: We subtract 1 from the number of items in the data set, because variance estimated on the basis of n-1 is said to be unbiased:

$$9 - 1 = 8$$

Step 6: By dividing the number in Step 4 by 8, the resulting number in Step 5, we obtain the variance:

$$2576 / 8 = 322$$

Step 7: The square root of the variance represents *s* the standard deviation figure:

$$s = \sqrt{219.778}$$

$$s = 17.94$$

The above example illustrates the pervasiveness of pure numeric treatment commonly undertaken in statistical analysis.

8.8. Numeric Focus in Probability Theory

Two central theorems in Probability Theory significantly contribute to the numeric focus of the science, notwithstanding their unquestionable accuracy. One is the Law of Large Numbers, and the other the Central Limit Theorem, both intimately related to one another.

8.8.1 Law of Large Numbers

The Law of Large Numbers expresses the fact that the numeric characteristics of a random sample become closer to the characteristics of the whole population as the cardinal size of the sample increases.

For example, a single roll of a fair six-sided dice produces one of the numbers 1, 2, 3, 4, 5, or 6, each with equal probability. Therefore, the expected or **mean** value of a single dice roll is

$$\mu = \frac{1+2+3+4+5+6}{6} = 3.5$$

Consonant with the Law of Large Numbers, if the number of rolls of the six-sided dice becomes large, the average value of the total sum of the top faces shown or the mean value is likely to be close to 3.5, while the precision increases as the number of throws or trials increases. By the same token, this law establishes a correlation between the variable space and the abscissa of the coordinate system underlying the Gaussian Distribution Function. Because each side of the dice has a probability of 1 or 100% to appear when the dice stops, there is correspondence between the probability value of 1 or 100% with the μ mean value for total number of trials: the mean value must coincide with the probability *density* axis.

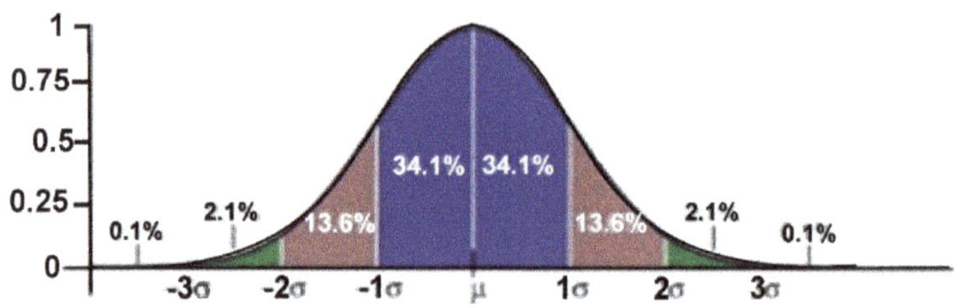

Fig. 8.3 Despite acknowledgment of a vertical coordinate axis by the Law of Large Numbers, the axis is not officially acknowledged in Probability Theory

In the empirical system of Probability Theory however, the vertical axis of the coordinate system is uncharacteristically not acknowledged, because the system technically privileges the area under the curve in the determination of probability values (Fig. 8.3).

8.8.2 Central Limit Theorem

The Central Limit Theorem establishes that, for the most commonly studied scenarios, when independent random variables are added, their sum tends toward a *normal* distribution even if the original variables themselves are not *normally* distributed. The foremost normal distribution is the commonly known Gaussian distribution. This theorem generalizes the law of distribution on account of trial counts as it stipulates that all distributions, no matter their original forms, tend to conform to the normal distribution as the trial counts reach a larger and larger number. On that account, probabilistic and statistical methods become universally applicable since they can be applied to many problems involving other types of distributions.

Just as the Law of Large Numbers principally correlates to the horizontal axis of the probability coordinate system, the Central Limit Theorem directly correlates to the putative vertical axis of the probability coordinate system in that it establishes the same to be the materialization of the scalar state of a normal distribution. The larger the scalar state of the sample, one should conclude, the more resolved the normal distribution. In that sense, what we normally call the probability density as it relates to the vertical axis of the Gaussian distribution function, must in fact be equated to a scalar density.

8.9. Toward the Physical Meaning Underlying Statistical Accounts

We are next going to undertake an inquest of the Gaussian distribution with the aims of uncovering analytic meaning beyond pure numeric accounts.

8.9.1 Re-Dimensioning the Standard Deviation Metric

Following the vein of predominance of the role played by σ the standard deviation figure, one should pay attention to curvature evolution throughout the rundown of the function. This undertaking reveals a different metric to be superimposed on the abscissa of the coordinate system. That is the metric arising from the curvature sectionals making up the curvilinear pattern of the function. A study of these sectionals in terms of Tensors and Operators has been undertaken in Chapter 4. Most importantly, the symmetry implications of this approach contribute to the most fundamental order of certainties ever revealed to be associated with the Gaussian distribution as Quanto-Geometrically re-dimensioned. Effectively every sectional corresponds to an ontological and phenomenological Norm characterizing both the structural and evolution modalities of every object in the universe.

It follows from the above that the distribution is best normalized transversally by catering to the curvature sectionals as strata across the q-axis, given that the discontinuity at $s = 0$ further squarely forbids application of the 99-95-68 % empirical rule.

This set of Nine Norms represents the highest order of certainties encapsulated in the Gaussian distribution. It incontestably projects to every group or topology of objects fit for statistical analysis. The sigma super-metric established over the s variable (Fig. 4.4) represents the symmetry metric grading the fabric of space in its continuum, while the values of the q variable establish the true first-order quantization basis applicable to fitting topologies. Note that we are expressly avoiding the characterization of quantization of space because the space variable shall always be visualized as a continuum and not a bundle, given that what essentially distinguishes one space spread from another is not a quantity but a quality, to be precise a geometric or symmetry quality.

We hope that the reader is fully taking stock of this development which is increasingly deflating the problem of applying analytic schemes to the macrocosm that are at the same time valid for the quantum realm. I cannot overstress that to the extent that statistical theory indeed contains the complete quantization basis and the geometric and symmetry elements characterizing the continuum of space, it brings to every observable object or subject of analysis in the macroscopic realm this quantum background long coveted for a unified view of physical dynamics.

8.9.2 Event Transformations in Probability Theory

In this interpretation of Statistical Theory, what we have arrived at thus far is a three-fold conclusion:

- The Theory implies a description based on a coordinate system with a horizontal axis for true coordinate space.
- The Theory implies a coordinate system with a vertical axis for the representation of scalar density.
- Statistical events may have physical meaning beyond the probability percentiles.

We have arrived at the above three-fold conclusion despite the fact that the Theory heavily leans toward a flat numeric treatment of distributions. Taking into account the physicality of the elements participating in a distribution, we may define statistical events as follows:

An event stack (or a phenomenon) is the larger frame of spatial transformations under which a variables set is preserved as an invariant agent to the action.

Under that vision, an event stack ultimately constitutes a new object based on two distinct variables. The randomness of the numerous iterations by the agent of action creates a new scalar state that is distributive in nature. That much becomes quite clear when we consider the throw of a dice not from the point of view of the thrower but from the point of view of the dice. Independently of how long it takes to repetitively throw the dice, or how long the intervals between the trials last, the dice as the point object of action transforms into a new scalar spread, if we allow ourselves to become blind to the throwing hand and if we run the clock on the event stack. The mass of the dice acquires a new state, of a distributive nature, that pertains to a **different scalar order** than the original and still invariant-compact state.

Likewise, if we give proper consideration to the position spread of the dice at every throw, it becomes manifest as an entity in its own right. Wherever the dice is being thrown, the immediate physical boundaries of the support medium constitute the *amplitude* of the position spread experienced by the dice. It must be understood as the *stationary limit* of the action which may be or must be visualized as the wavelength of a stationary wave developed by the very *position space* or position spread. This position space spread is real coordinate space in nature, not a fictitious mathematical space, the one to which the Quantum Theory has unfortunately accustomed us. The many different iterations of the throw within this position space spread creates the *harmonic character* of the action within the spread, independent of the direction of the momentum in development. From the point of view of the dice, as a point object it is subject to or develops on its own a (repetitive) harmonic motion constrained by a position space spread. We have just described nothing else than the physicality of the material *wavefunction*, as mysterious and controversial as it has always been. It is a harmonic motional artifact that springs from the very void of physical space.

The two new variables that we have uncovered thus far are ***distributive scalar state*** and ***position space spread***. They will be directly and straightforwardly modeled by a Cartesian coordinate system whereby the vertical axis incarnates the first as the dependent variable and the horizontal axis incarnates the second as the independent variable.

8.9.3 From Variance to Ontological Covariance

In the traditional probability graph, abscissa and ordinate axes are not correlated, simply due to the inexistence of the ordinate in the graph. Therefore, the variance expressed in terms of factors of σ on abscissa does not have any particular meaning in reference to a putative ordinate variable:

$$\sum(X-\mu)^2 = N\sigma^2$$

Hence the variance admittedly has very little meaning and only roughly gives an idea of spread. It is knowingly not used for much at all, except to compute the standard deviation figure, which is the main protagonist of the show. However, the concept behind the variance is a remote prelude to the very important concept of ontology or physical meaning ascribed to the abscissa variable of the Gaussian Distribution Function. As a descriptor of spread, the concept of variance relates to the notion of ***position space spread*** that we spoke of previously for the abscissa of the Quanto-Geometric coordinate system. It indeed refers to the inner variability of physical space spreads which *covariantly* relates to scalar density as the second tenet of the ontology subtending all statistical physical objects, not to say universal objects.

8.9.4 Parallel Between the Quanto-Geometric Covariant Spectrum and Statistical Percentiles

Suffice it to translate standard statistical accounts to the Quanto-Geometric canon for all the blanks of statistical analysis, not just to be filled in by causation, but to be replaced by an order of normed certainties representing the highest level of knowledge universally attainable. So it is despite the fact that the metric on the abscissa variable distinctly and more stringently goes up to 4σ in Quanto-Geometry, beyond the ordinary 3σ used in empirical analysis. Table 4.1 reflects the Primitives of the Quanto-Geometric Order of Certainties.

The referred symmetry primitives constitute the analytical categories that belie the empirical z-value calculations of gradients of probability involving σ and variance ordinarily undertaken in statistical theory. As previously highlighted, there is normally no meaning other than flat numeric cardinality to a percentile value of probability. What this science had ignored, having no means to otherwise inquire, is that there exists a transverse normative order to the probability distribution which is given by a different metric of σ over the variable. This analytic framework considers not the area or fractions of area under the curve but the curvilinear sectionals of the curve. It is there that statistical results acquire ontological and phenomenological qualities making up the deepest and most fundamental descriptive order of certainties in nature. In the strictest sense, the visualization of an area under the curve that spans across the probability density axis is illegitimate because the implicit expression of the function mandates discontinuity at $s = 0$. It is no wonder that there exists no natural or intrinsic meaning to the parcelization of the area under the curve and that the founders of this science had resorted to empirical methods of computation garnered in the z-tables in order to draw meaning and usefulness to the Function.

In a translation of the old methodology and its results to the new quanto-geometric dimension of the distribution, we would have to discard all z-values implicating transverse areas under the curve crossing the probability density axis. Only the individual tail z-values (involving external arms of the distribution) would remain valid.

For example, in educational settings, the bell curve is known to perfectly categorize test results of any class. Results are always such that the bulk of students will score the average or a C grade, while a smaller number of students will score a B or a D. Yet an even smaller percentage of students will score either an F or an A. The question is what is the cause of this distribution or what does it mean?

The Quanto-Geometric Eigenfunction, grandfather of the Gaussian distribution function, teaches the following:

A.- The position space spread axis mandates a division of students in two groups (Fig. 7.4). One group that positively interacts with the study material (positive s-axis spread), making an honest and consistent effort to apprehend the material. One other group that negatively interacts with the material (negative *s*-axis spread), engaging inconsistently with it. Per the axis, both groups are mirror images to one another.

B.- On the positive side, the 1% of students obtaining an A grade, experience the largest *dispersion* or *detachment* in their learning experience with the material. By exerting the ability to take *cognitive distance* from the material they are able to better prospect and visualize the material. This ability is a characteristic of the Quanto-Geometric asymptotic or transcendental layer where the *s* value set of the wavefunction is at its maximum.

C.- At the other end of the spectrum, we have the about 30% of students, the largest group, with the average C grade. These are the students in this group who could realize the least amount of cognitive detachment from the material. Their interaction with the material was too constraining or intrusive, or perhaps obsessive; they were unable to *elevate* over it for better visualization and thus better grasp. This ability or better say inability is typical of monocentric ontology experienced primarily at the monolithic layer of the Quanto-Geometric spectrum.

D.- The intermediate group of students in this rubrique, quantified in the distribution at about 13%, are those obtaining a B grade. Their B grade is due to a cognitive *distal* ability that is below the A group but above the C group. They are essentially marked by the Quadratic ontology experienced in the spectrum at level 4, where position space spread equals scalar density in covariant influence.

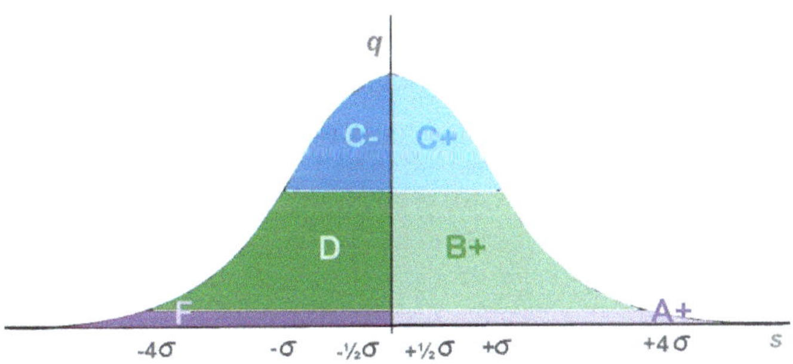

Fig. 8.4 Standard grade distribution in educational settings as a distribution function over a Quanto-Geometric coordinate system. The *s* variable grades student's cognitive distal ability and the *q* variable student count percentile.

One can see that, by catering to the ontological variables, the distribution acquires meaning much beyond the percentile figures. If we were to be more granular about the percentiles and taking into account the full set of expected grades (A+, A-, B+, B-, etc.), and further delimit them according to the σ super-metric, the quanto-geometric set of 9 transverse primitives would become pointedly manifest in the distribution for us to observe (Fig. 8.4).

Let us now examine the left side of the distribution, namely the distribution of grades F, D, and C$^-$. These students experienced the negative or least performing side of the ranking, which corresponds to the negative side of the position space spread axis. While on the positive side of the axis the position space spread showed the quality of cognitive *absorption*, on the negative side the distal variable equates cognitive *distraction* rather, two qualities that are concordantly antithetic to one another.

A.- The F students have equal amount of separation with the subject as the A students, but in this case separation has the negative quality of *distraction* or attention "away" from the subject, as opposed to attention "toward" the subject experienced by the A students.

B.- At the other end of the spectrum, the C- students could have spent equally long hours of study as the C+ student counterparts demonstrating to be equally studious, but their attention was directed away from the subject for the most part, running in a mode of distraction instead of absorption. Therefore, they score the lesser part of the average grade, a C$^-$ grade.

C.- The intermediate group of students in this rubrique are those obtaining the D grade, counterparts of the B grade students. Their D grade is due to a cognitive *distal distractive* quality that sits in between the C$^-$ group and the F group. They are essentially marked by the Quadratic ontology experienced in the spectrum at level 4, where position space spread equals scalar density in covariant influence.

The same remarks that we have advanced regarding the granularity of grade distribution on the positive side of the distal variable with respect to ontological qualities are valid here as well. In all, one ought to consider each individual student with his/her learning materials and the educational events occurring as interactions between the two as an event stack forming a *quanto-geometric* object, to be even more exact a *distributive* object. Consequently, the population of students shows the full spectrum of possible quanto-geometric correlations or covariance accessible to all material ontologies as described by the Grand Eigenfunction.

This example virtuously demonstrates **how to use the Grand Eigenfunction in the analysis of macroscopic statistical phenomena, in particular how to model the variant variable, whatever its nature, with the Quanto-Geometric space spread independent variable.** One must endeavor to identify in the sample the element that embodies the spatial spread inherent to the Quanto-Geometric visualization, which is not always quite obvious. Once that physicality is uncovered, we are a long way into unraveling the subjacent quanto-geometric qualities to the distribution that make it materially meaningful or causational beyond the mere probability percentiles.

We shall conclude that the above analysis has made it clear that cleverness and/or success in learning is not in direct proportion with compulsive learning behavior, but instead directly proportional to the learner's cognitive *distal* ability. It is thru the mental ability of detachment and elevation that one is able to scope a subject and establish its possible connections with other familiar physical elements, which then confers grasp and significant absorption of the subject matter. The web of symmetries that one is able to establish from scoping a subject constitutes the most important element of successful learning while setting the basis for deep memorization of the subject elements. At best realization, the learner becomes *transcendentally* one with the subject.

8.10. Classical Statistical Physics v. Statistical Quantum Mechanics

Classical Kinetic Theory, first formulated for gases or fluids and subsequently for solids, studies the behavior of large populations as single whole as well as how the properties of the whole relate to those of the constituent units of the whole. It is interesting to note that the formulation of the theory no longer privileges the probability distribution function directly but its functional derivative (Fig. 8.5).

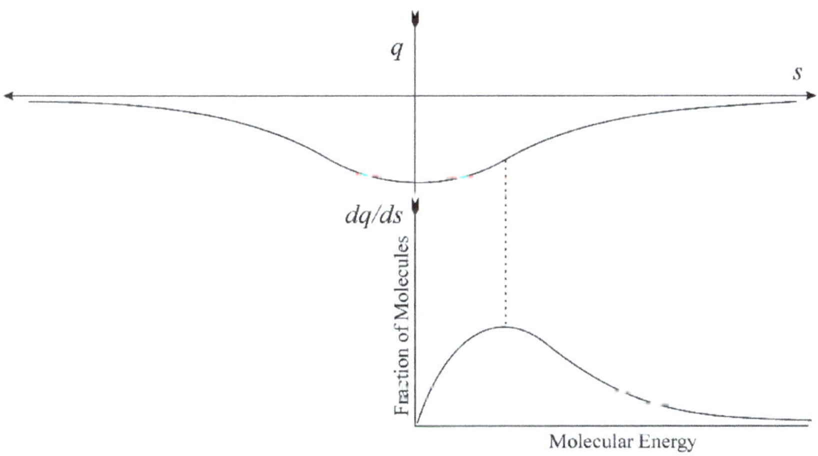

Fig. 8.5 Distribution of molecular energies in an ideal gas

An important fact to keep in mind is that in practice, I argue theoretically as well, it is almost impossible to study the evolution of the unit particles constituting the whole in a deterministic fashion, essentially along a timeline, that is. The proponents of the theory have argued that the behavior of the unary constituents do not bring pertinent information about the whole population or system, despite the fact that it is the sum total of their properties and behavior that give birth to the macroscopic behavior of the population as a whole. Given the eminent role played by the time variable in classical physics, the proponents of the Kinetic Theory (Maxwell, Boltzmann, etc.) have brushed aside the fact that the behavior of the unit constituents could not be studied deterministically.

We however find a quasi-similar scenario at the level of individual particles evolving in a stationary manner conducive to an atomic orbital. In this case the large number count of the population is surrogated by the large number of probabilistic iterations within the position space spread. Furthermore, in quantum mechanics we must assume that the particle is forced to behave as though restrained in a cubic box for the wavefunction to be extolled, which turns out to be a stationary wavefunction independent of time. Likewise, to study the dynamics of a population of gas molecules, we must admittedly assume a cubic container of any desirable volume. The container or its size determines the amplitude of the position space spread but not the space spread itself. Here the role of the position space spread is played by the velocities accessible to the molecules. **It is important to understand that it is the space spread that instantiates the motion which becomes apparent as molecular velocities. It is not the discrete molecule that creates the motion, but its space spread of inhabitation**. The discrete molecules only become submissive to innate spatial motion to be visualized as defoliation or the *un-folding* or *de-foliation* of physical coordinate space. The group of molecules accessible to the many different velocities (or wavelengths of the position-space-spread wavefunction) creates the co-relational discrete scalar states of the distribution. Those discrete groups of molecules are directly comparable to the scalar density states of an electron in orbitalization. To be exact, it is not the number of molecules but the percentile fractional number of molecules that compare to the scalar density states. Effectively the vertical axis of the distribution graph represents a derivative function, by nature a rate. Therefore, every velocity is a wavelength value of the material wavefunction of the ideal gas as a whole. The principal figure that characterizes an ideal gas in the Kinetic Theory is the product of the pressure P and volume V occupied by the gas, such that: $PV = \left(\frac{2}{3}N\right)\cdot\frac{mv^2}{2}$

where v^2 is the square of the average of all molecular velocities, N the total number of molecules and m their individual mass. Note how close v^2 is notionally to probabilistic variance. If we set the moment Q of one molecule to be:

$$Q = mv \text{ , then } PV = \left(\frac{2}{3}N\right)\cdot\left[\frac{Q^2}{2m}\right].$$

PV is thus an expression of the kinetic energy of the gas. In the Hamiltonian operator for the orbitalized or bound electron the core term is:

$$\left[-\frac{\hbar^2}{2m}\right]\cdot\left[\frac{d^2}{dx^2}\right].$$

The first part of the term expresses the kinetic energy displayed by the bound particle. Note its analogy with the same term in the ontological formula for the ideal gas. The difference between $\hbar$ and Q is simply the difference between quasi relativistic motion and non-relativistic classical motion. There is a factor of physical scale as well to account for the difference between the both phenomena visible in the mathematical basis for their formal treatment. One is in development in the atomic Shell of matter and the other in the Astral Shell, medium of inhabitation of the living where we setup our laboratories. Analysis conducted in Chapter 6 has shown the determinants to the formation of those Shells. However, our principal interest here is to make explicit the large similarities between the two phenomena in their statistical interpretation.

Once we understand that mass behaves as a density scalar state, that the fraction of gas molecules constitutes a discrete derivative scalar state, coupled with an understanding of the innate dynamical behavior of real position space spread, we will begin to understand a common ontological ultra-structure to all probabilistic behavior in all realms of matter, whether quantum or large-scale. What lies beyond is an equal understanding of the covariant relationship between those two ubiquitous variables in the serial development of Shells of matter, well developed throughout previous Chapters of this study.

8.11. Randomness is Relative

Randomness is incomplete in scope. Beyond randomness is the certainty or teleology of types otherwise expressed as normative typologies involving co-variance between the fundamental variables underpinning the ontology of physical objects or evolution of events. Beyond randomness lies the new physics or the physics of:

- Spatial wavefunction of a single object
- Symmetry roots belying all transformation groups
- Ontological covariant spectrum

The incompleteness of Quantum Mechanics does not altogether reside in Quantum Mechanics itself but essentially in the mathematical statistical theory that sustains its development. The required analytic categories for a physical interpretation of the ontological configuration of a bound electron simply falter in the probabilistic science of chances as known thus far.

A. Einstein famously conceded that *he cannot imagine the electron hoping like a bug* as implied in the quantum treatment. Could he have conceived that free will is not random either? There is nothing in existence, I repeat NOTHING, that can escape the Quanto-Geometric ontological covariant spectrum. Cognition of the living, bug or human, develops in covariant quantum-space phases according to the normal *quanto-geometric* distribution harbored by the Grand Eigenfunction, just the same way or in the same *modalities* that an electron evolves within an atomic orbital.

8.12. Conclusion

In the quest for a Unification Theory that single-handedly furnishes a formal and assertive description of the entire sweep of the successive Shells of matter, banking on Probability Theory is good intellectual investment. The universal applicability of the Theory, from quantum density states to rolling dice to the distribution of parameters characterizing the living, is an inestimable asset in that endeavor. The timelessness view implicated in the Theory applies across the full spectrum of Shells of matter, calling for a dramatic ban of the familiar but unphysical time variable in macroscopic ontologies and phenomenology as well. When re-dimensioned under the umbrella of the Quanto-Geometric Grand Eigenfunction, the Gaussian distribution function at the core of Probability Theory acquires discretisation properties and wavefunction descriptive capabilities that place the applicability of quantum science directly in the midst of macroscopic phenomena. Is this not a new day in Quantum Theory!

BOOK II

Chapter 9

How can it be that mathematics, being after all a product of human thought which is independent of experience, is so admirably appropriate to the objects of reality?
A. Einstein

CRITICAL ANALYSIS OF ABSTRACT REPRESENTATION THEORY

This second cast of our study specifically addresses the questions of inner fabric of universal space, theoretical derivation of masses and wavefunctions of condensed matter objects, as well as atomic architecture including a number of unsolved issues in nuclear quantum physics and quantum chemistry. We first engage a critical review of the main Theories in vogue in quantum physics, in particular Quantum Field Theory, its deriving Standard Model of particle physics, along with Abstract Representation Theory as its mathematical foundation. We critically review as well at the outset the Theory of General Relativity, first pillar of 20th Century physics, highlighting its contributions and limits.

Quanto-Geometric Theory being a very distinct vision of physical dynamic systems, it is important to establish the parallel between its analytical constructs and developments and those pertaining to the Theories footing modern physics, towards better appreciation of its contributions to resolution of pending problems. We must pointedly underscore the reservation that a critique to a Theory is not a personal critique to its proponents, all too aware as we are that the forthright aims of the entire academic establishment is scientific advancement for the good of all, despite serious drawbacks at times from other underlying interests.

Let's get started.

9.1. Quanto-Geometric Canon

The concepts that we discussed in Chapter 4 Book I of this publication arguably stand for the highest level of abstract experience accessible to mind. They expose the number stack and its properties as well as those of the geometric stack. Two paramount notions arise from the Quanto-Geometric vision:

1. Physical reality is laid upon a stack of nine typified Primitives as its universal building block.

2. The nine typified Primitives form a spectrum with a quintessential tri-partite partitioning.

Together those two axioms compose what we dub the Quanto-Geometric canon, namely the spectrum canon encapsulating the typified Primitives and the tri-valent canon encapsulating the differential partitioning of the primitive spectrum.

The Quanto-Geometric canon represents the foremost generic group theory ever possible, taking roots at the highest level of abstract experience. Nothing lies beyond the Quanto-Geometric canon, it is the beginning and the end of all things physical and abstract. In the following we aim to show that the Quanto-Geometric canon subsumes traditional Abstract Group Theory and its derivative: Representation Theory.

9.2. Group Theory and its Axioms

First a brief review of Group Theory. Abstract group theory is a system of representation of populations of tangibles and intangibles. It is based on 4 principal axioms which we describe next.

9.2.1 Closure Axiom

Per this norm, if two elements a and b pertain to a group, then element a combined to element b or $a \cdot b$ must also be part of the group. The rule of composition is whatever one wants to make of it, here represented by a dot, and constitutes once chosen the active operator inside the group.

9.2.2 Associativity Axiom

Per this norm, if elements a, b and c are part of the group then *a* combined to *b* *(a.b)* which is in turn combined with c, that is *(a.b).c*, must be as a whole equal to *a* in composition with combined *b* and *c*, that is *a . (b.c)*. In equation form:

$$(a \cdot b) \cdot c = a \cdot (b \cdot c)$$

This spells out the rule of association.

9.2.3 Identity Axiom

The norm of identity rules that the group must contain an element *e* such that when combined with any other element in any manner it must leave the element unaltered. That is:

$$e \cdot a = a \cdot e = a$$

9.2.4 Inverse Axiom

According to this axiom, any element *a* of the group must have an inverse a^{-1} such that:

$$a \cdot a^{-1} = e \text{ and } a^{-1} \cdot a = e$$

9.2.5 Abelian Axiom

The Abelian axiom is not a mandatory axiom for the existence of a group. It is a view of commutative behavior between two elements inside the group, whereby element *a* combined to element *b* or *(a.b)* can be equal to *(b.a)*. In equation form:

$$a \cdot b = b \cdot a$$

Group Theory expressly holds the commutative axiom as a facultative rule in the composition of groups.

9.3. Interpretation of Group Theory Axioms

Starting our examination with the Identity Axiom, one can easily see that this axiom is a condition for the existence of a monolith among group elements. Monoliths are elements that do not break up, and thus do not interact or combine. This is represented in type 1 Primitive of the Quanto-Geometric canon.

The term Identity in the denomination of this axiom is somewhat misleading insofar as it hides the true behavior of the identifying element and leads to the unintended notion of a defining element for the group. Most likely meant to signify that it does not change the Identity of other group elements when combining with them, it must preferably be construed as the all Unique and Monolithic element of the group in structure and behavior.

The Abelian or commutative axiom is a match to type 2 Primitive of the Quanto-Geometric spectrum canon. It is an axiom that formulates the condition for Interchangeability. Interestingly, Group Theoreticians have made it a facultative condition seemingly in an attempt to broaden the scope of application of the theory to the largest possible number of populations of objects.

The Associative axiom broadens the inclusion scope of the group in mandating the existence of a relationship between a set of three elements of the group, beyond the identity element. At the very least, this relationship is forcibly of distributive nature. The law of multiplication arises from the constraint brought about by the Associative axiom. It is clearly a match to type 3 Primitive of the Quanto-Geometric spectrum canon.

The Inverse axiom mandates the existence of fractional or imaginary elements within the group. If a is a real physical object, what can be the inverse of that element or a^{-1}? It can only be an *image* of a, an *imagin-ary* that is. The Inverse axiom is thus a match to the (type 5) Entropic Primitive of the Quanto-Geometric spectrum canon.

Lastly, the Closure axiom makes explicit the scope of the group thru the introduction of a rule of composition within the group. It means not closure as in an end or a boundary but instead it stipulates the breadth of a group topology thru a specific inclusion rule. The closure axiom posits the inner modality of composition within the group or the Operator proper. It defines the relationship between elements of the topology.

The equivalent of the Closure axiom in Quanto-Geometry is the Quanto-Geometric functional representation which defines the inner rule of composition between the variables (dependent and independent), or their Quanto-Geometric mutual articulation.

It is hence logical to conclude at this point that Abstract Group Theory is essentially laid upon a background canvas that is the Quanto-Geometric spectrum canon by surrogating the Monolithic Primitive, the Binary Primitive, the Associative Primitive, the Entropic Primitive as well as the Operator principle postulated by the Quanto-Geometric functional articulation. We will shortly show that Group Theory encapsulates as well the Enthalpic Primitive.

9.4. Interpretation of Group Theoretic Categories

Abstract Group Theory distinguishes three classes of Groups: Finite or Discrete Groups, Infinite or Continuous Groups and Semi-Continuous Groups. Finite or Discrete Groups have a number of properties that distinguishes them from Continuous Groups. These properties are gathered in Theorems that express their particular class behavior.

The Lagrange Theorem for instance establishes that elements of a Finite Group can only be evenly subdivided or else the topology is not divisible in sub-groups. The definition of the order of a Group, the cardinality of its topology, is a characterization that derives as well from the finite nature of Discrete Groups. From a Quanto-Geometric vantage, it is easy to recognize the primacy of the scalar in the configuration of the so-called Discrete Groups.

The notion of Continuous Groups forms a representation for addressing principally topologies of abstracts instead of strictly scalar elements. Continuous groups are concerned with elements undergoing changes, translations or transformations and garner the topology of these transformations which it identifies as symmetry operations. Elements of Continuous Groups are essentially differential equations and the analytic structures arising from differential geometry. In its physical applications, the notion of Continuous Groups pursues the description of motion thru changes and transformations.

For a set of elements R to form a continuous group, in addition to fulfilling the standard axiomatic requirements for a group, they must depend on a number of real parameters a such that $R(a) = R(a_1, a_2, \ldots a_r)$ and that a small change in one of the factors produces a corresponding small change in their product, resulting in a serial or continual suite for the topology of elements.

Under the regime of change and parameterization, symmetry transformations become a major focus for Continuous Group Theory. It is not difficult to see that this regime with its primacy for change, transformation and symmetry operations is a complete match to the Quanto-Geometric Hyper Space Operator within the Tri-Valent canon. Whereas the notion of Discrete Groups focusing on discrete or scalar topologies, and static algebraic equations when it comes to motion, is a match to the Hyper Quantum Operator within the Tri-Valent canon.

Is it now any surprise that Abstract Group Theory do envisage the existence of Semi-Continuous Groups as a sheaf with properties that sit midway between Discrete Groups and Continuous Groups? For Discrete Groups, the different operations of the same sheaf represent a single operation repeated once, twice, thrice or any order of times. For Continuous Transformation Groups different operations of the same sheaf correspond to different whole numbers, which may be commensurable or incommensurable. Whereas for Semi-Continuous Groups these operations correspond **only** to different commensurable numbers. This translates into the existence of a topological space that is an intersection of the topologies pertaining concurrently to both of the other two major Abstract Groups. Formally, a Semi-Continuous Group is a group G and a topology T on G making l_a and r_a continuous for all $a \in G$. Again, here the concept of Semi-Continuous Group undoubtedly shows a match to the Hyper Quadratic Operator within the Quintessential Tri-Valent canon. We must stress however that Semi-Continuous Groups have little presence if any in the physical science application of Abstract Group Theory.

If one were to wonder why it is that there exist three and only three broad categories of Abstract Groups, although Abstract Group Theory is patently a human view of nature, one ought to find the root cause of this categorization in the primacy of the Quanto-Geometric Quintessential (Tri-Valent) canon and nowhere else.

9.5. Interpretation of Representation Theory of Groups

Representation Theory of Abstract Groups starts with the recognition that matrix groups stand for a higher frame of generalization that subsumes basis axiomatic Group Theory. Theoretically, this capability derives essentially from the fact that the composition rule for matrices encapsulates the addition, subtraction and multiplication operators thru the dot product and the determinant methods at work in their organization and resolution.

Analysis of the regular representation of a symmetry group yields that every finite group bears a one-to-one relationship with a matrix group. Accordingly, general groups can be thoroughly studied by using matrix group representations. It stands to reason that if there is something about matrixes that make them fit for representation of general axiomatic groups, then that identity must stand higher than the axiomatic set itself including that of closure, and that a Matrixial Operator must be that identity.

When studying the numeric and geometric stacks (or in one word the quanto-geometric stack) in Chapter 4 of Book I we have identified type 6 Primitive, a promoter of Order and Match, as the Matrixial Operator. Conventional mathematics does not profess a Matrixial Operator by nomenclature, however the method in action for the configuration and resolution of a matrix is essentially based on a canvas of Order and the operation between matrix elements essentially develops thru Matches. Both of these features are the principal attributes of the Matrixial Operator deriving from the qualities of the quanto-geometric stack at typified layer 6.

One may wonder why something so specific as a matrix group may become a generalization that lies above the whole concept of Abstract Group. The reason to that assertion is the innate hierarchy of the quanto-geometric stack by which the Matrixial Operator lies above the Imaginary Operator, the Associative Operator, the Dissection or Subtraction

Operator, the Monolithic or Summation Operator, as well as the Root Operator, so far as mathematical operators are concerned.

9.6. Abstract Group Theory on the Quanto-Geometric Map

One quite novel idea posited in Quanto-Geometric Theory is the notion that Mathematical Operators bear more than arithmetic computational value and represent each an ontological and phenomenological proposition. We show how the nature of these Operators is reflected in the qualities of each of the Primitive or Normal layers of the primeval Quanto-Geometric stack. Each layer was shown to be in association with an Operator which informs the structure of its numeric elements and determines the configuration of its number set as a Group. When viewed Quanto-Geometrically, the number stack, as well as the geometric stack, portrays conceivably the most natural of Group Theories. In that condition the quanto-geometric stack will help us understand the origin of and the background to traditional Abstract Group Theory, indeed strictly a human development, and further assess its usefulness together with its limits.

Effectively Abstract Group Theory is an imperfect prelude to the Quanto-Geometric framework in that it amalgamates axioms, theorems and lemmas in a way that disparately reflects the perspective behind the number stack (or the quanto-geometric stack as a whole) as the highest of all Groups. The genericity of the rule of composition is an attempt at a generalization that nevertheless fails to constitute the proper framework for that role. The mapping between Group Theory categorization and the quanto-geometric stack is shown in Fig. 9.1.

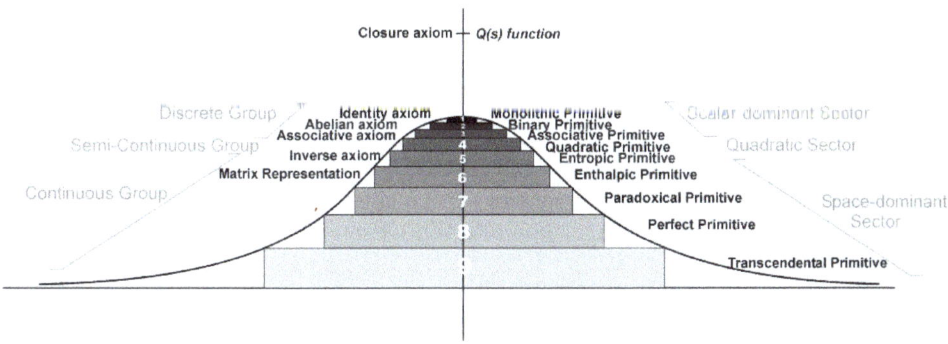

Fig. 9.1 Mapping between Abstract Group Theory categories and the layers of the Quanto-Geometric stack

Furthermore, we must recognize that a generic rule of composition cannot be unspecific, notwithstanding it high-level standing. We have learned that it must be instead of a class such that all the other rules of composition organically derive from it. The Quanto-

Geometric functional articulation reveals that this most generic rule must be of *transcendental* nature and that all the other rules of composition deriving from it must each exhibit a specific kind of typification as they indeed all do.

In the following we delve into the structural composition of some of the most common mathematical or arithmetic Operators with the aims of eliciting the affiliation of these Operators with specific number sets, beyond their traditional mathematical meaning.

9.6.1 The Monolithic Operator

The Monolithic Operator is one of Totalization or Summation. It is the Operator that foments Prime numbers and endows them with the quality of compactification or non-breakability as it were. One can make the statement that the scalar topology that makes up a Prime number is under the regency of the Monolithic Operator when it comes to its composition method. Elements of the topology cannot become explicit but are only of implicit nature because they are complete meta-level abstractions to the numeric domain. Summation as an operation applied over the entire range of elements pertaining to the number realm stands for a degenerate inheritance or projection of the Monolithic Operator to the number domain, one that lacks the ability to impose the law of Monolithisation or compactification in its fullness. Even adding Prime numbers together does not necessarily result in Monoliths.

Phenomenologically this Operator creates Autonomous behavior, Selfness or Solipsism. This feature further contributes to the fact that Primes do not necessarily sum up with each other to form Primes. What percolates into them from the Monolithic Operator informs them in a way that perpetrates their individual segregation from one another in relational experience.

9.6.2 The Parity Operator

The Parity Operator is one of Di-vision or Sub-traction or Sci-parity (Se-paration). The Parity Operator governs the scalar topology that makes up each Odd real number in their class. When it comes to its composition rule, this Operator is all about Separation or Bi-Polarization.

In the form of Subtraction, this Operator is about whether or not one may dissect at all a number in one group of another, at once. In the form of Division, this Operator poses the question about whether or not:

1. A number may be dis-sected in Groups of another, and
2. How many times the dissection may be done.

The Parity Operator promotes internal bi-polar segregation of the scalar topology that makes up each Odd real number. This may be easily understood by analyzing the geometry of an ovoid shape, which we may assimilate to a geometric representation of scalar sci-parity or separation.

Sweeping a vertical line across the egg-shape figure (Fig. 9.2) will always segregate it in two sheaves of symmetries about the line that are never equivalent to one another except when the line sits at the center position. The vertical line makes explicit Parity Separation or Bi-Polarization of the topology of elemental points making up the egg-shape figure. These results will remain invariant had we chosen to move the figure across the vertical line itself or even under rotation about the line.

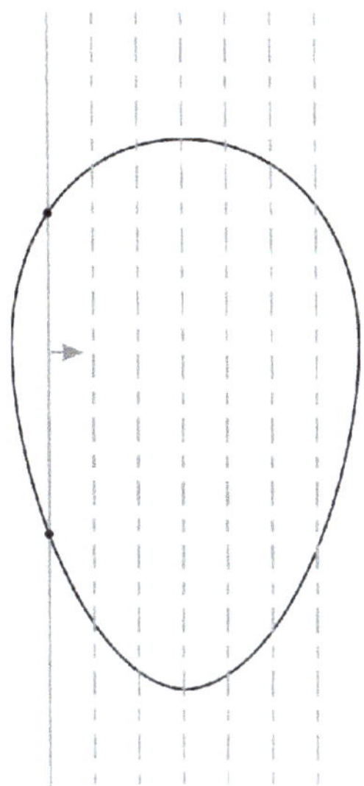

Fig. 9.2 A visualization of Bi-Polar symmetry

However, an element of Interchangeability does appear as a derivation of the transformation at the center position of the vertical partitioning line. If the object is C and A is one half of C, and B is the other half of C, then a lateral shift of the line or rotation of the figure about the vertical axis results in:

$$C = A \cdot B = B \cdot A \text{, } (\cdot) \text{ being the rule of Segregation}$$

Unlike the Monolithic Operator, the Parity Operator is heavily invested in the composition of numeric elements under its regency within the Odd Composite Group, beyond internal scalar state of individual elements. So it is because this Operator is less about compactification than it is about Spread. It does so thru the methods of Division and Subtraction and not thru the method of multiplication as traditional Number Theory has it. Suffice it to view the smaller or lower elements as Divisors of the larger or upper elements in the Group for this to become inarguably clear. In support of this view, one ought to remember as well that the scalar order of progression across the numeric stack is that along the line of breaking down the Monoliths.

9.6.3 The Associative Operator

The Associative Operator is one of Distribution thru Evenness. It is a close incarnation of the closure axiom in traditional Group Theory. It broadens the scalar scope of the composition method however, casting it within boundary or closure. While Division (of integers) may carry a remainder, showing thereby a form of the state of Evenness, Multiplication (of integers) always casts whole multiples. This specific form of the state of Evenness is a rendering of the multiplet state seminally native to the Associative Operator.

The scalar multiplet state is the modality by which the Operator creates the individual elements pertaining to the natural Even Composite Group of real numbers. Because a multiplet may involve not just two but any number of elements, it is in essence a Relational entity. The multiplet state innate to the Associative Operator is correctly expressed based on a minimum of three elements A, B and C such that:

$$(A \times B) \times C = A \times (B \times C)$$

It important to realize at this point that the scalar scope of our Operators is broadening as we move along their Quanto-Geometric hierarchy line, while their compactification accordingly diminishes.

9.6.4 Matrixial, Rational, Algebraic and Transcendental Operators

The three Operators above studied, the Monolithic, the Parity and the Associative, arguably stand for the most difficult to relate to specific Number domains. The next in line, the Power Operator, as it relates to square, cubic and power numbers in general, seem to be a straightforward proposition in that regard. As do the Imaginary Operator with its direct correlation with the complex number realm.

In that context, the structure of a complex number (a + b*i*) must be understood as a method of composition typical of the Imaginary or Complex Operator. The attribute of Imaginary for the Complex Operator is very well suited to the description of its phenomenological behavior. From a Quanto-Geometric standpoint, it reflects the significance acquired by the Intangible (spatial-geometric) with respect to the Tangible (scalar). Since a complex number may lack its real part but not its imaginary part, the preeminence of the intangible over the tangible posited in the Complex Operator is a clear representation of the Quanto-Geometric space dominant canon to which the Complex Norm belongs.

Likewise, we must talk about a *Matrixial* Operator, a *Rationalization* Operator , an *Algebraic* Operator and a *Transcendental* Operator as the key Norms of composition operating among the number sets represented by Matrices, Rational numbers, Irrational or Algebraic numbers and Transcendental numbers respectively. Again, the phenomenological attributes studied in Quanto-Geometric Number Theory fully apply to those entities. In that context, one must view functional representation, and even more so transcendental functions, as the highest forms of mathematical composition possible and their associated methods of differentiation, integration and differential expression as methods native to the Operation of *Transcendentalization.* In support of this claim, one must remember that most every function may be written as a Taylor series in accordance with the attribute of infinite serialization inherent to Transcendentalization. To that account one may further add the fact that the analytic derivative of every function represents a suite or sequence of infinite order.

9.6.5 Limits of Representation Theory of Groups

Representation Theory of Groups, in conjunction with operator mathematics, has now for long stood for the ultimate interpretative frontier of modern physics. Physicists have found it to yield an exceptional modeling tool for particle physics in particular. However, the Theory lacks the grander view that is required for a full grasp of the physics of matter in all of its deployments, not to mention its contending a series of views that are outright erroneous.

For all the importance acquired by the notion of symmetry in the Theory, it fails to comprehend the identity of the basis states of symmetry within their natural hierarchy. The Theory only witnesses symmetry transformations of objects in translation or rotation and attempts to establish an order of symmetry transformations that is ill ranked at the outset. One may read in any presentation textbook such assertions as: "The more symmetry transformations that an object admits, the more "symmetric" it is said to be. On this basis, the circle is "more symmetric" than the square, which, in turn, is "more symmetric" than the triangle". This assessment reflects plain misjudgment.

We have shown in the study of the geometric stack that the state of symmetry inherent to the circle is in fact the least prolific in symmetries due to the simple fact that this state of symmetry is inherently unitary. Because a triangle implies the 3 focal points (3 center points of symmetry) that sustain parabolic symmetry, it promotes a higher level of symmetry that lies above the circle in the non-linear class of symmetry. At the end, Representation Theory can only describe the behavior pattern of objects in motion, it cannot explain the origin or root cause of motion itself because it is not able to uncover the deep states of symmetry and the inner dynamics innate to the void of physical space. That also stands for the reason of its notable incapacity to explain the behavior of singlets for the most part but only that of population of objects.

Further, Representation Theory, despite its ability to model certain topologies of particles via special orthogonal and special unitary Groups (SO and SU), can barely describe the structure and dynamics of matter in the higher Shells beyond elementary particles and does not offer any sensible framework to account for the deployment of matter in distinct Shells. These Shell deployments are otherwise well studied in Book I from the standpoint of their building blocks, the mysterious but finally deciphered "fundamental" physical constants. Representation Theory has not helped us derive from any of its principles one single physical constant or mass of infinitesimal objects in all of the spread of matter (except for the questionable *g*-factor of the electron)! In fact, the Standard Model of Particles physics is fraught with problems and lacks that one ought to put on account of the very modeling tool.

Lastly Representation Theory has not at all contributed in clarifying the qualities of physical space and its possible internal architecture by obfuscating the matter in adding even more fictitious spaces to the list of artificial mathematical spaces in use in modern physics (i.e. parameter space, Hilbert space and others).

9.7. Limits to Physical Application of Hermitian Operators

Conventional Operators in mathematical science are quite distinctly defined in comparison with our Quanto-Geometric Operators. Very generically, a mathematical operator is a prescription, a procedure, a recipe or a rule. It generally applies to a function in an abstract function space. When applying the operator to the function, if the operator returns the same function but only modified by a coefficient, the function is considered proper to the operator and is designated an *eigenfunction* of the operator. The coefficient is in turn designated an eigenvalue of the operator. We can express the operation as follows:

$$Pf = af$$

where P represents the operator, f the function undergoing the action, and a the coefficient. We can readily see that the idea of a conventional operator is very narrowly related to functions with derivatives that are more or less equal to the original function. We may rewrite the above as follows to shed better light on the argument:

$$\frac{d}{dx} f(x) = a \cdot f(x) \,.$$

Considering the operator to be expressed by the *d/dx* notation, the question posed by the above equation is: What is the function which when differentiated (or under action by a derivative operator) returns the same functions times an *a* constant? The solution to the equation or answer to the question is:

$f(x) = A \cdot e^{ax}$, an exponential function with an *A* coefficient.

The differential operator, if not the only type or class of operators, is a prominent and widely used class of linear operators acting on function spaces. One of the problems that motivated the development of mathematical operators is the famous problem of determining the shape of a drum from its sound, or the shape of a standing wave in a string fixed at its boundaries. Applying the laws of mechanics to infinitesimal portions of the string, and letting *h(x,t)* be the functional expression of the lateral displacements of one stressed string of the musical instrument, then *h* satisfies the partial differential equation:

$$\frac{\partial^2 h}{\partial t^2} = c^2 \frac{\partial^2 h}{\partial x^2}$$

Applying the method of separation of variables in multi-variable functions, we obtain:

$$\frac{d^2}{dx^2} X = -\frac{\omega^2}{c} X \text{ and } \frac{d^2}{dx^2} T = -\omega^2 T$$

These eigenvalue equations are satisfied by the functions:

$$X(x) = \sin\left(\frac{\omega x}{c} + \varphi\right) \text{ and } T(t) = \sin(\omega t + \phi)$$

The final eigenfunction comes to be:

$$h(x,t) = \sin\left(\frac{n\pi x}{X_m}\right) \sin(\omega_n t)$$

where n is any integer and X_m a chosen length of the string. The eigenfunction is thus an entire family of standing waves with ω_n frequency or the frequency of the n-th harmonic. Fig. 9.3 gives a graphical sense of these eigenfunctions and eigenvalues.

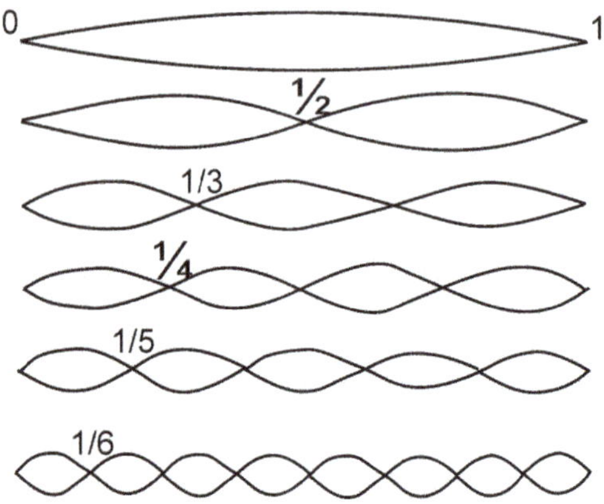

Fig. 9.3 Overtones of a vibrating string or drum as eigenfunctions and eigenvalues. The waveforms are the eigenfunctions of the vibrating element. The eigenvalues 1, 1/2, 1/3, 1/4, etc., are the periods of the eigenfunctions and make up the musical harmonic series.

Likewise, as we have seen in Book I, the question of a particle constrained to move within the confines of a virtual box and developing what appears to be standing waves, notably a bound electron, may be fruitfully treated from an operator viewpoint. When we apply an operator of kinetic and potential energy on the particle's apparent wavefunction, the famous Hamiltonian operator, we obtain eigenfunctions φ_k and a set of eigenvalues E_k for a resolution. The solution to the problem posed in those terms is a family of eigenfunctions and eigenvalues of the form:

$$\phi(t) = \sum_k e^{-\frac{iE_k t}{k}} \cdot \varphi_k$$

To further venture a comparison of the above constructs with what actually takes place in a bowling alley, the player is in the role of instigator or operator (the intelligence), the eigenfunction maps to a particular strike of the throw ball with respect to the pins sitting at end of lane. The throw that is able to hit the first pin and trigger a chain reaction that strikes down all the other pins, stake of the game, is comparable to the eigenfunction for that turn. It is important to realize that the arrangement of the pins stands for an important component of the coveted chain reaction resonance. There must be a suitable relationship between the pins for a particular throw to be successful. The pin set count and the structure of the pin set constitute in operator mathematics the basis of the function space, each pin being comparable to a basis function. There must be a par symmetry between the basis functions such that in any set of two among them one is *orthogonal* to the other when assimilated to a vector. They must obey thereby the dot product or inner product (a triple product between the magnitudes of two vectors and the cosine of the angle they form between one another). An orthogonal basis for a function space is expressed as below:

$$\int f_j \cdot \bar{f_i} = 0$$

where $\bar{f_i}$ is the conjugate of f_j. When the orthogonal basis condition is met the operator is said to be Hermitian, while the basis function space with its vector orthonormal condition becomes a Hilbert space. The Hamiltonian Operator perfectly meets the conditions of a Hermitian Operator, and on that account it has been able to deliver the very important family of Eigenfunctions and energy eigenvalues that unraveled the energy density states of the hydrogen atom. However, this scheme says nothing about the intimate nature of orbital wave functions. While the operator framework is perfectly suitable in the interpretation of waves cast by a vibrating string - since the solution eigenwaves and the waves forming the function space are perfectly and parsimoniously described mathematically under the concept of a wave frequency - there is by contrast no such match for the virtual or physically unobservable wave function of a bound particle. This makes for the very root of the tremendous interpretation problem that has hampered Quantum Mechanics to this day and sparked the famous controversy between the school of Copenhagen and the opposing school of determinism led by A. Einstein.

Alternatively, the Quanto-Geometric Operators lie high above this fray and this entire approach. They are not mathematical models awaiting physical application. They transcend both numeric abstraction and physical reality and are best thought as transcendent mathematical-physics entities that create physical reality.

9.8. Nomenclature of the Nine Quanto-Geometric Operators

At this stage we hope to have made it abundantly clear that the scalar-space Norm postulated in Quanto-Geometric Theory lies well beyond any number base spectrum of any Number system, as well as the different categories of number realms known in traditional mathematics. The Quanto-Geometric Operators are natural abstract entities that inform the entire base spectrum of any number system one pleases to uphold, in scalar and geometric terms, while they equally dictate the Group structures of all natural number realms. The Quanto-Geometric Function is a reunion of the studied set of 9 Operators. In that condition one may think of it as an Eigenfunction, nothing less than the Grand Eigenfunction beholding the expression of all of matter or the solutions to every physical problem posed by physical objects, the latter being in essence derivate objects across all Shells.

The effort to remain as close as possible to familiar terminology in the description of the Quanto-Geometric Eigenfunction and its Operators is one that seeks to provide an easy bridge for comprehending a new concept. Despite their meta-level mathematical nature however, the generic nomenclature of these Operators may be expressed as noted below.

$$\langle q_i \mid s_i \rangle = \int_b^a \sqrt{1+\left(\frac{d}{ds}\right)^2}\, ds,$$

where i is an ordinal of n and $[a,b] \in \{s_q\}$
and s_q is the primeval symmetry-breaking sequence

Specifically,

Monolithic Operator: $$\langle q_1 \mid s_1 \rangle = \int_0^{\sigma/2} \sqrt{1+\left(\frac{d}{ds}\right)^2}\, ds$$

Parity Operator: $$\langle q_2 \mid s_2 \rangle = \int_{\sigma/2}^{2\sigma/\sqrt{7}} \sqrt{1+\left(\frac{d}{ds}\right)^2}\, ds$$

Associative Operator: $$\langle q_3 \mid s_3 \rangle = \int_{2\sigma/\sqrt{7}}^{\sigma} \sqrt{1+\left(\frac{d}{ds}\right)^2}\, ds$$

Power Operator: $$\langle q_4 | s_4 \rangle = \int_{\sigma}^{\sigma\sqrt{8/5}} \sqrt{1+\left(\frac{d}{ds}\right)^2}\, ds$$

Imaginary Operator: $$\langle q_5 | s_5 \rangle = \int_{\sigma\sqrt{8/5}}^{\sigma\sqrt{10}/2} \sqrt{1+\left(\frac{d}{ds}\right)^2}\, ds$$

Matrixial Operator: $$\langle q_6 | s_6 \rangle = \int_{\sigma\sqrt{10}/2}^{2\sigma} \sqrt{1 + \left(\frac{d}{ds}\right)^2} \, ds$$

Rationalization Operator: $$\left\langle q_7 \mid s_7 \right\rangle = \int_{2\sigma b}^{\sigma\sqrt{7}} \sqrt{1 + \left(\frac{d}{ds}\right)^2} \, ds$$

Algebraic Operator: $$\left\langle q_8 \mid s_8 \right\rangle = \int_{\sigma\sqrt{7}}^{4\sigma} \sqrt{1 + \left(\frac{d}{ds}\right)^2} \, ds$$

Transcendental Operator: $$\left\langle q_9 \mid s_9 \right\rangle = \int_{4\sigma}^{\infty} \sqrt{1 + \left(\frac{d}{ds}\right)^2} \, ds$$

These expressions represent the segmented arc lengths of the Grand Eigenfunction within each of the normalized intervals (symmetry-breaking intervals). They portray *unified* quanto-geometric entities because they each contain both a scalar value and a spatial or geometric value thru the arc metric. Each arc segment thus represents a quanto-geometric composition (or unit-norm) under the umbrella of a typified symmetry within the class of nine. We apply an Operator over a set of observables in order to either query them for that qualified class of symmetry or to expressly inform an observable or set of observables with that particular class of symmetry. While the first part of this assertion bears resemblance with operator entitlement in conventional mathematical operators, the second part of the assertion entails a completely distinct vision of the phenomenology surrounding an Operator. In its application as a query of states of physical observables, one may read the norm of these Operators as follow:

For any observable subject to functional representation as *f* or *–f*:

If the square of the derivative of this function added to 1 and square rooted is integrable over a normal interval, then the observable conforms to the specific corresponding Ontology and Pheno-type of a Quanto-Geometric Operator.

The attributes that we have chosen for the nomenclature of these Operators in the above enumerated expressions reflect the type of mathematical composition associated with each of them. Whether one chooses mathematical attributes for the nomenclature of these Operators or ontological attributes (our so-called Primitives) is completely irrelevant from a taxonomic perspective. Only from an analytic standpoint, when the action of an

Operator is to be stressed or elicited, may such a distinction become meaningful. The intrinsic dual nature of these Operators constitutes the foundation of all mathematical-physics abstractions thru the Grand Eigenfunction.

From the form of the above expressions, one may intuitively project what the expression of the Quintessential Trivalent Operators will be. We specifically report them below with no further analytical comment:

Hyper Quantum Operator: $$\langle q_{\vec{1}} \mid s_{\vec{1}} \rangle = \int_0^{2\sigma/\sqrt{7}} \sqrt{1+\left(\frac{d}{ds}\right)^2}\, ds$$

Hyper Quadratic Operator: $$\langle q_{\vec{4}} | s_{\vec{4}} \rangle = \int_{\sigma}^{\sigma\sqrt{8/5}} \sqrt{1+\left(\frac{d}{ds}\right)^2}\, ds$$

Hyper Space Operator: $$\langle q_{\vec{9}} \mid s_{\vec{9}} \rangle = \int_{\sigma/\sqrt{8/5}}^{\infty} \sqrt{1+\left(\frac{d}{ds}\right)^2}\, ds$$

The arrow on top of the indices accompanying the entities within the brackets distinguishes these Hyper Operators from what would be their counter Operators. In the following Chapters of this text, we propose analysis based on the Grand Eigenfunction and these Operators on a number of unresolved questions pertaining to condensed matter physics (atomic and subatomic particles) as well as quantum chemistry.

9.9. Conclusion

It was physicist Murray Gell-Mann who by mid-20th Century called upon Abstract Representation Theory to provide for analytic structures and taxonomic methods for studying and grouping the 30 or so "fundamental" particles of matter then inventoried in the Quantum Theory. While the application of this suit to Quantum physics helped him opportunely predict the $\acute{\eta}$ neta particle, it is not clear from our standpoint whether or not the benefits provisioned by this tool outweigh its analytic drawbacks and limits. Nevertheless, the review conducted above of Abstract Representation Theory in the light of Quanto-Geometric constructs has endeavored to incontestably show that the Quanto-Geometric canon completely subsumes the former. It is in that comparatively superior analytic capacity that the Quanto-Geometric Theory has brought about the stupendous results exposed in Book I, and, going forward, is about to decisively shed light on a number of problems that have thus far defeated the Quantum Theory, as notably displayed in the last three Chapters of this study.

Chapter 10

The sound of One Hand clapping...
Zen Koan

MODELS OF SPACE VACUUM IN MODERN PHYSICS

One of the major caveats of all-time physics is its protracted inability to properly characterize the vacuum and adequately expose its behavior, notwithstanding the commendable effort of General Relativity in that regard. This inability takes its roots in culture which at this stage of human development is profoundly determined by quantum-dominant mentation. It is natural that minds primarily moved by the quantum may not find merits and qualities to the vacuum in any substantial or prolific manner, and that such minds tend to seek quantum contents of sorts within the state of vacuum in order to formulate a characterization. They have failed to comprehend that everything the vacuum is stands for the antithesis of the quantum-scalar, this by definition and by extension.

When physics had segregated from Natural Philosophy in the 18th Century to become experimental science, it has virtually lost all abilities to reflect on the intangible, a paramount requirement to understand the vacuum state. Henceforth physics had turned all questions relating to the physical intangible to philosophy. Experimental observables became reduced to the world of tangibles, i.e., the question of the acceleration of an apple fruit falling from a tree. By contrast, such questions as what the sound caused by a single human hand applauding is like, which is an intangible, became typified as non observables of no experimental value. Such interrogations were left or handed back to Natural Philosophy or in time simply philosophy, as it were in Eastern Zen philosophy. Even the question of a definition of mass or the very scalar was summarily turned to philosophy for its resolution, academic physics considering that its business is principally that of behavior of mass, which of course produces experimental observables, and not its definition. Such arguments were held as late as the 1970's.

Twentieth-century physics has seen a reconsideration of the stance of physical science vis-à-vis philosophy, some of the most prominent physicists, such as Niels Bohr, a figure of immense intellectual authority in the field in his time, formally engaging with Philosophy on the basis of an academically learned philosopher. Most notoriously, Wolfgang Pauli entertained a long-standing cooperation with renown psychologist Carl Gustav Yung in his pursuit of the unraveling of the mysterious number incarnated in the fine structure constant. In that regard, one knows the deep-seated philosophical roots and inter-connections between the science of psychology and philosophy. It is nowadays by a complete reversal of fortune that physicists are calling for theoretical inputs from contemporary philosophers, much beyond epistemological considerations, having clearly understood the significant role played by the intangible vacuum at all orders of the development of matter, from infinitesimal particles to the vast cosmic realms.

10.1. Inconsistent Philosophical Notions of the Vacuum

In the Quantum Theory only the interpretation of the wavefunction rivals in fuzziness the notion of the quantum vacuum. In Quantum Field Theory the vacuum or vacuum state is defined as the quantum state with the lowest possible energy. This definition leaves one to wonder whether the application of the term quantum refers to a generic attribute originating from the name of the theory or to a quantum or scalar content of the vacuum. In any event, the vacuum is **not** a quantum state. It is instead an anti-quantum state, should the term quantum be interpreted as the minimum scalar or scalar density as it is commonly understood.

10.1.1 Energetic Vacuum and Zero-Point State Vacuum

For sure Planck's original quantum of action is the minimum energy density, just as thermodynamic temperature represents an alternate non-quantum form of the energy. One would tend to believe that the vacuum state is an artifact that lies even below the scale of energy characterization since it participates by derivation in the buildup of that physical quantity we call energy, whether potential, kinetic or else. So therefore, it appears to be quite unfitting to use the figure of energy to characterize the state of vacuum, even if we were to hold the vacuum state at energy level 0. However, the contention is that the energy level of the vacuum is non-zero, more often than not, consonant with the so-called vacuum expectation value.

Another common characterization of the vacuum is that it is a zero-point state. Inasmuch as that characterization refers to a non-quantum state, it is exactly correct from the Quanto-Geometric perspective. However, it quickly becomes troublesome when that zero-point state is ranked within the scale of energy. Effectively all quantum vacua have zero-point energies and vacuum fluctuations, according to Peter W. Milonni. The energy of the quantum vacuum, or the zero-point field, is defined as a fluctuation or vacuum fluctuation. In P. W. Milonni's presentation of his vacuum model he further states that "the vacuum is regarded as truly a state with all physical properties equal to zero". This assertion is quite frankly pretty difficult to reconcile with the notion of inner energy harbored by the zero-point vacuum. While we do not at all object to the notion of intrinsic activity of the vacuum, the characterization of that activity as disbursement of any conventional form of energy is somewhat problematic.

10.1.2 Virtual-Particle-Ridden Vacuum

There are in existence several different quantum theories of the vacuum. The above however is sufficient to expose the difficulties that a characterization of the vacuum experiences in modern-day quantum physics. The difficulty essentially lies in the requirement of finding physical properties intrinsic to the vacuum outside of quantum content and behavior. The minds that have tackled this problem thus far have blatantly failed at it, since the consensus nowadays is that the zero-point fluctuations originate from *virtual particles* existent in the fabric of the vacuum. It would be interesting to further call for a definition of the term "virtual" in this context. In other words, in order for the vacuum to be active, some inner quantum objects have got to cause that activity, but because the vacuum is acknowledged to be patently void, the quantum objects must be *virtual!*

This has all the likes of the abstractions that caused A. Einstein to intractably oppose the interpretations of Quantum Mechanics, a theory in his mind misfit to properly and completely describe nature. We will later address the characterization of the vacuum that springs off A. Einstein's General Relativity for its own. Calling on particles, camouflaged under the terminology of virtual, to be the cause of action in the vacuum is at the very least ignoring that the particles themselves are made up of vacuum, arguably even more so if they are virtual! It is the vacuum that characterizes by derivation the particles, not the particles that determine the state of the vacuum from which they derive. There is in existence nothing *virtual* without a referential factor that determines the relationship between the two entities under consideration. Common experience as well as optical analysis have taught us that this referential must have the reflective properties of a mirror in one way of another for the virtualization to occur in the portrayal of a fictitious image. Accordingly, symmetric properties of some sorts would have to be attributed to the vacuum for it to convey phenomenological or kinematic properties of virtual particles, which could not then reside inside of it very symmetric fabric. At the end, the notion of "fluctuations" of the vacuum caused by endogenous virtual particles must be taxed quite unsound philosophically: there are no more virtual than there are real particles inside of the vacuum. I must add that the notion of virtual particles that are nevertheless real, which is firmly established in quantum chromodynamics, is a very dubious notion that is further offensive to plain and simple philosophy if not intellectual honesty.

The above objections equally apply to the notion of vacuum polarization, an effect that hypothetically results from the annihilation of *short-lived virtual* particle/anti-particle pairs residing inside of the QED vacuum. This effect is known to be allegedly responsible for the decrease in intensity of a background electromagnetic field. The contribution of a fermion/anti-fermion pair to the polarization of the vacuum is represented in a loop stream as illustrated below (Fig 9.1).

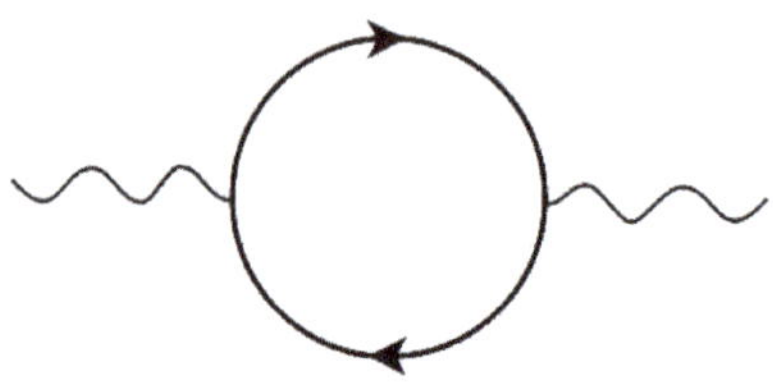

Fig. 9.1 Polarization of the vacuum

Given the significance of the time component in the phenomenology of the polarization effect, experimental physicists have attempted to provide a formal basis for this notion by calling upon the Heisenberg uncertainty principle in the form below:

$$\Delta E \cdot \Delta t \geq \hbar$$

Proponents of the idea theorize that the shorter-lived the particles, the stronger the polarization effect. However, not all physicists agree with the implied idea of a time budget against a physical quantum observable and tend to dismiss the whole idea altogether. Most importantly objections are principally on account that energy and time are not known in the Quantum Theory to form a canonical conjugate. But again, the very idea that particle phenomenology determines the properties of basis vacuum state is plain inconsistent.

Furthermore, it is indeed a very specious argument to conclude, as P. W. Milonni did, that the vacuum is not empty because Quantum Mechanics teaches that the most devoid vacuum chamber must harbor some particles or atoms by virtue of non localization of the particle wavefunction. This would result from the probability density function which grants any of them locality anywhere from the infinites to the inner chamber. If the particle wave function is an obstacle to the realization of perfect vacuum, by no means can dimensional phenomenology associated with particles become part of the definition of the vacuum state itself. All the more so when it is the very vacuum component of the particle (its wavefunction) that interferes with the chamber's state of vacuum.

In that same vein, Quantum Mechanics asserts, as previously reported in Book I, that the vacuum is not at all empty but contains fleeting electromagnetic waves as well as particles which constantly pop into and out of existence. This view further expands the idea of virtual particles within or at the other side of the vacuum. From a philosophical vantage, it requires though a narrow view of existence to exclude from the realm of reality all that exists virtually or potentially. Furthermore, the idea that *alter* objects erupting ex abrupto into a world of things contrary from nowhere seems to contradict the very laws of conservation, as much as the idea of things ceasing to exist ipso facto, just vanishing in an ex-nihilo fashion! In his 2008 paper *Creating the Physical World ex nihilo*, Luciano Boi reflects:

"More fundamentally, Quantum Mechanics predicts ... a correction to the energy called zero-point energy [which] consists of energies of virtual particles that have a brief existence."

Translating: non-matter particles come from somewhere beyond our physical vacuum (or from nowhere), interact with the physical vacuum ever so briefly in an action that causes them to become physically existent in our realm, and then suddenly disappear. To fill the obvious lacks of this picture, some physicists, such as Joseph Silk, cited in the previous volume, call instead for the intervention of no single particles, but virtual pairs of particles, in order to make grounds for their subsequent spontaneous mutual annihilation. The least one can say, from a critical philosophical viewpoint, is that this theory is still immature and has a lot more explaining to do.

10.1.3 QED Vacuum or Electromagnetic Vacuum

The Theory of Quantum Electro-Dynamics (QED) straightforwardly builds upon unified electro-magnetism by James C. Maxwell, in order to make inroads into the interpretation of the vacuum. The theory capitalizes on the fact that electromagnetic waves are the only known ones to propagate thru the vacuum, and thus asserts a parallelism or superposition of sorts between electromagnetic key defining properties and the properties of the vacuum. It therefore assigns electric permittivity as well as magnetic permeability to the vacuum , those values being slightly different than classical ε_0 and μ_0. The properties of QED vacuum are set to be analogous to the Quantum Mechanical properties of an (electromagnetic) harmonic oscillator. Accordingly, that construction endows the vacuum with a ground state with an "expectation value" for operators in its field. In other words, the QED vacuum is to a large extent a form of electromagnetic field very close to the classical electromagnetic radiation field. However, problems arise in the interpretation of the interaction of elementary particles with the field. The Casimir effect, an attraction force experimentally observed between uncharged conductive plates, is oftentimes proposed as a proof of the electromagnetic characterization of QED vacuum. Nevertheless, experimental physicists of the field acknowledge that the electromagnetic properties of QED vacuum alone do not and cannot account for the measured intensity of the Casimir force. Other contributions inherent to the interacting particles are found to be necessary, leaving the electromagnetic effects of the vacuum on a scale that is disappointedly too miniscule to experimentalists' expectations.

10.1.4 QCD Vacuum or Scalar Vacuum

The most blatantly objectionable characterization of the vacuum comes from the theory of Quantum Chromodynamics, which squarely places a scalar inside of the very fabric of the vacuum as a property of the vacuum itself. Although different models of the vacuum exist in QCD, they all portend to the scalar visualization of the vacuum to a very large degree. The Savvidy model of the vacuum contends that at its ground level QCD vacuum or the vacuum of sub-atomic particles cannot be understood as conventional Fock vacuum that is devoid of particles and fields. The particles that the model refers to are scalar points or knots that cause the vacuum at that scale to be unstable in its ground state and to express itself as scalar *condensates*. In other words, the state of the vacuum is a scalar condensation as opposed to a basis void state that could give rise to a tensored field.

The instanton would be one such scalar of inner QCD vacuum, as contended in the instanton ensemble model. Non-fundamental string, posited in string or dual resonance models, is another scalar variety of the topology of QCD vacuum. Even superconductor tubes come to be part of the fabric of QCD vacuum under a so-called superconducting model.

Some of these models were short-lived, as was the instanton model, while others among the survivors look more or less promising than their competitors. When it comes to the still active ones, it remains universally understood that they are far from maturity, given that none is able to facilitate a comprehensive description of the hadron spectrum or provide insight into confinement. In conclusion, it cannot be overstressed that ill-positioned abstractions and philosophically unsound theorization can never lead to any useful outcome with the likes of a solution. The vacuum can never be described as a scalar entity, however overwhelmed its expression and behavior might be at the sub-particle scale. What's more, an explanation to the factual scant expression of void in that vacuum remains a due and would have to be inherently derived from or justified by the theory.

10.1.5 The Higgs Scalar Field as Energy of the Vacuum

The momentum acquired by the Higgs research at the Large Hadron Collider in Geneva has apparently led some to speculate that the Higgs scalar field might be the correct representation of the vacuum energy density. The Higgs field in that scenario would be the original vacuum state from which all other forms of the vacuum originated by symmetry-breaking at phase transitions giving rise to the development of the vacuum. One can see that this scalar view of the vacuum is directed at seeking an interpretation of the vacuum structure, not even at QCD or sub-particle level scale, but at the macroscopic cosmic scale. It further attempts thereby to pave the way for a possible Grand Unified Theory. Problem with this view: the computed energy level of the vacuum which puts it extremely close to zero is not in a match at all with the Higgs field energy which is rated at multiple whole orders of magnitude higher. Moreover, these values each appear to be at the opposite end of the scale!

Hence the Higgs field does not seem to promise anything other than the possible genesis of mass in the universe, which is another controversial view from our perspective. Reconciling the energy scale of the Higgs field with that of basis vacuum state in the 3-dimensional universe remains an unfulfilled quest, with no resolution in sight at the moment.

10.1.6 Vacuum Energy Density and Cosmological Constant

Lastly, in the context of characterization of the vacuum, the idea of a cosmological constant, which has been floating around in physics ever since A. Einstein introduced it in the 1930's but shortly thereafter demoted it, has been gaining momentum once again after the discovery in 1998 that the universe is not only expanding but accelerating in its expansion. The notion of a cosmological constant is compatible with that of a static universe as the term "constant" might imply. It would be a non-null number, the value of Λ in A. Einstein's Relativity tensor equations. Einstein summarily demoted that notion, after the Hubble Telescope showed via the Hubble law based on electromagnetic red shift that the Universe, beyond the static stars of the Milky Way, is instead in expansion. He then assumed that ρ, the energy density of the vacuum was 0, which would make Λ equal 0.

However, physicists kept musing over this question for decades, promoting the constant just to subsequently demote it, as other explanations to observational results seemed more plausible. In the decade of the 1990's they started to revive the notion once again by attempting to capitalize on the fact that only the contribution of the electromagnetic force was taken into account in A. Einstein's equation. The hope was that perhaps the intervention of other yet-to-be discovered action parameter within the fabric of the vacuum would contribute to cancel out the value of the other terms to exactly 0. The discovery of the accelerating expansion of the Universe unfortunately shattered those hopes because it implied that Λ must be slightly non-zero, so that we are back to A. Einstein's original point of a non-zero value for this constant. The accelerating expansion further means that physicists are left with the vexing problem of accounting for the fact that positively contributing factors and negatively contributing factors cancel out only to 120 digits of the resulting number. In other words, the value of Ω would be a decimal 0 number up to the 120th decimal place. At the 121st decimal place, an integer appears! The number is of the form:

$$1.000\ldots000xxxx\ldots xxxx\ldots \times 10^{-120}$$

where x is a positive integer.

It so happened that physicist Steven Weinberg had argued in 1987 from basic Quantum Mechanics principles that the cosmological constant must be 0 to within one part in about 10^{120}. The vacuum energy density calculated on the basis of electromagnetic vacuum casts *10^{93} grams per cc* as an equivalent measure, yet a very large value. The calculated average mass density of the universe as a whole is just about *10^{-28} grams per cc*. There can never be a bigger disparity between prediction and fact in science, which prompted Stephen Hawking to tax this result as the most spectacular failure of any theory in the history of physics. In the words of physicist Leonard Susskind, the discovery of the acceleration of the universe and its implications constitutes a "stunning reversal of fortunes", while others acknowledge that it is literally causing a seismic turmoil in the very foundation of the entire field of theoretical physics. Physicist Sean Caroll further elaborates:

"To a good approximation, we believe that the vacuum is the same everywhere in the universe, so the vacuum energy density is a universal number which we call the cosmological constant. (More precisely, the conventionally defined cosmological constant lambda is proportional to the vacuum energy density Rho lambda.)... There are many effects which contribute to the total vacuum energy, including the potential energy of scalar fields and the energy in "vacuum fluctuations" as predicted by Quantum Mechanics, as well as any fundamental cosmological constant... Understanding the smallness of the cosmological constant is a primary goal of string theory and other approaches to quantum gravity."

To the end the smallness of this number puts the physics community in front of a conundrum of purely mathematical nature the likes of which it has probably never seen before. In the words of an exasperated Lee Smolin, this number is "ridiculously small"! Be that as it may however, the community believes that an outstanding requirement imposed on all potential Theories of Everything is that any legitimate and successful characterization of basis vacuum state of their creation must explain the physically observed cosmological constant. The Quanto-Geometric Theory has already made strides in that direction, as evidenced in Book I of this publication, with more contributions to come in the present one.

The above account was the state of affairs until the end of 2014, when in early 2015 the BICEP2 satellite program, commissioned to confirm the acceleration of the Universe, reduced all hopes to dust as it was found and announced that that all pieces of evidence collected thus far were actually due to galactic dust! The mood could not become more somber on any matter in the community!

10.2. Cosmological Constant: a Long-Running Convoluted Problem-Solution

The main reason why the storyline of the cosmological constant is so convoluted and the concept of no apparent good auspices is because the characterization of the nature and properties of the vacuum has been historically profoundly flawed in all of modern physics. We have attempted to highlight in the previous account some of the most striking of these conceptual flaws. The most important caveat confronted by the physics community in that regard is the overwhelming misunderstanding of the many different symmetric disciplines of 3-dimensional vacuum. This conclusion does not even spare the General Theory of Relativity, regarded perhaps as the most accomplished theory of modern physics.

To that effect we are going to briefly revisit Einstein's tensor or field equations (EFE) which provide the foundation for General Relativity. Below its summarized form:

$$R_{\mu\nu} - \frac{1}{2} R g_{\mu\nu} + \Lambda g_{\mu\nu} = \frac{8\pi G}{c^4} T_{\mu\nu} \quad (2\text{-}1)$$

where $R_{\mu\nu}$ is the Ricci curvature tensor, R the scalar curvature, $g_{\mu\nu}$ the metric tensor, Λ the cosmological constant, G the gravitational constant, $T\mu\nu$ the stress-energy tensor and c the speed of light. The left-side term of the equation is considered a description of the curvilinear fabric of spacetime, and the right-side term the matter-energy content of spacetime. Hence the equation relates the structure of spacetime to its energy content. The time component appears in the full expression of each of these terms which are in fact differential equations with time as the independent variable. The EFE are thus formulated in the classical context of a 4-dimensional physical background. In their full expression the EFE are a system of 10 coupled, non-linear, hyperbolic-elliptic partial differential equations. For this brief analysis we will only retain the summarized form in presented in *Eq. 2-1*.

10.2.1 Fabric of Space Modeled by Tensors

Relativity must enjoy the credit of the first Theory engaging a working description of the fabric of physical space in and of itself, in the exact terms stated by E. Schrödinger in his "Fabric of Spacetime" publication that we quoted at end of Book I. Despite its very audacious goal, and its revolutionary contributions to the understanding of physical matter dynamics, this attempt was nowhere near perfect. Since the very limit of the Theory was evident in the terms of the EFE themselves, as we shall later explain. Nonetheless what this framework has realized is an application of important concepts in differential geometry surrounding tensors and manifolds to a description of the dynamical behavior of real space in a way that is consistent with some aspects of natural scalar-space relationship in the physical universe.

It is important to stress that the EFE only apply to spatial phenomenology in the cosmic heavens, namely Shell 7 up. They are not an ontological description of generic physical space, nor do they account for what we know to be the many different regimes of spatial symmetry that we have otherwise dubbed Shells of matter. This question has remained an open question, not even answered by the Quantum Theory, and which contemporary Theoreticians are still grappling with under the quantum gravity approach. If the Relativity Theory had higher genericity, it would principally pose the question of reference frames in terms of relationships between the seminal constant-based Shells of matter and in that context the cosmological constant figure would acquire a firm and assertive characterization.

We have stated in previous discussions that the vacuum in the 3-D universe exhibits 9 basic symmetries at the heart of the formation of its Shells. Intra- and extra-galactic vacuum is not the same as solar system vacuum, and not at all the same as intra- and extra-particle vacuum, to just name these three instances of the vacuum. At each Shell of matter, the vacuum experiences a particular level of entropy or symmetric discipline which when coupled with the basis scalar of the specific Shell produces the physical constants that govern its particular physical structure and dynamical behavior overall.

So therefore: **no**, the vacuum conceived as being the same everywhere is not a "good approximation" to its authentic nature. No, the vacuum cannot be regarded as "a state with all physical properties equal to zero". There is a gradient for the symmetric discipline of the vacuum which takes different values for different Shells, as stated in Book I in a slightly different terminology. The transition or phase event that we have traditionally called in physics symmetry-breaking is principally what happens when the symmetry gradient of the vacuum takes special values along that scaling line of the symmetric discipline (Vol. I, Fig. 4.4).

10.2.2 Problematic Stress-Energy Tensor

Notwithstanding the correlation established by the EFE between the inner geometry of cosmic vacuum and its energy content, the symmetric discipline of the vacuum is a not an order of energy, nor is it an order of energy density. That order is more a state property of basis 3-D vacuum than it is related to energy, because energy is what it may become under a certain transformation, specifically the thermodynamic temperature form of energy. That order of the void lies below the field concept as well, as it is understood both in Relativity and Quantum Field Theory. So it is because that basis symmetric state of the vacuum strictly speaking becomes a physical 3-D field under intervention or transformation by an overarching Hyper Operator of one of three distinct possible norms. The field development occurs as an epi-phenomenon by the basis symmetric state which may be captured by functional differentiation analysis. In other words, before the gestalt vacuum can become or derive into a field, it is endowed with a transverse symmetric discipline beholding a particular gradient. This characterization springs off the Grand Mappa Mundi function.

The principal objective of General Relativity is to describe gravitational attraction between two bodies not as a force of inertial origin but as a result of the geometric structure of spacetime and its dynamical behavior. In that context spacetime does not mediate the influence between two massive objects but the relationship between a massive object and its accompanying surround spatial spread or field is what acquires significance. Under that approach the principle of *covariance* (equivalent to what we have called *quanto-*

geometric articulation) acquires utmost importance in laying down the Theory. That is why it was mathematically strictly expressed in terms of tensors that give the measure of the different forms of stress experienced by space in the mass-space correlation. It is only by virtue of this seminal compliance with natural canon that the Theory has been able to give account of a number of questions despite several serious flaws in perspective and mathematical formulation.

Effectively the two terms employed in the contracted expression for the metric tensor and for the stress-energy tensor have all but caused continuing headache to A. Einstein in the course of developing the Theory and were never properly resolved at the end. The generalization of the metric tensor to all vacua is a mischaracterization that acknowledges the vacuum to be the same everywhere. However, Quanto-Geometry prescribes that there is not one cosmic vacuum since, by our canon, there are at least 4 different cosmic vacua as previously indicated, each with its own physical constant or "cosmological" constant in the vague terms thus far employed in physics for the higher physical heavens. In the next few years after publishing the Theory of Relativity A. Einstein introduced a modulation to the metric tensor of spacetime with the associated figure of a cosmological constant Λ, in order to account for the static universe known at the time. The stars that experimental physics was then able to study all appeared fixed in the sky, the notion of a galactic entity was unknown, the universe hence appeared to be static.

Now even on that sole account, the concept would still be faulty because it would still inappropriately conflate planetary space, solar-system space, and interstellar space, each being laid under the regency of their own physical constants as we have reported in the Chart of coupling constants in Book I. I repeat and repeat: the farthest origin of gravitation is not something the likes of generic mass, Newton Gravitational G constant is borne out very specifically by Shell3 of matter, the Atom Shell.

Without Λ the EFE would allegedly describe at the very least an unstable universe, if not an expanding or a contracting one. Much later, the Hubble telescope had accumulated enough data to pave the way for a deeper understanding of the composition and structure of the cosmos in terms of larger-scale and self-dynamical objects. And so, catering to those facts, A. Einstein dismissed the concept of cosmological constant. He famously called it the greatest mistake of his life because he ostensibly realized that the original abstraction he started out with, deprived as it was from a notion of cosmological constant, was correct and was definitely ground for another revolutionary prediction about a dynamical cosmos.

But this trellis was a difficult one to thread, because a cosmological constant figure does not necessarily result in all-static cosmic objects, depending on one's visualization of the cosmos and what particular strain of solutions one cares to derive from the tensor equations under their very broad scope of solutions. At any rate, from our standpoint, the generalization implied in the figure is all but ill-fated. Modern-day cosmological physicists, however, have given a second lease on life to the notion when confronted with the evidence of not only an expanding universe but a universe in accelerated expansion.

Cosmologists have attempted to make the metric tensor with its adjoining cosmological constant in the contracted EFE an energy attribute by association with the stress-energy tensor. By doing so however, they have compounded a problem that was already quite serious. Effectively the stress-energy tensor is another pernicious imprecision of the EFE to say the least. A. Einstein himself could not precisely define the physical meaning of this term. He admitted this much:

"... in a more complete analysis, the energy tensor can be regarded only as a provisional means of representing matter. In reality, matter consists of electrically charged particles... It is only the circumstance that we have no sufficient knowledge of the electromagnetic field of concentrated charges that compels us, provisionally, to leave undetermined in presenting the theory, the true form of this tensor... The right-hand side [in the contracted general tensor equation] is a formal condensation of all things whose comprehension in the sense of a field theory is still problematic. Not for a moment... did I doubt that this formulation was merely a makeshift in order to give the general principle of relativity a preliminary closed-form expression. For it was essentially no more than a theory of the gravitational field, which was isolated somewhat artificially from a total field of as yet unknown structure."

No one to this day has unfortunately made good on A. Einstein's wishes to bring further precise physical meaning to the stress-energy tensor. The above assertion stands for unmitigated acknowledgment that this term of the EFE is, not to put it in crude vernacular terms, hocus pocus. A. Einstein was all too aware of his mass-energy equivalence principle for not disavowing this term for lack of a better visualization. He understood that there is not and cannot be any order of mass in the fabric of the vacuum, therefore no order of energy either. What modern-day physicists have done in equating the cosmological constant with vacuum energy density is not in keeping with this clarity. They have split the stress-energy tensor into two terms, one representing matter and one other representing the vacuum, according to the expression:

$$T_{\mu\nu} = T_{\mu\nu}^{matter} + T_{\mu\nu}^{vacuum}$$

They have further found that the vacuum can be modeled after a perfect fluid for which the well-known equation of state is the pressure over density ratio:

$$\omega = \frac{p}{\rho}$$

Assimilating the state of that perfect fluid to the stress-energy tensor of the vacuum:

$$T_{\mu\nu}^{vacuum} = -\frac{p}{\rho}$$

So that after several algebraic manipulations the general contracted EFE becomes:

$$R_{\mu\nu} - \frac{1}{2} R g_{\mu\nu} + \Lambda g_{\mu\nu} = 8\pi G T_{\mu\nu} \left(T_{\mu\nu}^{matter} - \rho_{vacuum} g_{\mu\nu} \right)$$

$$\text{and } \rho^{vacuum} = \frac{\Lambda}{8\pi G}.$$

While this effort has the merit of an intent, and the comparison with behavior of perfect fluids a consonant insight, it has yet to contribute in clarifying anything further at all and has in fact contributed instead to further obscure the matter. This approach, which is unable to visualize little more than an order of energy in the vacuum, has probably miscarried the development of the science of cosmology in no small part. Among other mis-portrayals, it has failed to understand that the relationship between the vacuum and mass is not always a gravitational relationship, and that matter-of-factly in the higher physical heavens several forces of a different nature underpin the dynamics therein in development. One ought to comprehend that the dynamical regimes of the higher physical heavens, otherwise called the large-scale cosmological structures, in the 3-D universe, well in the terms of specific physical coupling constants that define the correlation between their individual scalar objects and associated spatial spreads within their own Shell of matter. There is no hidden dark matter or dark energy within the fabric of space anywhere that is responsible for pulling celestial objects or group of objects away from one another against natural gravitational attraction.

10.3. Conclusion

The contentious matter that we have discussed above, along with a number of other problems confronted by cosmology today, will not be resolved within the framework of the EFE because of an important measure of conceptual inadequacy in the latter as well as imprecise mathematical formulation, specifically in relation to the metric tensor and the stress-energy tensor. Nowadays, there is manifest awareness in most academic spheres of physics that there may very well be no additional energy density component to spacetime but rather that the EFE themselves are in dire need of revision. This awareness comes as a result of the sheer number of problems accumulated in the field of cosmology, notably the so-called coincidence problem, the flatness problem, the galaxy formation problem, the anti-matter problem… not to mention the long unresolved problem of compatibility of General Relativity with Quantum Mechanics with the aims of recovering unity in physical law. Despite the distinct success scored by the field just a few months prior to this writing in original cast with the discovery of the long predicted gravitational waves, we agree with the prevailing sentiment that these problems call for a revision of the EFE and most definitely a new theoretical framework able to supersede them. While A. Einstein abstract insights were uncommonly prolific, his hope for greater simplicity in the granular mathematical formulation of the EFE, which he even turned into a rule, and in particular his wishes for greater elucidation of the role of energy relative to the dynamical behavior of spacetime, could not be met by subsequent generations of physicists, within the framework of General Relativity. Time to let go of space-*time* and embrace the symmetries of scalar-space correlations Quanto-Geometrically framed.

Chapter 11

Nevertheless, for the sake of his servant David,
the Lord was not willing to destroy Judah.
He had promised to maintain a lamp
for David and his descendants forever.
Star of David – 2 Kings 8-19

QUANTO-GEOMETRIC INTERNAL ARCHITECTURE OF SPACE VACUUM

At this stage of our study, we propose delving into the inner architecture of space vacuum that naturally springs from the Quanto-Geometric constructs thus far formulated. We will first further discuss the visualization of inner fabric of coordinate space in Quantum Field Theory and General Relativity and will lay down our own vision thereafter. The question of the seminal fabric of space is one of the most important questions posed to our global culture and our science altogether. There is no holding back credits to all those who have tackled this issue from the standpoint of the intangible and the formal profiling of possible structures intrinsic to the intangible, whether in science or in culture. There is no doubt that mastery in this architecture constitutes the key of the gate to the higher heavens, springboard of an imminent historic transmutation of human culture.

11.1. From Hilbert Space to Momentum Space

In fundamental physics, the Quantum Theory is inarguably the science with the least insight into the seminal architecture of physical space. The reasons for that deficiency, which we have substantially analyzed in the previous Chapter, are plentiful. The contentious and unsettled subject of the physical interpretation of the wavefunction stands for a historic testimony to that constitutional inability.

Not all mathematical tools, as intricate and resourceful as they may appear, necessarily result in blissful physical application. We believe that the idea of raising an intangible such as abstract mathematical spaces, to model another intangible such as physical space, will pose representation problems that are doomed to conspire against clear analytical interpretation, which at the end is the aim of the whole exercise. To make matters even worse, such an enterprise may lead to the compulsion to unfathomably virtualize reality itself, as we shall see in our critical review of the Quantum Theory next to come in this study.

Following that trend, the diverse actors in the Quantum Theory as a whole have chosen the mathematics of Hilbert space to primarily formulate Quantum Field Theory and to partially reformulate Quantum Mechanics. As a reminder, a Hilbert space, which becomes *momentum* space in its physical application, is defined as an abstract inner-product functional space with orthogonal basis, and with the expression:

$$\langle f \mid g \rangle = \int_b^a f(x) \cdot g(x) \cdot dx \text{, where } g(x) \text{ is a conjugate of } f(x)$$

The mathematics of Hilbert space is knowingly quite extensive and highly abstract, with multiple layers of compounded abstractions, all of which are brought to bear on the physical observables they attempt to model and interpret. Further, this approach is set to only unravel population behavior, given that it requires an orthonormal basis of functions, which leaves it with the crippling handicap of being unable to interpret the wavefunction of a single particle or a singlet. Its reach is thus limited to particle topologies.

The model of Hilbert space applies in the matrixial reformulation of Quantum Mechanics as undertaken by W. Heisenberg, which, many will agree, was a perfectly facultative exercise. E. Schrodinger, by using a direct Euclidean space formulation, had already resolved the states of bound electrons and elucidated the structure of the Mendeleev Periodic Table of elements. This development was achieved based on the construct of position space or real space as the background of a moving object or particle. Position space is the set of all position vectors r in real 3-D space. The position vector defines a point in space, in real coordinate space. E. Schrödinger had demonstrated that the motion of the particle was not traceable in time, which in our judgement was the second revolution of early 20th century physics.

The construct of momentum space is based on a visualization of the inverse of r and defining r as a wavelength, such that:

$$k = \frac{1}{r}.$$

By analogy with the classical relationship between frequency and wavelength:

$$f = \frac{1}{\lambda}$$

k becomes an equivalent to the frequency figure and takes the identity of a wavevector. Momentum space thus becomes an abstract frequency space or wave-vector space that is related to the original real or coordinate space by a transformation. This transformation is equivalent to a Fourier transform because it takes us from the Euclidian *time* domain (or so the proponents believe) to the abstract frequency domain. It is worth noting that in general the reason for utilizing Fourier transforms, particularly in electrical circuit analysis, is because it saves us from the complexities of exponential functions in the time domain by reformulating the problem in simple algebraic terms of *s*-space or frequency domain.

Momentum space or *k*-space (also understandably denoted reciprocal space) further uses the De Broglie wave-particle principle that makes the momentum of a particle a quantity equal to its wavelength times the Planck constant. The momentum space pivotal identity is expressed below:

$$p = \hbar k$$

where ***k*** is the wavevector (or wavenumber), ***p*** the momentum, and $\hbar$ the reduced Planck constant. At this point one may reasonably wonder what the purpose of this development is in taking us away from perfectly good Euclidean space into a world of abstract functional space. Where is the inner-structure proposal, one may ask. There is none indeed, because momentum space is not a model, but a tool that is only going to be useful when it comes to the analysis of crystals and the behavior of electrons inside a crystal lattice, aside of course from alternatively reformulating wave mechanics of the bound electron. And for the task of crystal lattice analysis, *p* and *k* have to be surrogated into defining quantities attached to the crystal as a topology of unit cells and no longer to an individual particle. Even on that account, the tool cannot describe what happens to the crystal *k*-space wavefunction within a single unit cell but only how the wave moves from one unit cell to the next.

We are not aware of any other seminal model for the void of space in Quantum Theory beyond the visualization of *k*-space. Modern-day models of Loop Quantum Gravity and String Theory primarily represent attempts to bridge the Quantum Theory and Relativity thru the development of models for the seminal fabric of spacetime. Since they have not matured yet by their own standards and only hold thus far the merits of a continued attempt, we will not discuss them in this study.

11.2. From Mathematical Tensors to Einstein Manifold

When Albert Einstein intently made the choice of modeling space with mathematical tensors, which are essentially represented by differential equations, he had started to crystallize the idea that the void has internal geometric structure. Since the tensor equally measures the correlation between a spread of spatial void and a scalar within that void, in terms of stress imposed on the former by the latter, one may at first conclude that the geometric structure of the void relates to the scalar imposing stress to its vacuum habitat, so to speak. One may further assert that when the scalar is far away from a vacuum spread or field of consideration, the tensor would then vanish, leaving the referred spatial void or field flat or devoid of curvature. However, A. Einstein had posed that the metric-gravitational tensor does not vanish with "empty" space or non-correlated space. In other words, in the absence of the Ricci tensor, or with $R_{\mu\nu} = 0$, which implies as well that the stress-energy tensor $T_{\mu\nu} = 0$, empty void still holds a ground or basis state of curvature, expressed by the so-called Einstein tensor:

$$G_{\mu\upsilon} = R_{\mu\upsilon} - \frac{1}{2} R g_{\mu\upsilon}$$

At that point, General Relativity is no longer simply a gravitational theory or a gravitational interaction theory but a theory of the structure of *space-time* as coined by A. Einstein coined after Minkowski. That is why the EFE in that particular circumstance of the vanishing of the two above-mentioned tensors are considered *vacuum field equations.* From the standpoint of our quest for the space model that springs from General Relativity, this brings us to the representation of the Einstein manifold or the manifold (Fig. 11.1) that corresponds to basis or flat Minkowski spacetime within the Theory.

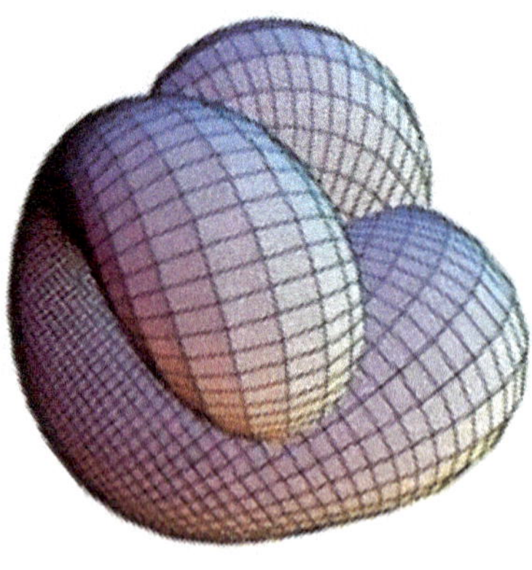

Fig. 11.1 An example of the construct of a manifold in differential geometry

We must recognize that the Einstein tensor ought to forthrightly take us to the notion of a geometrical constant, incarnated in a manifold of sorts, that sets the unit building block of the space component of space-time, if spacetime is geometrical in any way. One cannot imagine spacetime to be as homogenous as it is without that identity. Furthermore, the unit geodesics, a construction that A. Einstein coined in order to explain ensuing gravitational motion, cannot exist without the regularity of that seminal ground-state symmetry, even though he ultimately imputed the cause of motion in the Theory to irregularities or non-linearities of spacetime with the geodesic equations.

The minimal view offered in mathematical-physics science for an Einstein manifold is the requirement of constant sectional curvature *across the surface*. In that sense Euclidean space or a Euclidean portion of the void is an Einstein manifold, as well as a sphere in any number of dimensions one cares to represent or imagine it beyond 3. If this is the hierarchy in which curvature is graded, then the view of a manifold in Relativity falls short of a true and thorough description of the folds of the void of space, both in its basis state and excited state. We have shown well what this hierarchy looks like in the qualified symmetries that have led us to the Quanto-Geometric Grand Eigenfunction.

We fully understand that the domain of General Relativity is the cosmological heavens. The question of a cosmological minimal manifold, building block of that sector of the Void, does not seem to have even been perceived by the Theory. We understand as well that the necessity of an account for motion, ultimately rendered as the movement of the geodesics, contributed to make even more difficult the visualization of a *unit of action* for the vacuum in the Theory. That is the reason why A. Einstein formulated the Theory in two folds as it were, one for the geometrical structure of spacetime and one other for motion in spacetime. The most important obstacle that the Theory of General Relativity was unable to abate is the vision of time as a dimensional variable in the cosmological background. There lies another key to the limits of the Theory, as regards to its inability to uncover the origin of the metric, much coveted by A. Einstein, the identity of the stress-energy tensor with its evoked association with the Gravitational Constant, and very importantly its irreducible incompatibility with Quantum Mechanics a-posteriori.

Here is what we know:

With the derivation, from pure mathematical-physics abstraction, of the metric figure that quantifies the speed of light c, we have shown the true dimension of the metric tensor and the order of geo-metrical symmetry conducive to the conformation of this tensor.

By way of a re-formulation, the primeval symmetry-breaking sequence and the Grand Eigenfunction give us the following identities that we have well otherwise studied in Book I:

$$\text{for } n=3,\ s_3=\pm\frac{2\sigma}{\sqrt{7}}$$

$$\text{and } Q_3(s)=\frac{e^{-2/7}}{\sigma\sqrt{2\pi}}=\frac{0.299796065\cdot 10^{-9}}{\sigma}$$

$$\text{or } q_3\times\sigma=0.29979606510^{-9}$$

$$\text{or } q_3=0.29979606510^{-9}\sigma^{-1}$$

In summarized form (bis):

$$\boxed{\begin{aligned} n_3 &= 3 = D \\ s_3^{-1} &= 1.322875656\cdot\sigma^{-1} \\ q_3 &= 0.299796065\cdot 10^{-9}\sigma^{-1} \end{aligned}}$$

These three identities represent the highest characterization for 3-dimensional space. They crystallize in order of representation: the dimensional modulus of our physical universe *(D)*, the true dimension of reciprocal space, which is not an abstract space, and the seminal figure of the metric tensor in its quantification of both the distance metric and the kinematic metric. These figures lie far above the Relativity tensors and Quantum Theory's momentum space. We will show next what the intimate structure of the Void ensuing from this vision turns out to be. But first, a brief revisit to the old Devil of time.

11.3. Human Time: a Hindrance to Space and Motion

The notion of time only comes to translation in physical reality thru scaling. From abstract continuum it has to become a scaled unit for a physical incarnation. The only way to measure time is thru a transform of space motion. The abstract continuum has to be broken into stops or scaling units so that it can be measured, which is its sole purpose, that of measuring, that is. Duration, on that physical timeline or time circle, is the distance, linear or angular, between one stop or scale unit and another. Consequently, duration or time lapse is a function of a distance.

The only thing in the distance metric that may contribute to the notion of longevity incarnated in the concept of time lapse or duration is some kind of self-dynamism of the unit distance or metric unit of space. Space has to be auto-dynamic for any time lapse to ever make sense.

The inner variance of space, which obviously cannot be measured in time, is the only sensible anchor ever for time lapse. This space auto-dynamism of sorts is asserted in what we have called the propagation or deployment of space. Said act of deployment of space is universal and may be portrayed as a field of an infinite number of self-propagating lines in all directions. Best way to represent it is to liken it to a pressure field akin to earth's atmosphere where every point of the field receives pressure from all directions. The main difference is that in the case of space the representative vector is oriented outward because the origin of the push is internal. That constitutes the very act of split or deploy of space. Space in its texture is thus the inverse of a pressure field.

11.4. Unit of Action in the Fabric of Space

Space is propagating in all directions at speed c. In fact, c is the propagation speed of the void of space. In first approximation, the representation of a unit bundle of 3-dimensional space is a *cube of action* with an infinite number of diagonals crossing at the center point of the cube. The propagation originates from the center point and spans the integral of all paths or directions. These diagonals represent a trajectory, the trajectory of an intangible element, the void of space.

It is important to recall these two statements that we have made in Book I about the origin of 3-Dimensional space:

Space in U_3 is characterized as a three-dimensional form of space owing to its origin of parabolic symmetry. The embedded parabolic conic section within the geometry of the Quanto-Geometric function is its native realm. Within the work of creation of U_3, the Uni-Form Operator acts on the dimensional configuration of U_3 space in order to further modulate it, notwithstanding the constancy of its dimensional modulus, which remains inalterably 3.

Uni-Form operation by the Hyper Quantum Operator, whose signature is type 1 symmetry or mono-centric symmetry as studied in Quanto-Geometric Number Theory, compels the three dimensions of U_3 space to exhibit mono-centric symmetry, a constraint whereby all directions must exhibit same unit length, and all angles between them must lay equal to one another. The only possible configuration that responds to this constraint is U_3 space cast from an ortho-normed Euclidean referential, as shown in Fig. 4.3 (Book I).

The geometry of the fabric of 3-D space is a result of space-scalar articulation in the 3-dimensional domain of the Grand Mappa Mundi universe. Such a geometry materializes the stress imposed on 3-D space by the 3-D scalar, which is the total scalar mass of the 3-dimensional realm. The Hyper Quantum Operator further homogenizes the 3 dimensions by equalizing them to one another. A comparison, which is admittedly not always reason, will nevertheless help portray the entity: a matrixial line of 3-D space behaves like a river flowing at speed *c*. If by some special means we were able to freeze a small bit of water in the flow, the ice bit would flow along as a tensor imposed on a bit of the flowing water. That tensor may be equally construed as a perturbation of a bit of the water in the flow. If the perturbation or tensor is light enough (i.e., clear ice), it will propagate at the same propagation rate inherent to the flow. But if the perturbation or tensor is sufficiently strong or compact (i.e., hard ice), it will resist the flow and propagate at its own distinct rate. One may compare the first case to light in all its wavelengths, and the second to the motion permeable to most other objects in 3D. Note well that the above terminology is tantamount to an attempt at elucidating our vision by utilizing terms that have become common in the literature. However, make no mistake or confusion, this vision is in no way equivalent to QFT's *perturbation theory or* Relativity's *tensor framework!*

Therefore, in order to describe the fabric of space, one must pose its propagation speed *c* as the invariant reference frame and make all other motion deployed on that background not just relative to *c* but principally *degenerate* from c. In fact, one should not describe any other motion as relative to *c* because space is not discontinuous in its fabric despite its Shelling. A grander and more accurate view of motion and transformations is that Space in its ubiquity is one and that the Shelling it creates in order to inform and conform physical objects constitutes forms of regionalization of the intangible entity.

The above leads us to a clear understanding of the reason why no 3-D object may move at a speed faster than *c* in the 3-dimensional universe: the source of all motion is space itself. There is therefore no Minkowski *spacetime.* In the oblivious notion of time, all values of speed are or must be possible along a fictitious scaling timeline, a hypothetical dimension in and of itself. However, there is no time in the dimensional background (or foreground) but only infinite and unbound space of which every object is part of geometrically and dynamically by derivation. All translational and periodic motion experienced by or confined within objects are artifacts of universal space. The kinetic qualities of the fabric of universal space represent the back drop that sets the limits to the kinetic experience of all objects.

On may wish to critically tax our kinematic representation of the void of space as unphysical. And we would have to wholeheartedly agree that the characterization would be correct indeed, so long as the criticism means to oppose to our view that of a momentum behavior of sorts of the void of space (in the traditional meaning of momentum, a product of mass and velocity) or perhaps Relativity's view of geodesics.

To the criticism from a Quantum vantage, we must say that there is no scalar entity or mass, whether positive or negative, within the void of space. One may wish to utilize the concept of imaginary mass to consign the kinematic experience of the void of space. However, one must be careful to comprehend that imaginary mass is not quite the same as negative mass and that the suitable methods to render the experience of a dominant scalar may not reveal much if used to render the experience of a pre-dominant intangible.

On the other hand, the vision of moving geodesics in time is a remote approximation of the time-less motion of the intangible because the motion of the intangible is not subject to *timelines*. It requires new mentation along the lines described above to visualize the fabric of space devoid of all geodesics or tangible strings or quantum loops or scalar matrix but yet kinematically active. Furthermore, we must remind all those who may be quick to bring the charge of unphysical to the presented model that A. Einstein himself was seemingly apprehensive that his Theory was unphysical, as he wrote:

"... in the general theory of relativity, space and time cannot be defined in such a way that differences of the spatial coordinates can be directly measured by the unit measuring rod, or differences in the time coordinate by a standard clock... this requirement... takes away from space and time the last remnant of ***physical objectivity****."*

Our dismissal of time from the dimensional background is no more counter-intuitive today than the entanglement of time to space was in early 20th Century. Some may even justifiably advance that Relativity as a whole continues to be today a counter-intuitive vision to most, not only to the laymen but equally to the learned minds.

11.5. Relationship Between Parabolic and Cubic Symmetry

Consistent with our premise, the *cubic unit of action* in the fabric of space responds to parabolic symmetry. We endeavor to undertake in the following a formal demonstration of that assertion by analyzing the geometric features of a cornered cube. The illustration in Fig.11.2 will help us picture those divisions of the cubic figure.

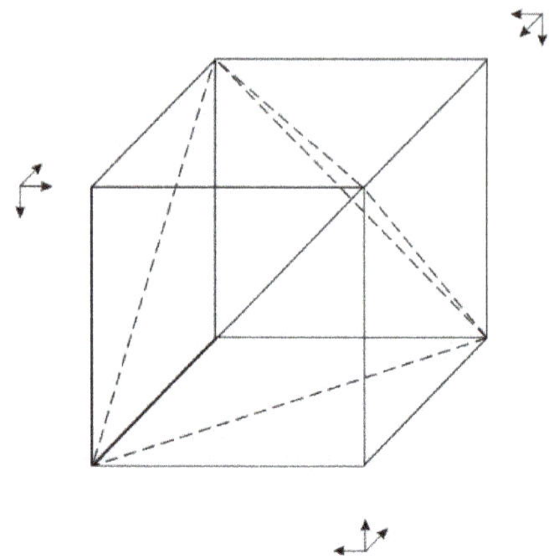

Fig. 11.2 Tetrahedral divisions of the cube: 4 corner cube tetrahedra and 1 internal isosceles tetrahedron

In demonstrating the relationship between the parabolic and cubic symmetry, the only certainty we have starting out is that ABCD, face of the cube, is a square. We first demonstrate that AH = HB from the face of the cube in order to prove that triangle AHB is right isosceles.

From Pythagorean theorem for ADB triangle, per illustration in Fig.11.3:

$$AB^2 + AD^2 = DB^2$$

$$AB^2 = DB^2 - AD^2 \quad (11\text{-}1)$$

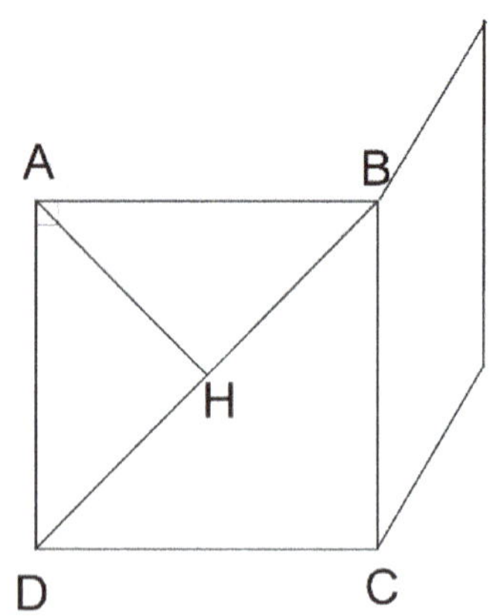

Fig. 11.3 Division of any face of the cube into two right isosceles triangles from one of the two diagonals

From Pythagorean theorem for HBA triangle:

$$AB^2 = AH^2 + HB^2$$

$$AH^2 = AB^2 - HB^2$$

By replacing AB^2 with its value obtained in *(11-1)*

$$AH^2 = DB^2 - AD^2 - HB^2 \quad (11\text{-}2)$$

Since the height of an isosceles triangle (ADB) divides its base in 2 equal parts:

$$HD = HB$$

And $DB^2 = (2HB)^2$

By replacing DB^2 in *(11-2)*:

$$AH^2 = 4HB^2 - HB^2 - AD^2$$

$$AH^2 = 3HB^2 - AD^2$$

Since $AD^2 = AH^2 + HD^2$, AHD being a right triangle:

$$AH^2 = 3HB^2 - (AH^2 + HD^2) = 3HB^2 - AH^2 - HD^2$$

$$2AH^2 = 3HB^2 - HD^2$$

Since $HD = HB$

$$2AH^2 = 3HB^2 - HB^2$$

$$2AH^2 = 2HB^2$$

Dividing each side by 2:

$$AH^2 = HB^2$$

So therefore, by taking the square root of each side: $AH = HB$

And triangle *AHB* is isosceles. Consequently, it must be a right isosceles triangle. The angle *AHB* being formed by the height and base of a right triangle, such as triangle *ADB*, must be a right angle. That condition added to the fact just proven that HA = HB, triangle *AHB* must be right isosceles. And so must be *AHD* by homology.

We now set out to demonstrate that a right isosceles triangle is a sufficient condition for a parabola (Fig. 11.4).

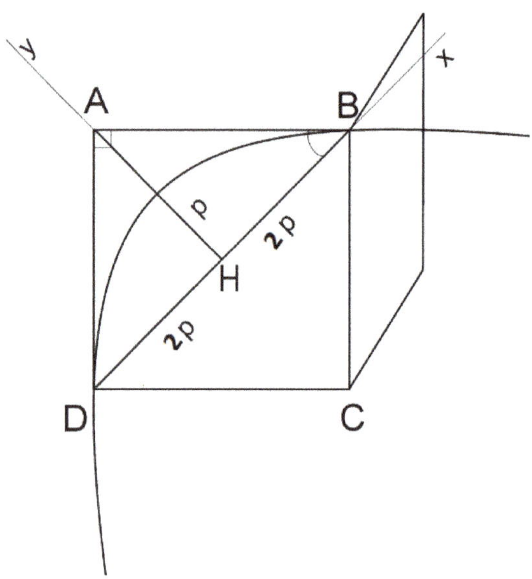

Fig. 11.4 Parabola inscribed within right isosceles triangle

For that we need to show that the base of the triangle *ADB* is equivalent to the latus rectum of a parabola.

The test: If angle *ABD* = 45° then triangle *DAB* is right isosceles, since from square *ABCD* angle *DAB* is right. Diagonal *DB* is assumed to be the latus rectum of a parabola.

Equation for a parabola is: $y = \frac{x^2}{4p}$

Taking the direction of *AH* as the y-axis of a Cartesian referential and the direction of *DB* as its x-axis, then the derivative of the function is:

$$\frac{dy}{dx} = \frac{2x}{4p} = \frac{x}{2p}$$

At $x = 2p$, the value of the derivative is:

$$\frac{dy}{dx} = \frac{2p}{2p} = 1$$

At that point the tangent to the curve must yield a 45° angle since the derivative is equal to 1, according to the relation:

$$\tan\theta = 1,\ \theta = \tan^{-1} 1,\ \theta = 45°$$

Since we already know that the angle θ is equal to 45° as the base angle of a right isosceles triangle, then the condition is verified. And the right isosceles triangle stands indeed as a sufficient condition for a parabola. This is a direct demonstration of the relationship between the square and the parabola. We may transpose this demonstration from 2-dimensional to 3-dimensional, by projecting the right isosceles triangle into a right tetrahedron and the parabola into a paraboloid. Right tetrahedra are also designated corner cubes.

We recall that 3-D space is a progeny of space of s_3 of parabolic symmetry while being constrained by a Quintessential Quantum Dominant Operator which homogenizes its structure. The original symmetric quality of 3-D space under the constraint of the

Operator conforms the building block of its fabric into a basis cube. The cube substantiates both the *multiplet* quality imparted by type 3 Primitive of the Quanto-Geometric function and the monocentric symmetry imparted by the Quantum Dominant Operator to universal 3-D space. In other words, the void of 3-D space is a tetrahedral population of cubes. However, these corner-cubes are intangible and cannot be represented by any scalar artifact.

The spatial cubic tetrahedra are not *virtual*, they are entirely **real**, but non-scalar. In addition, they are *fleeting* entities, which qualifies the unit cube as the universal *cube of action* as we have called it. It is important to understand that the unit cube is not structurally crystalline or compact but *composite tetrahedral.* That property is paramount to the kinetic behavior of space s_3. Further, the *cubic tetrahedron* stands in a sense for the unit manifold of the 3-Dimensional realm.

Consequently 3-D space cannot be visualized as a fabric of points but preferably of lines because the building block is a linearly structured entity. These lines are not strings or string objects or even geodesics. There is no scalar object at the multiple apexes of the basis cube because it is entirely intangible. A line in universal 3-D space is an alignment of fleeting basis tetrahedral cubes along a determined direction.

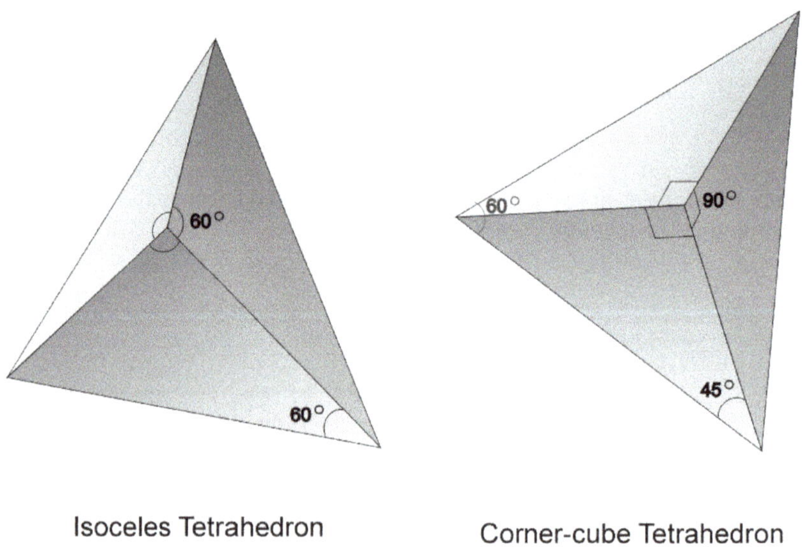

Fig. 11.5 Complete representation of the external corner cube and the internal isosceles tetrahedron

We must stress that the alignment of the basis cubes is cast in all directions. However, the fleeting basis cubes are not discrete or separable from one another in their alignment.

The spatial fabric is not interstitial in any ways because there is no void in the void. In this architecture what would be a void is still the same void because the void of space is completely homogenous within its imponderability. If we represent the imponderable in a discrete manner, then the imponderable is no longer such. That also stands for one of the reasons why we shall preferably not use the term *quantized* to describe the partitioning of the void.

To be precise, the *s*-line in universal 3-D space is an alignment of fleeting basis elongated tetrahedra along a determined direction. The basis elongated tetrahedron is made up by the inner-core isosceles tetrahedron of the cube flanked at the bottom by the corner cube along one of the four diagonal directions of the cube. The unit cube of action consists of 4 corner-cube tetrahedron and one inner isosceles tetrahedron (Fig. 11.5)

Fleeting lines of tetrahedral cubes cross each other in all directions. One tetrahedral apex of any basis cube constitutes the translational moving front of the cube, the direction of propagation always being that of aligned diagonals of successive cubes, which is the line of elongated tetrahedra. Because the whole topology is inherently dynamic and omni-directional over the entire tapestry, one cannot quite legitimately make a *lattice* representation of the situation. The motion is rated at *c* in basis state and degenerates from *c* in all objects and all transverse Shells of its composition. Because every direction in space is dynamic, a lattice representation would imply an infinite number of lattices in superposition so that the integral of all possible paths is covered. At the end a lattice view of the fabric of space in the dimensional background amounts to a static view that significantly departs from its very intimate and original nature, although one may help oneself with that representation if unable to embrace a seminal vision of the imponderable tapestry. The familiar lattice geometry is indeed a derivative of the primal fabric at a lower stage of the history of deployment of the identity, the Material Shell as we have coined it.

If this architecture seems unphysical, it is... because it is. It is the architecture of the imponderable. It requires an entirely new form of mentation in order to represent spatial identity. One cannot understand spatial identity from a scalar point of view. One may only understand spatial identity thru contemplation of the imponderable and by visualizing an order of intangible content. What will guide us in that instantiation and keep us far and away from fiction is the verifiability of the model, to which we will devote our attention shortly.

11.6. Homogenous Tetrahedral Cubic Matrix

An important quality of the unit tetrahedral cube that supports the mandated homogenous or isotropic quality of the void of space is its infinite reducibility. We may easily trace that quality in the geometry of the unit tetrahedral cube which consists of 4 corner-cube tetrahedra and one inner-core regular tetrahedron. Suffice it to show that a corner cube and the equilateral tetrahedron are infinitely reducible for that quality to be verified.

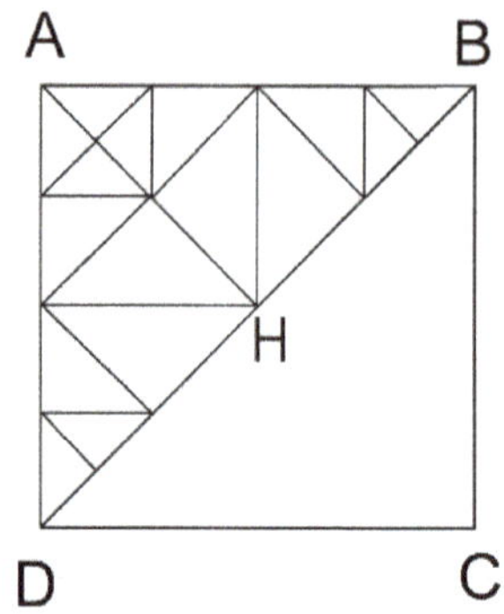

Fig. 11.6 Sweeping subdivisions of right isosceles triangles into the same figure down to the infinitesimal

By reducing the two figures to their 2-D counterparts, we obtain for the corner cube an isosceles right triangle and for the regular tetrahedron an equilateral triangle. We show in Fig. 11.6 how a right isosceles triangle can be infinitely broken into smaller similar triangles by forming the height of every newly obtained triangle. We could continue this operation until the pictured square becomes completely and uniformly dark.

We show in Fig. 11.7 how an equilateral triangle can be broken into smaller identical figures by successively forming the parallel to one side inside the figure, an operation that can be infinitely repeated.

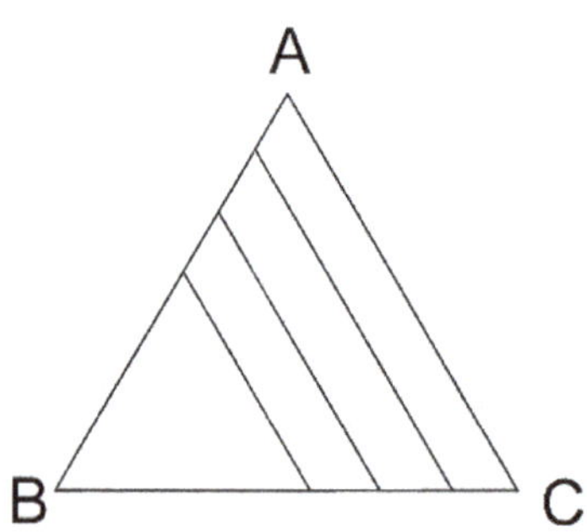

Fig. 11.7 Sweeping subdivisions of equilateral triangles into the same figure down to the infinitesimal

Consequently, by visualizing the continual deconstruction (or construction) of the two figures in step with one another in their 3-D layout, one can easily realize the property of reducibility of the whole 3-dimensional matrix.

Furthermore, this property supports the dynamic quality of the identity by providing the basis for uniform omni-directional propagation, in the understanding that the unit entity is not just the simple 6-face cube but the *tetrahedral cube*.

11.7. Spatial Cubic Matrix as Background to Laws of Geometric Optics

The laws of geometrical optics that describe the propagation of light thru material media were first proposed by Danish physicist Christian Huygens and were never unseated even by the soon-to-come momentous Maxwell's Electromagnetic Theory. Huygens's principle undertakes to describe the propagation of light as a wave artifact thru any medium by stipulating the modality of formation of a wavefront. This principle prescribes that all points of a light wavefront can be represented as a plane perpendicular to the direction of propagation and are to be each taken as the center of a circle of propagation (Fig. 11.8). The tangential plane to the ensemble of circles at any distance $c \cdot \Delta t$ from the original wavefront represents the new wavefront.

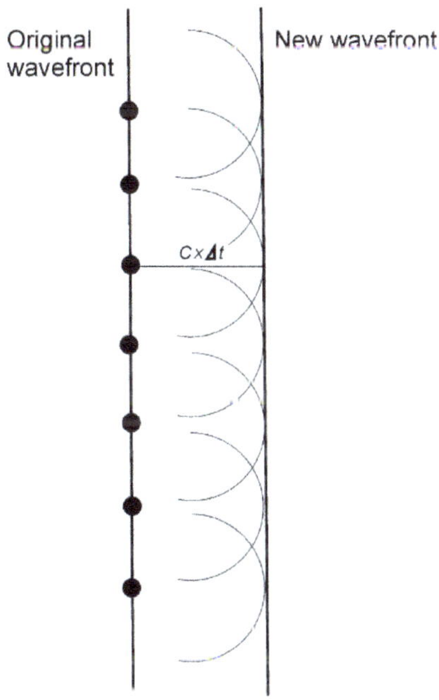

Fig. 11.8 Huygens principle of propagation of a wavefront

The point to circle representation is not physical but simply a heuristic artifact aiming at modeling the displacement. Yet this notion of a wavefront is powerful enough to model with simple trigonometric formulae the following:

- The phenomenon of reflection when the light ray hits a reflective surface.
- The phenomenon of refraction when the light rays transits from one medium to another.
- Even the phenomenon of diffraction when the light traverses an aperture with length or diameter in the same range as the wavelength.

So therefore, the Huygens principle reaches so far as the Compton Wavelength figure, expression for the wavefunction inherent to quantum states of particles in Quantum Theory. If we assimilate the visualization of points making the planar wavefront of light waves to that of apex points of the diagonals of the cubes of action along the direction of propagation in the spatial matrix, one can see that the Huygens principle reduces to tetrahedral cubic space in homogenous translation. And because the cubes definitely bear a set metric, one can readily see how they contribute to wavelength and become the cause of such phenomena as slit diffraction or interference.

Light waves have historically been regarded as a curious phenomenon, principally because we have visualized them as an emanation of some sorts that is able to propagate in empty space, a property that makes them unlike any other form of wave phenomenon we know of. So long as we fail to understand that beyond geometry the vacuum is *structured* and that the structure is based on intangible *fleeting* tetrahedral cubes, we will be unable to understand light waves as a result of the propagation of the vacuum and not as an object within the vacuum. In the next Chapter we will delve even deeper in the innate kinematic of wavefunctions based on the visualization of tetrahedral cubic matrix of space.

11.8. Adjusting Shell Nomenclature

The concept of Shell of Matter, rooted on super-structural action over the tetrahedral cubic matrix of spatial void, is critical and bears as much significance in Quanto-Geometry as reference frames do in General Relativity. We are going to make a slight adjustment to their nomenclature applicable to all prior publications by making the number associated with each an integral part of the descriptor they represent, no longer letting them be just a flat ranking ordinal. The 9 Shells that make up 3-dimensional matter each respond to a Quanto-Geometric Primitive, albeit not belonging to vertical hierarchy. The signature of that Primitive constitutes the most towering imprint over the ontological bundle represented by the Shell as a superstructure. The new nomenclature follows:

- Shell1 – Shell of Quarks and Gluons
- Shell2 – Shell of Elementary Particles
- Shell3 – Shell of Atoms
- Shell4 – Shell of Molecules
- Shell5 – Shell of Materials
- Shell6 – Shell of Karyotes
- Shell7 – Shell of Astral Bodies or Singlets

- Shell8 – Shell of Solar Systems
- Shell9 – Shell of Galaxies and Galactic Groups

11.9. Conclusion

Tensor-based manifolds do not evaluate to the ultimate representation for the void of space, whether in the quantum foreground or the cosmological background. The reason to that deficit is their inability to construct a working unit of action for the void that is consistent with what we know to be the values certain of the fundamental physical constants and the kinematic behavior of space given in wave mechanics.

The presence of *c*, the speed of light constant, and *G*, the Newton Gravitational constant, in the role of coefficient of terms in the Einstein Field Equations indubitably downgrades the Theory of General Relativity, as regards to the long coveted ultimate description of nature without bounds. So does the presence of *h*, the Planck energy constant, in its reduced form, in the high-level momentum-space identity that underpins Quantum Field Theory. Both Theories so reveal in their respective hallmark mathematical formulations a constitutional inability to uncover the mathematical-physics dimension behind these pivotal physical constants. Said inability exposes their lack of elevation for the seminal description of the inner structure of the space matrix. The task of reaching a Grand Unified Theory will not be fulfilled by producing extensions to the framework of General Relativity or that of the Standard Model, poster child of Quantum Field Theory. To achieve that goal of long, both Theories must be peremptorily superseded. That said, we must acknowledge the enormous analytical contributions of the Theory of General Relativity, for its own, to our understanding of the identity of the physical intangible.

Chapter 12

It is important to realize
that in physics today we have
*no knowledge of what **energy** is.*
Richard Feynman

THEORETICAL DERIVATION OF PARTICLES' MASSES AND WAVEFUNCTIONS

Ever since particles of matter were discovered in first haft 20th century, one interesting but ongoing quest in modern physics had been the question of determination of elementary particle status, in many a mind a qualifier for the ultimate building block of matter in the sub-atomic realm. As it happens, at every turn when candidates for irreducible components of the infinitesimal have surfaced, shortly thereafter other candidates of finer granularity have taken the stage to ouster the former from the prime role. While the sentiment nowadays tends to grant the Higgs particle this much coveted role, some are already hypothesizing the possible existence of a whole class of Higgs particles, in response to some of the serious impasses that the Higgs has aroused. Is the search for the ultimate building block of matter a sensible quest? The principal task we have set for this stage of our study is that of unraveling the deep principles that govern the distribution and allocation of mass to elementary particles of matter and the mathematical derivation of these masses, a goal of a very tall order, thus far out of reach to physical science. This treatment will bring unequivocal answers as well to the question of the identity of elementary particles, along with the theoretical derivation of their wavefunctions.

12.1. Particle's Genesis and Units of Measurement in Physics

At the point where the quark model emerged from the Standard Model of particle physics, physicists had already come to terms with the idea that no one particle seemed to be more fundamental than others and that every single one of them emerges from the existence of the others. The implication of this view, peremptory and inescapable though it was, did not deter physicists from hypothesizing a 13-billion-year history of the Universe where particles were cast into existence in the first moment following the gestalt of a Big Bang, with subsequent structures sequentially arising from the previous along a scaling timeline. Effectively, if no particle is more fundamental than any other, the Universe (3-D) must be in deployment over a timeless background for the one part, and, for the other, it must be that everything had *always* been there. Hence the necessary conclusion that there is no singularity to be the single cause of the existence of matter. In this context, all forms of ordinalization in phenomenology must be strictly local, just as all continuum is only apparent in the macroscope just because deep underneath lies a world of quantization, per insight apportioned by the Quantum Theory.

Accordingly, interdependence in the world of infinitesimal matter does not imply a process of phylogenesis. This principle lies deeply embedded in the Quanto-Geometric Primitive canon. We will promptly show that group topology of the world of real or *on-shell* particles severally obey a Quanto-Geometric distribution and that the *generations* of particles contented in the Standard Model lack all taxonomic legitimacy with regards to all applicable theoretical and overarching principles. In effect, vertical hierarchy within the Quanto-Geometric spectrum shows a decreasing scalar order from the Monolithic layer to the Transcendental layer. Should one wish one normal layer to beget another, which hierarchy shall we choose to achieve that form of phylogenesis? A transcendental entity would have to play the role of finest unitary quantum to serve as building block of all other entities in the spectrum. Whereas the largest monolithic entity would have to play the role of unitary wavefunction from which all others are built. Since every particle is simultaneously both a scalar and a spatial wavefunction, phylogenetical construction of a particle becomes conceivably impossible. A particle would have to be there even before its precursors have come into existence. In consequence, the construction and creation terms in this study, whenever they arise, should always be taken in a timeless context and solely qualify as an artifice of the narrative.

The above considerations show at a stroke the limits of the notion of physical units or traditional units of measurement. We asserted σ, a multivariate-symmetry parameter, and its many different dimensions, to be the single most natural unit scale. By the same token, from this perspective, numbers acquire full phenomenological dimension whereby their inner scalar state expresses something of significance about the physical entity that they quantify. Very specifically, it is no longer a man-made physical unit of measurement that scales the physical variables but the very number that expresses the value of the variable in conjunction with the σ qualifier. The measurement scale thus becomes intrinsic to physical quantities, revealing that true human knowledge and realization of matter can only result from the depth of a mathematical-physics perception. The few among us that are able to live sensorially within the realm of pure mathematics, such as the late physicist Richard Feynman with his known synesthesia, are those who have crossed this door leading to the evoked mathematical-physics experience in its most authentic dimension. Interesting though this might be, we are not going to delve into such an analysis in this study. Suffice it to underscore that scalar states of numbers along with space parameter σ represent a natural inner scale for the representation and gauging of physical identities.

12.2. Quanto-Geometric Spectrum of Elementary Particles

From the dichotomic classical view that a particle is strictly a scalar and a wave strictly an oscillatory field, physics has come to coalesce around the idea that all particles have an associated wavefunction, and that all harmonic oscillators or fields have an associated scalar component. Louis Victor de Broglie and A. Compton have firmly contributed to establish this unifying view of granular matter, with A. Einstein's contribution to the matter of no lesser significance.

The notion of a Compton wavelength of a specific wavefunction inherent to each and every (fermionic) particle inherits from the De Broglie general notion of a wavelength associated to all physical objects in motion, even though historically A. Compton had first discovered a momentum to be associated with pure photonic radiation. The Quanto-Geometric primitive spectrum represents the canvas that shapes the distribution and internal structure of all sub-atomic particles. What we have called the space tenet in a Quanto-Geometric object is a characterization of the wavefunction experimentally known to be inherent to all sub-atomic particles down to the quarks. Whereas what we have called the quantum or scalar tenet stands for the mass they are all known to exhibit.

Further down we will delve even deeper in the details of the internal architecture of sub-atomic particles of matter. Suffice it to underscore for now that they are a faithful embodiment of the Quanto-Geometric Primitive canon.

12.3. Empirical Estimation and Measurement of Masses

Interestingly enough, the heavy focus put on scalarity and its mass derivates in modern physics does not typically come with expert or clever rationalization and experimental measurements of mass, let alone the question of its definition. There is a good half dozen different formal approaches to the subject of mass in all-time physics that we shall briefly review subsequently. Short of being able to define mass, we have traditionally approached it in terms of its behavior and effects relative to other physical quantities as we shall see below.

12.3.1 Inertial Mass

Because an object naturally resists motion when it is subject to the application of a force, we rightfully believe that mass naturally leans toward a state of inertia. This form of behavior points to the existence of inertial mass inside of physical objects. The greater the inertial mass of an object, the greater its resistance to imparted motion by application of a force.

12.3.2 Active Gravitational Mass

The Newtonian view of mass of an object further identifies it as that element which causes the object to induce kinematic behavior in other surrounding objects because of an attraction force it exerts on those objects, its gravitational pull. The inner element causing the pull we call active gravitational mass of the object.

12.3.3 Passive Gravitational Mass

All objects respond to gravitational pull in a gravity field because of the mass element they enclose. Within a gravitational field two different objects will experience the same acceleration or gravitational pull, but they will experience a force constraint of different magnitude. The experience of that force constraint is what we call the weight of an object, the larger object experiencing the larger weight. Hence, weight is a result of passive gravitational mass of an object.

The physics of inertial and gravitational mass principally relies on weight scales for the experimental measurement of mass. It is important to realize that these are different views of the same quantity resulting from examination and analysis of exogenous behavior of objects. The quantity in question is a scalar entity native to physical objects despite the fact that it is utterly difficult to define in and of itself.

12.3.4 Curvature of Spacetime

Mass has the ability to exert pressure on spacetime and curb its very fabric according to the theory of General Relativity. The difference in ticking rates of clocks on Earth and in space is considered support to this view. The difference in elapsed time in both of these cases is conceivably the manifestation of a form of curvature denoted "gravitational time dilation". Useless to say that we do not quite subscribe to those views, we simply report them here to summarize the different constructions surrounding the notion of mass in modern physics.

12.3.5 Quantum Mass

Quantum mass is intimately related to the wavefunction native to a quantum scalar, its Compton wavelength. Quantum mass is a derivative of the principle of Mass-Energy equivalence in that the latter stipulates that the rest-mass of a particle is equivalent to the amount of energy harbored by its wave function. Quantum physics models particle masses as one of the irreducible representations of the Poincaré group in Representation Theory.

It follows that quantum physics heavily relies on spectroscopy for the experimental determination and measurements of particle masses.

12.3.6 Polemical Mass-Energy Equivalence

A long-standing controversy in physics has been the question of appropriate interpretation of the principle of Mass-Energy equivalence ever since Einstein published what turned out to be likely the most popular and celebrated physics formula ever:

$$E = mc^2$$

From laymen to learned, we all commonly talk about energy without anyone able to satisfactorily and accurately define the ontological meaning of these terms, although the relativistic energy formula itself, strongly supported by experimental evidence, has never raised any objection whatsoever.

A. Einstein had his own interpretation of the principle of mass-energy equivalence, both of these tenets being in his view a reflection of the exact same property of matter. Another school of physicists had historically countered that they are instead two distinct properties of matter. Such opposing arguments were held as late as 2002 by physicists the like of Hermann Bondi, C.B. Spurgin and Mark Lange who believe that the equivalence principle is essentially a statement about material state transformation and not ontology.

This controversy is not likely to be resolved to anyone's satisfaction unless a clear definition of the properties of mass and energy comes to fruition and gather sufficient consensus. We have already invoked the philosophical barriers that keep most concerned minds away from achieving a theoretical resolution to the nature of these tenets. From the classical viewpoint, the two substances that make up matter is the tangible stuff we call mass, and an intangible stuff we call field, an example of which is the electromagnetic field. For A. Einstein the tangible part of matter harbors both mass and energy while the physical field only has or is energy. But because of the preponderance of the principle of equivalence, he came to think, the field comes to acquire mass and end up having both mass and energy. Physicists from the opposing school have begged to differ from this view.

The Quanto-Geometric view of mass has it as a scalar, to be exact an idealized scalar. A scalar is a pure mathematical abstraction that cannot come to physical reality. The Grand

Mappa Mundi function, we otherwise dub the Grand Eigenfunction, inherently forbids the physical inception of a pure scalar and so do all images of this function. Therefore, all physical scalar in any material topology must coexist with an individual space field or void spread in more proper terms. Furthermore, in the portrayal of this function even as a bare Gaussian function, deprived of all Quanto-Geometric properties, we know the same to be a density function. Consequently $Q(s)$ or q_i in its representation of a scalar can only depict a *spatial or dimensional scalar*. This scalar is **inertial Mass**, the tangible stuff of matter. But because it intrinsically contains spatial spread as well, the resulting internal scalar-space articulation constitutes the inner Energy rendered by A. Einstein's formula. Energy is conceptually associated with motion; in this case it is the spatial spread that constitutes the dynamical source which moves the scalar within its own boundaries and produces the energetic articulation. This ontology remains true no matter how far up we wish to integrate the physical scalar in order to recover its monolithisation or pure compactness: it will always end up being a spatial scalar, of a diminished dimension true, but nevertheless a spatial scalar.

The above readily explains why an object acquires mass by implosion and not by explosion. Mass being a density function, it increases with increase in scalar density. As it occurs with particle decay mediated by the so-called weak nuclear force, a **Down quark** particle for instance is able to emit a $\mathbf{W}^-$ boson, something that is about 40,000 times more massive than the very quark! Where was all this mass hidden (within the quark), one may ask, in the ignorance that the nature of the quark's mass was *dimensional scalarity*? It was always there as scalar in dimensional spread. Contrary to common sense assessment, a more massive object is not at all a larger object but a smaller object.

Likewise, the explosion of a particle, which is a drastic drop in its scalar density, leads to a less massive but bigger entity. It is well-known that the annihilation of an anti-symmetric particle pair, i.e., an electron and a positron, leads to the production of pure radiation elements according to the reaction:

$$e^- + e^+ \rightarrow \gamma + \gamma$$

The particle and anti-particle catastrophically destroy each other to release two enormous gamma photons. This is not a non-homogenous explosion whereby the original fermions are destroyed and converted into a pure cloud-like energetic radiation. This example is often but inappropriately cited to support the notion that mass can be converted into pure energy, a notion held by the school of thought led by A Einstein. Both particles in the pair are completely conserved in status thru the reaction. This is a circumstance where two fermionic particles of equal and opposite spin (+1/2 and -1/2) are forced to collide, an event that would cancel both of their spin to 0. But because spin 0 is forbidden by the Grand Eigenfunction, each particle explodes into a huge photon. The dimensional scalar in each particle does not disappear but instead experiences a density drop and a change in state or class whereby it becomes subjugated or eclipsed by the accompanying spatial wave field, the second ontological tenet of the particle, in a transformation that has now turned the original fermion into a bosonic photon. It cannot be overstated that a scalar per se can never become an energy field if one envisions an energy field to be a wavefunction, nor can a spatial wavefunction become a scalar, as we have well emphasized before. The transformations that are accessible to the dimensional scalar can only be executed along the line of density or dimensional changes of its own.

The assertion sustained above about the dimensional nature of the scalar equally explains why all objects, even at rest, have an energy budget. This energy or form of energy duly resides with the inertial mass. Note well that the relativistic energy of an object is not the classical potential energy, which is the energy associated with an object (or its inertial mass) as a result of its elevation in a force field, such as a gravity field. The relativistic energy in A. Einstein's formula is strictly intrinsic to the inertial mass independently of all exogenous conditions to the object.

The relativistic energy formula is a consignment that in U_3 the primary operating property of the internal space spread of mass in all objects is the native propagation speed of space s_3, universal and invariant c. One is free to call its composition with the scalar (mc^2) Energy or whatever else one please to denote the term. When adopting the denomination of Energy for this term of the relationship, one has to be absolutely mindful that it is of a distinct nature from the space wavefunction (or any of its artifacts) that accompanies the mass tenet in an object, which we most commonly refer to as energy as well. At the very least, in terminology they ought to be considered different forms of the energy.

12.4. Inner Structure of Condensed Matter Wavefunctions

In the context of a discussion about mass one can hardly avoid addressing ontological contributions of the wavefunction to physical objects. One may wonder why the relativistic mass formula above discussed does not take the form of the Newtonian momentum, which is one of the prime statements of state for a physical object. It would cast $p = mc$ instead and $E = mc^2/2$. One element of answer to this interrogation lies in the nature of c. The so-called speed of light c is a peculiar form of velocity that is not accessible to fractional or overt mass. The only explicit mass identity that can move at speed c is the integral mass of the U_3 universe by reason of $\langle q_3|s_3\rangle$, the trinary Operator directly originating from the Quanto-Geometric Gran Mappa Mundi.

All fractional masses (mass of common objects) are a result of a composition between an image function of the Grand Eigenfunction and the Quintessential Quantum Operator. All masses are distributed in accordance with the Quanto-Geometric primitive canon whereby a companion space spread is a mandatory composition tenet, and allocated by mandate of the Hyper Quantum Operator. It is only when the space wavefunction (or void spread) inherent to an object jugulates or eclipses its mass partner within their quanto-geometric articulation (or coupling) that the object as a whole is able to acquire the degree of freedom intrinsic to basis space s_3, its propagation speed, because the object's wavefunction and basis space s_3 become integrally and absolutely conflated with one another. Thus far, we have used the term *dominant* in this study to qualify the ontological state of mass and space in quintessence. Henceforth we will use as well the terms *overt* and *covert*, a syntax that may facilitate better grasp of the analysis in the context of the specificities of mass and wavefunction. We have stipulated that the first 3 normal layers of the Primitive spectrum constitute states of articulation between the scalar and its space partner whereby the partnering scalar jugulates or dominates the space spread. This preponderance of the scalar over the space spread means that the latter is in *covert* and not able to manifest itself phenomenologically. Likewise, we have indicated that the 5 last normal layers of the Primitive spectrum constitute states of articulation between the two primal ontological partners whereby the space partner eclipses or dominates its scalar mate. This preponderance of the space spread over the dimensional scalar (or mass) means that the latter is in *covert* and finds itself unable of outright phenomenological expression. By contrast, in the previous case the phenomenology of the scalar was *overt*, and in this latter case the phenomenology of the space spread is the *overt* one.

Covert mass is **absolutely not** equivalent to mass $m = 0$. In this situation the mass entity is not able to manifest itself beyond the boundaries of the object but it remains fully existent and integral inside of the object. The covert state of the mass in those objects is what prevents the onset of a momentum figure like $p = mc$. In other words, at speed c, all masses are in a covert state, as if there were no mass in the object. Consequently, the integral of p, which would be $E = mc^2/2$, may not exist either. However, not because the momentum of these objects may not exist as a product between mass and speed is it altogether forbidden. The object's momentum may perfectly be visualized completely on the side of the wavefunction, which plays a preeminent ontological role in the configuration of these sorts of U_3 objects.

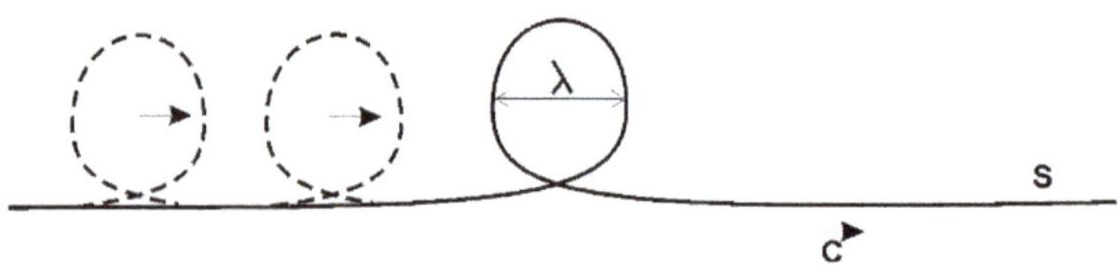

Fig. 12.1 Dynamic model of the wavefunction impulsed by the fabric of spatial void in U_3.

If the mass-energy equivalence had sparked a long-standing controversy in physics, the wavefunction had been for its own an area of feverous disagreement and even discord, to this day far from settled. Furthermore, a complete treatment and elucidation of the notion of wavefunction associated with physical objects is mandatory to any effort aiming at the theoretical derivation of particles' masses. In the following we are thus going to further build on the analysis initiated on this topic in Book I of this study. We start by posing once again that every outward regionalization of a dimensional scalar, its restriction to a limited space spread, that is, creates a wavefunction proper. Space, s_3 space in our case, reacts to the presence of a scalar by dropping a de-ploy shed or an un-fold vis-à-vis the constraining scalar, while continuing its fleeting race at *c* along a particular direction. The boundaries of this fold delimit the object's wavefunction and the number of harmonic cycles by the scalar inside the fold per unit distance along the propagation path characterizes for us the frequency of the wavefunction (Fig. 12.1). We find that the quantity which sometimes bears the designation of *wavenumber* is in close correspondence to that identity. Thus, our frequency or wavenumber is not rated in time but in the distance metric. The wavenumber line is *quasi* homomorphic to our *s* line. The wall-to-wall boundary of the fold represents the oscillatory unit of the wavefunction, for us its wavelength. It is the distance over which the dimensional scalar is forced to repeat a motion cycle of any kind inside the fold.

The waveform of the wavefunction may be as simple as a sinusoid of monocentric symmetry, in which case the object takes the form of a sphere, or as complex as a hyperboloid of linear (asymptotic) symmetry, in which case the object takes the form of the typical Gaussian distribution (or hyperboloid morphology).

This entire characterization emanates from the image Grand Eigenfunction instantiated by the work of creation for the generation of condensed matter in U_3, in composition with the Hyper Quantum Operator. The transformation that underpins our visualization of the wavefunction in U_3 follows:

$$T_{19} \rightarrow \upsilon = \frac{q_3^{-1}}{\left[\dfrac{dQ(s)}{ds}\right]_{s=1}}$$

where *v* is the frequency characterizing the wavefunction, q_3^{-1} is the parabolic quantum from the Grand Eigenfunction or its inverse to be precise, and *dQ(s)/ds*, the derivative of the image grand Eigenfunction.

The derivative expresses the relationship between the dimensional scalar and it space partner. We evaluate it at a specific value of *s* in the relation above. When evaluated at *s = 1* or *1o*, it yields the whole derivative to be equal to *1*, which sets *v* to be equal to *1/q*₃ or:

$$\upsilon = \frac{1}{q_3}$$

The only entity that can be equal to q_3^{-1} or $1/q_3$ is s_3 according to the primal Quanto-Geometric relation:

$$q \times s = 1 \text{ or } s = \frac{1}{q}$$

In that instance therefore the frequency v is equal to c or

$$\upsilon = c,$$

the maximum value of any wavefunction's frequency in U_3. The Transform above grandfathers a well-known relation that stands for the precursor of the momentous Planck-Einstein relation ($E = h\,v$), which is:

$$\lambda_i \upsilon_i = c \quad \textit{(12-1)}$$

Rewritten in a more revealing fashion, the relation becomes:

$$\upsilon_i = c \cdot \frac{1}{\lambda_i}$$

If λ is normalized to *1* in that relation, we obtain the identical outcome of :

$$\upsilon = c \quad \textit{(12-1b)}$$

So therefore relation *(12-1)* is only a particular case within the more general descriptive framework of the wavefunction for condensed matter in U_3 made explicit in the T_{19} Transform. From a propagation constant characterizing basis 3-D space, c becomes a frequency constant in the organization of 3-D condensed matter objects. One might affirm that the Transform has *ported* c into the spectrum of 3-D condensed matter objects.

The parabolic quantum positioned in the T_{19} Transform, as one might expect, plays overall a key role in the configuration of objects in the condensed matter realm of U_3. We will shortly see that not only the reciprocal quantity but q_3 itself, the direct figure of the quantity, plays an instrumental role as well in the configuration of condensed matter in U_3. The parabolic quantum represents the great allocator of scalarity in the 3-D Universe, under modulation of the Quintessential Trivalent Operator. It represents the seminal source from which all dimensional scalarity is provisioned within the spectrum of 3-D condensed matter objects.

12.5. Theoretical Derivation of the Electron-Volt Constant

Pursuant to the above assertion in relation to the role played by the parabolic quantum in the configuration of sub-atomic objects, the transform of creation further undertakes to *rationalize* q_3 into an energy figure for these objects. The direct parabolic quantum q_3 now becomes, under mandate of Hyper Quadratic Operator, the point source for the generation of the entity by rationalization or fragmentation (Fig.12.2), with the first-order derivative of the image Grand Eigenfunction being the modulus of that fragmentation.

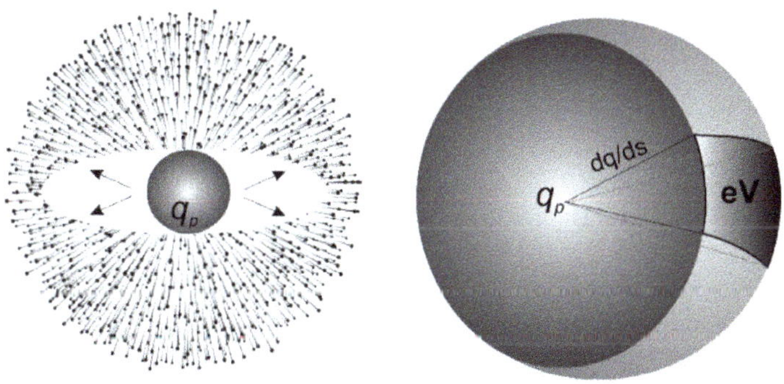

Fig. 12.2 Generation model of the ***eV*** Energy figure

The expression of the condensed matter Energy transform follows:

$$T_{14} \rightarrow E = \frac{q_3}{\left[\dfrac{dQ(s)}{ds}\right]_{s=1}} \quad (12.2) \quad \longrightarrow \quad E = \frac{q_3}{\left(\dfrac{-s}{\sigma^3\sqrt{2\pi}} \cdot e^{\frac{-s^2}{2\sigma^2}}\right)_{s=1}}$$

$$E = \frac{\dfrac{0.299796065}{\sigma}}{\dfrac{0.2419707245}{\sigma^2}}$$

$$E = 1.238976598 \cdot \sigma \quad (12\text{-}3)$$

where E is the energy or first statement of state for the treated object. To be clear it means a Quanto-Geometrically articulated *condensed* object under the composition by the two Operators in question. We have evaluated the derivative $dQ(s)/ds$ in the denominator of the above relation at $s = 1\sigma$ once again because that unitary point is the anchor for the generation of elements appealed to play the role of building blocks.

Although arguably insufficiently credited the importance that it deserves as a constant, the quantity above derived is well known in modern physics, and represents the electron-volt *(eV)* constant also expressed as hc. Its known value is: **1.239841930 · 10⁻⁶** with a valuation of being a lot more than just the plain product of h and c. Given the large uncertainty known to the measurement of h which is part of the latter quantity, the number is in a good match with the Quanto-Geometrically derived figure. Beyond the mere product of h and c, the electron-volt is a turnkey unit in condensed matter physics, particularly in high-energy physics, where it grades Mass, relativistic Energy, Momentum and even Temperature. When grading other physical quantities than Energy, its dimension is compounded with another unit such as in eV/c^2 for Mass, eV/c for Momentum or eV/k_B for Temperature (Table 4.1). Yet the practice of dropping the associated unit by setting it equal to 1, such as is done with $c = 1$ or $\hbar = 1$, is widespread in particle physics owing to a search for "natural units" in the expression of the physical variables germane to this domain. And so, this practice often renders all these figures simply as eV as well. Very particularly, particle scattering lengths, which are very closely related to particle's wavelengths, are oftentimes presented in units of inverse particle masses. For these reasons, some physicists have called the electron-volt constant the conversion constant, honoring thereby the fact that it always shows up from one unit to another in dimensional quantity conversions.

Should one opt not to take into account the relativistic principle of mass-energy equivalence, in evaluating this constant derivation outcome, one should not wonder why a scalar point source, our parabolic scalar q_3, radiates an element graded as energy in our physics. The key is in the modulus of rationalization used by the transform which is a quantum-space derivative, a point that we will later further clarify.

The electron-volt is formally defined as the energy acquired by *an electron* that has been accelerated across a potential difference of 1 volt. Here again we see the electron in the role of base unit! The anchoring of this figure at $s = 1$ in our relation is in conformity with other foundational energy parameters such as the electron Compton wavelength anchored as well at that point and obviously found tightly associated with the electron, not to mention the very Planck constant h.

The above T_{14} Transform grandfathers the paramount relativistic energy relation:

$$E = \frac{hc}{\lambda}$$

Rewritten in a more revealing fashion, the relation becomes:

$$E = hc \cdot \frac{1}{\lambda_i}$$

If λ is normalized to *1* in that relation, we obtain the outcome of:

$E = hc$, an equivalent to *(12.1b)*

VARIABLE	UNIT	SI VALUE
Energy	eV	$1.602176565(35) \times 10^{-19}$ J
Mass	eV/c^2	1.782662×10^{-36} kg
Momentum	eV/c	5.344286×10^{-28} kg · m/s
Temperature	eV/k_B	11604.505(35) K
Time	h/eV	6.582119×10^{-16} s
Distance	ħc/eV	1.97327×10^{-7} m

Table 4.1 Physical variables quantified in *eV* units

The electron-volt constant (*hc*) is indeed a true physical constant in its own right. It is a *transformation constant*, akin to the porting of *c* from basis space s_3 to the spectrum of condensed matter. What the transform T_{14} reveals is that when we apply an energy transformation to a condensed matter object under the influence of the *electron-Volt* constant, the internal quanto-geometric tenets of the object, its constitutional mass and wavefunction, will be transformed into mutual equivalents and will become *ambi-valent* akin to two homozygotic twins, quasi but only quasi-indistinguishable from one another. That assessment sets the most solid physical basis for the much-debated relativistic mass-energy equivalence and constitutes an extension to our previous analysis of the matter in Book I. This homogenization transform is quite well exemplified in the possible and quite correct application of the *eV* unit to almost all forms of physical characterization of parameters in condensed matter objects. Despite the fact that these units are not officially adopted by the SI board, the practice of their use is unofficially well accepted worldwide and is additionally fully reported by CODATA.

Finally, the T_{14} Transform stands for a firm contributor to the momentous Planck-Einstein relation:

$$E = h\upsilon$$

which by virtue of the T_{19} Transform yielding $\lambda_i \upsilon_i = c$ finally becomes after replacement of v:

$$E = \frac{hc}{\lambda}.$$

These simple but impactful relativistic relations all originate from these two transforms which are able to elucidate at the same time the origin of the physical constants these relations are towered on.

12.6. Mass Measurement Units and Techniques

In general, several different units are in use for the specific quantification of mass in physics as a matter of standard. These units are utilized according to their suitability for the particular realm of matter being treated. Chiefs among them are the familiar gram or kilogram *(Kg)*, the atomic mass unit *(u)* and the electron-volt *(eV/c^2or eV)* above discussed.

The kilogram is the standard for measuring weight of macroscopic objects per the SI metric system. The etalon for the kilogram is a solid metal object stored at the International Bureau of Weights and Measures in Paris, France, to the great dissatisfaction of metrologists because unavoidable microscopic contaminants are slowly but continually chipping off its mass.

The atomic mass unit *(u)* taken to be *1/12* of the mass of *carbon-12* represents the unit of preference to express the masses of atoms and molecules. It is important to point out that in the system of atomic units, the atomic unit of mass *(u)*, counter-intuitively enough, does not stand outright for the unit of mass. In that context the unit of mass is m_e, the mass of the electron. This choice is further Quanto-Geometrically justified by a paramount reason we have already alluded to and which we shall further discuss shortly.

Common in particle physics is the electron-volt for the measurement of mass as indicated previously. The electron-volt is initially a unit of energy. However due to the mass-energy equivalence principle, its conversion to a unit of mass would be straightforward and consequently leads to its direct use as a unit of mass. The mass value in this context is originally obtained in eV/c^2.

Of singular importance in the physics of mass measurement is the electron rest mass m_e. Considered to be one of the fundamental constants of physics, it enjoys great consideration in chemistry due to its relationship with the Avogadro constant. Since no particle, including the electron, can be decelerated to stationary state, the idea of weighing a particle "at rest" becomes fictional, forcing the estimation of rest mass of particles from mathematical formulae or by deduction. The relation that serves as a barometer for the determination of the mass of the electron is the Rydberg constant, while empirical measurement of that mass is conducted in the context of atomic edifices.

$$m_e = \frac{2R_\infty h}{c\alpha^2}$$

One measurement technique uses the subterfuge of replacing an electron in a helium atom shell by an anti-proton and to subsequently apply manageable measurements in atomic mass unit to deduce the mass of the substituted electron. This technique leads to the determination of the mass in u and the application of a conversion formula to cast that mass in Kg. The electron's relative atomic mass is an adjusted parameter in the CODATA set of fundamental constants published by the NIST. It is not quite the same however as the rest mass of the electron which is computed from the guiding Rydberg constant formula. The electron's relative atomic mass unit participates in the calculation of all other relative atomic masses which are estimated as ratios.

Many corrections must be applied in the process of determination of the electron's atomic mass, which entail a rather long cascade of approximations and iterations until a stable value is reached. The value obtained at the end is 5.4857991111 x 10^{-4} u with relative uncertainty of 2.1x 10^{-9}. One must consider that value as an approximation to the guiding value in eV/c^2 from the Rydberg constant formula. Below is the corresponding computation.

$$m_e = \frac{2 \times 10973731 m^{-1} \times 4.135667516 \cdot 10^{-15} eV \cdot s}{299792458 m/s \times (7.29735256 \cdot 10^{-3})^2}$$

$$m_e = 5.6856629865 \cdot 10^{-12} eV/c^2$$

We have made sure to enter the quantities in dimensions that would cast the final value in eV/c^2. Had we chosen to obtain the final value in Kg rather , with quantities entered in their appropriate dimensions, the final value would be: 9.109383 x 10-31 *Kg*. We will however give precedence to the eV/c^2 value since it allows direct comparison with the value obtained from another formula involving directly the electron's Compton wavelength:

$$m_e = \frac{h}{\lambda_e c}$$

This formula yields:

$$m_e = \frac{4.135667516 \cdot 10^{-15} eV \cdot s}{0.242631058 \cdot 10^{-11} \times 299792458 m/s}$$

Hence:

$$m_e = 5.685628003701 \cdot 10^{-12} eV/c^2$$

The computed values from both formulae are in appreciable match with one another. Should one want to derive the value in eV from measurement, applying the conversion factor of $1eV$ for 1.78266270 x 10^{-3}Kg, the mass quantity comes to be:

$$m_e = 5.109986875 \cdot 10^{-28} eV$$

This eV value is the one most often reported officially for the rest mass of the electron. Whether one wants to utilize one or the other is a matter of circumstantial convenience. From the Quanto-Geometric perspective however, the official SI kilogram unit is the least significant and least telling of all, given its detachment from the deep parameters that govern the allocation of mass in condensed matter physics.

12.7. Operators' Composition and the Inverse Square Law

We shall now address another form of composition between the Hyper Quantum Operator and the Eigenfunction instantiated by the work of Creation in the distribution of dimensional scalarity or mass in U_3. We recall that one of the primary effects of the intervention of the Hyper Quantum Operator was the homogenization of the ontological tenets of U_3, in particular the space tenet, by laying down the symmetries buoyed in the dimensional background of U_3. An important realization of the Operators in composition for the construction of the scalar foundation of U_3, condensed matter physics below the atomic scale that is, is what has been known as the ***Inverse Square Law*** (Fig. 12.3). It will be no surprise that that law shows an implementation of the derivation from a point source as a key feature of the Operator composition that we have seen in action in the two previous cases.

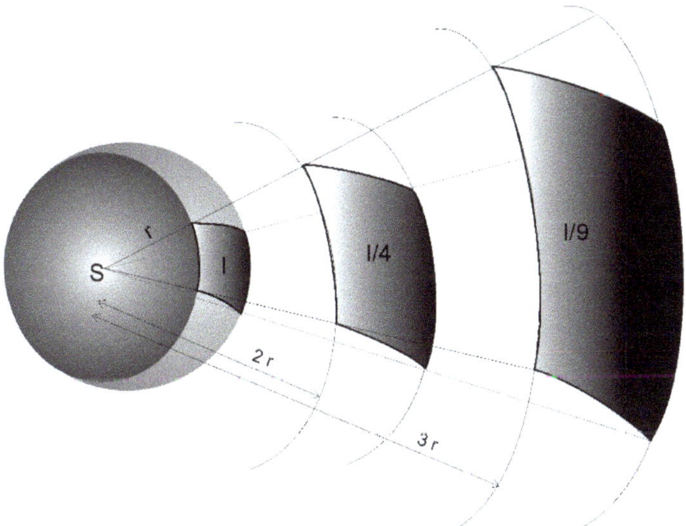

Fig. 12.3 The inverse square law by definition

The inverse square law states that if P is the point of radiating power at center of a sphere, S the emissive power of the point source and I the radiated intensity over a cross section of the surface of the sphere, then the intensity of radiated power from the source over the cross section is given by:

$$I = \frac{S}{4\pi r^2}$$

In fact, the formula is best expressed with I as a function of r, because it is not just about the sphere but most importantly how the sphere volume varies parabolically with r^2, a clear dictate of the composition between trinary symmetry from normal layer 3 of the Grand Eigenfunction (or its image) and monocentric symmetry keenly underscored by the Hyper Quantum Operator. So that:

$$I(r) = \frac{S}{4\pi r^2}$$

As r varies, so does the radiated intensity in inverse proportion to r^2. The radiated intensity on the cross section drops with the square of its radial distance from the point source. The radius r here does not function simply as a constant of spherical symmetry but as the independent variable of a parabolic function.

The origin of the inverse square law is undoubtedly the composition between the Quanto-Geometric Operators mentioned above as they lay down the foundation for our 3-D Universe.

12.8. Derivation of the Mass of the Electron

Common wisdom in modern physics thus far has it that the absolute value of particle masses may not sensibly be obtained from first-principle deduction, given that units of measurements are completely man-made. Hence tackling the issue on the side of finding the relative value of masses with respect to one another, or in terms of their ratios, apparently stood to better reason. Therefore, in particle physics, current mass quantities are heavily based on ratios, an approach credited with the ability to produce the dimensionless quantities for mass thought to be derivable.

The long-known Koide formula is one such precursor of theoretical derivation of mass. Yet, because it has no solid connection with Theory, it remained to this day no more than an empirical curiosity from which nothing further could be built. More recently there has also been interesting work by physicist Sergey Efimov who came up with empirical formulae for the calculation of magnetic moments of proton and neutron particles. They generated little interest however due to the lack of connection with Theory.

A fact that is not well understood and much less accounted for by principle in particle physics is the inseparable character of the electron, although the singularity of the element is well recognized. In passing, one must highlight the fact that likewise the electric charge quantity is rooted on that of the electron, so conferring to the charge of the electron the value of universal unit of charge.

Our task of the moment is to derive from Quanto-Geometric principles the value of the mass of the electron, an entity with a key position if not a paramount role in the configuration of the spectrum of elementary particles. We have mentioned in Book I that the derivation of the Electron Compton Wavelength already stood for one important step in that undertaking. We are now going beyond that prelude to show the deep origin of the allocation of scalar density or mass in the configuration of the U_3 universe. For sure, a theoretical system with the ability to derive the Compton Wavelength of most elementary particles would be a great contribution to the physics of masses, by further counting with the agency of the relativistic principle of mass energy-equivalence to arrive at the final quantities. Quanto-Geometric Theory however is able to bring more to the table.

One important constraint to keep in mind in this development is that the quark or sub-particle Shell is not rooted on a physical constant to the same extent as all the other Shells because of the indetermination of the Quanto-Geometric function at $s = 0$. Akin to the generation of the previously studied variables in Book 1, the generation of mass in U_3 proceeds in accordance with a composition between a sub-Operator from Quintessential Tri-Valent Operator and the derivative of an image function of the Grand Mappa Mundi function. In this case the sub-Operator is the Hyper Quantum Operator, obviously the one best positioned for the task of mass generation.

The transform presiding on the generation or allocation of mass to particle objects is a composition between monocentric symmetry and parabolic symmetry, which is materialized in the so-called Inverse Square Law, just as the electron-volt constant is a form of that composition. The mandated transpose by the Hyper Quantum Operator is here again that the radius of the sphere must stand for the derivative of the image function of the Grand Mappa Mundi, but that the point source of mass is to be the *inverse* of the parabolic quantum (q_p^{-1}) and the cross section thus obtained for the mass of the electron to be the unit mass for the world in construction (Fig. 12.4).

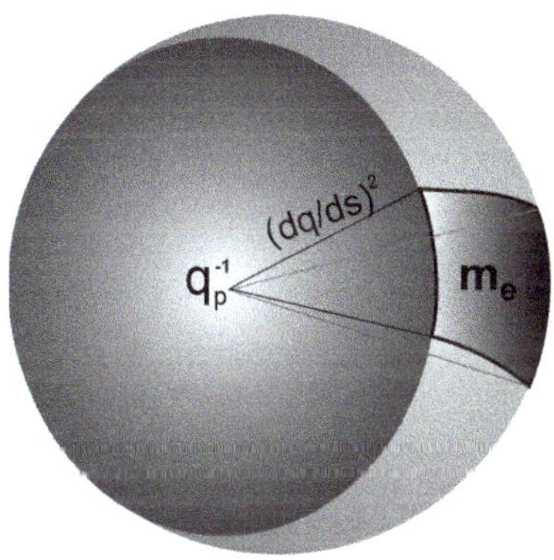

Fig. 12.4 Mass of the electron from the parabolic quantum as the radiating point source scalar

One more reason for this chosen path in the composition of mass in U_3 is the impossibility to generate mass from a $Q(s)$ scalar constant, given that the point $Q(s) = 0$ in the image function is indeterminate. In the process of construction, the reason for the mass of the electron to be further specifically construed as the unit mass instead of that of any other particle stands in conformity with the very parabolic nature of the world in construction. Reciprocally, the particle of type 3 must acquire singularity in the construction because of the qualified superposition of its ontological type with that of the world under generation as the superstructure. This translates into a boost in its standing. There lies the proper reason of the inseparability of the electron, a character that should be devolved instead to the neutron as the elementary particle spectrum's monolith.

The Transform for the rendition of mass for the unit object under the Inverse Square Law Quanto-Geometrically dimensioned comes to be:

$$T_{11} \rightarrow m_e = \frac{q_p^{-1}}{\left[\dfrac{dQ(s)}{ds}\right]^2_{s=1}}$$

The electron is already spotted in the construction at s = 1 for the generation of its defining Compton wavelength as indicated in Book I, as well as the electron-volt constant above studied. Its mass allocation is done at that same seminal point, to yield:

$$m_e = \frac{1}{\dfrac{0.2419707245^2}{(\sigma^2)^2} \bullet \dfrac{299796065 \cdot 10^{-9}}{\sigma}} \qquad (12\text{-}4)$$

One may readily see in the above equation a clear representation of the inverse square law, whereby the unitary numerator stands for totality of the scalar source, the squared factor in the denominator surrogates r^2 and the other factor the constant. From relation *12-4*:

$$m_e = \frac{1}{\dfrac{0.17553009 \cdot 10^{-9}}{\sigma^5}} = 56.97028895 \cdot 10^9 \sigma^{-5}$$

$$\boxed{m_e = 5.697028895 \cdot 10^{10} \sigma^{-5}}$$

It is easy to appreciate the close figurate match between this theoretically derived quantity and the empirically obtained quantity in (x). Given all the convoluted steps and readjustments it took to arrive at the cited empirical value, we will conceivably stick with and give preeminence to the theoretically derived value from the Quanto-Geometric system. Furthermore, it is interesting to note the new dimension of σ^{-5} in our figure, which stands for the unit for mass in our system, indeed the most natural dimension for mass.

Going back to the Compton wavelength relation, we may rearrange the formula to show:

$$m = \frac{h}{c} \cdot \frac{1}{\lambda}$$

If we normalize λ to *1* in that relation, we obtain the outcome of:

$$m = \frac{h}{c}$$

which stands for the smallest quantity of overt mass or universal unit of mass. The constant (h/c) is known as the inverse meter atomic-mass unit and valued at: **1331.02250394(19) $\times 10^{-15}$ u**. Interestingly, we do not arrive at this quantity from the T_{11} Transform, but in a different way from what will transpire later. We therefore hold and uphold the value of **$5.697028895 \cdot 10^{10}$ σ^{-5}** for the smallest universal unit of overt mass, an exhibit of the electron particle.

12.9. Primitive Spectrum Configuration of Elementary Particles

The epithet of elementary applied to particles of matter has covered a continually different set of entities throughout the history of modern physics. The choice of the epithet is attributed to internal structure of those elements, mandated to meet the test of a monolith in order to merit that qualification. Their stability against decay into other elements, which one would expect to be a contributing factor to the qualification, is surprisingly not a factor in the official assessment for the status of elementary.

Furthermore, if an elementary particle ought to demonstrate the structure of a monolith, the concept could only conceivably apply to a class of particles forming a distinct topology of their own, but not heterogeneously as it is to particles across distinct families or groups. The Quanto-Geometric distribution emanating from the higher number stack abstraction does not support this notion of the elementary.

Is there not something quite inadequate with a classification that grants the status of elementary to a particle such as the muon which naturally decays into electrons and neutrinos, whereas it taxes the proton as non-elementary or composite despite its absolute stability and its pivotal participation in the organization of higher Shells of matter like the atom?

What's more, in the current model of classification in particle physics, there appears to be no functional consequence to the status of elementary or composite, both types of particles being part of the most heterogeneous groups and families as they are. It would be a more compelling assumption and determination if the taxonomy resulted from theoretical principles able to uncover the specific reasons that make a particle one of either a composite nature or a stable nature. We are touching here one of the most important problems of the Standard Model of particle physics, which is a large taxonomic deficiency due to the ignorance of the deep laws that govern the internal configuration of particles of matter as well as the distribution and allocation of mass and wavefunctions to those entities.

Consequently, we are going to subsequently dismiss the qualifiers of elementary and composite as elements of classification and will strictly cater to the guidelines cast by the Quanto-Geometric vision as evidenced in the ontological study thus far conducted.

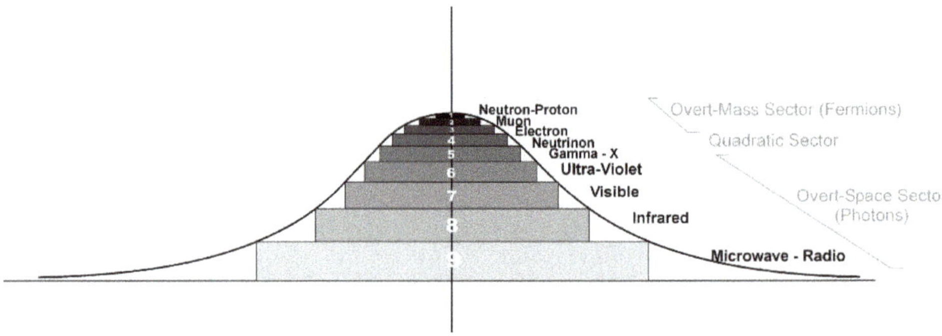

Fig. 4.5 Primitive Spectrum of Shell2 Particles

The study just conducted above already profiles the spectrum that underpins the topology of Shell2's condensed matter objects. The three Transforms hereby laid down, which fully reflect the towering quanto- relativistic relations above reported and analyzed, thoroughly capture the profile of that taxonomic spectrum. It is indeed a Quanto-Geometric Primitive spectrum (Fig. 4.5), though not a pure one because it is the result of a composition between two sets of Operators. One might tax it as a skewed spectrum because it is essentially the product of a Hyper Quantum Operator intervening over a pure image of the Grand Eigenfunction, which it duly redimensions. One can readily see that the T_{19} Transform principally lays down the precursory range of Overt-Space particles (Layers 5 to 9), the T_{11} Transform principally lays down the precursory range of Overt-Mass particles (Layers 1 to 3), while the T_{14} Transform principally lays down the precursory range of Equalized Mass-Space particles (Layer 4). We have grouped below the corresponding quanto-relativistic relations.

$$\upsilon_i = c \cdot \frac{1}{\lambda_i}$$

$$E_i = hc \cdot \frac{1}{\lambda_i}$$

$$m_i = \frac{h}{c} \cdot \frac{1}{\lambda_i}$$

For the sake of one-to-one comparison, we have subsequently grouped the Transforms themselves.

$$T_{19} \rightarrow \upsilon = \frac{q_p^{-1}}{\left[\frac{dQ(s)}{ds}\right]_{s=1}}$$

$$T_{14} \rightarrow E = \frac{q_p}{\left[\frac{dQ(s)}{ds}\right]_{s=1}}$$

$$T_{11} \rightarrow m_e = \frac{q_p^{-1}}{\left[\frac{dQ(s)}{ds}\right]_{s=1}^2}$$

The main realization of the Hyper Quantum Trivalent Operator here at work is to force the generation of objects from a point source naturally chosen to be the basis parabolic dimensional scalar. It operates on the overt-space layers of the image function (normal layers 5 to 9) thru the T_{19} Transform and yields to the frequency characterization of objects under treatment, naturally imposed by overt space ontology in that region of the Primitive spectrum. The Hyper Quantum Trivalent Operator further acts on the overt-scalar layers of the image function (normal layers 1 to 3) thru the T_{11} transform in order to underscore or boost the mass characterization of objects in that region of the spectrum, naturally imposed by overt-scalar ontology in that region of the Primitive spectrum. Lastly, the Hyper Quantum Operator composes with the Quadratic layer of the image function thru the T_{14}Transform in order to yield the Equalization characterization of objects in that region of the spectrum termed Energy, a natural form of expression of the ontology in orchestration in that region of the Primitive spectrum. The subscripts in the symbol of the Transforms mean to typify both the nature of the Hyper Operator (first subscript) and the region of the image function the latter purports to effect (second subscript).

In comparing the quanto-relativistic relations with their Quanto-Geometric Transform counterparts, one can see that the notion of material wavefunction, when characterized by wavelength λ, roughly approximates the derivative of the Quanto-Geometric Grand Eigenfunction. In that sense Frequency, Energy and Mass are best expressed as true functions of the wavelength λ, such as $v(\lambda)$, $E(\lambda)$ and $m(\lambda)$ and λ is itself a complex function which takes its variable in Quanto-Geometric coordinate space. Further, the Quanto-Geometric Transform relations teach us that, whenever we choose to characterize the wavefunction by its wavelength, it is a distinct entity for overt-mass condensed matter objects as it is for overt-space and equalized space-mass objects. A parallel between the two sets of relations show that the wavelength λ is almost on par with the first-order derivative of the image function $dQ(s)/ds$, but that for overt-mass objects *the derivative is distinctly squared*. Modern physics does distinguish between the De Broglie wavelength essentially associated with photonic radiation and the Compton wavelength strictly associated with fermionic particles, however they are both represented by the same Greek letter symbol λ, which reflects the squint visibility that field theory science thus far has had over the intimate physical and mathematical nature of this variable. The common notion that all objects, condensed matter objects or else, have an associated De Broglie wavelength adds more evidence to this record of misevaluation. It is however of strict universal admittance that only overt-mass objects or fermionic particles exhibit a Compton wavelength, even though the Compton wavelength of a particle is evaluated and defined as the equivalent of the wavelength of a photon harboring the same amount of energy as the rest-mass of the particle. The record on this matter is definitely mixed, to say the least.

12.10. Quantitative Origin of Compton Wavelength and Mass of Fermions

Once the mass of the electron has been defined and positioned to play the role of unitary element in the portion of the spectrum of particles of matter with overt-scalar ontology, the mass of all other particles is apportioned from that of the electron. In fact, what we see is that the idealized quantum or scalar parcels of the image Eigenfunction are set to compose with the value of the mass of the electron in order to generate the distribution of masses in that sector of the new spectrum. We dub these pilot scalar parcels ***gestalt scalars.***

From computation exposed in Book I of this study, we have:

$$q_1 = \frac{0.39894228}{\sigma}$$

$$q_2 = \frac{0.352065327}{\sigma}$$

$$q_3 = \frac{0.299796065}{\sigma}$$

These three elements represent the gestalt scalar parcels pertaining to the first three normal layers of the image Grand Eigenfunction. They are appealed to play a crucial role in the generation of the mass tenet of overt-mass objects, and to a certain extent their wavefunction as well in accordance with the mass-energy equivalence principle.

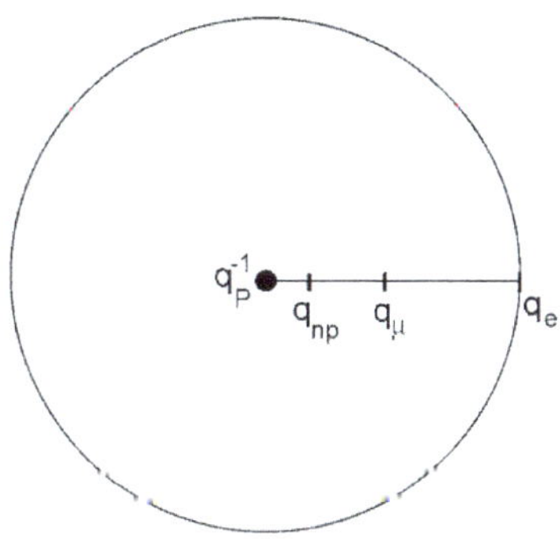

Fig. 4.6 Generation of the masses of overt-mass particles from gestalt scalars

The ratios q_1/q_3 and q_2/q_3 are of paramount significance within this hierarchy, in the understanding that q_3 at normal layer 3 is surrogating the scalar ranking of the electron within the new ported spectrum. The first ratio should tell us something about a particle or particles with monolithic signature, while the second ratio should tell us something about a particle or particles with binary signature. Effectively, the q_1/q_3 ratio determines mass and wavefunction characterizations for the neutron and the proton, whereas the q_2/q_3 ratio determines the mass and wavefunction characterizations of the muon particle. Computation of the first ratio follows.

$$\frac{q_1}{q_3} = \frac{0.39894228}{0.299796065}$$

$$\frac{q_1}{q_3} = 1.330712196$$

This ratio is obviously a dimensionless number in our context, free from the σ parameter, that is. This number is known as the *inverse meter-atomic mass unit relationship*, with the value of ***1.33102505120 × 10⁻¹⁵ u***. In evaluating the official quantity, it will be interesting to look further into what the atomic mass unit ***u*** stands for. The official definition of one atomic mass unit ***u*** sets it to be the approximate mass value of one atomic nucleon, either a single proton or a single neutron. Modern physics has always been in a bind with regards to exactly defining a universal unit quantity for mass. The best that has ever been achieved is an approximation by approaching the matter from several different angles. When it comes to the evaluation of the mass of one proton or one neutron, the difficulty in evaluating the mass of these particles in free state was no more significant than that of evaluating their mass in atomic bound state as nucleons, because of the so-called mass defect of atoms, resulting from the nucleons having to shed some of their mass in contribution to binding energy for the nucleus. The existence of isotopes with slightly different number of nucleons does no more than further complicating the matter. In the end the atomic mass has been officially set to be the 12th part of the nuclear mass of the carbon atom ^{12}C, with a count of 12 nucleons, yielding the mass of one proton to be equal to ***1.007276466812 u*** and the mass of one neutron to be equal to ***1.00866491588 u***. To a two-digit decimal point approximation they both equate unity at **1.00** for a 1 to 1 relationship between the unit and either the proton or the neutron. Thus, any quantity expressed in atomic mass unit ***u*** directly compares to or is referenced to the mass of the neutron or the proton. By that measure, the empirical quantity for the *inverse meter-atomic mass unit relationship* expressed in ***u***, well in close figurate match with the Quanto-Geometrically derived quantity, is unquestionably telling us something about the proton and the neutron, if ever the closeness in their physical characterizations.

The inverse of this quantity is a relevant and well-known quantity as well:

$$\frac{q_3}{q_1} = 0.75147729$$

The corresponding quantity by CODATA is $\mathbf{0.75130066166 \times 10^{13}\ m^{-1}}$, the *atomic mass unit-inverse meter relationship.*

12.10.1 Derivation of Neutron Wavelength and Mass of Muon

Quite straightforwardly, the product of the q_1/q_3 gestalt scalar ratio with the derived mass of the electron yields the neutron Compton wavelength according to the relation:

$$\lambda_n = \frac{1}{m_e \cdot \frac{q_1}{q_3}}$$

$$\lambda_n = \frac{1}{5.6970042820\ 5\sigma \cdot 1.330712196}$$

$$\lambda_n = 1.3199007448 \cdot 10^{-1}\sigma^{-1}$$

The denominator of the relation is a scalar quantity and quite fittingly the inverse of this quantity (the whole rational term of the relation) yields an anti-mass dimension, that is to say a wavefunction characterization, in accordance with the primal Quanto-Geometric axiom stating that: $Q \times S = 1$, which implicates that one tenet is the inverse or reciprocal of the other. The official value of the neutron Compton wavelength is ***1.319 590 9068*** $\times$ ***10*** $^{-15}$ ***m***, a solid 4-digit figurate match with the theoretically derived quantity.

The gestalt scalar ratio q_2/q_3 for its own should yield the mass or Compton wavelength characterization of the muon particle, according to the root ratio:

$$\frac{q_2}{q_3} = \frac{0.352065327}{0.299796065} \quad \rightarrow \quad \frac{q_2}{q_3} = 1.1743493939$$

This quantity, however, remains completely unreported if at all known. What we currently know for a comparable quantity is an expression of the mass of the muon in atomic mass unit, an expression of its mass referenced to that of the neutron (or proton), that is. This quantity is instead rendered in our system by the q_1/q_2 gestalt scalar ratio:

$$\frac{q_1}{q_2} = \frac{0.39894228}{0.352065327}$$

$$\frac{q_1}{q_2} = 0.113314845$$

The known comparable quantity for the mass of the muon is ***0.1134289267 u***. Note well that in this case the unit digit is part of the figurate match, for a total of 4-digit match between the two quantities. It is important to underscore once again that all measurements in ***u*** stand for an explicit reference of the measured or computed quantity to the neutron or proton parcel. Because the physics of sub-atomic particles has turned out to be that of high-energy physics, and despite its unitary boost in the spectrum, the electron has been relegated to the back seat when it comes to the unit of measurement of mass and energy in favor of the neutron and the proton, which are located higher up in the energy scale. It is common practice to express the mass of particles in terms of their ratio with that of the proton as reflected in most physics textbooks and CODATA official publications. We recognize that the practice is not quite theoretically illegitimate however, given that those two particles represent the monolithic identities of the spectrum. This equivocation is certain to stem from the difficulties mind experiences in establishing referential or origination points in a naturally timeless system, which we have evoked at the outset of this part of our study.

12.11. Origin of the Proton Compton Wavelength

The foregoing developments have shown that the combination of the gestalt scalar ratios directly involve the electron, the muon and the neutron particles. How does the creation function deal then with the proton, an important integrant of the spectrum?

Physicists of the great generation of early 20th century have long considered the proton and the neutron to be the same particle, and that only a minor shift in some manner of an associated vector characterization would determine whether the particle ontologically shows up under one facet or the other. The isospin doublet concept, arguably an inappropriate choice of nomenclature, partially captures this characterization when it comes to the close values of their masses. The wavelength differential between a free neutron and a free proton is very minimal as well. The differentials between both of their masses and wavelengths are certainly not supported by any combination of the gestalt scalar ratios in that portion of the spectrum.

How then does the creation function allocate the parameters that characterize the proton particle specifically? The Hyper Quantum Operator simply directly *ports* the value of reciprocal multivariate-symmetry of basis space s_3 reflected in the image function to the proton identity in the new (composite) spectrum under development as a last resort! From the Grand Mappa Mundi function:

$$s_3^{-1} = 1.322875656 \cdot \sigma^{-1}$$

The projected wavefunction for the proton so acquires the conflated and slightly redimensioned wavelength value of $1/s_3$. One may formally construe this operation as a conflation transform to be expressed as follows:

$$[T_{11}] \otimes [s_3^{-1}] \rightarrow \lambda_p = s_3^{-1} \cdot 10^{-1}$$

or

$$\boxed{\lambda_p = 1.322875656 \cdot 10^{-1} \sigma^{-1}}$$

The official value of the proton Compton wavelength is ***1.321409856*** × ***10***$^{-15}$ ***m***, a quantity with an appreciable figurate match with the theoretical conflated value arising from the transform. As to the mass of the proton, having shown the deep Quanto-Geometric origin of the quantum relativistic relations, one may legitimately use in a theoretical exercise the mass relation m_i to compute the mass of the proton from its derived wavelength from the procedure exposed above.

12.12. Ambivalent Neutrinos and Origin of their Mass Quantity

One of the most fascinating particles in modern physics came out to be the neutrino which exhibits very peculiar properties, mostly uncommon to the rest of the flourished world of condensed matter particles. Many of its properties remain misunderstood if unknown while feverish investigation about this particle continues.

12.12.1 Ambivalent Phenomenology of the Neutrino

In Book I of this study, we have seen how the Hyper Quadratic Operator adjudicated two particular values of multivariate-symmetry space from the derivative of the image

Grand Eigenfunction to the electron in order to establish the unitary basis for the development of U_3's Shell1. These s values were: $\frac{1}{2}\sigma$ and 1σ. Said Operator adjudication gave us the *Electron Compton Wavelength* at $s = 1\sigma$, and the *Electron Gyromagnetic Ratio* at $s = \frac{1}{2}\sigma$. On that basis, the distribution of Shell2 Particle objects becomes essentially configured into two topological ensembles under the stamp of each of these two multivariate-symmetry space values, namely the world of overt-mass particles under $\frac{1}{2}\sigma$ and that of overt-space particles under 1σ. Following a different theoretical path, statistical quantum mechanics has been driven to the same conclusion through the determination of what has been called Fermi-Dirac statistics for fermionic topologies or fermion particles and Bose-Einstein statistics for bosonic topologies or boson particles. Fermions are commonly said to be *massive* and bosons to be *mass-less*. The correspondences are so clear they need not be addressed.

Within the Quanto-Geometric system one must understand this consignment as an overloading action onto the image function by the Hyper Operator in order to modulate the topological qualities of the distribution. This leaves the quadratic layer of the image function supporting the distribution with no possible adjudication of a distinct electron *s-value* wrapper, other than a conflation of both of the values of s in action, if ever possible, as a stamp to the new element. It is important to keep in mind that the element sitting at that level of the distribution is already under an ambivalent topological stamp due to the very quadratic quality of the layer. The topological object emerges from the composition doubling down on its ambivalent qualities.

Have we been talking about the neutrino? A resounding yes is the answer. Known to be a massless particle ever since 1950, more recent discoveries in 1998 pointed to the existence of some level of *massive* behavior implying the particle does have mass, however squint, by experimental observation; it is evaluated in the minuscule *0.2 to 3.5 eV* range. The exact mass scale of the neutrino and the exact amount of its different flavors or types still remain unknown. Yet since the particle does not have a Compton Wavelength it is fair to distinguish the nature of this mass from the common mass of fermionic particles. The travel speed of the neutrino is experimentally known to be extremely close to the speed of light, which puts its wavefunction in the category of a bosonic or De Broglie wavelength instead. However, because the particle does show mass behavior, it is acknowledged that the nature of its wavefunction cannot be exactly identical to that of vector bosons. Therefore, the sensible conclusion about the mass and wavefunction properties of the neutrino is that the particle both has something close to *massive* and something close to a vector boson wavefunction. This is well characterized within the Quanto-Geometric framework by the inherent qualities of the Primitive spectrum at the quadratic normal layer, with the attributes of equalized scalarity and wavefunction, which one may justly translate into *free* (or uneclipsed) mass and *free* (or uneclipsed) wavefunction. The two tenets of the articulation appear to be *free* because their mutual influence equates one another.

The discovery that the neutrino does have massive behavior, most often spelled out as simply "having mass", has called for a conceptual adjustment in relation to the nature of the different versions or species of these particles. A departure from the generation of 3 particles postulated in the Standard Model, peremptorily imposed by evidence of the existence of right-chiral eigenstates (to be further discussed), became unavoidable. The hypothesized new neutrino version has been coined *sterile* or *inert* neutrino. There is a number of conjectures still floating around in attempting to come to terms with the new reality of the neutrino, however sterile neutrinos have borne out something more than a plausible status at this stage and may be considered a universally accepted hypothesis, despite the fact that it implies a complete departure from the traditional theoretical framework represented by the Standard Model. We have identified sterile or inert neutrinos to be the representatives of quadratic particles homing on the Quanto-Geometric quadratic normal layer of the Spectrum. The attribute of sterile or inert has been chosen to signify that these neutrino types do not interact at all with matter in general, except possibly gravitational mass, in contrast with the other neutrino species of the triplet generation, namely the tau neutrino, the muon neutrino and the electron neutrino.

This notion is fully supported in Quanto-Geometric Theory when one takes care to look at propensity to interaction thru the prism of the relationship between mass tenet and wavefunction tenet of physical objects. Potential for interaction in that view results from the regions of unbalance between both tenets within the Primitive spectrum canon that makes the object couple with the environment as if in search of thoroughness thru a conjugate partner. Accordingly, a mass-dominant object would tend to couple to a space-dominant object, and reciprocally an overt-space object portends to couple with an overt-mass conjugate partner. Because of equalization between the two ontological tenets in a quadratic object, the element "feels" internally thorough and becomes unable to interact with surrounding elements. This is reflected in the functional derivative as the point where $dQ(s)/ds$ is maximal.

Therefore, the mass condition of an object is unquestionably a significant variable that must be treated conceptually with highest accuracy. The common notion that bosonic particles *do not have mass* is vastly inaccurate because experiments such as Compton scattering clearly shows that photon particles do carry a momentum, therefore must structurally harbor mass in their particulate composition. Just as we recognize all fermionic particles to carry an associated wavefunction, any theoretical approach aiming at faithfully capturing the physics of condensed matter parcels, must incorporate the notion that bosonic particles do have mass. The specificity about such mass quantity in bosons is simply that it is overwhelmed and its phenomenological expression therefore severely constrained. Whereas the peculiarity about the mass of the neutrino, an ambivalent element, is that it is neither overwhelmed nor underwhelmed – being simply covariantly *free*. Strictly speaking, the neutrino is not a fermion. Consequently, if the inert neutrino shows *massive behavior* or does *have mass* its mass expression is not indistinguishable from that of Dirac fermions. Because it is not a fermion, the natural point-source driven allocation of scalarity to the particulate structure of the inert neutrino substantially differs from that of the other fermions of the spectrum, for one, and for the other part, even with a smaller mass than the electron, inert neutrino mass does not hold unitary or scaling value in any way within the spectrum.

12.12.2 Quantitative Origin of Mass of Inert Neutrino

Consistent with the above, we are going to coin the term *neutrinon* to designate the inert neutrino, since within our framework as well as that of the Standard Model extensions it is a completely different particle than the twins pertaining to the triplet generation. Hence, for the 9 generations of particles of this Spectrum (Shell2 Particles) the naming nomenclature is: Protons – Neutrons – Muons – Electrons – *Neutrinons* and Photons. The symbol for the *neutrinon* will be v_n. The motivation for this slight change in nomenclature is strictly to appropriately designate a particle of its own class on the one hand, and on the other to facilitate the linguistics of this analysis. In some quarters, the inert neutrino has been tentatively classified as a Neutral Heavy Lepton, however we take exception with the term Neutral in this context and will wittingly discard that nomenclature.

Once the question of typification of the mass of the neutrinon is resolved, the remaining question to elucidate is that of the quantity of that mass. We do expect this quantity to be smaller than the mass of the electron, simply because mass within the hierarchy of the Spectrum diminishes from one normal layer to the next in order of progression toward the asymptotic layer.

Being a very elusive particle, the most plausible projection of the value of the neutrino's mass has required data from different sources to be combined together. Although the determination of difference between the values of the mass of each flavored neutrino does not pose a problem, exact determination of the value of the mass of each neutrino particle still faces serious challenges. The mass of all flavored versions of this particle are estimated very nearly equal to one another, the order of the difference between them being in the *meV* (milli eV) range. The KATRIN experiment, pending completion as of this writing, will be searching for a mass quantity down to *0.2 eV* and up to *2.3 eV*. By combining cosmic microwave background data with other data from galaxy surveys and baryon acoustic oscillations, an upper limit on the value of the conflated mass of the neutrinos has been obtained and evaluated to be:

$$\sum_i m_i < 0.23eV$$

When it very specifically comes to the mass of the sterile neutrino, this quantity remains totally unknown, only wildly approximated in conjectures by incipient extensions to the Standard Model aiming at tackling the neutrino problem. In Quanto-Geometric Theory however, the mass scale of particles is ultimately represented by the *q*-axis scalar scale of the Quanto-Geometric coordinate system, which maps the mass value of each normal layer of objects. According to the *q*-axis scalar scale the mass of the neutrinon can only be smaller than that of the electron, which is the immediately preceding layer of particles. Furthermore, the phenomenon of neutrino oscillation, discussed further down, which sees neutrinos spontaneously changing flavors, can only occur in our view when degeneracy currents hit the last tally of the scalar scale for overt or free mass. Therefore, the Quanto-Geometric framework puts the flavored neutrino masses on the same footing as the neutrinon. Consequently, the quantitative estimation of that mass at no more than $0.23eV$ validly includes that of the neutrinon for us.

Turning to our computational methodology, we note that in the context of the generation of this Spectrum the Operator constraints spelled out at the start of this section place a serious impediment on a naturally transparent allocation of mass to this particle. In addition to those constraints, because the neutrino is not a fermion or overt-mass particle as mentioned before, the computational method associated with allocation of mass from a radiant point-source scalar does not quite apply to the determination of its *free* mass. Hence the scalar value of the mass is left to be directly ported from the quadratic scalar q_4 from the image of the Grand Mappa Mundi to the particle's configuration within the 3-dimensional Spectrum under genesis.

The value of the quadratic scalar in the Grand Mappa Mundi function is:

$$q_4 \times \sigma = \frac{e^{-1/2}}{\sqrt{2\pi}} = 0.241970725 \text{ or}$$

$$q_4 = \frac{e^{-1/2}}{\sqrt{2\pi}} = 0.241970725 \cdot \sigma^{-1} \quad (4\text{-}6)$$

A transform T_ν of the kind that we have already seen for the proton wavelength is likewise hereby required in order to transmute σ^{-1} into the σ^{-5} unit so far at work for the scaling of masses:

$$[T_\upsilon] \otimes [q_4] \rightarrow m_\upsilon - q_4 \cdot \sigma^{-4}$$

Which yields, from relation *4-6*, the ported mass of the neutrino noted below, a real tiny mass in comparison with the mass of the four fermions of the spectrum:

$$\boxed{m_\upsilon = 0.241970725 \cdot \sigma^{-5}}$$

This quantity presents a non-trivial figurate match with the *0.23 eV/c²* reported above.

12.13. Vector Bosons

The bulk of our deliberations on the wavefunction in this study thus far has endeavored to show, and eloquently so we hope, that the wave function is not an etherish entity with only a life in abstract mathematical spaces but an artifact of real coordinate space, understood and visualized thru the prism of keen Quanto-Geometric constructs.

In the context of the three-dimensional universe, we have described its first characterization as multivariate-symmetry space s_3, which is as a member of the Operator $<q_3|s_3>$ deriving from the Grand Eigenfunction. This particular wavefunction, of which the Grand Eigenfunction yields at the same time the associated propagation rate, represents universal basis space for $U_3$and provides the encompassing matrix for the entire range of phenomena involved in the ontology of U_3.

Subsequently the Transform T_{19} has put forth for us the clearest expression of what the least amount of intervention over basis space s_3 amounts to, in the form of frequency (or wavenumber) characterization of the wavefunction of condensed matter particles. By giving us a generation of overt-space objects, the transformation reveals to us the innate qualities and behavior of space s_3 itself, one of which is chiefly its symmetric qualities.

Effectively the corresponding generation of objects demonstrates that property at a very pronounced level, captured in modern physics under the construct of Bose-Einstein statistics. In a population of such objects, their group behavior shows them to be absolutely transparent to one another, one element being able to superimpose onto another or any number of others. If one expresses the wavefunction of every particle in terms of its position, i.e. $\psi(r_1)\ \psi(r_2)\ \psi(r_3)\ \psi(r_4)\ldots\ \psi(r_n)$, the value of the grand product of these wavefunctions remains the same no matter how many times we permute them within the product. We exemplify this fact in the equations below for a number of 6 particles:

$$\psi(r_1) \times \psi(r_2) \times \psi(r_3) \times \psi(r_4) \times \psi(r_5) \times \psi(r_6) = a$$

$$\psi(r_2) \times \psi(r_1) \times \psi(r_4) \times \psi(r_3) \times \psi(r_6) \times \psi(r_5) = a$$

$$\psi(r_2) \times \psi(r_1) \times \psi(r_3) \times \psi(r_4) \times \psi(r_6) \times \psi(r_5) = a$$

$$\psi(r_6) \times \psi(r_1) \times \psi(r_4) \times \psi(r_3) \times \psi(r_2) \times \psi(r_5) = a$$

There exists one and only one other alternative form of arrangement to these objects, the matrixial arrangement, and it would yield a different outcome for each topology. So it is, because permuting two terms of a matrix changes the sign of its determinant. Note well that the final wavefunction figure for the whole physically observable population is in fact given by $|\psi|^2$ in this mathematics. This entails that even if two individual wavefunctions were to have their arguments of same value and different signs, the resulting value of the final wavefunction $|\psi|^2$ for the whole population would still be the same for all topologies. For this reason, and because in addition multiplication is commutative, the topology characterized by the wavefunction of these objects is said to be *symmetric*: the mathematics that yield this result is the Bose-Einstein population statistics which so describes group behavior of **photons** or more generally vector bosons.

Modern physics knows not how to specifically define the wavefunction of a single photon (not the same as the wavelength) and has engaged at such effects, among other driving reasons, in the long-protracted enterprise of the construction of Quantum Field Theory. The QFT approach has elected to grapple with the problem from the vantage of global topologies or fields. By contrast, the parameterizable relation underscored by our T_{19} Transform succinctly constitutes the expression of this elusive singular entity.

12.14. Field Theories' Erroneous Notion of Electric Charge

For 19th and early 20th century physicists, the question of defining and computing the mass of a particle necessarily entailed the inclusion of the portion of mass that resides with the particle in the form of energy subsumed in its electric field. They presumed that the mass of a charged particle should include the mass-energy in its electrostatic field. The electric field mass of a charged spherical-shell particle was then equal to:

$$m_{em} = \frac{q^2}{8\pi r_e},$$

where q is the particle's charge and r_e its radius. For the electron, the most commonly known particle at the time, its so-called classical radius, is given by:

$$r_e = \alpha \frac{\hbar}{m_e c},$$

where α is the fine-structure constant and the rational part of the relation stands for the reduced Compton wavelength of the electron. Since the mass m_e is measurable experimentally r_e is easily computed and substituted in the m_{em} relation for final evaluation of the latter. In summary the total effective mass of the particle comes to be the sum of its actual bare mass in the spherical shell and that of its electromagnetic mass equivalent. Furthermore, in attempting to determine the finite size of the electron as a point particle the question was posed as to what non-electromagnetic stresses would cause the electron to hold together against the mutual Coulombic repulsion of its finitely sized “parts”, which must presumably be negatively charged.

This was the beginning of a string of aggravatingly vexing problems that made their way from classical physics into quantum physics and severely hindered the progress of physical science until the 1980’s. These notions are altogether erroneous in many ways:

1. The idea of an electric field surrounding the particle as part of its own individual ontology is unsound and driven by early interpretation of observable behavior of the electron in 19th century physics.

2. The idea that the electric field surrounding the particle contributes to its ontological mass is flatly unsound from a Quanto-Geometric perspective as we have previously explained in relation with the space tenet in condensed matter objects.

3. The notion that the ontological extension of an entity would interactively react back over the entity to the same extent as a completely foreign element to the entity would is completely unsound from a philosophical viewpoint.

4. The notion of internal sub-components to the electron particle that are negatively charged and would make it necessary for a counteracting force to overcome such charges so that the electron could stand whole was another ungrounded hypothesis.

The mathematics developed to support those ideas notably ran into the unsolvable problem of the infinities, if not the unsustainable non-causal pre-acceleration of particles in the theory which portrays them starting to move even before application of any force! Before exposing the details of how Quantum Field Theory experienced the problem of the infinities, let us emphasize that in their approach to a quantum field theory in 1925, W. Heisenberg, M. Born and P. Jordan did develop a framework completely free of all notion of electric charges or current, a “free field theory” as commonly known today, which was free from the problematic infinities as it were. Unfortunately, two years later, P. M. Dirac reverted this trend by coming up with a Lagrangian that included the electric charge of the electron and the electromagnetic field as quantum mechanical objects. We have evoked in Book I of this study his bitter complaints toward his peers who appeared to be selling the problem of the infinities as no problem. It was only because P. M. Dirac’s subsequent work produced a successful response to the Stern-Gerlack experiment by uncovering the spin angular momentum of the electron that the strain he introduced in the development of Quantum Field Theory had been able to survive.

The electron-electron interaction and the electron-photon interaction are mediated by the attributed static electric field of the electron which becomes an electromagnetic field with the particle motion involved in the interaction. This self-energy of the electron creates a distortion in the form of an energy loop within the kinematic of the interaction. A pictorial method perfected by R. Feynman, which was to become extremely popular in modern physics under the designation of Feynman diagrams, firmly contributed to a clear visual representation of these interactions and their systematic treatment. Based on a mix of directional lines for particles and wavy lines for wavefunctions, they portray how two particles enter an interaction, exchange energy and leave the interaction. The diagram in Fig. 12.1 shows an electron-electron interaction with a distorting ultra-violet loop in the interaction.

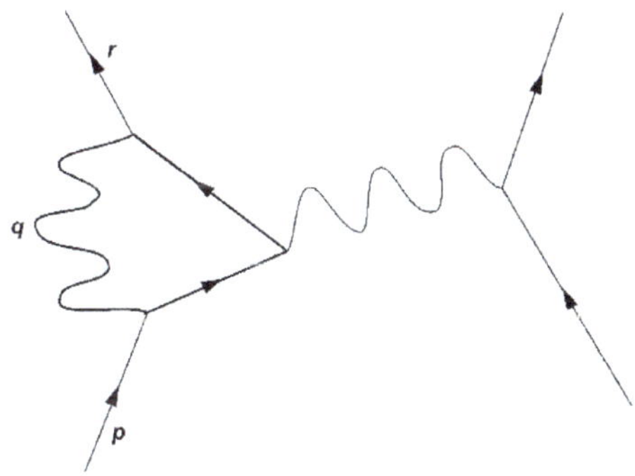

Fig. 12.1 Electron-Electron interaction with ultraviolet loop

The theory prescribes that in an interaction particles typically outgrow their shell due to this circumventing self-electromagnetic-energy field. For initial energies of both particles entering the interaction to be conserved, distorting self-energies from both particles must be equal and opposite to one another in order to cancel each other out as the interaction develops. Therefore, in order to determine the momentum-energy amplitude of the whole loop, it is necessary to integrate over all possible combinations of energy and momentum in travel around the loop. This is where vernacular language would state that "all hell breaks loose": these integrals are divergent and yield an infinite number of solutions. Below one such specimen:

$$e^3\gamma^\mu\gamma^\alpha\gamma^\rho\gamma^\beta\gamma_\mu\int\frac{d^4q}{(2\pi)^4}\frac{q_\alpha q_\beta}{(r-q)^2(p-q)^2q^2}$$

which long rang the toll for this theory. There were other problems with the theory aside from the ultra-violet loop divergence, in particular a second class of divergence called infrared divergence resulting from theorized kinematics of mass-less particles in the interaction. However, the most troublesome and unassailable one remained the ultra-violet divergence described above. Before QFT ever did, classical physics had run into these dead ends essentially originating from the theorized electric field around particles and attempted to invoke a procedure destined to tame those divergent integrals called re-normalization. Quantum Field theoreticians followed that vein and simply attempted in turn to arbitrarily manipulate these divergent integrals by introducing domains or ranges of validity for the troublesome variables. This much decried renormalization procedure was consummated in the early 1950's to the dismay of many, among them P. M. Dirac and even R. Feynman who termed it mathematical "hocus pocus", due to the heavy-handed suspect methods employed. It consisted in setting arbitrary points of validity for the value of the theorized electron's electric charge and off-shell mass, which were declared to not be *real* variables or, to be exact, *"not correspond to the physical constants measured in the laboratory"!* The Theory's starting formulae along with the normalization of the very quantum fields were directly and summarily modified in a call to *"constrain the theory to make contact with reality"!* Kenneth Wilson was one instrumental physicist in shaping the Renormalization Group Theory, a theory about a theory, or to be more precise, a theory about how to interpret and peruse a theory!

What's more, the Renormalization Group Theory did not even complete the job, because the unease about using infinities to cancel infinities ($\infty - \infty$, a mathematically forbidden operation) persisted until 1995 when Steven Weinberg, Felix Villars and others undertook to develop independently several different methods, under the name of Regularization, in order to put a more formal wrapper over this unsightly and unorthodox procedure. While R. Feynman came to fully accept QFT only in its incarnation as Quantum Chromodynamics (QCD) around 1974, the serious hurdle of the Landau Pole (read the infinities) being absent from this outcome, P. M. Dirac and many other theoreticians from the older generation never withdrew their critical stance toward QFT in general, in their eyes a framework that does not meet the test of good and honest science. A notable and telling absence through the course of this development, which nevertheless spanned several decades of his own productive years, is famed physicist A. Einstein: he had completely rebuffed and quarreled the "abstract and fictitious" quantum mechanics wavefunction concept for its own, busy as he was in his repeated but unfruited attempts to unveil an alternative deterministic description of condensed matter's wavefunction. Had physicists' mathematical sense prevailed during the pioneering decades of QFT, they would have dismissed this approach by virtue of the mathematical formalism known as *reductio ad absurdum*: starting with a premise that may or may not be true, you develop tight logical argumentation from that statement which brings you instead to a patently false outcome, then you have thereby proven that the premise is untrue! The premise of a cloud of photons making up an electromagnetic self-energy field around elementary particles in ground state had been mathematically proven purely and simply untrue by the absurd infinities!

In the end, one must underscore two salient facts out of this journey:

- Early on, W. Heisenberg and others had shown that the concept of electric field germane to the electron and by extension to all fermions in unnecessary to the incipient formulation of a Quantum Field Theory.

- Once the electric field artifact and its electromagnetic consequences make it into the Theory, the only way out of the ensuing intractable impasses is to proceed to remove them as premises from the framework, or else the theory stalls and falls apart.

When it comes to Quanto-Geometric Theory, we have shown in the first cast of this study two important facts worth highlighting here:

- On the one hand, the electric charge constant is not a particle-Shell physical constant but a molecular-Shell constant.

- On the other hand, the transforms that we have hereby uncovered, which are endowed with the unprecedented ability to derive the values of masses and wavefunctions of a whole set of key condensed matter particles, do not imply or have recourse to any electric charge property inherent to these particles in any way.

The electric charge and field are properties attributable to the molecular medium, reflecting on particle parcels when they are brought into this medium. One can never underestimate or undervalue the properties of the immediate spatial medium of an observable, a paramount flaw that has traversed all of physical science in the ignorance of the very consequential properties of the vacuum other than its so-called *polarization* and *zero-point energy*.

Classical physics had found that rubbed objects may behave in an attracting manner with regards to other objects, and has thus defined positively and negatively charged objects. It ruled that electrical charges of the same kind, whether negative or positive, repel each other but that charges of different types attract each other. Later the electron parcel was discovered and found to be responsible for the negative charge of objects. The knowledge of Electricity evolved into circuits and the science established that electrical currents flowed from the positive pole of a circuit to its negative pole. The discovery of atomic parcels soon prompted the identification of the proton as the source of positive charges. The loss of electrons in an atom results in increase in proton charges and contributes to make the material positively charged. At some point, the science understood that electrical current does not at all flow from the positive pole of a circuit but from the negative pole instead, because it was a result of the flow of electrons which are plenty at the cathode end. Further down the road, with the onset of quantum mechanics, we understood that in solid materials, there is no electron flow at all, in terms of parcels travelling from anode to cathode, but that the electric current in solids is essentially a travel wave. When the neutron particle was discovered, the idea of the electron being the source of the negative electric charge and the proton the source of the positive electric charge, had already solidified with Maxwell theory of electromagnetism and its derivative laws, such as Lenz's law, well established. The neutron, recognized to be an uncharged parcel, demonstrated however a magnetic moment, a heresy against electromagnetism. This was the beginning of troubles with this notion. However, physicists never questioned this notion and unsuccessfully attempted to fit the numerous observations of electrically neutral particles with yet a magnetic moment into current theories or relegated them to the stack of back burner problems. Quantum Field Theory however could not get around this problem which unknowingly defeated it time and time again.

Furthermore, our development has shown that the association of the photon particle with a presumed electromagnetic field is unnecessary to an understanding of the ontology of those types of condensed matter parcels. The categorization of fermions as "matter" particles and bosons as force carriers, prescribed by the Standard Model of particle physics, is not quite supported by Quanto-Geometric Theory. What we have shown is that fermions, photons and neutrinos belong to an authentic Quanto-Geometric Primitive spectrum. Fermions do not enjoy the status of particles any more than photons do. We have not arrived at this description thru the notion of perturbative fields akin to QFT's approach. We are not describing a field, whether of quantum or electromagnetic nature, with the photon and particle interaction arising as perturbations of said field. We have framed the photon in terms of its individual ontology in the T_{19} Transform. Besides, nowadays it is apparently well understood that the photon may be perfectly described without invoking an electromagnetic field, notwithstanding the fact that the parcel nature of the photon had unmistakably transpired thru the photoelectric effect and Compton scattering which were manifestly radiation phenomena. Lastly, it is noteworthy that Quanto-Geometric Theory returns that the most natural way of framing the ontology a fermion is thru its mass tenet, while the most natural way of framing the ontology of a photon is thru its wavefunction tenet, although all these expressions dually include a scalar quantity and a wavefunction derivative by canon.

I shall conclude this section by agreeing with René Descartes that the mark of great thinkers is their ability to put everything they know on the table in front of mounting evidence pointing to the contrary or overwhelmingly objecting to their dearly held truths. If Max Plank and Niels Bohr were demonstrably of that class, how many others in the field could or can make this claim? As to Albert Einstein, he patently showed so much heart that he openly conceded his blunders!

12.15. Quantum Field Theory's Particle Conjugation

The principal business of Quantum Field Theory is to formally describe the wavefunction associated with condensed matter parcels in terms of a mathematical function. However, proponents of the theory sought as well to make it render the observations of elegant Quantum Mechanics which had by then solidified. The third pursuit in developing the Theory was to achieve unification between relativity and Quantum Mechanics in order to recover undoubted physical unity in nature.

This triple goal, however laudable and ambitious, was a tall order to accomplish in one stroke. Early on P. M. Dirac was able to develop a Lagrangian for a single particle, akin to E. Schrödinger's Hamiltonian, which satisfied both Lorentz invariance of Special Relativity and the quantized ½-spin figure of the electron particle posited in Quantum Mechanics. Yet P. M. Dirac's equation ruled by extension the existence of "negative energy" and thus "negative mass" for known particles, which demanded physical interpretation. P. M. Dirac proceeded with the invention of Hole Theory, postulating that space is initially filled up with a sea of negative-energy particles that are invisible and undetectable despite their very real existence. These particles would occupy all the lower energy states, so that by preponderance of the Pauli exclusion principle all other particles are prevented from sharing those states of the vacuum. At times however, a negative-energy particle would escape the sea and leave behind a hole, which would act as a positive energy particle with a reversed charge. The latter would be the known protons.

Since critics found this portrayal to imply infinite negative charge of the universe, P. M. Dirac countered unconvincingly that the perception of universal negative charge was in fact the normal state of zero charge that we commonly perceive. Further Herman Weyl proved that this theory was inconsistent with the factual difference in masses between protons and electrons and that in fact the theory was completely symmetric with regards to particles' positive and negative charges. P. M. Dirac later argued that a sufficient number of negative-energy particles had been extracted from the sea, with their mass becoming thereby *positive*. Hence, they became the ordinary positive-mass particles that we see around and about. A sufficiently large number of them, he argued, had undergone that conversion so that the observable universe was possible. The holes left behind would be filled back up and endure as anti-particles. R. Oppenheimer and I. Tamm undertook to prove however that the theorized annihilation between positively charged and negatively charged particles present in the universe would cause an accelerated disappearance of physical matter altogether. In front of these serious objections, P. M. Dirac modified his Hole Theory again postulating this time the existence of the positron, a new particle of positive charge and identical mass to the electron.

He scored some success when a year later in 1931 the positron, the positively charged electron (or anti-electron), was discovered. However, the problem of infinite universal charge remained whole, as well as that of bosons not fitting into the picture because they do not obey the Pauli exclusion principle. Posterior developments in Quantum Field Theory by V. Fock, W. Furry and R. Oppenheimer specifically addressed these issues in a firm attempt to perfect P. M. Dirac's theory as best they could. They reiteratively started out with the mission of quantizing the hypothetical field surrounding an electron, that is to say set to have two and only two orientations when referenced to the momentum of the particle, in accordance with the Pauli exclusion principle. They incarnated this notion in the following composite exponential time function:

$$\varphi(s) = \sum_k u_k(s) a_k e^{-iE(k)t}$$

where k denotes the momentum and the spin of a charged particle in the coordinate z-direction, $E(k)$ the energy of the particle and a_k associated creation-annihilation operators. This function requires the following expression for the Hamiltonian of the particle's energy field:

$$H = \sum_{k_+} E_k a_k^\tau a_k + \sum_{k_-} |E(k)| b_k^\tau b_k + E_0$$

where H is the Hamiltonian for total energy of the particle, a_k^τ and b_k^τ the two exponential time functions from the premise, E the energy of the field and E_0 the negative energy of the vacuum. The first term at the right is a sum over positive energy states of the field and the second a sum over negative energy states. Analysis of the exponential functions a_k and b_k shows that one ought to be a creation Operator and the other an annihilation Operator. This is at best a mitigated success since, on the one hand, we do not obtain eigenvalues or eigenfunctions independent of time from this equation, hallmark of Quantum Mechanics in our judgement, and on the other hand E_0 is an infinite negative quantity that has to be removed by hand from the equation, so renormalizing it, for the total value of the Energy to make sense. Furthermore, this approach does not yield the value of +/- ½ for the spin angular momentum of the field, a quantity certain to characterize it.

What we learn from this extension of Dirac's model is that the quantum states of every particle-antiparticle pair can be interchanged in:

- electric charge by applying a Charge Conjugation Operator,
- chirality (left-right symmetry) by applying a Parity Operator,
- and in time by applying a Time Reversal Operator.

This translates into an abstract **CPT** space with the requirements of same mass m, same spin state J but opposite electric charges q and $-q$ for any particle-antiparticle pair. It also implies that only fermionic particles may have distinct anti-particle correspondents, further entailing that both bosons and electrically neutral particles cannot have distinct anti-particles. In the abstract **CPT** space, only a hybrid position of sorts is accessible to the latter, which sets them to have themselves as their very own anti-particles.

What is most remarkable from all this development is that the presumed value of the electric charge q of 1.60 x 10^{-15} Coulombs, in this theory a paramount figure quantifying this electron field of Coulombic nature, never partook of the calculations in any ways and was never accounted for. At the very least this Theory should justify the pertinence of fixing the charge of an electron to -1 by convention instead of just perusing it. There are good reasons to that lack which we shall visit later.

12.16. Quantum Field Theory's Time Inversion

The introduction of the time variable in the parity basis of the interaction forces studied in Quantum Electrodynamics is a force-fed element in the Theory that ultimately lacks physical support. The Feynman diagrams frequently show the arrow of time pointing forward for the matter particles while the anti-particles are shown moving opposite to the arrow of time. The underlying concept is that under time inversion transformation the dynamical properties of particles are inverted, as opposed to the interpretation of antimatter particles moving backward in time.

Whatever the proposed interpretation, the invocation of a time arrow in the parity basis remits to an inversion character dependent on the orientation of the time arrow. Because there is no known physical phenomenon, object or parameter assumed to be consonant with a negative time flow or time arrow, the referential parity of $+t$ and $-t$ becomes truly nonsensical and outright inapplicable. Entropy is generally the physical entity called upon when approaching the question of direction of time flow or time arrow. We must emphasize that all antimatter elements duly constitute an integral part of the real physical world: they are not "fleeting objects" that pop in and pop out of existence" as an all too frequent characterization has it. Case in point: atomic antimatter objects have perfectly been formed in the laboratory. Substitution of a proton by a positron in a hydrogen atom in order to form an onium atom has been realized, in a clear demonstration of coexistence of matter/antimatter in the real physical universe. An onium atom unobjectionably shows that the properties of an antimatter positron particle remain invariant in the timeless world of an atomic edifice of energy density states, even though this particular edifice is understandably unstable.

12.17. Particles in Coordinate Anti-Symmetric Space

Quanto-Geometric Theory approaches the question of particle-antiparticle pairing from a completely different perspective. Quite aside from the concept of time inversion, the notion of electric charge of particles is central to QFT treatment of particle conjugate pairs. Quanto-Geometry for its own provides a foundation for re-dimensioning that notion before applying it to the treatment of conjugate pairs. While it is perfectly reasonable to modify a theoretical model, preferably in a non-artificial manner, in order to make it yield known experimental observables, as we have seen with QFT in regards to particle conjugate pairs, the Quanto-Geometric model *naturally* incorporates most if not all analytical elements applicable to the matter under scrutiny.

12.17.1 Quanto-Geometric Coordinate Transformation

When exposing the model at the beginning of the first cast of this study, we have described a feature of the model labeled anti-symmetry gathered in the trigonometric or azimuthal plane of the model. We have stipulated under that umbrella three distinct forms of symmetry exposed in Fig. 2.13 Book I:

1. Anti-symmetry between adjacent quarters.
2. Transverse symmetry between matching threads of contiguous quarters.
3. Group or Family symmetry throughout a thread.

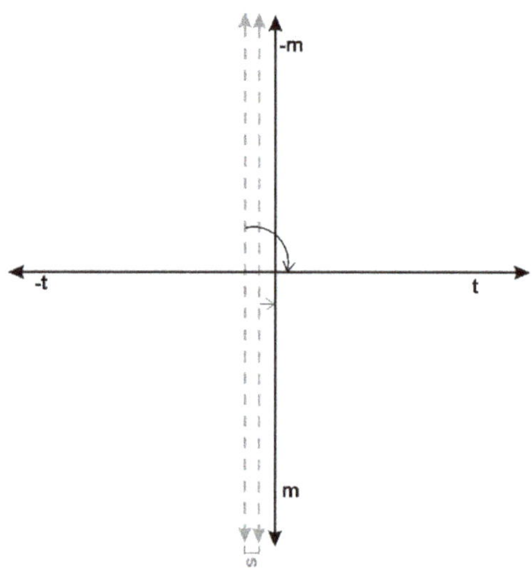

Fig. 12.2 One-Dimensional to Bi-Dimensional *s*-axis transformation

One may choose any of the 4 hemi-hyperboloids to represent the 3-dimensional universe hereby under scrutiny. Once chosen the azimuthal plane of that section finds its exact depiction as represented in Fig. 2.14 Book I. We find however that the hemi-hyperboloid centered on the positive *m*-axis uniquely positioned to provide a faithful representation of 3-dimensional reality (Fig. 12.2). There is indeed an authentic *transformation* beyond the split of the one-dimensional *s*-axis to the two-dimensional *m-t* coordinate plane. While this transformation splits the original *s*-axis in two and repositions the *m*-axis engender along the same direction as the former *s*-axis, it rotates the *t*-axis engender 90° from the original direction.

As previously explained axes *m* and *t* making up the new azimuthal plane represents the splitting of the two essential components of physical space: *symmetry* and *multivariance* (or *motility)* under the Quanto-Geometric approach . To that extent, the organization that takes place in the azimuthal plane is no more than an extension of the stakes implanted in the one-dimensional *s* variable. For that matter, one may further construe this transformation as a spontaneous act whereby the *motility t*-axis assumes the multivariance phenomenon intrinsic to *s* space which it translates into motion away from the original direction, while the *symmetry*-specialized *m*-axis does no more than statically conserving the original direction of the *s*-axis generator. Notwithstanding this spontaneous movement, the irrotation form of the *m*-axis transformation is due to a higher-level determinant in the context of the formation of U_3, as we will see later.

Further, the azimuthal space that results from the coordinate transformation is accompanied by a translation or projection of the multivariate symmetry values from the *s*-line or *s*-axis into the radii of the azimuthal circles on the *m-t* plane, whereas the azimuthal bands reflect the former normal layers in the graph of the *Q(s)* single-variable function.

For the hemi-hyperboloid centered on the positive *m*-axis, the one of interest to us thus far, let's note the following:

- Both of its quadrants yield equal values of the *m* (or geometric gestalt) variable for any inner point.
- One of the quadrants (left in Fig. 2.13 Book I) yields negative or lesser values of motility, given that they stem from the negative *t*-axis, than the other quadrant does, since these correspond to the positive *m*-axis.

This intrinsic unbalance of the quarter duplet making up the hemi-azimuthal plane *(-t/+t/+m*) translates into a qualified specialization of one of the two quadrants in *symmetric* or *geometric* valuation and the other in *multivariance* valuation. This characterization of the model establishes the foundation of what has transpired as polarization of the vacuum in Quantum Field Theory or even simply the Coulombic charge of the vacuum.

12.17.2 Depictors of Matter-Antimatter Symmetry and Asymmetry

A positive charge +*t* (would-be equivalent to the conventional electric + symbol) is a quality that establishes an object in azimuthal space with a multivariate-symmetry latency graded by the couple *(+t, +m)*. A negative charge -*t* (would-be equivalent to the conventional electric – symbol) is a quality that establishes an object in azimuthal space with a multivariate-symmetry latency graded by the couple *(-t, +m)*. Therefore, an electric poten tial is a complex or polar number, even though we may elect to drop its symbol to one only element of the representative couple because the second element *(+m)* is always the same. These charges are not a field that can be represented by lines of force going in and/or out of the object while circumventing it, as depicted in the usual electric field in classical physics or quantum theory. The charge latency is an intrinsic quality of the space making up the object determined by its azimuthal quadrangular origin. Quanto-Geometric azimuthal charge is therefore a very different concept than classical charge.

Attraction between those charges is a direct result of the motility latency component and only that latency type. Further, at azimuth $m = 1$, which is the positive m-axis itself, the Grand Eigenfunction is perfectly defined. So therefore, there exists a form of azimuthal charge laying half way between the two quadrants, which we call *neutral* latency, in essence a *hybrid* of both types above described. In this sense neutral does not mean 0 (i.e., 0 electric charge), but a cross in the true meaning of the term hybrid. This is a very important distinctive point to keep in mind.

The next characterization inherent to azimuthal space is what we might call *laterality,* a structural category that we have termed *group* or *family symmetry* in the exposition of the model in Book 1. These categories appear in Fig. 2.13 Book I as slices or threads delimited by the radii of the azimuthal semi-circles, and they are graded by values of the azimuthal arcs themselves. In the context of condensed matter physics, these threads typify the quality of *handedness* of condensed matter parcels. This quality is a direct reflection of the global character of chiral symmetry originating from the left-quadrant/right-quadrant division over the inner spindle partitioning of the quadrants themselves. To that extent, left/right categorization (mirror image symmetry or chirality) becomes part of the basis that establishes these sub-divisions.

We shall note, before proceeding further in that direction, that both the charge and laterality characterization in the azimuthal plane directly stem from the Grand Eigenfunction. We shall further add however that, with regards to the formation of U_3, they must be viewed thru the wider prism of the composition of Operators which intervene in the formation of basis space s_3, which we briefly discuss.

We have seen that the Hyper Quantum Operator is uniquely responsible for homogenizing (in other words applying unitary or monocentric symmetry to) basis elements of U_3 $(\langle q_3|s_3\rangle)$, giving us the totality of known laws of conservation. By the same token, it is also responsible for the azimuthal characterization of space s_3 by dictating a rotational motion to the *m*-axis transformation earlier studied for the coordinate background of U_3. While one may think that this symmetry may be of any kind, the Hyper Quantum Operator indeed expressly intervenes to specifically define that symmetry, setting it up to be spherical or *monocentric* symmetry in faithful rendition of its very nature. The Hyper Quantum Operator not only imposes a 90° irrotation to the *m*-axis transformation but further homogenizes the two azimuthal space quadrants of U_3 so created by *transforming* each of these two spaces in a physical state consistent with the nomenclature below:

$$(+t, +m) \rightarrow +T$$

and

$$(-t, +m) \rightarrow -T$$

In other words, instead of the many different values for positive or negative charges that would otherwise be possible for these couples of values inside the two quadrants, we end up with one single value for each. What's more, this constraint further gives rise to a hybrid spread of space stemming from the azimuthal intersection between $+T$ and $-T$, because the point $m = 1$ is a well-defined point. The boost *transforms* this hybrid intersecting space into a spindle spread we denote T_0.

Consequently, the $+T$ quadrant represents a form of polar space *unitarily* geared to exhibit mobility latency, while the $-T$ quadrant represents a form of space *unitarily* geared to exhibit geometric latency (because of its original deficit in motility). It means not that the other tenet $(+m)$ is completely absent from the expression of the specific quadrant, it rather means that it is eclipsed. It is important to underscore that these spaces do not refer to abstract mathematical spaces, such as those in QFT and gauge theories in general, these descriptors apply to real coordinate space in U_3. They are indeed the most towering descriptors for the asymmetry of the vacuum, which has been characterized in terms of the conjugate particle/anti-particle vacuum. P. M. Dirac had struggled unsuccessfully to give a plausible description of the vacuum in these terms through his many different relativistic interpretations of the vacuum. We do not believe that the creation and annihilation Operators are wholesale faithful descriptors of the vacuum because, among other reasons, we view as completely unnatural the resulting restriction that two completely existent elements in their full rights, such as bosons and anti-bosons, altogether do no more than annihilating each other constantly as if with no other ontological mission.

If this model faithfully portrays real coordinate space, we should expect the following four important characterizations:

1. two grand families of parcel objects spread across two broad states of background basis space lending them their intrinsic properties of symmetry, which are antithetic to each other.

2. a thriving family of parcels with an expanding multiplicity of objects corresponding to the very *motile* or *interactive* $+T$ spread vs. a pronounced rarity of objects if at all corresponding to the $-T$ spread due to its deep *motile* or *interactive deficit.*

3. a propensity for particulate objects inheriting the specific type of handedness grandfathered by the motile quadrant to experience particularly more intense interactions in comparison with their chiral conjugates.

4. a dissociative latency associated with the hybrid space spindle (T_0) as a consequence of its adjoining two opposite properties ($+T$ and $-T$) thereby creating a fault line for inherent segregation.

The great virtue of the above framework lies in its ability to *naturally* explain experimentally observed behavior of condensed matter objects consistent with principled guidelines, but absolutely not in terms of *violations* to ill-devised principles or *expectations* that find nature rebellious and incompliant, a complete theoretical aberration the like of gyromagnetic anomalies. Indeed, we are uncovering here the physics that physicists justly suspected all along must exist beyond **CPT** (Charge Conjugate – Parity – Time) symmetries and their violations, a conviction particularly reinforced by the late discovery of neutrino mass. The detailed analysis of the referred asymmetries we will undertake in due course.

We want to take the opportunity to further clarify and expand on item 4 above because it represents a characterization of azimuthal space of far- reaching consequences. To put it otherwise, the propensity to mobilize adjoined to an equal deficit in mobilization potential in the same element creates a structural under-current of instability. As a consequence, the mid-radial division across the hybrid spindle constitutes a latent fault line between $+T$ and $-T$ spaces. This categorization, symbolized by $\mp T_L$ and $\mp T_R$ informs hybrid azimuthal space with a second layer of internal symmetry, in fact of transversal asymmetry. This sub-discipline of symmetry subsumes both motile asymmetry and chiral asymmetry as gathered in the symbols. On chirality we shall expand later.

This sub-division of symmetry within the hybrid azimuthal intersection is indeed quite complex of an affair. While its universal class is that of hybridity, it harbors the sub-classes of juxtaposed association, overlapping association and neutral association. We will not however delve in this exposé into said ultra-structural level of the analysis and will leave that symmetry at a relative equivalence between the notion or class of hybrid and that of neutral.

The general formula that captures the symmetries buoyed in the azimuthal plane as proposed in Book I follows:

$$\widehat{Z}_{\vartheta} = \frac{\pi}{2|N|+1} \qquad (12\text{-}1c)$$

where Z_θ is any arc subtended by the basis azimuthal angle of distribution θ, N is any rational number. The N quantity represents an index of global symmetry for the azimuthal plane or its modulus of transverse symmetry and sets what we might call the azimuthal anti-symmetry degree of freedom. This formula constitutes the paramount numerical expression that crystallizes the scope and structure of the symmetry genomic underpinning azimuthal transverse symmetry.

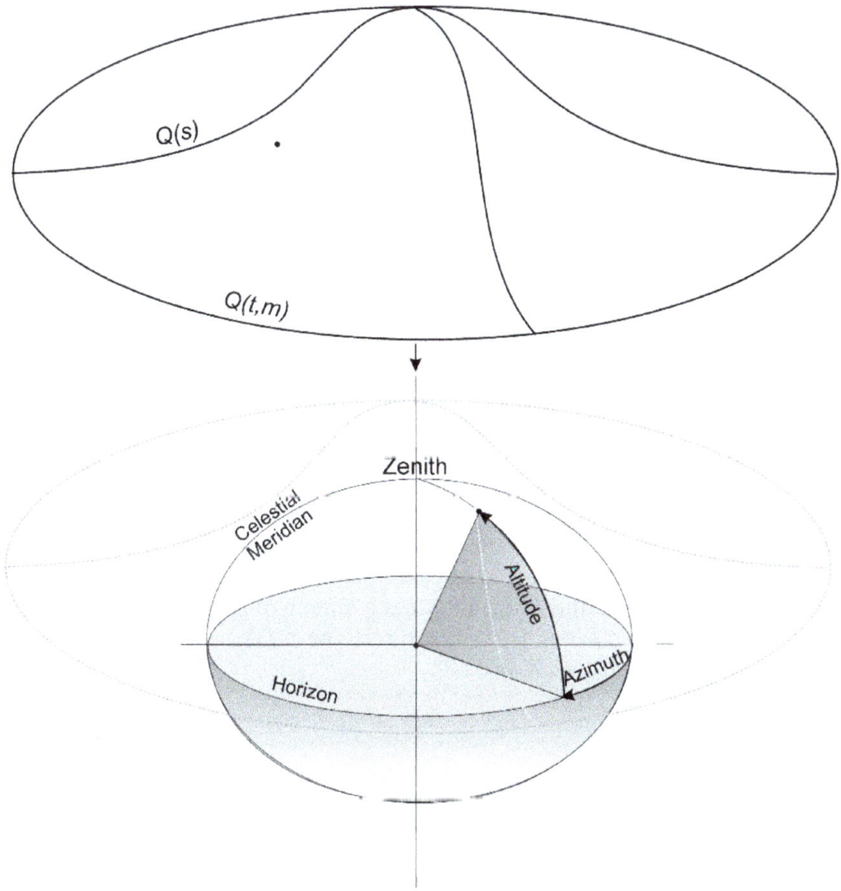

Fig. 12.3 Zenithal and azimuthal symmetry dimensions of the Quanto-Geometric model

It represents the quantization basis, in particular the Quanto-Geometric quantum number N_i for azimuthal symmetry, akin to the primeval symmetry-breaking sequence attached to the zenithal order (function of one variable s) of the Quanto-Geometric Model (Fig. 12.3). We will come back to this relation further down in the light of other critical categories of symmetry once studied.

12.18. Spin in Quantum Theory v. Quanto-Geometric Theory

Spin angular momentum is an important descriptor of quantum states that has arguably not received best possible interpretation in the Quantum Theory as a whole. Early proponents of the Quantum Theory did not quite know how to formally incorporate this entity, of unequivocal experimental evidence, in their equations. Spin angular momentum of a particle was originally expressed as a replicate of the orbital angular momentum of a bound electron, which is nevertheless an entity of a different nature. Under the impetus of Wolfgang Pauli, spin angular momentum of the electron was modeled after the orbital angular momentum with the Operators $\acute{L}^2$ and $\acute{L}_z$. Below reported these operators for the orbital angular momentum:

$$L^2\psi_{n,l,m} = l(l+1)\hbar^2\psi_{n,l,m}$$

$$L_z\psi_{n,l,m} = m\hbar\psi_{n,l,m}$$

The mathematical treatment found that the x and y components of the orbital angular momentum Operator (Lx and Ly) in the three-dimensional space of an atom must be equal to 0, leaving only the Lz component as actual descriptor of the motion. This outcome implicated major consequences in the physicality of the motion of a bound electron, since it meant a preference of the electron for certain directions in the 3-dimensional space but not travel paths traceable by vector analysis.

The experimental evidence provided by the Zeeman effect and the Stern-Gerlach experiment that the bound electron had intrinsic irrotational motion demanded an incorporation of that reality both in the Hamiltonian of the Hydrogen atom and in the Field Theory of free electrons or particles. Although these two forms of motion are quite different from one another, aside from their common angular nature, the adopted solution was to transpose these angular momentum relations to spin angular momentum. W. Pauli then proposed to restrict to two values the z-axis component of these operators, the only meaningful one, in accordance with experimental evidence, so that the resulting characterization for the spin angular momentum became:

- Magnitude of the spin angular momentum pseudo-vector also *called spin quantum number* or just *spin*:

$$|S| = \sqrt{s(s+1)}\hbar, \text{ with } s = \frac{\sqrt{3}}{2}$$

- *Z*-axis component of the spin pseudo-vector also called the *spin projection quantum number:*

$$S_z = m_z \hbar \text{, with } m_z = \pm 1/2$$

Posterior treatment attempted a more formal mathematical rendition of the spin thru the use of a different Operator deemed to represent a *generator of rotations*, the so-called Rotation Operator. Clearly an operator that attempts to capture the angular artifact of rotational motion of an object about its own axis, without a translation component, that is, has to seize what occurs in the neighborhood of 0-value for the translational angle (not the rotational angle). If $\check{n}$ is the rotation axis and Φ the translational angle, one may define a Rotation Operator $R(\check{n}, \Phi)$ relating to the spin angular momentum operator in t he form of the following expression:

$$J_n \equiv i\hbar \lim_{\phi \to 0} \frac{R(n,\phi) - 1}{\phi},$$

where $J_{\check{n}}$ is the angular momentum operator, $R(\check{n}, \Phi)$ the rotation operator, *1* the identity operator, the very rotation axis, that is. By extracting $R(\check{n}, \Phi)$ the relation becomes:

$$R(n,\phi) = \exp(-i\phi J_n / \hbar)$$, where *exp* is matrix exponential.

In that manner one creates an abstract space of rotations as the basis to grade the spin or helical rotations of condensed matter objects. Sub-operators such as *L* or *S* in that space become *modified partial rotation operators*. The most important consequence of this approach is that rotations are not necessarily 3-dimensional, with a 360° rotation not necessarily representing a full turn. Generally, a 360° rotation in that abstract space represents less than a 3-dimensional full turn , only small rotations tending to equate their 3-dimensional image. This is the *SU(2)* symmetry group of rotations of the Standard Model.

While this treatment is completely valid, the resulting physical interpretation commonly contended is completely inappropriate. Quantum physicists generally extrapolate from this treatment that particle spin is only a metaphor and that spin is unrelated to any sort of motion in space! They have lost sight of the fact that a theoretical model is about *interpreting* physical reality from a virtual construct but never about *equating* the construct to physical reality itself. It is no wonder that the profound theoretical deficits and heavy-handed habits of confounding theory with reality long in use in QFT had led to those deviations of sorts.

Another erroneous assessment from these developments is the stipulation that electrons, and in general all fermions, are *point-like particles that do not have a size*. This notion is both philosophically and physically unsound. We have reiteratively asserted in this study that the ontology of all physical objects is based on the mass-space doublet. If an object were to lack the space component which gives rise to its wavefunction, it would have to be "absolutely" *massive* and would as such vanish ipso facto leaving no trace behind. The dimension of an object is a direct extension of its Quanto-Geometric wavefunction or space component. Further this quality pertains for the most part to the *symmetric* or *geometric* tenet of the space variable, its *m* sub-variable, since all dimensional construct possesses an intrinsic geometric quality.

If an object has "no size", it can never yield a form of poly-composite of its own that is itself endowed with size. It stands to good reason and mathematical sense that 3 or 4 times 0 is still 0. However, the phenomenology of a *bound* electron is that of an object with measurable size and observable shape, the radial probability density cloud that an atomic orbital turns out to be. The radial probability density contour exists because the angular spread that the contour stands for springs not only from the mobility of the inner object in action but very importantly from the dimensions of that object as well.

One other related misrepresentation in QFT to be put on the same account of wrongful assessment is the notion that the muon particle has no internal structure, akin to the electron, and of which it is presumably a surrogate. How can a particle that spontaneously decays into others lack at all internal structure, not experience structure at the very least in a potential manner, or otherwise not possess an *ultra-structure* which would conceivably dictate the main fault lines (so to speak) of its decay process? The physics that produce this assessment is patently at odds with principled philosophical constructs.

Coming back to spin, it turns out that particle spin is an artifact of the space tenet and its wavefunction. Condensed matter objects spin not because of their mass, but because of their spatial wavefunction. Their ontological space component is indeed the primal *generator of rotations*. The angular momentum operators constitute, as they should, abstract models, but they model real 3-dimensional objects, despite quantum physicists' contention to the contrary. Quanto-Geometric Theory may endorse these angular momentum operators only in that condition. If all one can do with a Theory is to stage another Theory as the end product of the exercise, instead of producing clear and unequivocal interpretations of physical observables, then we shall agree with A. Einstein that the Quantum Theory is vastly and badly incomplete and would have to be put to rest at this stage, I might add. Every time that the mind loses visibility into the logic behind a span of observables, legitimately or not it tends to seek refuge in abstract spaces for interpretative schemes. The abstraction however must map back to the physical reality along clear lines of correspondences in order to be of any usefulness. Yes, one may use *SO(3)* or *SU(2)* abstract theoretic groups to decipher the meaning of spin, however all constructs in that context must map back to *SO(3)*, the group of 3-dimensional rotations which is the real physical universe of objects, with clear explanations of 3-dimensional behavior of intervening elements, and not stop short at the fictitious "internal" spaces as if they were the final stake of the exercise. Any interpretative outcome that fails to address and make explicit 3-dimensional behavior of the observables remains insufficient, if at all acceptable, no matter how formal the mathematical treatment.

Notwithstanding the fact that the spin angular momentum with symbol *s* in the Quantum Theory is not quite interchangeable with the independent *s* variable in Quanto-Geometric Theory, the values both variables take in both theories make for an interesting parallel, when it comes to physical interpretation of the spin. Quanto-Geometric Theory distinguishes between a range of semi-integer values of which ½ is promoted to highest significance by Operator composition, and another range of fractional and integer values of which *1* is promoted to highest significance by Operator composition as well. In the Quantum Theory common values for the spin angular momentum are 0, ½, 1 and 2 with semi-integers that are multiples of ½ and integers that are multiples of *1* (or simply integers) , all of which is to a large extent a subset of our symmetry-breaking *s*-values of the space wavefunction. Quantum spin $1/2\hbar$ or Quanto-Geometric *s* degree of $1/2\sigma$ is the unit of motility in U_3, the smallest possible modulus of motion, which is a representation of the helical de-ploy or un-folding of space s_3 (read *closed* translational motion). Quantum spin $1\hbar$ or Quanto-Geometric *s* degree of 1σ is the unit of motility for overt-space objects in U_3, the smallest possible modulus of *open* translational motion which is a measure of linear de-ploy or un-folding of space s_3.

12.19. Spin in Euclidean Space

Other properties associated with spin will no doubt be interesting to analyze. First let us briefly revisit the Stern-Gerlach experiment which puts particle spin under a quite shiny light (Fig. 12.4). The experiment essentially channels a beam of silver atoms with outer-shell solitary electrons, so to speak, across an inhomogeneous magnetic field which in turn constrains them to horizontally spread over a target screen. One would expect the orientation of the axis of helical rotation of the particles to be randomly distributed in the 3-dimensional space of the laboratory. As a result, given the non-uniform shape of the magnet, they would create a dark spot on the screen at the location indicated in the illustration, because the particles would arrive at the screen with all kinds of inclinations of their helical axis of rotation. However, they unexpectedly hit the screen forming two and only two up and down lines indicating that they arrive at the screen only in two angles of inclination of their helical rotational planes. Since there is no other constraining factor in the space of the laboratory, one must conclude that the "up and down" state of the spin plane is a property inherent to the ontology of the particles.

Now it is important to realize that even if one were to turn around the whole apparatus, the electrons would keep forming the same outlining pattern on the screen, insensitively to the orientation or rotation of the apparatus. If you rotate the magnet a bit, the double-line pattern on the screen would take a different orientation, but the doublet character of the lines will persist. Therefore, we may positively conclude that the spin orientation doublet of the particles constitutes a symmetry that remains invariant under 3-dimensional rotations. This is group theory parlance in use in the Quantum Theory and its Standard Model.

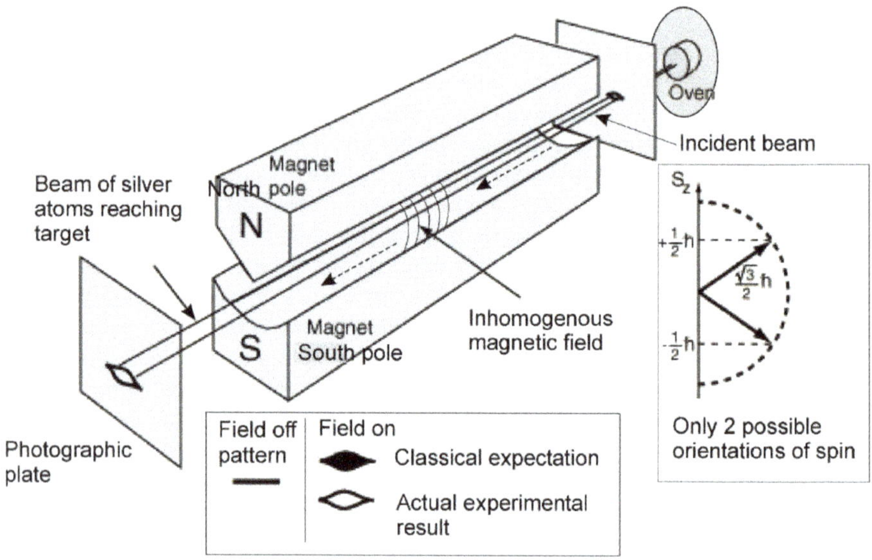

Fig. 12.4 Stern-Gerlach experiment putting particles' spin in evidence

In fact, what one should firstly conclude is that the space tenet of the particle has demonstrated holding a life of its own that is independent of the *space spread* in extension in the laboratory. In Quanto-Geometric parlance, the spatial spread that gives rise to the spin character attributable to the wavefunction of the particle is a Shell2 type, while the spatial spread in extension in the laboratory is at the very least a Shell5 type, the astral Shell. Quanto-Geometrically we do not expect Shell2 space to behave exactly the same way as or to be tightly dependent on Shell5 space, because we know that their local gauge or tensor level if you will, which transpires in the values of the physical coupling constants that sustain both Shells, have a different value for each. Further the fine Zeeman wavelength lines showing from the electronic orbital in hydrogen spectroscopy cast clear evidence as well that the outer shell electron of the hydrogen atom does conserve its spin orientation doublet, despite the incessant dance of the whole atom in all directions. This effect, from an electron that does have orbital angular momentum, unlike that of the silver atoms in the Stern-Gerlach experiment, manifestly advocates for the spin orientation doublet of the electron to be positively and thoroughly conserved as well within Shell2 space, the Atom Shell. The electron spin doublet is obviously equally conserved in Shell4 space, the Molecular Shell.

12.19.1 Classical Analysis of Spin

The most common representation of the origin of particle spin in modern physics is the relationship between mass and electric charge of the particle. If we conceive the spin angular momentum as an entity created by the spinning mass around an axis laid across any diameter of the more or less spherical particle ball, then the electric charge associated with the particle will generate a circulating current in the rotation plane, which in turns creates a magnetic field with a dipole moment in opposite direction to the angular momentum vector.

Classically, the strength of the ensuing magnetic field is given by:

$$\mu_s = -\frac{e}{m} S,$$

where μ_s is the magnetic moment, e the electric charge, m the mass and S the spin angular momentum. S however cannot be measured, only a component along an axis of a coordinate system. Measuring the spin along a z-axis, the above relation becomes:

$$\mu_{s,z} = -\frac{e}{m} S_z \quad (12\text{-}5)$$

Because the spin is quantized in two values of direction and a single value in amplitude, the expression that captures this behavior is:

$$S_s = m_s \frac{h}{2\pi}, \quad \text{for } m_s = \pm 1/2$$

where h is the Planck constant (the smallest quantity of action or energy) and m_s the *spin magnetic quantum number*. When the spin pseudo vector is parallel to the z-axis, m_s is positive and the spin is said to be *up*, while the latter is said to be *down* when the spin pseudo vector is anti-parallel to the z-axis. Consequently, by replacing S_z in Eq. *12-5*:

$$\mu_{s,z} = \pm \frac{eh}{4\pi m}$$

We so arrive at the quantity defined as the Bohr magneton, the smallest possible magnitude of a magnetic field, which is that generated by the spinning electron:

$$\mu_B = \pm \frac{eh}{4\pi m} = 9.27 \times 10^{-24} J/T$$

It became common practice to establish μ_B as the unit of spin magnetic dipole moment and to express the spin of all condensed matter particles in terms of μ_B. So therefore, for the electron:

$$\mu_{s,z} = 1\mu_B$$

12.19.2 Quantum Analysis of Spin

The above is all well and good except that the measured magnitude of the electron's magnetic field does not match that value. It is in fact **twice** as large! The quantum physics of the electron denoted Quantum Electrodynamics, in several different formulations here already evoked, had wrestled with this problem for three decades. But this analysis exposes the reason why the magnetic moment of elementary particles had been dubbed *anomalous*. It is a misnomer that more justly applies to the analysis itself: the reason why the result is so badly inaccurate is because the premises of the analysis are wrong! Therefore, anomalous are the notions behind the analysis and not the natural physical quantity itself!

A *g* correction factor to the spin magnetic dipole moment had to be introduced in order to prompt the Theory to make contact with reality. The resulting new formulation of the spin magnetic moment for the electron follows:

$$m_S = -g_S \cdot \frac{\mu_B S}{\hbar} \qquad (12\text{-}6)$$

where S is the angular momentum, g_s the g-factor correction of 2 and $\hbar$ the reduced Planck constant. If the g-factor along with the Bohr magneton and the reduced Planck constant are taken as a single quantity, this formula expresses the spin magnetic moment as a negative value of the spin angular momentum, which means that for the electron the the spin magnetic moment is anti-parallel to the spin angular momentum (Fig. 12.5).

These developments took place during the period known as the era of older Quantum Theory when classical thinking was still very fresh and vivid among physicists. Yet despite the correction factor introduced by George Ulenhbeck and Samuel Goodsmidt in the new formulation, the situation was unsatisfactory to most physicists, notably W. Pauli and Lorentz. W. Pauli objected to the fact that such g-factor equal to 2 is much unknown to classical physics and Lorentz computed the speed at which the electron charge would have to be spinning in order to produce the experimental value of the magnetic moment, which puts it significantly beyond the maximum limit of the speed of light, a relativistic paradigm.

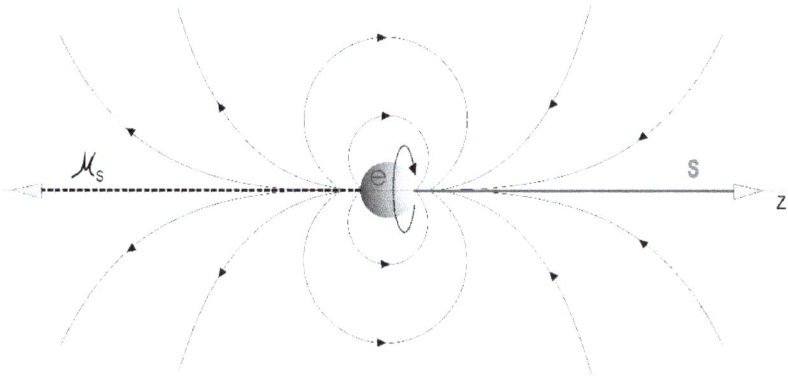

Fig. 12.5 Spin angular momentum and magnetic moment of the electron

Hence the Theory of Quantum Electrodynamics that followed attempted, under the influence of W. Pauli and P. M. Dirac, to accommodate an anomalous magnetic moment *a* for the electron particle related to the *g* correction factor, according to the following formula:

$$a = \frac{g - 2}{2}.$$

Generalizing, the spin magnetic moment anomaly a_i for every particle correlates to a particle-specific correction factor g_i according to the relation:

$$a_i = \frac{g_i - 2}{2}$$

The *g*-factor for the electron in accord with experimentation is: 2.00231930436. The Theory was adjusted to produce this outcome by invoking quantum electrodynamical effects concerning loops and vertex functions represented in Feynman diagrams. Following one such diagram (Fig. 12.6).

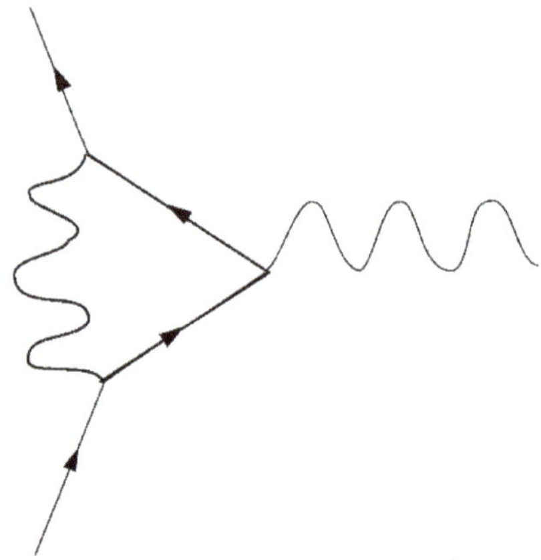

Fig. 12.6 A typical Feynman interaction loop diagram

We have already explained the thinking behind the nature of these loops in terms of a self-interaction of the electron via the supposed electromagnetic cloud surrounding the particle, and how these led to the undesired divergences. It took several more decades to arrive at seemingly acceptable calculation procedures known as renormalization and regularization initiated by Julian Swinger in 1948. It is the same J. Swinger who found a relation between the value of the anomaly a and α, the fine structure constant, as reported below:

$$a = \frac{\alpha}{2\pi} = 0.001158 \text{ , where } \alpha = 729 \times 10^{-3}$$

So impactful was this finding that in recognition of J. Swinger's entire contribution to physics this specific formula was later inscribed on this tombstone at his passing. It led to the QED expansion for the g_e correction factor for the electron:

$$\cdot\ g_e = 2(1 + \frac{\alpha}{2\pi} + ...) \text{ or } g_e = 2(1 + a + a^2 + ...)$$

The experimental value is 2.0023193043617(15). Dirac's Theory, despite all its complex mathematics, has only been able to find g_e equal to an even *2*. It is noteworthy that the gyromagnetic anomaly value of *2* may equally be obtained by linearizing the Schrödinger Hamiltonian equation, as well as the Klein-Gordon equation leading to Dirac's. It took the Theory of Quantum Electrodynamics a long journey of many decades (from the 1920's to the 1970's), in a game where several notable voices had to blow the foul whistle, to finally arrive at numerical results approximating the experimental value in a satisfactory manner. The Theory claims the ability to have been able to "predict" these values to an unprecedented degree of precision in the history of modern physics! Note well however that the famous QED precision tests rely on the determination of the empirical value of α, which makes the claimed precision almost a circular argument. For the claim of precision to legitimately hold, the Theory ought to have derived the fine structure constant as well. Further, the term "prediction" in this case has to be toned down in favor of **inference**, which is more descriptive of the historical process, given that the line between first principles posited for the analysis and the target experimental values of the observables has been at best tortuous, not to say broken. Finally, to put it simply, there is no way to be absolutely certain that the computed "12-digit precision" number is the real number, and this is what QED precision advocates acknowledge when they caution that "the precision of the computed *g*-factor depend upon extreme precision on the value of α, which can only be obtained from another QED experiment".

If the premises and principles invoked by the Theory at the outset were correct and legitimate, it would not have needed the introduction of an arbitrary correction factor based on experimental evidence, to start with. We identify the problem to be essentially in the claimed dependency of the spin magnetic moment upon the *c/m* ratio. As previously stated, the mass of the particle has next to nothing to do with the spin magnetic moment, the spin being directly a function of the space variable or the space wavefunction. If QED had a good handle on participation of mass in the numeric analysis, it would have expectedly shown the straightforward ability to compute g_μ, the spin correction factor for the muon particle, to the extent that the latter is deemed identical to the electron except for its much higher mass. The g_μ value cast by QED from its numerical analysis is patently not in agreement with the experimental value, admittedly a very long-standing problem in QED.

Secondly, QED remains completely moot in regards to the question of why the neutron, an electrically neutral particle, does show a spin magnetic moment, that is furthermore similar to that of the electron in direction, with its own g_n correction factor. Quantum Chromodynamics (QCD) was born as an attempt to resolve this conundrum, but its response remains unsatisfactory to many, despite its contribution in other areas.

Thirdly, if the electron bears a superficial charge in circulation inducing a current that creates the circumventing magnetic field, a QED visualization owed to J. Swinger, what does the photon bear in order to create its spin component of value 1? The photon particle is knowingly electrically neutral as well. One is left to wonder, by contrast with the electron visualized with a cloud of circumventing photons, what would the photon itself be featuring in its immediacy as a particle in order to produce its observable magnetic moment?

Fourth, the electric charge is a building-block constant of a higher Shell of matter, totally different from α, the fine structure constant, which rules the architecture of the

Particle Shell. In fact, the Quanto-Geometric expression of α transpires at every turn when the *quantum electrodynamic* corrections are applied to the anomalous moments of elementary particles in order to recover their true experimental values.

The principal problem confronted by QED throughout its development is the reliance on 18th century classical concepts in the interpretation of quantum phenomena. Classical physics is chiefly about higher Shell phenomena such as those taking place in the Material Shell and the Astral Shell, the human living medium. Exporting higher Shell variables to the Shells of the infinitesimal is no more and no less than a risky proposition that can only be successful in full knowledge of the absolute energy scale of matter.

Very telling is the fact that QED was at the end able to propose a valid equation of the electron magnetic anomaly in terms of α, while no conclusive results were achieved directly and strictly under the assumption of the electron charge in circulation or in association with its mass. From our perspective, the fine structure constant is the very physical constant playing the role of building block of the Particle Shell. It is that unfortunate conceptual inheritance from classical physics that dwarfs the interpretation of the mathematical-physical outcome turning it into abstract reality, because of its departure from classical conceptualization. The notion that the electron is not moving at all irrotationally, because of the dissonance of QED's classically-laden mathematical development at addressing the issue, is plainly improper.

Undoubtedly, if modern physicists understood the true nature of the fine structure constant as the central quantity to the configuration of all Shell2 condensed matter particles, they would long have adopted its proposed expression by J. Swinger, though not at such effects, as its most authentic numeric expression. Effectively the reduced version of the empirical quantity (729×10^{-3}), which is $\alpha/2n$, gives it a figuration of the Quanto-Geometrically derived quantity. By the same token, one can see that it is the pluri-composition of that quantity which produces for the most part the anomaly relation that quantifies the correction for the true value of the spin. What this plainly means is that in its reduced expression (or its Quanto-Geometric expression) α represents an authentic scale for the spin and that a formulation outside of the charge and mass constructs is fully conceivable. This also means that even when one attempts to express lower Shell phenomenology thru higher Shell variables, one still ought to go thru the local symmetry central to the lower Shell.

The Quanto-Geometric approach to the question of particle spin is as definite as it is resolute. First let us note one other interesting twist in the QED treatment. Going back to the *4-3* relation, we may choose to express it in terms of the ratio of the spin magnetic dipole moment to the spin angular momentum as follows:

$$\frac{m_S}{S} = -g_S \cdot \frac{\mu_B}{\hbar}$$

This gives us the well-known Electron Gyromagnetic Ratio constant, which we have well derived in the course of the study exposed in Book I (See Table 4.3 Book I)) with the expression: $0.17603266/\sigma^2$. The par value in modern physics is: rad/s·T, a dimension of an angular velocity per unit of thermodynamic temperature. There is nothing abstract about this quantity. The Quanto-Geometric derivation of its equivalent instructs us that the nature of massive particle spin is the **derivative**-wavefunction of the Grand Eigen-function, evaluated at the particular value of s = ½ of the multi-variate symmetry line of U_3 space. Several interesting facts arise from this prescription:

1. Spin is an irrotational wavefunction of a mathematical derivative nature.

2. Particle spin wavefunction is an expression of probability density distribution of their internal scalar parcel(s).

3. Massive particle spin wavefunction is parameterized on the quanto-geometric wavenumber of ½.

There is no cloud of any nature surrounding condensed matter particles, just as there is knowingly none surrounding an atomic orbital taken as an object, i.e that of an ionized hydrogen atom which truly appears as a stand-alone proton-based distributive object. What the atomic orbital object purports to be is a spatial probability density distribution of a smaller or inner scalar object which happens to be the massive or overt-mass particle. The space spread beyond the immediate probability density contour is still the object. It is strictly the quality of this space spread that mediates interactions of the orbital object with its environment as we shall study later, it is never something that reacts back onto the object. Nor is there anything in the interior of the maximal density contour , such as a charge, that projects outside or outward as a field. Likewise, particle Shell objects are regions of probability density distribution with sub-particle(s) in circulation, in the terms that we have previously exposed in the prelude to the quanto-relativistic analysis devoted to them.

In the broader analysis, it is worth highlighting once again that the quantities systematically reported in Table 4.2 Book I expose the quantization foundation for all U_3 measurables beyond $\langle q_3|s_3\rangle$. Those quantities expose the scale for motion momenta (both angular and linear) in the incarnation of σ^*, which occur by degeneracy from c as well, the scale for thermodynamic temperature with , the scale for energy packet with h and the scale for energy radiation with λ_C. At the top of this canvas are the basis units for space parameterization or parameter space. However, this space, far from being an abstraction, is all but real because it provides the scale for the symmetry-breaking wavenumbers.

Combining these foundational units with one another in order to gauge physical measurables, although quite feasible, may nevertheless not turn out to be useful and revealing, because the units are already unified on the σ scale. On that account, singling out $\hbar$ as the natural unit for spin did not reveal much about the nature of spin, quite aside from the fact that this determination is not Quanto-Geometrically supported.

I shall conclude this section on the note that if the Electron Gyromagnetic Ratio, or its spin magnetic dipole moment, does not characterize a real-world entity in motion, as commonly contented in the Quantum Theory, then none of the other three physical constants of the group, namely the Planck constant, the Compton Wavelength and the Stephan-Boltzmann constant is real either. With that very few if any physicist would agree.

12.20. Chirality of Condensed Matter Particles

If the structure of condensed matter parcels is deemed uniform, despite their possible various harmonic shapes, one would tend to believe any one particle can be from its peers strictly from their superficial layout. In general, forms of primal symmetry such as a sphere or an ellipsoid are invariant from the point of view of their laterality or chirality: they look the same left or right. Yet, for good reasons, the Standard Model of particle physics, regards the aspect of laterality of condensed matter parcels of utmost significance. Some unfavorably argue that this feature was artificially implanted in the Standard Model.

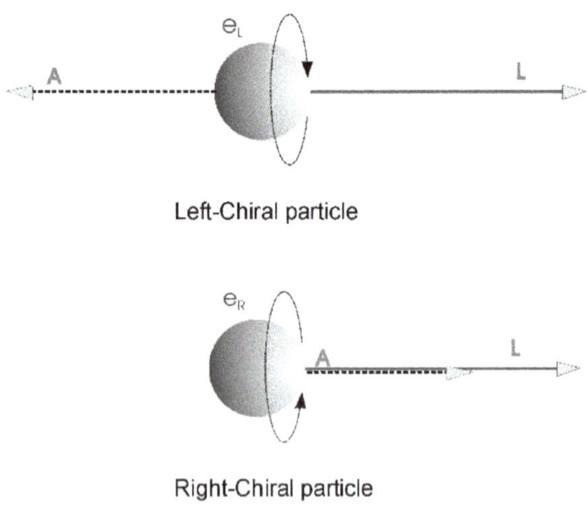

Fig. 12.7 Definition of particle's chirality property

We argue however that chirality is in large part a consequence of the azimuthal quadrant subdivision in of the Quanto-Geometric model. This high-level division, which is a left-right category in the graphism of the model, reflects back on the actual composition or topology of each quadrant.

Chirality is based on spin orientation and is intrinsic to all infinitesimal matter parcels we call particles. It is a property that is independent from the larger observational reference frame, even though the reference frame may contribute to make its detection more or less difficult to achieve. The nature of the wavefunction inherent to the particle, of which the spin is a direct expression, may equally render its detection from a higher reference frame more or less difficult depending on whether the wavefunction is overt or covert.

The orientation of the spinning motion of the particle is used to define its chirality or handedness. Chirality comes to be formally based on the orientation of the spin angular momentum vector relative to the linear momentum of the particle (Fig. 12.7). This characterization makes use of the right-hand rule for vector angular momentum, according to which if the sense of right-hand curled fingers is in accord with the sense of the spinning, then the thumb points to the direction of the linear momentum. When thumb points in same orientation as the linear momentum vector, then the particle is right-handed, else it is left-handed. In order to handle chirality on a numeric basis, the qualifier of helicity has been defined and set to be the ratio m_s/s between the z-component of the spin and the magnitude of the spin.

It is important to stress that the spin orientation of the particle, which defines its chirality or handedness, is a property inherent to the particle whether this physical attribute is detectable or not from a specific reference frame of observation. When the motion of a right-handed *massive* particle is lesser than that at of the reference frame from which it is being observed, its linear momentum vector would then point in the opposite direction of its actual motion, which would then falsify the observation of the chirality of the particle. This situation cannot arise for bosons, which are particles naturally always moving at the highest possible speed in U_3, the speed of light, owing to their overt-space nature. Thus, the chirality of bosons, because of its frame independence, is always an undistorted observable.

If an electron is moving to the right and its spin oriented for right-handedness (in reference to its angular momentum) , then the magnitude of the magnetic dipole moment is parallel to its spin angular momentum. Hence the particle is right-handed with spin up and may be symbolized: e_R^U .

If an electron is moving to the left and its pin oriented for left-handedness, then the magnetic dipole moment is parallel to its spin angular momentum. Hence the particle is left-handed with spin up and symbolized: e_L^U.

If an electron is moving to the right and its spin oriented for left-handedness, then the magnitude of its magnetic dipole moment is anti-parallel to its spin angular momentum. Hence the particle is left-handed with spin down and symbolized: e_L^D.

If an electron is moving to the left and its spin oriented for right-handedness, then the magnetic dipole moment is anti-parallel to its spin angular momentum. Then the particle is and symbolized: e_R^D.

In other words, the spin may be either clockwise or counterclockwise with respect to the center axis of rotation, while in either case the particle may be travelling either along the sense of the positive axis of rotation, or along the sense of negative axis of rotation, per axis orientation of the quanto-geometric coordinate system. This results in a 4-state spinor in multivariate symmetry space s (not Minkowski spacetime as the Dirac spinor). As a consequence of the splitting transformation from the unitary dimension s of the space variable to the explicit multivariate symmetry plane (*m-t* plane), the 4-state spinor becomes explicit and conserved within the azimuthal plane, so casting its very basis layout in 4 main partitions per quadrant, and inside of the hybrid spindle equally. The 4-state spinor laid on the azimuthal platform constitutes the foundation of particle genesis and distribution in the 3-dimensional realm. The graphical symmetric display portrayed in Fig acc represents the foundational symmetry genomic behind the symmetry patterns, transformations and interactions studied in the Standard Model of Particle Physics, which we address in further detail down below. Again, the reader ought not to take the term genesis here in its strict meaning of origin or primal source, but only in the sense of a re-production, given that all the particles currently present in the universe have always been there! They can only experience annihilation and re-birth as solitons or integrants of a group.

In summary, the above characterization integrally shows three main hierarchies or divisions of rotational symmetry informing the configuration of condensed matter particles. The first division is that of quadrant-based anti-symmetry essentially pivoting around the motility quality of multivariate space s_3, with values $+T$ and $-T$. Pictorially, these quadrants are reduced to two large spindles in their graphism due to the emergence of the second division of symmetry. The second division of azimuthal symmetry is that of hybrid motility with value $\mp T$ (read plus **and** minus T or neutral T). Should one prefer the term *neutral* T for this item of the nomenclature, one shall keep in mind that neutral here is not and means not 0 but (+) **and** (-) concomitantly, in the acception of hybridity. Lastly the third division of azimuthal symmetry is represented in the 4-state spinor, a symmetry that cuts across both the first and the second division.

12.21. Periodic Distribution of Condensed Matter Particles

Based on the divisions of azimuthal symmetry above analyzed, the distribution of condensed matter particles proceeds in a periodic pattern very similar to the familiar periodic table of atomic elements in many respects. This should not come as a surprise given the overwhelming role played by spinors and the Pauli exclusion principle in atom distribution based on electron quantum energy shells. What may be tantamount to a pleasant surprise is the quasi revelation of the foundational symmetry genomic behind the constantly arising symmetric patterns in particle physics, which physicists knew for long must exist, but could never uncover since the inception of the Standard Model by Glashow, Weinberg and Salam in the 1970's.

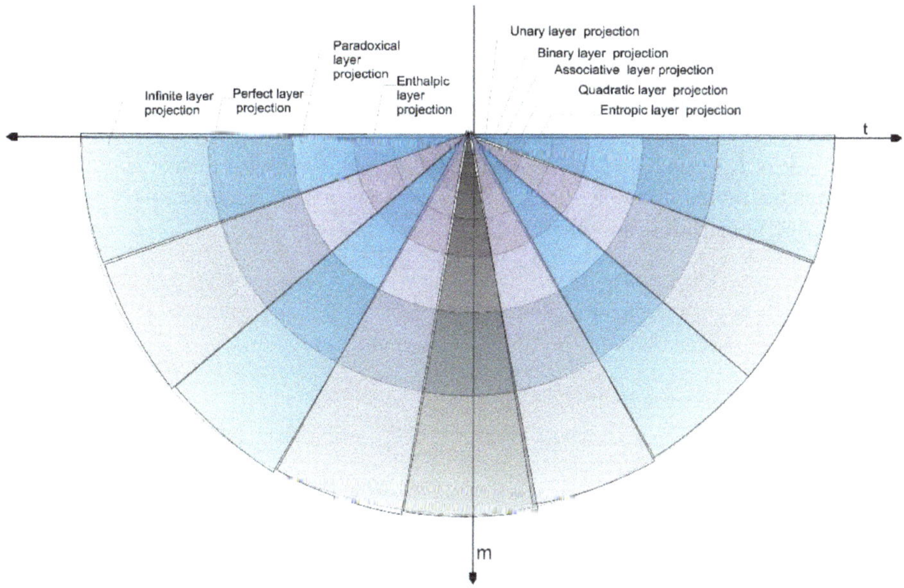

Fig. 12.8 Azimuthal partitioning of the Quanto-Geometric model with overlaid Primitive layers

The charts in Fig. 12.14 - 12.19 do not represent all the particles inventoried in the Standard Model since the quarks and other singular hadrons are not hereby portrayed. The latter do not belong to the Primitive spectrum of Shell2, but their own, the Shell of proto particles or Shell1. Nevertheless, they are subject to the same symmetry genomic here in display, as we shall see in due course. The 4-spinors are represented on both the left azimuthal quadrant and the right quadrant with 4 spindles each. The neutral or hybrid spindle lies in-between the two groups of 4 on the m-axis. The linear sides of these spindles, radii of the azimuthal circles, each represent the primal quanto-geometric wave-number line. Each azimuthal band represents a particle generation or a particulate Primitive. The trapezoid-like partition represented by the cross between a spindle and an azimuthal band stands for a real particle object, very specifically one of the 4 subdivisions within the trapezoid.

An expression of the *(12-1c)* relation, the general formula that captures the encompassing symmetries buoyed in the azimuthal plane, in angular measure is:

$$\theta = \frac{180^o}{2|N|+1} \qquad (12\text{-}7)$$

where θ is the basis azimuthal angle of distribution, N is the integer that sets the anti-symmetry degree of degree freedom.

N is a variable per Primitive Spectrum depending on other intervening factors when the whole spectrum participates in meta-level interactions, which is almost always the case. At the end, it is the number of spindles in one quadrant, which for the current Primitive Spectrum is the number of spin states accessible to particles. Consequently, for the current Primitive Spectrum $N = 4$. So therefore, from relation *12-7*:

For $N = 4,\ \theta = 20°$.

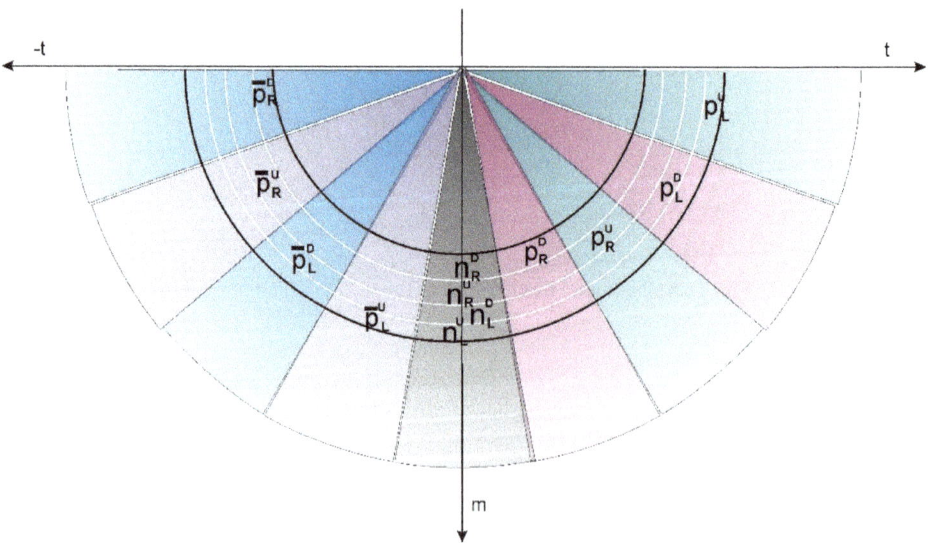

Fig. 12.9 Primitive Period 1 of Shell2 Particle Spectrum as an azimuthal band with spinor subdivisions. This period is formed by protons and neutrons.

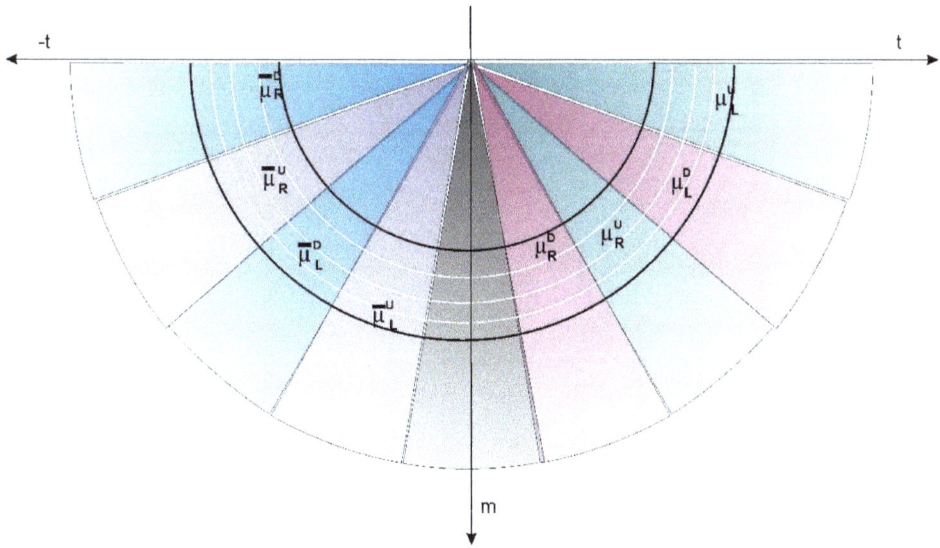

Fig. 12.10 Primitive Period 2 of Shell2 Particle Spectrum as an azimuthal band with spinor subdivisions. This period is formed by muons.

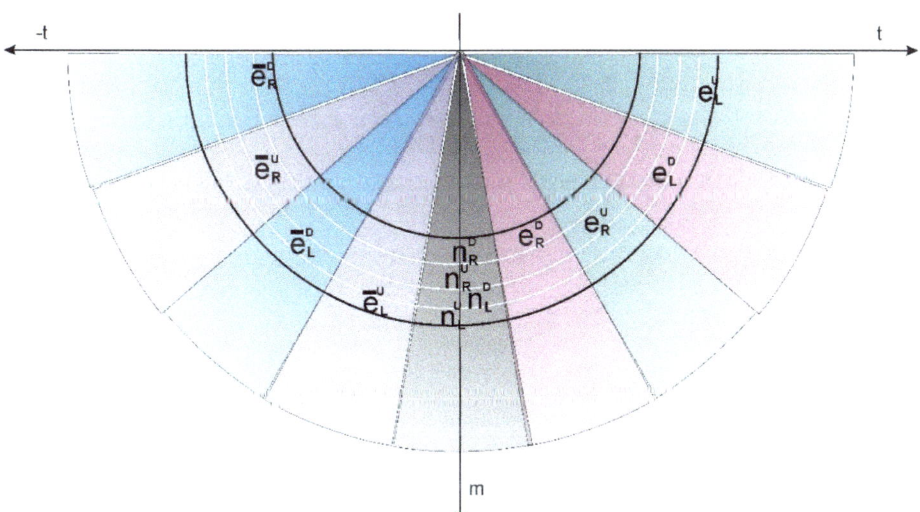

Fig. 12.11 Primitive Period 3 of Shell2 Particle Spectrum as an azimuthal band with spinor subdivisions. The period is formed by electrons.

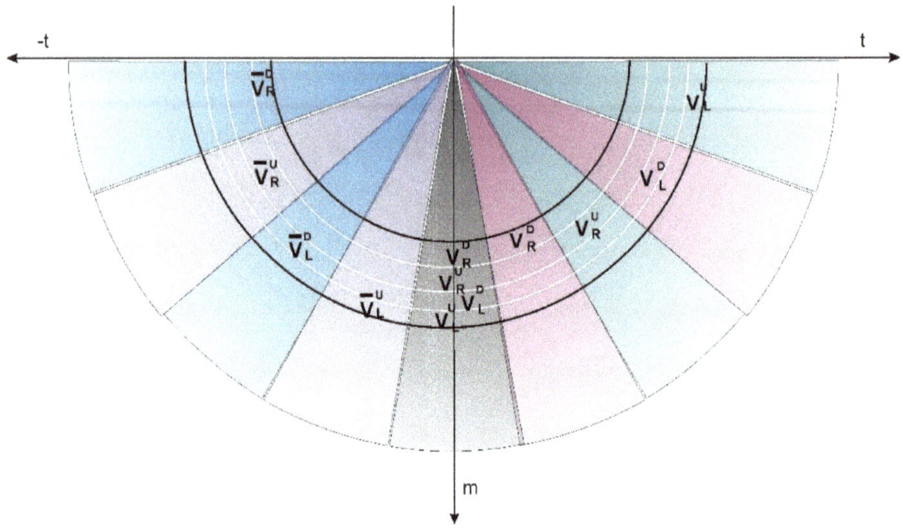

Fig. 12.12 Primitive Period 4 of Shell2 Particle Spectrum as an azimuthal band with spinor subdivisions. This Period is formed by neutrinos.

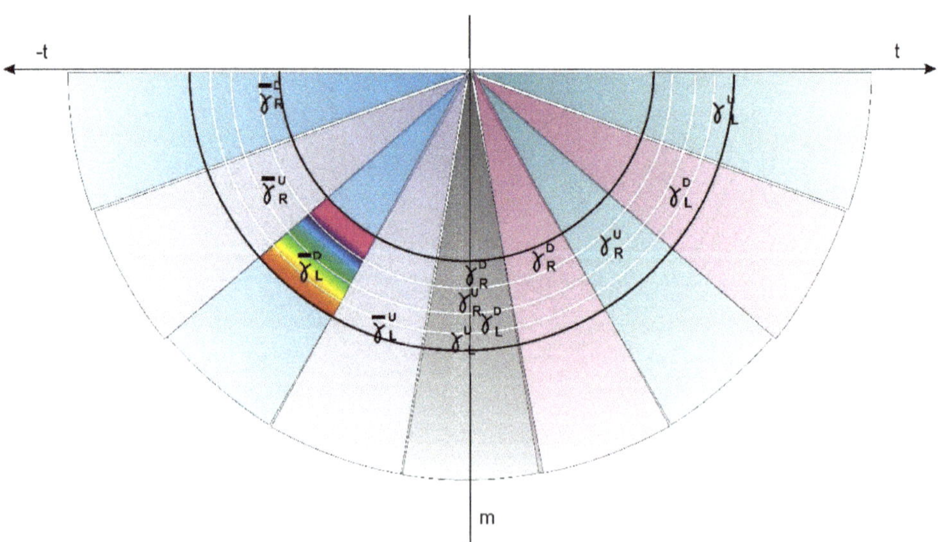

Fig. 12.13 An overt-space Primitive Period of the Shell2 Particle Spectrum as an azimuthal band with spinor subdivisions. The Spectrum comprises 5 photonic azimuthal bands: Primitive Period 5 up to Primitive Period 9.

12.22. Majorana Statistics Under the Quanto-Geometric Model

The discovery in 1998 that neutrino particles have massive behavior caused widespread commotion in the worldwide physics community because for the first time the Standard Model of particle physics received an arguably fatal blow stricken at the heart of its axiomatic. Chirality is so entrenched in the Theory that some have justly dubbed it a chiral theory. For long Quantum Electrodynamics has been developed around the idea that neutrinos are uncompromisingly massless and that only two chiral states exist for neutrino conjugates, left-chiral for neutrinos and right-chiral for anti-neutrinos, in accordance with all experimental observation hitherto conducted. Concerns were that if neutrinos are massive, then in regards to the neutrino triplet **each** of the two matter-antimatter flavors must have **both** left- and right-chiral states and the Higgs mechanism by which particles acquire mass, one of the towering postulates of the Standard Model, suddenly becomes groundless. Effectively according to the postulated mechanism, if a left-handed neutrino attempted a collision with a Higgs particle, it would have to become right-handed, because the spin-0 Higgs changes the handedness of every colliding particle while funneling mass into them. Because the right-handed neutrino does not admittedly exist, then its collision with the Higgs is a strictly forbidden event in the Theory: hence neutrinos *cannot acquire mass*. In that manner, mass-less state for the neutrino in the Standard Model goes hand in hand with the real-world absence of right-handed neutrinos and left-handed antineutrinos.

The serious blow stricken to the Theory with the discovery of massive behavior of neutrinos has created the need for some form of extension of the Standard Model that may still credibly account for neutrino physics without raising other insurmountable problems. This unsolved question remains whole to this day. We have seen that Quanto-Geometric analysis thoroughly endorses the hypothetical inert neutrino contended by these extensions of the Standard Model because this neutrino species finds natural fit in the Quanto-Geometric model. This leaves the status of the traditional Standard Model triplet neutrinos to be clarified in terms of their classification.

After now almost two decades of musing, the extensions of the Standard Model addressing the neutrino mass issue are still immature and raise on one side contradictions extremely difficult to resolve and on the other quite farfetched hypotheses that are utterly difficult to falsify. At the heart of the problem is a lack of *representational* basis in the Quantum Electrodynamics theory that would allow seizing the neutrino particle's peculiarities in their full dimension. Effectively in order to theoretically harness the neutrino, one needs a scheme that is able to fully embrace spaces of a quadratic nature because there are not one but two layers of intermediacies or crosses in the ontology of this particle as we will shortly see.

Let us first note that in our azimuthal model, all particles are accessible to the full 4-state moduli of motion as each normal layer subdivides into 4 sub spaces, each corresponding to one of the 4 spinors. On that account, the neutrino as all other particles occupies all 4 spindles in both quadrants of space representation. This leads us to a much-needed clarification on matter and anti-matter with particles denoted to be their own anti-particles. This terminology wrongly lends itself to the interpretation that there are out there particles that spend their lives in the company of their anti-particles in an endless annihilation struggle. Most particles are observed as beams, a pure beam of kindred particles being a polarized one. But since photons of both polarizations can be emitted in a general interaction process, it becomes pointless to distinguish between a photon and an anti-photon. Then one generically says *a photon* in reference to the elements of the beam or the process, in the understanding that the term applies to both the photon and its anti-particle. This leads us to make the statement that the anti-particle of a photon is a photon. In the case of the neutrino, the conflation could not be made because a neutrino was considered to have only one-handedness, not the two possible ones, and it was necessary to explicitly refer to the neutrino (a left-chiral element) when one meant to designate the matter-like element of the species, and to the anti-neutrino, the right-chiral element of the species, whenever one meant to designate the antimatter-like element.

Now that *massive* or mass-prone behavior of the neutrino has been observed and attributed to the inert neutrino (or *neutrinon* in our own designation), the question posed within the spirit of the Standard Model is this: can neutrinons or antineutrinons be differentiated only by chirality? In other words, do right-handed neutrinons and left-handed anti-neutrinons exist as separate particles? Contemporary physicists are still musing over the answer to this question because, among other reasons, their referential framework is heavily based on the attribution of an intrinsic electric charge (positive, negative or neutral) to elementary particles, which plays an important role thus far in the determination of matter versus anti-matter in most cases. The Quanto-Geometric answer to the question posed is a resounding yes. Let this be universally clear: all matter-like particles are all independently accessible to the full 4-state spinors (right chirality and left chirality) as all antimatter-like particles are. As shown in Fig. 4.18 the quadratic layer of distribution, home to the neutrinons, shows the 4-state spinors for them to occupy spindle after spindle, akin to all other layers.

12.22.1 Majorana Distribution

In relation to the characterization of a particle being its own antiparticle, it is important to note that no physical law dictates that wherever there is a boson particle, like a photon for instance, there must be as well an anti-boson such that their fate is to constantly annihilate each other. The common statement that photons are their own antiparticles leads to this erroneous conclusion. It is not unfair to say that in the Theory of Quantum Electrodynamics this view appears to be officially stamped, even though some quarters of physics push back against it with the conceptual adjustment that we have posted above. Be all that as it may, there is no more truth in this visualization of boson and anti-bosons obligatorily and perpetually annihilating each other than there is in giving a mandatory and perpetual antimatter companion of the same kindred to any and every fermion and having them annihilate each other as their sole destiny.

This exception having been acknowledged, there is indeed a dimension that seizes the neutrino with the possibility of being its own anti-particle, in fact this dimension applies to all elementary particles. This is where Majorana statistics come handily into play. In the Quanto-Geometric vision, a Majorana state is simply the hybrid or intermediate azimuthal state (T_0) between the two quadrangular sub-spaces making up real coordinate space. The Majorana state, Quanto-Geometrically assessed, is not a property that requires neutral electric charge and deemed only accessible to chargeless fermions and the chargeless bosons. In fact, the Majorana state, as a view of fermions being their own anti-particles as originally postulated by E. Majorana, must be understood within the context of Quanto-Geometric hybrid azimuthal space.

Let us briefly revisit the Majorana postulate. The Majorana equation is a relativistic wave equation akin to the Dirac equation sustaining fermion topologies with the difference that it includes the charge conjugate ψ_c of a spinor ψ. The equation follows:

$$-i\delta\varphi + m\varphi_c = 0$$

with the derivative operator including the γ matrices along with a summation over the spinor components, ψ_c the charge conjugate of ψ.and m the Majorana mass. The quanta contemplated in this equation are two particle species of which one is the neutral matter particle and the other its neutral anti-matter conjugate. One can immediately see that this characterization reflects the premises that we have set forth for Quanto-Geometric hybrid azimuthal space, reflecting even the axial fault line subdividing the hybrid azimuthal spindle. Effectively the Majorana equation is predominantly an equation of state. Frequently the condition that $\psi = \psi_c$ is introduced so that the depiction portrays a single particle with a single multi-state Majorana spinor. One must however take this restriction as a particularity applied to the generality stated in the original Majorana's equation. Majorana's approach to frame the possible existence of this specific topology within the spectrum of condensed matter particles narrowly approximates a faithful mathematical representation of Quanto-Geometric Theory's hybrid azimuthal spatial states. There is a number of experiments currently underway attempting to probe whether or not the triplet neutrinos are Majorana particles. It is our hope that the conceptual adjustments brought about in this exposé will positively impact the conditions and targets of experimental neutrino physics going forward.

Our contention is that all generations (Spectrum's normal layers) of particles, fermions, neutrinos and bosons, exhibit Majorana particles. The graphical depictions above displayed (**Fig.**12.12 to 12.15) clearly reflect hybrid azimuthal space to be accessible to not only inert neutrino particles but universally to all generations of particles within the Spectrum, since the hybrid spindle cuts across all normal layers. However, while the quadrangular spindles yield only one spinor state of choice for occupation at a time, the hybrid spindle at mid-azimuth gathers all 4 spinors for concurrent occupation by a specific particle in each generation throughout the Spectrum. *The occupation of all 4 spinor states* *represents a constraining circumstance which lends a boost to the background of dissociative latency inherent to the hybrid nature of mid-azimuth space, forcing its explicit expression as particle degeneration.* There lies the deep origin and true dimension of subatomic particles' **spontaneous** degeneration known as decay or radio-activity, otherwise called the electroweak force. (We are specifically referring here to spontaneous degeneracy outside of all outright Shell3 atomic context). We must add however that final expression of degeneracy results as a composition between the dissociative latency and spinor constraints of the hybrid spindle on one hand, and on the other, the very Primitive typology characterizing the layer of distribution.

12.22.2 Neutron Particle as a Hybrid Monolith

On the monolithic layer of distribution, at hybrid azimuthal space we find the neutron particle per Fig. 4.15. The neutron is registered as an electrically neutral particle. When asked the question about the electric charge of the observed anti-neutron particle, which looms as the main differentiating factor between matter and antimatter particles, physicists invoke the quark engenders to which the Standard Model has arbitrarily assigned fractional charges. The assignment of these charges is done in a manner to have them cancel each other out in order to account for the no net electric charge of the particle. Quantum Chromodynamics has it that the three quarks making up the neutron particle have one fractional charge of +2/3e and the other two set at -1/3e . The charge distribution is identical among the the three anti-quark making up the antineutron. There is two problems with this hypothesis, in fact three.

1. The quarks cannot manifest as free particles so that the charge assignment become impossible to falsify.

2. If the electron is acknowledged to provide the unitary basis for the configuration of particles pertaining to this spectrum, only multiple but not fractional charges of the electron can be construed as defining values of charge.

3. The verifiable magnetic moment of the neutron constitutes a challenge to the non net charge status of the particle since long standing theory recognizes that any charge in motion produces a magnetic moment and any magnetic field in motion concomitantly produces a (non-zero) electric field.

Many physicists remain uneasy about the arbitrary assignment of fractional charges to the quark constituents of the neutron and acknowledge that this notion has been parachuted within the Standard Model and lack principled support. There is also something to be said about how and why the properties of inner constituents of a composite object come to directly make up observable physical properties of the very object. By comparison, the interactive properties of a molecule are germane to the identity of the molecule, i.e., the water molecule H_2O, but do not result from a direct transplant of the properties of the constituents, whether the hydrogen or the oxygen atom in cited example. This is very ostensibly the foundation of the entire science of chemistry. It could very well be that, despite the internal fractional charges cancelling each other out, the composite object still exhibits a charge possibly for reasons germane to a higher ontological order.

The uneasiness created by fractional quark charges responsible for the zero net charge of the neutron, joined to the observable but yet inexplicable magnetic moment of the particle, has prompted some physicists, as late as 2007, to venture a theory about uneven distribution of charge throughout the internal structure of the particle in order to account for the particle electric charge status and magnetic moment. The theory has gained no momentum.

If we discard this inauspicious notion of quark fractional charges, then the question becomes on what basis does the antineutron become a conjugate particle to the neutron? Consistent with the point-source methodology of the derivation of the masses and Compton wavelengths of the proton and neutron particles previously discussed, which showed both particles on the same leveled footing, the azimuthal characterization finds the proton particle on quadrangular space $+T$, its antiparticle, the antiproton on quadrangular space $-T$ and the neutron particle within hybrid azimuthal space $\mp T_L$ of the same period. Therefore, the hybrid neutron is under the mandate of exhibiting concurrently all 4 spinor states in one single particle. However ,the Primitive-monolithic nature of the particle intervenes to impose the rule of monocentric symmetry associated with the Primitive and constrains the particle to exhibit *one* singlet per spin state. Nonetheless, the dissociative latency inherent to hybrid azimuthal state becomes manifest and causes the particles to spontaneously disintegrate at ground state. Hence, unlike its proton partner, the neutron status is that of an unstable particle with the lowest level of volatility (some would say longest lifetime) and with 4 distinct particles per conjugate on the basis of their spin states.

To understand the very significant differences between the proton particle and the neutron, and yet at the same time their close isomass (or isospin) relationship, is to invoke these determinants and no other.

Lastly, is it any surprise that an element informed by an ontological Primitive of Monolithisation (or self-containment) exhibits the least intense course of segregation among all existing unstable particles? In effect the neutron is the only particle whose rate of spontaneous segregation is long enough to be graded in minutes, totaling exactly 12 minutes, as opposed to fractions of milliseconds experienced by other unstable particles. By the same token one can see that degenerate current intensity is a better and more precise scale to grade particle instability than the time scale which maintains a relativistic dependency on space frames and gravitation if credited any value whatsoever.

12.22.3 Hybrid Bipolar Muon

On period 2 of the Spectrum, in consonance with our premises, one would expect the existence of a hybrid muon particle μ_0 which would be unstable if allowed by the Primitive informing the period. Well, the nature of said Primitive is already Bipolar-Dissociative, reason why not only μ_0 would be volatile but the muons in quadrangular spaces on both sides of the period would be themselves degenerate as well. The muon is indeed observed to be a spontaneously degenerate particle.

I shall add that within the framework of Shell generation, the whole Shell2, the Shell of Particles, which we are hereby analyzing, bears the stamp of the Dissociative Primitive within its own domain, and so contributes a meta-level boost to the dissociative phenomenology of the Muon particle. This boots causes an enhancement of the degenerative behavior throughout the entire period.

When it comes to the hybrid muon particle specifically, theorized since mid-20th century, it has never been observed. Is this lack of experimental observation any wonder given the high level of volatility imparted to the particle by the conjunction of aforementioned determinants? R. E. Marshak and E. C. G. Sudarshan had argued since 1957 in a short paper the possible existence of the Neutral Muon. In their view, the most probable reaction that would yield a neutral muon is:

$$K^+ \rightarrow \mu^+ + \mu^0, \text{ where } K \text{ is the unstable Kaon particle.}$$

If the hybrid muon μ_0 were not to be so volatile and ephemeral, could we expect it to live to show all 4 spinor states in 4 distinct elements for each of matter and antimatter particle as the hybrid neutron does? Because the muon is not ontologically anchored on a monolithisation mandate such as the neutron is and exclusively so, it would not be able to counteract the hybrid-dichotomic mandate in the same compacted manner, and would most likely have to drop some of the spinor states. In fact, because the ontology of the period is geared to effect di-chotomization, the construction would be led to show only two possible chiral states for the matter particle ($\mp\bar{\mu}_L^D, \mp\bar{\mu}_L^D$) and only two possible chiral states for the antimatter particle ($\mp\mu_R^U, \pm\mu_R^D$), all represented in the same spindle as for the hybrid neutron. This hypothesis can be construed as an attempt to di-chotomize the spinor block of this hybrid element by availing two and only two opposite spinor sets. The meta-level symmetry of the hybrid spindle gives us a side (right side) that is predominantly matter-prone and another (left-side) that is predominantly antimatter-prone. The would-be asymmetry would then be naturally constructed across that line, the center line of the spindle. The end representation comes down to the depiction in Fig. 12.15.

Can we in good justice regard the muon neutrino as the Neutral Muon μ_0 since the latter remains unknown? The description above fits the muon neutrino on almost all counts, very importantly it explains the critical feature of the muon neutrino, and for that matter all other flavored neutrinos, to demonstrate marked preference for one or the other chiral state. Many experiments have proven that matter flavored neutrinos are overwhelmingly left-chiral, while anti-matter flavored neutrinos are overwhelmingly right-chiral, to the point that this distinction became the construction cornerstone of the Standard Model of particle physics. Obviously, it explains as well why flavored neutrinos are, by reason of their hybridity, neutral elements and as such interact very minimally with matter. The only discrepancy remains the mass quantity attributable to the Neutral Muon which would have to be more or less similar to that of the other elements in the period by reason of *isomass* (otherwise called isospin), the reference mass being very specifically the computed mass of the muon particle. This hypothetical value for the mass of μ_0, ***0.1134289267 u*** that is, is manifestly not attributable to the muon neutrino which exhibits a mass quantity so small that they were long believed to lacked all mass. However, one may justifiably argue that the muon neutrino represents a degeneracy of the very volatile Neutral Muon. Effectively, in examining the nature of the affiliation between the muon particle and the so-called muon neutrino one finds that this affiliation is grounded in the natural allowance of many reactions that associate the Muon with the muon neutrino while others, where all parities are conserved but not a straight Muon/Muon-Neutrino relationship, are disallowed. For instance, the following reactions are all allowed:

$\mu^- \rightarrow W^- + \nu_\mu$, where W is the "unstable" W boson.

$$\bar{\nu}_\mu + p^+ \rightarrow \mu^+ + n^0$$

$$\pi^- \rightarrow \mu^- + \bar{\nu}_\mu$$

$$\bar{\nu}_\mu + p^+ \rightarrow \mu^+ + n^0$$

Whereas the following ones which should be possible are disallowed or unobserved, simply because the muonic path of degeneracy is privileged:

$$\pi^- \rightarrow e^- + \bar{\nu}_e$$

$$\bar{\nu}_\mu + p^+ \rightarrow e^+ + n^0$$

One may then safely conclude that the relationship between the Muon and the Muon Neutrino is one of identity or parenthood, the second representing a broken or dwarf version of the first. We argue that the broken link between the two is the volatile hybrid Muon μ_0.

The instability inherent to the Neutral Muon by hybridity is further passed down to its poster child the muon neutrino, which we find able to transmute into other neutrinos. Those other two we will examine shortly. The transmutation of the muon neutrino into an electron neutrino or a tau neutrino, in fact any flavored neutrino into any other flavored neutrino, represents the phenomenon known as "neutrino oscillation". Because this transmutation involves a change into another neutral particle, no matter the modality, it is best understood as a form of current taking place within the hybrid spindle. This current is of the same strain as what is traditionally called a neutral current. Such a form of current is more affine or comparable to the type shown in neutron decay for instance than it is to the quadrangular muon decay, generally denoted *charged* current. One might argue that the flavored neutrino has no other type of neutral fermion to decay down into, so it reverts back to its flavored kindred in a continual counter dance. Azimuthal hybridity stands for the prime provenance of these types of particle transformations.

We will take this opportunity to vehemently object to the characterization that the flavored neutrinos are stable particles as displayed in official literature. If stability means remaining unrelentingly wholesome without end, then when a particle exhibits state changes from one kind to another, then the particle cannot be considered stable. When a muon neutrino mutates into an electron neutrino or a tau neutrino, or an electron neutrino mutates into a tau neutrino or vice versa, even if one were to ignore by way of unawareness the change of taxonomic period occurring during these mutational oscillations, one can never in good logic ignore the change of state that patently occurs during these transformations, else the leptonic affiliation of flavored neutrinos is to be acknowledged artificial and makes no sense. There are however undeniable proofs that flavored neutrinos are distinct from one another. Considering for instance the process that produces the two reactions below

$$\bar{\nu} + p \rightarrow \mu^{+} + n$$

$$\bar{\nu} + p \rightarrow e^{+} + n$$

one finds that if the two intervening neutrinos were the same in both reactions, then the percentage rate of occurrence of each of these reaction would have to be a split 50%. However, the second one is never produced in that process, justifying the identification of the generic antineutrino in the first reaction to the muon antineutrino specifically, and the generic antineutrino in the second reaction to the electron antineutrino specifically.

Now the change of period that occurs *de facto* throughout the above discussed flavored transformations puts a complete ban on the epithet of stable for the flavored neutrino particles, because it reflects changes at the ontological level, so begging for profound conceptual corrections to be brought to the relatively young neutrino physics. That said, we acknowledge that neutrino physics is indeed very complex, intricate and difficult, as well as plagued with many a pitfall.

12.22.4 Hybrid Distributive Electron

Consistent with the methodology being applied in this analysis, we postulate the existence of a Hybrid or neutral Electron e^0 to occupy this block of the period. However, of all known particles the electron is by far the oldest and yet no neutral element in its class has ever been found, much less an unstable one. Nevertheless, the validity of this hypothesis remains as thorough as it was in the two previous cases. The ontology and phenomenology of the period, stated to be multiplet-Distributive, offer a permeable background to the hybrid dissociative propensity at work at mid-azimuth. This squarely puts the fate of e^0 closer to that of μ^0 than n^0. Hence understandably e^0 closely shares the dubious distinction of being remarkably elusive by virtue of utter volatility.

Therefore, here again one may justifiably regard the well-known electron neutrino, a neutral particle, as a direct degeneracy of the volatile Neutral Electron e^0. The affiliation between the electron particle and the so-called electron neutrino is patently grounded in the natural allowance of several reactions that associate the Electron with the electron neutrino. Some of these reactions follow:

$$\overline{\nu}_e + p^+ \rightarrow n^0 + e^+$$

$$n \rightarrow p^+ + e^- + \overline{\nu}_e$$

The last reaction is the one that mediates beta decay of neutron particles whether in unbound or bound states, the phenomenon that prompted the development of the Electroweak Theory where physicist Enrico Fermi so deservedly distinguished himself.

Lastly to the account of inherited unstable behavior, one must add flavor oscillation which the electron neutrino particle is accessible to, to the same extent as all other flavored neutrinos. It may knowingly oscillate between muon neutrino state and tau neutrino state.

12.22.5 Hybrid Quadratic Neutrinos

The question of the identity of the hybrid neutrino on the quadratic period of distribution we are now set to address. The tau neutrino, third and last flavored neutrino to be addressed in this analysis, unmistakably arises as a very plausible candidate. Is there yet any sound theoretical ground for considering the tau neutrino the hybrid quadratic neutrino?

The unequivocal answer to the question is yes, despite the official affiliation of the tau neutrino with the Tau particle. We are hereby switching the affiliation of this particle to the Inert Neutrino or Neutrinon in the hope that further experimental data will contribute to cement this view. Short of plentiful knowledge in the intrinsic properties of the tau neutrino, there are a few phenomenological artifacts that strongly point in the direction of said proposed affiliation with the Inert Neutrino.

12.22.6 Reassessment of Tau Neutrino's Leptonic Nature

First off, the Tau is **not** a *bona fide* particle. It is not a *real* particle. The Tau has not the same standing as the neutron, the muon, the inert neutrino, the three flavored neutrinos or the photons, all of which appear as natural and legitimate entities belonging to this Particle Spectrum within the Quanto-Geometric calculational framework. Unlike all of these, the Tau is a virtual entity, at the most a parcel resonance, or what has become common terminology in QFT an *off-shell* particle: in other words, a particle with no substance of its own.

QFT's notion of virtual particles, resulting from its perturbative-field approach of particles' ontology, universally perused in the Theory in order to explain interaction between condensed matter objects, has widespread debilitating consequences. One of these aberrant consequences is the blurring of lines between what is real, material, shall we say physically tangible, and what is mere hypothesis, unreal, theorized, imaginary or virtual. A credentialed physicist at the head of a public institution publishes, quite unjokingly, on its cyberspace portal the following statement: *"Virtual particles are indeed real particles."* (sic). Nobody thinks that their virtual image in a mirror is *real*, which is as physical as anything virtual will ever be. Apparently, much in vain did Steven Weinberg, one of the three artifices of the Standard Model, undertake to carefully and expressly avoid mentioning this notion of virtual particles throughout his entire publication *The Quantum Theory of Fields*. Yet many find evidence of the existence of virtual particles in their representation as the internal wavy lines of Feynman's diagrams, sometimes following the infamous loops, forgetting that such diagrams are only meant to support a mental representation of particle interaction. Those otherwise lucid minds do forget that it requires an integration of an infinite number of these diagrams to minimally approximate what is taking place in one single interaction. The question of the infinite divergences that we have extensively discussed previously falls squarely at the heart of this issue. We are going to avoid engaging at this stage in an all-out critique of the notion of virtual particles per se, deferring it for later treatment if ever. We will only remind the reader that there are in existence many Quantum Field Theories out there other than the official Standard Model QFT, where these particles are not at all invoked. Most importantly, the reason why such theories exist is because their promoting physicists are well aware of the historical development of QFT and thoroughly understand that these virtual particles have no other life than that of hypothetical entities theorized within the framework of QFT's perturbative vision of particle physics.

The implications of QFT's misrepresentations of state and internal structure of condensed matter particles, whether official or unofficial, are consequently widespread and potentially damaging to the advancement of the science. Even probing physical experimentation becomes a tall order feat under these conditions, since the experimental constraints under QFT's regime become enormous, having to establish truth on the basis of results that are just a few dozen counts out of billions, along with correlations of signature decay channels that say something about the presence of a particle instead of direct measurable effects from interaction with matter. The fact is that the Tau particle was not directly detected, but its existence was inferred by Martin Perl in 1995from events that appeared to be anomalous. The theorized reaction was:

$$e^{+} + e^{-} \rightarrow \tau^{+} + \tau^{-} \rightarrow e^{\pm} + \mu^{\mp} + 4\nu$$

In order for a matter-antimatter electron pair to produce an observed matter-antimatter pair of electron and muon, there would have to be a few neutrinos produced as well for conservation of energy to be observed. In addition, for lepton number to be conserved the produced neutrino would have to be of a new flavor, which implies their affiliation with a new Tau particle. A pair of such Tau particles must conceivably mediate the reaction. The so-called third generation of leptons was born. Except that unlike the electron and the Muon leptons, the Tau remained a virtual creature, and that as later theorized it can even decay into hadrons, being the only lepton with that exceptional capability! It may not only decay into hadrons, but into low-level hadrons such as quarks (**down** and **anti-up** quarks), which is even more exceptional! Trans-generational activity experienced by the Tau particle in the theory, posited but not derived, is indeed an exceptional problem. That the existence of a particle was inferred is not at all a problem and responds to our long-standing method of scientific discovery, it is the corroboration of its existence and its ontology as a real physical creature that are problematic.

By contrast, the tau neutrino has garnered more traditional experimental evidence. A solid evidence of its stature as a thorough and real particle is the Cerenkov radiation. Cerenkov radiation is a bluish white light in the visible spectrum that is caused by particles entering a highly purified water medium at a greater speed than their speed of deceleration as they enter and travel through the water medium. This radiation, resulting from interaction between the incident particle ray and atomic nuclei, will be in effect as long as the speed of propagation of the particle within the medium is greater than the speed of light thru the same medium. Cerenkov radiation has come to play a key role in the experimental determination of distinct flavors between neutrinos, as specific tones taken by the radiation constitute a key pointer to the identity of its indirect generator.

12.22.7 Inhomogeneous Leptonic Symmetry

With the discovery of the charge-less antineutrino, the decried notion of electric charges at the particle level entered in crisis mode toward mid-20th century. The claim of an opposite electric charge to distinguish the neutrino from its newly discovered antiparticle, everything else being equal, could not possibly hold since the particle is electrically neutral. Physicists of the day, specifically E.J. Konopinski and H.M. Mahmoud, postulated a new quantum number called the Lepton Number L in order to provide support for the mandatory distinction between the two particles. The new quantum Lepton Number was simply modeled after the electric charge concept with two possible values for L: $+1$ and -1. The value of $L = +1$ was assigned to e^-, μ^- and ν altogether, while $L = -1$ was assigned to e^+, μ^+ and $\overline{\nu}$ altogether. The lepton number concept later underwent some level of refinement with the introduction of an individual lepton number for the Muon and the Electron, the Tau being still unknown at the time. By the 1970's the concept of the lepton number was fully incorporated in the Electroweak Theory and subsequently in the Standard Model, which further endeavored to provide support for the observed restriction in handedness by the two conjugate neutrino particles thru the notion of Parity Violation.

It is clear that azimuthal symmetry constitutes the larger symmetry discipline that the lepton number rule, in fact all the postulated conservation rules of particle interaction, attempts to describe. Yet the leptonic notion appears under a traumatic light when it purports a separate distinct requirement for each of the three established sets of lepton particles: the Electron, the Muon and the Tau. The leptonic symmetry discipline is already broken here and support for the three generations simply vanishes. The order of family symmetry loses support because one global rule to characterize them does not exist but only three different rules apply to each. In effect, the lepton rule later came to encompass an *electronic number* L_e for the Electron generation or doublet, a *muonic number* L_μ for the Muon generation or doublet and a *taunic number* L_T for the Tau generation or doublet. A statement of state by which all three generations would be related to one another falters altogether in the Standard Model. Although these empiric rules may explain why we see certain interactions but not others, the phenomenon of neutrino oscillation, where particle transformation patently crosses two generations in any direction, stands in flagrant opposition to the Lepton generation concept. This is further reflected in the decay formula for each generation which shows a different putative invariant anchoring its mathematics. For the typical Muon decay according to

$$\mu^- \rightarrow e^- + \overline{\nu}_e + \nu_\mu$$

the decay rate Γ is

$$\Gamma = K_1 G_F^2 m_\mu^5$$

where K_1 is a putative muonic degenerative constant. For the Tau decaying according to

$$\tau^- \rightarrow e^- + \overline{v}_e + v_\tau$$

the decay rate Γ is

$$\Gamma = K_2 G_F^2 m_\tau^5$$

where K_2 is a putative taunic degeneracy constant. Matter-of-factly K_1 and K_2 are not legitimate physical constants because their values depend on the mass of their related lepton. The relation is not an Eigensolution of any known or postulated operator equation. And, to close the argument, the decay rate of the Electron as the 3rd lepton generation is obviously null (with $\Gamma = 0$).

We hope the above argumentation establishes firmly enough the permissive grounds that assist us in hereby summarily dissolving the lepton class and reassigning the affiliation of the Tau neutrino to the v Neutrinon. One additional quality of the Tau neutrino, shortly addressed below, warrants this classification.

We recognize the quadratic layer of the Quanto-Geometric Primitive Spectrum one of exception because it is the one and only one that shows equalized covariant freedom between the two internal composition tenets. Because of this equalization, the identity emerges in the context of interaction as a non-polar or saturated element. It is against this ontological backdrop that the Neutrinon on the hybrid spindle of the layer, the putative v^0 particle, develops its imparted azimuthal hybrid character.

12.23. Quark Particles at the Top of the Mass Scale

The exposition in Book I of this Treatise has put in evidence the primordial role played by the derivative of the Grand Eigenfunction, or its image, in the construction of several hierarchies of U_3's physical properties, very particularly its coupling constants or building blocks. We have seen how the symmetry-breaking intervals of the Quanto-Geometric wavefunction line determines the many different Shells of matter. We have examined in depth the developments of Shell2 to Shell8, which occur based on particular values of the wavefunction line, specifically those initiating the intervals.

Shell1 remained somewhat of a conundrum due to the absence of an initial value for its symmetry-breaking interval. Effectively the derivative of the Grand Eigenfunction or its image at $s = 0$ does not exist, since the function is undefined at that point. This interval is the first on the s variable line, delimited by 0 and ½:]0, ½[. The contour of this interval is key to the establishment of the primal invariant destined to model all the key configuration quantities for the Shell1 edifice. This interval is uncharacteristically open on both ends.

However, before we turn our attention to the details of interval]0, ½[, let us highlight the fact that this entire space sweep represents normal layer 1 of the grander image Eigenfuncion presiding on the development of U_3. By virtue of the monolithic typology defining such a normal layer, one should expect an encompassing over-densification trend applied to the corresponding elements. This stands for a mandate applied to the development of the relevant objects, which translates into a monolithisation boost.

Secondly, the particulate objects within the cited interval evolve along the line of their own Primitive Spectrum, just as we saw neutron to photon particles develop along the line of their own Primitive Spectrum. In this context the Grand Eigenfunction represents a meta-level umbrella that directs and composes with the suite of particular Primitive Spectra ultimately shaping up the U_3 objects in construction. This composition produces a correlation between layer 1 of the meta-level Primitive Spectrum with all nine levels of the lower Primitive Spectrum. We expect that correlation to result in an extra mass boost to the objects occupying the overt-mass sector of the lower Spectrum. Therefore, two densification functions are de facto acting over the overt-mass sector of the lower Spectrum producing a disproportionate unbalance of mass with respect to space within the ontology of those objects.

We have identified the quark-gluon family to be the infinitesimal particulate elements of Shell1. These objects are studied in the extension of QFT called Quantum Chromodynamics, so named due to the so-called color charges at play in this region of particle physics. Effectively, we do witness the high mass quantity exhibited by these objects and their associates to be the highest known, per Table 5.1. This goes in concordance with our view that mass is a density function, and only as such is it gradable in terms of energy. So it is, because these elements are located at the top of the $Q(s)$ density function in both of the Spectra where they are involved. In effect, consistent with QCD, the Quanto-Geometric Theory puts the quark family of objects at the highest level of the mass scale, even though experimental QCD is notoriously inefficient at measuring the mass quantities of these elements. Particle physicists agree that the reported values for these different masses are not to be taken too seriously and, in several cases, an exact quantity cannot be obtained but only a range of most probable values instead. The mass of the up quark for example is specified in terms of a range of *1.7-3.3 MeV*. The mass of down quark is given in terms of a range of *4.1-5.8 MeV*. There are different experimental reports for the mass of the bottom quark, which in some cases is measured at *4.19 GeV* and in some others at *4.67 GeV*. Perhaps the most contentious of these inaccuracies dwells in the very features of QCD whence it postulates different mass values for the same combination of quarks! The pion particle, formed by a combination of up and anti-down quark, has a mass rated at *139.6 MeV*, whilst the rho meson formed by the same combination of quarks exhibits a mass about 4 times larger that is rated at *770 MeV!* The Theory does not invoke any possible internal binding feature that might contribute to the discrepancy. Many think therefore with good reason that QCD is still an immature Theory despite its long-running development. To that account one must add that the predicted glue-balls by QCD, which should occur at energy levels well within reach of our common particle accelerators, have never been observed.

12.23.1 Origin of the Phenomenon of Quark Confinement

The aforementioned accrued densification mandate contributes to a dramatic effect on the development of the quark-gluon family of elements, beyond their positioning at the top of U_3's mass scale. In a trend to over-massify the quark particles located in the overt-mass sector of their Primitive Spectrum, the scalar amplification mandate forces their development in duplet agglutinations or condensates in their own Shell, thereby denying them a topological extension based on singlets or single particles!

Consequently, no free quark may be observed at ground state. It would take an amount of energy that is beyond the U_3 physical scale to reverse the accrual densification constraint and individually free up the quarks. All quarks are physically manifest as mesons, bound states of two quarks. These condensates are essentially matter-antimatter quark particles. QCD has inventoried more than 200 of these condensates, all of which are unstable and ultimately decay (or explode) thru the weak interaction into particles of the previous (lower) mass scale or Primitive Spectrum. However, this decay activity is admittedly not to be qualified as radioactive decay, the latter being typically the breakup of atomic nuclei. This decay is rather best catalogued as a phenomenon originating from the top of the physical mass scale, ignited by energies quantified at that level whether artificially in particle colliders or in nature as very high-energy cosmic jets of interacting neutrons and protons. Such energy levels are experimentally far above the weak energy scale.

In QCD the phenomenon of agglutination of quarks in duplets is known as confinement. We may mathematically consign this behavior as the product of a Transformation experienced by the $Q_{1\text{-}4}(s)$ sector in the quark Primitive Spectrum from the application of a similar functional sweep $Q_{1\text{-}4}(s)$ from the overarching image Grand Eigenfunction.

In QCD quarks agglutinate as well in triplets to form baryons. Be that as it may, the nature of quarks' association in triplets inside baryons, such as neutrons and protons, is different from the glue that binds them in duplets. In baryons quarks are bound by the fine-structure constant α, a true Invariant applicable to Shell2 particle Spectrum. It is a profound analytical error and a lack of grander view on the part of QCD to have established the running strong force as the force that subtends the structure of both baryons and entire atomic nuclei, the latter being cores of Shell3 objects. We will address in due course the true nature of the drivers for the configuration of atomic nuclei.

12.23.2 Mathematical Origin of the Phenomenon of Asymptotic Freedom

The peculiarities of the]0, ½[interval on the *s* line for the image Grand Eigenfunction for U_3 gives rise to the second most dramatic effect in the development of quark-gluons physics, namely the phenomenon of asymptotic freedom. From Book I, the primary expression of the coupling constant generator is:

$$M_3(s) = 0.66114378278 \bullet \frac{-s}{\sigma^4\sqrt{2\pi}} \cdot e^{\frac{-s^2}{2\sigma^2}}$$

The symmetry-breaking sequence of the *s* variable generates the sweep of values of $M_3(s)$ that represent the physical coupling constants. This function is not allowed to take a value at $s = 0$. So therefore, the interval]0, ½[., interval under the stamp of rotational symmetry, is left with no down anchor for the generation of the coupling constant destined to support the physics of U_3's Shell1, Shell of the uttermost infinitesimal particulates under consideration. To the limit, in defect of a set lower bound, the upper bound of the cited interval could be set as an inverse anchor for the desired dependency; however, by definition that point does not belong to the interval either. Instead, it represents the initial point of the next interval.

For our target coupling constant what we are looking for is a composite quanto-geometric invariant at the beginning of the symmetry-breaking interval for Shell1. Short of the application of the full derivative figure resulting in a value for $M_3(s)$, we may choose to explore the limit to the *Q(s)* Function as *s* approaches 0^+ or from the right side. As a reminder, the Function would be differentiable at that point if the limit of the function in the form of affine *s* increments $(s + \Delta s)$ as *s* approaches 0^- (or from the left side) and the limit as s approaches 0^+ (or from the right side) both exist in the real number realm and are equal to one another. That is unfortunately not the case for the Grand Eigenfunction.

Taking the analysis down to the limit figure, we may elect to examine the most affine numeric variance possible past 0 along the positive number line, or the positive Quanto-Geometric wavenumber line, in an attempt to determine an invariant.

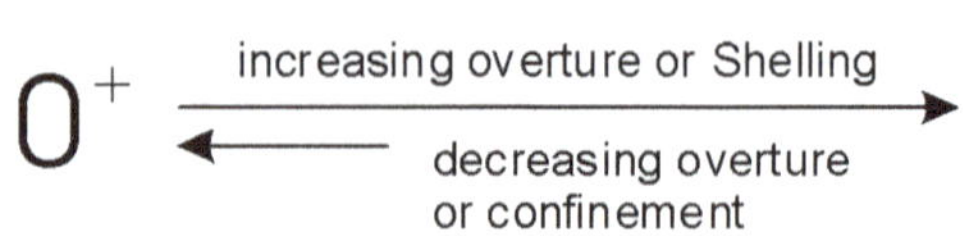

One may readily note that analytical limit considerations clearly spell out the fact that as the limits are considered for the series of symmetry-breaking points, their values thru $M_3(s)$ subtend or configure increasingly larger Shells of matter. Whereas for point 0 the only sensible limit figure is that one running contrary to the previous ones, which leads to the representation of the most affine possible quantity approaching 0 only from the right side. In the absolute this quantity conceivably exists for $Q(s)$, but it is a laxed figure, not a set one. In other words, this quantity is a variant, not an invariant. We know of no mathematical procedure to compute it. Its expression follows:

$$\lim_{s \to 0^+} Q(s) = 0.\overline{0}X$$

This quantity is an irrational with 0 for unity, followed by *any* number of 0 decimals followed by *any* amount of non-zero integers. So therefore, if analytical limit considerations lead to the formation of Shells of increasing space Overture for the other wavenumbers of the wavenumber line, for the 0 wavenumber they instead lead to the inverse of Overture, a down trending or running to a state of ever smaller spread coupled to increasingly smaller scalar, in a manner akin to an asymptote. Zero is thus the asymptote at origin of $M_3(s)$ that determines the quanto-geometric coupling subtending the ontology of quarks and gluons within their Primitive Spectrum. The eventual values of $M_3(s)$ continually diminish as s approaches 0 from the right side. The coupling force is not only running but it is running toward 0, without ever reaching it however. The internal quanto-geometry standing of these entities features a race to the bottom that qualifies them as *stateless* particles. This qualification applies to their derivates in duplets as well. It is only because of the down-run of the coupling that these particles, despite the tremendous densification accrual mandate on their ontology, are still unable to solidify in a *stateful* stable physical configuration. Consequently, the resulting binding is ever smaller as s approaches 0, or as the wavelength approaches 0.

The above sufficiently clarifies, we hope, the origin of the phenomenon of *asymptotic freedom* in vogue in the QCD nomenclature for these particles. QCD however does not consider internal quark structure and applies this qualification to the binding between quarks in duplets and even triplets. Unquestionably, all existing physical entities must have determinants for their own internal ontology, without which they cannot subsequently exist in associations with other elements from exogenous phenomenology. Asymptotic freedom between quarks in mesons is only a derivate of asymptotic freedom in quarks' internal structure. In QCD visualization, the binding between quarks is stronger at larger distances than at shorter distances separating them. The beta function effectively describes this behavior in sufficiently close agreement with experimental observation. However, one must remain mindful that the beta function does not explain the physical origin of the negative coupling, and most importantly, it does not represent the fundamental coupling for the Shell (or quark structure if you will). QCD's beta function follows:

$$\beta(g) = \frac{-g^3}{16\pi^2}\left(\frac{11}{3}N_C - \frac{4}{3}\frac{N_F}{2}\right)$$

where g is the quark-quark coupling constant, N_C the number of color charges and N_F the number of quarks in the association.

Despite the visualization of a gluon cloud around quarks causing self-interactions, wrongful in our view, the mathematical formulation of QCD has miraculously escaped the quagmire of insolvable infinite series of integrals that has plagued and hampered Quantum Electrodynamics for long.

The above characterization equally resolves at a stroke the paradox of *virtual* particles that are nevertheless *real* in QFT. It is the very absence of the coupling constant (not the coupling or coupling force) that creates the condition for a state that cannot be fully characterized from a scalar or even a field point of view. This condition is best described as a *stateless* configuration instead of *virtual.* Said condition may only appear at very high-energy physics, completely disappearing with Shell3 objects (atomic realm) where energies are much lower, thereby giving rise to exclusively *stateful* objects. The condition of the existence of a variant instead of a physical invariant is paramount to the symmetries of quark scale physics. If one is adamant in finding a coupling constant where it naturally does not exist, then one is left to describe nature in terms of unsustainably paradoxical nomenclature.

12.23.3 Quark and Gluon Distribution Under the Primitive Spectrum Canon

The distribution of particles in the quark family proposed by the Standard Model exactly fits the Quanto-Geometric Primitive Spectrum canon. The quarks proper are known to obey the Fermi-Dirac statistics, which makes them spin-½ particles or fermions. Therefore, the group of up, down, bottom, strange, charm and top quarks make up the Hyper Quantum sector of the Spectrum. We remind the reader that the Hyper Quantum sector is that sector of the Primitive Spectrum with objects featuring overt-mass/covert-wavefunction ontology. They correspond to baryons and leptons of the Shell2 Primitive Spectrum.

Gluons associated with quark particles are known to be boson objects. The Standard Model has them in the incarnation of a field or cloud of particles surrounding the quarks. They are otherwise called force carriers for the strong force. By obeying the Bose-Einstein statistics (spin-1 objects), they map to the Hyper Space sector of the Quanto-Geometric Primitive canon. As a reminder this is the sector with objects featuring overt-wavefunction/covert-mass ontology. They correspond to the photon group of the Shell2 Primitive Spectrum (Fig. 12.14).

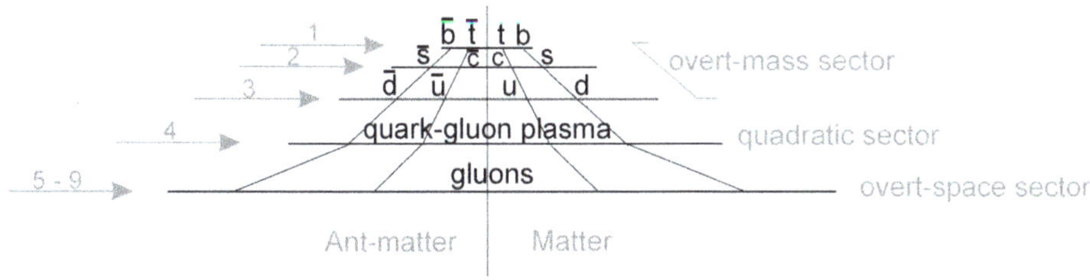

Fig. 12.14 Primitive Spectrum of Quark and Gluon particles

Lastly, the Standard Model contemplates the existence of a quantum state called quark-gluon plasma as an aspect of the phenomenon of *asymptotic freedom.* This ontological phase of the quark-gluon family is a faithful representation of the Quadratic sector of the Quanto-Geometric Primitive or Trivalent Canon, whereby the quanta and wavefunction of an object exert equal influence on one another or couple in a *covariantly free* manner. This covariantly-free primary coupling, which results from equated reciprocal influence, creates a state of interchangeability between the two tenets of the articulation, as previously explained.

The Quanto-Geometric Model reproduces QCD's taxonomic generation in its normal layers of distribution. That arrangement shows the top and bottom quark at normal layer 1, the charm and strange quark at normal layer 2 and the up and down quark at normal layer 3, consistent with QCD's generations. What is definitely not conserved in the Quanto-Geometric Model is this notion of generations that gather both leptons and quarks in each of the three families, on one hand, and the grouping, on the other hand, of all the so-called force carriers (photons, gluons, Z and W bosons) in the same family. We will later delve into more particle properties that overall confer more accuracy to the Quanto-Geometric Model. Let's note for now that the mass quantities known to the quarks, despite their inaccuracies, are consistent with the scaling within the Quanto-Geometric Primitive Spectrum. Not enough is known about quark and gluon ontology to allow a comparison of their structural properties with the typologies arising from the Quanto-Geometric Model, reason why we will forego that aspect of the analysis.

Likewise, the properties inherent to the azimuthal dimension of each period we will examine later in this study in the context of a review of Quantum Field Theory.

12.24. Conclusion

In this stage of our study, we have analyzed the parameters conducive to the theoretical computation of condensed matter particles' masses. The analysis has put in evidence that the computation of mass requires sound understanding of the behavior of the void of universal space, as well as the necessary derivation of the wavefunctions themselves. We have shown that the order of things in void is so prolific that the term wavefunction may have even become too vague for a descriptor. However, the momentous triplet Transforms *(T_{19}, T_{14} and T_{11})* that we have uncovered are key to the understanding of condensed matter physics and extol the natural descriptors emerging from behaviors and structures of objects pertaining to those two Shells of U_3. The time-free **frequency** characterization, the **energy** characterization and the **scalar-mass** characterization all emerge naturally from the inner workings of physical matter and bear an intimate relationship with the primal metric that is the compass for U_3. It is well worth stressing that the Transformations that have realized the development of U_3 are completely consistent with the model of the fabric of the spatial void that we have advanced in the previous Chapter. For certain, there is more to this seminal spatial architecture than we have exposed. Yet enough of this ultra-structure have we elucidated for the ensuing derivation of particle masses and wavefunctions to successfully cast the known empirical figures for those quantities. Finally let me loudly and vehemently state that in Quanto-Geometry we do not invoke or peruse fictitious abstract spaces because they do not exist in nature, are of little usefulness in the task of deciphering matter, and have proven to be instead quite obtuse in nature. We use real coordinate space and real image functions to unveil the hidden structures and phenomenologies of the physical realm on a mathematical-physics basis. We believe this standard should be enforced.

Chapter 13

The first principle is
that you must not fool yourself -
and you are the easiest person to fool.
Richard Feynman

TO DECONSTRUCT QUANTUM FIELD THEORY AND ITS STANDARD MODEL

There is no doubt that the onset of the Quantum Theory in early 20th Century had contributed to deepen the revolutionary change in the visualization of physical matter dynamics in course in that period. The Quantum Theory came as a response to a challenge posed to common-sense or classical devising of the dynamics in the infinitesimal realm. The intellectual capital invested in its early development by pioneers such as Max Planck, Erwin Schrödinger, Werner Heisenberg and Niels Bohr, was truly uncommon, only rivaled by visionary Albert Einstein with his audacious attempt to uncover the geometry and dynamics of the Void. When early Quantum Mechanics yielded the stage to the Quantum Field Theory vision, many of the new thinkers that joined the ranks did not always keep with the unwavering intellectual probity and rigor that characterized the science thus far. We have already mentioned in Book I the discomfort felt and publicly voiced by P. M. Dirac and R. Feynman in that regard. In this Chapter, we are going to expand this critique, in light of the grander view apportioned by the Quanto-Geometric Theory when it comes to analytic structures best appropriate for elucidating wave and quantum mechanics. By the same token, we will shed light on a number of the concealed symmetries that have mystified the theoretical physics for long and in so many different ways.

13.1. Misrepresentation of Unitary Groups

The three artifices of the Standard Model, Sheldon L. Glashow, Julian Weinberg and Abdus Salam, found mathematical basis for the model in the rotational symmetry groups:

$$\boldsymbol{ISpin(3,1) \times SU(3) \times SU(2) \times U(1)} \quad (13\text{-}1)$$

or $\boldsymbol{ISpin(3,1) \times G}$

where G is *SU(3) x SU(2) x U(1),* a tensor product of the irreducible representations of the contained groups.

In the *(13-1)* formula, ISpin(3,1) stands for the geometrical symmetries of space-time in the role of the grander frame where the rest of the symmetries ***SU(3) x SU(2) x U(1)*** play out as local or internal symmetries exhibited by elementary particles inhabiting that spacetime. It is an attempt to frame physical matter in the most encompassing manner, notwithstanding the notable absence of Gravity. In the internal symmetry group ***SU(3) x SU(2) x U(1)*** each partial group respectively represents the strong force, the weak force and the electromagnetic force. This approach is consistent with Quantum Field Theory visualization of field perturbations which sees particles as perturbations of a field with all particle interactions directly mediated by field entities. Thus, specifying an irreducible representation of *U(1)* is to specify how it interacts with the electromagnetic force. The irreducible representations (particles) of *U(1)* are labeled by the integer called **electric charge** *q*. This *U(1)* however is not the *U(1)* unitary representation pertaining to the subgroup *U(2) x U(1)* as it technically is in pure Representation Theory mathematics! It is another *U(1)* standing on its own. This is no more than an arbitrary tweak which physicists justify by invoking entanglement between the weak force and the electromagnetic force. They contend that *U(2) x U(1)* describes how an elementary particle interacts with the "electro-weak" force. If that still gives the theory legitimate right to the claims that Representation Theory formalism will produce at the end of the treatment is anyone's guess. We suspect an orthodox mathematician would hold a *no* response, just as he or she would object to the arbitrary cancellations in the procedures afore-reported in the previous Chapter.

13.2. Misappropriation of Particle Electric Charge in the Standard Model

Considering the group *SU(2)* of the tensor product, one notes that its Lie Algebra consists of skew-adjoint 2x2 complex matrices, with basis matrices $\frac{i}{2}\sigma_j$, where

$$\sigma_1 = \begin{pmatrix} 0 & 1 \\ 1 & 0 \end{pmatrix}, \sigma_2 = \begin{pmatrix} 0 & -i \\ i & 0 \end{pmatrix} \text{ and } \sigma_3 = \begin{pmatrix} 1 & 0 \\ 0 & -1 \end{pmatrix}$$

For the group *SU(1)* of the tensor product, the Lie Algebra consists of skew-adjoint *1x1* complex matrices which are the same as imaginary numbers. The chosen basis vector is *i/3*. The Quantum Theory prescribes that physical observables map to self-adjoint operators. So therefore, given any unitary representation ρ of *SU(2)* on some Hilbert space, the above-mentioned basis for *SU(2)* is consequently mapped by $d\rho$ to self-adjoint operators on that space. Division by *i* yields self-adjoint operators I_1, I_2 and I_3, summarized below:

$$I_j = d\rho\left(\frac{1}{2}\rho_j\right)$$

The three I_j components are called components of the **weak isospin**. However, in order to obtain I_j in the form shown in the equation we must have arbitrarily canceled some factors of *i* in a completely heavy-handed manner. The isospin figure represents a par symmetry between doublets of the same mass.

Following the same reasoning and procedure shown above for the group *U(1)* we obtain the following self-adjoint operator, so called **hypercharge** *Y* by applying the same arbitrary cancelation of some factors of *i:*

$$Y = d\rho\left(\frac{1}{3}\right)$$

It will be no surprise that the representation for the electric charge will utilize the same math and the same arbitrary cancelation of some factors in the equation of the self-adjoint operator in order to obtain the desired equation for the electric charge *Q*. Let's see how that works. So, the irreducible representation of the Lie Algebra for the electric charge is

$$d\rho(i)\psi = iq\psi$$

where ψ is a vector on the Hilbert space of representation. In order to obtain the self-adjoint operator from the skew-adjoint operator *dρ(i)*, we divide the equation above by *i* resulting in:

$$Q = i^{-1} d\rho(i)$$

We have to arbitrarily remove some factors of *i* once again to obtain the coveted identity below:

$$Q = d\rho(1)$$

which implies that

$$Q\psi = q\psi$$

The *Q* observable in the representation bears the name of **electric charge**. *U(1)* is so entrusted to be the irreducible representation for the notion of electric charge of particles to which it tributes a semblance of packetization or quantization nature, realized in the portrayal of the smallest electric charge *q*. Note well however that the Model is completely unable to extol the value of that charge quantum, which it purports to be the familiar ***e***, charge of the electron particle. The Model further treats this charge *q* with the customary value of *-1* instead.

When it comes to the quark sub-component of hadron particles, the Standard Model relies on the same Representation Theory math above reported, somewhat tweaked to the needs of the cause, in order to formally support its standing as mediator of the strong nuclear force. A major inclusion in the representation basis is the notion of fractional charge valued at *1/3e*, without any prior theoretical justification. Here is how it is done.

For any number *y* with *3y* belonging to *Z*, there is a unitary irreducible representation ρ of *U(1)* denoted the hypercharge *y* irreducible given by:

$$\rho(e^{i\theta)}\psi = e^{3iy\theta}\psi$$

By differentiating with respect to θ as we did before we obtain an irreproducible representation *dρ* of *U(1)* as a skew-adjoint operator such as:

$$d\sigma(\frac{i}{3})\psi = iy\psi$$

To obtain a self-adjoint operator we divide *dρ* by *i:*

$$Y = i^{-1} d\sigma(\frac{i}{3})$$

which in short notation is the coveted *dρ(1/3)*. This is how fractional charge of *1/3e* has been forced to represent a basis for the charge of quark particles. This type of irreducible representation historically had first been used by Heisenberg in order to establish basis for similarity between neutron and proton phenomenology. He proposed *SU(2)* as a symmetry group for the strong nuclear force whereby proton and neutron are unitarily identified as nucleons (integrants of the nucleus). Under this view protons and neutrons represent two different states of one same atomic nucleon, a change of spin orientation (in fact a change of charge) undergone by the particle establishing the identity it takes following the transformation. It is important to highlight that Heisenberg was only following the analytical model of electron spin, in the understanding that there was no such spin in the nucleon states he sought to describe. He called the new entity *isospin*, after the word isotopic applying to nuclei with different number of neutrons but same number of protons, the objects of his study. That the same math may be used to characterize other physical variables is not necessarily objectionable. We have seen in Book I how the physical constants called coupling constants have similar computational formulation despite the difference in the nature of the physical quantities involved in concerning identities. On account of such similarity in the formulation is it that the term coupling righteously applies to the physical phenomenology they develop. Should one close eyes for a moment on the clear procedural faults that undermines the development of the Group *G*, a glimpse of a higher truth may come to light. That the same irreducible representation procedure, from a class or classification Theory such as Representation Theory, may categorize three different physical variables strongly advocates for a class commonality between these variables. Prospected thru Quanto-Geometric lens, one finds that all three groups under the overarching tensor product are in fact attempting to depict a symmetry class that is no less than the par or transverse symmetry embodied in the Quanto-Geometric azimuthal model. This trend is very clear in the characterization of electric charge and hypercharge. It is no less visible in the isospin doublet symmetry, which depicts a class of symmetry between two particles of the same mass and same interaction condition. Through the irreducible representation method, the Standard Model is indeed proposing a charge class to which the electric charge of the electromagnetic force and the hypercharge of the weak force integrally belong. That is the reason why most physicists acknowledge that the electric charge in the context of the Standard Model is no longer a fundamental figure, favoring the generic charge figure or charge class itself to hold that role. To that account one must add the color charge borne out by quarks in the Model.

Because the Standard Model lives in the ignorance of the bigger symmetry picture scaffolding condensed matter physics, it is not able to appropriately and thoroughly describe all the symmetries implied in the scaffolding, nor is it able to uncover its origin. That also is the reason why the Model was deeply shaken by the discovery of neutrino mass implicating the existence of more chiral symmetry than contemplated.

13.3. Triplet, Octet and Decuplet Symmetries in the Standard Model

Physicists have long recognized the paramount role played by symmetries in the development of condensed matter parcels. Beyond the symmetry axioms discovered by Emmy Noether, early on physicists had realized that symmetry constraints even take the form of geometric patterns shaping topologies of matter particles. Since the early 1960's Murray Gell-Man, a pioneer of the Standard Model, endeavored to uncover those patterns which helped him predict the existence of the η particle. In so doing he uncovered the very important pattern called the Eightfold way, which presides over the topologies of baryons and mesons. He also uncovered the Decuplet pattern which applies to spin-3/2 baryons. The decuplet pattern helped Murray Gell-Man theorize the existence of the Ω^- particle, at amiss to complete the pattern, and shortly found thereafter. We show below (Fig. 13.1) the baryon octet pattern.

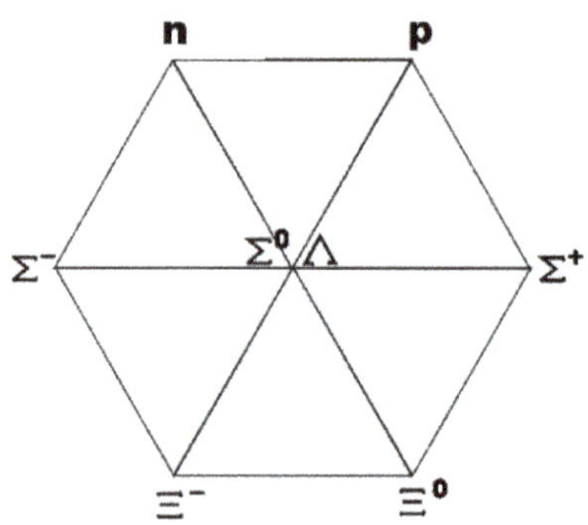

Fig. 13.1 Baryon octet pattern

The meson octet topology follows in the illustration in Fig. 13.2.

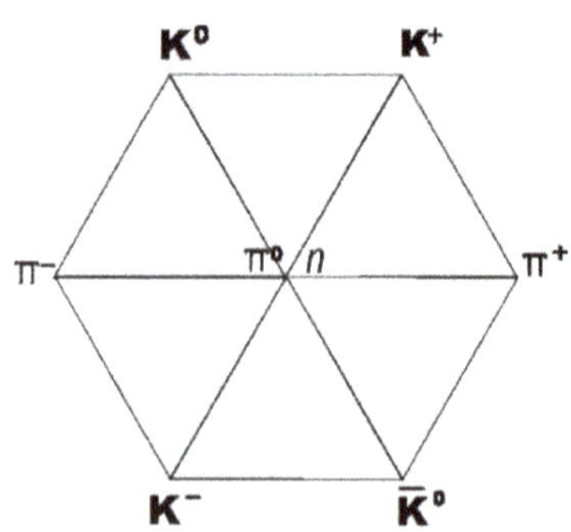

Fig. 5.2 Meson octet pattern

Finally, the decuplet pattern that also underlies the topology of other meson particles is displayed in Fig. 13.3:

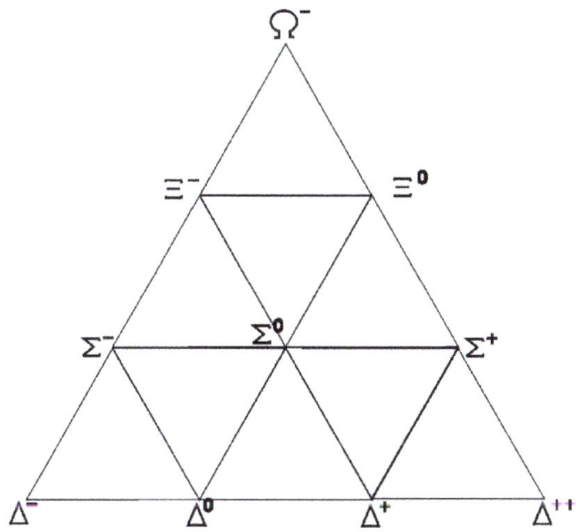

Fig. 13.3 Meson decuplet pattern

The name and nature of the particles are immaterial to the analysis. Note rather the alignment of charges of same nature across all patterns. In considering the geometry of these patterns one would tend to make the equilateral triangle the building block of such patterns. In fact, the building block is the partitioned trapezoid (Fig. 13.4), as illustrated below.

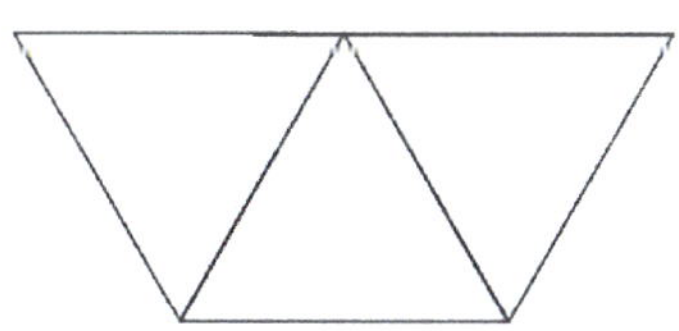

Fig. 13.4 Partitioned trapezoid as building block of Shell1 particle topologies

This trapezoid is no more and no less than a representation of Quanto-Geometric azimuthal symmetry in ground state. We obtain this pattern by setting $N = 1$ in the azimuthal genic relation. N with value of 1 yields one single spindle per quadrant, each flanked by the central hybrid spindle, with resulting basis azimuthal angle of 60°. The schematic in Fig. 13.5 puts that contention in a bright light:

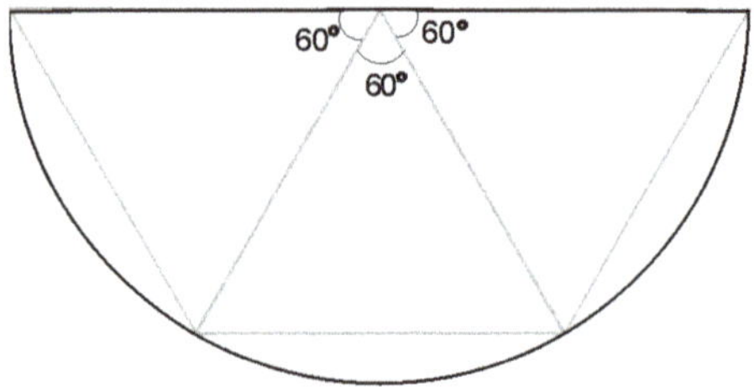

Fig. 13.5 Basis azimuthal symmetry pattern with $N = 1$

From relation *4-2*, replacing N by 1:

$$\theta_Z = \frac{180^o}{2|1|+1} \text{ or } \theta_Z = 60°$$

For the octet figure, if the topological symmetry was based on the triangular triplet, one would notice a flat value for the character typifying the element (in this case the generic charge) because of the invariance of the triangular figure under half or full rotation. Instead, one sees on the horizontal axis of the octuplet a triplet formed by two opposite characters (+ and -) with a neutral character in between (^{0}). For the meson octet, the azimuthal symmetry is even repeated across the diagonal $K^+ - \eta - K^-$.

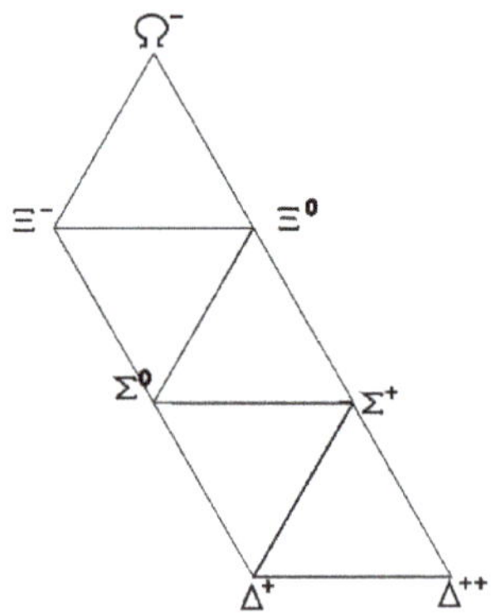

Fig. 13.6 A slice from the decuplet pattern

One finds the triplet symmetry therein both across the horizontal axis and the two diagonals. This clearly suggests that the symmetry modulus of representation is the full trapezoidal triplet as described by azimuthal symmetry genomic and the relation that crystallizes it.

This symmetry is all over the decuplet pattern as well in any direction one cares to look. This pattern can be decomposed in a number of azimuthal trapezoids featuring the aforementioned charge line (+ : 0 : –).

In the slice in Fig. 13.6, we can see the axial line Ω^- : Ξ^0 : Σ^+ with its 3 spindles. But we can also see the line Ξ^0 : Σ^+: Δ^{++} with its own 3 spindles. These two azimuthal genic blocks overlap or inter-merge with one another precisely because the unit of symmetry here is the 3-spindle scaffold. In order to make sense of the complete scaffold, one ought to visualize two azimuthal genic blocks separate from one another and then bring into superposition, with the end result being the observed overlap or inter-merge. We can also see the embedded 3-spindle scaffold anchored on the axial line Ξ^- : Σ^0 : Δ^+ in an even more faithful anti-symmetrical charge distribution than the block anchored on Σ^+ for the hybrid spindle. We may repeat the same analysis with the slice of the decuplet topology shown in Fig. 13.7.

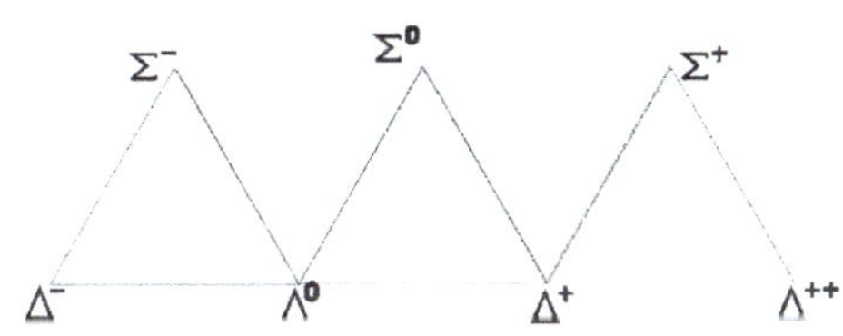

Fig. 13.7 Horizontal slice from the decuplet pattern

In some other instances we see a triplet alignment of charges that are apparently not quite anti-symmetric, and with a neutral median, i.e Δ^0: Δ^+ : Δ^{++}. However, when we look deeper into the pattern, we see that these indices correspond to a quantitative description of the qualitative artifact of the nature of generic charge. We thus see a quality depicted in a 0-1-2 quantitative progression. This depiction fully retains the azimuthal intermediacy concept with 1 flanked by 0 and 2 at each side.

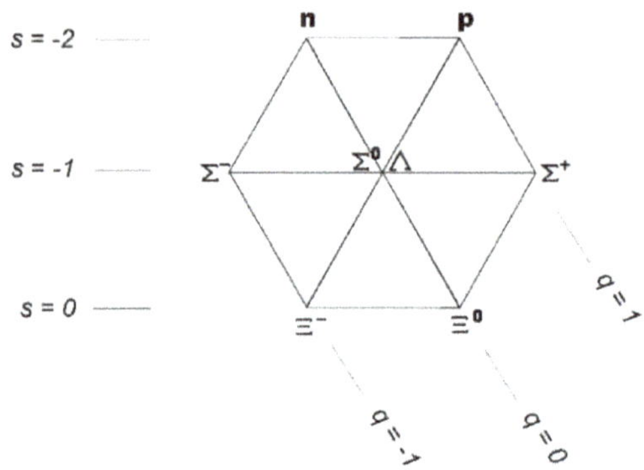

Fig. 13.8a Electric charge, hypercharge, strangeness, baryon number and hadron number from symmetry pattern underpinning Shell1 Particles: Octet format.

So it is, because the imperative of the 3-spindle symmetry bears preponderance over the generic charges that populate them, in a clear showing once again that the azimuthal triplet spindle represents the very template underpinning the generation of these topologies. The azimuthal symmetric raster interweaving those patterns amounts to a tantalizing symmetry language best understood thru the vantage of superposition of 3-spindle rasters.

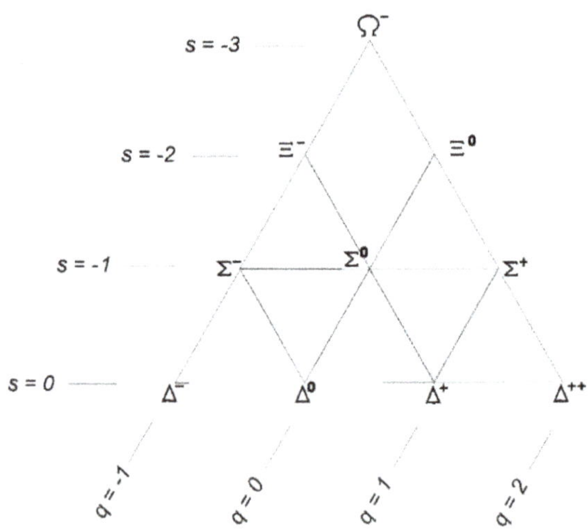

Fig. 13.8b Electric charge, hypercharge, strangeness, baryon number and hadron number from symmetry pattern underpinning Shell1 Particles: Decuplet format

The several different facets of the elements of the charge class portrayed by these rasters, specifically the electric charge, the hypercharge, the strangeness, the baryon number, the hadron number, are all progenies of the azimuthal transverse symmetry as mathematically formulated in previous discussion. The conservation rules constraining the charge class are all an expression of the infallible observance of the geometric layout implicated in the azimuthal symmetry genomic.

Why nature prefers to work in that very specific manner based on octet and decuplet patterns, physicists have wondered generation after generation? The answer is the global azimuthal symmetry presiding over the development of U_3, and its local reflection on particulate topologies as a constraint.

See illustrations in Fig.13.8a and 13.8b for the graphical representation of these symmetry patterns.

13.4. QFT Energy Scale and the Hierarchy Problem

It is not without consequence that a theory may uphold an inaccurate view of the structure of matter and consign faulty mathematics featuring unorthodox procedures of cancellation and adjustments to support it. Renormalization, the procedure of heavy-handedly curing the equations with infinite integrals, may have provided some relief in terms of taming those integrals and may have culminated in the techniques known as Effective Field Theories, largely viewed as a success. However, one can never pretend to trick nature, which will recover its solemn rights sooner or later in one way or another. Consequently, expect more problems down the road.

The view that condensed matter objects are surrounded by a field, which ends up giving rise to self-interactions by the bearing particle, even before exogenous interactions with other particles can take place, leads as we have seen to a mathematical formulation with no resolution within the realm of well-established mathematical procedures. This visualization has led to a skewed view of what the coupling constants characterizing condensed matter objects essentially are.

By not addressing directly the question of internal structure of a particle, and pretending to unravel the phenomenological properties of the particle exclusively from its field behavior, Quantum Electrodynamics has completely missed the nature of the fine-structure constant. By the way, by being notably unable to derive the constant from contended principles, the Theory exclusively manipulates the constant as an empirical quantity. Considering a particle like the electron for instance, QED has it that the fine-structure constant stands for a dimension-less constant which determines the strength of the electromagnetic force on the electron particle. This QED's quantum relativistic view predicates that the calculation of an electron interaction with another is hindered by that infinite series of possible virtual particle interactions in a surrounding cloud of quantum relativistic fluctuations. Such a cloud of so-called quantum relativistic fluctuations, the Theory further believes, changes the measured value of the coupling constant and sets it as a dependent on the energy scale at which the observation is being carried out.

While α may influence the strength of the interaction taking place between the particles, it remains an endogenous physical entity to the particles that is germane to their individual ontology. Furthermore, the least that a quantity considered as a physical constant can do is to remain constant or invariant under conditions of its most immediate reference frame. As a consequence, when one attempts to measure the coupling strength between objects in interaction mediated by a known force, aka the intensity of their interaction, one should not expect that measurement to stand for the very physical coupling constant possibly implicated in the interaction. The strength of the interaction is at most a function of the coupling constant but not the coupling constant itself. So therefore, the idea that the value of the physical coupling constant changes with the energy imparted to the particle is plainly unsound. The Theory predicates that the electromagnetic (fine-structure) coupling constant increases with energy while the strong and weak nuclear coupling constants decrease with energy. In order to give the idea a formal cover, QFT goes on to invoke a certain beta function with the coupling parameter as the variable. If g is the coupling variable and μ the energy level imparted to the particle or its "energy scale", the running of the coupling is given by the value taken by the β function as expressed below:

$$\beta(g) = \mu \frac{\partial g}{\partial \mu} = \frac{\partial g}{\partial \ln \mu}$$

The theory finds that at low energies, ground state of the electron, $\beta(g)$ or $\alpha \approx 1/137$ and at high energies, i.e., energies approaching that of the weak scale *(90 GeV)*, $\alpha \approx 1/127$. The encoding of the β function shows that as the energy scale μ increases so does $\beta(g)$ or α, and it does so in a logarithmic fashion. Hence the coupling becomes infinite at some finite energy as Lev Landau noted. However, physicists acknowledge the abnormality of this situation and choose to reason that the Landau pole situation seemingly arises because one is attempting to apply perturbation treatment in a situation of high energy where it no longer applies. To top it all, the Theory holds that the true scaling behavior of the fine-structure constant at high energies remains unknown. It is not a mischaracterization to call this end result a dead end.

13.5. QED's Blindness to Particles' Mass Scales

Quantum Electrodynamics directly associates the observed value of alpha with the energy scale of the electron mass. Because the electron is considered the lightest charged object whose quantum loops may contribute to the running of alpha, the electron particle (whether in its matter or anti-matter version) becomes the lower bound for the energy level or scale associated with the observation or measurement of alpha. Commonly the referred energy level is the kinetic energy acquired by the particle by acceleration in a particle collider. QED thus considers *1/137.036* to be the value of the fine-structure constant at *zero energy*, or level zero of the energy scale that tallies condensed matter objects.

This assessment has seemingly led the Theory to lay all condensed matter particles on a single mass or energy scale, including atomic nuclei, the configuration of which the "strong force" very particularly attempts to characterize. Despite the differences in the irreducible representations classifying the three forces with their associated causal particles, the Standard Model represents a flat basis that amalgamates all particles on a single mass or energy scale, with a single universal cover which is the class of three-dimensional rotations. QED assesses that as the energy scale increases, the strength of the electromagnetic interaction not only increases but eventually approaches that of the other two fundamental interactions, suggesting a single energy scale for all involved elementary particles at least in the upper region before the electromagnetic coupling diverges.

Particles	*MeV/c²*	*u*	*Kg*	
top quark t	173,070	186.3	3.093	10^{-25}
Higgs particle	125,900	135	2.25	10^{-25}
Z	91,187,6	97.8393	1.62557	10^{-25}
W	80385	86.312	1.433	10^{-25}
bottom quark b	4420	4.75	7.88	10^{-27}
tau	1776.82	1.90749	3.16747	10^{-27}
charm quark c	1275	1.369	2.273	10^{-27}
neutron	939.5653	1.0080649	1.6749273	10^{-27}
proton	938.272	1.00707646	1.67262177	10^{-27}
muon	105.6584	0.113428927	1.883531	10^{-28}
strange quark s	95	0.102	1.7	10^{-28}
down quark d	4.8	0.0052	8.6	10^{-30}
up quark u	2.3	0.0025	4.1	10^{-30}
electron	0.5109989	0.00054858	9.1093829	10^{-31}
tau neutrino	0.24 10^{-6}	2.58 10^{-10}	4.289	10^{-37}
muon neutrino	0.24 10^{-6}	2.58 10^{-10}	4.289	10^{-37}
electron neutrino	0.24 10^{-6}	2.58 10^{-10}	4.289	10^{-37}

Table 13.1 Masses of Elementary and Composite Particles inventoried in the Standard Model

The Representation Theory of Groups that belies Quantum Electrodynamics and the Standard Model does not capture particles' ontology but only a putative aspect of their phenomenology, which is their immersing field and the perturbation of such fields. Representation Theory is essentially a mathematical taxonomic theory that arguably holds diminished intrinsic computational power (beyond operations inside and over matrices). The computation of the most sensible physical variables must be conducted at the periphery of the Representations, principally in accompanying Lagrangians, and subsequently transplanted onto the Representational constructs. While Representation Theory of Groups may be useful to study particles from the point of view of their topologies, it cannot uncover the overarching hierarchy or hierarchies to which they belong because the Theory is not able to engage into the computation of the mass and wavefunctions of particles. Therefore, QFT's Standard Model has run into many tracks to nowhere and faced stiff walls at every turn. The question of particles' mass scale, which transpires in the so-called *Hierarchy Problem*, is one such major problem it is notably unable to resolve. Accounting for particles' masses is a heavy burden that the Theory is left to tackle in a clueless manner, given that QFT starts out with the construct that all particles are initially energy fields and at the outset provides no computational framework for mass derivation from primal principles. All particles are initially altogether massless in the Standard Model since there is not even a Dirac mass term in the accompanying Lagrangian. The fact that Dirac mass terms could not be incorporated in the Lagrangian contributed to deprive the theory of a sense of mass scale for condensed matter physics. Effectively the incorporation of mass terms to acquire definite values in a sliding modality is in direct contradiction with gauge group invariance underpinning the Theory. None of the Standard Model fermions or bosons can begin their lives with built-in mass except by abandoning the requisite of gauge invariance, cornerstone of the Theory.

Once left with no tool to explain particle's masses, an obvious element of their composition, physicists turned to the electroweak force for clues and possible mechanisms of mass provisioning to these objects. They came to believe that the physics of mass apportionment is inextricably linked to the mass scale of the mediators of that force, the two W scalar bosons and the Z scalar boson. Table 13.1 below reports the known exact or approximate masses of the most important elementary and composite particles studied in the Standard Model. One of the key reasons for querying the weak force for the source and apportionment of particle masses is that, among the three forces studied in the Standard Model, the weak force is the only one that witnesses crosses of particle generation within the diverse interactions it mediates. Degeneracy from tau to quark and the phenomenon of neutrino oscillation previously studied are two examples of such crosses. In this context the Higgs boson had been postulated as the master purveyor of mass and the Higgs mechanism of spontaneous symmetry breaking, previously evoked, proposed to account for the modality of mass apportionment.

From Table 5.1 one can already pinpoint some of the problems that this view faces. One is that the mass of the Higgs particle, a particle long prospected but finally discovered recently, is smaller than that of the *top quark t* of its supposed gestation. Most importantly the symmetry for the four scalar bosons of the weak force, the Higgs, the two W's and Z bosons, which must be known for the symmetry breaking mechanism to apply, has remained completely elusive. Hence the involved mass scale cannot be adequately determined. Worse, the Theory projects the mass scale at an enormously high quantity that is larger in many orders of magnitude than what is observed in nature. Again, only if several mass terms contributing to the total mass value in the Theory are heavy-handedly removed or artificially caused to cancel, can this obstacle be resolved.

Physicists have insistently attempted to determine the absolute ruling mass scale for the physics of condensed matter particles, because they harbor no doubt that nature must hold such a referential axis in the distribution and allocation of mass to infinitesimal particles. That is why they have turned to the fundamental physical constants in order to determine that scale. In this endeavor, the Planck's constant h, the speed of light c and the gravitational constant G, considered key fundamental physical constants, have been combined with one another in order to produce a term whose units would be those of energy.

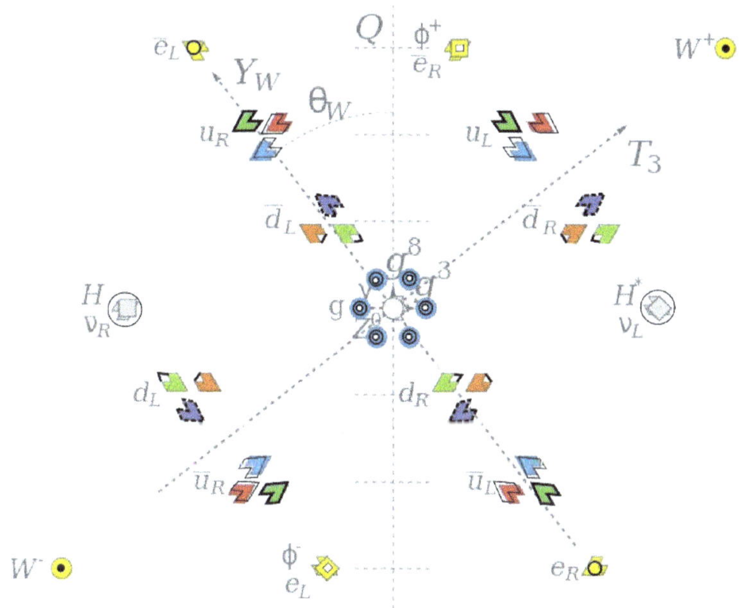

Fig. 13.9 Standard Model's symmetry-breaking mechanism on a flat sheet

This scale of energy, the so-called Planck scale of energy, is ruled the fundamental scale of energy in physics. Aside from energy, the mentioned physical constants have been combined in units to produce as well a Planck length, a Planck mass and a Planck time. The Theory considers the Planck length for instance to be the natural length scale for quantum gravity, of utmost significance to any visualization of unification of Relativity to the Quantum Theory. The Planck length computation follows:

$$L_{Planck} = \sqrt{\frac{\hbar G}{c^3}} = 1.6 \times 10^{-35}\, m$$

The computed Planck time is in the order of 10^{-43}sec, while the calculated Planck mass is in the order of 10^{19} proton masses. Physicists believe these quantities to be too disparate to reflect nature however. They are not only exceedingly removed from the macroscopic scale, but most significantly, they are too far removed from the known order of magnitude of the weak scale, to which modern physicists tribute the role of grand

Generator. And so, the question becomes, if the Planck scale mirrors the fundamental scale of physics, why is the weak scale, the chosen candidate for that role, comparatively so low? This spells out in a large measure the terms of the long-standing *Hierarchy Problem* in quantum physics.

From the Quanto-Geometric vantage, one can already see the resulting compounded problems with that visualization. Yet there is more to Table 5.1 that is further revealing in that regard. The table reflects a single hierarchy of mass for all inventoried particles, which is exactly reflected in the graphical view of the postulated symmetry-breaking mechanism in the Standard Model (Fig. 13.9). All the particles, from all classes and generations, are therein represented together on a single flat sheet. However, the reader who has been following closely our development may already point out that there is in fact two distinct mass scales for condensed matter particles: one represented in the quantum axis sustaining the Quark Primitive Spectrum (Shell1 of matter) and the other represented in the quantum axis sustaining the Primitive Spectrum sustaining Shell2 of matter. We have arrived at these two scales not by an arbitrary measure but consistent with the Quanto-Geometric axiomatic. Furthermore, even the classes and generations within each of these two main groups are not to be inventoried on a flat sheet. This product of the Standard Model lacks the crucial zenithal dimension that stands for the primary spectral representation giving rise to the most significant physical quantities characterizing 3-dimensional objects. In other words, the Standard Model of classification of sub-atomic particles only captures part of the azimuthal dimension of the Quanto-Geometric Model and altogether misses the zenithal dimension of the latter, which is keenly represented in the Quanto-Geometric function of single-variable s.

13.6. Particle Mass Allocation by Symmetry-Breaking

Physicists have long known that three-dimensional rotations stand for a key artifact in the phenomenology of condensed matter particles. It is not surprising that they have looked into this dimension of the phenomenology when confronted with the requisite of accounting for the very real artifact of particle masses, given that they have started from an analytical perspective that exempts the latter.

Physicists' tool of preference in that endeavor is the Higgs mechanism, already hereby invoked, along with the weak force interactions. Let's note at the outset that while the Higgs has been experimentally observed, the Higgs mechanism itself has yet to fall under experimental observation. Some aspects of these constructs are supported by the Quanto-Geometric Model, we must say, but others are strictly not. The notion that a Higgs scalar field permeates all space and that particles acquire mass from the universal 3-D spatial field is an utterly awkward notion that is totally at odds with the Quanto-Geometric axiomatic. Proponents of this construct explain that the mass contained in the Higgs field is originally in the form of potential energy. This proposition is quite frankly tantamount to a euphemism since it is only a quantum form of the energy that can be transformed into mass as we have seen in the quanto-relativistic analysis. Proponents of the model do further admit however that the quantum of the Higgs field is indeed the Higgs scalar boson. It is worth noting that within the framework of the Theory, other important particles such as the neutrinos remain with no mass acquisition, since the Theory has been built to make them massless. Gluons' masses are not accounted for by the Standard Model either, all of which peremptorily discounts the quality of universal to the construct of space scalar field, unless more is to come now after a long quasi 100-year wait period. The proponents of the Higgs field have lost sight that the construct of space scalar field has arisen from a Theory that has specifically elected to start out with a field representation of condensed matter particles. Nature has not started out in the incarnation of "fields" which acquire mass by perturbation. At least not so in other Theories. In the Quanto-Geometric view of nature, all objects are concomitantly and dually space and scalar parcels. From that construct we have been able to derive an unprecedented series of parameters, constants, variables and mass quantities with very good approximation to their known empirical values, in a clear demonstration that the Quanto-Geometric Theory deserves the stature of closest representation of physical nature ever postulated, thereby standing on a more assertive and credible footing.

This notion of scalar fields embedded in the fabric of space shows many more caveats as one ponders even more over its implications. Should one not ask the question "Where does the scalar field come from at the first place, or what is the makeup of spatial potential energy that can turn scalar?" In the real-world potential energy is a derivative of spatial distance. Likewise, what is the nature of the dimension in space that is "potentially" scalar or can turn scalar? Or should one limit oneself to not ask the question "What is the origin of the very spatial field that originally and single-handedly initiated the life of the particle, before it needed to acquire mass?" Even more realistically, one must ask of the Theory the ability to derive the proper wavefunctions of particles themselves, given that in all visualization this entity must be closely related to the notion of spatial field. We know those answers altogether falter.

13.6.1 Mass Acquisition by Spontaneous Symmetry Breaking

The details of how particles interact with the Higgs field in order to acquire mass are very interesting to explore. It is important to remind the reader that to this point said mechanism qualifies as no more than a speculation, given that it has not yet been experimentally verified. Most interesting is the fact that the famous Mexican-hat shape of the field had not been in any way derived from the Theory but arbitrarily posited (Fig. 13.10).

The Higgs symmetry-breaking effect theoretically occurs when a gauge boson particle enters a Higgs Field which subsequently causes it to naturally transition into its lowest energy states. This process breaks the gauge symmetry of the field allowing the intervening particle to acquire new degrees of freedom by which it absorbs mass from the Higgs bosons of the field. The Higgs bosons play the role of quanta of the Higgs field. It is important to stress that the Higgs mechanism very specifically applies to the generation of masses for the two W gauge bosons and the Z boson only and no other particles. Its application domain is strictly the electroweak force.

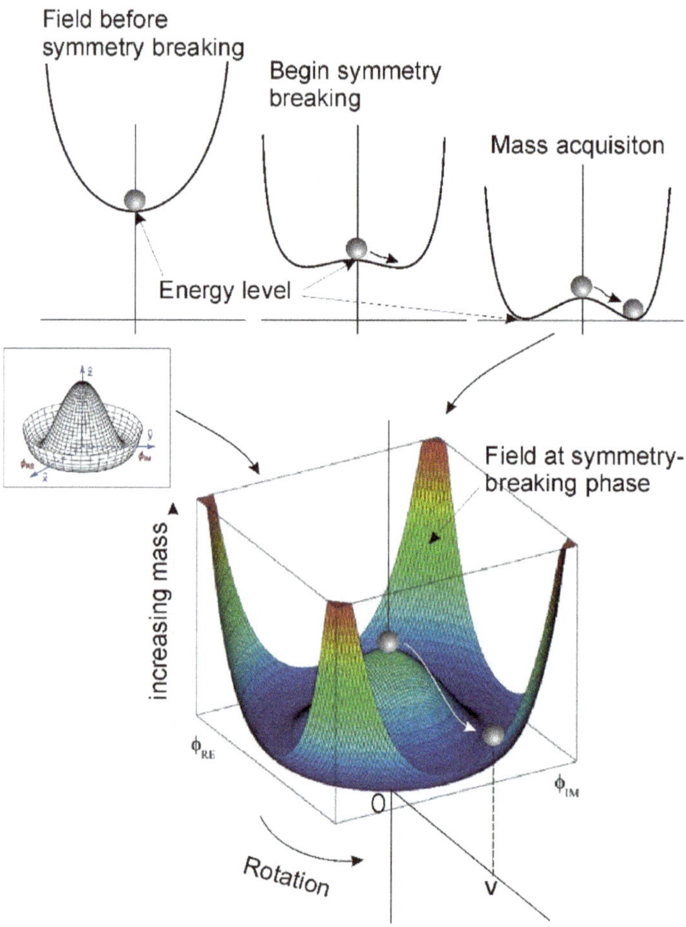

Fig. 13.10 Flow of the Standard Model's symmetry-breaking mechanism

The field has the form of a paraboloid before symmetry breaking, with the particle at its origin. At the symmetry breaking phase the shape of the field evolves into the characteristic Mexican hat. When the particle transitions from origin of the field to its quadratic point v by the transformation $\Phi' = \Phi - v$ the fields W and B recombine to reemerge as the photon field A, a neutral massive vector particle Z and the massive doublet W^+ and W^- particles. The transformation from (B, W) to (A, Z) may be considered as well a rotation of the mixing angle θ_W. The empirical formulas below give the value of the acquired mass of the new massive particles:

$m_W = \frac{gv}{2} = 78.4 \pm 0.7 GeV$, mass of the $W^{\pm}$ gauge bosons

$m_Z = \frac{m_W}{\cos\theta_W} = 89.1 \pm 0.7 GeV$ *(5-2)*, mass of the neutral Z gauge boson

$m_A = 0$, mass of the photon

These values obtained by replacement of :

$v = \pm\sqrt{\frac{-2m^2}{\lambda}}$, complex expression of the minimum energy value v of the field at the quadratic point.

$\theta = \tan^{-1}\frac{g'}{g}$, the weak mixing angle.

Let us first note that from origin up to the quadratic point this field at symmetry breaking phase is eerily reminiscent of the Quanto-Geometric hyperboloid. The proponents of the Standard Model have never explained why the Higgs field must feature this very particular shape or the mathematical origin of the posited potential function, which has long remained an item on the to-do list of the Theory.

An examination of the mechanism from the Quanto-Geometric vantage quickly reveals that it is only a remote approximation of allocation of mass to sub-atomic particles in that framework. In Quanto-Geometric Theory mass is directly computed from the zenithal dimension of the model, the function of one variable s, from a construct of point source derivation of all implicated physical entities in composition with the canonical scalar hierarchy. The mass quantity of every particle is essentially a value taken by the function on its quantum axis, the natural mass scale. Most tellingly every particle is represented by a normal Quanto-Geometric layer enthused with a particular form of symmetry within 9 distinct states of symmetry contemplated in the Model. This primary discipline of symmetry, delimited by the primeval *symmetry-breaking sequence* on s, represents the symmetry stage that breaks and transmutes at every layer, at the same time that it corresponds to a specific mass value on the scalar axis at every step. The symmetry not only breaks but reappears into a new symmetry in the borne-out object. The evolution of the shape of the Higgs field, from parabolic to Mexican-hat, is a remote reflection of the artifact of that Quanto-Geometric transmutation. The order of symmetry is a built-in feature of the Quanto-Geometric Theory which not only explains what it means for the symmetry to get broken yielding mass quantities but why the symmetry was there in the first place. It critically reveals the identities of the natural distinctive states of symmetry subsumed in the physics of nature. Those questions, long posed by physicists but with no answers in sight to date, find clear resolution in the Quanto-Geometric framework.

13.6.2 Origin of the Symmetry-Breaking Seesaw Mechanism

There is significant level of Quanto-Geometric support as well for the seesaw mechanism inherent to the apportionment of mass to the leptons of the Standard Model, although this mechanism at this stage remains a hypothesis, no more than the generality of the Higgs mechanism itself. The Higgs seesaw mechanism is one whereby the Higgs particles change the chirality of the interacting lepton particle as they funnel mass into the lepton thru its travel course across the scalar field (Fig. 13.11).

As we have seen, neutrinos were formerly exempt from participation in this mechanism in the Standard Model. However, the experimental observation of neutrino oscillation has notably created a crisis in the life of the Model, prompting the late creation of extensions to the Standard Model contemplating modalities of participation of neutrinos in the seesaw mechanism without dismantling the rest of the Theory.

In support of this hypothesis, and quite fittingly, in the Quanto-Geometric azimuthal model, one may notice that each subdivision of every azimuthal band corresponds to both a scalar gradient in the zenithal dimension of the model (a value on the quantum axis) and a specific chiral state. If we imagine a particle travelling thru the azimuthal band in any direction as represented, it will transition longitudinally from one scalar level of the specific azimuthal band to the next scalar level, changing chirality at every step. Not quite a seesaw pattern but definitely a chiral oscillation with mass transformation if not acquisition. This would be an isomass (isospin) contribution to the particle's already existing mass otherwise acquired by the entity, but definitely not the main contributor.

Clearly among all neutrinos, the inert neutrino is the one that would fully experience this longitudinal transformation. The flavored neutrinos, being hybrid-spindle creatures, can possibly experience such state transformations only through zenithal (top down) currents across the hybrid spindle.

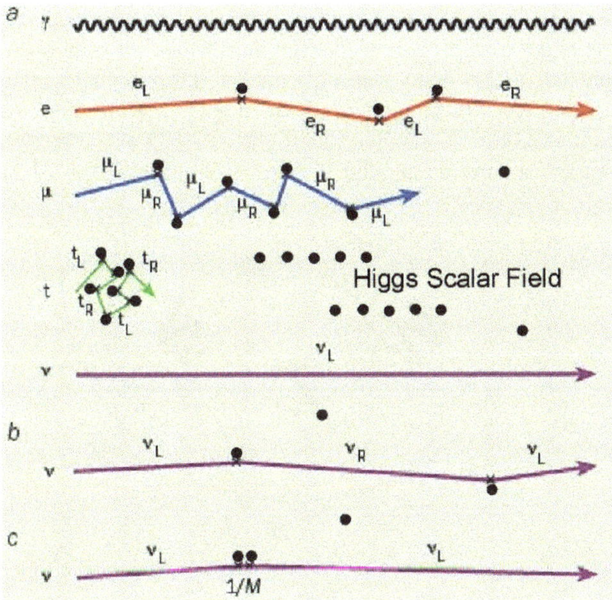

Fig. 13.11 Symmetry-breaking seesaw mechanism

When it comes to the question of interpretation, it is important to stress that the Quanto-Geometric model is a theoretical description, in the suit of a mathematical functional formalism. This is to mean that the azimuthal band is not a physical scalar field and it is not phenomenological in and of itself. Therefore, the analysis above does not describe *acquisition* of mass by particles, no more than it describes *acquisition* of wavefunction. It is a structural mathematical-physics description of 3-D objects in U_3. In that sense it is totally different from the Higgs proposition, despite the expansive view and support that it provides to the same.

13.6.3 Theoretical Derivation of the Weak Mixing Angle

The weak mixing angle otherwise known as the Weinberg angle is a rotation parameter representing the angle by which the original W^0 and B^0 vector boson plane must spontaneously rotate under symmetry breaking constraints in order for the mass-less neutral photon and principally the massive neutral Z^0 boson particle to arise. It is coded as the tangent of the ratio formed by the weak force running coupling constant g and the electromagnetic running coupling constant g'. As such it is a running parameter that depends on the energy level at which the interactions are taking place. At energy level 91.2GeV corresponding to the mass of the Z boson, the weak mixing angle approximates the valued of *28.75°*, from $sin^2\theta_W = 0.2312$. The value of this angle is not quite settled yet, different experiments at different physics laboratories casting a slightly different value for the angle under the sine square figure that is most frequently used. An approximate value of *30°* is admittedly the best consensus at this point. It is acknowledged that the specific value of the angle is *not* a prediction of the Standard Model: it is an open, unfixed parameter. To date there has not been any generally accepted theory that explains the specific value of the measured weak mixing angle. Well, an examination of the nature of this angle from the Quanto-Geometric vantage forthrightly casts it as a direct representation of the Quanto-Geometric azimuthal angular symmetry captured in the *12-2* relation.

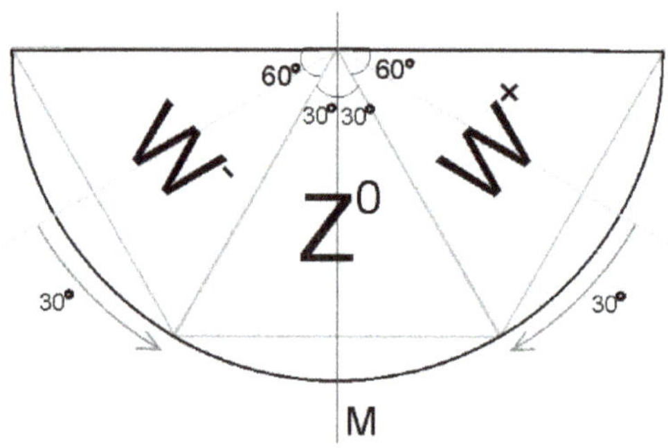

Fig. 13.12 Generation of mass by angle mixing

QFT's Higgs symmetry-breaking theory prescribes that the gauged vacuum spread at the origin of the mechanism must have a phase, that is to say it must be "charged", not electrically but generically charged, we say in the exact sense of what we have called bipolar transverse symmetry. The Theory further mandates that for a scalar field to have a phase, it must be canonically dual, which confirms our interpretation thus far. It must be either complex or show two fields with a symmetry that rotates them into each other! Are we witnessing here a description of the Quanto-Geometric azimuthal symmetry? No doubt we are. The vector potential, the description continues, that is the gauge level of the field dictates how much the real and imaginary part of the two partitions of the field spontaneously rotate into one another at symmetry breaking. The Theory finds that the above requirements come to incarnate into the Mexican-hat field with the phenomenologies that we have already exposed.

We find the above theoretical requirements to be a pretty faithful representation of our definition of azimuthal transverse symmetry (Fig 13.12). Taking the azimuthal symmetry genic relation with a value of $N = 1$ for ground state, we obtain the triplet scaffold featuring a median hybrid spindle flanked by two lateral spindles. At ground state the unit spindle angle is equal to 60° from the relation. If the side spindles are a representation of the matter/antimatter W doublet, by rotating them into one another as prescribed we obtain the Z meson, because the rotation places the right-side half of the W^- meson spindle and the left-side half of the W^+ meson spindle right into the location of the mid-spindle, so forming the neutral or hybrid character of the Z meson. The rotation span that accomplishes this symmetry or the breakdown of this symmetry is a *30°* angle, which is half the value of the azimuthal angle for ground state of $N = 1$ from the azimuthal genic relation. The equitable translation of the W doublet into one hybrid or mixing spot justifies the alternative name of weak *mixing* angle to the Weinberg angle. The relation showing how the weak mixing angle derives from the azimuthal angle follows:

$$\theta_W = \frac{\theta_Z}{2} = \frac{\left[\dfrac{180°}{2|N|+1}\right]_{N=\pm 1}}{2} = 30°$$

The preceding analysis undoubtedly uncovers in the most precise manner the hitherto unknown origin of the weak mixing angle, empirically estimated at *30°*. This analysis very narrowly concurs with the topological symmetries of the Octuplet pattern of other mesons previously studied.

In the larger picture, the rotation scaled by the weak mixing angle on the azimuthal plane portrays an isospin (or isomass) transformation. In the Quanto-Geometric azimuthal model, every particle on a normal layer (the azimuthal band) occupies a specific chiral sub-division of the layer, which corresponds to a specific scalar gradient on the scalar axis within the global scalar range attributable to the layer. Consequently, an azimuthal rotation may perfectly add complementary allotted mass to mass acquired by zenithal symmetry breaking. This provides theoretical support to the empirical formula expressed in *(12-2)* for the mass of the Z boson.

13.6.4 Representation of the Higgs Scalar Field

All of the above considered, we firmly assert that the Higgs field of the Standard Model is no less than a representation of the Quanto-Geometric normal layer within a canonical Primitive Spectrum. Because Quantum Field Theory is unable to frame an assertive mathematical description of particles' ontology, it is left with a phenomenological story line whereby it projects the buildup and allocation of mass to individual entities from scalar point sources within a theorized scalar field. This narrative is comparable to gazing at an object far away located on the horizon line and being able to only vaguely perceive some of the elements of its shape and structure. If the Higgs particle is going to be an apportionment source, it ought to be visualized apportioning both verifiable wavefunctions and scalarity to the particles. The object that surrogates that role in the Quanto-Geometric framework is the $<q_3|\ s_3>$ entity. But again, we have cautioned against adopting a timeline narrative based on single-point origin in describing the configuration of condensed matter objects in the physical universe. We have already explained the reasons that assist us in holding this view. Further, the conceptual distance between the scalar source (or the Higgs if you will) and the condensed particulate objects is truly enormous; because, in order to have a World or a World brane, you don't just need to funnel mass into particles, you ostensibly need to have the foundational quantization basis on which to erect the objects destined to populate the dense region of that construction. You need overarching modeling and scaling Operators to act over the primal quanto-relativistic $<q_3| \, s_3>$ entity in order to build the laws of symmetry, as well as an all-encompassing quadruplet quantization basis *(h, λ, ρ, T)* for the new World, before you can build up or funnel mass into a single particle!

The spacetime model conveyed by the Higgs Scalar Field is that of a lattice of some kind with nodes that internally contain a potential distribution in the form of a Mexican hat (second-order polynomial of two variables). Objects foundationally take origin from that field for the configuration of their condensed matter plane. The Quanto-Geometric model instead visualizes the fabric of space as being always void, with a population of scalar parcels in quantum states in direct quanto-geometric correlation with their immediate space spread, thereby forming a wider topological spread and a distributive Shell of matter under the regency of pivotal physical coupling constants.

Our vision of mass in Quanto-Geometry is that it is irrevocably a density function, at all scales of matter. Our mathematical representation of this construct is the image Grand Eigenfunction presiding over the development of the 3-dimensional Universe, in the values taken by the q-coordinate of the Function. If the concept of acquisition of mass by phenomenology is going to be plausible and useful, it ought to represent how mass is acquired not by additive scalar juxtaposition, which seems to be the modality of the Higgs mechanism, but strictly by scalar density accrual. We have demonstrated the same by ontology in the method of theoretical derivation of the mass quantities of a number of elementary particles in previous Chapter of this exposé.

13.6.5 Identity of Scalar Bosons

This examination also tells us that the terminology of *scalar boson* in use in the Standard Model is probably impertinent and at best imprecise. Arising from the Goldstone treatment of the vacuum which permeates all of Quantum Field Theory, this notion is central to the Higgs mechanism. It further remits to a bosonic particle which plays the role of a scalar or acts as one, or reciprocally a massive particle that behaves as a boson. Again, the phenomenological narrative is that the Higgs mechanism occurs because the scalar components of the Higgs field are absorbed by the boson of mesonic nature as degrees of freedom. The Goldstone boson particle under transformation interacts with the Higgs field in order to acquire its mass thru Yukawa coupling as an end result. Why the new massive meson remains a boson subsequent to the transformation is unclear, other than the fact that the Theory is intent on making a fit for the new particle as a so-called force carrier for the electroweak force. The measure by which the sum total of all the processes involved in the symmetry breakdown correctly allocate mass to the particles according to their statistical identity remains unclear. The statement that fermions interact more or less with the Higgs bosons depending on their to-be mass level is not quite satisfactory. Since the Theory's entire ambition is to explain the mass quantities observed to elementary particles, one is in absolute good right to demand a computational account of a lot more of these elements than just simply two, and especially those with a clearly defined statistical nature involving their mass, such as the fermions. A variable-dependent scale that sets the amount of allotted mass in the spontaneous breakdown of a symmetry with incidence on the ontological nature of condensed matter particles, or with calculable influence on their spin angular momentum, seems to falter altogether in the Theory. In our view the only dimension that makes a *scalar boson* a sensible taxonomic concept is the quadratic dimension of the Quanto-Geometric Primitive canon. We have seen that in both Shell1 and Shell2 Spectra, this role is respectively and exclusively played by the Quark-Gluon plasma and the Inert Neutrino.

Any theory pretending to treat particles' ontology from a unilateral component of their structures is doomed to fail. We cannot overstress that any successful treatment of physical objects, whether macroscopic or infinitesimal, must address both their scalar component and their wavefunction component, and not only in a concomitant manner but very importantly as orthogonal elements to one another.

13.7. Unified Theory of Condensed Matter Physics

By realizing back projections to origin to the known or so-called running physical constants, physicists seek to find a common mass or energy scale that might help decipher the physical evolution of the universe and better understand the pattern of masses and distribution of particles present in the Universe. A single mass value from that projection of the three Standard Model's running coupling constants constitutes the starting point for that coveted Grand Unified Theory. This enterprise ran into two problems. Firstly, the computed unified mass is a huge $10^{14}GeV$, a quantity far too high for any conceivable probing particle accelerator. Secondly, short of having to directly probe the out-of-reach computed mass scale, in a Grand Unified Theory everything should decay from origin in order to give rise to today's observable universe and so the proton, like the neutron, should undergo structural decay. Only that such a decay has never been observed experimentally. Thanks, God, for its inexistence, because the artifices of the experiment, human individuals or whole civilization cycles, might not survive decay of their own protons to get to know the result!

A Grand Unified Theory cannot exist on the basis of a single mass sheet or a single hierarchy as implied in the original unification concept. The backward projection of the coupling constants makes virtually no sense because first of all, except for the quark family's coupling, those constants are not running, they are fixed invariants. The running quantities from those constants are the actual forces arising from their application. The only true variant, applicable to the quark-gluon family, runs down asymptotically to zero as it happens. This spells out the second aspect of the so-called Hierarchy Problem. It is a problem that in no small part results from a mistaken visualization of the nature of the fundamental physical constants. To put it otherwise, it is because modern physics has not been able to theoretically derive the fundamental physical constants, an enterprise that would reveal the true origin of those constants and their actual role in the configuration of condensed matter physics, that the mistaken projections from these quantities become plausible.

On the other hand, at high energies the Standard Model falls in complete disarray, requiring the intervention of other frameworks that would harmoniously yield all the observed symmetries while safeguarding the successful interpretative features of the Model. A Grand Unified Theory does not arise from projections carried out inside of the Standard Model. These projections ultimately only offer the Model means to probe itself and to identify problems with its own features and structures. They do not necessarily set the path for the resolution of these caveats. This implies that not only the Standard Model ought to be superseded but in the process all the background theories that have subtended its development as well. The Hierarchy Problem, we must say, is formal nomenclature for prosaic language that would otherwise state this: “We know there is a major problem with our system, but we just don’t know what the problem is!”

13.7.1 Gauge Invariance across Dimensional Boundaries

When viewed thru the angle of gauge invariance, the Hierarchy Problem is to be seen, to some extent, of modern physicists’ own making. In their pressing quest for an accurate interpretation of the dynamics operating in the infinitesimal realm, they have summarily obliterated all the boundaries between the various dimensions of condensed matter physics.

It requires no prior definition, due of the bulk of the evidence, to hold the quark particles and their relational dynamics as one dimension of matter, as much as baryons such as neutrons, protons, etc., which are made up of three quarks and appear manifest in their full rights, as another dimension of matter; just as atoms, consisting of specific baryons, represent another very specific dimension of matter. Everyone can agree unreservedly with this characterization, if anything just because every layer of this genesis showcases objects that are made up of a small number of elements from the previous layer (except for the quarks perhaps), at least from the scalar point of view. Hence, taken together they form three different physical dimensions. These individual dimensions are what is called in Gauge Theory a gauge symmetry. A gauge symmetry is a reunion of all the physical laws that interrelate a certain set of physical objects at the same layer of expression of matter. These laws create or result from a state of invariance or symmetry. The most pivotal element at the basis of this invariance is the physical coupling constant figure. This pivotal symmetry gives constitutionality and independence to the physical layer of matter of its application. If any layer of 3-dimensional matter were to lose constitutional independence of its symmetry, it would collapse under the transformations occurring in the next layer up or down, which means the Universe ultimately would tend to collapse or could matter-of-factly collapse extemporaneously. The conservation of gauge invariance of one dimension, under transformations occurring whether in another larger or more encompassing dimension or a lower-level one, is paramount to the organization and stability of the entire 3-dimensional Universe. Exceptions to this rule, if ever, should be extensively and peremptorily scrutinized.

On that account, one should not look into the physics of one symmetry for laws that explain the physics of another symmetry as a matter of course. Unfortunately, modern physics routinely incurs this misguided exercise. Reiteratively, the notion of appropriating fractional electric charges to quarks so that the baryon particles could end up bearing their known electric charges, hence conveniently setting the neutron with no net electric charge, is one example of these cross-boundary infringements that have caused great harm to modern physics. Most physicists remain uneasy with the characterization of quark's fractional charges, contending that it has not been theoretically justified. Beyond this lack however, it is a wrongful view to make the charge property of a baryon a direct result of the charge property of its inner constituents as if a baryon is constitutionally on the same playing field as the quark or the meson. One would have to address the reason why baryonic symmetry allows this transmigration of charge or charge effects into its own province for the notion to acquire any legitimacy.

13.7.2 Another Dimension of Symmetry Violations by the Weak and Strong Forces

The literature also contemplates the notion that the weak force and the strong force are *nuclear* forces. The epithet suggests that these forces originate from the dynamics of atomic nuclei proper. Correction: they are instead *sub-nuclear* forces. There are many different contributors to the weak force, several being of different origin. However, if we strictly consider the nature of its main mediators, the W and Z mesons, one ought to keenly comprehend that the weak force is strictly a quark-dimension phenomenon. Considering a radioactive object such as a molecule with a Nobelium atom for instance, the disruption of the object by the weak force, is no less of a molecular nature than it would be of a nuclear-atomic nature. There is nothing within the molecular structure that orchestrates the disruption, nor is there anything inside of the atom proper that orchestrates the disruption. The origin of the breakdown is indeed in the mesonic dimension of the object. Hence, the term weak *nuclear* force for all types of radioactive activity is misguided and should be discouraged.

When it comes to the strong force, cross-dimensional determinants become even more prominent in the Theory. Gauge Theory has the strong force with two different facettes: the *fundamental* strong force which binds quarks together in baryons and the *residual* strong force which binds nucleons (atomic protons and neutrons) together. In other words, the coupling between quark triplets in baryons is not only responsible for their internal structure, but percolates residually outside of these objects into the next dimension up of matter in order to provide the glue that binds the whole edifice of an atomic nucleus! To say the least, this is quite of an awkward characterization whereby all sense of harmony (or symmetry) has completely been thrown away. According to this picture overall, condensed matter physics seems to be ruled by one single symmetry, the quark symmetry, since mesons and baryons are internally structured by quarks, which give them their electrical properties, and the quark-quark force providing the glue that binds them also provides the glue that binds the much larger edifice called the atom. Yet three different layers of matter, with distinctly structured objects that are integral to their individual configuration, loom undeniably before our eyes. Is it any wonder that there is a gauge hierarchy problem with this picture and that modern physics has had a long-standing Hierarchy Problem?

Evidently, the three different dimensions of condensed matter physics are each integrally ruled by their own symmetry, whether the Standard Model likes it or not. Surprisingly enough, there is as we can see a different dimension than the customary one to the notion of Symmetry Violation in use in Gauge Theory and it is at the heart of the infamous Hierarchy Problem. By comparison, the Quanto-Geometric notion of Shell of Matter is as inescapable as effective a visualization for an appropriate treatment of all gauge symmetries.

It is of utmost significance to realize that there is more to the organization of matter than invariant behavior under three-dimensional rotations or even Lorentz transformations. Beyond the realization that the expectation value of the vacuum is not always a state of zero-point energy, one must come to the full realization that the vacuum is not at all the same everywhere. While the view that privileges the tracing of invariant interaction or behavior under three-dimensional transformations may yield quite useful descriptions of the dynamics at play, it unfortunately contributes to obscure the very ontology of the underlying vacuum, which are in fact, most often at least, two different vacua at different dimensions of the involved physical dynamics.

One of the reasons why the properties of the vacuum have remained imperceptible for the most part thus far is very much because we have historically developed the physical sciences essentially on the basis of phenomenological analysis. The importance of seizing the many different dimensions of matter as it lays before us did not transpire. As a result, our physical description of nature took in large part to the practice of mathematically expressing physical variables in terms of any other variable the description finds suitable, regardless of its origin. The sense of hierarchy, essential to an encompassing comprehension of nature and an accurate formulation of its dynamics, remained ever more absent throughout the development of the science. The recurrent example of how we want the figure of electric charge, an almost macroscopic phenomenon discovered at the molecular level, to be part of the life of every lower-dimension object in the condensed matter realm, even though the actual value of the elementary charge can never find clear applicable circumstance in that realm, is a case in point. Virtually every formula in our physical science is testimony to this cross-dimensional transgression. Another case in point, this one formula expressing the fine structure constant:

$$\alpha = \frac{e^2}{\hbar c} = \frac{1}{137.04}$$

Viewed thru Quanto-Geometric lens, it involves three different dimensions of condensed matter physics represented in each of the variables at the right side of the relation.

In that context, the unit-dimensions stood to acquire foremost importance, because the science would quickly crumble under deficient house-keeping of those units or loss of concordance between them from one numerical expression to the next. Unit-dimension reconciliation has for good reason remained a subject of great scrutiny among physicists, given that the new unit-dimensions that emerge are oftentimes quite obscure and difficult to visualize. It is interesting and even ironic to note that it is by probing the possible meaning of combinations of the most significant of these cross-dimension variables and their projections that modern physics has come to the awareness of a Hierarchy Problem in its structure. Only on account of good unit-dimension house-keeping as well as the inherent similarities or homomorphisms between the symmetries pertaining to different layers of matter, has the science been able to develop with sufficient numerical accuracy. However, it is no surprise that dimensional analysis itself has proven that a significant Hierarchy Problem does exist in the midst of the Standard Model. This simply stands for one more aspect of this problem.

13.8. Conclusion

In this Chapter and the previous we have formulated detailed mathematical basis for the symmetric discipline of azimuthal symmetry, of paramount importance in the configuration of ontologies and topologies accessible to condensed matter particles. This symmetric discipline represents the key allocator of particles' spinor states and the matter-antimatter division of the entire 3-dimensional domain. Despite the relationship of the former with the elements of mass configuration, it is clear that the symmetric discipline pertaining to vertical hierarchy or zenithal symmetry remains a closer effector to mass and wavefunction in its direct relationship with the gestalt scalar *Q(s)* of the image Grand Eigenfunction, which is the key allocator of mass to condensed matter particles. Nevertheless, the importance of azimuthal symmetric discipline will only grow in the development of the next Shell up, Shell3 of atomic elements, as it will later transpire. Once those paramount analytic structures are unraveled and understood, the various flaws and mires of the Quantum Theory, single-handedly reflected in the Hierarch Problem, can be cured and happily overcome, because the principal problem of the Quantum Theory to date is a blatantly incomplete comprehension of symmetry in physical law.

Chapter 14

...[It is possible] that Nature is the same damn thing over and over again... that nature has only one way of doing things, and She repeats her story from time to time.
Richard Feynman

PERIODICITY OF ELECTRONIC SHELLS IN UNIFIED ATOMIC ARCHITECTURE

Following the momentous discovery of the geometric nature of the force of gravity by Albert Einstein, the subsequent discovery by Ernest Schrödinger of the energy density states that subtend the organization and dynamics of electron particles within atomic elements marked a tremendously impacting milestone in the historic development of the physical sciences. His famous Hamiltonian equation independent of time was able to virtuously cast the famous transcendental eigensolutions representing these energy density states for the electrons spread around the atom's nucleus. It is no surprise that the basic dual structure of the quantum Hamiltonian, a canonical structure in Quanto-Geometric Theory, bore the necessary features to ascend to the transcendental Eigensolution.

Shortly thereafter, Werner Heisenberg found a matrixial formulation of atomic (electronic) quantum mechanics. However, it was only when E. Schrödinger ventured to show that quantum mechanics could explain the already known distribution of elements in the so-called Mendeleev periodic table of elements, that all the guards came down and the new quantum mechanics was acclaimed. Even later, P.M. Dirac reformulated quantum mechanics in the terms of a Lagrangian in order to account for the spin of electrons in atomic orbitals as well as the free transitioning of electrons throughout the complete orbital spread thru non-zero minima between the saturated zones.

Quanto-Geometric Theory grandfathers all these characterizations, albeit not pretending to substitute their adopted mathematical formalism. Our objective is rather to put in evidence analytical symmetries between the treatment under Quanto-Geometric Theory, insofar as the organization of electron particles within atomic architecture, and that of the other mentioned Theories. Most importantly we will endeavor to show that the same laws governing the buildup and dynamics of electron particles within the edifice are the same, lest a few variants, that dictate the organization of the other two particles making up the edifice: protons and neutrons.

14.1. Characteristic Structure of Atomic Energy Density States

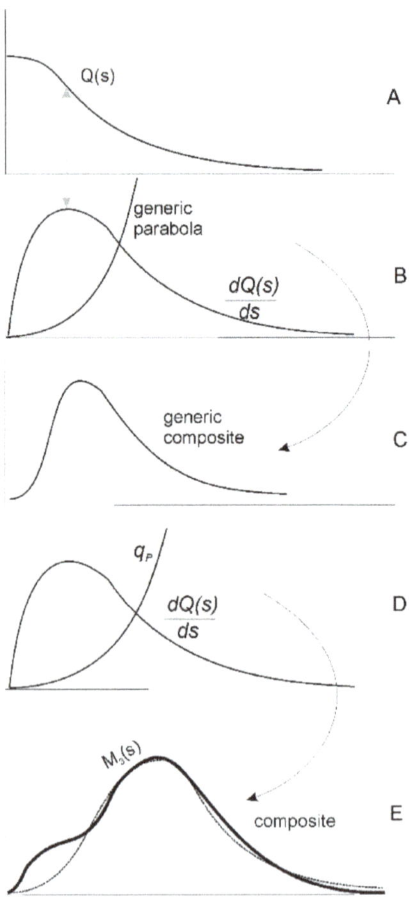

Within the framework of the development of U_3 on the basis of a Primitive Spectrum, the atomic Shell or Shell3 bears the *trinary* signature typified by Layer 3 of that Spectrum. Since U_3 is itself under that same signature within the Spectrum of the Grand Mappa Mundi, the atomic Shell so receives a trinary boost, which translates into a forceful trend to develop according to the same directives dictating the gestalt of U_3 by homomorphism. Under said mandate, a composition between a generic parabolic function and the derivative of an image Grand Eigenfunction stands to govern the gestalt of Shell3.

The illustration in Fig. 14.1 attempts to highlight the signature of this phylogenesis, accomplished essentially by composition between a first-order derivative of an image Grand Eigenfunction and a parabolic function. Some would further specify that this composition is Abelian distributive. And we would hasten to further stress that it is time-less or time-free.

Fig. 14.1 Composition between the two functions giving rise to atomic energy states (B to C and D to E)

14.1.1 Primitive Basis of the Atomic Orbital Map

Below are the four root M_{xi} composite functions that represent the irreducible symmetries apportioning the foundation of the entire atomic edifice followed by a graphical representation of the same.

$$M_{x_1} = s^2\left(-s \cdot e^{\frac{-s^2}{2}} \cdot \frac{1}{\sqrt{2\pi}}\right)^2$$

$$M_{x_2} = \left(s^2 - 1\right)^2\left(-s \cdot e^{\frac{-s^2}{2}} \cdot \frac{1}{\sqrt{2\pi}}\right)^2$$

$$M_{x_3} = \left(s^2\right)\left((s-1)^2 - 1\right)^2\left((s-1) \cdot e^{\frac{-(s-1)^2}{2}} \cdot \frac{1}{\sqrt{2\pi}}\right)^2$$

$$M_{x_4} = \left(s^2\right)\left((s-2)^2 - 1\right)^2\left((s-2) \cdot e^{\frac{-(s-2)^2}{2}} \cdot \frac{1}{\sqrt{2\pi}}\right)^2$$

The first composite function represents a product of a generic parabola and a derivative of an image Grand Eigenfunction elevated to a square. This function occupies Primitive layer 1 of the construction.

The second composite function represents a product of a composite parabola and derivative of an image Grand Eigenfunction elevated to a square. This function occupies Primitive layer 2 of the construction.

The third composite function portrays a product between a generic parabola, the previous composite parabola at layer 2 shifted by 1 unit of the independent variable and the derivative of the image Grand Eigenfunction shifted by 1 unit as well, all elevated to a square. This whole function occupies Primitive layer 3 of the construction.

The fourth composite function features a product between a generic parabola, the previous composite parabola at layer 3 shifted by 1 unit and the derivative of the image Grand

Eigenfunction shifted by 1 unit as well, all elevated to a square. This entire function occupies Primitive layer 4 of the construction.

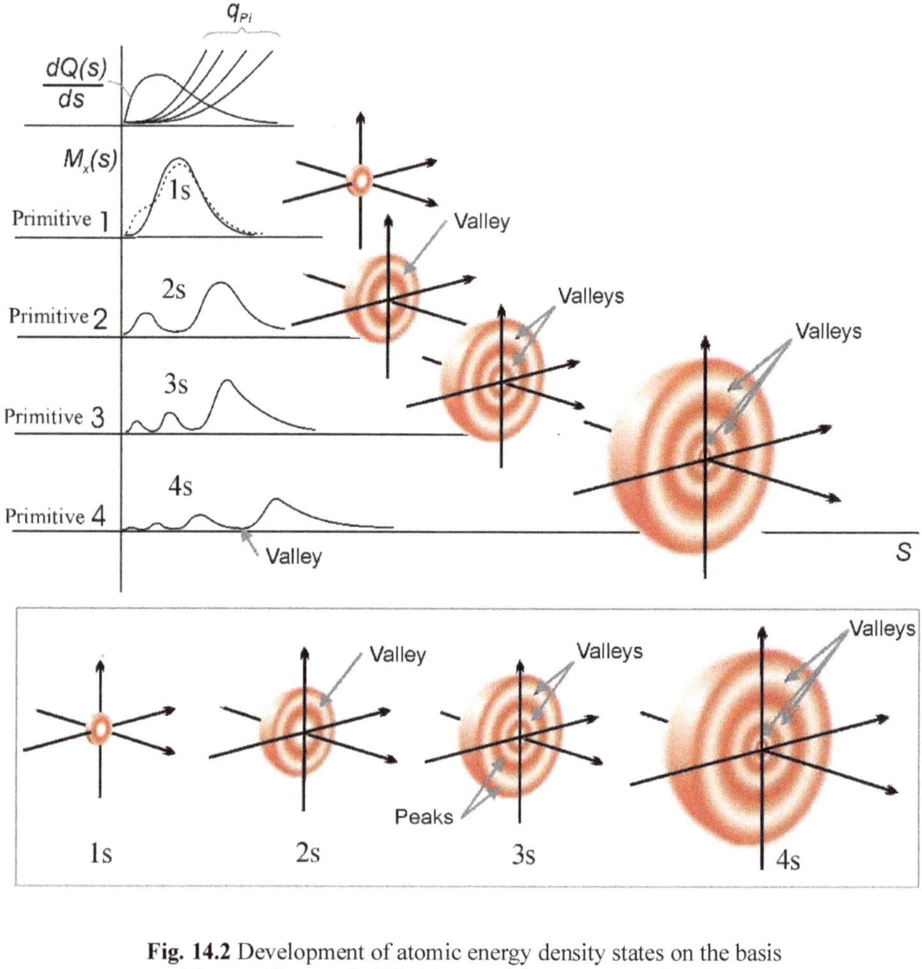

Fig. 14.2 Development of atomic energy density states on the basis of Quanto-Geometric Primitives and the composite functions

The shifts experienced by the independent space variable *s* mean to express the amplification of the spatial spread in the density state of the point element of action from one Primitive layer to the next. The above-mentioned energy density states are represented in illustration in Fig. 14.2. This analysis will attempt to retain as much as might be feasible the well-known nomenclature utilized in Quantum Mechanics for the description of atomic quantum states. As such our analysis simply proposes a parallel path to the derivation of these states, at least for bound electrons. We will also make good on the promise to demonstrably exhibit those same states as origin of the distribution of neutrons and protons within the edifice.

We have expressly chosen to take into account only the positive (or real) side of these graphs. Furthermore, we have also assumed that the generation of real physical matter is based on two equitable or homomorphic space variables, as much as we conform to the Quantum Mechanics representation whereby these independent variables behold an x-y representation, for a z independent variable. The graphism is based on the x-y-z coordinate system.

$$M_{[s,x]} = \left[(s^2 + x^2) - 1\right]^2 \cdot \left(-(s + x) \cdot e^{\frac{-(s^2+x^2)}{2}} \cdot \frac{1}{\sqrt{2\pi}} \right)^2$$

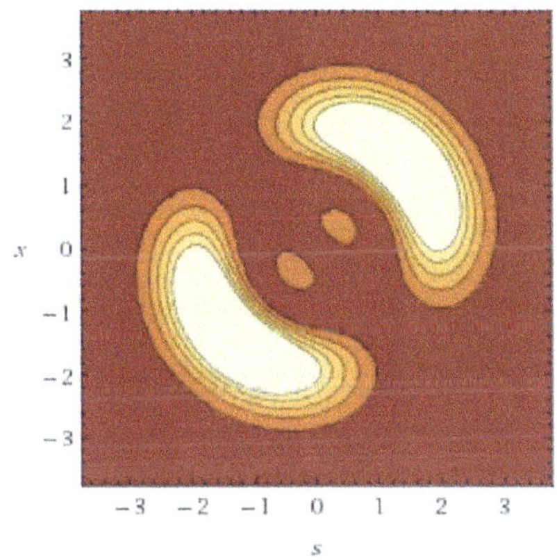

Fig. 14.2b Example of a 3-D density contour plot of the M_{x2} function by a popular math program with x and s as symmetric independent variables. See Appendix.

The above energy density states (Fig. 14.2b) represent to a close approximation the expression of the radial function of Schrödinger's Hamiltonian in Quantum Mechanics. From the Quanto-Geometric vantage, these states are distributed in a manner to form a Primitive Spectrum as determined by the prime parabolic-differential Function presiding over this development. Accordingly, the scalar states and geometric states of symmetry intrinsic to the Primitive canon become patently explicit in these formations.

14.1.2 Inner Quanto-Geometric Symmetry States in Orbital Morphology

In the illustration at Fig. 6.2 we show how the geometric trends emanating from the inner states of symmetry germane to each Primitive Layer inform the radial density states previously discussed. The spherical state of symmetry arising from Primitive 1 meets a spherical geometry already formed by the composition in the radial pattern and leaves it geometrically unchanged. At Primitive layer 2, the inner symmetry state of elliptic geometry purports to imprint an ellipsoid shape to the radial pattern. At Primitive Layer 3, the inner parabolic state of symmetry informs the radial pattern with a paraboloid shape very characteristic of the egg shape which is a reunion of two parabola of different foci parameters. Finally at Primitive Layer 4, the innate hyperbolic state of symmetry imprints the radial pattern of the orbital lobe with a hyperboloid shape, which we attempt to highlight in the illustration with the dotted line. Effectively, the internal saucer shape of the mid density spreads does possess a hyperboloid shape that might otherwise remain unapparent, as much as the lower part of the density lobes exhibit that same shape.

There is no objecting that any further Primitive Layer of development of these density states would likewise develop according to their inner geometric state of symmetry. In Quantum Mechanics, it is the intervention of the angular function of the Hamiltonian which, by contributing to geometrical restrictions on the radial spread of these energy density states, determines their final morphology. What we have seen so far is the manner in which basis state of symmetry of the angular hierarchy informs the creation of one density state or distributive object. However, the geometric expression innate to the Primitive doubles down on their geometric directives by imposing replication of these density objects according to the linear hierarchy of symmetry. We capture this trend in Fig. 14.3.

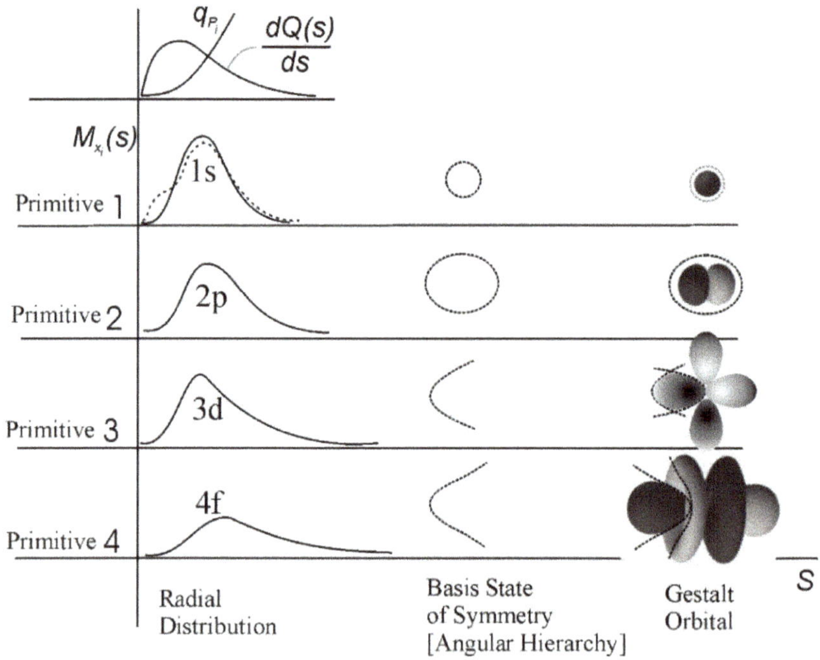

Fig. 14.3 Geometric trends emanating from the Primitives which inform the development of the basis energy states

In the seminal study of states of symmetry originating from the higher number stack in Book I, we have shown how symmetry states correlate to scalar states. One alternative way of looking at the expression of these symmetries is to consider that the circular or rotational state of symmetry emerges from **one point**; the elliptical state of symmetry emerges from **two focal points** or a line; the parabolic state of symmetry emerges from a focal center made up by a directrix line (2 points) and one point, which together define a plane (**3 points**); the hyperbolic state of symmetry emerges from 2 focal lines intersecting each other, but two non-intersecting focal lines (**4 random points**) determine or create the cubic space.

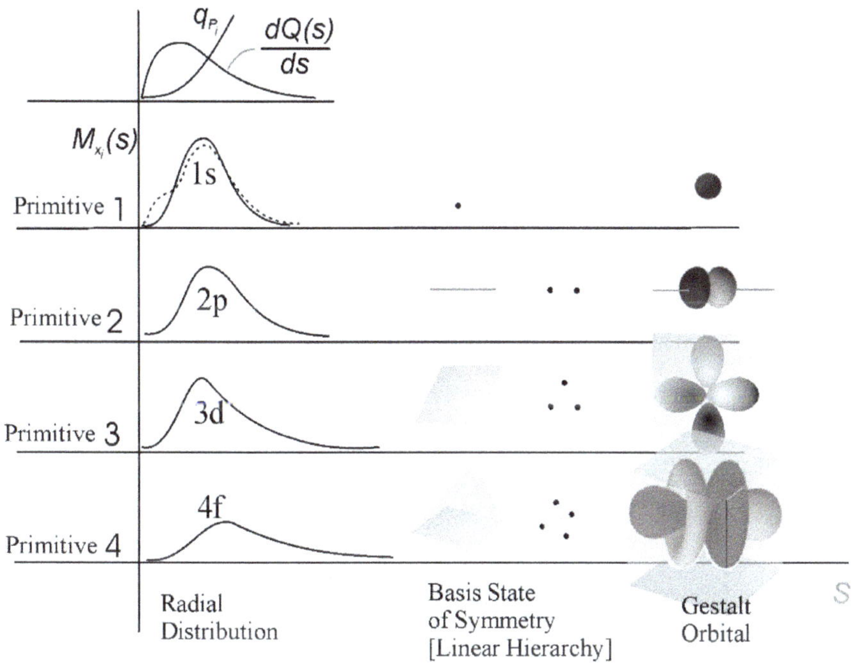

Fig. 14.4 How the basis states of symmetry from Primitives' Linear Hierarchy informs the development of the basis energy states

When this quanto-geometric hierarchy of symmetry innate to the Primitive layers of the canon intervenes in the development of the distributive objects it forces their multiplication with the aims of informing their expression in accordance with these spatial trends (Fig. 14.4). Consequently, we see object 1s, already a radial object constrained to retain its spatial symmetry with respect to one point, in the morphology of a sphere (Fig. 14.5).

Fig. 14.5 Orbital 1s

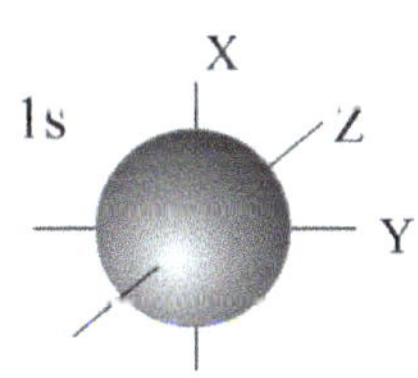

Next, we see object 2p multiplying in 3distinct objects that take each the orientation of the 3 possible directions of 3-Dspace, per display in Fig. 14.6.

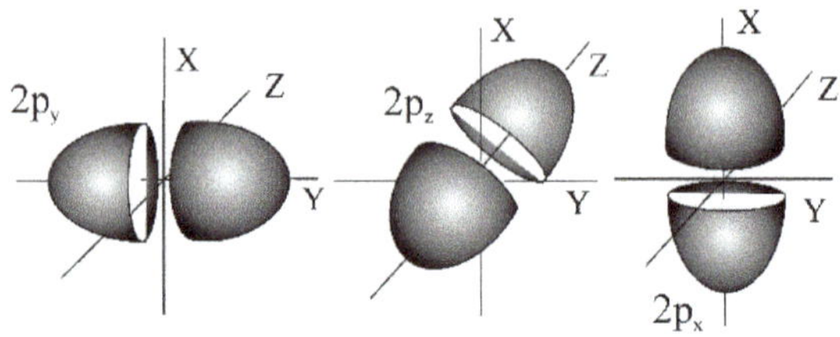

Fig. 14.6 Orbital 2p

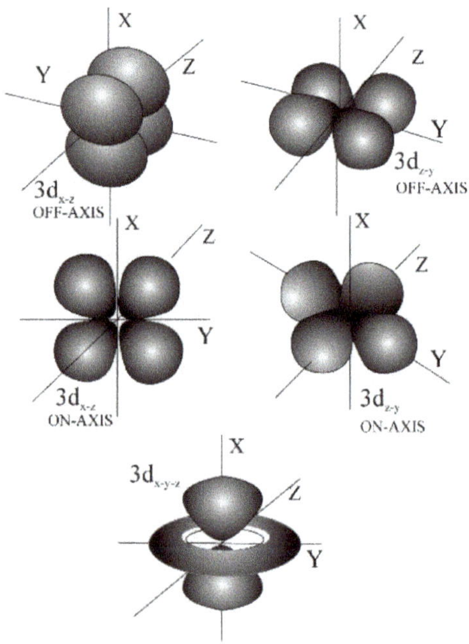

Fig. 14.7 Orbital 3d

Object 3d for its own undergoes multiplication as well in order to describe all 5 possible planar occupation modalities of 3-D space as shown in Fig. 14.7. These objects occupy plane X-Z off axis, plane Y-Z off axis, plane X-Y off axis, plane X-Y off- axis, and finally, a state of putative occupation of all planes as seen in the $3d_{X\text{-}Z\text{-}Y}$ distributive object. Effectively the latter object features a torus going all around plane Z-Y and two lobes across the X axis of the 3-D coordinate system.

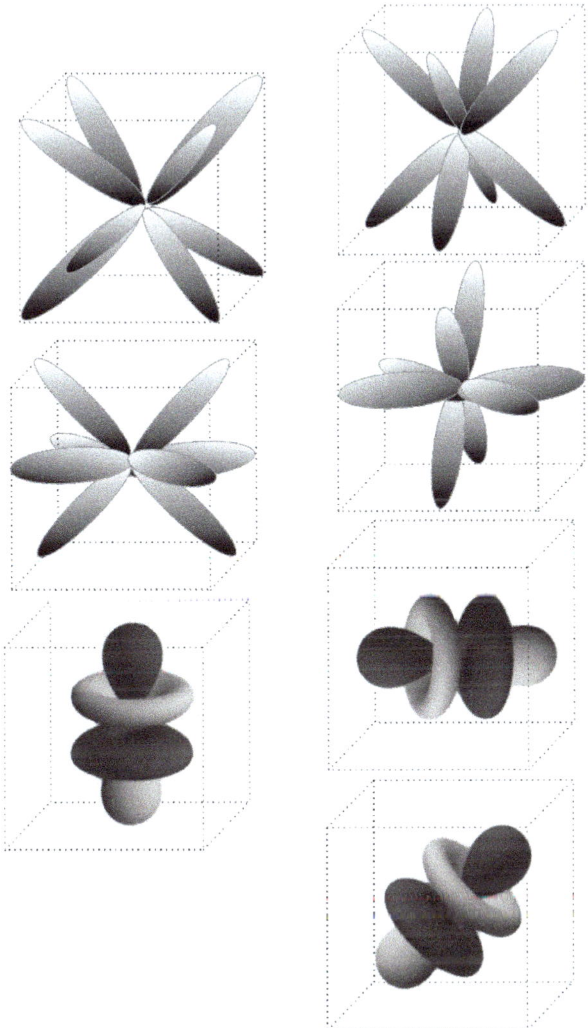

Fig. 14.8 Orbital 4f

The torus may be viewed as two regular lobes sweeping rotationally plane Z-Y in an attempt to exhibit both $3d_{Z\text{-}Y}$ objects by conflation. Whereas the axis-centered vertical lobes of the $3d_{X\text{-}Z\text{-}Y}$ object attempts to emulate both $3d_{X\text{-}Z}$ objects concomitantly in their occupation of the only two vertical planes. At the end, one may view object$3d_{X\text{-}Z\text{-}Y}$ as if executing a directive of conflating all the other 4 objects or, which is the same, occupying all 3 coordinate planes.

Finally, when it comes to distributive *f* objects, it takes little scrutiny to realize from the graphs in Fig. 14.8 that *f* objects are a concretion of an emulation of cubic space from a 4-point focal symmetry standpoint. The lobe packets of each *f* distributive object attempt to form, occupy or create a spatial cube in different ways. It is important to realize that we are dealing with a 3-D object that is attempting to simulate 3-D space. But it is a 3-D object of quadri-focal symmetry in the linear or dot hierarchy. This directive reveals 7 possible forms of density distribution in reference to an X-Y-Z standstill coordinate system.

It is worth emphasizing that all these distributive objects, from *s* to *f*, are already 3-D objects. What they are doing here is an over description or over emulation of 3-D space deploys forced by the quanto-geometric directives inherent to the Primitives they are subject to, simply because they are acting under an overall mandate of retracing the phylogenetical process of U_3 itself. We remind the reader that this mandate originates from the trinary boost underlying the development.

14.1.3 Scalar States of the Quantum Radial Map of Orbitals

This development so much exhibits a homomorphic replication of the directives conducive to the very genesis of U_3 that even the primal inner scalar states are reproduced. Hence, we see that the number of states developed by each layer conforms to the cardinality (or inner quantum state) native to its Primitive, per Table in Fig. 14.9.

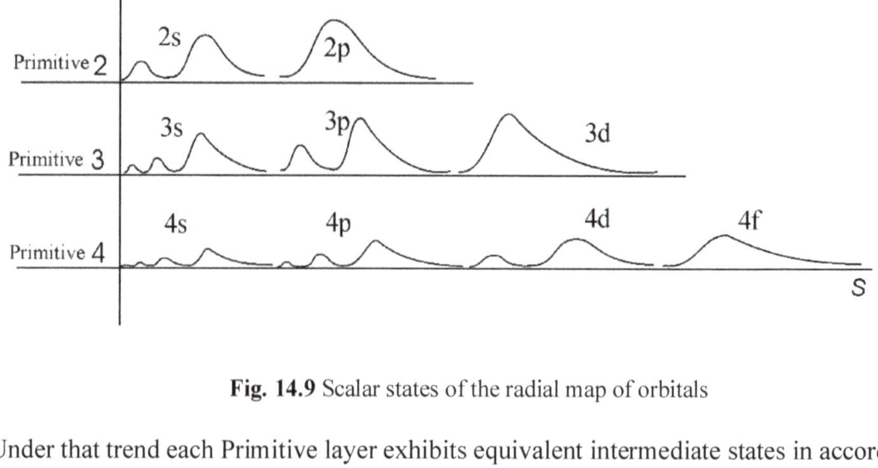

Fig. 14.9 Scalar states of the radial map of orbitals

Under that trend each Primitive layer exhibits equivalent intermediate states in accordance with the numeric hierarchy of the number realm in order to realize its cardinality.

Those equivalent states are such because, for one Primitive layer, they feature the same integral space-quantum distribution, although their number of nodes and peaks may be different from one state to the next. This means that the areas under all curves in a Primitive layer, which is the integral of their functions, are equal and represent the very scalar state subtending the layer. This map of the radial density states is an exact representation or emulation of the quanto-geometric number stack.

14.1.4 Contextual Symmetry Background of the Hamiltonian

The Quantum Mechanics *quantum number n*, setting the longitudinal spread of the radial density state, corresponds roughly to the Primitive layer in our analysis. It is sometimes equally referred to as typifying an *energy shell* or *level*.

The Quantum Mechanics *quantum number l*, setting the shape of orbitals, roughly corresponds to the geometric symmetry that largely determines the morphology of orbitals in our analysis. This number is equally designated the *azimuthal quantum number*. The symmetry subdivisions of the quantum number *l* are the *s* state for sharp, the *p* state for principal, the *d* state for diffuse and the *f* state for fundamental. One must concur that the attributes of the *s*, *p*, *d* and *f* labels make little sense in terms of their own defining qualities for atomic orbitals within the Quantum Theory. In fact, they only came about as a result of the historical development of the Theory in typifying the spectral line series observed in atomic spectroscopy. They acquire full and consistent meaning in the Quanto-Geometric analysis in terms of the symmetric qualities they tribute to the wavefunction.

The Quantum Mechanics magnetic *quantum number m*, setting the orientation taken by orbitals in a magnetic field, roughly matches an aspect of the directional space affinity of orbitals in our analysis.

The Quantum Mechanics spin *quantum number s,* which sets every orbital with two and only two possible values for the spin of the bound electron, is a direct match to transverse **azimuthal** symmetry in Quanto-Geometric analysis. It is exercised here by emulation as much as the other key symmetries.

The *Pauli exclusion principle* is another example of emulation that directly springs from higher order azimuthal transverse symmetry in the matter-antimatter characterization studied in previous Chapter. According to this principle, no set of bound electrons can reside concomitantly in the same exact atomic orbital by experiencing the exact same set of quantum numbers, but two and only two bound electrons which then must have their spin *quantum number s* with opposite values (+½, -½). This is a direct transposition of the *[(+t,m),(-t,m)]* characterization of matter-antimatter in the 3-D physical universe implemented by the emulation function.

The bulk of the symmetries and associated directives that we have uncovered in the above analysis represent the contextual background to the radial and angular functions of the Hamiltonian. The specificities of these determinants are ultimately lost through the resolution via the election of real-valued terms at every turn of the solution process, which stand for the only path to the resolution of the Hamiltonian in the real realm. It is a path chosen for the practical reasons that it is not known how to resolve the probability eigenfunction invoked by the Hamiltonian as far as the imaginary components of that function:

$$\phi_m(\varphi) = \frac{1}{\sqrt{2\pi}} \cdot e^{im\phi}$$

In that regard, it is well known that the resolution of the Hamiltonian strictly thru the real-valued terms of its derivates, which are the only workable ones, causes loss of information about the probability wavefunctions. However, the Quanto-Geometric symmetries and associated directives, all pertaining to the Grand Eigenfunction and its phenomenologies, integrally spell out the diversity of artifacts and minutia arguably called for by Quantum Mechanics' unfathomable imaginary eigensolutions. So it is, for not only the entire map of electronic orbitals but for those of bound protons and neutrons as well, as we shall later see.

14.1.5 Order of Uncertainty v. Certainty

One of the cornerstone principles of Quantum Mechanics is the Heisenberg uncertainty principle which states that no two quantum variables about a particle may be determined simultaneously, yielding that if one variable such as momentum is known, the other, such as position, remains intractably insoluble or uncertain. This so-called uncertainty is admittedly at the core of atomic orbital phenomenology.

While this view remains absolutely correct as long as one considers the point element of action in a quantum state in itself, it becomes inappropriate to say the least when one considers the quantum state or orbital itself. The quantum state is no longer a particle element, it becomes a higher order object, what we have called a distributive object. In that new incarnation of the particle element, the order of uncertainty binding the element yields to an order of certainty embodied in the Quantum Mechanics time-less operators. Said operators are able to bring about a whole slew of information about the new distributive object.

Quanto-Geometric Theory takes this higher order of certainties to a whole new level by bringing to bear its Operator hierarchies, the Primitive Norms and the Symmetry disciplines in the analysis of condensed matter physics which duly includes the atomic realm. One must comprehend the seminal difference there is indeed between a free electron and a bound electron. The latter is no longer uniquely an electron particle essentially but a new spatial object within a new Shell of matter. The Shell or Shelling view of matter implies, as we have seen, a distinctly independent treatment to the very void of space and stands for one of the most important characterizations of the Quanto-Geometric framework. This view with all the uncovered associated Symmetries constitutes a higher order of certainties that overwhelms all uncertainty dictates any physical element may rightfully encapsulate in its profiling within the scale of deployment of 3-dimensional matter.

At the end of his intellectual voyage thru the quantum realm, W. Heisenberg reportedly broke the news to his peers in great sadness about his discovery of an unredeemable limitation in the knowledge concerning the behavior of physical quantum variables. We may now conclude that he surely was making thereby an understatement about the qualities of the Quantum Operators, let alone representing a misperception of the revelatory Higher Order of Symmetries lying beyond. We believe this higher order to be integrally orchestrated in the Quanto-Geometric framework.

14.2. First Principles Underlying the Periodic Table of Elements

Despite its foundational role in the material development of atomic elements, the matrix of basis energy density states suffices not on its own to the formation of natural elements. Other high order imperatives in the development of those orbitals must be invoked in order to account for the actual development of atoms as we find them in the physical universe. One consequence of the lack of these intervening causal elements in Quantum Mechanics' formalism is its regrettable inability to describe any atom from that vast family of objects other than solely the Hydrogen atom. Hence a few additional theories and empirical rules, some more suitable than others, have been proposed to explicate the proliferation of artifacts encountered in the orbital structure of these elements.

The most towering Quanto-Geometric determinant that intervenes in the establishment of the global configuration of electronic orbitals as integral ensembles is the Hyper Space Operator pertaining to the Tri-Valent Canon. Effectively the Hyper Space Operator arbitrates this process at the highest level by enforcing the electron in the role of Spatial propagator. It establishes the operation of Overture or Exteriorization as the primary driver in the formation of electronic orbitals. By that measure, a definite preponderance in the formation of external orbitals as compared to the internal ones in an electronic orbital packet marks the formation of the latter from one element to the next.

Additional intervention of the following three secondary principles is no less instrumental to the genesis of all the elements making up the familiar periodic table of elements, not only in their electronic configuration, but in their neutron and proton configurations as well:

1. An overruling Quanto-Geometric Primitive Spectrum
2. The Quanto-Geometric transverse azimuthal symmetry
3. Continual phylogenetical emulation of the seminal genesis of U_3.
4.

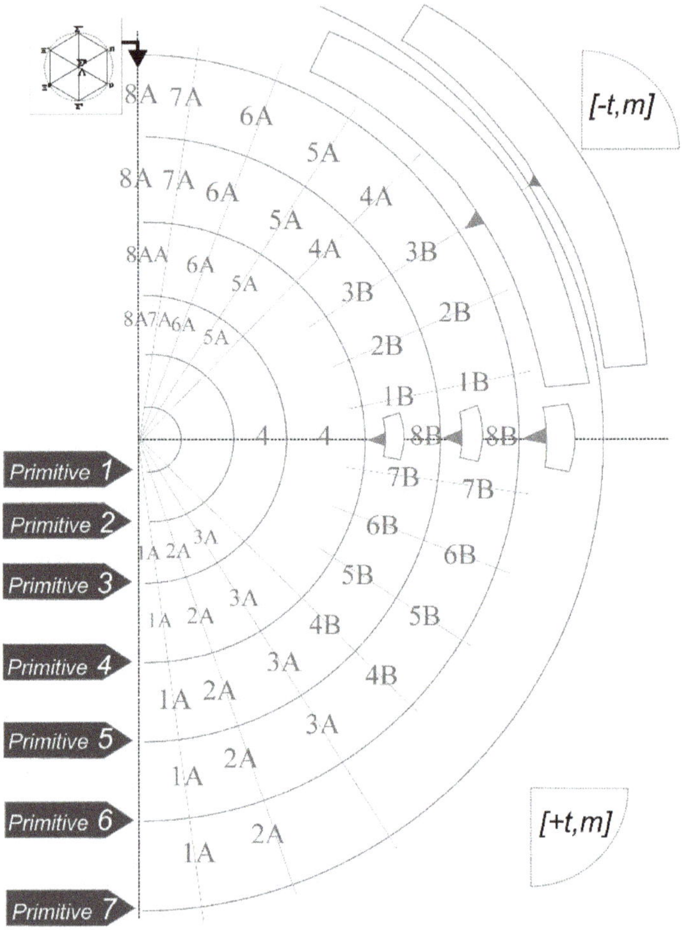

Fig. 14.10 A compendium of all the symmetries underpinning the common Periodic Table of Elements

We have formerly seen how some aspects of two of these imperatives already contribute to the makeup of the basis energy density states, notably the emulation function and the Primitive canon. However, they are here redeveloped on a new or higher encompassing order of genesis which treats the basis states as an encapsulation. It is a rather arduous enterprise to express the interplay between these four imperatives in the new order in terms of a single formal functional expression. For these reasons we will limit the analysis here to nomenclature and graphism for the most part.

Illustration in Fig x purports to represent a compendium of the set of four above-mentioned imperatives as they shape the universe of actual atomic elements in their electronic shells. We report the familiar Periodic Table of Elements as well making sure to annotate a 0 for the relevant undeveloped orbitals in order to better support the analysis.

14.2.1 Origin and Nature of the Symmetries Underpinning the Periodic Table

The curious fact about a large number of known empirical rules to explain the very diverse artifacts subtending the development of electronic orbitals is that they are never absolutely true and quickly break down within their applicable scope, leaving the analysis with significant exceptions to deal with. When there are significant or too many exceptions to a rule, we must conclude that the rule does not capture the true nature of the underlying trends it purports to model. The soundness of any rule, and indeed in no small part its very usefulness, precludes the existence of an excessive number of exception cases.

The so-called construction principle, of which N. Bohr was one of the artifices, stands for one of these paramount rules expected to shed light on the formation of atomic orbitals in polyelectronic elements, beyond the reach of the Hamiltonian. It rules a progressive approach to the buildup of orbitals whereby the states with lower energy level have precedence in their formation over those with higher energy level. As natural as this rule may appear, it is only true until the 20th element of the Table, the Scandium atom (Sc), is formed. It fails to give proper account of the construction of the vast remaining number of elements, where we repeatedly see states with higher energy exercising precedence in the construction order, specifically in what has been called the transition Groups (B family).

The concurrent application of the three towering principles posited above are poised to uncover the most important symmetries behind the periodic Table, which the very necessary construction principle is patently unable to capture. We are going to attempt a description of their origin and the modality of their application to the formation of electronic orbitals in natural polyelectronic atomic elements. To do so we will expressly refer to the symmetry scaffolding illustrated in Fig. 14.10.

Perhaps the most important and inarguable classificatory element of the Periodic Table is the "periodic" feature in its structure. What seems to stamp the periodicity of rows in the Table is its cardinality based on the octet quantity. One has to go back to the symmetries studied in the analysis of Shell1 particle's ontology to exactly trace back the origin of octet cardinality of the periods in the Table. The Octuplet symmetry pattern that underpins the development of baryons and mesons represents the farthest precursor of the periods' octet cardinality in the development of polyelectronic atoms' orbitals.

We have shown that the 3-tier scaffold, pictured at the top inlet of Fig. 6.10, is the embodiment of the Quanto-Geometric azimuthal symmetry axiomatic. It is the coupling or juxtaposition of two of these figures to one another that creates the Octuplet symmetry pattern, mimicking thereby the full province of physical matter the Model calls for. On the basis of that grid do we get the 8 elements, baryons or mesons, that construe the symmetry, which by extension serves as the core of the Decuplet pattern as we have seen. The encapsulation of azimuthal transverse symmetry in the Octuplet pattern represents the high mark of completion of this symmetry discipline.

What we witness in the Periodic Table is on, the one hand, the layout of azimuthal symmetry scaffolding as the premier trellis of genesis of natural elements, and on the other hand, the punctuation or completion of each azimuth with Octuplet cardinality by "remembering" or emulating the genesis modality of U_3's Shell1. At the same time, the signature of the period responds to the qualities of their corresponding Primitives, which are either inherited from or overplayed on the underlying basis of gestalt orbitals. Thus, all three posited principles converge in the creation of natural polyelectronic atoms.

At first glance, the first and perhaps only Period in the entire Table (Fig. 6.11) that appears to be completely well-behaved, judging by minimally flat rules, is the 2nd Period. Exclusively based on *p* orbitals, which bears the generic signature of Primitive 2, it also represents Period 2 of the construction cycle replicating the Binary signature of its basis orbital constituents. It further completes in accordance with the Octet cardinality since it exhibits a total of 8 orbitals (2 *s* orbitals and 6 *p* orbitals). It does concord with the progressive construction rule as well, since the 8 constituting orbitals are progressively formed in an orderly fashion with regards to their energy levels.

If this Period were to set the pace for all the other Periods of the Table, one would expect the next Period, Period 3, to be strictly based on *principal quantum number* 3 with orbitals *3s*, *3p* and *3d* in formation, for a total of 18 orbitals to be realized. However, this Period experiences de facto a severe breakdown of the progressive rule and can only be understood thru the prism of the triplet principle that we have raised for this analysis. The imperative of Octuplet cardinality as the mark of completion of a Period truncates the development of the *d* orbitals. Once the last of the *3p* orbitals is formed, the Octuplet cardinality is realized so that the element holding that configuration must be located at end of azimuth. The 10 *d* orbitals that are due to follow must be then dropped because they have run out of azimuthal space on this period.

The above constraint sets the prelude to the *3d-4s* incongruence much debated in the literature before the notion of "total energy" versus "ionization energy" had been expressly developed for its elucidation. Effectively Period 4 sees the orbital *4s* in formation before

3d, which exhibits a lower energy level, thereby openly breaking the progressive rule. The two *4s* orbitals so get formed before the *3d* ones even start developing. This comes truly about as a direct result of the imposition of azimuthal transverse symmetry imperative along with the standard of Octuplet completion at end of azimuth in regards to the genesis of polyelectronic atoms. In other words, the construction norm that produces this result is not the "total energy" versus "ionization energy" gathered in the parallel theories addressing this sort of incongruency encountered in the transition metals, but principally the azimuthal imperatives as represented in the Fig. 6.10. Pervading Quantum Chemistry science, the Hartree-Fock method has been developed for the computation of sum total of atomic orbital energy in polyelectronic atoms, and surely deserves credit from utilitarian considerations. However, it cannot uncover the true active background imperatives and symmetries commanding these formations. In fact, it only has the virtue of making explicit results mandated by the imperatives above cited. This method, along with others of mitigated and controversial success, are based to a large extent on the notions of electrostatic repulsions and attractions between bound electrons mutually and between bound electrons and atomic nucleus.

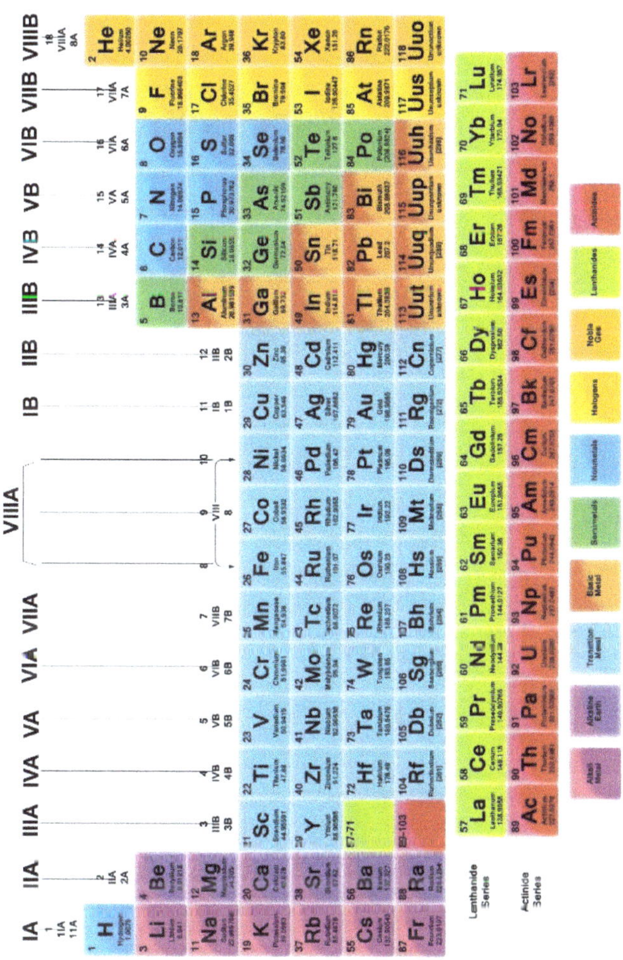

Fig. 14.11 Periodic Table of Elements based on electronic orbitals. With family grouping nomenclature by both schools of thoughts.

These views can only prosper in the ignorance that the orbital configurations and formation trends that apply to bound electrons equally apply to neutrons and protons in the nucleus core where putative considerations of electrostatic interaction with another inexistent core manifestly cannot apply.

The atomic construction issue represents a central matter in Quantum Chemistry because of its implications regarding both physical and chemical properties of natural elements. The significant breakdown of the progressive construction rule gives rise to a whole series of elements, the so-called transition metals, which chemists recognize form a whole family of their own. However, the question of how to tabulate these elements in a way that is consistent with their orbital structures and their chemical properties has been for long a matter of continued controversy.

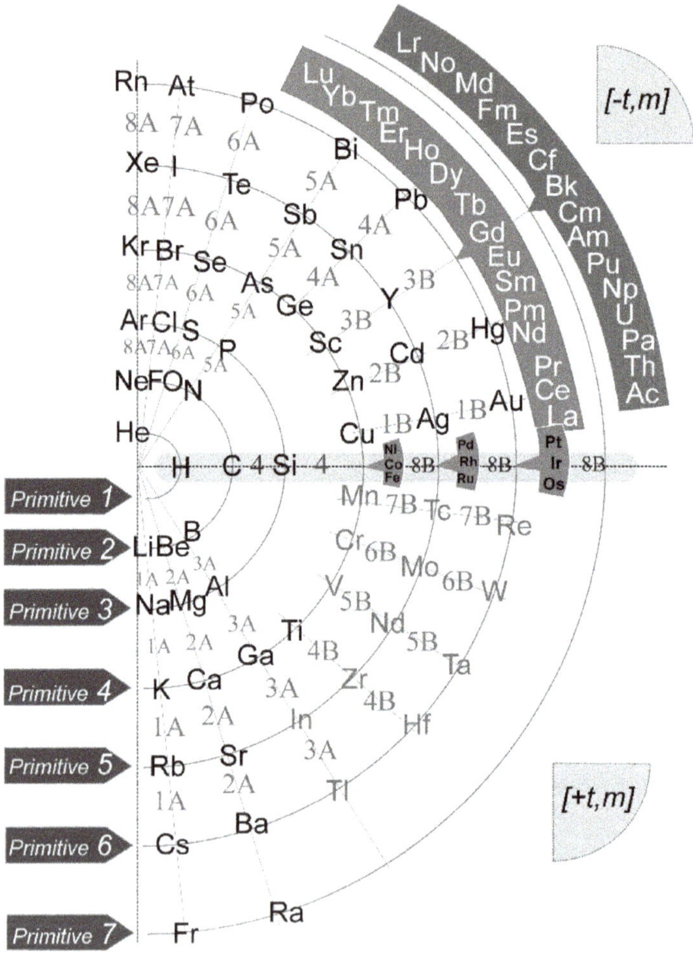

Fig. 14.12 Periodic Scaffolding of Elements based on azimuthal symmetry

Whilst most chemists agree that nature profiles two large distinct families of elements, they disagree on proper classification for each in the Periodic Table. Many find that the line after elements Ni, Pd and Pt should be the dividing line for two main families of elements, which they designated B family for the groups at the right of that line and A family for the groups laying to the left side of that line. The A family starts with group 1A which aligns elements H, Li, Na, K, Rb, Cs and Fr, ending with Ni, Pd and Pt as group 8A. The B family starts with group 1B embodied by elements Cu, Ag, Au, ending with group 8B incarnated in the noble gases. The view that has seemingly prevailed is that of a B family starting in an awkwardly backward manner with group 1B in the embodiment of elements Cu, Ag, Au and ending with a group 8B in the embodiment of Ni, Pd and Pt. The latter classification obliterates the expected order of progression based on phenomenology taking place in external orbitals and strictly caters to the total number of orbitals reflected in the Atomic Number.

Truth to this matter is that unwelcomed incongruent outcomes arise in both views leaving us to make the claim that no tabular graphism may adequately yield the construction taxonomy of natural elements due to their inability to reflect the true symmetries belying atomic construction. The two trends that are at odds in the distribution are, on the one hand, core formation naturally resulting from successive addition of electron quanta to the shell, and on the other hand, normed superficial formation orchestrated by the outer part of the whole shell packet.

By contrast with the usual methods, the azimuthal symmetry scaffolding garners in the most natural way the layout of the two large families of elements and their group partitioning because it is the symmetry par excellence that governs the construction. And that is the reason why, as a first order proponent of spatial imperative from its parenthood with the Hyper Space Operator in command, it grants primacy to superficial formation over core formation in the construction order.

In this partitioning, each of the two families represents an azimuthal quarter. Note well the very important azimuthal hybrid line that is a paramount feature of this symmetry. Note further the radial lines that incarnate the very Group definition, as well as their mirror symmetry across quarters. Azimuthal mirror symmetry is as we know responsible for antithetic properties, which we expect to be reflected in structural, physical and chemical properties of corresponding elements. As a consequence, a period is not a tabular row but an azimuthal π arc. A group is not a tabular column but an isotropic radial line. This architecture has the virtue of reconciling both aforementioned taxonomic views by gathering the principal facts or artifacts that they each attempt to capture, while shedding light on the very origin of other obscure phenomena appearing in the construction (Fig. 14.12). This first order of azimuthal symmetry in the distribution develops on the basis of partition parameter $N = 8$ in the azimuthal genic formula so that:

$$\theta_Z = \frac{180^o}{2|8|+1} = 10.5°$$

Azimuthal transverse symmetry constitutes the foundation of the paramount concepts of Ionization Energy and Electron Affinity long established in Quantum Chemistry science, otherwise formulated as the octet rule. Ionization energy is an expression of the physics of external orbitals or valence electrons by which they are all the more bound the closer they are to saturation to 8 parts, the configuration of next noble gas or next middle transition triad in each period. Electron affinity determines how prone external orbitals or peripheral electrons are to mobilize outward or leave the parent atom. This trend sees the packets all the closer to the previous noble gases or the transition triad in a period the more prone they are to leave the parent atom, thereby showing the saturation configuration of the previous noble element to each of them respectively. The octet rule sums it all by stating that elements to the right of the mid group tend to capture or attract more electrons toward accomplishing octet configuration proper to the next noble elements, while elements to the left side of the mid group tend to shed their bound external electrons toward accomplishing the octet configuration proper to the previous noble element.

The octet characterization is a replicate of azimuthal transverse symmetry norm strictly based on affinity to mobility. One might typify right side elements as those featuring valence electrons or orbitals registering a deficit in mobility that becomes deeper as we get closer to the octet position, the maximum grade of this deficit. Whereas one might characterize left-side elements as those experiencing a propensity to mobility that becomes broader as we move to the left of the table or the azimuthal chart. The fact that there is no element on the down-side octet line (the negative m-axis) in the azimuthal layout is a testimony to the accuracy of this characterization, given that the full octet line incarnated in the full m-axis represents completion state or stability in saturation. Consequently, what we commonly view as octet stability is a rendition of maximal motile deficit.

Faithful to the azimuthal symmetry norm, the mid line embodied in Group 4 shows elements that are, again not neutral from an Electron affinity standpoint or Ionization energy standpoint, but ostensibly hybrid in the sense that these elements behave in both ways with regards to one trending characteristic. The Carbon (C) element for instance may both acquire 4 electrons in its chemical behavior and alternatively indulge in “losing”, as we commonly say, its 4 valence electrons as the other facet of its chemical behavior. It is important to properly comprehend the essential trait of azimuthal hybridity in terms of a conflation of two dissimilar or anti-symmetric properties and not as the zero-sum phenomenon advocated by the term *neutral*. Without that intimate understanding one will gain no insight into the principal trend occurring on the hybrid line of the distribution and for all that matter the complete family of transition elements.

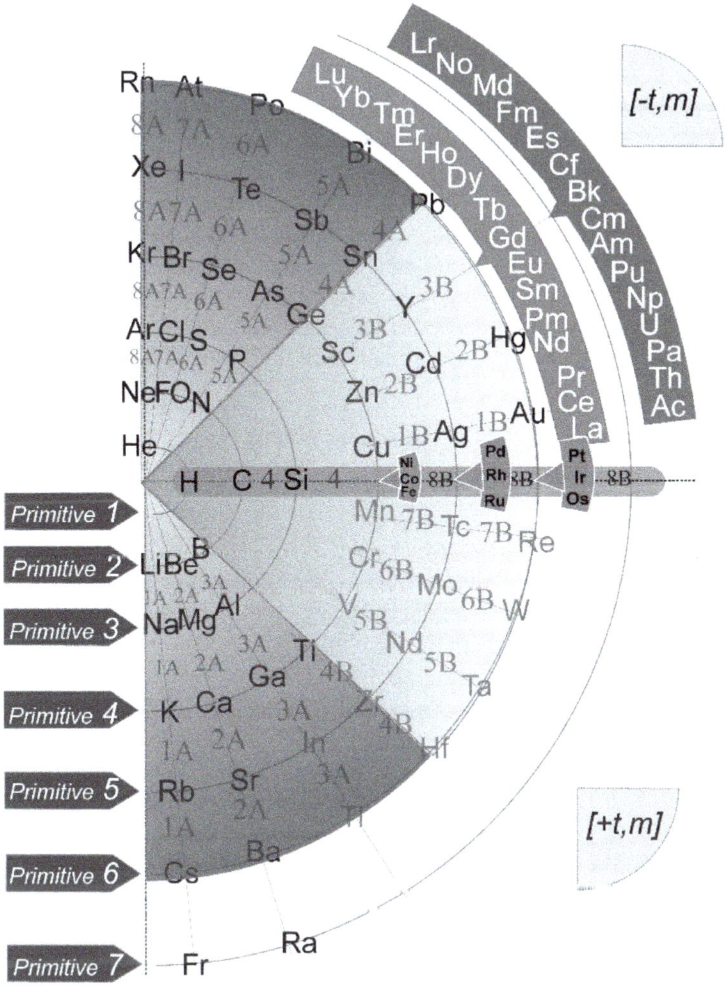

Fig. 14.13 Periodic Scaffolding based on azimuthal symmetry parameter *N* = *3*

The Carbon (C) element and the Silicon (Si) element are well known to both exhibit this hybrid or ambivalent behavior in their chemical properties. They are so to speak well-behaved on the hybrid line. The whole A family is generally well-behaved because their external *p* orbitals with a maximum possible number of 8 states seamlessly fit within the octet trending imperative by the azimuthal symmetry norm. The B family faces the problem of realizing the octet trending imperative based on external *d* orbitals with a natural maximum count of 10 orbitals. Therefore 2 excess valence orbitals must be taken care of or else the construction according to the higher norms falls apart. The construction would have to show the violating Group 9 and Group 10 proper by reason of the excess of *d* orbitals beyond 8. The azimuthal symmetry imperative however enforces its norm proper by conflating the two excess Groups into the single Group 8 on the hybrid line. We will shortly see why a Group 8 or an octet Group can be laid elsewhere than on its legitimate *m*-axis line. We will concentrate for now on elucidating the nature of the hybrid Groups.

The most legitimate transition Group 8 elements are Iron (Fe), Rhodium (Ru), Osmium (Os) and Hassim (Hs) which feature 8 valence orbitals. The hybrid conflation experienced by the other sets of elements with 9 and 10 valence orbitals forces a homogenization of those elements with the characters attributable to the legitimate Group 8 elements. The most telling evidence of this characterization is that valence-9 and valence-10 Groups show the same oxidation number as their natural Group 8 element in the period. In point of fact, they all show the oxidation numbers pertaining to Iron (Fe), namely +2, +3 and +4. The bulk of all the physical properties of these elements are exactly identical, reason why they have received the common denomination of "rare earth" elements.

Due to the imposed hybrid conflation, all the rare earth elements share the following additional physical and chemical properties:

- They are silver or silvery-white or gray.
- Metals among them have a high luster and quickly tarnish in open air.
- They demonstrate high electrical conductivity.
- They have almost identical gradient of solubility.
- They generally occur together in minerals.
- Sameness in their physical identity is so strong that they are difficult to separate or even to distinguish from one another.

Our azimuthal chart (Fig. 14.13) shows the position that each of these triads occupies on the hybrid line as a conflated single element. One other telling evidence of the hybrid azimuthal characterization for these elements is that despite the octet state of their valence orbitals, which should stamp them with chemical inactivity or stability, they instead demonstrate sheer chemical interactive properties.

Perhaps the most dramatic concretion of hybrid azimuthal symmetry in this distribution is given in the emulation of Shell1 particulate azimuthal symmetry, portrayed in Fig. 14.13. It develops a symmetry based on partition parameter $N = 3$ in the azimuthal genic relation so that:

$$\theta_Z = \frac{180^o}{2|3|+1} = 45°$$

This symmetry engulfs the entire transition family (B family) into a single middle or hybrid 90° spindle while regrouping the main family (A family) across two opposite 45° spindles. This second order azimuthal symmetry mandate is superimposed over the first order layout and contributes to the formation of physical and chemical properties of the periodic elements in an appreciably large measure. In particular it fully justifies the epithet of "transition" associated with the B family, a term with flavor of intermediacy, quite frankly hybridity as well. One arm of the angle is the radial line 4A, the other arm is the radial line 4B, while the bisector of the angle is the original hybrid line 4A/8B. It is principally by reason of this superimposed overwhelming hybrid character that we see the development of two similar conflated groups of elements, the lanthanides and the actinides along the 3B radial line. They borrow the hybrid expression of conflation experienced by the original hybrid line where, on the one hand, three triad groups are each conflated into one element, and, on the other hand, they borrow the very usual +3 oxidation number of the original hybrid triads in their preferential ranking over the 3B radial line.

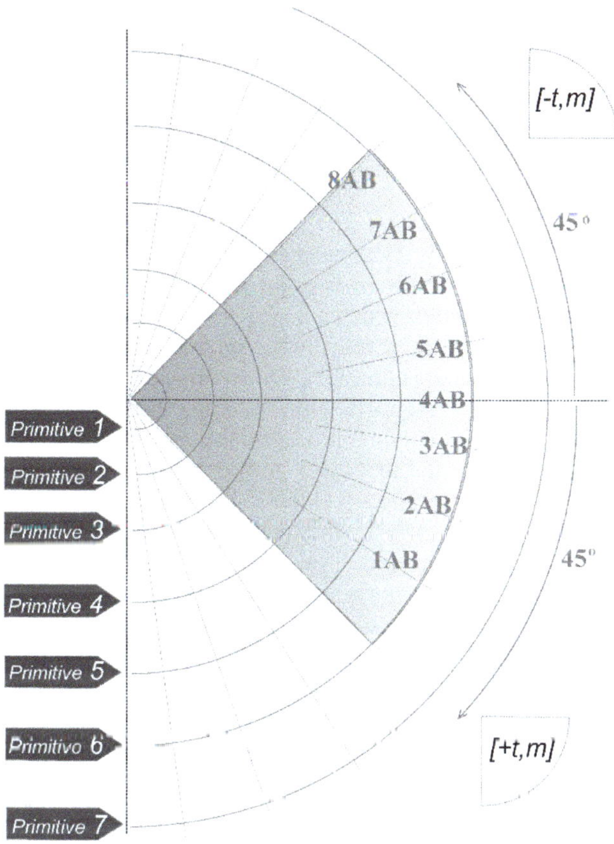

Fig. 13.14 *Mixing* conflation of A and B element spindles

Traditional Quantum Chemistry attempts to explain the very effectual contraction phenomenon experienced by the sub-groups of actinides and lanthanides in terms of an effective electrostatic nuclear force of attraction exerted on their valence orbitals. The true determinant here at work is hybrid conflation which forces homogenization of all the elements in a group into one element that would presumably occupy the matching position over their radial line in the construction. This element would have physical and chemical characters determined by its exact position in the construction, in particular a size ranking in-between the element at the left and that one at the right. The conflated elements are simply forced to emulate those properties.

There exists another azimuthal feature, of vast consequence from the standpoint of construction of this distribution, that we will briefly visit at this end term of the analysis. One will recall the intervention of a mixing transformation in the ontology of Shell1 particulate objects in previous Chapter, the hybrid nature of which we have amply elucidated. The intervention here of an *inverse* mixing or segregating transformation constitutes another azimuthal imperative undeniably at work in this one ontology.

Starting with the representation in Fig. 13.14, we first proceed to conflate the bottom spindle of A elements with the top spindle of B elements. Both spindles have their radial line sets similarly numbered from 1 to 4, which obviously reflects the cardinality of their valence orbitals. We follow suit with a conflation of the top spindle of A elements with the bottom spindles of B elements. Both have as well the same number of radial lines similarly numbered from 4 to 8. This gives us the central 90° spindle illustrated in Fig. 6.14 of conflated elements, each of which is fused with both anti-symmetric A and B properties. By accomplishing a single transformation that rotates these elements simultaneously and equally at a 45° angle northward and southward, we take the first step toward the creation of the real-world distribution of these elements in the numbering of valence orbitals reflected in the radial lines modeling the latter. We so obtain the 1 to 8 block at the bottom as well as the 1 to 8 block at the top. To obtain the exact division of A-B groups in each block, further mixing in the A-B property indexing at the original hybrid spindle level is required. However, we will not pursue in this exposé further analysis in that direction. Nevertheless, one may firmly conclude from the above that the inverse mixing artifact virtually obliterates the notion of temporal and individual successive steps in the construction (translate creation of one element after another) if ever conceded and further takes the very artifact of sequencing to a whole new level.

14.3. Conclusion

In concluding this Chapter, we wish to stress the importance of azimuthal symmetry in the configuration of Shell3 objects in U_3. This class of symmetry, thus far unknown in the Quantum Theory for the most part, is responsible for a whole host of phenomenologies in the entire infinitesimal quantum realm, from Shell1 to Shell3, and beyond as well. The hybrid dimension of this symmetry is one of the most consequential effectors, and perhaps the most complex configuration element, in all of 3-dimensional matter. Without this symmetry, the large variety of facts and artifacts known to the distribution of natural electronic elements cannot be properly interpreted in their fullness and peculiarities. In our judgement, the ignorance of the analytic components of this symmetry has prompted the creation of the string of approximative methods, such as the Hartree-Foch method, for the interpretation of unruly behavior in the distribution. When it comes to the very modality of the construction, while zenithal symmetry presumably yields a bottom-up movement or a top-down movement in the construction, at anyone's election, azimuthal symmetry dictates a transverse bi-lateral movement from a conflated or mixed hybrid state, at all 7 Primitive levels. Both orders of symmetries taken together summarily obliterate all notion of a timeline sustaining the construction. A remarkable fact that should be of great consideration to cosmological physicists.

PERIODIC TABLE OF ELECTRONIC ELEMENTS
TABLE I

Atomic Number	Element	(1)s	(2)s	(2)p	(3)s	(3)p	(3)d	(4)s	(4)p	(4)d	(4)f	(5)s	(5)p	(5)d	(5)f
1	H	1													
2	**He**	2													
3	Li		**1**												
4	Be		2												
5	B		2	1											
6	C		2	2											
7	N	**He*	2	3											
8	O		2	4											
9	F		2	5											
10	**Ne**		**2**	**6**											
11	Na				1										
12	Mg				2										
13	Al				2	1									
14	Si		*Ne*		2	2									
15	P				2	3									
16	S				2	4									
17	Cl				2	5									
18	**Ar**				**2**	**6**									
19	K						0	1							
20	Ca						0	2							
21	Sc						1	2							
22	Ti						2	2							
23	V						3	2							
24	Cr						5	1							
25	Mn						5	2							
26	Fe						6	2							
27	Co			*Ar*			7	2							
28	Ni						8	2							
29	Cu						10	1							
30	Zn						10	2							
31	Ga						10	2	1						
32	Ge						10	2	2						
33	As						10	2	3						
34	Se						10	2	4						
35	Br						10	2	5						
36	**Kr**						**10**	**2**	**6**						

PERIODIC TABLE OF ELECTRONIC ELEMENTS
TABLE II

Atomic Number	Element		(4)d	(4)f	(5)s	(5)p	5(d)	5(f)	(6)s	(6)p	(6)d	(6)f	(7)s
37	Rb		0	0	1								
38	Sr		0	0	2								
39	Y		1	0	2								
40	Zr		2	0	2								
41	Nb		4	0	1								
42	Mo		5	0	1								
43	Tc		5	0	2								
44	Ru	*Kr*	7	0	1								
45	Rh		8	0	1								
46	Pd		10	0	0								
47	Ag		10	0	1								
48	Cd		10	0	2								
49	In		10	0	2	1							
50	Sn		10	0	2	2							
51	Sb		10	0	2	3							
52	Te		10	0	2	4							
53	I		10	0	2	5							
54	**Xe**		**10**	**0**	**2**	**6**							
55	Cs	*Xe*					0	0	1				
56	Ba						0	0	2				
57	La			0			1	0	2				
58	Ce			1			1	0	2				
59	Pr			2			1	0	2				
60	Nd			4			0	0	2				
61	Pm			5			1	0	2				
62	Sm			6			0	0	2				
63	Eu			7			0	0	2				
64	Gd			7			1	0	2				
65	Tb			8			1	0	2				
66	Dy			10			0	0	2				
67	Ho			11			0	0	2				
68	Er			12			0	0	2				
69	Tm			13			0	0	2				
70	Yb			14			0	0	2				
71	Lu			14			1	0	2				
72	Hf			14			2	0	2				
73	Ta			14			3	0	2				
74	W			14			4	0	2				

PERIODIC TABLE OF ELECTRONIC ELEMENTS
TABLE III

Atomic Number	Element		(4)d	(4)f	(5)s	(5)p	5(d)	5(f)	(6)s	(6)p	(6)d	(6)f	(7)s
75	Re			14			5	0	2				
76	Os			14			6	0	2				
77	Ir			14			9	0	0				
78	Pt			14			9	0	1				
79	Au			14			10	0	1				
80	Hg			14			10	0	2				
81	Tl	*Kr*		14			10	0	2	1			
82	Pb			14			10	0	2	2			
83	Bi			14			10	0	2	3			
84	Po			14			10	0	2	4			
85	At			14			10	0	2	5			
86	**Rn**			**14**			**10**	**0**	**2**	**6**			
87	Fr	*Rn*									0	0	1
88	Ra										0	0	2
89	Ac										1	0	2
90	Th								0		2	0	2
91	Pa								2		1	0	2
92	U								3		1	0	2
93	Np								5		0	0	2
94	Pu								6		0	0	2
95	Am								7		0	0	2
96	Cw								7		1	0	2
97	Bk								8		1	0	2
98	Cf								9		1	0	2
99	Es								10		1	0	2
100	Fm								11		1	0	2
101	Md								12		1	0	2
102	No								13		1	0	2

Chapter 15

When we have found how the nucleus
of atoms is built up we shall have found
the greatest secret of all – except life.
Ernest Rutherford

ARCHITECTURE AND PERIODICITY OF ATOMIC ELEMENTS' NUCLIDES

Efforts to unravel the possible quantum energy states subtending the behavior of nucleons in atomic nuclides had been ongoing in nuclear physics as early as the 1930's. To this day the analysis, albeit successful in deriving the identity of many of these states and their match with the so-called "magic numbers", has yet to culminate in the presentation of the full development of their periodic configuration. In particular, presentation of the universe of known nuclides in a classification scheme similar to the familiar Mendeleev's Periodic Table of Elements, has yet to be achieved. We believe that a deeper understanding of the higher laws governing nuclear properties of atoms is a necessary requisite to the elucidation of the big picture with the potential to facilitate said periodic classification. We propose in the following an analysis of shell periodicity of Atomic Elements' Nuclei within the context of global Atomic Architecture. We present a Quanto-Geometric theory of such physical laws and culminate the analysis with the presentation of the anticipated Periodic Table of Atomic Nuclides.

15.1. Early Shell Model of Atomic Nuclei

A Shell Model of atomic elements' nuclei had been suggested since 1930 by J. H. Bartlett who noted the occurrence of two forms of periodicity in atomic nuclides: n + p + n + p and n + n + p + p + n + n. Later work by María Goeppert-Mayer, and by Hands D. Jensen independently, showed that nucleons experienced spin-orbit angular momentum, and on that basis provided a successful interpretative background to the distribution of nucleons' "magic numbers". The latter term was coined by Eugene Wigner to reflect the high-level stability observed for certain nuclides with a specific number of protons or neutrons, namely 2, 8, 20, 50, 82 and 126. The shell model is known to accurately predict or interpret certain properties of normal nuclei, such as their specific angular momentum. Yet for nuclei in highly unstable states, the shell model is considered inadequate and must be modified and connected with or replaced by other models, such as the liquid-drop model, the collective model, the compound-nucleus model or the optical model. Nevertheless, the sum of all these methods or even attempts to integrate them, such as realized in the collective model by unifying the shell model predicates to those of the liquid-drop model, remain altogether unable to deliver this coveted all-encompassing interpretative framework consistent with first principles.

2	$1s_{1/2}$
8	$1p_{3/2}$, $1p_{1/2}$
20	$1d_{5/2}$, $2s_{1/2}$, $1d_{3/2}$
28	$1f_{7/2}$
50	$2p_{3/2}$, $1f_{5/2}$, $2p_{1/2}$, $1g_{9/2}$
82	$1g_{7/2}$, $2d_{5/2}$, $1h_{11/2}$, $2d_{3/2}$, $3s_{1/2}$
126	$2f_{7/2}$, $1h_{9/2}$, $1i_{13/2}$, $3p_{3/2}$, $2f_{5/2}$, $3p_{1/2}$
184 (?)	$2g_{9/2}$, $1i_{11/2}$, $1j_{15/2}$, $3d_{5/2}$, $2g_{7/2}$, $4s_{1/2}$, $3d_{3/2}$

Table 7.1 Quantum states and orbital magic numbers by the old nuclear shell model.

Effectively despite continued attempts to provide the shell model with as much of a respectable background as could be possible, it did not develop as one might expect from the elevated principle of Schrödinger's Hamiltonian. It is rather more of a nemesis of the Hartree-Fock method, surrogating the Hamiltonian in order to come up with an explanation of the sheer diversity of facts encountered, in one case, in the distribution of elements' electronic quantum states, and, in the other case, that of nuclide's quantum states. The shell-model method uses the same nomenclature utilized for the labeling of electronic orbitals: s, *p, d, f, g, h, i*, etc. It further postulates the formation of orbitals in the successive shells or orbital packets by so-called nucleons which might be either protons or neutrons indistinctly. We show Table 7.1 the postulated quantum states, with each term incorporating the total angular momentum as a subscript associated with the letter indicating the orbital type. Each orbital can accommodate a maximum of $2j+1$ nucleons, with j being the radial (principal) quantum number and the exact number of orbitals within a shell varying for neutrons and protons.

Despite the obvious and unsuitable conceptual amalgamation, on the basis of that scheme the octet-like completion stability encountered in the distribution, as the "magic number" of orbitals in different shells is realized, may be properly accounted for. While this distribution may even account for accrued stability in nuclides with the magic numbers realized both for protons and for neutrons (so-called doubly magic), for nuclides that are removed from those positions, this spherical shell model simply falls apart. From our standpoint, the principal problem confronted by nuclear physics in general, not only the shell model or its substitutes, is essentially that of deriving nuclear properties from the laws that govern global configuration of atoms and, in particular, that of the nucleus in the context of that global architecture. Evoked at times in the previous Chapter, we make this quest our subject matter of the treatment to follow, by showing the true dimension of nuclear quantum state shelling within the Quanto-Geometric framework.

15.2. Hyper Tri-Valent Operators as Chief Modelers of the Atom

The Tri-Valent Hyper Operators account for the presence of three and only three particles in the makeup of atomic elements, which they lay along the elemental axial line of their own hierarchy.

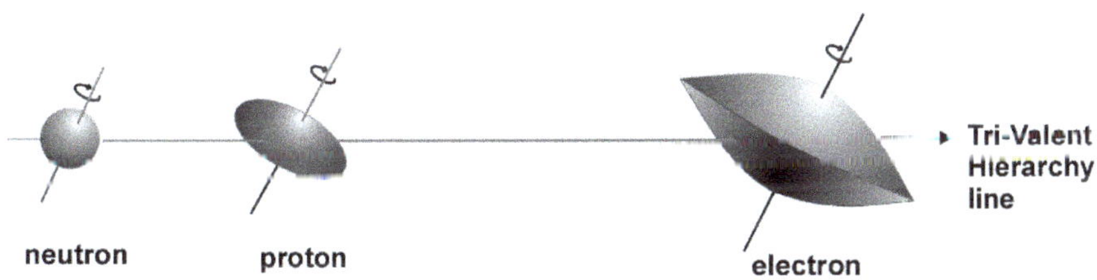

Fig. 15.1 The 3 particles that make up the atom along the hierarchy line of the Tri-Valent Operator

That line gives us the following key elements of the architecture in terms of the scalar or quantum aspect of the hierarchy (Fig. 15.1):

1. A neutron particle of highest mass at top of the hierarchical line.
2. A proton particle of high but lesser mass ranking at second position in the hierarchy.
3. An electron particle of comparatively very low mass in rank 3 of the hierarchy.

Furthermore, the natural wave functions of these particles (their Compton wavelengths) are in a match as well with spatial hierarchy directed by the modeling Tri-Valent Operators. The Compton wavelength of the free electron is the longest *(2.4263102389* 10^{-12} *m)* and lies longer than that of the free proton *(1.321409856* 10^{-15} *m),* which is even longer than that of the free neutron *(1.31959090* 10^{-15} *m).* So therefore, at higher mass shorter wavelength in ground state of these particles, which is a tenure reflected in the hierarchy of the Hyper Operators that had elected them for the makeup of the atomic edifice.

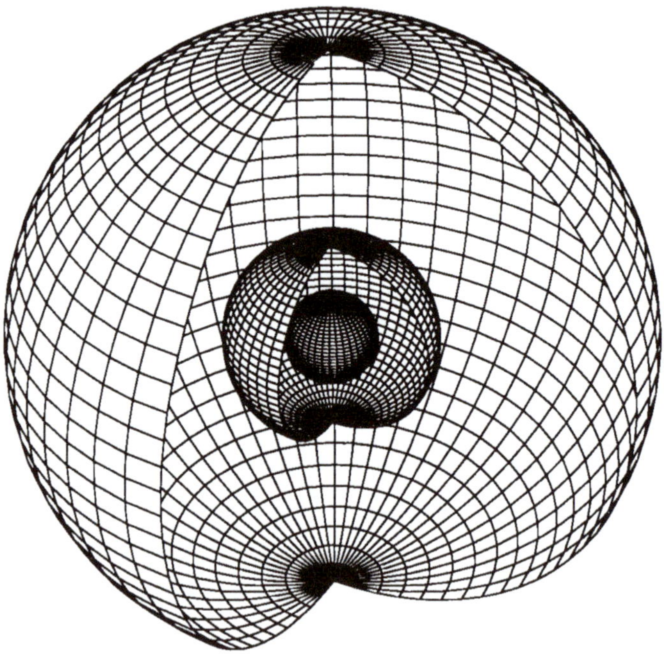

Fig. 15.2 The 3 embedded super-shells of the atom where the nucleus portrays an aggregate of 2 radial cores: neutron core embedded within a proton core

In accordance with Quanto-Geometric canon, each of these particles falls under the direct regency of one of the Hyper Operators within atomic architecture. Specifically, the Hyper Quantum Operator naturally orchestrates the positioning and kinematics of the neutron particle, heaviest of all. As formerly indicated, the Hyper Space Operator materializes the positioning and Overture-stamped kinematics of the electron particle, the one with the widest spatial wave function. Whereas the Hyper-Quadratic Operator dictates the phenomenologies associated with the proton particle, the one with intermediate or covariantly-free scalar/wavefunction value in the context of the atomic edifice.

This framework produces a three-tier atom per illustration in Fig. 15.2. The accrued densification Norm by the Hyper Quantum Operator organizes the neutron or neutrons as an inner core to the whole edifice.

The norm of accrued Spatial Overture by the Hyper Space Operator organizes the electron or electrons as the outermost shell (a super-shell) of the edifice. Whereas the Ambi-Valent Norm of the Hyper Quadratic Operator positions the proton or protons at the intermediate quadratic level of the edifice. It is important to note that the quadratic level, the locus of the proton core, is not a mid-distance but truly a quadratic point in accordance with the scalar/space hierarchy line at the origin of the Tri-Valent Operator itself. In summary, the Hyper Tri-Valent Operators model the atom in 3 super-shells spread as depicted above.

15.3. Bi-Focal Atomic Nucleus

The concept of an atomic nucleus in the form of an amalgam of neutrons and protons, or even as an agglutination of alpha particles, holds very thin no matter how useful the liquid drop model of atomic nuclei, the model that notably posits this architecture, may have proved to be. Atomic elements truly hold **two distinct nuclei** embedded in one another. That structure stands for one of the key reasons why the distribution of neutrons' and protons' energy shells is possible, akin to the electronic shells, as we shall see later.

15.4. Universality of Energy Density States

The atomic orbital map stemming from the four M_{xi} composite differential functions discussed in previous Chapter defines and orchestrates the low-level energy density shelling of the whole architecture. Orbitalization, based on energy density states *s, p, d* and *f*, is universal in atomic architecture and subtends all three spreads of super-shells, namely the neutron core, the proton semi-core and the electronic sheaf. The differences in specific developments of these sheaves of basis density states as platforms to each of the three super-shells are caused by the directives and modulations imposed onto them by the Hyper Tri-Valent Operators individually. Consequently, the final realization of the atom is a composition between, on one side, the M_{xi} composite differential functions that provide the base energy density states resulting in the conventional shells or orbitals, and, on the other side, the Hyper Tri-Valent Operators which over modulate the spread and development of these shells according to their own generic Norms.

We remind the reader that the known base energy density states *s, p, d* and *f* in Quantum Mechanics, are pre-defined by the primal Quanto-Geometric function from its inherent Primitives which give rise to spatial states of symmetry. By that measure, state *s (1)* is an expression of *mono*-centric symmetry and adopts spherical geometry, state *p (2)* is an expression of *bi-polar* symmetry and prescribes elliptical geometrization, state *d (3)* is an expression of *parabolic* symmetry and promotes parabolic geometry, while state *f (4)* is an expression of *hyperbolic* symmetry and dictates hyperbolic geometry.

15.5. Rules Governing Nuclear Energy Density States

There exist 4 important rules mandated by the Hyper Quantum and the Hyper Quadratic Operators in the development of core shelling implicating both the neutron core and the proton core.

1. The development of energy density shells in cores is subject, from their ground state, to a level of accrued compression mandated by the Hyper Quantum Operator whereby the near-end (to core center) of the distribution undergoes a shrinking effect in a more pronounced manner than the far-end of the distribution, akin to a non-uniform spring.

2. An exclusion principle is prescribed by the Hyper Quantum Operator whereby if, as a result of compression, two energy states happen to have any of their maxima at the same radial distance from core center, the generation of the whole distribution for one of the states is prohibited. Only the first one to be generated in the quantum order of construction (or the more "localized" one) is allowed. This principle manifestly seeks to obtain broadest possible stability for the edifice.

3. An exception to the Rule 2 applies when the aggregation function is satisfied with a high level of nucleon accumulation, allowing thereby the formerly prohibited quantum states to be occupied. However, this concession runs at the expense of stability since the saturation of these states *potentially* triggers an act of disintegration of the edifice.

4. The Hyper Quadratic Operator, per its accrued Ambi-Valence Norm, obligates the proton(s) to relate to both electron(s) and neutron(s) in a double pairing manner, which amounts to *ambi-valence* overloading. Per this norm, a proton ought to concomitantly relate to one electron and to one neutron. A proton ought to relate to only one electron; if it has to relate to more than one neutron, an inflationary presence that may and will occur because neutrons are in the business of atomic densification, it does so at the cost of stability.

In the aggregate, the bulk of the densification constraints exerted by these two Operators are embodied in a fraction of nuclides' binding energy and accounted for by the so-called *mass-defect* of atomic nuclides. In other words, asked about how we know all of the above to be true, our simple answer is: *mass defect!* Effectively the mass deficit of nuclides, a qualified difference between the total mass of all integrants as free particles and their total mass as nucleons in a nuclide, testifies to this investment in binding energy at the expense of mass. This act is veritably effected by the core densification function intrinsic to these Operators, notably the Hyper Quantum Operator. The systematic of the binding embodied in the facilitating rules above described seeks not only the creation of the radial core structure but its post-creation conservation. The formula for nuclear binding energy is reported below for reference:

$$BE = \Delta m \cdot c^2$$

where BE is the Binding Energy, Δm is the change in mass after formation or mass defect, and c the speed of light.

15.6. Specificities of Nuclides' Energy Density States

All attempts to describe nuclide organization primarily based on interaction between piled-up nucleons are doomed to fail, similarly to the notorious failure of yesteryear in classical physics with regards to its attempt to describe the electronic configuration of atoms, well portrayed in the Mendeleev Table, based on electromagnetic energy dynamics between electrons and protons. The modern notion that quarks' gluons, whose mission is to primarily bind them together, may carry their influence across the baryon particle, neutron or proton, and out into the atom environment in the incarnation of a so-called *residual strong force* as inter-nucleon binding force for nuclides, we have formerly decried and taxed at the very least as plainly unsound. This notion portrays a convoluted, if not abusive, picture which violates the necessary independence of Shells of Matter (or Symmetry Groups), each known to be built on their own physical constants.

It is only when the classical school of thought was overcome by the notion of a particle (the electron) constrained to move *freely* in a virtual *box* and experiencing discrete energy states that a Hamiltonian and later a Lagrangian could be reached which ultimately unraveled the distribution of energy density states for the electron. By doing so, they were able to give at least satisfactory account to the then well-known Mendeleev Table of [Electronic] Elements. If this construct was successful in leading to an adequate interpretation of the dynamics of the bound electron, even more so will it be in attempting to describe the dynamics of bound protons and neutrons at the core of atomic edifices. The "box" effect, nothing else in fact than a quantum tensor applied to a space void spread, must patently be even stronger in that region of atomic elements than at their periphery. The following point is a pivotal argument to this proposal:

Were it not for the proposed two radially separate core formations in atoms' nuclei, their proton and neutron integrants could not have the periodic sets of energy density shells factually known to them since 1963.

Should one be tenacious in abiding by the older concepts, the analysis underway here in a sense amounts to providing a visualization of embedded radial nuclei as *two virtual boxes* that make the older quantum dynamical equations (Hamiltonian and Lagrangian) directly applicable to these objects.

The distribution put forth by the Shell Model is flawed and far insufficient only because it is unable to recognize the additional densification imperatives that constrain the development of the basis energy states for nuclides. Once those imperatives are understood and properly taken into account in the modeling, we are well on our way to successfully draw the real-world distribution of natural nuclides. But again, only from the Quanto-Geometric perspective of nature's dynamics can these imperatives arise.

As usual, the impactful intervention of azimuthal symmetry imperatives over the formation of orbital packets puts the completion seal over the construction. In effect, it is the imperative of octet completion at end of azimuth that leads to the "magic numbers" by staging the realization of saturated shells or sub-shells of orbitals at that particular position of the distribution. Furthermore, the following fact cannot be overstressed: the set of known "magic numbers" which confers accrued stability to the nuclides of their count, apply to both neutrons and protons independently. *Consequently, if these magic numbers are in fact the total count of integrants of a nuclide* ***whose energy density states have reached shell saturation****, then there must be one independent set of quantum state distributions for neutrons and one other for neutrons.* By exactly exemplifying that portrayal, our distribution Tables consistently reconcile the majority of the known pieces of this puzzle.

15.7. Ontological-Flavor Field

Something to be said about Rule 4 above is that it supplants the concept of classical and QED/QCD electrostatic charges presumably at work within the atomic edifice with that of ontological-flavor field. We have formerly amply demonstrated that the construct of a field exogenous to particles of matter as a separate spread to their internal wave harmonics is totally unjustified and unjustifiable. Indeed, no object in the universe projects such an identity. Tensored vacuum around an object is veritably the ontological vacuum out of which the object is constitutionally built. It is a rarefied vacuum with a tensor or tension that falls exponentially with radial distance.

At the most the qualities of the ontological field, should one insist in utilizing the notion of field, may be viewed thru the prism of phenomenology, which goes in accord with the signature of the relevant ontology. If we were to put forth a Quanto-Geometric field theory, we would have to retain that: 1) all interaction fields are a derivate of the wave function of particles and 2) the phenomenological type deriving from the Quanto-Geometric primitive that defines the object reflects onto its interaction field as its inherent flavor or phenotypical form of interaction. However, in our judgment, we do better by completely doing away with the notion of field as a separate entity, principally because it is as unjustified as conducive to confusion.

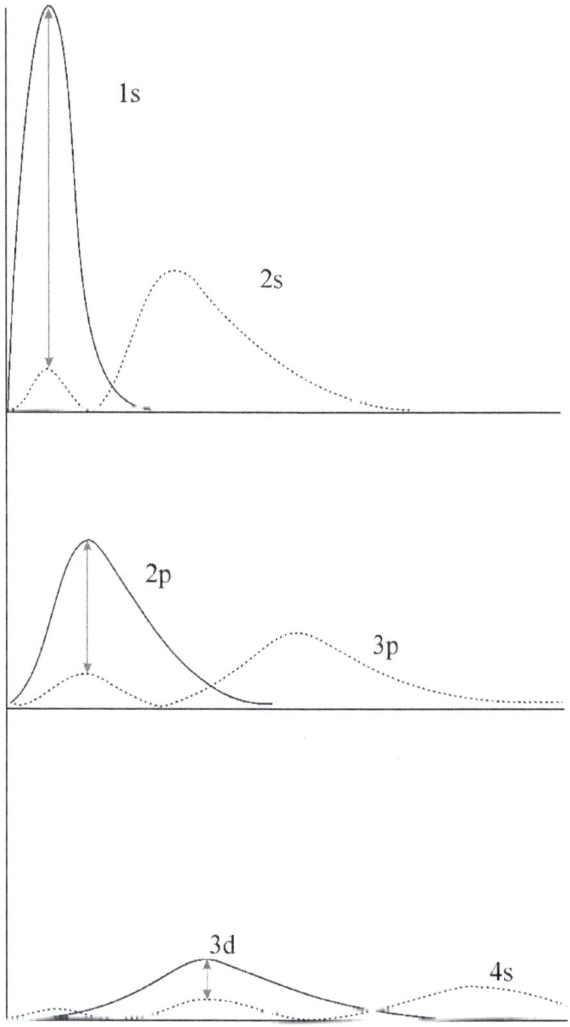

Fig. 15.3 Quantum energy density states under the compression effect resulting in peak coincidence between several states

Let me further state that it is only because the classical concepts of + and – poles or charges remit to some level of ontological-flavor types that they have been able to assist us historically in some limited measure in the interpretation of physical matter dynamics. Accordingly, the field germane to a free neutron would be of *autonomic* flavor. The action of this field or charge is to *isolate* or to *stop*. Whereas the *dichotomous* action of a free-proton field *repels* as much as the *associative* action of a free-electron field or charge *attracts*. Those characteristics putatively apply to particles that are on-shell (in their own Shell2 of matter). When overtaken in higher-level ontology, i.e., as bound particles (in orbitalization), these characteristics no longer directly apply because the imperatives of the new Shell take over, dictating ontologies and phenomenologies consistent with signatures of the new Primitives.

Due to its familiarity however, we will allow ourselves to continue perusing the field term, but in complete ban of any implicated self-interaction. We do so in the firm and intractable understanding that the notion of field means not exogenous phenomenology but endogenous ontology with respect to its integrant.

Accordingly, it is against the backdrop of ontological-flavor field interaction that particle pairing and over-pairing invoked in Rule 4 takes place, in substitution of electrostatic charge constructs. It is simply an interaction between spatial wavefunction spreads native to particulate orbitals. A towering view of atomic construction must embrace it in a global dimension which therefore implies that all three involved particles, neutrons, protons and electrons, are being concomitantly added to the edifice, per the Tri-Valent Operator norm. The Tri-Valent Operator is the absolute owner of this pairing action which it executes or mediates as described in the terms of Rule 4.

15.8. Genesis of Atomic Nuclides

As previously and copiously mentioned, energy density states of atomic nuclides are based on the same distribution known for electron energy density states in the manner previously derived in this study and from the Hamiltonian that our framework fully endorses. This includes the few basis states initially discarded by the exclusion rule afore stated. We graphically depict in Fig. 15.3 how the compression effect shrinks the individual spread of these states and brings them to the level of superposition where certain density peaks in some states coincide with other peaks in others. Consequently, per Rule 2, state *2s*, posterior to state *1s* in quantum order of construction, is prohibited because its first peak coincides with the one single peak of state *1s*. State *3p*, posterior in quantum order to state *2p*, is prohibited because its first peak coincides with the one single peak of *2p*. And finally state *4s*, posterior to state *3d*, is prohibited because its second peak comes to coincide with the peak of *3d*.

We finally note that the Pauli exclusion principle fully applies here as well, just as it does for electron energy state distributions: every energy density state only allows two particles, one in spin up and one in spin down.

15.9. Periodic Construction of Atomic Nuclides

In order to generate the Periodic Tables of Atomic Nuclides, we call upon the concepts of *isotopy* and *isotony* to assist us in that task. We make each of them an intrinsic property of nuclides consequential to the aggregation and ambivalent Operator Norms imposed onto them and the ontological-flavor fields of their nucleons as well. We thus gather those concepts in the figures we coin *isotopic* index or *valence* and *isotonic* index or *valence*. As a reminder, isotopes are nuclides with same number of protons but distinct number of neutrons and isotones are nuclides with same number of neutrons but distinct number of protons.

These indices play a somewhat similar role to electronic valences, without quantifying however orbitals in external shells. Thus, an isotone with a valence of *4 means that it has net neutron charge of 4 or 4 surplus neutrons (read neutron orbitals). The * symbol surrogates customary electrostatic charge and typifies the *autonomic* flavor of the

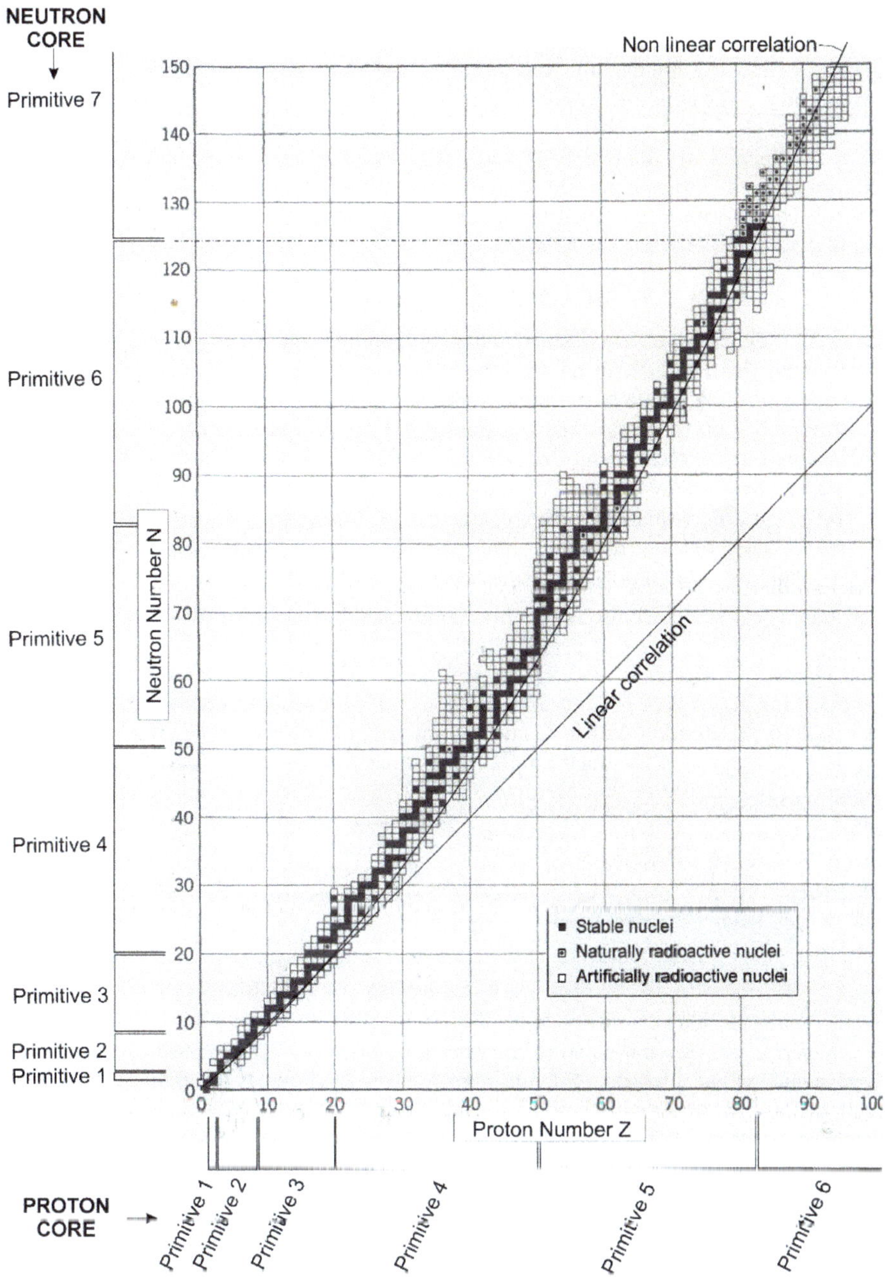

Fig. 15.4 Chart of known nuclides displaying Neutron Cores v. Proton Cores

neutron's charge in this context per previous discussion. This character arises from particulate-level ontology and principally from Hyper-Quantum Tri-Valent imperative imposed on neutron orbital ontology. The isotopic valence of a nuclide element is simply the difference between its number of neutrons and its number of protons. This number is a characterization for neutron cores in nuclides. It is expressed in the formula below:

$$\textit{Isotonic Valence} = N - Z$$

where N is the number of neutrons, Z the number of protons. When the A Mass Number, the sum total of neutrons and protons in a nuclide, is the known variable, the Isotonic Valence may equally be computed this way:

$$\textit{Isotonic Valence} = A - 2Z$$

On the other hand, a nuclide with an *isotopic valence* of +1 means that it has a deficit of 1 proton for a proton-neutron match. This number is a characterization for proton cores in nuclides. The formula coding the isotopic valence follows:

$$\textit{Isotopic Valence} = |Z - N|$$

The nuclide chart above which reports all known nuclides up to element 149, with their neutron N number vs. their proton Z number, provides the numeric basis for the analysis. We study the evolution of neutron cores by progressive aggregation of neutrons to the core's shells, which leads us to represent the amounts of nuclides with same number of neutrons or isotones in the valence column of the corresponding tables. Likewise, we study the evolution of proton cores by progressive aggregation of protons to the core's shells, which leads us to represent the number of nuclides with same number of protons or isotopes in the valence column of the corresponding tables. In the chart reported in Fig. 15.4, we read isotopes by looking at a Z number and examining the number of squares along the vertical for that number. Each square along a vertical represents an isotope. We read isotones in the Table by looking at an N number and examining the number of squares along the horizontal for that Number. Each square along a horizontal represents an isotone.

Let us illustrate with the case of $N = 50$. There exist a total of 6 stable and naturally radioactive nuclei with $N = 50$:

- 50 Neutrons – 36 Protons (stable)
- 50 Neutrons – 37 Protons (naturally instable)
- 50 Neutrons – 38 Protons (stable)
- 50 Neutrons – 39 Protons (stable)

- 50 Neutrons – 40 Protons (stable)
- 50 Neutrons – 42 Protons (stable)

We compute the difference between number of Neutrons and Protons to yield from last to first: 8, 10, 11, 12, (13), 14. These numbers represent the surplus number of neutrons over protons in the various isotones. We note them in the orbital distribution Tables with a *symbol to represent the surplus (ontological-flavor) neutron charge: *8, *10, *11, *12, *(13), *14. We can readily see that such numbers have the value of a valence, though not with the same identity of electronic valence because they do not represent peripheral elements to the orbital shell.

The Nuclide Chart readily spells out that the most primal directive in the construction is that of aggregation or accumulation in terms of a many-to-one function, in accordance with the first-order role played by the Hyper Quantum Operator in nuclear architecture. As the number of proton progresses, the number of possible neutrons per Proton Core in a nuclide progresses even more according to a binomial function the likes of a parabola (non –linear correlation curve in illustration). Starting from Proton Core 20, there is no more equity between the number of neutrons and protons in nuclides in general (no more elements on the line portraying linear correlation in illustration). Incorporating data from the Nuclide Chart, we report two sets of Periodic Nuclide Tables: one for Neutron Cores and another for Proton Cores.

In Quantum Chemistry, elements take their names, and therefore their identity, after the number of protons (or electrons) they bear. This practice only responds to the historical development of the physical sciences which firstly got acquainted to electrons, and subsequently protons, among the many particles of matter. Nothing forbids naming nomenclature accounting for number of neutrons instead in identifying atomic elements. Physical law underlying their development would remain unaltered. The Neutron Core Table reports the N number or N numerals for identification of listed elements, by contrast with the Proton Core Table which lists them by the familiar conventionally adopted names.

15.9.1 Extraordinary Density of Atomic Nuclides

Quantum Physics literature oftentimes states that atomic nuclei become unstable when they incorporate too many neutrons for their included number of protons. However, a distinct threshold ratio between the nucleons beyond which instability would occur is never to be found in the literature. The absence of such a figure makes the assertion unfounded and even discredits it as an empirical rule. The question of stability vs. instability is not based on the number of nucleons in nuclides. What there is an aggregation or densification function that agglutinates elements at the very core of the atom. Because the neutron core constitutes the true heart of the atom, the Hyper Quantum Operator, here the operating function, disproportionately funnels neutrons, a particle naturally heavier than the proton, into that sector of the edifice. In other words, neutron constituency within the nucleus is in the business of densification, compactification or monolithisation, just as bound electrons are in the business of spatial overture or exteriorization.

Effectively, computation based on the liquid-drop model, which considers the nucleus an amalgam of neutrons and protons, yields an extraordinarily high figure of nuclear density, namely $2.44 \times 10^{14}\ g/m^3$. One cubic centimeter of nuclear matter, comparable to a small end piece of a human finger, weighs about 250 million tons! Without a doubt an eminently monolithic function is at work within atomic nuclei. If we further consider that neutrons and protons are segregated into two distinct populations within the nucleus, per our contention, then an unimaginably dense nucleus looms before us.

15.9.2 Nuclides' Periodic Distribution Based on Quanto-Geometric Primitives

In examining the reported Tables of proton and neutron cores, note well the natural progressive realization of neutron and proton orbitals that lead to completion status at the exact numbers known as "magic numbers". It is well worth taking stock of the details of this development by reviewing the natural progression in the realization of orbitals and sub-shells. It is well known that the nuclides that experience any of these numbers in their total count of neutrons or protons experience markedly increase stability in relation to their immediate pars. The stability in question is spelled out in terms of high binding energy, high resistance to fission and fusion, as well as abundance in nature.

This progression gives us 7 Primitive periods for neutrons and 7 Primitive periods for protons. For Neutrons, namely:

- Primitive 1 period spanning Element 1 to 2
- Primitive 2 period spanning Element 3 to 8
- Primitive 3 period spanning Element 4 to 20
- Primitive 4 period spanning Element 21 to 50
- Primitive 5 period spanning Element 51 to 82
- Primitive 6 period spanning Element 83 to 124
- Primitive 7 period spanning Element 125 to 146

For Protons, we have:

- Primitive 1 period spanning Element 1 (H) to Element 2 (He)
- Primitive 2 period spanning Element 3 (Li) to Element 8 (O)
- Primitive 3 period spanning Element 4 (F) to Element 20 (Ca)
- Primitive 4 period spanning Element 21 Sc) to Element 50 (Sn)
- Primitive 5 period spanning Element 51 (Sb) to 82 Element (Pb)
- Primitive 6 period spanning Element 83 (Bi) to Element 124

- Primitive 7 period spanning Element 125 to Element 146

Notice the absence of the prohibited *2s, 3d* and *4s* orbitals in the initial movement of the construction, shown in the Tables with 0 orbitals, and note how they intervene thereafter in the construction starting with Neutron Core 85 and Proton Core 85, in accordance with the prescription spelled out in Rule 3. Note well the correlation between the odd-even-number oscillation from one orbital of action to the next (or one core to the next) and the singular-plural oscillation in valence indices that indicates binding energy, spin angular momentum and even abundance in nature.

These Periodic Distribution Tables represent the one single backdrop to the very prolific series of facts known about atomic nuclides which the collection of nuclear models individually attempts to address. We will not address here the granular details of this reconciliation, limiting ourselves to expose the foundational structures that lead to this minutia. In that regard, one shall understand well that, on the one hand, a nuclide is a reunion of two cores, a proton core and a neutron core, and, on the other hand, certain features in nuclear properties, such as the super-enhanced stability of "doubly magic" nuclides, arise from correlation between both cores. Other aspects of this distribution where the trends seem to break as exception cases pertain to phenomenological behavior attributable to the very signatures of the Primitives, in conjunction with Azimuthal Symmetry underlying the distinct periods. However, this characterization pertains to Quanto-Geometric ultrastructural analysis, which lay outside of the intended scope of this exposé. Spontaneous yield to fission and fusion demonstrated by certain nuclides, as part of the binding energy picture, leads to one of the most dramatic effects known to atomic nuclides, the impactful nuclear radioactive phenomenon. That one phenomenon we will devote particular attention to next.

Let's now allow the Tables to speak for themselves. Valence numbers in parenthesis indicate unstable cores. An empty cell for the valence figure indicates no existent stable core for the element. End-of-period elements are highlighted so that the periods may be readily observed.

PERIODIC TABLE OF NEUTRON CORES
TABLE 1A

Element	1s	2p	3s	3d	4p	4d	4f	Isotone	Isotonic Valence
1	1							2	0, +2
2	**2**							**1**	**0**
3	2	1						1	0
4	2	2						1	-1
5	2	3						2	0, -1
6	2	4						2	0, -1
7	2	5						2	0, -1
8	**2**	**6**						**2**	**0, -1**
9	2	6	1					1	-1
10	2	6	2					3	0, -1, -2
11	2	6	2	1				1	-1
12	2	6	2	2				3	0, -1, -2
13	2	6	2	3				1	-1
14	2	6	2	4				3	0, -1, -2
15	2	6	2	5				1	-1
16	2	6	2	6				3	0, -1, -2
17	2	6	2	7				1	-1
18	2	6	2	8				3	0, -1, -2
19	2	6	2	9				0	
20	**2**	**6**	**2**	**10**				**5**	**0, -1, -2, -3, -4**
22					2			3	-2, -3, -4
23					3			1	-3
24					4			3	-2, -3, -4
25					5			1	-3
26					6			3	-2, -4, -6
27					6	1		1	-5
28	*Element 20*				6	2		5	-2, -4, -5, -6, -8
29					6	3		1	-5
30					6	4		4	-2, -4, -5, -6
31					6	5		1	-4
32					6	6		3	-4, -5, -6
33					6	7		1	-5
34					6	8		3	-4, -5, -6

PERIODIC TABLE OF NEUTRON CORES
TABLE 1B

Element	*1s-2p-3s-3d*	4p	4d	4f	5s	5p	5d	5f	Isotone	Isotonic Valence
35		6	9						0	
36		6	10						3	-6, -7, -8
37		6	10	1					1	-7
38		6	10	2					3	-6, -7, -8
39		6	10	3					0	
40		6	10	4					4	-6, -8, -9, -10
41		6	10	5					1	-9
42		6	10	6					4	-6, -8, -9, -10
43	*Element 20*	6	10	7					1	-10
44		6	10	8					4	-8, -9, -10, -12
45		6	10	9					0	
46		6	10	10					4	-8, -10, -11, -12
47		6	10	11					1	-11
48		6	10	12					4	-10, -11, -12, -14
49		6	10	13					1	-11
50		**6**	**10**	**14**					**6**	**-8, -10, -11, -12, (13), -14**
51					1				1	-11
52					2				4	-8, -10, -11, -12
53					2	1			1	-11
54					2	2			3	-10, -12, -14
55					2	3			2	-11, -13
56					2	4			4	-10, -12, -14, -16
57					2	5			1	-13
58					2	6			5	-10, -12, -13, -14, -16
59					2	6	1		1	-13
60					2	6	2		4	12, 13, -14, -16
61					2	6	3		0	
62	*Element 50*				2	6	4		4	(-12), 14, -15, -16
63					2	6	5		1	-15
64					2	6	6		4	-14, -15, -16, -18
65					2	6	7		1	-15
66					2	6	8		1	-16
67					2	6	9		1	-17
68					2	6	10		1	-18
69					2	6	10	1	1	-19
70					2	6	10	2	4	-16, -18, -19, -20
71					2	6	10	3	1	-19

PERIODIC TABLE OF NEUTRON CORES
TABLE 1C

Element	*1s–3d-4p-4d-4f*	5s	5p	5d	5f	6s	6p	6d	6f	Isotone	Isotonic Valence
72	*Element 50*	2	6	10	4					4	-18, -20, -21, -22
73		2	6	10	5					1	-21
74		2	6	10	6					5	-18, -20, -21, -22, -24
75		2	6	10	7					1	-21
76		2	6	10	8					3	-20, -22, -24
77		2	6	10	9					1	-23
78		2	6	10	10					5	-20, -22, -23, -24, -26
79		2	6	10	11					1	-23
80		2	6	10	12					2	-22, -24
81		2	6	10	13					2	(-24), -25
82		**2**	**6**	**10**	**14**					**7**	**-20, -22, -23, -24, -25, -26, -27, -29**

PERIODIC TABLE OF NEUTRON CORES
TABLE 1D

Element	2s	3p	4s	5s2 5p6 5d10 5 f14	6s	6p	6d	6f	Isotone	Isotonic Valence
83					1				1	-23
84				*Element 82*	2				2	-24, -26
85	1				2				2	(-23), -25
86	2				2				2	-24, -26
87	2	1			2				1	-25
88	2	2			2				4	-24, -25, -26, -28
89	2	3			2				0	
90	2	4			2				4	-26, -27, -28, -30
91	2	5			2				1	-27
92	2	6			2				2	-26, -28
93	2	6	1		2				1	-29
94	2	6	2		2				4	-26, -28, -29, -30
95	2	6	2		2	1			1	-29
96	2	6	2		2	2			2	-28, -30
97	2	6	2		2	3			1	-31
98	2	6	2		2	4			4	-28, -30, -31, -32
99	2	6	2		2	5			1	-31
100	2	6	2		2	6			3	-30, -31, -32
101	2	6	2		2	6	1		1	-31
102	2	6	2		2	6	2		3	-30, -32, -34
103	2	6	2		2	6	3		1	-33
104	2	6	2		2	6	4		3	-32, -33, -34
105	2	6	2		2	6	5		1	-33

PERIODIC TABLE OF NEUTRON CORES
TABLE 1E

Element	2s	3p	4s	$5s^2\ 5p^6\ 5d^{10}\ 5\ f^{14}$	6s	6p	6d	6f	7s	7p	7d	7f	Isotone	Valence
106	2	6	2		2	6	6						3	-32, -34, -36
107	2	6	2		2	6	7						1	-35
108	2	6	2		2	6	8						4	-32, -34, -35, -36
109	2	6	2		2	6	9						1	-35
110	2	6	2		2	6	10						3	-32, -34, -36
111	2	6	2		2	6	10	1					1	-35
112	2	6	2		2	6	10	2					3	-36, -37, -38
113	2	6	2		2	6	10	3					1	-37
114	2	6	2		2	6	10	4					3	-36, -37, -38
115	2	6	2		2	6	10	5					0	
116	2	6	2	*Element 82*	2	6	10	6					4	-36, -38, -39, -40
117	2	6	2		2	6	10	7					1	-39
118	2	6	2		2	6	10	8					3	-38, -39, -40,
119	2	6	2		2	6	10	9					1	-39
120	2	6	2		2	6	10	10					2	-40, -42
121	2	6	2		2	6	10	11					1	-41
122	2	6	2		2	6	10	12					3	-40, -41, -42
123	2	6	2		2	6	10	13					0	
124	**2**	**6**	**2**		**2**	**6**	**10**	**14**					**3**	**-41, -42, -43**
125									1				2	-43, (-44)
126				*Element 124*					2				4	-43, -44, -45, (-46)
127									2	1			3	(-43), (-44), (-46)
128									2	2			3	(-44), (-45), (-46)
129									2	3			3	(-46), (-47), (-48)
130									2	4			2	(-46), (-48)
131									2	5			3	(-46), (-47), (-48)
132									2	6			2	(-48), (-50)
133									2	6	1		2	(-47), (-48)
134									2	6	2		2	(-48), (-50)
135									2	6	3		1	(-47)
136									2	6	4		3	(-48), (-49), (-50)
137									2	6	5		1	(-47)
138									2	6	6		3	(-48), (-49), (-50)
139									2	6	7		1	(-50)
140									2	6	8		3	(-49), (-50), (-52)
141									2	6	9		1	(-51)
142									2	6	10		2	(-50), (-52)
143									2	6	10	1	2	(-51), (-52)
144									2	6	10	2	1	(-54)
145									2	6	10	3	0	
146									2	6	10	4	1	(-54)

PERIODIC TABLE OF PROTON CORES
TABLE 2 A

Z	Element	1s	2s	2p	3s	3p	3d	4s	4p	4d	4f	Isotope	Isotopic Valence
1	H	1										1	0
2	**He**	**2**										**2**	**-1, 0**
3	Li	2	0	1								2	0, +1
4	Be	2	0	2								1	1
5	B	2	0	3								2	0, +1
6	C	2	0	4								2	0, +1
7	N	2	0	5								2	0, +1
8	**O**	**2**	**0**	**6**								**3**	**0, +1, +2**
9	F	2	0	6	1							1	1
10	Ne	2	0	6	2							3	0, +1, +2
11	Na	2	0	6	2	0	1					1	1
12	Mg	2	0	6	2	0	2					3	0, +1, +2
13	Al	2	0	6	2	0	3					1	1
14	Si	2	0	6	2	0	4					3	+1, +1, +3
15	P	2	0	6	2	0	5					1	2
16	S	2	0	6	2	0	6					4	0, +1, +2, +4
17	Cl	2	0	6	2	0	7					2	+1, +3
18	Ar	2	0	6	2	0	8					3	0, +2, +5
19	K	2	0	6	2	0	9					3	+1, (+2), +5
20	**Ca**	**2**	**0**	**6**	**2**	**0**	**10**					**6**	**0, +2, +3,+4, +6, +8**
21	Sc							0	1				3
22	Ti							0	2				+2, +3, +4, +5, +6
23	V							0	3				5
24	Cr							0	4				+2, +4, +6, +6
25	Mn	*20 Ca*						0	5				5
26	Fe							0	6				+2, +4, +5, +6
27	Co							0	6	1			5
28	Nu							0	6	2			+2, +4, +5, +6, +8
29	Cu							0	6	3			+5, +7

PERIODIC TABLE OF PROTON CORES
TABLE 2B

Z	Element	1s2 2s2 2p6 3s2 3p6 3d10	4s	4p	4d	4f	5s	5p	5d	5 f	Isotope	Isotopic Valence
30	Zn		0	6	4						5	+4,+6,+7,+8,+10
31	Ga		0	6	5						2	+7,+9
32	Ge		0	6	6						5	+6,+8,+9,+10,+12
33	As		0	6	7						1	+9
34	Sc		0	6	8						6	+6,+8,+9,+10,+12,+14
35	Br		0	6	9						2	+9,+11
36	Kr		0	6	10						6	+6,+8,+10,+11,+12,+14
37	Rb		0	6	10	1					2	+11,(+13)
38	Sr		0	6	10	2					4	+8,+10,+11,+12
39	Y		0	6	10	3					1	+11
40	Zr	*20 Ca*	0	6	10	4					5	+10,+11,+12,+14,+16
41	Mb		0	6	10	5					1	+11
42	Mo		0	6	10	6					7	+11,+8,+10,+11,+12,+13,+14,+ 16
43	Tc		0	6	10	7					0	
44	Ru		0	6	10	8					7	+8,+10,+12,+14,+15,+17,+19
45	Rh		0	6	10	9					1	+13
46	Pd		0	6	10	10					6	+10,+12,+13,+15,+17,+19
47	Ag		0	6	10	11					2	+13,+15
48	Cd		0	6	10	12					5	+10,+12,14,+15,+16
49	Ln		0	6	10	13					1	+15
50	**Sn**		**0**	**6**	**10**	**14**					**10**	**(+12),+14,+15,+16,+17+18,+19, +20,+22+, +24**
51	Sb						1				2	+20,+22
52	Te						2				7	+18,+19,+20,+21,+22,+24,+26
53	I						2	1			1	+21
54	Xe						2	2			8	+16,+18,+20,+21,+22,+23+24,+ 28
55	Cs						2	3			1	+23
56	Ba	*50 Sn*					2	4			7	+18,+20,+22,+23,+25,+26,+27
57	La						2	5			2	(+24),+25
58	Ce						2	6			4	+20,+22,+24,+26
59	Pr						2	6	1		1	+23
60	Nd						2	6	2		7	+22,+23,+24,+25,+26,+27,+28, +29

PERIODIC TABLE OF PROTON CORES
TABLE 2C

Z	Elmt	2s	3p	4s	4p6 4d10 4f14	5s	5p	5d	5f	6s	6p	6d	6f	Isotope	Isotopic Valence
61	Pm					2	6	3						0	
62	Sm					2	6	4						6	+20,(+23),+24,+25,+26,+28
63	Eu					2	6	5						2	+25,+27
64	Gd					2	6	6						6	+24,+26,+27,+28,+29,+30
65	Tb					2	6	7						1	+29
66	Dy					2	6	8						6	+26,+28,+29,+30,+31,+32
67	Ho					2	6	9						1	+31
68	Er					2	6	10						6	+26,+28,+30,+31,+32,+34
69	Tm					2	6	10	1					1	+31
70	Yb					2	6	10	2					7	+28,+30,+31,+32,+33,+34,+36,
71	Lu					2	6	10	3					1	+33
72	Hf					2	6	10	4					6	+30,+32,+33,+34,+35,+36
73	Ta					2	6	10	5					1	+35
74	W				*50 Sn*	2	6	10	6					5	+32,+34,+35,+36,+38
75	Re					2	6	10	7					2	(+35),+37
76	Os					2	6	10	8					7	+32,+34,+35,+36,+37,+38,+40
77	Ir					2	6	10	9					2	+37,+39
78	Pt					2	6	10	10					5	+36,+38,+39,+40,+42
79	Au					2	6	10	11					1	+39
80	Hg					2	6	10	12					7	+36, +38, +39,+40, +41, +42, +44
81	Tl					2	6	10	13					6	+41,+43, (+44), (+45), (+46), (+48)
82	**Pb**					**2**	**6**	**10**	**14**					**8**	**+40,+42,+43,+44, (+46), (+47), (+48), (+50)**
83	Bi									1				5	+43, (+44), (+45), (+46), (+48)
84	Po									2				7	+42, (+43), (+44), (+46), (+47), (+48), (+50)
85	At	*1*								2				2	(+46), (+48)
86	Rn	*2*								2				3	(+47), (+48), (+50)
87	Fr	*2*	*1*		*82 Pb*					2				1	(+49)
88	Ra	*2*	*2*							2				4	(+47), (+48), (+50), (+52)
89	Ac	*2*	*3*							2				2	(+49), (+50)
90	Th	*2*	*4*							2				6	(+47), (+48), (+50), (+55), (+56), (+58)

PERIODIC TABLE OF PROTON CORES
TABLE 2D

Z	Element	2s	3p	4s	5s2 5p6 5d10 5 f14	6s	6p	6d	6f	Isotope	Isotopic Valence
91	Pa	2	5			2				2	(+49), (+52)
92	U	2	6			2				3	(+50), (+51), (+54)
93	Np	2	6	1		2					
94	Pu	2	6	2		2					
95	Am	2	6	2		2	1				
96	Cw	2	6	2		2	2				
97	Bk	2	6	2		2	3				
98	Cf	2	6	2		2	4				
99	Es	2	6	2		2	5				
100	Fm	2	6	2		2	6				
101	Md	2	6	2		2	6	1			
102	Lw	2	6	2		2	6	2			
103		2	6	2		2	6	3			
104		2	6	2		2	6	4			
105		2	6	2		2	6	5			
106		2	6	2		2	6	6			
107		2	6	2	*82 PB*	2	6	7			
108		2	6	2		2	6	8			
109		2	6	2		2	6	9			
110		2	6	2		2	6	10			
111		2	6	2		2	6	10	1		
112		2	6	2		2	6	10	2		
113		2	6	2		2	6	10	3		
114		2	6	2		2	6	10	4		
115		2	6	2		2	6	10	5		
116		2	6	2		2	6	10	6		
117		2	6	2		2	6	10	7		
118		2	6	2		2	6	10	8		
119		2	6	2		2	6	10	9		
120		2	6	2		2	6	10	10		
121		2	6	2		2	6	10	11		
122		2	6	2		2	6	10	12		
123		2	6	2		2	6	10	13		
124		**2**	**6**	**2**		**2**	**6**	**10**	**14**		

PERIODIC TABLE OF PROTON CORES
TABLE 2E

Z	Element	$2s^2\ 3p^6\ 4s^2\ 5s^2\ 5p^6\ 5d^{10}\ 5f^{14}\ 6s^2\ 6p^6\ 6d^{10}\ 6f^{14}$	7s	7p	7d	7f	Isotope	Isotopic Valence
125		*Element 124*	1					
126			2					
127			2	1				
128			2	2				
129			2	3				
130			2	4				
131			2	5				
132			2	6				
133			2	6	1			
134			2	6	2			
135			2	6	3			
136			2	6	4			
137			2	6	5			
138			2	6	6			
139			2	6	7			
140			2	6	8			
141			2	6	9			
142			2	6	10			
143			2	6	10	1		
144			2	6	10	2		
145			2	6	10	3		
146			2	6	10	4		

15.10. Deep-Seated Roots of Radioactive Decay in Nuclides' Shell Structure

The isotopic and isotonic valence numbers noted in parenthesis in these tables mean to flag the nuclides that are known to be radioactively unstable. Consistent with our premises and particularly Rule 3, one may note that the vast majority of radioactive nuclides occur after the full state of saturation of the initially prohibited states *(2s, 3p* and *4s)* along with that of the immediately following outer shells within the heavy nuclides where the prohibited states become allowed. We see radioactivity starting to occur even within the saturated end-of-period elements such as *82 Pb*, with 4 of its eight proton cores showing radioactive instability. All radioactive neutron cores in that condition, for their own, start after core 124, the saturated end-of-period element.

Radioactive instability is a phenomenon primarily driven by neutron core behavior in submission to Rule 3. It is clearly initiated immediately after the genesis of neutron core 124, an end-of-period element. Two major decay trends take place, namely decay by core fission, and decay by particle emission, both consistent with our theoretical premises.

15.10.1 Orbital-Compression Provenance of Radioactive Disintegration

Stability of generated nuclides in the construction become compromised from the very moment that neutron sub-shells *2s, 3p* and *4s* come into play in the construction with element 85. The other neutron orbital sets they correlate with under the compression effect, namely *1s, 2p* and *3d,* experience as well the undercurrent of instability (Fig. 7.3). However, the strength of the Hyper Quantum Operator in command instills a sufficiently high level of compactification on constructed elements to keep them stable all the way up to the subsequently-filled outer shells *6s-6p-6d-6f,* the last of which at saturation receive a natural boost in stability (Neutron Core 124). Immediately thereafter however, the next orbital sets in the neutron core the *7s-7p-7d-7f* shell, come to experience too much deficit in stability to keep whole and exert effective continual correlation with their pars in the proton core. This correlation is normally exerted thru the space component of orbital spatial spreads of both pars (read ontological-flavor field interaction if you will). The new neutron cores becoming explicitly unstable drag along in their breakup the associated proton elements with the higher core. The latter therefore peremptorily disintegrates as well. This fact represents the veritable trigger for the phenomenon of radioactive disintegration, and not at all the alleged excessive neutron/proton ratio oftentimes cited in the literature but not theoretically justified.

Consequently, proton cores noted as unstable in the Proton Core Table, starting with element 81 (Tl) to the end, are not inherently unstable but only become so by association with unstable neutron cores in the nuclides they both correlatively make up. There is no noticeable construction particular with element 81 that would in itself trigger any manner of break up event. Yet radioactive disintegration begins with the innocuous Thallium (Tl) element and even proceeds with the supposedly stable end-of-period element 82 (Pb)! So it is only because these proton cores are in correlation with neutron cores already unstable.

Quite contrastively, we notice that neutron core 124, an end-of-period element, with both sub-shell groups *2s-3p-4s* and *6s-6p-6d-6f* at saturation do not demonstrate any level of instability. The increasing undercurrent of instability due to concretion of formerly prohibited orbital sets and their matches is simply overcome by the boost in stability granted by azimuthal end-of-period completion of shell 6 in neutron core 124. In all, the absence of such radioactive isotones for neutron core 124 translates indeed the boost in stability received by its 3 isotones in exempting them from their radioactive fate. This gives us the opportunity to address the question of whether or not element 124 is indeed the end-of-period element. Admittedly, nuclear physicists have attributed the condition of "magic" or endowed with exclusively saturated shells to element 126 instead, in all likelihood mainly because of its larger abundance with 4 surrogates, the largest in the period. However, the above argumentation keenly demonstrates this contention to be a mis-qualification, as regards to the subjacent dynamics that are actually taking place with shell configuration of these nuclides. Despite its mitigated level of binding energy, neutron core 124 is indeed the rightful end-of-period element by naturally progressive orbital construction. All the cores that follow are understandably yielding to radioactive decay (including neutron core 126), having no particular to contravene the forceful concretion of the "excessively localized" inner orbital sets we have pinpointed above.

15.10.2 Specific Orbital Shell Origin of Radioactive Series

Consistent with all of the above, all shell-*7* orbitals in the neutron cores subsequent to core 124 are going to detach from their core's nucleus but retain their correlated linkage with their par in the proton core. While there is a current to disjoint the neutron orbitals, nothing plays against their proton correlation mandated by the Hyper Quadratic Operator. Consequently, all filled-up neutron orbitals adjoined to their corresponding proton orbitals in all sub-shells of this period are going to leave the nucleus in the form of packets of 2 neutrons with 2 bound protons, the so-called alpha particles. Alpha particles are admittedly equivalent to Helium nuclides in composition and structure, except for one feature that we later discuss.

Therefore, starting from the last naturally generated element, neutron core 146, all paired orbitals in neutron core's period *7* are going to leave the nucleus in the form of alpha particles in a cascading manner from *7f* orbitals down to *7s* orbitals. Likewise, all single neutron orbitals are going to leave the nucleus in the form of single particles, however these neutron particles must necessarily do so by entering proton core space where they explode into a proton, an electron and a neutrino, all mass/energy relation conserved.

NEUTRON CORE PERIOD 7
TABLE 1F

Element	2s	3p	4s	$5s^2\ 5p^6\ 5d^{10}\ 5f^{14}$	6s	6p	6d	6f	7s	7p	7d	7f	Isotone	Isotonic Valence
124	**2**	**6**	**2**		**2**	**6**	**10**	**14**					**3**	**-41, -42, -43**
125									**1**				2	-43, (-44)
126									2				4	-43, -44, -45, (-46)
127									2	**1**			3	(-43), (-44), (-46)
128									2	2			3	(-44), (-45), (-46)
129									2	**3**			3	(-46), (-47), (-48)
130									2	4			2	(-46), (-48)
131									2	**5**			3	(-46), (-47), (-48)
132				*Element 124*					2	6			2	(-48), (-50)
133									2	6	**1**		2	(-47), (-48)
134									2	6	2		2	(-48), (-50)
135									2	6	**3**		1	(-47)
136									2	6	4		3	(-48), (-49), (-50)
137									2	6	**5**		1	(-47)
138									2	6	6		3	(-48), (-49), (-50)
139									2	6	**7**		1	(-50)
140									2	6	8		3	(-49), (-50), (-52)
141									2	6	**9**		1	(-51)
142				**4n + 0**					2	6	10		2	(-50), (-52)
143									2	6	10	**1**	2	(-51), (-52)
144				**4n + 1**					2	6	10	2	1	(-54)
145				**4n + 3**					2	6	10	**3**	0	
146				**4n + 2**					2	6	10	4	1	(-54)

Nuclear radiochemistry science recognizes 4 distinct radioactive series, namely:

1. Series **4n** (or 4n + 0) designated the Thorium series
2. Series **4n + 1** denoted the Neptunium series
3. Series **4n + 2** designated the Uranium series
4. Series **4n + 3** denoted the Actinium series

Each of these series either takes their name after their firstborn element or the longest-lived element. The associated mathematical formulae that attempt to rigorously capture their disintegrative behavior lacks however proper mathematical legitimacy. The formulae exactly mean to express that by dividing the nucleon number of a radioactive nuclide by 4, one would obtain a remainder of either 0, 1, 2 or 3. They so suggest that by giving a certain value to the n parameter, which stands for the nucleon number of the nuclide, we might recover a defining feature or the typical element of a specific series. In fact, to properly describe behavior, n can only take the two values of 59 or 58, which is a match for the mass number $A = 236$ *or* 232 respectively. That makes it not a parameter but a label, therefore a figure with diminished descriptive value.

For the **4n + 2** series that starts with Uranium-238, we decompose the mass number 238 as follows:

$$238 = 236 + 2 = (4 \times 59) + 2 \rightarrow 4n + 2, \text{ if } n = 59$$

This Uranium isotope is known to have 92 protons. With a mass number of 238, the resulting number of neutrons is 146. Accordingly, we note the 4n + 2 series in our Table in neutron core 146.

For the **4n + 3** series that starts with Plutonium-239, we decompose the mass number 239 as follows:

$$239 = 236 + 3 = (4 \times 59) + 3 \rightarrow 4n + 3, \text{ if } n = 59$$

This Plutonium isotope is known to have 94 protons. With a mass number of 239, the resulting number of neutrons is 145. Accordingly, we note the 4n + 3 series in our Table in neutron core 145.

For the **4n + 1** series that starts with Neptunium-237, we decompose the mass number 237 as follows:

$$237 = 236 + 1 = (4 \times 59) + 1 \rightarrow 4n + 1, \text{ if } n = 59$$

This Neptunium isotope is known to have 93 protons. With a mass number of 237, the resulting number of neutrons is 144. Accordingly, we note the 4n + 1 series in our Table in neutron core 144.

For the **4n** series that starts with Thorium-232, we decompose the mass number 232 as follows:

$$232 = 232 + 0 = (4 \times 58) + 0 \rightarrow 4n + 0, \text{ if } n = 58$$

This Thorium isotope is known to have 90 protons. With a mass number of 232, the resulting number of neutrons is 142. Accordingly, we note the 4n series in our Table in neutron core 142.

We immediately notice a gap at neutron core 143. In order for this core to start a series, its proton *Z* number would have to be 143-52 = 91 for 91 protons. Therefore, its mass number *A* would have to be 234 which lies between 232 and 236. Because the 4n + x expression is already complete with these two A values, there is no more room for another one of one such expressions. We also note the inversion of order for the expression between core 145 and core 146. All this is to mean that this expression is plainly and simply inadequate to describe the symmetries belying radioactive behavior experienced by nuclides below core 124. This behavior is best modeled by *7s-to-7f* orbital ontology.

We commonly know two key facts regarding radioactive disintegration: radioactive nuclides disintegrate either by releasing alpha particles (2 pairs of proton-neutrons orbitals) in a rundown cascade spanning one element after another until a stable nuclide is reached, or by what is called negative beta decay whereby a neutron transmutes into a proton releasing an electron and a neutrino particle. The examination of *7f* orbitals readily explains why beta decay is overwhelming present in all forms of radioactive decays: the sheer greater number of paired orbitals must definitely give rise to a copious number of alpha particles from one step of the ladder to the next until stable end-of-period neutron core 124 is reached back, which is known as Pb-206, Pb-207 or Pb-208. These are the 3 stable elements noted with valence 41, 42 and 43 in row 124. The table readily shows that elements with exclusively paired orbitals have an even neutron number, which knowingly participate very heavily in alpha decay.

Elements with unpaired neutron orbitals provide the singlets that give rise to decay by transmutation of single neutrons. These odd numbers are highlighted in the Table. The lesser amount of these orbitals explains why in all forms of decay beta decay demonstrate a secondary stream status. An examination of the 4 tables commonly illustrating the four decay series is quick to show that this stream is much less numerous than the alpha-particle stream. At the same time, the Table clarifies for us why beta decay, rooted on single neutron particle disintegration, involve nuclides with odd neutron number. Perhaps most importantly, the ontology of *7s-to-7f* orbitals demonstrates why all decay cascade is always both a stream of alpha decay and beta decay in a zigzagging line, never demonstrating a clear-cut between the two. Effectively the deconstruction of these orbitals indistinctly involves both unpaired orbitals in the outer shells and paired orbitals in the inner shells as the disintegration cascade proceeds from one element to the previous.

15.10.3 Intimate Orbital-Shell Origin of Gamma Radioactive Decay

The ontology of orbital 7 objects equally clarifies the intimate origin of gamma radiation in unstable nuclides, the third known form of radioactive decay. While there is a variety of different sources of gamma radiation including Shell2 3-D objects, radioactive gamma decay is a phenomenon that alpha-decay particles are primarily responsible for. Gamma radiation essentially consists of photons with energy ranging from a few dozen *KeV* (kilo-electron-volts) to several hundred to several millions *KeV*. Common nuclear nomenclature indicates that alpha decay occurs when a heavy nucleus releases a helium-4 nucleus in a process that leaves the original nucleus with excess energy that has to be emitted in the form of a gamma ray. In a context where every one nuclide obtained at any step in a decay chain will degrade into the next, one or two levels down, there is not a clear standard for what is called "excess" energy. This statement is no more precise than that one ruling disintegration to be unavoidable result of nuclei having too many neutrons for their number of protons. On the basis of ontology made explicit in Table 1E, what we can see is that all alpha packets springing off the nucleus are located at radial level 7. They are very excited orbital objects, with an important diffusion component, that is. Before they can truly become equivalent to a helium nucleus, two neutron-proton orbital spreads at $1s^2$, they must transition from radial level $7s^2$ to $6s^2$ to $5s^2$... all the way down to $2s^2$ before reaching what might be considered the stable ground level of $1s^2$. At each transition the alpha particle loses energy which is radiated in the form of concomitant gamma radiation. Unlike its cited predecessor in the construction (or deconstruction) context, the $1s^2$ orbital is indeed an end-of-period element, which grants it the necessary enhanced stability to make it the end product of the chain reaction experienced by the orbital packet.

15.11. A Few More Nails in the Coffin of Unphysical Time

We hope to have made it eloquently clear at this point that orbital-7 ontology constitutes the foundation for all nuclear radioactive phenomena. Nuclear physicists have found an equation based on the time variable to account for this process. The proposed radioactive disintegration rate equation follows:

$$N = N_i e^{-\lambda t}$$

where N_i is the initial number of nuclei in a sample, N is the number of nuclei present at time t in the sample and λ the disintegration rate constant for the material. They have further put in practice the half-life figure $t_{1/2}$ to characterize the time necessary for half of the amount of substance initially present to undergo complete decay. Some may even further argue that the practical benefits garnered by this approach from a technological point of view further justify it. Strictly speaking however, this time-driven formalism is simply an interpretation, one must guard against the wild and erroneous view that it represents a natural foundation for time as a dimensional variable.

First of all, there is a narrow correlation between half-lives and disintegration energies of alpha emitting nuclides. The shortest-lived nuclides emit the most energetic alpha particles, and the longest-lived nuclides emit the least energetic alpha particles. This artifact summarily puts any measurement or gradation framework for the phenomenon on the side of process intensity or energy scale rather than the notion of time duration. It follows that any time characterization, should one insist in the peruse of the defunct variable, becomes then fully dependent on energy scale for the energy level transitions underscoring the entire phenomenon, and not the other way around. All surrounding velocity derivates should be graded over or against c.

Secondly, once one has endeavored to uncover the true energy density states behind this phenomenon, one finds that it is entirely based on orbital-7 ontologies, which are statistical objects in nature and not time-driven phenomena. If a time-less orbital object is going to give life to something to be called *time*, then time is at most an epiphenomenon related to deconstruction of objects already cast to life, would only behold local value, and can never harbor the purported value of a dimensional background variable.

No one has ever argued however against time-free quantum analysis of molecular objects or crystal-based materials since the onset of the Quantum Theory. No less than they should find better characterization for nuclear radioactive phenomena than time-driven formalism, now that it is demonstrated that radioactive decay is based on the same orbital map as the well-known electronic orbital shelling.

Indeed, the regularities or symmetries put in evidence in nuclear radioactive decay are entirely based on the wealth of artifacts protagonized by orbital-7 ontologies in distinct core distributions for neutrons and protons. We do not delve into further details in that regard, because the original scope of this publication did not contemplate such studies. The reader must remain fully convinced however that the above study has raised a sufficiently ample umbrella toward this granular treatment. No, we do not need orbitals g, i and other engenders to fully understand the dynamics of atomic nuclei. We now know better. The idea of a time budget for energy are hereby completely dismissed and no one should ever conceive, nor propose, suggest or write that:

$$\Delta E \cdot \Delta t \geq \hbar$$

It is an aberration and an insult to the Quantum Theory.

15.12. Conclusion

The standard that we have set in Quanto-Geometric Theory is that of the *numbers*. To those with core objections to the Theory who wish to propose other new and better ideas, we say: show me your *numbers*. Henceforth all ideas for the advancement of the Quantum Theory shall demonstrate, as we have done throughout this entire proposal, a consistent path to the universal physical quantities known as physical constants. In meeting this standard, we have peremptorily removed from the Quantum Theory all the abstract spaces that have given rise, since its inception, to an endless series of misconceptions and exotic but useless visualizations of physical matter. By the same token we have cured its theoretical weaknesses and infringements against good conceptual practice. An eternal truth: there is no *anomaly* in all of Creation but only in the minds that do not ascertain a broader grasp of its ever-increasing scope. There will be no end to knowledge, even around and about a single object, and life at all levels is forever.

Chapter 16

The difference between something good
and something great is attention to detail.
Charles R. Swindoll

A QUANTUM OPTICS DECONSTRUCTION OF MAXWELL'S THEORY OF ELECTROMAGNETISM AND THE KINETIC THEORY

Both Maxwell's Theory of Electromagnetism and the Kinetic Theory have stood the test of time in mathematical physics and remained consummated exemplars of successful mathematical modeling of physics. Despite the fact that these theories could not give way at any rate to the Quantum revolution in physics, we have not sufficiently identified thus far the caveats that prevented them from paving the way to Quantum Physics, notwithstanding the statistical grounds that the Kinetic Theory in particular share with Quantum Mechanics. Undoubtedly the continuum perspective that surrounds all of classical physics stands for a major element of incompatibility between the Classical vision and the Quantum vision. I show in this Chapter, from an unrelenting quantum perspective, that in a Unified Theory of all physical dynamic systems, the continuum vision is as inappropriate as untrue to nature, in an all-out deconstruction of both Maxwell's Theory of Electromagnetism and the Kinetic Theory. I argue that further advances in photon theory and quantum optics toward complete unravel of photon behavior, firmly debuted in Chapter 4, require this profound deconstructive revision of our understanding of Energy distributions and State topologies as inherited from Maxwell's Electromagnetism and the Kinetic Theory respectively.

16.1. Brief Revisit of Electrostatic Laws

Let us first revisit the basics of electricity and magnetism, as established in classical theory, starting with the first.

16.1.1 First Law of Electrostatics

Since 600 BC, the Greek philosopher Thales had stated the observation of natural physical attraction between substances and had particularly noted this effect to be very notable in the *Elektrum*, a resinous amber substance.

Electricity took its name from the Elektrum and was established as the science of electrization, a phenomenology resulting from putting two different substances in static or rubbing contact with one another, which thereby confers to them the artifact of intangible charges, the so-called electric charges.

The first law of Electrostatic thus stipulated the existence of positive and negative electric charges in all substances, whether in a balanced or unbalanced fashion.

16.1.2 Second Law of Electrostatics

According to the second law of Electrostatics, the equalized number of opposite charges in a neutral substance remains constant in isolation of the substance.

This principle of conservation of charges further implies that the algebraic sum of charges in any neutral substance remains constant under any interaction or transformation that may affect the distribution of charges in the substance. Thus, electrical neutrality is independent of the distribution of opposite charges in a substance.

As a consequence, electricity or the property of attraction or repulsion by a physical substance, otherwise viewed as electrization, is not a property that can be altogether transfused into an object but attributed only by manipulation of the electric charges inherent to the object.

16.1.3 Third Law of Electrostatics

Having identified the existence of two different types of electricity, physicists have defined a convention to characterize the nature of the two discovered forms of charges based on the rubbing substance used to produce *electrization*.

What can be construed as the 3rd Law of Electrostatics stipulates that like charges repel each other and that charges of opposite nature attract each other (Fig. 8.1).

16.1.4 Coulomb's Law

It is natural to think that the total amount of electricity or charge Q possessed by a substance is quantified by the algebraic sum of all charges, positive and negative, in that body. It is as much natural to consider that the strength of the electric interaction between two bodies, whether in the form of attraction or repulsion, is quantifiable and dependent on the number of net charges experienced by each body (Fig. 8.1).

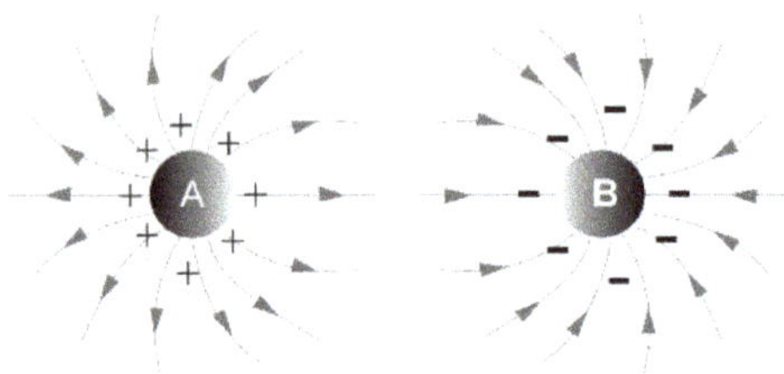

Fig.8.1 Electric monopoles

Coulomb's Law numerically describes the electrostatic force F between two bodies as follows:

$$F = k\frac{QQ'}{s^2} \quad (16\text{-}1)$$

where k is primarily an electric constant characterizing the electric permittance of the medium between the bodies, or its dielectric property, Q and Q' the charges on the two bodies, and s the distance between them.

The similarity in the numeric expression between Coulomb's law and Newton Second Law had not evaded physicists' attention throughout the history of physics. There are good reasons to this comparison, although physicists have been thus far notably unable to resolve it theoretically.

16.1.5 Gauss Law of Electric Field

The notion of a force being exercised between two charged bodies naturally led to the idea of that force being materialized in the form of lines of force or electric flux in the immediate region circumventing the charged body (Fig. 16.1). Hence Carl G. Gauss came to formulate a relation between the net electric flux circumventing a charged body and its enclosed electric charge as follows:

$$\oint E \cdot dA = \frac{q}{\varepsilon_0} \quad (16\text{-}2)$$

where ε_0 is the permittivity of the medium surrounding the object, most commonly empty space or air. This relation, known as the *Gauss' Law for electricity*, expresses equality between the integral of the electric flux within the spherical circumventing area to the charge and the ratio of the charge and permittivity of the medium. In other words, Gauss' law states that the total number of lines of flux composing the electric field radially from a closed charged surface is equal to the total positive charge enclosed in that surface.

Indeed, more than a natural law this theorem is a convention that seeks to set the basis for a numeric treatment of the physics of electric charges and their fields. This convention is accompanied by another of dimensional nature which assigns an arbitrary value to ε_0. According to that convention, the intensity of an electric field of *1 Newton/Coulomb* is represented by 8.85×10^{-12} lines of force that are normal to a surface of $1 m^2$ on the basis of the MKS unit system. The intensity of the electric field or its flux in a spatial point is thus represented in the number of lines of force per unit surface traversing a surface normal to the field in the vicinity of that point.

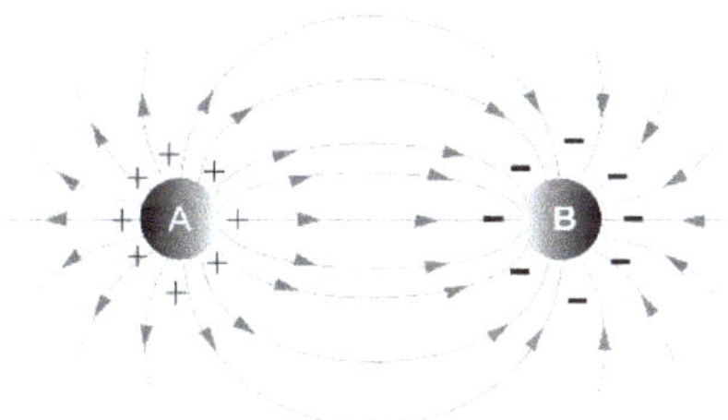

Fig. 16.1 Electric fields created by two proximate equal but opposite charges

In the Maxwell's formalism of the Gauss Law for Electricity, he attributes the $\nabla \cdot$ operator to the divergent radial electric flux generated by the charged surface and proposes the following nomenclature:

$$\nabla \cdot E = \frac{1}{\varepsilon_0} q \quad (8\text{-}3)$$

where $\nabla \cdot$ is a differential divergent operator.

16.2. Empiricism of Gauss's Law for Magnetism

The rules or laws above analyzed for the phenomenologies surrounding the electric charge result from experimental observables which in theory came to be shaped numerically for computational purposes. While one ought to acknowledge that Gauss law for electricity is a mere and complete convention, the Coulomb law takes the shape of a more natural formalism and its similarity with the Newtonian gravitational relation certainly adds to its naturalness as it were.

When it comes to magnetism, experimental observation yields differing phenomenologies from those cast by charge and electric fields, contrarily to what one might expect of a phenomenon that shows the likes of similarity with that of charge and electric fields. When we examine the electric field created by two opposite charges that are equal in magnitude, one sees an identical pattern to the one developed by a bar magnet, in terms of the distribution of lines of force around the body of the magnet. Accordingly, one would expect that the laws governing the phenomenon of magnetism would be a faithful replicate of the laws governing electricity. However, such replicates do not exist.

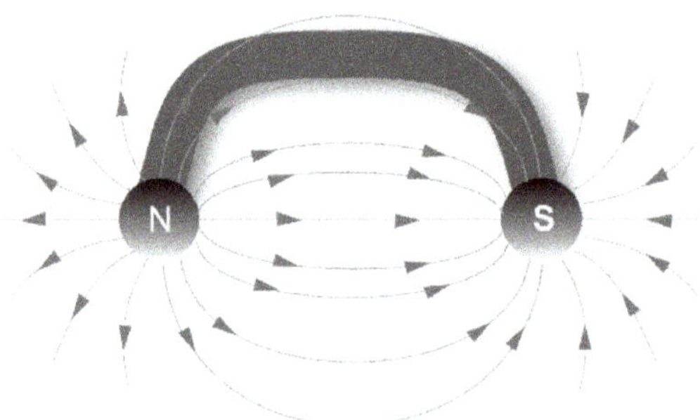

Fig. 16.2 Dipole of a horseshoe magnet

Remarkably enough, the simplest magnetic structure that can be enclosed by a Gaussian surface is not a unipole akin to the case of charges, but a dipole (Fig.16.2). For instance, we can perfectly have an entire spherical body with a distribution of the same charges, all positive or all negative, over its surface. However, no such magnetic singlet charges in a distribution over an entire closed surface has ever been found. In other words, there does not exist any magnetic monopole. Magnetism only exists in a bipolar form or a dipole.

Gauss's Law for magnetism asserts this fact by stating that the net magnetic flux Φ_B thru any closed Gaussian surface is zero:

$$\phi_B = \oint B \cdot dA = 0$$

In the Maxwell formalism, it is expressed with the differential $\nabla\times$ operator as follows:

$$\nabla \times B = 0$$

While this statement remains in the form of a law in Maxwell's formalism, it must not deter us from raising the question as to why this law is anti-symmetric to the Gauss' Law for electricity. Although hitherto never offered in theoretical physics yet, a satisfactory answer to this question is not only relevant but mandatory if the Maxwellian formalism impersonated in the $\nabla\times$(curl) operator, as much as the $\nabla\cdot$ (dispersion) operator, must be upheld.

Maxwell-Gauss second law amounts to an insufficient interpretation of magnetism because it does not address the reasons why magnetic monopoles do not exist, as do electric monopoles. If one is going to find a parallel between electricity and magnetism as proposed in the unifying Maxwell Theory, then a plausible answer to that question must be peremptorily furnished. Such a formalism ought to make abundantly clear why magnetic monopoles are not in existence whilst electric monopoles do exist.

Between mid-1800's and early 20th Century, in an attempt to define the best unit system to properly express electric and magnetic phenomena, and in a way that is compatible with the units long is use in mechanics, physicists in the Standard Bodies have invented many things to that end, one more unfortunate than the next. The concepts of *permittivity* and *permeability* of the vacuum had been invented to offer a numeric rendition of how the vacuum reacts to the presence of the fields. Further the product of these constants has been arbitrarily related in such a way *(8-3b)* to produce the then known value of the speed of light, since the electromagnetic wave was measured to travel at that same speed. Changes in the experimentally measured value of the speed of light, which has a long history of measurements as a function of technological improvement and refinements in optical science, prompted corresponding adjustments in the values of these constants. It is my unreserved opinion that physicists tend to overvalue Maxwell's equations in their inquest for physical structures and kinematic trends in dynamic systems in general.

Maxwell's formalism seizes these two faux concepts, the permittivity and permeability of the vacuum, to build on, making good on the artificial relation:

$$c = \frac{1}{\sqrt{\mu_0 \varepsilon_0}} \quad (16\text{-}3).$$

More than helpers to the Gaussian unit system, these artificial classical parameters became an intricate part of the theory, to the point that many a modern theorist unjustifiably believe that the speed of light had *naturally* dropped off Maxwell's equations! It is well recognized however in modern theory that these constants say nothing about the intimate structure of the vacuum and do not at all constitute depictors of the vacuum.

In all, if Maxwell's theory of electromagnetism had the towering value that physicists have sanctimoniously credited it with historically, it would have to have been built in such a way to instead derive the fine structure constant α and notably the unit of charge e, these quantities having been widely associated with particles of matter and electromagnetism. The following cannot be overstated: *the dynamics of the magnetic field and the electric field maintain only a concomitance/correspondence relationship, not a direct causal or covariant one, and do not belong to the same gauge group assumed in Maxwell's theory of "electro-magnetism".* Given the overwhelming influence of Maxwell's field equations in physics, which had even cleverly been re-written later as a single equation in tensor form, the above assertion spells out one of the major causes of the long-running stagnation in theoretical physics! Field theory, which decisively peruses the artificial constants of permittivity and permeability of the vacuum, is not the best way to approach the dynamics of the vacuum if the objective is to uncover its hidden fundamental symmetries.

Manifestly, Maxwell's equations only constitute a phenomenological description which fails to comprehend the underlying physics of the vacuum and ignores the dynamics of the hitherto unknown quanta that are being played over its tapestry. That is why the theory cannot explain the reason why the void of the vacuum can transport the so-called electromagnetic wave, a phenomenon to this day unexplained.

16.3. Maxwell's Vision of Material Displacement

Maxwell had further introduced the somewhat etherish notion of *displacement current* to provide for a broader understanding of electro-magnetic dynamics. Specific laws for specific applications by J. C. Maxwell's predecessors could not predict the existence of magnetic wave effects due to the lack of this apparently instrumental ingredient.

The central idea in the physics encapsulated in Maxwell's Theory of Electromagnetism is that of motion being the intimate cause of magnetism as well as qualified electric current and since that motion is undertaken by the electric charge, electromagnetism or the electromagnetic wave ensues as a consequence of displacement. That motile displacement, which may be cursive or dispersive, he encapsulated in the ∇ operators. It is fair to note that these operators have transcended the framework of electricity and magnetism and unduly became models in physics for motion and the visualization of other physical phenomena quite distinct from the former.

16.4. Maxwell's Triple Merger of Electricity, Magnetism and Optics

That Electromagnetic energy travels in free space at the speed of light provided Maxwell the basis for unifying the theory of electromagnetism to well-known geometric optics. In the eyes of generations of theoreticians, the triple merger of electricity, magnetism and optics constitutes a triumph and the epitome of the Maxwell Theory. However, the inability to provide an intimate understanding of the structural elements making up the diverse fields involved, beyond their phenomenologies, was already an indicator of the limits of the Theory, which became quite clear with the transformational advent of Quantum Theory, the latter of great success in elucidating the **discrete** spectral distribution of light (or so-called electromagnetic radiation) experimentally observed.

The lack of this critical assessment has led modern theoreticians to model gauge theories after the electromagnetic field representation formalized by Maxwell. The idea of the invariance of electromagnetic laws within any chosen potential differential within the field, from which gauge invariance is modeled, led to the erroneous idea that the vacuum is the same everywhere, and that the laws of physics remain invariant over the tapestry of any vacuum. A divorce with those views is for certain necessary if theoretical physics must gain insight into and assertiveness about the structure and activity of the vacuum. Gauge theories inspired by Maxwell classical view have recently led to one of the most grueling failures in physics insofar as the attempt to computationally characterize the vacuum, with the calculation of the cosmological constant, aka the vacuum energy density. The unfathomable computed number has been characterized as a **"cataclysmic failure for physics"** by one very authorized voice and by another **"the most notable predictive failure in the history of physics"**.

16.5. The Electromagnetic Wave: an Artificial Construction

The electromagnetic wave is an artificial human artifact of a different nature than light and even more so than the photon particle, both of which are principally native entities in Nature. It is important to understand that Maxwell's classical view of radiation leads to the visualization of wave mechanics being developed from the tapestry of a continuum. That stands for the preeminent reason why it could not decipher the radiation spectrum emitted by black bodies and subsequently energy density distributions in atomic elements as experimentally revealed in distributed optical spectral lines.

Key to the energy structure in those distributions is the quantum packet well encapsulated in the photon parcel visualization. The photon particle is not known to bear an electric charge, nor does it diffuse an electric field. It is known to be an electrically neutral particle, estranged to all electric artifacts structurally. Physics of the photon cast it as the quantum of light radiation and the sole point constituent of light rays or beams. Hence light is not an intrinsically **electro**-magnetic phenomenon per Maxwell's Theory of Electromagnetism.

The above arguments must be taken as a scathing counter-truth to Maxwell's assessment that light is of electromagnetic nature. Because the electromagnetic wave propagates at the speed of light does not automatically make light an electromagnetic wave, no more than a balloon that moves with the wind at its speed means that the balloon is the wind or a form of wind. The electric field that may or may not accompany the production of our artificial sources of light are not nascent to the beam and do not constitute an intrinsic part of the light beam. Despite the seduction of Maxwell's mathematical treatment of these concepts, we must endeavor to theoretically extricate ourselves from these erratic views.

16.6. Why Do We Have a Natural Unit of Charge e But Not a Natural Unit of Magnetic Flux?

The absence of the discovery of a natural unit of magnetic flux had seemingly led to the creation of both of the artificial constants of μ_0 and ε_0, the vacuum permeability and permittivity constant respectively. If magnetism and electricity are codependent or covariant phenomena, then the existence of a natural unit of charge should be naturally matched with the existence of a natural unit of magnetic flux. There is no reason why the same vacuum spread should not behave symmetrically in reaction to the presence of the two forces or even ontologically if the two forces are nascent to the vacuum itself. Because Maxwell's Theory pre-supposes this symmetry which does not exist empirically, it has been forced to adopt and consolidate the use of the two artificial constants, which it cleverly relates to one another in the dubious relation, above reported *(8-3b),* involving *c*.

Anti-parity in the laws of electricity with respect to magnetism strongly suggests that the two phenomena do not belong to the same physical scale, even though they may otherwise be inter-dependent.

16.7. Significance of Shell Hierarchy in the Development of Matter

It is a regrettable fact, heavy in consequences, that throughout the development of physical science, physicists have been impervious to the Shelling evolution of matter, which is nonetheless an ostensible observable.

As I have explained in Book I with a profusion of numeric details on axiomatic basis, from galaxy groups, to galaxies, to solar systems, to planets and stars, to lunar satellites, to materials lattices, to karyotes, to molecules, to atoms, to elementary particles and to quarks, the universe is organized in Shells in a way that an association between the previous elements of the hierarchical suite comes to form the building block of the next element in the suite (Fig. 16.3 bis). The two most basic elements of this notable development have never been understood, which is the evolution of mass, defined as scalar density, and the evolution of the vacuum, unfortunately always disregarded for nothingness, throughout the hierarchy. One should not attempt to understand behavior and properties of interactive elements in one Shell from the properties of its inner constituents, because these must remain ***invariant*** as a condition for the structural stability of the very Shell. Furthermore, the qualities and properties of the (upper) Shell element are always very different from those of the inner constituents because this is how mother Nature creates its own evolution (thru symmetry breaking). The vacuum plays a seminal role in this transformation, while this overarching Shell symmetry frames the notorious symmetry axiomatized by Emmy Noether (Noether's theorem) and the laws of conservation in total.

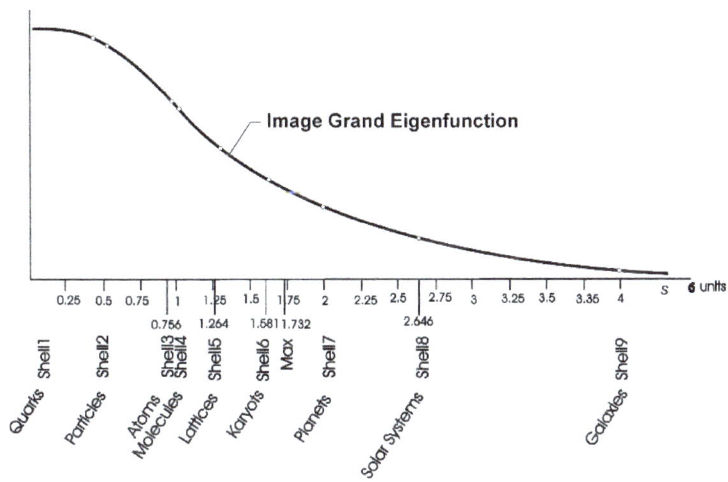

Fig. 16.3 A view of the evolving suite of Shells of 3-D matter. Units in σ.

It turns out, from theoretical numeric derivations emanating from the Quanto-Geometric framework exposed in Book I, that magnetic phenomena are native to the elementary particle Shell of matter, whereas electricity is native to the molecular Shell of matter, surprising though this might be to most theoreticians. It is the misunderstanding of the role played by the vacuum in the hierarchical Shelling development of matter, in terms of fashioning the fundamental symmetries on which each Shell is built, that is wholly responsible for the erroneous perception of the modality in which both magnetic and electric phenomena develop in Nature. Thus far, it has remained common wisdom in physical science to consider the vacuum to be the same everywhere, throughout all Shells of matter that is, and that laboratory experiments concerning any Shell element can be conclusively conducted from any other Shell indistinctly, such as the Astral Shell which is where we and our laboratory experiments live. A certain sense of the Shelling has only started to develop recently with the new experimental science of nanotechnology. The ignorance of the specific tensorial properties of the vacuum in each Shell and the fundamental scalar-space covariant relationship on which each Shell is built had resulted in the attribution of physical properties natively belonging to one or the other of the higher Shells to elements germane to the lower Shells under investigation. We have amply and severally discussed those symmetries. Suffice it to reiterate here that what we call the *fundamental physical constants* are no other than the invariants that sustain the transformations taking place in a Shell in the role of the very building block of the Shell.

In accordance with the above, and as derives from the Quanto-Geometric functional framework, the fine structure constant α represents the building block of the Elementary-Particle Shell, namely the primary suite formed by neutrons, protons, muons, electrons, neutrinos and photons. Whereas the fundamental physical constant represented in the Coulomb unit of charge e stands for the building block of the Molecular Shell.

16.8. The Fine Structure Constant α as the Magnetic Constant

From the Quanto-Geometric functional framework, the fine structure constant arises derivatively as the scalar-space invariant that sustains the Particle Shell of matter. What we have denoted the Particle Shell exhibits, as previously mentioned, a parcel spectrum composed of neutrons, protons, muons, electrons, neutrinos and photons, indeed a first-order Quanto-Geometric Spectrum.

The value naturally cast by the framework for this fundamental invariant is: $\boldsymbol{0.1164346625 \cdot \sigma^{-3}}$.

If we evaluate the value of α from the following formulae among the many known to yield this quantity:

$\alpha = \dfrac{e^2/\hbar c}{4\pi\varepsilon_0}$ or by substitution of *(3b)* $\alpha = \dfrac{\mu_0 c e^2}{2h}$, where h is expressed in eV-s, we obtain:

$$\alpha = 0.116916654 \cdot 10^{-20}.$$

Given that that μ_0 and ε_0 are imprecise and largely arbitrary quantities, it is not surprising that the numeric match falls off after the 4th decimal in the figure. Most physicists are familiar with the value of $7.2973525698 \cdot 10^{-3}$ for α. Yet if we divide that value by 2π we obtain:

$$\frac{\alpha}{2\pi} = 0.11614097 \cdot 10^{-2} = a$$

The interesting tidbit about that *a* value is that it is known to be the electron *anomalous magnetic moment*. Keep in mind that despite the epithet there is nothing anomalous about this magnetic moment, if not the fact that it does not correspond to the theoretical or expectation value that is computed under a classical definition of a magnetic moment for the electron. The correction, denoted the *g*-factor or *g* correction factor, was necessary for the theoretical value to match the experimentally measured value. Nonetheless, it is a very probing fact that a simple division by 2π, a mathematical scalar, transforms the nominal figure into a **magnetic** quantity, in an irrefutable proof that the fine structure constant is a seminal magnetic constant in nature. This figure is attached to every elementary fermionic particle, taking values that are quite close to the same quantity per experimental measurement. Indeed, α is the building block underlying the symmetry upon which the entire spectrum of elementary particles, namely neutrons, protons, muons, electrons, neutrinos and photons, develops.

In that regard, it is worth stressing that the figure of the magnetic moment does fully apply as well to the neutron particle, which is however known to be an **electrically neutral** element! This fact, while confirming our hypothesis, further advocates against the view of direct covariance between magnetic phenomena and electric phenomena. Physical science strongly anchored on Maxwell Theory of Electromagnetism has never been able to furnish a satisfactory explanation to the paradoxical magnetic moment of the electrically neutral neutron particle, except for the dubious conjecture of fractional charges attributed to the quark sub-constituents of baryonic particles in general in the Standard Model, which is far from consensual among theoreticians.

16.9. Parenthood Between the Fine Structure Constant and the Sackur-Tetrode Constant

The figure of the Sackur-Tetrode constant is important in the context of this study for two main reasons. First it shows an even broader numeric match with the corresponding axiomatically derived figure in the Quanto-Geometric formalism. The experimental value of this constant is: ***0.11648**709·10* in conditions of pressure $p = 101.305\ kPa$ (or *1 atm*) and temperature $T = 1K$, with the following formula, already reported in Book I:

$$\frac{S_0}{kN} = \ln\left[\frac{V}{N}\left(\frac{4\pi m}{3h^2}\frac{U}{N}\right)^{\frac{3}{2}}\right] + \frac{5}{2},$$

where S_0 is the entropy of the gas, V is the volume of the gas, k the Boltzmann constant, N the number of gas particles, m the mass of each gas particle in atomic-mass unit and U the internal energy of the gas. In the portrayal of the ratio S_0/kN, it expresses the entropy of a monoatomic ideal gas with sufficient accuracy even at temperatures that can be quite high. It is clear from the formula that if the term of the logarithm *ln* becomes equal to 1, then the logarithm will take a 0 value, falsifying the entire quantity. This will occur when N is very large with respect to U or V in particular. This spells out the condition that the average inter-atomic distance within the gas become significant with respect to the size of the atom, or what is denoted the thermal De Broglie wavelength. The latter has the following expression:

$$\frac{1}{\Lambda^3} = \sqrt[2]{\left(\frac{4\pi m U}{3h^2 N}\right)^3},$$

where Λ designates the thermal De Broglie wavelength. One can further note, due the presence of h in the expression, that this wavelength quantity is of a quantum nature. Despite the fact that in the quantum regime, the Sackur-Tetrode equation as a whole becomes invalid, remarkably enough it remains the root of the new expression that applies in those cases. The quantum regime contemplates two states, one where the gas particles are indistinguishable from one another, yielding the Bose-Einstein type of population (bosons), and the other where the gas particles are distinguishable from one another, yielding the Fermi-Dirac type of population (fermions).

The above considerations lead us to the second reason why an analysis of the Sackur-Tetrode constant is noteworthy to this study. When attempting to provide an explanation of what drives the entropy of a population of quantum-scale particles, the notion of thermal De Broglie wavelength acquires capital significance. Effectively, notwithstanding the utility and effectiveness of statistical physics as a model, it does not provide an ontological interpretation of entropy or the determinants behind entropic behavior of dynamic systems. The notion of thermal De Broglie wavelength takes us one step closer to the Quanto-Geometric wavefunction, broadly discussed throughout this entire publication. In an ideal gas, the reason why particles move in all three spatial directions is not because of an innate kinematic engine of their own, but because that kinematic activity is innate to the surround spatial void to each, which drives and determines the whereabouts of the particle. The spatial void associated with the particle is both external and internal to the particle. As stipulated in the Quanto-Geometric axiomatic, the generic tapestry of spatial void spreads is based on a wavelength metric that expresses the innate kinematic of the entity.

16.10. Dimensional Analysis of the Classical Frequency Figure in Relation to the Planck Constant

When it comes to the nature of the so-called electromagnetic spectrum, it is important to make a clear distinction between, on one side, natural light such as combustion light, phosphorescent and plasmic light, artificial light produced with the means of quantum optics (i.e., lasers), and, on the other side, man-made light resulting from alternating currents thru resistive components or gaseous materials.

The latter in the above enumeration is likely to be the one and only one proper to the characterization of *electromagnetic radiation*. Much more than the very J. C. Maxwell, champion of this development was Nikola Tesla, who notably discovered the means to produce artificial optical radiation (or artificial Light) and radio transmission alike from alternating currents and **Tesla** coils, and further demonstrated prodigious genius in inventing a string of deriving applications. Credit to Maxwell must be granted only to his successful formal attribution of similarity in propagation properties between Light waves and man-made radio radiation. It is clear that, except when alternating currents are used in the production of light, there is no electromagnetic nature to the source of artificial radio waves from which the ensuing radiation could possibly inherit this quality. Irrefutable proof of the latter is that direct current thru a wire, which **in its flow** is devoid of electric potentials or voltage gradients and creates no travel electric fields, can produce light in the same devices that alternating currents can. Only a static magnetic field void of oscillation or frequency feature is created in this case. Therefore, the least of permissible conclusions is that while the electromagnetic radiation may resemble light radiation in certain aspects, light is generically **not** an electromagnetic radiation, contrarily to the Maxwellian proposition.

The electromagnetic radiation is irrevocably associated with oscillation or *frequency* defined on the basis of the eternal sinusoid. Not so when it comes to any of the aforementioned other types of light. Combustion light, phosphorescent or fluorescent light, plasmic light, laser light, or chemically induced light are all quantum optics phenomena deriving from energy level transitions experienced by particles or within atoms, molecules or material lattices. These entities know of no time flow and develop dynamical behavior only statistically construed. This remains all true, even though the mathematical formalism used for this treatment in technology, in particular laser technology, unfortunately remains impervious to that important fact.

16.11. From Electromagnetic Frequency to the Planck Constant

One of the most important unidentified problems that exists at the foundation of the Quantum Theory is the expression of the primordial quantum constant, the Planck constant h, in units of time, as *Joules-sec* or *eV-sec*. If it is true that an atomic orbital as an energy density distribution is enacted by an electron with no momentum (with motion that is not a function of time), then the fundamental unit that scales that energy density distribution cannot conceivably be a time-driven engender. Dimensional analysis starting with the towering A. Einstein's quantum relativity formula, also denoted the Planck-Einstein relation, will be keen in demonstrating that contention.

$$E = h\nu \quad (8\text{-}3c) \quad \text{or} \quad h = \frac{E}{\nu} \quad (16\text{-}4)$$

where ν if the frequency. Expression *(16-4)* means that there is an invariant relationship between the energy E of a radiation mode and the particular frequency of the radiation mode ν housing that energy. That makes h a constant of "Energy per unit of frequency".

A more encompassing view from that vantage will cast h as a constant of Energy distribution per frequency in a radiation spread with defined boundaries. In other words, this view spells out how energy is distributed over a semi-finite range of frequencies, which is at most $\left[a, \infty\right[$[6]. Planck's formula for energy density of the radiation [E(λ)] per wavelength follows:

$$E(\lambda) = \frac{2\pi c^2 h}{\lambda^5 (e^{ch/k\lambda T} - 1)} \qquad (16\text{-}5)$$

In practice the aforementioned intervals are even more reduced at the upper end because of the limits imposed by thermodynamic temperature (Fig. 16.4 bis).

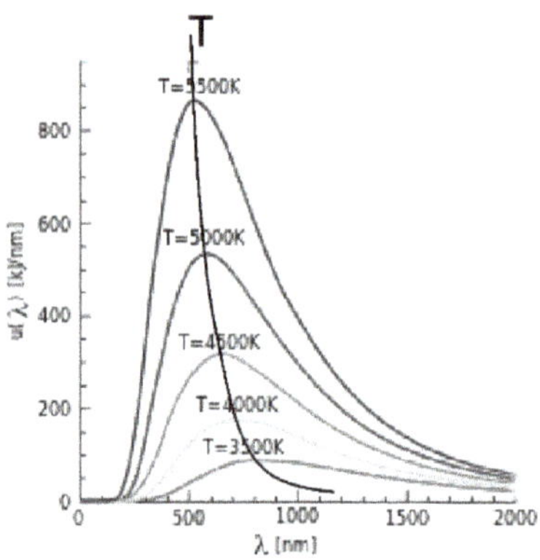

Fig.16.4 Plots of radiation energy density v. wavelength

So therefore, h stands for Energy per frequency or Energy Density over the frequency scale. From frequency defined as oscillation in unit of time *(sec)* a Period T ensues such that:

---

[6] Note that there is always a higher limiting value of frequency much before the neighborhood of infinity in the spectrum range. The lower limit a always exists in all spectra as a non-zero value.

$$\nu = \frac{1}{T} \quad (16\text{-}7)$$

By replacement of *v* in *(16.7)* we obtain

$$h = E \times T \quad (16\text{-}8)$$

We may elect to express *E* either in *Joules* or *eV* (by conversion) while *T* is expressed in *seconds*. That is how and why the dimension of *h* is borne out in *Joules – sec* or *eV – sec* as officially mandated. However, one can see that despite the fact that its unit-dimension is a product *(Joules-s* or *eV-s)*, *h* is in fact a constant of **Energy Density**. In spite of A. Einstein's original attribution of the frequency figure to the radiation mode that knocked the electron off the metal surface in the photoelectric effect, for which he composed this formula, the ambiguity long remained in physics about the application of the frequency figure to both the radiation and Einstein's hypothesized photon as the quantum unit of said radiation. Modern physics nowadays tend to apply the formula to the photon of the radiation. Nevertheless, in one way or another, there are problems arising from any of the two attributions.

Applying the formula to the photon as the quantum energy packet of the radiation is a rather abusive exercise because, thus far, we do not know at all what is oscillating within the particle or the nature of its harmonic behavior: the quantum theory is to this day still devoid of a morphology definition for the photon as much as a representation of its ultra-structure.

Should we instead decide to apply the frequency figure to the radiation mode, we cannot use relation *(8-4b)* for the expression of the Period any longer but we ought to define the Period in relation to the propagation speed of the radiation and express the Period as a wavelength in space instead, per the expression below:

$$f = \frac{c}{\lambda} \quad (16\text{-}9)$$

where λ is the wavelength or harmonic unit of space. By replacement of *(16-9)* in *(16-8)*, keeping mind that *v* is another symbol for *f*, we obtain:

$$h = \frac{1}{c} \cdot E\lambda \quad (16\text{-}10)$$

Since c is expressed in *m/s* and λ is expressed in *m*, the unit of h remains *E-sec* or *Joule-s* and *eV-s*. However, this form of the equation *(16-10)* means that h is dependent on Lorenz invariance and that it is a product of discrete Energy and harmonic unit of space. Expression *(16-10)* thus explains that for h to remain a constant, there must not only be a co-variant relationship between the energy E of a photon and its wavelength λ since c is a constant, but the expression imposes a finite wavelength spectrum to the definition of h such that $0 < \lambda < 1$. This finite wavelength spectrum is incompatible with the quasi continuum of applicable frequencies left open with expression *(16-4)* of the Planck-Einstein relation. The latter expression only forbids $v = 0$, since the ratio h/v becomes undefined in that case.

It is important to stress at this point that the very notion of h as distribution of energy density over the harmonic or oscillatory scale is incompatible with that of frequency continuum implicated in the generalization of electromagnetic oscillations. From Maxwell classical vantage, any electromagnetic wave of any frequency value can be generated without any restriction other than empirical means, and the distribution of energy over that generalized electromagnetic spectrum must be furthermore linear. Importing the actual notion of frequency to the analysis of observed light radiation, defined as previously stipulated, faces the challenge of bound frequency spreads or finite spectra, which ultimately means that the associated oscillations do not develop in the continuum of time but in units of space or on the space scale. So therefore, when it comes to the quanta of the electromagnetic wave, these constituent energy packets (photons), which undergo a finite distribution throughout the oscillatory tapestry of the radiation, shall not and cannot be individually described in terms of oscillation per second or frequency (*Hz*) because the distribution is squarely independent of time.

One of the most significant supportive arguments to this result is the specific observation of the photoelectric effect whereby the emission of electrons caused by incident radiation with the correct harmonic measure is immediate, and does not require any build up time for energy E to be accumulated in the material before the effect takes place, as wrongly expected in classical Theory. In addition, and as a reminder, in the classical view, for most dynamic scenarios, Energy at first order is a function of time. Classical theory had long bucked at the striking immediacy effect, well observed empirically in photoelectric interaction, without ever being able to explain it in the least amount.

A further confirmation of our view is offered by the Compton effect which the formula in *(16-4)* assertively and thoroughly interprets, if the harmonic characterization is applied to the photon as the quantum of *X*–ray radiation instead of a putative electromagnetic *X*-ray wave. Only if we attribute an equivalent mass to the photon that pairs with its quantized Energy as modeled by *(16-4)* can we then assume a momentum for the *X*-photon particle which ultimately explains the angular dispersion and wavelength drift observed to the photon particle after colliding with the electron particle (Fig. 8-5b). We set the mass-laden Energy of the particle to be equal to its harmonic-driven Energy as shown below:

$$E_m = mc^2 \text{ and } E_\lambda = \frac{hc}{\lambda}.$$

Setting $E_m = E_\lambda$, we obtain the momentum p of the travel photon particle:

$$p = mc = \frac{h}{\lambda} \quad (16\text{-}11)$$

Note that the photon does not have an explicit mass, therefore the *mc* product is heuristic and only meant to lead us to the expression of h/λ for the momentum. Once we arrive at the momentum characterization for the particle, we may then invoke an elastic collision with the electron, which ultimately leads us to the final relation:

$$\lambda' = \lambda + \frac{h}{m_0 c}(1 - \cos\theta) \quad (16\text{-}12)$$

where λ' is the observed drift wavelength of the photon particle and θ its observed receding angle[7].

Unequivocally, no momentum figure can be applied to a wave function such as that characterizing an electromagnetic mode. Therefore, it is clear that the harmonicity invoked in the Planck-Einstein relation with ν is not your classical electromagnetic frequency but a wavefunction applicable to the energy packet making up the quantum of that radiation. Furthermore, that harmonicity is proven best described with the spatial wavelength figure λ as a descriptor of a material wavefunction animating the particle. This visualization has proved to be so useful that it has been extended to qualify the wavefunction of quantum parcels that are manifestly principally particles (fermions) but exhibit wavefunction properties as a minor. All quantum parcels or particles have been experimentally found to exhibit wavefunction properties that have been characterized as their Compton wavelengths. The Compton wavelength of a fermionic particle is defined to be *equal or equivalent to the wavelength of a photon exhibiting an amount of energy equivalent to the mass of the particle.* Despite these advances, we have continued to ambiguously use in physics the Maxwellian figure of *f* or ν to qualify harmonicity, undoubtedly due to our intractable inability to extol the seminal nature of the material wavefunction characterized with λ.

---

7 A complete demonstration may be reviewed in most physics textbooks.

16.12. Empirical Derivation of h from X-ray Radiation Spectra

It is noteworthy that the Planck constant may be derived from *X*-ray spectra (Fig. 8.5a) instead of visible optical spectra, given the fact that *X*-ray harmonicity is only amenable to the concept of the material wavefunction characterizable in units of space and not the classical frequency figure. This exercise further provides a direct means to express the unit-dimension of the constant in a completely time-free fashion by eliminating the use of the *Joule* figure.

16.12.1 Experimental Setup for Obtaining X-Rays

X-rays are very energetic radiation with wavelengths situated between 10^{-8} and 10^{-11}m. One way to obtain X-ray waves is to accelerate a flow of electrons in an evacuated chamber onto an angled metal plate thru the application of a high electric potential, per Fig.16.6. These fast electrons collide with the atoms of the metal target for the most part and lose energy via thermal release.

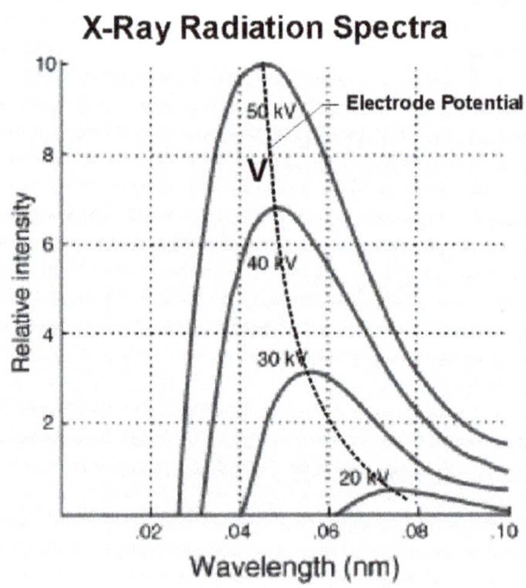

Fig. 16.5 X-ray radiation spectra

For those electrons that are instead decelerated, there is a classical expectation of a release of electromagnetic energy in all directions, with the maximum frequency depending on the electric potential. It is experimentally found, however, that there is an uncharacteristic limit on the high frequency side of the emanating X-ray radiation. The semi-finite spectrum (Fig. 16.5 bis) can only be explained on the basis of the hypothesis of the photon in this way: an electron can be decelerated several times while approximating the atoms of the anode material so that an X-photon is emitted every time with an energy: $E = h\nu$ or more exactly $E = h/\lambda$.

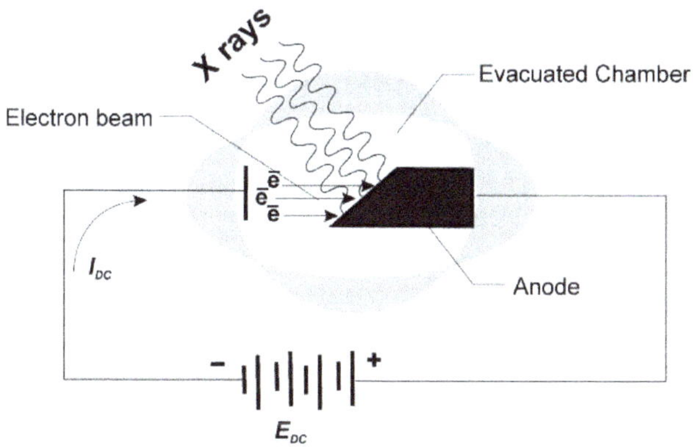

Fig. 16.6 Production of X-rays from a continuous electron beam (DC current)

In fact, this Energy corresponds to the decrement in the kinetic energy of the particle experiencing deceleration, expressible as:

$$\Delta E_k = \frac{h}{\lambda}.$$

The most energetic photon possible out of the emission is the one that is produced by a complete conversion of the drop in the kinetic energy of the electron. Since the electrons hit the metal target with energy $E = eV$ the energy of the most energetic photon is captured in the following relation:

$$eV = \frac{h}{\lambda_{max}} \quad (16\text{-}13) \quad \text{or} \quad \frac{e}{h} = \frac{1}{V\lambda_{max}} \quad (16\text{-}14)$$

where e is the electric charge of the electron and V the accelerating potential. This relation *(16-14)* translates what had been denoted the Duane-Hunt law. The quantity e/h may be calculated directly from the distribution plots and since e is well known, h can thus be derived, yielding a value that perfectly accords with that obtained by other methods. The Duane-Hunt law, little evoked in the literature, has the additional important virtue of consigning h with unit-dimensions totally free from time dependency as:

$$h = eV \cdot \lambda \quad (16\text{-}15),$$

thereby showing h in eV-m dimension, a more authentic dimension for the constant in the light of the premises of this discussion.

16.12.2 Electromagnetic Continuum is a Misrepresentation

This demonstration must cement the proposition that the very notion of the photon contravenes the idea of electromagnetic continuum of frequency applicable to optical radiation, despite the perception entertained by the Quantum Theory's pioneers.

Fig. 16.7 Cyclical structure of the tonal character of the radiation spectrum, clearly observable in the visible sector

When considering a spectrum of optical radiation, the artifact of harmonicity must be applied to the quanta of the radiation spread and that harmonic behavior shall not be consigned in frequency units as measured in *Hertz*. **There does not exist any electromagnetic continuum that joins the artificial man-made electro-magnetic radio alternations to the spectrum of natural radiation spreads**. The latter never come to be observed in nature as a continuum but as finite energy spectra. To start with, if one were to simply cater to the visible section of the optical spectrum, it is manifestly evident that the tonal character of the radiation is not linear but cyclic, since the near infra-red wavelength region and the near UV wavelength region inarguably relate to one another thru a tonal reddish (or purplish) character to the radiation that calls for cyclicity throughout that finite span of radiation. This fact is very well known in colorimetry but not sufficiently interrogated parallely on the radiation plane (Fig. 16.7). We argue that this cyclical structure must be extended to the entire wavelength spectrum of natural radiation.

16.12.3 No Electromagnetic Phenomenology in X-Ray Radiation

It is opportune to point out that the famous X-ray experiment is one that obliterates the idea of electromagnetism in the nature of radiation, owing to the important fact that the accelerating voltage which ultimately causes the radiation is not of alternating nature. Consequently, there is subsequently no possible presence of an oscillating envelope to the emitted X-ray radiation that could be construed as the modality of the radiation, as is otherwise the case in the man-made antenna-based or cavity-resonator-based radiation. All harmonic behavior can only be brought to bear on the quanta of the X-ray radiation. Arguably, if this feature is true for a finite spectrum, so it must be for the entire domain to which all partial spectra belong, lest the entire domain is disjoint.

16.12.4 Inexistence of a Joint Double Scale of Harmonicity

Furthermore, and from a generalized viewpoint, it is not possible to create a coherent double scale for radio radiation in terms of a frequency metric and a parallel wavelength metric. In the frequency metric, the scale must start with 1, per mandate emanating from the very definition of harmonicity in the frequency characterization, 1Hz that is. Frequency is classically defined to be the number of cyclic events or complete alternations per unit of time or *second* by Standard bodies. In the wavelength metric, there is no classical restriction on the spatial extent of a wavelength, a quasi infinitely long wavelength being a possibility. That extremely long wavelength, with minimalist energy, would have to correspond to *1Hz* (or a quasi *0 Hz* at the very least) in the frequency metric. Both of these wavelength and frequency points cannot be virtuously represented as a match on a double scale line (Fig. 16.8). One can obtain a sense of the disjunction between the two scales by considering that the scale of frequency is limited to interval $[1,\infty[$. Frequency is undefined for any value less than 1$(or\, f < 1)$ or for any Period $T < 1\ sec$. If f or T takes any fractional value, i.e., $f = 1/3\ Hz$, then from expression *(1)*:

$$E = h\upsilon \rightarrow E = \frac{h}{3},$$

which unacceptably implicates that h is not the smallest energy quantum possible in a radiation spectrum, the value $h/3$ being possible. Fractional amount of frequencies with extremely low energy would have to be possible, however, in order to match low energy from the extremely long wavelengths putatively localizable in that same sector of the spectrum. This irreconcilable spells out a definite disjunction between the two scales and makes these pictorial representations of the "electromagnetic spectrum" in both frequency and wavelength real dubious concepts.

Arguably, the formula proposing:

$$f = \frac{c}{\lambda}$$

may be to blame for entertaining in physics the idea of compatibility between frequency and wavelength, by showing a conversion path between the two quantities. It is not always perceived, if at all, that f and λ in the formula do not belong to the same order of harmonicity and instead pertain to two different harmonic domains or metrics. It is uncertain whether the formula is viewed at any extent as a conversion tool between two differing domains or not.

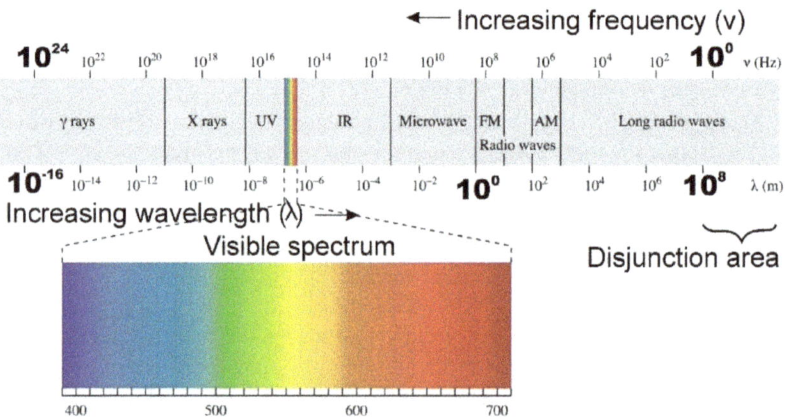

Fig. 16.8 Putative double scale in frequency and wavelength of the radiation spectrum

What's more, the sub-divisions in frequency groups that traditionally qualify the universal radiation spectrum cannot be justified from a flat frequency point of view, but only on the basis of the wavelength characterization. Notwithstanding the admittedly artificial and arbitrary spectrum partitioning in the frequency order, these sub-divisions respond to well-known distinctive behavior of radiation spectra in their individual interaction with matter. The wavelength spatial order is the only one that may sensibly account for those phenomenological interaction specificities, which are of both physical and chemical nature, as well as other radiation properties such as cyclicity above mentioned, but not the putative frequency continuum.

16.12.5 Universal Reciprocal Space Spectrum

Let us rewrite relation *(8-8)* in the manner that follows in an attempt to sensibly express the Planck constant in its best unit-dimension:

$$h = \frac{eV}{\lambda^{-1}} \quad (16\text{-}16).$$

That form of the expression renders the Planck constant h as Energy Density over the scale of reciprocal space. It represents indeed the most authentic and legitimate representation of h. As a thoroughly time-free form of the expression of h, it accords with its foundational predicates and opens the way to the most assertive visualization of the material wavefunction, herein stated as reciprocal space. We refer the reader to Book I for a mathematical expression of the material wavefunction represented by λ^{-1}, as well as a detailed description of its physicality, long debated and hitherto only hypothesized in physics. See Book I as well for a discussion of reciprocal space as harmonic basis of universal space with numeric support.

16.12.6 Statistical Derivation of the Planck Constant

We will not conclude this section of the study without reporting one important yet neglected form of derivation of the Planck constant, pertaining to the Old Quantum Theory. Such a derivation is of statistical nature, and thereby attests to the time-free nature of the constant as much as its estrangement to the classical notion of frequency and electromagnetic oscillation. The Planck constant must be viewed as one of the most important descriptors of radiation and stands even above the electromagnetic frequency characterization when allowed, because this constant describes the ultra-structure of radiation as it relates its internal energy content to its harmonicity. As such the Planck constant is more fundamental than the Hertz as a unit of electromagnetic radiation and stands as a perfect vantage to inquest the notion of radiation frequency.

That the constant can be derived from pure statistical methods via the study of distribution of other quantum objects strongly argues in favor of time free harmonicity by contrast to harmonicity on the time scale.

Effectively what we see in the statistics of gaseous states, subsequently evoked in this study in the context of an analysis of the Kinetic Theory, is how scalar density is distributed over the scale of motional speed of point particles composing the state. Both H. Tetrode and O. Sackur had independently demonstrated, even before the advent of Quantum Mechanics, that h could be derived from the distribution of mercury vapor molecules in gaseous state. Tetrode put forth the concept of discretization of **phase space** encapsulated in "elementary volume domains" in order to model the gas entropy of mercury vapor molecules. He expressed the concept with the equation:

$$\delta p_1 \delta q_1 \ldots \delta p_n \delta q_n - \sigma - (zh)^n$$

where q and p are conjugate variables, z a parameter that here must be equal to 1 in accordance with gathered data relating to mercury vapor pressure, and n the number of degrees of freedom. In this case, n must be equal to $3N$ due to the *3-D* elementary volume domains in entropy phase space, N being the number of point objects in the population. Although their original purpose was to explain numerically the entropy of an ideal monoatomic gas, the two authors anchored their demonstration on the Planck constant by showing that their theoretical results implicated the value of the Planck constant as known, since the expressions that they had independently reached prominently incorporated the Planck constant. Specifically, from the satisfactory numeric comparison of the value of h, which they obtained from thermodynamic data on mercury, with the then known value of the Planck constant from black-body radiation, they inferred the adequacy and accuracy of their equations and developments. This chapter in the evolution of the Quantum Theory has unfortunately fallen prey to oversight.

The fact that absolute entropy is a dependency of h should never be overlooked when it comes to analysis of the distribution of energy *v.* harmonicity. The common thread in the various applications of h is the harmonic behavior of the variables scale over which the energy distribution is taking place, which is always ultimately an attribute of coordinate space, whether the variable is the speed of atoms or molecules in the Materials Shell or wavelength of radiation quanta in the Particle Shell. When it comes to molecular or atomic speeds in state distributions, we must remember that according to De Broglie all motion involves a material wavefunction to be depicted by a wavelength quantity. Again, we have accumulated this many different harmonic concepts in physics (De Broglie wavelength, thermal De Broglie wavelength, Compton wavelength, etc.), because of or our inability to formulate a single unified scale for harmonicity and the legitimate connection of coordinate space with motion as the originator and propagator of action. Evidently, all because we have remained prisoners of the faux concept of universal electromagnetism coined by J. C. Maxwell as well.

16.13. Establishing State Physics on a Spectrum of Nine Typified Degrees of Freedom

The above discussion directly brings us to the proposed critical analysis of the Kinetic Theory, in light of the effective discrete perspective brought about by the quantum vision in photonic physics. The most important formula in statistical physics is inarguably the Boltzmann relation that numerically characterizes entropy:

$$S = k_B \cdot \ln W \quad \textit{(16-17)},$$

where S is the entropy of the ideal gas, k_B the Boltzmann constant and W the state multiplicity of the gas particles. The multiplicity for single monolithic gas particles, such as atoms and elementary particles, is equivalent to the 3 dimensions of 3-dimensional space or its 3 degrees of freedom. These degrees of freedom are the 3 possible basis orientations accessible to the state of any one gas particle.

By transforming the expression into a ratio akin to the Sackur-Tetrode constant, we gain better insight into its meaning and implications:

$$\frac{S}{k_B} = \ln W \cdot$$

This form of the expression instructs us that gaseous state or the gas scale of matter, which is typified by the ratio, cannot exist with 1 degree of freedom but a minimum of 2 degrees of freedom since $lnW = 0$ at $W = 1$. Because the 3 fundamental dimensions of 3-D space cannot be further extended, $D = 3$ being a foundational physical constant for our Universe, it requires a different order of expansion within 3-D space for any epi-phenomenon taking place on its tapestry to implement an order of expansion that can stretch beyond its 3 basis degrees of freedom. As the theory of statistical mechanics has it, many molecular gases may implement as many as 6 degrees of freedom, the extra ones being rotational and vibrational. One may possibly argue that the nature of the gas monolith may account for the difference; the argument however does not hold because **molecular** gas entropy is treated on the same numeric foundation as the **monoatomic** gas (i.e., He, Ne, Kr, etc.), and indeed such gases as water vapor (H_2O), oxygen (O_2), nitrogen (N_2), etc. are experimentally observed to conform to the numeric descriptions of the Theory.

It is critically important for the analyst to comprehend the gist of the above discussion. In the classical Maxwell-Boltzmann interpretation, the Theory goes on to statistically give an account of the distribution of energy or kinetic energy throughout the gas per fraction of molecules making up the entire gas spread (Fig. 8.8). This account, though appropriate and necessary as might be, should not mask, as it always had, the inner activity that matter-of-factly produces the energy and how this kinetic energy belongs to or produces a different foliate of the space spread as an overlaid phenomenon on the basis 3-D space spread. The notion of Thermal De Broglie Wavelength for its own, when materially visualized, unmistakably remits to this physical characterization. Quantum statistical physics has not remedied this lack of vision and in fact has spread such lack further to the analysis of the microscopic realm. Undoubtedly, there exist many different stages of the meta-molecular vacuum that both the classical Maxwell-Boltzmann statistical description and the quantum Boltzmann statistical description remain oblivious to.

1.	***Amorphous Solids***
2.	***Crystalline Solids***
3.	***Powders***
4.	***Waxes***
5.	***Colloids***
6.	***Oils***
7.	***Aqueous Liquids***
8.	***Condensed Vapors/Clouds***
9.	***Volatile Gases***

Although statistical physics only offer hitherto an account of principally 3 different states, there exists in fact a spectrum of 9 individual well-defined states in the Shell of Materials, according to the ranking below:

The Theory mainly accounts for: States 1 and 2 which it conflates into **solids**, State 7 which it qualifies as a **liquid** and State 9 which it qualifies as a **gas**. The above suite surrogates or mimics what we have called a Quanto-Geometric Primitive Spectrum which defines the degrees of freedom that characterize the void spread typical of each of these states. One can readily see the progression from amorphous solid to gaseous or the gradation in the gauge or ruggedness of the successive elements from 1 to 9. From Stratum 1 to 9, compactification as a state descriptor diminishes, while the state descriptor of fluidity increases in the same order. The surrogation of the 9 Quanto-Geometric primordial degrees of freedom constitutes the basis of the development of the epi-formations of the materials realm only partially contemplated in classical statistical physics. The very fact that the Theory is unable to furnish a convenient numeric description for seven of the nine manifest states speaks of its limitation in scope and the contrivance of its numeric prescriptions.

16.14. The Kinetic Theory in a New Light

Despite the wealth of variables developed in the Kinetic Theory such as gas pressure, specific heat, mean free path, molar gas constant, etc., it is unable to explain the reason for such fundamental invariants as the mole, the molar volume of an ideal gas, entropy of an ideal gas, let alone the molecular speed distribution function. In one word, the Kinetic Theory represents descriptive or empirical physics, inapt at elucidating the remote causes of molecular kinetics originating from an applicable fundamental Symmetry of circumstance in Nature. The towering gas law states that:

$$PV = nRT \quad (16\text{-}18),$$

where P is Pressure, V Volume of the gas, n the number of moles, R the gas constant and T temperature.

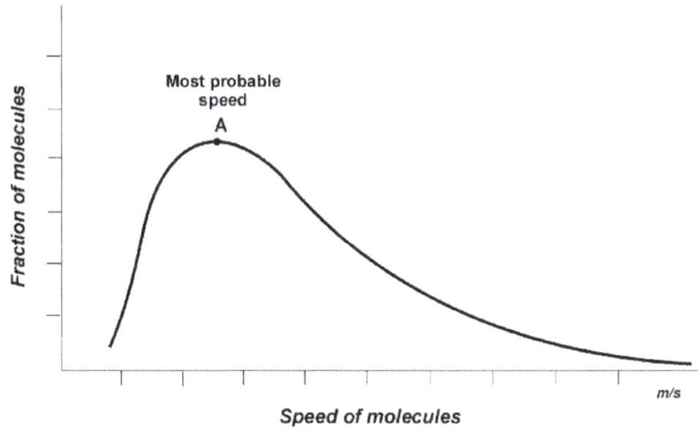

Fig. 16.9 Plot of molecular speed distribution

One of the most important lacks of the Theory is a foundational explanation to the cause of the motion of molecules in the gaseous state. If we acknowledge, as the Theory does in the ideal gas law, that the size of molecules in the distribution is negligible with respect to inter-molecular distances and that molecular collisions are random and do not experience any particular order, then the entire order of the distribution, which is homogenous and well-behaved per its graphical analytic layout, becomes unexpected and must be seminally explained (Fig. 16.9).

What the Kinetic Theory does not recognize is that the gaseous state forms or belongs to a new Shell of matter that cannot be interpreted exclusively on the basis of physical molecular properties. A *new level* of 3-dimensional coordinate space arises beyond the molecule, with new degrees of freedom. In that context, molecules agglutinate in order to create the scalarity (or scalar density) of the new Shell, while the symmetry of molecular coordinate space breaks or explodes in order to create gaseous coordinate space. It is the inner activity of this new space Shell that couples with the new scalar state to covariantly create the fundamental invariant that structurally sustains the new Domain, which we have dubbed the Materials or Lattice Shell in the Quanto-Geometric formalism. The primal fundamental invariant that grounds the new Materials Shell is embodied in the *Van der Vaals* ***a*** constant, with the value of $\boldsymbol{0.2441197/\sigma^{-3}}$ as derived from the Quanto-Geometric axiomatic system in the condition of a functional derivative essentially. The reason that motivated **Diderick Van der Vaals** to produce an equation beyond the previously well-known equation of state for ideal gas *(8-11)* was the need to produce an interpretation that could include a broad spectrum of inter-molecular conditions common to the gaseous state in general. Effectively, at low pressures real gases may be more compressible than an ideal gas, and at high pressure they are less compressible. Further, at high pressure the volume of the molecules is no longer negligible and become a significant part of the total volume, because at higher pressure the total gas volume is smaller. The Van der Vaals' general equation, which virtuously explains the physics of both ideal and non ideal gases, follows:

$$nRT = \left(P + \frac{n^2 a}{V^2} \right) = (V - nb) \quad \textit{(16-19)}$$

where V is the volume occupied by n moles of gas, and ***a*** an experimental gas constant with some level of theoretical support. The Kinetic Theory blindly attempts to explain the state of Materials in terms of a force of attraction between molecules at the molecular scale from properties innate to molecules, without recognizing the new degrees of freedom creating the coordinate space of Materials beyond molecular space. It is on that account that the Kinetic Theory had failed to perceive the importance of the Van der Vaals ***a*** constant for the H_2-H_2 gas (Hydrogen gas), the simplest of all Materials, as the fundamental physical constant for the Materials Shell or gauge group. The measured value of the ***a*** constant for the Hydrogen gas is: ***0.2444*** $\boldsymbol{lit^2}$***-atm*** $\boldsymbol{mol^{-2}}$, well in conformity with the previously noted theoretically derived value within the Quanto-Geometric framework[8]. Also note that it is not possible nor necessary to invoke so-called inter-molecular forces to explain gaseous state for the Hydrogen gas, as postulated in the Kinetic Theory, because its behavior responds to the "ideal" condition, given that for that Material $PV/RT = 1$ at 0°C. Therefore, the value of the Van der Valls ***a*** constant for Hydrogen gas becomes manifest as the universal *scalar-space* coupling constant for the entire Materials Shell. On that account, the ***a*** constant, as prescribed by its Quanto-Geometric correspondent or precursor, already encapsulates the definition of a mole or the Avogadro Number. Note that in the Kinetic Theory this constant remains an empirical quantity or an empirically-driven quantity, given that its attempted derivation proceeds from a functional expression that is inapplicable to state transition physics as discussed further in this study.

Nevertheless, the Van der Vaals ***a*** constant is not only valid for gases but for the entire Primitive Spectrum of Materials that includes the aforementioned series of 9 Primitive states. In support of our view, note that the critical term in *(8-11b)* is the term $\boldsymbol{n^2a/V^2}$ or $\boldsymbol{a(n/V)^2}$, which is known to be equally accountable for condensation of gases to liquids.

The idea of an "ideal monoatomic gas", central to the Kinetic Theory, indeed is a faux concept that mistakes an apparent exception to the rule for the rule itself, given that the pivotal invariant on which Materials state transformations as a Shell are based is rooted on ***mole-cular*** building blocks, not atomic building blocks. Only when a molecular topology reaches the quantity of a *mole* is the new Materials state created on the basis of a mole as the new *scalar density* element. Accordingly, the new scalar is no longer the molecular scalar (i.e. weight or a.m.u of 16 for O_2 or 12 for ^{12}C) but a multiple of the **mole**-cular scalar conforming the Material. It is that new scalar form that covariantly couples

---

[8] This value is the previously offered quantity by the Standard bodies, before they chose to offer a new value of: *0.247 liter* atm^{-2} mol^{-1}. We have chosen to retain the older value in catering to the many misconceptions in state physics tending to affect experimentation and experimental calculations and because the mentioned value is closer to the independently derived value by a purely theoretical axiomatic framework.

with the transmuted Materials coordinate space transpiring in the pressure variable to create the invariant experimentally perceived as the Van der Vaals ***a*** constant. In that regard, the reader must remain keenly aware that in the Quanto-Geometric axiomatic framework, there is no direct intervention of empirical physical entities computationally, such as atoms or molecules or even outright physical variables, but only mathematical entities modeling physical variables, which so discloses their intimate nature and remote origin. Within our framework, the Van der Vaals ***a*** constant arises as the following functional derivative (Book. I):

$$k \cdot \left[\frac{dQ(s)}{ds} \right]_{s=l} = 0.2441197 \cdot \sigma^{-3} \qquad (16\text{-}20)$$

The broad match of this quantity with the Van der Vaals constant (for Hydrogen gas) sheds a bright light on the nature and origin of the experimental constant as well as its relationship with other fundamental quantities, this on account of the generic abstractions upheld in the Quanto-Geometric Theory.

In particular, the Kinetic Theory relates pressure P, volume V and Temperature T to codify state as if these three variables existed on the same footing. In the ignorance of the origin of these fundamental variables, the Theory only seeks with these variables to give order to empirical data while constructing a tool for the numeric manipulation of state behavior of Materials. One of the most important drawbacks of this approach is the masquerading of the material wavefunction under the Pressure figure, and to some extent the Volume figure alike when it comes to the PV product. Except when a gas is bound by a container, its expansion is free running with no physical obstacle to the action on all paths. Thus, pressure can only be invoked heuristically as a substitute for the hitherto incomprehensible intangible that creates the expansion. It is interesting to note that the construct normally invoked in the Theory for heuristic pressure is the representation of a virtual box bounding the molar volume of the ideal gas whereby pressure is asserted for a gas molecule by its longitudinal back and forth travel and repetitive impacting of one wall of the box. One molecule moving toward that chosen wall will hit it every $2 \times l$ *cm* of its trajectory[9], because, subsequent to the first impact, it must go the opposite way to hit the opposite (imaginary) wall before the next impact on the first wall can re-occur. While this representation is thereby invoked to explain gas pressure, it closely mimics the required representation for the visualization of repetitiveness or harmonicity that defines the wavefunction inherent to coordinate space within the Quanto-Geometric framework (Book I – Chapter 8). As such it is a close parent to probability amplitude as discussed in the referred Chapter of Book I. Assuredly, coordinate space, the true agent of pressure, is

[9] If the box is a cube, l is the length of its side, which is also equal to the wall-to-wall longitudinal travel path of the molecule.

moving in all directions, and appears as reciprocal space with a defining wavelength or wavenumber for its harmonicity.

The key element advanced in the Kinetic Theory to account for the behavior of the *PV* product of an ideal gas is a putative inter-molecular force of a polar nature resulting from attraction between any set of two molecules in the gas. It cannot be overstressed that state covariance that explains the Materials Shell as a symmetry is between a scalar of molar nature represented in the *Q(s)* function as shown in *(16-20)* and coordinate space *s*, the function's dependent variable. The chemical bonding force, essentially based on *e* the electric constant, only exists at the level of a flat population of molecules that has not reached in quantity the Avogadro number or one mole. Such reduced populations are circumscribed in the Molecular Shell. At one mole count in the topology, spontaneous symmetry breaking occurs that transmutes molecular coordinate space into Materials coordinate space, which creates the fundamental invariant known as the principal Van Der Vaals constant ***a***. So therefore, it is not an inter-molecular force of a kind that creates Materials states, whether gaseous, liquid or solid. This significant error in vision, ultimately due to the ignorance of the nature of the material wavefunction and its history, may be to blame for the squint numeric development of the Kinetic Theory when it comes to liquid and solid state. The most compelling support to our view is that its implicated primitives straightforwardly yield the second most important quantity pivoting the Kinetic Theory, the molar volume of the ideal gas empirically measured to be:

$$[PV]_{P=1at}^{T=0} = 0.0224146m^3 - at / mol \quad (16\text{-}20).$$

This quantity expresses the Pressure-Volume product for one mole of a gas at 0° and infinitesimally low pressure trending to *0*. From the Quanto-Geometric axiomatic formalism (Vol. I), this quantity naturally arises as the square of *(16-20)* at a specific symmetry-breaking point:

$$\left[k \cdot \left[\frac{dQ(s)}{ds}\right]_{s=m}\right]^2 = 0.02246942 \cdot \sigma^6 \quad (16\text{-}21).$$

A comparison with the square of the $|\Psi|$ function that defines an atomic electronic orbital will quickly help us understand the true meaning of expression *(8-14)*. The square of the $|\Psi|$ function ($|\Psi|^2$) in Quantum Mechanics is credited to provide a representation of the probable whereabouts of the bound electron around the nucleus. Accordingly, what we call the volume of a gas is indeed the probability amplitude of locating the mole of the Material from a center point of radial expansion. Inasmuch as a derivative is also a function, the comparison is fully justified, however the first order functional expression ***Q(s)*** remains the primordial entity that informs the nature and remote origin of the physical variables. It is clear that what we have come to designate the wavefunction in modern physics is restricted at first order to the coordinate space component of the Quanto-Geometric root Function (its independent variable) which the Theory expresses as a wave-number on the basis of a harmonic distance metric. The derivative expressions of that function in its many different forms do conserve the variable ***s*** as independent variable, however one should always keep in mind that these functional expressions always instruct us on both scalar density and coordinate space with its internal degrees of freedom and inner kinematic.

From a numeric standpoint, the Kinetic Theory is at any rate not sufficiently prolific in developments addressing solid state of Materials as it is the gaseous state. The main problem the Kinetic Theory faces in addressing solid state is the inability to properly account for states formed by population of electrically neutral molecules, given that it confers to intermolecular forces of electric nature the role of bonding agency within solids. The Theory acknowledges however that in these particular cases the introduction of the Van der Vaals ***a*** constant is necessary in order to account for Material state in the crystals it designates "Van der Vaals's crystals". Examples of such crystals are carbon dioxide, carbon tetrachloride or the vast majority of organic compounds. Note that the other forms of crystalline solids classified by the Theory are ionic crystals, metallic crystals, covalent crystals and Hydrogen-bonded crystals, based on the proposition of inter-molecular chemical bonds.

The main problem with these definitions is their patent inability to explain why the distribution of average speeds of molecules, or their rotational or vibrational kinetic energy, all follow the same type of distribution per fraction of molecules (read per molar density) whether the state is gaseous, liquid or solid. These distributions are well known as the Maxwell-Boltzmann distributions, exactly identical for all three considered states (Fig. 8.8). The Kinetic Theory is unable to uncover the fundamental invariant that provides for the structural basis to those states, which would by the same token elucidate the reason for their similarity. In that regard, it is important to firmly highlight that the mole is the single most important scalar variable in the study of how molecules react with each other to form compounds or compound elements. It is an important variable in Chemistry and Physical Chemistry when it comes to balancing chemical reactions at STP, whereby when applied to a gaseous element, it transpires thru the figure of $\boldsymbol{0.0224146\ m^3}$ and when attached to a solid or liquid element, a mole of that substance must be admittedly taken to be the number of grams equal to its molecular weight in order to topologically satisfy the chemical reaction.

Let us examine the structure of the Van der Vaals ***a*** constant toward solidifying it as a universal state constant for the Materials Shell. In expression *(8-11b)* the Van der Vaals term $\boldsymbol{a\,n^2/V^2}$ captures ***a*** as a constant of proportionality between the mole which is a scalar and *Volume* which is a physical variable. At first it may be difficult to understand the meaning of a ratio between a scalar and a physical variable. It is convenient to understand, however, that this scalar is how nature creates *mass* for this Shell; this scalar is not a pure mathematical quantity but a mathematical-physics quantity in that it is the count or cardinality of a group of physical elements. And so, the mass unit becomes an agglutination of dimensional elements with cardinality of 1. It is the form taken by the scalar density function in this Shell, specifically molar mass. That unit of mass naturally appears to be that of one mole of Hydrogen H_2-H_2, in recognition of the Shell-wide unitary value of that molar singlet. So therefore, the state function here is not about how a molecule relates to another, the common analytic view proposed by the Kinetic Theory, but about how a molar bulk of molecules relates to another molar bulk of molecules. If we were to account for this relationship as a *Force*, we would have to express it in terms of a Molar Force such that:

$$F_M = a\frac{n \times n'}{V^2} \qquad \textit{(16-22)}$$

where n and n' are two identical molar bulks of molecules within the topology or any set thereof. Expression *(16-22)* simply represents another way of writing the Van der Vaals term. Compare this expression with that of the electrostatic force relating two charges:

$$F_E = k\frac{Q \times Q'}{r^2}$$

where Q and Q' are two electric charges of either the same polarity or opposite polarity, and r the distance separating them. We may even compare with the expression of the Gravitational force that directly relates two Masses, with a certain proportionality constant that is the Gravitational constant G, and inversely to the square of their distance r. Should the two masses be equal, then the expression of the Gravitational force would be exactly identical to the two expressions above as a reduced form of the same:

$$F_G = G\frac{M \times M'}{r^2}$$

The nominal figure of “Van der Vaals force”, with a totally different numeric formulation, had been introduced in physical chemistry as early as 1939 by Boris Derjaguin. Other theories of the same were developed in mid-20th century by Evgeny M. Lifshitz, Fritz London, Willem Keesom and others. They are all unassertive theories that have left this aspect of the Kinetic Theory in disarray since different texts including more modern literature present different and even opposite views about the nature of the Van der Vaals force. A repulsive force for some, it is given as an attractive force by others, yet some other authors seem to ascribe to the notion the totality of forces between every interacting pair of singlets in the topology, with no distinction whatsoever in the nature of these topological singlets, whether microscopic particulates or macroscopic objects. In the literature this force is also denoted the Van der Vaals pressure or Casimir pressure. We will forego a discussion of these theories at the numeric level on account of their total unassertiveness in characterizing state discriminatingly, on the heels of their complete misunderstanding of the Shelling evolution of the vacuum.

16.15. Material State Topologies

To the statistical distributions mathematically codified in the Kinetic Theory, there is more than just probable average speeds taken by groups of molecules. The most towering equation in the Theory is the Boltzmann distribution codified as follows:

$$f(v)=\sqrt{\frac{m}{2\pi kT}}\cdot e^{-mv^2/2kT} \quad (16\text{-}23)$$

where $f(v)$ is the fraction of molecules and v the average velocity of any molecule in the gas in one direction. This expression is nothing else than the normal probability density function incorporating a few physical variables of circumstance. Only when one is able to grant this expression its proper quanto-geometric attributions is one prepared to uncover the quantum abstractions subsumed in its dependent and independent variables. The same remains as much true when this function becomes the Maxwell distribution thru incorporation of the 3 spatial dimensions accessible to molecular speed:

$$f(v)=4\pi\left[\sqrt{\frac{m}{2\pi kT}}\right]^{3/2} v^2\cdot e^{-mv^2/2kT} \quad (16\text{-}24)$$

which is the one majorly represented in textbooks with "Fraction of Molecules" as the independent variable, and under the adopted designation of Maxwell-Boltzmann distribution of molecular speeds (Fig.8.8). The Maxwell-Boltzmann mathematical-physics contribution essentially consists in cleverly substituting the generic x variable in the normal probability density function with the kinetic energy expression (in fact the entire term $x^2/2$), with v as the independent variable of the function. In the ignorance of the quanto-geometric covariance regime internal to that function, their attempt has the merit of at least describing the observed empirical motion of particles in the gaseous state. The higher truth, however, is that when re-dimensioned with quantum and geometric abstraction definitions, the normal probability density function becomes an Eigenfunction that **implicitly** incorporates the Kinetic Energy expression as part of the scalar component as well as an expression of the wave dynamics that together make up Materials state, as shown below:

$$\|\psi\|_{a_i b_i} = \left[\frac{-Q^2}{2q}\right] \bullet \left[\frac{2}{a_i e^{\left(\frac{2a_i + b_i^2}{4}\right)} + b_i e^{\left(\frac{2a_i + b_i^2}{4}\right)} + c e^{\left(\frac{2a_i + b_i^2}{4}\right)}}\right] \qquad (16\text{-}25)$$

In this implicit expression of the Quanto-Geometric Eigenfunction as a re-dimensioned Gaussian probability density function, the first term is a representative of the Kinetic Energy and the second the intangible wave component. It is the axiomatic centered on this Eigenfunction, primal cause or root to ***Everything***, that both the Van der Vaals ***a*** constant and the molar volume of ideal gas, above presented, have been theoretically derived (Vol. I). Therefore, we do not need to artificially incorporate the classical Kinetic Energy expression to the function in order extol the capability to predict distributions of molecular speeds, that exercise being in fact counter-productive.

Some have attempted to establish a topological standard in aligning the Maxwell-Boltzmann distribution, the Bose-Einstein distribution and the Fermi-Dirac distribution as portrayed below:

$$f(E) = \frac{1}{Ae^{E/kT}} \rightarrow \text{classical Maxwell-Boltzmann}$$

$$f(E) = \frac{1}{Ae^{E/kT} - 1} \rightarrow \text{Bose-Einstein}$$

$$f(E) = \frac{1}{Ae^{E/kT} + 1} \rightarrow \text{Fermi-Dirac.}$$

As seductive as this classification might be, it is outright misleading since, on the one hand, the Maxwell-Boltzmann distribution is not necessarily distinguishable, that property being only a presumption, and, on the other hand, it does not belong to the same topological class as the other two as previously discussed. The Bose-Einstein and Fermi-Dirac topologies are not molar in nature, because the mole quantity or the Avogadro Number does not yet inform the scalarity of the occupants of that Shell, the Particle Shell, nor the Molecular Shell. Topologies exist from the perspective of observation from an upper Shell in the downward direction. An element of a Shell looking in all directions within the Shell will always perceive the entire Shell as infinite, ultimately due to the very kinematic nature of coordinate space. To convince ourselves of this universal truth, let us remember that all things in existence in the 3-D Domain exist thru a spectrum, akin to the state spectra that we are analyzing here, which implicates extent trending to infinity at the tail of the spectrum; yet the entire 3-D Domain is one finite Domain among others of higher order within the Universe or the *Omniverse*. Which all means incidentally that the physical boundary between infinite and finite is quite a complex affair.

The true dimension of Material state topologies is given in the 9 Quanto-Geometric normal layers as described in the suite severally evoked here which goes from Amorphous Solids to Waxes to Volatile Gases. Refer to Book I of this publication for an expression of the numeric representation of this suite. Thus, a significant caveat of the Kinetic Theory is to make state definitions a function of properties of molecules. It is an erroneous proposition because, to start with, every topological molar assembly of molecules is accessible to all states, not only the three that are familiar to the Kinetic Theory but all nine postulated in the Quanto-Geometric Theory. The *Pressure-Volume* product *PV,* which pervades the classical Theory, is not fundamental in and of itself, being only an observable that provides no assistance whatsoever in the interrogation of the reason and origin of molar topology.

The topological molar definitions are the ones that determine the associative elements that create physical state of Materials and not the topological singlets themselves. That is why a molar topology of like molecules can be subject or made to become accessible to any of the nine possible states. What's more, at one point of temperature (known as the triple point of water), three different topologies of the same assembly of molecules may coexist spatially, an empirical fact that is an important driver of both the Kinetic Theory and Thermodynamic physics while they prove unable to address the reason for that artifact. The triplet coexistence of states obliterates the putative importance of the *Pressure-Volume* product and its role in the formation of physical states. That the same molar assembly of molecules could adopt different states in different spreads of coordinate space and very particularly under different temperature points could have understandably fouled the analyst, but that these various states of identical molecular population can coexist under the same temperature point leaves the conspicuous analyst with no other election than looking for different drivers to materials state phenomenology than the *Pressure-Volume* product or even the *Pressure-Volume/Temperature ratio* as eponymously posited in the Gas constant *R*:

$$R = \frac{[PV]_{T_0}^{P=0}}{T_0} \qquad (16\text{-}26)$$

The condition of triple point of water is generalized in what the Theory calls *phase equilibrium*, whereby at a certain *critical* temperature any molecular molar topology may simultaneously coexist in all three "phases", solid, liquid and gaseous that is, or any set of two of those phases. In this same context, of greater significance is the fact that all gases tend to behave identically and in the condition of "ideal" expansion, when let free to themselves by eliminating all external constraints such as pressure. In that situation, expression *(16-20)* is what applies to model their behavior. They then tend to occupy the largest volume, however the fundamental second-order invariant expressed in the ratio between molar density and volume, as stated in *(8-13)* and even more clearly in *(8-14)*, is strictly conserved. That gives us a primitive basis for a topological state normative, allowing us to break free from the chicken-and-egg situation between expression *(8-13)* and *(8-18)* in terms of which is first. Being largely an empirical theory, the Kinetic Theory has no way to discern origin and co-dependency of these variables.

While the aforementioned empirical figures are acceptable for practical purposes in applied physics, they are unacceptable for theoretical physics and should all have long been noted as unresolved problems awaiting elucidation. Sadly, a call this important has never been raised in all likelihood in mathematical-physics.

Beyond expression *(8-13)*, expression *(8-12)* constitutes the first-order primitive basis for topological state normative. Effectively, with all the preeminent role played by the *Pressure-Volume* product in the Kinetic Theory, it has proved unable to sufficiently cater to the dissimilar qualities of monoatomic gases in that order so as to remove them from the topological state classes mainly addressed in the Theory. These gases, otherwise known as noble gases, are mono-atomic at STP, unlike all others which are based on molecules; they uncharacteristically feature extremely low *melting* and *boiling points*. Very importantly, even with those elements with large atomic masses that are superior to the masses shown by other elements that would otherwise be solid with those levels of scalarity, these atomic singlets remain in the gaseous-like state. Among them, Helium presents several singular qualities that are worth noting when compared with other elements:

- Its boiling and melting points are lower than those of any other known substance.
- It is the only element that cannot be solidified by cooling under standard conditions: a Pressure of *25 at (2,500 kPa* or *370 psi)* must be applied at a temperature of *0.95 K (−272.200 °C* or *−457.960 °F)* to convert it to solid-like state!
- It is the only element known to demonstrate a so-called state of *superfluidity!*
- Its putative Van der Vaals' ***a*** constant is 7 times lower *(0.03412 liter^2-at mole^{-2})* than that of the simplest multi-molecular state topology, the H_2-H_2 topology.

16.16. Erroneous Continuum Perspective in the Kinetic Theory

This aforementioned string of dissimilar qualities is decisively telling us that the related elements (monoatomic gases) do not belong to the Materials Shell because of the fundamental reason that their topologies do not hold the required molar quality, nor does their wavefunction hold the proper harmonic measure for that Shell. They simply do not belong to this Shell, no more than a population of electrons or photons do, no matter the count level reached by their singlet populations. It requires a ***molecular mole count*** in covariance with coordinate space such as to satisfy expression *(8-12)* to so create the invariant that forms the basis of the 9 Materials states accessible **to poly-molecular** topologies. In those conditions, the Van der Vaals ***a*** constant for Hydrogen arises as the fundamental invariant quantity that characterizes the state of all Materials which behave akin to Hydrogen molar topology when let free to themselves or completely unconstrained.

As previously reviewed, the Kinetic Theory has been heavily centered on the study of three states exhibited by Materials, solid, liquid and gaseous that is, to the point that they come to appear as the three primordial states accessible to Materials. Accordingly, the transitions experienced between one of these states and any other of the three have been extensively studied and numerically codified thru a vast number of collected empirical data covering the variables of circumstance. Along that line, *state* has become *phase*, specifically when it comes to accounting for topological transitions of a substance. The role played by the thermodynamic variable of Temperature is unmistakably another key element in the Kinetic Theory construction. The most encompassing view that emerges from this intellectual enterprise is that the organization of materials state is played out on a background of absolute *continuum*. The development of an absolute scale of Temperature subtending the reduced Centigrade or Fahrenheit scale has been decisive in cementing the *continuum* perspective.

Nonetheless, in the physics of mother nature, including the realm of solid, liquid and gaseous materials, *discreteness* is the de facto currency in their physical state evolution. Reflecting the true distribution of state existent in nature, the reported quanto-geometric suite stands erected on discrete partitioning with the Quanto-Geometric Grand Eigenfunction representing the primal cause or provenance of that partitioning. Refer to Book I of this publication for an extensive analysis of the 9 universal Tensor partitions of that Eigenfunction and the function spaces that subtend them. The very fact that the triplet solid-liquid-gaseous state may exist in overlap of one another over the same thermodynamic temperature background essentially means that these states are not continuous or granularly contiguous to one another in their defining qualities. Effectively we experimentally find that the physical transition between these states or phases is abrupt or discontinuous. The best transformation applicable to the Van der Vaals equation *(8-11b)*, which formally turns it into a 3rd-degree polynomial with the molar volume as the independent variable, is patently unable to properly account for these transitions, wrongly casting them as if continuous. We operate this transformation simply by multiplying the entire equation by V, the molar volume, to obtain:

$$V^3 - V^2\left(b + \frac{RT}{P}\right) + V\frac{a}{P} - \frac{ab}{P} \qquad (8\text{-}19).$$

where V is the molar volume, ***a*** and ***b*** the Van der Vaals constants.

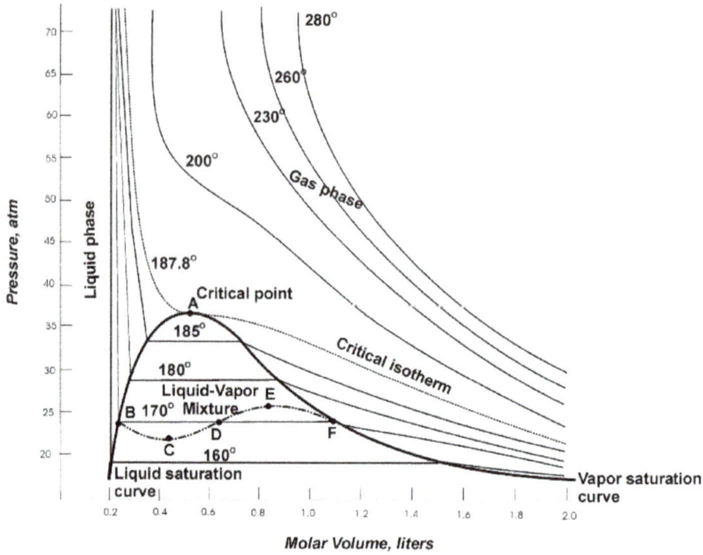

Fig. 16.10 Phase transition map between gas and liquid experienced by molar Materials

The chart in Fig.16.10 shows the root points of the polynomial ***(B-C-D-F-E)*** on a line that falsely suggests progressive state-to-state transitioning, whereas the underlying horizontal pressure line is known to reflect the true existence of these various states at the same pressure value. One should not be surprised though that this equation, despite its inadequacy in general to characterize state transitions, is able to yield an expression for the ***a*** constant that is correct thru resolution of its first-order and second-order derivatives as a set:

$$a = 3P_c V_c^2 \quad \text{or} \quad a = 3P\left(\frac{v}{n}\right)^2 \qquad \textit{(16-27)}$$

where v is the normal volume and n the number of moles of substance. What this all means is that in order to characterize generically Materials state, more than one equation is needed, no matter how complex and prolific in variables one wants to make that one single equation. We have a similar situation in Quantum Mechanics which has never been overcome in more than a century of its existence, whereby the Schrodinger's Hamiltonian Operator, in spite of its successes, can only explain the energy density distribution for one single atom, the simple Hydrogen atom. The rest of the Atomic Shell is populated with polyelectronic atoms whose behavior as experimentally known significantly surpasses the framework of the Hamiltonian Operator.

The Quanto-Geometric Eigenfunction offers a set of 9 expressions as the structural basis for that characterization. In that way it is well positioned to capture the discontinuous nature of states (or "phases") as they relate to one another, and further shows that the eponymous *solid-liquid-gas* triplet state is unnatural. If one wants to find a truly natural triplet partition in state topologies (Table 1), we propose the Quanto-Geometric *Tri-Valent* partitioning which originates ultra-structurally from the Grand Eigenfunction.

1	Amorphous Solids	Solids
2	Crystal Lattices	
3	Powders	
4	Waxes	Semi-Solids
5	Colloids	Fluids
6	Oils	
7	Aqueous Liquids	
8	Condensed Vapors	
9	Volatile Gases	

Table 8.1. Natural triplet partitioning of states

The *Tri-Valent* partitioning arises as a second-order functional derivative property of the Grand Eigenfunction which delimits three distinct sectors of the geometric rundown of the Function, as previously studied. Suffice it to note here as a reminder that the first sector (comprising Layers 1, 2 and 3) characterizes states or objects as a composition between scalar density and coordinate space whereby the scalar component overwhelms the space component. The second sector (comprising Layers 5, 6, 7, 8 and 9) characterizes states or objects as a composition between scalar density and coordinate space whereby the wavefunction attribute of coordinate space overwhelms the scalar component. The third sector is an intermediate Layer (Layer 4) whereby covariance between scalar density and coordinate space is equalized. Fundamental differential geometry inherent to the Grand Eigenfunction so gives us a topological map with 9 representation groups and 3 group covers identified in *Solids*, *Semi-solids* and *Fluids* as depicted in Table 8.1, the most natural of state triplets. In addition, all the physical-chemistry of solubility, fusion and reactants as well as the physics of entropy and enthalpy is generically contained within that framework, which is able to seminally uncover the intrinsic discrete nature of state and kinematic of objects in dynamics systems in all realms.

16.17. Discontinuous Quality of Probability Distributions

To consider the well-behaved curve of distribution of velocities per fraction of molecules as a continuous distribution, per common interpretation in the Kinetic Theory, amounts to a major mistake. Although the graph shows no discontinuity of the function at any point, this curve represents a derivate of the Grand Eigenfunction (either a straight first-order derivative or a composite derivative function) which incorporates deep properties of discreteness emanating from the quantum abstraction Norms at its origin. Just as **M. Planck** had to call upon discrete ***h*** quanta of energy to make sense of the energy distribution in black-body radiation spectra, there is here a discrete inner composition between physical variables that create the state distribution. For a better grasp of this concept, one has to translate the independent variable of speed *(m/s)* to the wavefunction metric and make of it a wavenumber, in the profound understanding that it is the coordinate space spread in the surrounds of a molecule that creates its kinematic experience within the molar state. Its identity is that one relationship between the molar scalar and coordinate space which is expressed in *(8-12)* as the molar ***a*** constant, or the Materials constant, otherwise known in the Kinetic Theory as the Van der Vaals ***a*** constant. Moreover, the visualization of the state wavefunction is the only one that can account for the broad similarity between the distribution of gases and that of liquids and solids, despite the fact that the independent variables are molecular speed in the first case and in the second of vibrational nature instead, according to their identification in the Kinetic Theory.

Furthermore, the Kinetic Theory acknowledges that at equilibrium *the distribution of instantaneous velocities is independent of* ***time***, as reflected verbatim in the literature. This realization is antithetic to the very notion of molecular speed incarnated in the abscissa of the graphical representations, the independent variable, because, absent a global or universal time scale for the entire molar population, every molecule cannot be separately time-driven or hold its own time scale. The Theory therefore plainly acknowledges that the latter does not exist. Nonetheless, theoretically speaking, the ultimate reference for the *continuum* notion is ***time***; in the admitted absence of time all phenomenologies must be stochastic and forcibly fall under the regency of the Quanto-Geometric Grand Eigenfunction. In Book I, we have shown what the fundamental orders of certainties existing beyond randomness come to be under the umbrella of the Grand Eigenfunction.

16.18. Expression of Discreteness in Classical Thermodynamic Physics

The nexus between the Kinetic Theory and Thermodynamic physics is indissoluble within the framework of classical physics which had reached its maximum expression at the height of the 19th Century. Hence a critical review of the Kinetic Theory could not go without addressing related caveats in Thermodynamic constructs. We are next going to address the most important Thermodynamic variable, absolute Temperature.

For us, temperature is a coordinate space property whose mission is to promote unity in Shell topology in the 3-D Domain and across all Shells. It is represented as a tensor imposed on coordinate space and not as a physical variable akin to Pressure or Volume for example. To provide *integrality*, it acts isotropically owing to the Tensor of Quadratic nature it is invested with. Isotropy is a consequence of *equalization* which is the defining

Norm of the Quadratic Tensor within the Quanto-Geometric axiomatic framework. These preambles are what Absolute Temperature brings to the background of transformations occurring within a Shell at all stages of said transformation. It obviously tends to bring theses transformations to the *critical* "point of equilibrium", which is the fullest expression of *equalization.*

Because of its mission of integrality, one could misconstrue Thermodynamic Temperature to be a *continuum* parameter. To make that case one could further argue that it can take any mathematical value possible among positive real numbers, *0* being the only exception case. In the Quanto-Geometric construction aforementioned in Book I, Absolute Temperature theoretically arises by derivation on a scale class based on the Stephan-Boltzmann constant in the role of its basis unit:

$$\left[\frac{dQ(s)}{ds}\right]^{-1}_{s=q} = 5.680763 \cdot \sigma^2 ,$$

while the officially measured value of the constant is:

$$5.6705 \cdot 10^1 Wm^{-2}K^{-4} .$$

Therefore, despite its naturally assigned normative basis of *equalization*, Absolute Temperature is itself granularly scaled. Because temperature is an attribute of coordinate space (or specifically of its wavefunction property), a rise in temperature corresponds to a metric expansion of coordinate space or an increase in its wavenumber. The concept of heat capacity $\bar{C}_V$ of a substance, defined to be the amount of heat required to change the temperature of the substance from T_1 to T_2, is a classical attempt at capturing the behavior of Temperature from a continuum perspective that is overtly deficient, because this view does not comprehend that all differential values *[*T_2 *-* T_1*]* are not indeed possible as experimentally known. If not all values *[*T_2 *-* T_1*]* are possible, Temperature as a physical parameter must develop on a quantization basis and, if it must behave isotropically (same everywhere in a spread), then there must exist a physical constant of thermodynamic temperature, akin to *h*, the constant of discrete distribution of Energy.

Effectively, there is a known limit to heat capacity for molar populations which is **6.96 cal deg^{-1} mol^{-1}**. This limit is approached, but not reached, at very high temperatures, since the very molecules in the population (the topological singlets) would dissociate un der any temperature increment from the cited limit. The reason for this behavior cannot admittedly be explained in the classical continuum vision. Nor can the latter explain the reason why different topologies of **diatomic molecules** have their heat capacity values restricted between ***4.97*** and ***6.96 cal deg^{-1} mol^{-1}***, an acceptable explanation coming only from Quantum Mechanics on the basis of an intervention of ***hv*** in the expression of heat capacity.

$$\bar{C}_{V,vib} = \frac{R(hv/kT)^2 e^{-hv/kT}}{(1-e^{-hv/kT})^2}$$

The development of an *absolute* scale of Temperature has created the false impression that the scale is granularly continuous to the finest desirable degree point. However, the radiant power j or heat capacity that Temperature is thermodynamically related to follows the relation prescribed by Joseph Stefan:

$$j = \sigma T^{-4}$$

Therefore, the Stephan-Boltzmann constant σ is a proportionality constant for Temperature while the scale is defined to be of a quadratic class. What that means is that the way that Temperature behaves within the evolution of dynamic systems is that for any positive real number of choice from the Number realm, Temperature will materialize as the inverse of the 4th power of that number $(1/T^4)$ and from there a change from T_1 to T_2 is regulated by the σ constant and will always be a multiple of the constant value. Further, the principled reason why Temperature acts on the quadratic point is because its mission is to promote *equilibrium* or *equalization* as mandated by the *Quadratic Tensor* from which it originated. Note that in all Thermodynamic distributions the isotherms all go thru the quadratic point (the point where the second-order derivative of the distribution equals *0*). Should we rewrite the relation with E_T instead representing heat capacity or intensity of radiation in the form below:

$$\sigma = \frac{E_T}{T^4},$$

it would be directly comparable to expression *(8-9)* for the definition of ***h***, both expressions showing a ratio between Energy and a coordinate space variable (λ and T). It is well fitting that the denominators in both ratios are in form reciprocal to one another, because the two relations are altogether orthogonal to one another, since the Planck constant is invested with the mission of *discretization* whilst the Stephan-Boltzmann constant is fulfilling a mission of *equalization.* Both constants occupy a towering mutual-conjugate position in the development of all physics. On that account, it should be no surprise that the Stephan-Boltzmann constant can be used to measure the amount of heat emitted by a black body, which absorbs and re-emits all incident radiant Energy thru a distribution regulated by ***h***, an isotherm being a component of all radiation spectra, as shown in Fig. 9.

16.19. How to Cure the Caveats in Quantification of Mass for Advancement of Solid-State Physics

One of the major problems that have traversed all-time physics, albeit well mitigated in the literature, concerns the lack of an ontological definition of mass in order to match the preeminent role it has been granted in our understanding of the development of matter. As a result of this deficiency, there exist an unwelcome long suite of mass abstractions in physics arising from perception of its behavior in the many different phenomenologies where it is engaged.

The best theoretical rendition of our common perception of a mass irreducible is the vision of a scalar, a monolith or perhaps a scalar monolith. This vision implies the existence of an irreducible scalar monolith that may quantify mass in all of its very diverse forms of manifestation. In physics, we have searched long and large, albeit to no avail, for a unitary expression for this universal mass irreducible. A universal mass irreducible based on the notion of a mass unit as an integrable monolith has long remained an elusive pursuit, since a single mass scale under the irreducible monolithic abstraction resulted unsuitable for unified quantification of mass in all physical domains. Beyond the question of convenience or suitability, this caveat amounts to a deep conceptual problem, because it spells out an inability to capture the Norm of distribution of mass in nature.

To be specific, the *Kilogram*, the eV/c^2 energy unit, the *atomic mass unit a.m.u.* or *unified atomic mass unit u*, the *relative atomic mass unit* $A_r()$ and others, are all mass units that demand constant monitoring and revisions and even deprecations by the Standard Bodies in order to keep physical science in synchrony with the expression of mass in nature. The kilogram is an attempt to qualify or characterize mass as a gravitational force or weight or a result thereof. Because the gravitational force does not seem to operate in an effective manner at the quantum level, the kilogram has come to lack applicability in quark-scale and elementary-particle-scale physics at the very least. We take advantage of the principle of energy equivalence to apply the energy scale to the quantification of mass in those two realms. We should keep in mind however that equivalence means not equality or sameness, and energy, while equivalent to mass in some respects, is definitely **not** mass.

We primarily apply the *Kilogram* to molecular physics, Materials or state physics and above. Modern physics incorporate the mole quantity for the definition of mass in molecular scale physics and the above realms. Although the atomic mass unit(s) have been primarily defined for quantification of atomic masses, the mole, which is an experimentally valid quantity for poly-molecular materials, has been retrofit to the definition of the atomic mass units. So therefore, from the primary definition of the (unified) atomic mass unit u, which has been centered on the amount of neutrons in a ^{12}C carbon atom nucleus as codified below:

$$1u = 1m_u = \frac{1}{12} m(^{12}C),$$

we also find an extension of the definition to include the mole as shown below:

$$1u = 1m_u = \frac{\frac{1}{12} 12g(^{12}C) \cdot mol^{-1}}{N_A},$$

where N_A is the Avogadro number and *mol* is the mole. This effort attempts to include the broadest number of factors influencing mass in order to circumscribe the most widely applicable irreducible for mass. Nevertheless, the effort is still insufficient because of the existence of isotopes in all samples of materials, reason why the Relative Atomic Mass $A_r(x)$ had been defined as well, a dimensionless ratio between mass of the standard naturally occurring element and that of its isotopes. This effort further results in the inconsistency that the mass of the electron m_e, the natural monolith at the atomic scale due to its indefectible unbreakability, does not function in the SI system as the atomic mass unit. Being an even smaller fermionic element than the neutron or the proton, one would expect the electron, in conformity with modern physics' perspective of mass, to play the role of mass irreducible. However, by all accounts, it does not appear to be naturally invested of the role of conferring mass to the atom, but solely to mediate its wavefunction component instead.

In all, modern physics ought to abandon the perspective of integrable monolith in its attempt to define mass, if it should become successful in establishing a single mass scale. Secondly, physical science must come to terms with the fact that mass is ontologically about a densification trend beyond pure monolithisation. Note that the notion of *mass density* defined as Kg/m^3 does not at all respond to this characterization despite the appearance, if anything because the *Kg* is not a fundamental unit. Thirdly, the bulk of the evidence shows that there exist not one mass scale but a suite of fractional mass scales corresponding to the suite of Shells of matter, notwithstanding the fact that those different ranges and qualities of mass may obey a single root function the likes of an Eigenfunction.

The Quanto-Geometric Eigenfunction faithfully meets these constraints in the definition and quantification of mass, because it is only from theoretical abstraction that a mass function can be determined along with the elements by which mass should be quantified as the theoretical system meets empiricism. In previous Chapters, we have developed an axiomatic program that has derived with a large degree of accuracy, on one hand, the mass quantities pertaining to a significant number of elementary particles, based on the notion of scalar density largely discussed in this text, and the Shell-to-Shell transmutation of mass, on the other hand.

In this section of the study in course, we specifically aim to redimension the notion of the mole which has been deeply misunderstood in physical science, along with its contribution to quantification of mass. First, we cannot overstate the importance of visualizing mass not as a sum of monoliths but as scalar density, in the understanding that it is a densification function that instantiates mass. This function is manifest in Shell1 of matter, the quark-gluon Shell, which mandates their nuclearization by disallowing their free-space manifestation, in the so-called phenomenon of "asymptotic freedom", almost a misnomer in our appreciation. Quark behavior responds to an aggregation or densification function which at its highest value materializes a highly accrued level of densification or compactification. This is well modeled by the Quanto-Geometric normal density Eigenfunction at the point where coordinate space s (the independent variable) takes near 0 value because $s = 0$ is forbidden. In the neighborhood of that point, the function as a representative of scalar density takes its highest possible value. It is because the s component of the matching entities diminishes toward 0, making them *quasi* devoid of a wavefunction component, that they "nuclearize" or tightly agglutinate.

At the other side of the span, the distribution of the scalar within its immediate coordinate space spread experiences a natural threshold, below which the macroscopic body will implode under inter-gravitational forces between its scalar parts. Effectively, what is known to be the **Schwarzschild radius** is in fact a scalar density threshold describing the minimum radial spread of any macroscopic object required for the object to remain integral. The Schwarzschild radius relation follows:

$$r_S = 2G\frac{M}{c^2},$$

where r_S is the Schwarzschild radius, G the gravitational constant, M the mass of the object and c the speed of light.

When the object implodes in the below-threshold condition of the radius, it does not become less massive, but more massive, precisely because its scalar density has prodigiously increased. It is by virtue of this tremendously accrued *mass* that the object, which will have become a so-called **black hole**, may experience the new and unusual ability to even attract and capture light rays. This is known to be true for all macroscopic objects starting with those pertaining to the Astral Shell of matter.

In between those two realms are several others, among which the Materials Shell based on the molar scalar as one form of scalar density. The Avogadro Number, together with the **Van der Vaals *a*** constant, give rise to the organization of polymolecular blocks and should never be regarded just as a mathematical number akin to any other real number. It is in essence a **rational** quantity because it is by nature a scalar abstraction germane to a functional derivative. Stress here the syntactic root term **ratio**. Again, even though the notion of molar mass is defined within the SI system and in terms of *Kg/mol*, it does not quite characterize the notion of molar density for mass that we are advocating here. The SI system offers a very sui generis definition of molar mass in the following expression:

Molar mass M is mass of a ***substance*** *divided by its* ***amount of substance****:* M = m/n

where m is the mass of the substance and n, the Avogadro Number, newly baptized ***amount of substance***.

We instead advocate giving the mole primacy in the definition of scalarity of mass, and in this case restricting the generic term substance to the bodies populating the Materials Shell, or first-order poly-molecular blocks. We understand that it is only when the blocks reach the Avogadro Number in quantity that a unit of mass is born for that particular Shell. In doing so we are only making good on the Van der Vaals ***a*** constant, the fundamental invariant for this Shell, per expression *(8-11b)*. So therefore, the mass of a Material in any of the nine state forms, is not the same as the mass of a molecule, or the mass of an atom, the mass of an elementary particle, or the mass of karyotic cells or that of astrophysical bodies, and quite importantly requires at least an Avogadro Number of units in quantity. A Material behavior starts with the collective behavior of an Avogadro Number or amount of poly-molecular units taken as **one**. The Avogadro Number quantity forms the core scalar that instantiates the mass of the material. This scalar incorporates a minimal expression of intangible coordinate space, because in the Quanto-Geometric vision there cannot materially be any pure scalar in existence. It is not necessary to quantify this squint element because it is eclipsed by the "weight" of the scalar and it remains endogenous to the expression of the scalar. It is not to be confused with the coordinate space that accompanies the scalar bulk, which composes the degrees of freedom and harmonic spread as the basis conjugate element to the scalar bulk. This molar mass is unique to itself and it is its relationship with its conjugate coordinate space component that is given in the fundamental invariant ***a*** for the Shell as a proportionality constant between the Avogadro or molar mass and the coordinate space spread (or harmonic volume) covariantly articulated to the mass bulk. In that sense the mass of one mole of molecular Hydrogen (H_2), as already discussed, represents the most natural unit of mass or mass etalon for all elements germane to the Materials Shell.

In this context, the physical variable of *volume* can no longer be considered nonspecifically because there is a volume constraint that comes with the molar mass emanating from its coordinate space conjugate. At this Shell level, the minimum volume that makes sense is $0.0224m^3$ because of the explosion in the metric of coordinate space in creating Materials space spread, which is the spread mandatorily associated with one molar mass. Accordingly, the molar volume, which is of harmonic nature from the wavefunction at its heart, is equivalent to an atomic orbital, because for one it defines the whereabouts of the molecular objects that form the mole in a stochastic fashion, and secondly it is numerically quantized at the mentioned quantity in m^3. As a result, the notion of physical density acquires a new dimension, because the scale for the topological spread is distinctly quantized at this Shell level. To be clear, the scalar units in the topology are made up by moles or molar groups and the spatial mosaic is scaled in volumetric pockets of $0.0224m^3$, irrespective of state. With higher priority over the density function per our usual definition of density is the interstitial organization of the topology resulting in lattice geometry or otherwise across all 9 stratified topological states for the Shell. This organization is essentially dictated by covariant coupling between scalar density and space wavefunction as defined by the Grand Eigenfunction on the basis of function spaces (and *eigenforms*). By that measure, one must consider the nine different states in the quality of energy density levels accessible to any molar topology, relatively akin to atomic energy density shelling. This view is firmly supported by the high energy level evidenced in the internal molar mobility of molecules in the volatile gaseous phase or state, the highest state stratum, which puts them at highest energy scale in contrast to molecules in the solid phase or state, the lowest stratum, which are stationary and only experience rotational or vibrational degrees of freedom. We do not offer in this writing a numeric treatment of this view in fullness, that exposé not being a part of the original motivation for this text. However, we hope to have transparently shown throughout this writing the elements for a thorough quantum treatment of molar state or state physics and now the broad lines along which a list of nine basis *quantum numbers* can be efficiently applied to this development.

Common wisdom among modern-day physicists has it that state physics in physical chemistry represents the intersection between microscopic physics and macroscopic physics, a key role being played by the Boltzmann constant k in that assessment. This view is erroneous on several counts. First of all, there is no complete science so long as there is one physical science for the interpretation of the macrocosm and another for the interpretation of the microscopic world. Because natural physical laws have created a single universe that appears to us in a unified and well-integrated fashion, a thorough understanding of natural physical laws cannot lead to disjoint science as it currently stands in theoretical physics. Consequently, we should never have at any juncture two differing bodies of interpretation that need reconciliation. Complete science resides in a single description of all physics. Secondly, only quantum physics has the potential to fully describe the macrocosm thru its natural ban of the fictitious variable of time and unraveling of the certainties behind stochasticity in macroscopic dynamical regimes. I fiercely and uncompromisingly advocate the view that, because time **does not exist** in nature, all physics is necessarily of quantum descent, making quantum physics and its evolution universal and totalitarian.

16.20. Multi-Level Structures in Molar Topologies

There is an interesting intersection between X-ray radiation physics and Materials state physics that is all worth re-examining in support of the analytic perspective we have advocated throughout this paper.

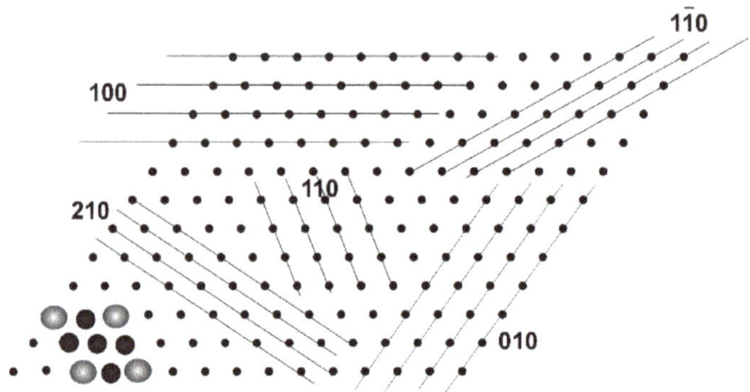

Fig. 16.11 Planes thru the lattice points of a crystal as seen from an axis perpendicular to the page

X-Ray radiation represents the tool par excellence utilized in the analysis of internal structure of crystallographic materials. X-Ray prospection has revealed that internal geometric distribution of molar units inside solids can be quite varied despite the idea of a unit-cell that may be central to the organization. Along any axis thru the solid material, there exist a number of different planes made up of lattice points that betray organizational constraints experienced by the Material during the formation process of the topology (Fig. 16.11).

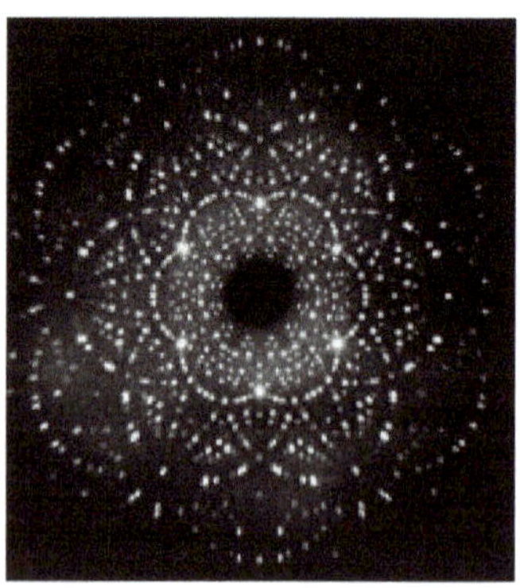

Fig. 16.12 Two-fold symmetry in the lattice structure of a crystalline material

The complex internal geometric structure of a crystal is illustrated in Fig.16.12, which portrays a clearly visible two-fold symmetry.

William L. Bragg, whose work had been pivotal in structural analysis in optical science, established the following formula that relates the radiation wavelength (reflected ray) to the planar arrangement of the lattice:

$$\lambda = 2\left(\frac{d}{n}\right)\sin\theta,$$

where d is the distance between two successive planes, n an integer defining the reflection order (first order, second order, etc.) and Θ the ray reflective angle from the lattice plane. A typification of the unit-cell of lattices had been developed which proposes fourteen different types for the so called Bravais lattices. One element of categorization is attentive to the manner in which the lattice units occupy corners, faces and internal spatial spread of the cubic outline delimiting the unit-cell and thereby distinguishes between simple cubic, face-centered cubic, body-centered cubic and end-centered cubic.

The most important aspect of lattice organization of molar topologies, which has never received the special attention it deserves, is the solid structure based on interlaced or interpenetrating lattices. For instance, in the sodium chloride salt, there are two interpenetrating lattices, one consisting of sodium atoms or ions and one other consisting of chlorine atoms or ions. An example of face-centered lattice is shown in Fig. 16.13A and a view of how the interpenetration between two lattices is realized in Fig. 16.13B.

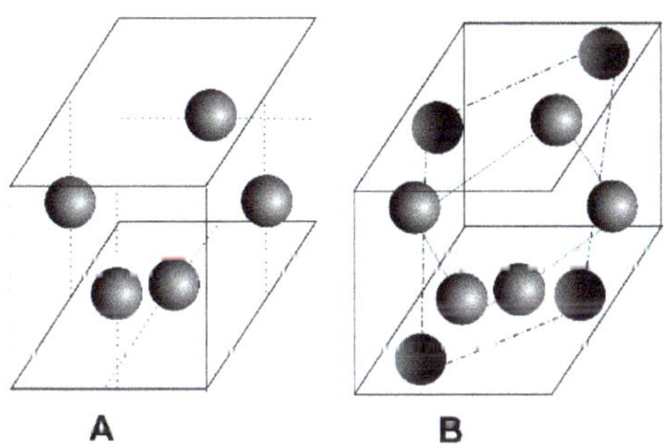

Fig. 16.13 A) Example of a face-centered Unit-cell
B) Single unit-cell view of interlacing between 2 face-centered lattices

The illustration in Fig.16.14 shows complete interlacing of not just 2 but 4 lattices in a topology, in a demonstration that the general rule in lattice organization is not flat one-dimensional repetition of unit-cells, whatever their type, but inter-penetration of paired lattices.

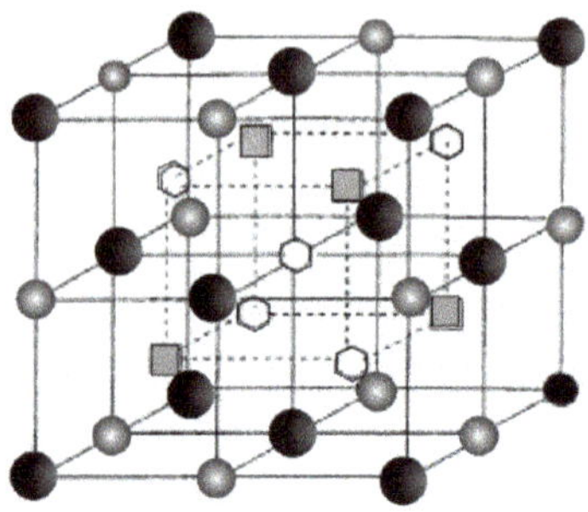

Fig. 16.14 View of a cut of 4 face-centered interpenetrating lattices

This structural pattern does represent the most fundamental level of organization of a large number of the 9 molar state strata and subsumes internal molecular bonding or chemical bonding. It directly derives from expression *(8-14b),* which captures molar state as a result of two molar topologies in interaction creating the Van der Vaals binding force. More generally, the Van der Vaals constant, invested with the Quanto-Geometric scalar/wavefunction definitions, in particular their first-order and second-order covariance as partially expressed in *(8-12),* the axiomatic rendition of *(8-14b),* is majorly responsible for the interaction between molar units that creates both the lattice and lattice interlacing arrangement inherent to lattice geometry.

In an even larger context, paired molar interlacing portrays a topology at the Materials Shell level that corresponds to the bi-localization of bound electrons in atomic orbitals at the Atom Shell. This interlacing molar topology is best viewed as an expression of energy density state akin to atomic orbitals. However, the Pauli exclusion principle, which limits atomic orbitals to two electrons, does not apply here because of the higher degree of freedom acquired by coordinate space at this Shell level. It is the evolution of this arrangement thru symmetry-breaking of the vacuum component that is going to create at the Karyot Shell level, next one in line, the populated box that viral and cellular organization is based on, thru the pivotal element of a *physical life constant.*

Regrettably, X-ray crystallographic analysis has not exploited in due fullness the potentials of this development, likely due to the prevalence of classical visualization of empirical facts. Among the benefits traditionally drawn from the effort stood the capability to compute crystal (mass) densities, or to determine X-ray wavelengths when instead the inter-planar spacing of the lattice is known. The computation of the Avogadro Number is the third benefit drawn by the Kinetic Theory from crystallographic analysis, so achieved by measuring the angles of X-ray reflections from the internal lattice planes. Beyond these representations, it is clear that *molar energy states* thru paired lattice interlacing organization represents the most fundamental levels of energy distribution in the composition of Materials. Again, to compare with atomic energy distribution, the spectrum of 9 states represents a suite of quantized energy density levels, each higher than the previous, which sets the stage for energy density transitions from topological molar behavior, much beyond rotational and vibronic degrees of freedom traditionally attributed to molecules within the lattice. As previously mentioned, this vision may incipiently be perfectly formalized under a *principal quantum number n* with integer values that are eigenvalues to the Quanto-Geometric Grand Eigenfunction, such that $1 \leq n \leq 9$ for each successive energy level or stratum, a development that is not to be presented however in this exposé.

16.21. State Physics within Quantum Optics

The sciences of state physics and quantum optics come together in crystallography to expose the rift that exists between quantum optics and classical physics, if one takes the time to deconstruct a series of notions long taken for granted. In particular, the survival in laser physics of a few classical notions surrounding radiation must be reconsidered, and ultimately demoted, if the coveted Theory of the Single Photon must firmly come to light.

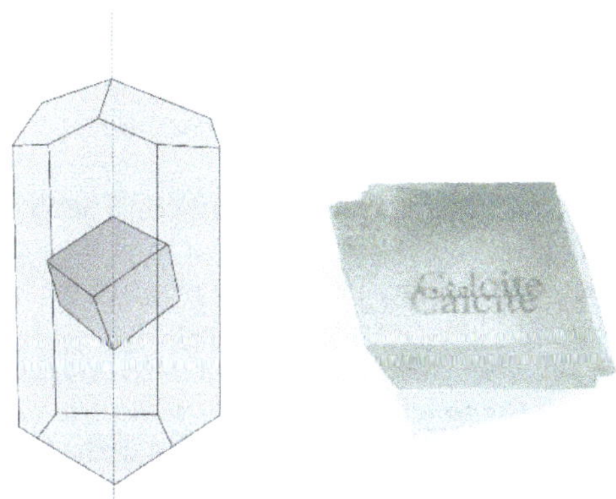

a.- Calcite crystal lattice **b.- Calcite crystal block**

Fig. 16.15 Bi-refringent property of calcite crystal blocks due to interlacing lattice populations

One of the questions in need of re-examination is the idea of the influence of electric fields on light beams, putative proof of which is found in the well-known Kerr effects. To some extent, one may put on that account as well the so-called Pockels electro-optic effect, in which an electric field presumably alters the propagation of a light beam travelling in a particular crystal lattice block. In the DC Kerr effect, light beams are diffracted when travelling thru liquid or solid dielectrics exposed to an electrostatic field. It is easy to misinterpret these results as direct influence of the electrostatic field on light beams, overlooking the role played by the propagation medium, although it is well admitted that the Pockels effect chiefly results from an alteration of the refractive properties of the medium by the electric field. Throughout the 19th Century when these notions emerged, atomic, molecular and intimate molecule-based lattice structures were unknown. In any valid theory of presumed electro-optical effects on light beams, it is necessary to properly relate the topological structure of the light beam to its constituent photon parcels, which contrastingly are electrically neutral particles. That has never been done following the advent of quantum photonic science by A. Einstein, let alone 19th Century classical physics.

One of the key properties known to light beams that ought to play an important role in any Theory of electric interaction with light beams is that of polarization. The electromagnetic theory of light posits that it is composed of two distinct sinusoidal waves, one of electric nature and the other of magnetic nature, and which are planar in their layout as well as orthogonal to one another (Fig. 16.16). However, it does not require interaction with any electrified material for the polarizing effects to be observed. We shall highlight two important observations to that effect:

- The creation and subsequent rotation of a polarized light beam only requires propagation thru a neutral bi-refringent material to take place, which is offered by interlacing crystal lattices such as in calcite (Fig. 16.15), for one.
- On the other hand, all reflective surfaces of neutral materials will reflect incident light beams with a polarized component.

In transparent materials, or materials based on lattices, a second component of polarized light, aside from surface reflection, will be refracted from the incident light beam thru the material in a manner that is parallel to the incident plane of the beam. These effects do not require the intervention of any electric field whatsoever or the material itself to be electrically active or poled. Therefore, it is difficult to reasonably relate the polarization property of light beams strictly to interacting fields of electric nature, but to lattice structure of materials, and not only in the Pockels effect but in the Kerr effect as well. On that account, the term electro-optic effect to qualify those interactions is as inaccurate as misleading, the most legitimate application of this qualifier being to the *photoelectric* effect.

In the so-called AC Kerr effect (aka optical Kerr effect), a laser light beam at extremely high intensity (in the order of GW/cm^2) finds itself unable to readily traverse the material because of interaction with the lattice, which it structurally alters initially causing a drastic change in the refractive index of the material. The change in the refractive index profile causes the beam to focus itself, a phenomenon known as *self-focusing*. Subsequently, the medium ionizes lowering the index of refraction, which allows propagation to proceed. This process is repeated in a series of successive focusing and defocusing steps. Because of the belief that an alternating electric field inherent to the beam is responsible for the ionization of the medium, the effect is alternatively designated optical or AC.

However, nothing prevents us from attributing the ionization effect to interaction between incident photons and electrons pertaining to the lattice molecular blocks, as it occurs in the photoelectric effect. Because of the very high optical intensity that is required for the optical Kerr effect to occur, one may reasonably hypothesize that the photon stream gets to significantly interact with the electronic orbitals germane to the molecular units constituting the nodes of the lattice.

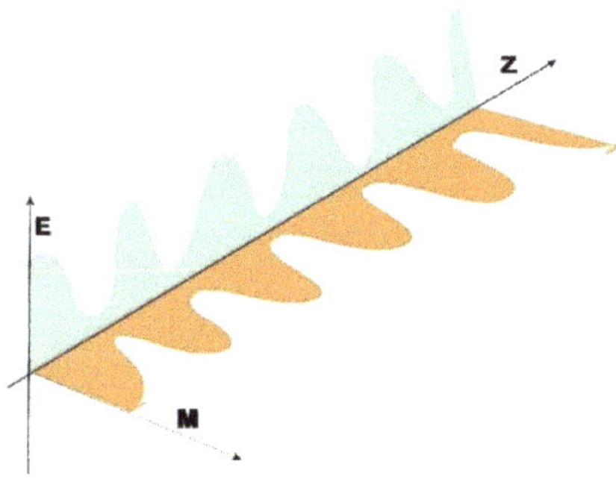

Fig. 16.16 A representation of the electric field and the magnetic field that make up electromagnetic radiation.

The idea that laser light beams are identical to alternating electromagnetic waves is a distraught notion that is incompatible with the fundamentals of quantum theory. In quantum laser physics, we hold that laser light is strictly due to transitions in energy levels experienced by atomic electrons when a population inversion has been achieved in terms of a higher number of atoms in the excited state than atoms in the ground state. Electro-negativity of atoms or molecules or lattices, or even possible intervening incident electric fields play no role whatsoever in stimulated emission of radiation. We refer the reader to the analysis of *X-ray* emission to support this contention (Fig. 16.16). Hence, the formalism that calls for an external electric field E_0 interacting with an inherent electric field to the light wave of frequency ω such that:

$E = E_0 + E_\omega \cos(\omega t)$, where E_ω is the vector amplitude of the wave, is simply unfounded. Furthermore, expressing the wavefunction of the laser light, which derives from energy density shells in spatial transition, in terms of the eternal sinusoid is an act of faith, a hypothesis at best. The very nature of the quantum wavefunction has simply remained unknown to this point, until we explicitly proposed a clear formulation of the same as a result of our research.

In a future writing, we will further soundly substantiate the above contention, while continuing to edify the Theory of the single photon, by integrating a significant number of facts known to photonic radiation both in classical physics and quantum physics, contributing to clarify at the same time the true polarizing nature of radiation.

16.22. Road Ahead for Quantum Optics

The primary motivation of this writing is to uncover important drivers in the quantum realm of mother Nature (which traditionally encompasses the Particle Shell, the Atom Shell, the Molecular Shell and the Materials Shell) thus far misunderstood, yet all appealed to play a significant role in furthering the development of quantum optics, flower of quantum science, at this stage.

These four Shells constitute the natural theater of the quanta. A detailed understanding of the foundations on which they are erected is a necessary prelude to the development of an all-encompassing Photon Model as well as the ensuing phenomenologies the dynamics of the photon throughout those realms may orchestrate. Modern physics' understanding of Light radiation coupled with our level of understanding of the internal dynamics of crystalline solid state principally has made it possible to stir quantum optics in the direction of laser optics technology. The science is new and still going thru its infancy stage. The perspective that we offer here is placed at the intersection where incremental changes, being no longer able to guarantee progress, must yield the stage to transformational changes in vision so that the journey can forge ahead. At the same time, this rendez-vous is also the place where the long-standing unresolved questions of a diverse nature in fundamental physics are being fervently reviewed, debated and resolved. Perhaps the problem of tallest order that we have forever faced in physics is the question of the man-made time variable as an inherent element to the background metric of coordinate space, which relativists have traditionally and wrongly qualified as *spacetime.*

The broadly discussed quantum formulation of Materials molar state represents a transformational stage in the theater of continual development of quantum optics science. It promises to be spectacular on the experimental side.

16.23. Conclusion

We believe the continuum vision in all-time physics stands for the most formidable obstacle to a unified mathematical physics formulation of Nature universal in the fullness of its physicality. Paramount to the continuum visualization is the false notion of time, which has played the infamous role of its single most significant descriptive variable.

This stubborn continuum vantage on nature's physics has prevented us from fully exploiting the seminal Planck constant insofar as reaching a generalized expression of harmonicity for all wave mechanics, under the umbrella of a clearly defined wavefunction, in substitution of the largely artificial notion of frequency in general and electromagnetic frequency in particular.

The continuum vantage has further barred our physics from deciphering the arguably most crucial element of the development of matter, which is the Shelling symmetry that provisions its stability and profiles its evolution alike. The continuum vision has led us to falsely believe that the vacuum is the same everywhere, by obscuring internal behavior of the vacuum that instantiates gauge symmetry or differential Shelling.

Lastly, the continuum vision is equally responsible for the classical formulation of state physics in the Kinetic Theory, whereby the Pressure-Volume product and the Pressure-Volume/Temperature ratio unduly play a preeminent yet unsatisfactory role. This continuum vision has thwarted the possibility of achieving a quantum delivery of state physics that would thoroughly and granularly account for all 9 states in existence and not just 3, let alone the ultra-structural identity and composition of lattice topology.

The tensorial and geometric qualities of the vacuum upheld in Einstein's Theories of Relativity addressing interaction in the higher astrophysical Shells of mother Nature are absolutely not at all incompatible with the Quantum vision. Once we have well understood the internal behavior of the vacuum that consistently instantiates its serial Shelling evolution, starting from its quantum realm, that otherwise intractable incompatibility vanishes.

Nonetheless, the above does not mean to argue that there is no place whatsoever in physics for the continuum perspective. Its authentic dimension simply lies elsewhere, where it is not an artifact of misperception of scope and granularity, as it has unfortunately been thus far in physics.

BOOK III

Chapter 17

The universe is driven by a very simple force
– Symmetry.
Mike Hockney

FOUNDATIONS OF SYMMETRY IN GEOMETRY

In this volume of the series, we make it our objective to unravel the foundations of Life through a particular biophysics vision along the mathematical guidelines set forth in the Quanto-Geometric Theory. As we have made clear at the end of Book II of this publication, the visualization of the origin of the symmetric disciplines belying the formation of the suite of U_3 physical Shells is of capital significance and becomes paramount in the analysis of the origin and development of Life about to take place. To that end, in this introductory Chapter, we are going to studiously devote our attention to the treatment of Symmetry in Geometry and mathematical physics. Our objective is not only to revisit this construct as commonly understood in science but to unearth the many significant shortcomings associated with the notion and properly re-dimension the same in preparation for the biophysics study of Life we are embarking on.

Symmetry stands for a central notion in Mathematics with paramount applications in Theoretical Physics. In 1915, mathematician Emmy Noether famously axiomatized this notion which resoundingly made its way in fundamental physics in its chapters of Lagrangian mechanics and Hamiltonian mechanics. However, there is only one dimension to symmetry in mathematics and mathematical physics: an invariance that is preserved under a particular transformation. Contrary to this approach, there exist many other dimensions to the artifact of symmetry, which, if correctly prospected, could have helped accurately answer the long-running question in mathematical physics concerning the very origin of those physical symmetries and the reason for the physical mechanism of spontaneous symmetry-breaking. The ignorance of origin of observable symmetries as well as the existence of other categories of symmetry is accompanied by a misunderstanding of the very states of symmetry, which is reflected in an inexact assessment of the ultra-structures subtending mathematical symmetry. In this writing, we aim to expose the caveats of symmetry specifically in Geometry, as viewed from the overarching Quanto-Geometric perspective, and to show their resolution with the application of special Differential Geometry principles emanating from the Quanto-Geometric framework.

17.1. Inexistent Definition of a Point in Geometry

Perhaps the most encompassing problem facing all known Geometries is that of the definition of an abstract point. Clarifying the nature of a point in all Geometries has never been successfully achieved, albeit relentlessly attempted. As trivial as this question may seem at the outset, it is of enormous consequences in the application of Geometry to the natural world and its dynamic systems. By the same token, the absence of that definition may very well be one of the major reasons why the question of a "*minimal Geometry*" has never been satisfactorily resolved in mathematics. Euclidean geometers long wrestled with the question of the definition of a point and, at the end, desisted from the quest for an answer. Hence in Euclidean Geometry, there is no formal definition for a point, whether abstract or physical, because the philosophy subjacent to Euclidean Geometry finds itself utterly unable to define it and therefore simply states that there is no prior construct from which it may derive. Since no other Geometry had arrived at a formal definition either, all Geometries have so been left to adopt the latter viewpoint. Nevertheless, the abstraction of a point remains a central basis figure to all Geometries, which all extensively peruse it, one might say, in an "empirical" manner.

If a point cannot be defined in Geometry, does it belong to Geometry as a metric category? Is it legitimate for mathematicians to build *"minimal Geometries"* based on the construct of points? Our resolute answer to these questions is that if a point cannot be defined by the philosophy subjacent to universal mathematical logic, then a point as an abstraction does not forthrightly belong to the geo-*metric* order. It belongs elsewhere and can only be used in Geometry as an import or the result of interaction between other mathematical categories and the geometric order. In other words, a point is simply not a primitive of Geometry or a legitimate geometric irreducible. Consequently, one is in good stead to expect any Geometry constructed on the basis of points as primitives, as all of known ones are, to face more or less serious difficulties when attempting to faithfully render the intimate physicality of mother nature in the 3-Dimensional arena.

17.2. Notions of Points and Lines in Geometry

Geometry, in our view, is primarily about spatial spread or dimensionality, on one hand, and form or shape, on the other hand. Ever since Geometry was axiomatized at the turn of the 19th Century, it treats these entities as abstract categories. If a point is dimension-less, and cannot have a defined shape, as considered by conventional mathematical wisdom, then it cannot be a primitive of space or Geometry. It can only be considered as a second-order category in Geometry.

Veritably, the construct of points does not belong to Geometry but the Number realm as scalar integrants of a number hierarchy. Hence, the most irreducible primitive of pure Geometry is left to be a **line**. So therefore, it is improper to state that a duplet of points define a line in pure Geometry, but that a line can be ultimately reduced to a minimum of two points. Simply put, two points may sufficiently represent a line but do not define the abstraction of a line.

Euclid attempted to define *"a point as that which has no part"*. This definition however is inaccurate and doubles as wrong. Something that "has no part" is a wrong construct in this case because inarguably a point is an entity with a single part. A single part is a "part" and not "no part". Hence, Euclid's definition makes the point a monolith rather, which complies with its true nature as a scalar.

The problems associated with this definition had steadily driven mathematicians away from it. They had since preferred to accept the point as is, claiming, as previously mentioned, that there is no prior cause or antecedent to a point in Geometry. I would venture that Euclid's definition of the point is probably most applicable to space instead, since one could apply the expression "*absence of part*" as a descriptor for true physical realization of the vacuum as a spread. A point, we firmly conclude, is not a legitimate element of Geometry, but an import of Geometry from the scalar order. It is now fair to concede that the use of the construct of points in Euclidean Geometry, even under either a mistaken or undefined identity, does not necessarily invalidate Euclidean Geometry altogether, or the other Geometries which equally peruse the entity. However, the equivocation in the identity of the element entails non-trivial consequences.

17.3. Elements of a Minimal Geometry

The notion of a "**minimal Geometry**" represents a paramount yet unachieved construction in mathematics. When it comes to Number Theory, it has clearly identified the minimal scalar hierarchy to be upheld for the construction of number systems and their associated operators, in the embodiment of the number base of a specific radix. Said base rests on the notions of cardinality and ordinality for its principal underlying principles. By contrast, we do not have a similar hierarchy in Geometry. We do not have a match to the construct of a number base in Geometry, an important and necessary development for the representation of the long sought "minimal Geometry". In Quanto-Geometric Theory, we have however undertaken recently that development and shown the elements of philosophical logic upon which a "minimal Geometry" is naturally erected. We will only discuss the issue here from a critical perspective relative to the views hitherto entertained in Geometry concerning its basic elements. A minimal Geometry must be visualized as the most primal or most irreducible geometric order and as such ought to elicit the seminal elements composing pure Geometry. We have already dismissed the construct of points, whether singular or in coalition, in the role of first-class integrants of the pure geometric order.

For us, a line, or specifically a line segment, ought to represent the most irreducible geometric element in a purely abstract representation of primal Geometry. Because, for us, Geometry is primarily about space or more generically *spread*, its most irreducible basis element must be a primitive of spread. Such an element is firmly given in a line, or its reduced form: a line segment. Note that spread is also a generator of dimension or size; the latter must be understood as the metric expression of spread when the notion of spread is applied to physical space. So therefore, all Geometry may be reduced to a coalition of lines, just as all and any numeric construction may be reduced to a coalition of scalars. Consequently, when modeling physical space, pure Geometry must populate the tapestry of space or physical vacuum with the dimensional element of the line and make sure that a point cannot exist monolithically in any generic spread or patch of space of its representation. Points can only exist within lines or line segments. If they ever come to be represented outside of lines, points cannot and shall not define spread or any physical vacuum tapestry but only infer attributes of line constructs and their derivates. For instance, what defines a circle is its radius but not its origin. The point origin is germane to the radius as a finite line or a line segment in the condition of a stop. *One shall not look at the origin of a circle and view it as an independent element to the circle with its own monolithic and autonomous existence. There is no agency, abstract or tangible, by which the point origin can relate to the circle other than through the radius element.*

These definitions are of paramount significance and derive from the particular Quanto-Geometric perspective that we bring to the analysis of abstract mathematical constructions. They are meant to provision a higher placed vantage point from which one can assess and cure the lacks and misgivings of the variously developed Geometries in mathematics, in particular Euclidean Geometry and Differential Geometry. In linear algebra, a line is given to be a one-dimensional vector space, the set of scalar multiples of a given "guiding vector" i.e. This definition of a line is not fundamental, however, because a vector is a construct that already assumes the existence or hypothesis of a reduced form of a line, which is the line segment represented in the length of the vector. In other words, the length element of the vector already implies a line segment or a finite line. Secondly, perhaps the most important flaw of that definition, insofar as pure Geometry is concerned, is its calling upon scalar elements, which do not seminally belong to the geometric order but what we have called the scalar or numeric order. While the line is the primary "space" or spread where a vector or all vectors live, we cannot turn around and make the vector the origin of the line. A vector is instead a progeny of the notion of a line.

Likewise, it is not proper to state that any set of 2 points determine a unique line, as posed in Euclidean Geometry. In pure Geometry, the line has preponderance and precedence over point elements. A space or spread has to be pre-determined for 2 ad hoc points taken within its tapestry to mean anything geometrically, because spread or space is primordial in pure Geometry. If the space is flat, 2 points mean or reflect a straight line or linear distance within the space, when the latter has been normed. If the space or spread is curved, 2 points mean or reflect a curvilinear distance within the space or what has been designated a geodesic. Applying the attributes of *affine* (curved space) or *metric* (flat space) to the tapestry is not defined or determined by the points per se but by both the spread and nature of the space patch.

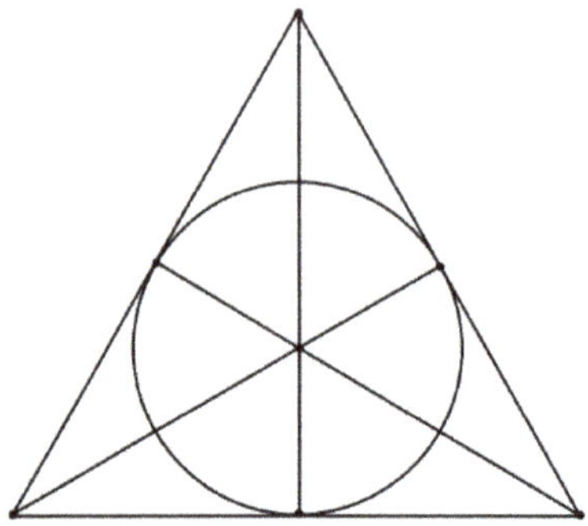

Fig. 17.1 Fano plane

Some have attempted to turn such figures as the Fano plane, with its alleged 7 points and 7 3-point lines, into the basis of a "minimal Geometry" (Fig.17.1), whereas some others sought to make of the Mobius strip (Fig.3) the irreducible element of pure Geometry. In fact, such figures represent superstructures that are very much subsequent to the prime irreducible level of any pure geometric construction. We will further discuss the Mobius strip later.

17.4. Parallelism in Geometry

The problems with the definitions of points and lines in Geometry have led to the other intractable problem of a formal definition for *parallelism*. Euclid had admittedly well struggled with this question and remained dissatisfied with his 5th postulate which only offered a side view of parallelism:

If a line segment intersects two straight lines forming two interior angles on the same side that sum to less than two right angles, then the two lines, if extended indefinitely, meet on that side on which the angles sum to less than two right angles.

Scottish mathematician John Playfair, offered in turn this axiom relative to parallelism:

In a plane, given a line and a point outside of it, at most one line parallel to the given line can be drawn through the point.

There have been many other propositions of parallelism in the history of mathematics, one as dissatisfying as the next. For instance:

Two lines are parallel if and only if they admit a common guiding vector, indeed a dissatisfying proposition despite its appealing succinct or reduced expression. Because of the prime role played by points and the unassertive view of the nature of a line, a straightforward accomplished definition of parallelism has yet to be achieved in Geometry. An angle, so much as the vector construct, is a geometric element that is posterior to the line construct in pure or absolute Geometry. They should not be used to define the very first order of two-line ontology, and if they are, the resulting axioms might not strictly obey the hierarchy of mathematical logic, or else it might fall short of compliance with Occam's law of parsimony (Occam's razor), though they may very well be of practical utility in applied Geometry. Following the Quanto-Geometric constructs of linear ontology, one may define parallelism between two lines in first-order Geometry as follows (Fig.17.2):

Given two straight lines T and S in a plane, and an equal reduction of each in terms of two segments of equal length on each line ($[AB]_T = [CD]_S$), if by joining the end points of each segment without intercross (BD and AC) the resulting interlinking segments are equal ($BD = AC$), then the two lines must be parallel to one another. If the interlinking segments are unequal ($BD \neq AC$), then the two lines must intersect with one another.

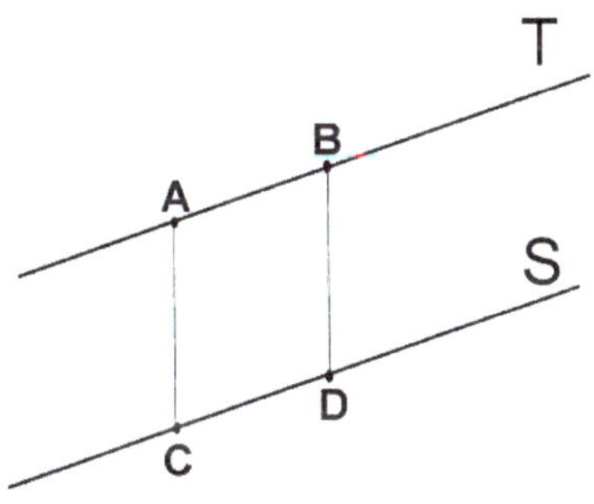

Fig.17.2 Definition of parallelism between two lines

This definition remains within the boundaries of the ontology of the line construct and has no recourse to angles, vectors or triangles, which are all derivate constructs. Note that if, alternatively, we were to have recourse to the figure of distance between the two lines for the definition, we would be otherwise referring to a 90° angle formed with a third line, which would represent an equally improper irreducible definition.

17.5. States of Symmetry in Affine Space v. Metric Space

If non-Euclidean Geometry had emerged only when Gauss, Bolyai, Lobachevsky and other Differential Geometry pioneers decided to drop Euclid's 5th postulate, in the fundamental question of linear parallelism the new Geometry did not demonstrate more assertiveness. Parallelism may well be the very first expression of explicit symmetry in primal Geometry, as later understood in mathematical physics. Effectively:

Considering a line as a non-finite element, should one view two lines as a progression of their reduced or finite forms denoted segments, parallelism is thus the invariant equality between the interlinking non-intersecting segments defined in the above axiom ***(BD and AC)****, as the finite co-linear segments* ***($[AB]_T = [CD]_S$) unfold*** *in an infinite manner.*

Here the *unfolding* is the transform.

Most non-Euclidean geometries are built on the notion of an opposition between *affine spread* and *metric spread*, owing to a profound misunderstanding of the very intrinsic nature of a geometric spread. At the heart of all geometric spreads is what in **Quanto-Geometric Theory** we have identified to be a *state of symmetry*. It is an intangible that only becomes observable by association or covariance with the scalar order, under the representation of a *form*. Parallelism, which is incarnate in pure or absolute Geometry, gives us evidence of the intangible state of symmetry within the geometric order. The state of symmetry is subjacent to linear ontology and exists both within the rectilinear construct of a line and the curvilinear construct of an arc or a curvature sectional. What makes an isometry[10] possible is the existence of that state of symmetry within geometric spreads.

Because of the way that we arrive at a granular description of curvatures, we have designated what is curved symmetry an *affine* isometry while we reserve the epithet of *metric* to rectilinear symmetry. It is clear that an "affine" spread is going to be more "resolved" than a "metric" spread because affine is the result of curvilinear behavior or tensorial curvature experienced by the spread. Whereas Euclidean metric has no rectilinear constraint because the directional spread in a line element is instead non-finite. To better understand this, consider a line and a circle. There is only one way that an arc segment, a curvature sectional of the circle, can be *con-cordant* to another, which is thru the reference of the one radius of the circle. Most importantly, the amount of metric concordances in the circle is a finite quantity because it is a closed spread, which spans from 0^0 to 360^0. In contrast, the amount of metric concordances in a line is infinite, because it is an open ended element, subject to an infinite amount of segmentation along a direction that is always preserved.

It is important to highlight that metric concordances and isometries exist because of the existence of states of symmetries which become observable as forms or extrinsic forms. One could construe a state of symmetry as Plato's *forms*, which to him were counter-intuitively intangible elements. His vision, however, was too metaphysical to be satisfactorily descriptive of the physical world, and furthermore could not identify the fundamental irreducible forms, indeed a large part of the essence of Geometry.

---

[10] Among the many definitions of isometry, it is also defined as a function from one metric space onto a second metric space, having the property that the length of a line segment in the first space is equal to the length of the image segment in the second space.

17.6. Affine Description of Foliates

What Gauss' *Theorema Egregium* recognizes, perhaps not so remarkably as he had hoped and once one is able to adopt the Quanto-Geometric perspective, is the existence of these states of symmetries immanent to spreads in two dimensions, which are already due elements of one-dimensional linear Geometry.

Theorema Egregium only reflects the fact that true *3D* or *2D* physical space are isotropic and homogenous as an application of spreads in pure Geometry. A Gaussian curvature is truly not a real or fundamental geometric curvature category in and of itself. We speak of a Gaussian curvature only because of the specific manner in which we arrive at the description or identification of the curvature, which is essentially from Gauss' curvature constant *K*, in which derivatives play an essential role. The elements that make up the curvature are, however, totally Euclidian in nature. They are the natural basic forms treated in Euclidean Geometry (circle, ellipse, parabola, hyperbola), to which we have to add other special forms of (secant) *hyperbolic* nature or unit-forms well captured in Analytic Geometry, albeit never typified or normalized. So therefore, it is only because of the universality of this spectrum of unit-forms, and because all forms are made up of these irreducible unit-forms, that *Theorema Egregium* is true. We offer further down a detailed critical analysis of the representation of a Gaussian curvature.

Consequently, it is not at all surprising that "local" curvature described by functional tangential behavior (derivatives) is directly integrable into a "global" Euclidean space. In that sense, the connection between Euclidean Geometry and Differential Geometry is one of complementarity but not hierarchical superiority by the latter. Inarguably, the notion of *minimal surface* in Geometry does not meet the standard of fundamental form, no matter how it has been defined in non-Euclidian Geometry. For instance, the local least-area definition:

A surface $M \subset \mathbf{R}^3$ is minimal if and only if every point $p \in M$ has a neighborhood with least-area relative to its boundary.

This definition is deficient by many regards, not the least of which is the determination of making spreads a primary composition of points. Furthermore, there exists no minimal surface other than the unit–form that a spread is reducible to, since a composition or aggregation of unit-forms can never be construed as a minimal spread. Beyond the unit-form, a manifold or foliate patch is necessarily a composite spread. No patch of spread that is not typifiable within the fundamental spectrum of unit-forms may exist and no spread may vanish in any region of a foliate patch. It may be that a physical foliate patch lacks a part of its fabric, but the intangible spread of which the physical patch is a descent can never vanish.

Arguably, the best proposition in Differential Geometry may be that of morphism and morphic evolution of surfaces, sheets, surface sheets or foliates. Say that you wake up in the morning and look back at the blanket on your bed. You observe therein a mosaic of interesting patterns in terms of folds and shapes and intertwining folds, and want to communicate a description of this interesting mosaic to the rest of the world. Should you try to describe them directly with discrete Euclidean elements in a suite, you realize the enormity of the task over which you would most likely have to spend the rest of your life. You quickly realize the urgency of a more sophisticated methodology enabling you to explain shapes and folds in terms of their curvilinear evolution throughout this "blanket" manifold. Differential Geometry makes this possible by studying all shapes of the foliate through their curvilinear evolution from an inner functional perspective of the curvature sectionals. The entire toolset of Riemannian Geometry is geared to that end, that is to say theorems or axiomatic definitions, morphic relationships, mapping methodologies, etc.

However, what the many artifices of Differential Geometry and their successors may not have quite realized or sufficiently pondered is the fullness of the following:

1. A tangent vector space implies a **functional** view of the geometric space under study (this, we concede, they probably did not miss).
2. The implied function(s) or *functional* contains the very geometric design of the **curvilinear evolution** of the foliate.
3. The curvilinear rundown of all functions in their graphs is a **serial suite** of Euclidean sectionals and other typified (secant) hyperbolic sectionals (in 2-space or 3-space).
4. Euclidean Geometry with the inclusion of those (secant) hyperbolic normals represent and present the **Eigensolution** to Differential-Geometry-laden geometric problems. One can view this to be a result of the hierarchy that exists between all three Geometries.

In all, in terms of a relationship between all three Geometries, it is fair to conclude that they are strictly complementary to one another in their approach to curvature.

17.7. Misgivings of Topology in Riemannian Geometry

Perhaps, the most important flaw in Riemannian Geometry dwells in the foundational representation of topology whereby a manifold is considered to be an assembly of points each with a Euclidean neighborhood. What we traditionally consider to be the basic elements of Euclidean Geometry are dots, lines, angles, surfaces and volumes. In Differential Geometry of surface sheets, foliates have been studied from two principal perspectives: *extrinsically*, as relates to their embedding in Euclidean space, and *intrinsically*, reflecting properties determined solely by distance within the surface as measured along curves on the surface, or entire isometries within the surface sheet. Within that representation, the topological map of the foliate is based on a coalition of points which must have a neighborhood more evolved geometrically than the point so that such population of dots can relate to one another.

However, the only way, in our sense, that a Riemannian point can have a neighborhood is for the dot it represents to be intrinsically at least bi-dimensional (bi-linear) or somehow endowed with further Euclidean faculties. In other words, since a dot can never be even minimally bi-linear as such, it is only as a result of *dimensional indiscernibility* that a manifold irreducible set to be a dot can encapsulate meta-Euclidean geometric elements, which are the only other elements possibly making up the neighborhood of dots, at any level. If the concept is inclusive and makes the neighborhood a part of the dot, the above argument ostensibly shows that it is tantamount to a flawed concept. If instead the concept is expansive and makes the neighborhood a periphery of the dot, then it is no less flawed because the neighborhood would have to spread over the surrounding dots, since the theory does not propose an interstitial arrangement within the fabric of curved surfaces.

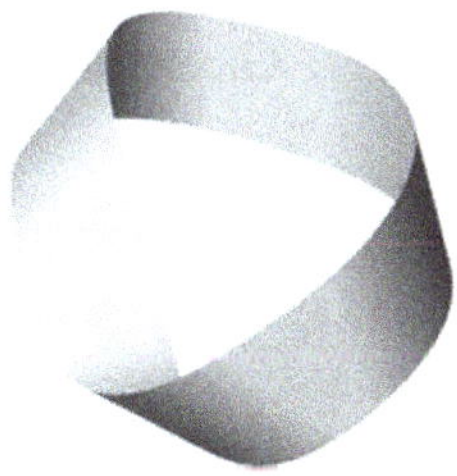

Fig.17.3 The Mobius strip

Only if one imagines that the ground state of a physical surface sheet is pre-populated with pre-formed geometric elements in each point that a point can become multi-dimensional in and of itself when the space is excited or no longer at ground level. For instance, the point designs (or miniaturized designs) on a balloon that are invisible, or whose inner punctual patterns are indistinguishable, when the balloon is deflated, and then become visible when the balloon is inflated. The theory, however, offers no such proposition, and furthermore one may well argue that such a space is not a purely abstract entity but already a physical surface sheet. As previously argued, in pure abstraction, a dot must be one-dimensional. Any way you look at a dot, for it to be so it must be visualized as an absolutely **monolithic** element. There is in abstraction no sub-order over which a dot would acquire dimensionality or multi-dimensionality in and of itself. It must be monolithic or atomic onto itself or it is not. A point that encloses a neighborhood is thus a skewed abstraction, whatever the nature of the neighborhood. Hence the notion of points in a manifold that each contains a Euclidean or metric neighborhood is definitely no less than a troubling notion.

How about a point with a differential topological neighborhood? Should points of the surface sheet be based on a differential irreducible, like a Mobius strip for instance (Fig.17.3), one still runs into serious conceptual problems. If the Mobius strip irreducible must describe a full form as acknowledged, then there is no lower geometric dimension to that entity, simply because the form must always be considered in total. So much so that it is sustained by one invariant, which, if broken, vanishes the form. So therefore, a Mobius strip is its own irreducible. From there, we can only consider composite topologies made up of these strip units. One may probably not know what a point on a Mobius strip means geometrically, because the point alone makes no sense if it does not relate to ALL the other points of the strip at once. That is why in topology structure is knowingly all that matters. And that is why a manifold based on irreducible topological patches or points each endowed with a topological neighborhood would equally amount to a troubling notion in higher-order canonical logic.

I shall add that the ultimate reason why in Differential Geometry points on curved surface sheets are accorded a neighborhood is simply so that the Theory affords itself means to connect them to each other through the device of morphism, which ultimately results in the Differential Geometry map. Seemingly, a foundation not too solid.

17.8. Order of Dimensions in Geometry

Some have argued that it may seem unnatural, when studying a curved surface sheet or manifold, to begin by embedding it into a flat Euclidean space, as primally posed by Gauss for a foundation of Differential Geometry. Let us dwell for a moment on the question of dimensionality in Geometry.

17.8.1 An Improper Account of Dimensions

If one considers a surface sheet to be a bundle of manifolding patches and attempts to describe the entire foliate in terms of a tangent vector space, one will find that the foliate order (otherwise designated Gaussian curvature) must be composed of irreducible Euclidean conics and conic sections as well as secant hyperbolic curvatures, all in the role of primitive spectrum of the geometric curvature order. That principle stands for the essential reason why *any differential (Riemannian) manifold can be embedded within a Euclidian manifold* $\boldsymbol{R}^N$ *for N large enough.* The above also means, and counter-intuitively so, that curved space can be demonstrated to be a derivation of flat space (and not the other way around), in the understanding that Euclidean metric is the mother of all metrics.

In other words, if it is true per Gauss' *Theorema Egregium* that curvature in a surface can be described intrinsically with no recourse to any referential element outside of the surface sheet, while the surface still necessarily and intimately relates to the extrinsic Euclidean space containing it, that is only because there is a common geometric order to both local or intrinsic space and global or extrinsic space. Effectively the following two questions are very legitimate interrogations to provide an answer to in this investigation:

1. What is the boundary between the "intrinsic space" inside the foliate and "extrinsic space" outside of it within the "global space"?

2. What physical property of "intrinsic space" is the generator of the curvilinear or curviplanar behavior of the foliate?

Such answers have never been provided by the pioneers of the Theory. In abstraction, extrinsic space (higher dimensional order) must be made up of intrinsic space elements (states of symmetry). The distinction between intrinsic space and extrinsic space in Differential Geometry has the misfortune of tending to mask the transcendent question of the nature of the internal tapestry of physical space, owing to the preponderance attributed to the relationship between 2-dimensional space and 3-dimensional space. To some extent, one may consider the 3-dimensional curved foliate a morphic realization or the explosion of the states of symmetry germane to the 2-dimensional Euclidean spread.

This representational problem is further exacerbated with the $\boldsymbol{R}^N$ representation of manifolds which suggests a possibly infinite number of dimensions for manifolds. Strictly speaking there are only 3 physical dimensions for physical manifolds which should be represented as: $\boldsymbol{R}^N$ such that $1 \leq N \leq 3$. So should equally be represented the higher up abstract spread which models physical space as well as the physically tangible foliates. Any physical dimension $N \geq 4$ remains outside of the scope of traditional Differential Geometry, because the Theory has never established the abstract formalism to support their existence or representation. Consequently, all $N \geq 4$ do not refer to *3-D* coordinate space but to simulations of higher order degrees of freedom duly rooted in *3-D* space, but not estranged to *3-D* space. There is no *extrinsic* meta-space beyond 3-space in all traditionally known Geometries, such abstractions being representationally vacuous in Differential Geometry as relates to coordinate 3-space and only useful for virtualizations to be implemented over 3-space elements. Nevertheless, we do not explicitly reject the possible physicality of meta-spaces beyond 3-space, which is a core subject in Quanto-Geometry.

17.8.2 Inappropriate Modeling of Physical Dimensions

For sure, Mathematics is a world of abstractions, and all mathematical spaces are abstract spaces, including the concept of Euclidean space. Physical space however is a real entity because we existentially experience its 3 degrees of freedom, and so our mathematical Euclidean space becomes an abstraction that we arrive at by inference from observed *3D* space. It is not clear that the portrayal of *3D* space in mathematics is properly meant to be abstract. When we speak of a vector space, we are clearly speaking of an abstract space, but when we speak of a volume or an area, those Euclidean notions unduly convey a representation of true space. This may be due to Geometry having been originally physical science, until its axiomatization by the generation of early 20th Century mathematicians. It is important to firmly uphold the principle that all mathematical spaces are indistinctively of abstract nature and it is only in that quality or the quality of a model that they relate to actual physical space. In all known Geometries, the conflation of abstract modeling space with real space leads to faulty constructions attempting to erroneously depict physical reality. Said conflation has fortunately left Euclidean Geometry intact due to the aforementioned reasons, whereas it has had deleterious consequences in Differential Geometry because the latter seeks a more intimate description of the intangibility of coordinate space.

No more than physical space, the notion of time is an estrangement to mathematics since it is essentially a physical concept. Does it not represent a direct qualifier for an intrinsic dimension of the natural physical world? It is, however, a bankrupt concept that we humans compulsively inject into reality whereas the natural world remains totally stranger to it. There is simply no timeline or flowing time to be observed or somehow witnessed anywhere. That also makes the representation of a "spacetime" dimension a false and corrupt concept, because it pretends to model a reality of its own creation. Consequently, the notion of Lorentzian manifold (Minkowski spacetime) qualifies as a false concept, no matter the mathematical formalism it has been given (+++,-). Even in pure abstraction, that representation is anti-canonical, since one may justifiably wonder why should **one** negative be associated with **three** positive. By which logical Norm? Further, both the 3 dimensions of space and that of time are quintessentially and equally indiscernibles; time is not "naturally" orthogonal to the 3 dimensions of space. So, while the mathematical mind may want to abstract out the 3 dimensions of space in stating that in the modeling manifold those 3 dimensions are "space-like", there is no support for the "time-like" dimension, logical or otherwise. And one can also see that creating an abstract generic entity to model space and characterizing it as "space-like" is already defeating the model or its purpose because the required distance between the model and its object is plainly not there. We will come back later to Makowski's notion of a *spacetime*.

Therefore, the fullness of the metrics projected thus far in mathematical physics, which universally incorporate both a time dimension and non-abstract space, are ex-abrupto mathematical constructions devoid of fundamental mathematical logic. Further elaboration on this point later on.

17.9. Why Gauss' First and Second Generic Forms are Not Fundamental

In the quest for a generic description of foliates, Gauss considered how a two-dimensional surface is intrinsically embedded in a *3-D* global space. He first assigned coordinate values to points on the local surface *(u,v)* from the global coordinate system such as:

$$x = x(u,v),\, y = (u,v),\, z = z(u,v).$$

He then considered a line element on the surface and generalized the Pythagorean Theorem to the length of that line segment:

$$ds^2 = dx^2 + dy^2 + dz^2.$$

He further considered the derivative functions *E*, *F* and *G*, intrinsic quantities independent form the global space:

$$E_{(u,v)} = \frac{\partial r}{\partial u} \cdot \frac{\partial r}{\partial u} \quad F_{(u,v)} = \frac{\partial r}{\partial u} \cdot \frac{\partial r}{\partial v} \quad G_{(u,v)} = \frac{\partial r}{\partial v} \cdot \frac{\partial r}{\partial v} \qquad (17\text{-}1)$$

which yields by replacement:

$$ds^2 = E(u,v)du^2 + 2F(u,v)du \cdot dv + G(u,v)dr^2.$$

Gauss called equations *(1-1)* the *first fundamental form*, which is endowed with all the properties required for Euclidean Geometry in terms of computation of lengths of surface curves and areas of surface patches.

Subsequently, Gauss considered the second order derivatives to *E, F* and *G* by extension, whereby *E* becomes *L, F* becomes *M* and *G* becomes *N*, such that:

$$L = n \cdot \frac{\partial^2 r}{du^2} \quad M = n \cdot \frac{\partial^2 r}{du \cdot dv} \quad N = n \cdot \frac{\partial^2 r}{\partial v^2}$$

He posited that these derivatives are an extrinsic characterization of the surface depending upon the embedding of the surface within the global space and proposed the expression:

$$d\sigma^2 = L(u,v)du^2 + 2M(u,v)du \cdot dv + N(u,v)dv^2 \quad \textit{(17-2)}$$

where $d\sigma^2$ is the expression of the *second fundamental form.*

Gauss finally related the two forms to one another in order to obtain a single figure allowing the categorization of the total surface curvature:

$$K(u,v) = \frac{LN-M^2}{EG-F^2} \quad \textit{(17-3)}$$

What he found remarkable or *egregium* was that in the final expression of the curvature constant *K* in the case of an orthonormal coordinate system where $F = 0$, in the expression of the constant, the functions *L, N* and *M* vanish leaving only *E* and *G*, so that:

$$K = -\frac{1}{\sqrt{EG}}\left[\frac{\partial}{\partial u}\left(\frac{1}{\sqrt{E}}\frac{\partial\sqrt{G}}{\partial u}\right) + \left(\frac{\partial}{\partial u}\frac{\partial\sqrt{E}}{\partial u}\right)\right] \quad \textit{(17-4)}$$

Gauss finally established that for $K = 0$ the geometry of the embedded form is *Euclidean*, for $K > 0$ the geometry is *elliptic* and for $K < 0$ the Geometry is *hyperbolic* (Fig.17.4).

What one may conclude after a rigorous examination of this construction is that it is fraught with arbitrary definitions, no matter how central it has been to the science of Riemannian Geometry to this day. First of all, it is not clear why we should stop differentiating the functions after the second derivation. If the functions are legitimate, then we should be able to pursue their differentiation to any possible order. We do not see any characteristic deriving from the dimensional order of the global space, which is *3D*, requiring only 2 orders of differentiation of the posed functions. Such restriction puts in question the authenticity of the order of forms posited in the construction as much as the foundational stature of the forms advanced in the Theory.

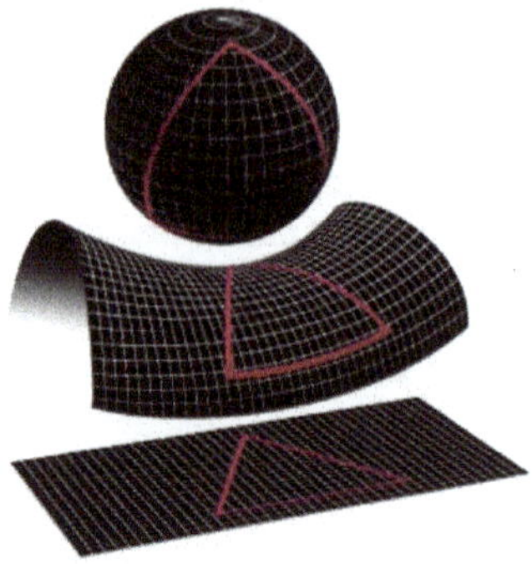

Fig.17.4 The three basis manifolds emanating from Riemannian Geometry

Secondly, in its condition as a ratio, the curvature constant K is not a natural constant but reflects one among many other possible ways Gauss could have chosen to express it algebraically. If Gauss had chosen to pursue a 3rd order derivative, the resulting *third fundamental form* would potentially offer more cases for the curvature characterization typified by the constant K. As it stands in Gauss' original theory, When $K = 1$ the curvature is said to be closed or elliptic, when $K = -1$ the curvature is open or parabolic, and when $K = 0$ the curvature is vanishing or planar. So therefore, one can see that the curvature hierarchy of *closed, open* and *vanishing* does not find a concordant numeric match in the progression of the value of K, which raises doubts about the legitimacy or accuracy of the construction.

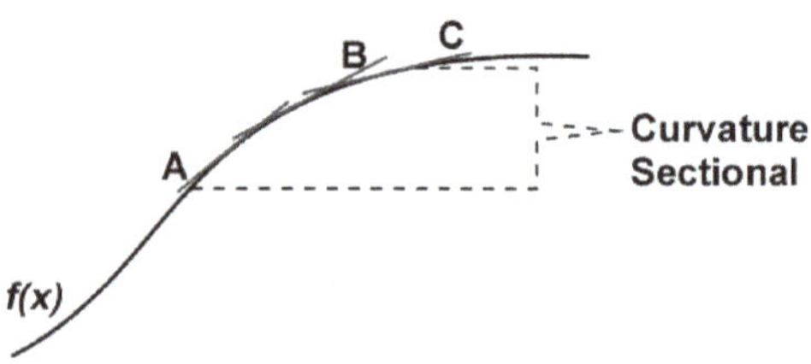

Fig. 17.5 Curvature sectional with curvature defined by the tangent bundle *[A, B, C]*

The truth resulting from an overarching perspective, which is even more elevated than the Gaussian perspective, is that the very order of differentiation in Analytic Geometry already contains the premises of the order of fundamental forms. Effectively, we know that a derivative is the evolution of a secant to a curve by affine variations until it becomes a tangent to the curve. Consequently, the method of differentiation is tantamount to the study of geometric curvature in curves, since the slope of the tangent at one point on the curve is an indication of its degree of curvature at that location. By extension, a two-dimensional tangent bundle is therefore a thorough descriptor of a curvature sectional within the graphical evolution of a Function (Fig.17.5). This view leads us to the discovery of the natural fundamental or irreducible forms deriving from the geometric stack that constitutes Absolute or Minimal Geometry afore discussed. In Book I of the series, we carried out an ample discussion of this view.

17.10. A New Dimension to Differentiation

A derivative to a curve is traditionally defined in Analytic Geometry to be the limiting value taken by the function following the affine approximations of a secant *[AB]* to the curve as point *B* approaches *A*. However, it is important to realize that the condition is stated to be as point *B* approaches *A* infinitesimally, or alternatively as *dx* → *0*, *dx* being the affine variation of the independent variable *(17-5)*.

$$\frac{df(x)}{dx} = \lim_{dx \to 0} \frac{dy}{dx} \quad (17\text{-}5)$$

Consequently, the tangent is **never** what the secant line segment becomes at **one** point on the curve, as commonly assumed. Point *B* approaching *A* or *dx* never vanishes, hence **the secant *[AB]* never vanishes.**

The tangent with the value of a derivative is best defined as what the secant becomes when **point *B* becomes the single most consecutive point to point *A*** on the curve at one side, which equates the absolute minimal value of *dx*. As a result, the suite of minimal values of the slope of all the secants $\{[AB]_i\}$ over a sectional of the curve potentially describes the curvature of that sectional. The question is: how does one extol the fundamental form embedded in the curvature from the minimal secants? Theoretically, the evolution of the slope of the minimal secants gives us at once the evolution of the curvature. If an orthogonal line is drawn at each end point of the one secant, and the resulting orthogonals intersect with one another, then the curvature must be closed. If the orthogonal lines are parallel to one another instead, then the curvature must be vanishing or planar.

We may also look at the angle formed by two consecutive minimal secant lines to infer the nature of the curvature encircled in the delimited sectional. Given two minimal secant lines *[AB]* and *[BC]*, the entire set of possible irreducible curvatures between *A* and *C* is determined by the value of the angle $A\hat{B}C$, with the possible set of values being (Fig.17.6):

$$90° \leq A\hat{B}C \leq 180°.$$

When $A\hat{B}C = 90°$, the curvature is circular. When $A\hat{B}C = 180°$, the curvature is vanishing or planar. For all values $90° < A\hat{B}C < 180°$, we run the gamut of curvatures from elliptic to secant hyperbolic according to the hierarchy of fundamental forms or *eigen-forms* established in Quanto-Geometric analysis of differential manifolds in Book I and II. The notion of an intrinsically negative curvature is not germane to that analysis and hence discarded.

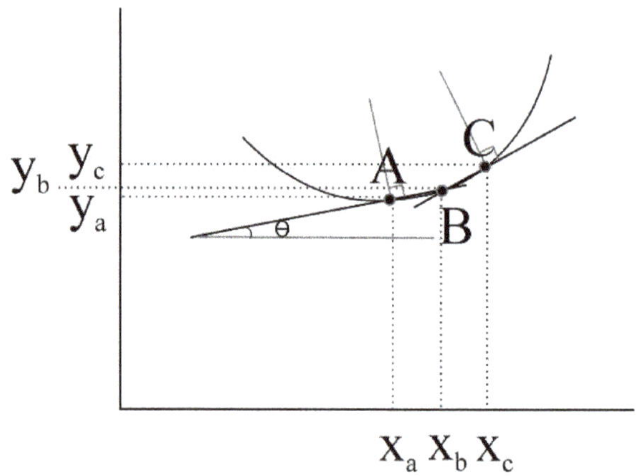

Fig. 17.6 Minimalist secant lines [AB] [BC] surrogating the curvature. This is an exploded view for sake of clarity.

In order to draw a curve from its functional expression, one may of course execute the approach of finding the maximum practicable number of points for the graph from the *(x,y)* values yielded by the function. By flooding the coordinate system with points, a continuous curvilinear evolution of the function will automatically show. Only possible with computing technology for a reasonably expedient execution. Another more economical approach is to draw a lower number of points and join them one to one in segments representing the secants. If a sufficient number of points is represented to sufficiently approximate the minimally proximate two-dot requisite, and depending on the resolution of the medium, a continuous curvilinear evolution of the function will equally show.

Further analysis shows that if one were to apply the tangent or tangent vector approach, the minimal secant approach would comparably trump the former, whilst showing a greater representational value as to curvilinear patterns. A set of tangents drawn from *(x,y)* points extracted from the function lacks the virtue of rendering a clear and exact representation of curvilinear patterns of the same. If we choose an individual point on the coordinate system and, assuming that point belongs to the function, we draw a tangent line at that point with a slope given by the function, we still do not obtain any knowledge of the nature of the curvature of the function at and in the neighborhood of that point. Note that any form of curvature will meet a tangent line at a single point, not only a circular curvature will. Further, for one tangent line, there will be many different circular curvatures that will meet the tangent line at the same single point (Fig.17.7). Let alone that not one curvature sectional may be drawn to the tangent line but two, one at each side of the line, causing even more indefinition.

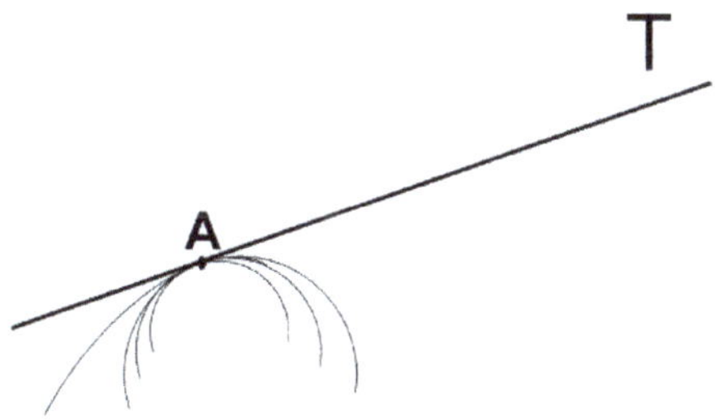

Fig. 17.7 A tangent to multiple curvatures at the same point

At best, applying multiple tangents, we obtain a sequence of straight lines across singular points, which we do not know how to relate to one another to precisely form the curve. The points alone would be even more telling and useful for the representation of the function's graph. Should we choose to give the tangent lines a vector representation, the function does not contain any information transferable to the modulus quantity of these vectors other than their slope or angular inclination. That stands for the main reason why in Riemannian Geometry so many different developments over abstract tangent spaces had been necessary, as much as the necessity to appeal to extra geometric elements as materialized in the differential chart, in order to approximate a real description of innate curvature.

In other words, differential calculus, largely based on a misunderstanding of the role of points in Geometry and ignorance of the preeminence of lines for its irreducible element, does not contain enough information, or the adequate sort of information, to provide for an appropriate and thorough description of curvatures. To achieve that goal, it takes a significant re-dimensioning of differential analysis, which we have completed in Quanto-Geometry Book I and II.

The above analysis also gives us a sense of the reason why the majority of differential equations do not have a solution in closed form. Under the definition of the derivative being a tangent line to the curvature (at one point), if the starting information is given in affine states *(dx, dy, dx²...)* which ultimately equate tangent lines by composition, one can see why the road to the antiderivative function will be steep, arduous and more often than not just impossible.

17.11. Primacy of the Minimal Secant Line Approach in All of Calculus

We must be convinced that the most important theorems and chapters of Calculus are forthrightly due to the primacy of the minimal secant approach being hereby advocated.

The so-called ***Fundamental Theorem of Calculus*** which establishes the basis relationship between differentiation and definite integration is a direct result of this view.

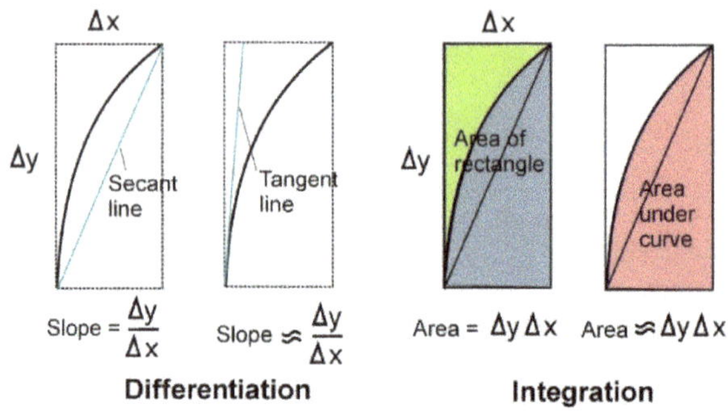

Fig. 17.8 Foundational approximations to Differential Calculus

The quotient $\Delta y/\Delta x$ emanating from the secant line is very much at work in order to project the affine approximation dy/dx, which leads to the tangent line, as previously discussed (Fig.17.8). Likewise, it is the product $\Delta y \Delta x$ that is used in order to approximate the minimal area under or above a curvature sectional of the Function, which leads to the integration technique in its very diverse methods.

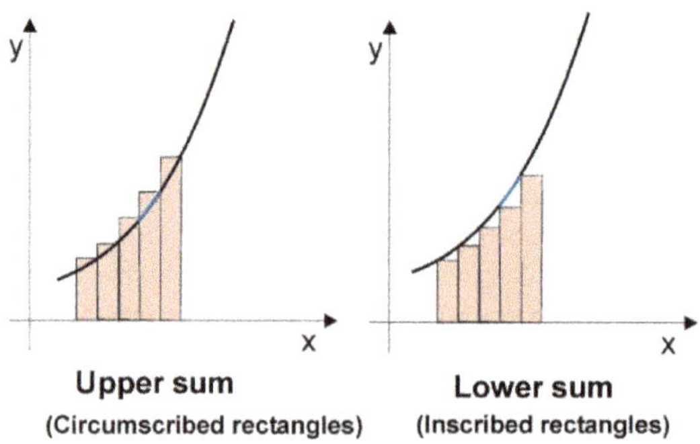

Fig.17.9 Approximations of area under a curve in Calculus through sums of circumscribed and inscribed rectangles

The secant line remains indefectibly at the heart of the minimal incremental ratio or product in differentiation and integration respectively. When areas below the curve are computed through approximation methods (Fig.9), the secant line determines width and height of rectangles above the curve (circumscribed rectangles), just as it determines width and height of rectangles below the curve (inscribed rectangles):

$Lower\_sum = \sum_{i=1}^{i=n} f(m_i)\Delta x$ (area of inscribed rectangles)

$Upper\_sum = \sum_{i=1}^{i=n} f(M_i)\Delta x$ (area of circumscribed rectangles)

where m and M are the two different heights of the rectangles and Δx is the conserved width of the rectangles (Fig.17.9). In this approximation case, the true value of the area is estimated lying between the value of the upper sum and that of the lower sum.

The above method is generalized in the definition of *Area of a Region in the Plane*, for an exact value of the delimited area under the curve, which states that:

If f is continuous and nonnegative on the interval [a, b], the area of the region bounded by the graph of f, the x-axis and the vertical lines x = a and s = b, is equal to:

$$Area = \lim_{n\to\infty} \sum_{i=1}^{n} f(c_i)\Delta x \quad x_{i-1} \le c_i \le x_i$$

where $\Delta x = \frac{(b-a)}{n}$.

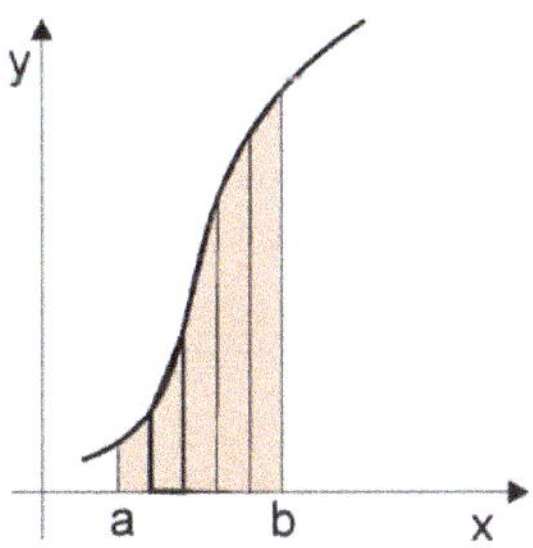

Fig. 17.10 A better approximation of area under a curve through affine trapezoids

The secant line determination intervenes just as ostensibly in the so-called *Trapezoidal Rule* [5] where the approximation quadrilateral is no longer a rectangle but a direct trapezoid with the secant as one of its sides (Fig.17.10). The rule is derived from the principle that:

If f is continuous on [a, b], the area of the region bounded by the graph of f, the x-axis and the vertical lines x = a and s = b, is equal to:

$$Area = \lim_{n\to\infty} \frac{[f(a) - f(b)]\Delta x}{2n} + \sum_{i=1}^{n} f(x_i)\Delta x$$

It is interesting to note that the Trapezoidal Rule, which is a very direct implementation of the secant line in the computation of area under the curve, leads to the very definition of indefinite integration because as $n \to \infty$ the above expression approaches:

$$\int_a^b f(x) \cdot dx$$

So it is, because the term with the limit is in fact equal to 0, making the second term the only non-null numeric expression for the Area. The second term directly translates the very definition of integration.

It is only because all of the above is true, in particular the primitive to the Trapezoidal Rule, that the **Fundamental Theorem of Calculus**, in its condition of liaison between differentiation and integration, plays a preeminent role in Calculus. The Theorem states that:

If a function f is continuous on the closed interval [a, b], and F is an antiderivative of f on the interval [a, b], then

$$\int_a^b f(x) \cdot dx = F(b) - F(a).$$

The most important advantage of this Theorem consists in allowing a direct means to accurately evaluate or compute a definite integral without recourse to the limit of a sum of quadrilaterals, provided of course that the antiderivative of the integration function exists. Furthermore, and very importantly, the indefectible reference to the interval [a, b] is always governed by the primacy of the minimal secant line primitive, not by the notion of the tangent line to a point on the graph.

17.12. Foundations of Symmetry in Geometry

If one ought to honor the syntactic root of the term *symmetry*, which essentially means sameness, the first and simplest expression of symmetry in Geometry is equality between segment lengths metrically measured. The next most significant expression of symmetry is parallelism between lines, earlier discussed.

In Quanto-Geometric Theory, states of symmetry and motion are two distinct inner properties of spreads. In mathematical physics they become related to one another under the figures of invariance and transformations through the notion of symmetry. Considering an isometry to be a distance-preserving transformation between metric spaces, commonly assumed to be bijective, a primitive form of isometry already appears in our definition of parallelism, each of the two lines playing the role of a metric space.

However, what a higher symmetry is defined to be, nominally an invariance that is preserved under a specific type of transformation, finds its roots in Quanto-Geometry in the identity of what results from covariance between the Scalar Order and the Geometric Order. At second order, this covariance elicits what we have called Quanto-Geometric Quintessence, an inner correlation between the two primeval existential tenets, intangible spreads and tangible scalars. Quintessence is expressed as a first-order derivative of the Quanto-Geometric Grand Eigenfunction. The function portrays three grand moments or phases (Fig.17.11 bis):

- One where the scalar tenet overwhelms its spread partner, expressed by $\frac{dQ(s)}{ds} > 1$.
- One where the spread tenet overwhelms its scalar partner, expressed by $\frac{dQ(s)}{ds} < 1$.
- One where the mutual influence of both tenets is equalized, expressed by $\frac{dQ(s)}{ds} = 1$.

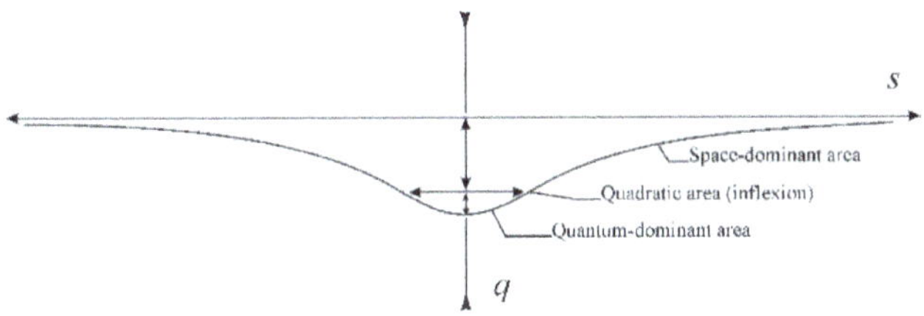

Fig. 17.11 The three sectors of the Quanto-Geometric Function *Q(s)* that grandfathers Quintessential Symmetry

In the Quanto-Geometric towering perspective, spread is of *auto-dynamic nature*, a feature which has never been understood thus far in science and mathematics. Further, spread is to be conceived as the generic abstraction for physical or coordinate space. Whereas the scalar is of static nature and constitutes the farthest notional origin of physical mass. In this context, Quintessence is the first cause of all mathematical- physics symmetry. Therefore, the invariance in a mathematical-physics symmetry is germane to the scalar component of physical elements or mass, whereas transformations are germane to the spread element or space.

Everything in nature complies with this fundamental prescription, including human artificial constructions. For instance, we directly observe a solar system in the form of:

1. a constellation centralized on a dense Star, endowed only with rotational but no translational motion,
2. with far out in its periphery a plethora of comets, with tiny masses, exercising tremendous degrees of freedom from the vicinity of the central Star to the outskirts of the system, and
3. in-between a set of planetary bodies each with a respectable amount of mass and translational orbitals around the Sun Star, all restricted within a plane (the ecliptic).

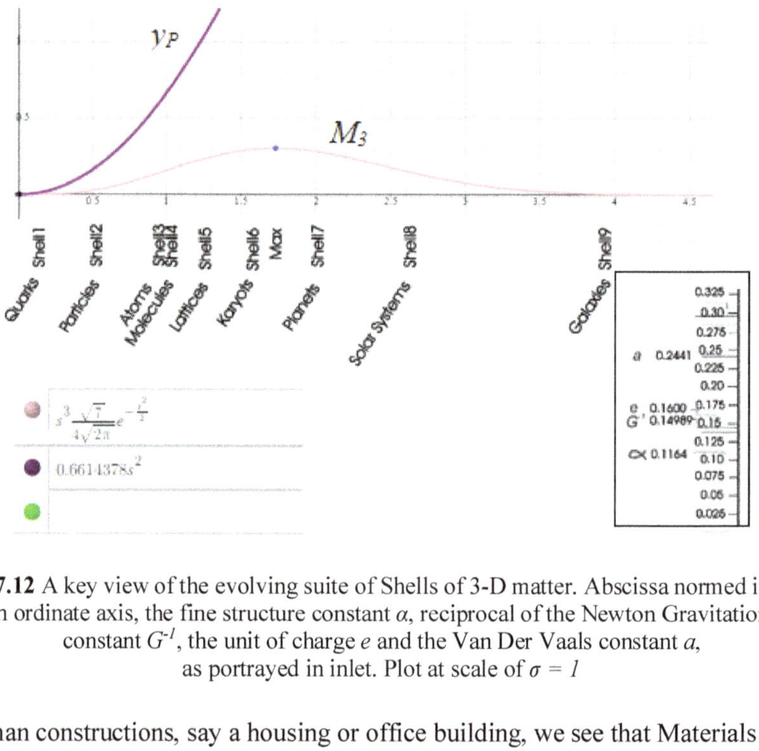

Fig. 17.12 A key view of the evolving suite of Shells of 3-D matter. Abscissa normed in σ units. On ordinate axis, the fine structure constant α, reciprocal of the Newton Gravitational constant G^{-1}, the unit of charge e and the Van Der Vaals constant a, as portrayed in inlet. Plot at scale of $\sigma = 1$

In human constructions, say a housing or office building, we see that Materials mass densities go from very dense for the foundations to lightest in the roofs. Foundations are intended to be and remain as invariant as might be structurally. Nobody is going to happily redo the walls of their home basement, any structural problem in basement walls and floor is considered a major if not catastrophic problem. The mid-section of our dwelling constructions is generally designed for living, which is where a tremendous number of human transformations is going to take place daily. Those structures may be readily renovated for better suitability to those transformations. Construction materials employed for those sections are lighter than those used in the foundations. Top parts such as attics and roofs are made of light materials and roofs are accommodated to deal with the elements of nature at large which may subject them to even greater transformations, i.e., winds, hurricanes, rain and snow storms, etc. They will be the first to suffer destruction under excessive natural turmoil those sorts.

Quintessential symmetry is applicable to all physical laws and the Laws of Conservation in particular. All physical laws are built around and about fundamental physical constants, which are indeed mathematical constants, as we have amply demonstrated in Book I and II. The dimensionless physical constants are directly derivable from first Quanto-Geometric principles that are completely independent of time and time evolution. We have seen that the dimensional fundamental constants are derivable from the Quanto-Geometric framework in the form of precursors that present a striking numeric figurate match with their counterparts (Fig.17.12). The fundamental physical constants represent in fullness the invariants that sustain the transformations taking place in dynamic systems as studied in the physical sciences. The following is worth a full repeat here:

"As I have explained in Book I and II with a profusion of numeric details on axiomatic basis, from galaxy groups, to galaxies, to solar systems, to planets and stars, to lunar satellites, to materials lattices, to karyotes, to molecules, to atoms, to elementary particles and to quarks, the universe is organized in Shells in a way that an association between a few of the previous elements of the hierarchical suite comes to form the building block of the next element in the suite (Fig.3b). The two most basic elements of this notable development have never been understood, which is the evolution of mass, defined as scalar density, and the evolution of the vacuum, unfortunately always disregarded for nothingness, throughout the hierarchy. One should not attempt to understand behavior and properties of interactive elements in one Shell strictly from the properties of their inner constituents, because these must remain ***invariant*** *as a condition for the structural stability of the very Shell. Furthermore, the ontological qualities and properties of the (upper) Shell element are always quite different from those of the inner constituents because this is how mother Nature creates its own evolution (through symmetry breaking)."*

The vacuum plays a seminal role in this transformation by creating the very impetus of motion and motorizing the transformation, while this overarching Shell symmetry frames the notorious symmetry axiomatized by Emmy Noether (Noether's theorem) and the laws of conservation in total. From our perspective, Noether's 1915 proven Theorem becomes very understandable:

Every differentiable symmetry of the action of a physical system has a corresponding conservation law.

The theorem had been numerically expressed in the category of the integral of an action in time, the latter being translated in a Lagrangian representation *(I),* such that:

$I = \int L(q, \bar{q}, t) \cdot dt$, where q is the coordinate set, $\bar{q} = \frac{dq}{dt}$ and L the Lagrangian notation.

The theorem in itself means that the conservation law represents the invariant of the symmetry, whilst the action of the physical system represents the transformation taking place. If the action is continuous, the symmetry may be represented by a differentiable function that is defined within the boundaries of the action or the domain of its variables. Noether's statement is principally an expression of Quanto-Geometric Quintessence. Under Quintessence as the model for symmetry, every global gauge group we otherwise call Shell of Matter forms a symmetry. Unlike Noether's symmetry however, Quintessence is a time-free symmetry that cannot be codified with the time variable, nor does it require such a variable.

The *mathematical proof* of Quanto-Geometric Quintessence is in the accurate computation of the fundamental physical constants representing the invariants that remain preserved under the Universe's Global gauge group transformations or the actions within the many different Shells of Matter.

As assertive as Noether's Theorem turned out to be, it fell in the pitfall of expressing motion, indeed a mathematical entity at its source, with the physical and false variable of time, through the Lagrangian code and the Euler-Lagrange equations. The reasoning is that the Symmetry *I* follows the principle of least variation or action according to:

$\frac{d}{dt}\left(\frac{\partial L}{\partial \bar{q}}\right) = \frac{\partial L}{\partial q}$ and if one coordinate q_T does not appear in the Lagrangian, then $\frac{\partial L}{\partial q} = 0$, necessarily implicating that $\frac{d}{dt}\left(\frac{\partial L}{\partial \bar{q}}\right) = 0$, which in turn imputes that the momentum $m_T = \frac{\partial L}{\partial \bar{q}_T}$ on the physical path of motion is conserved.

The problem with this description is that there is no motion in time along a trajectory when it comes to the natural Gauge groups. That stands for one of the principal reasons why our conventional numeric consignment of Symmetry in mathematical-physics has been irredeemably inapt at deriving the fundamental physical invariants of nature from first principles, an unreachable goal until Quanto-Geometric Theory took on the challenge. In the context of mathematical Symmetry, these fundamental physical invariants are called the *constants of motion.* In this case, they are the *most fundamental constants of motion.* As proven in Quanto-Geometric Theory, they consist in dimension-less, time-free and pure mathematical scalar quantities, very much similar to *e* or π (Fig.1.13).

Never to be overstressed is the importance of intellectually representing how:

1. Quarks inter-associate to form elementary particles.
2. Elementary particles inter-associate to form atoms.
3. Atoms inter-associate to form molecules.
4. Molecules inter-associate to form lattice-based (crystalline) materials.
5. Interlacing lattice groups inter-associate to form karyotic cells.
6. **Higher Karyotes inter-associate to form singlet celestial bodies.**
7. Singlet celestial bodies inter-associate to form solar systems.

8. Solar systems inter-associate to form galaxies.
9. Galaxies inter-associate to form galaxy groups.

The connection between the intimate endogenous association of sub-constituents based on an invariant bond and the galvanized exogenous Shell formation or transform, created through variance, constitutes the basis of phenomenal Symmetry.

17.13. Affine Secant Basis to the Principle of Least Action

It is worth highlighting hereby that the theory of minimal secant lines as primitives of curvature constitutes the primal justifying cause of the *principle of least action*, so instrumental in the granular analysis of dynamic systems in all-time physics. No unit of action is reducible to a point but to a minimalist linear segment which directs the infinitesimal variation of space spreads. If the state of symmetry innate to the space spread is vanishing (type 9), the least action is rectilinear in projection; any other typology of the innate state of symmetry (type 1 to 8) will induce a direction of least action that is curvilinear as determined by the secant as a vector. If we were to describe the tapestry of all space spreads, they would be stated to be based on intangible affine secants, *not points*. However, space spreads are not intimately static but auto-dynamic, therefore a spatial tapestry per se cannot exist.

17.13.1 Current Formulation of the Principle of Least Action

The principle states that:

The time integral between two points along a trajectory takes a stationary value with respect to the same integral taken over any other allowed group of trajectories.

The action and Lagrangian both contain the dynamics of the system for all times. The path taken by the system between times t_1 and t_2 and configuration instances q_1 and q_2 is a path for which the action is stationary to first order. Statement and definition of action are generally taken together in physical applications according to the following expression:

$$\delta \int_{t_1}^{t_2} L(q, \bar{q}, t)dt = 0,$$

where δ (lowercase delta) is the action conceived to be a *small change*. The term "*path*" refers to a curve traced out by the evolving system in terms of the coordinates in the configuration space, i.e., the curve $\boldsymbol{q(t)}$. The value of the path integral itself δq for a particular trajectory is minimal compared to its value along any other trajectory (since the path integral is a composite of them all - Fig.17.13).

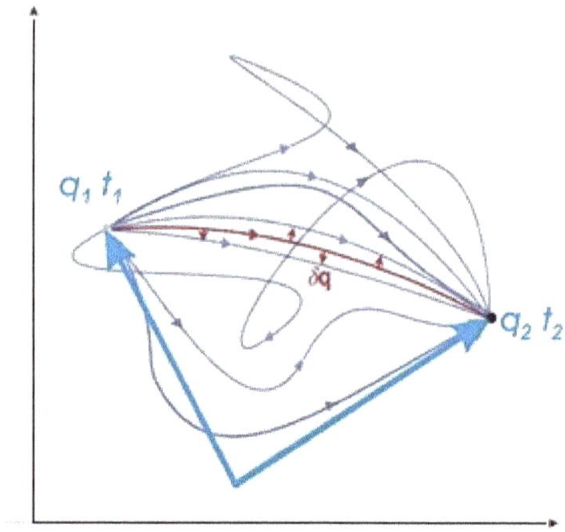

Fig. 17.13 Common description of motion as the path integral over time

The omnipresence of the time variable in the diverse formulations of the principle of least action constitutes undoubtedly one of the principal reasons for their inability to truly compute the so-called fundamental constants of motion. In all, this principle is an approximation to the cause of motion by employing the fictitious variable of time and the notion of action, since it is unable to represent the evolution of motion based on the innate state of symmetry to the spatial vacuum that vehicles the motion. Clearly what remains invariant here is the product of elements that make up the action, notably its consignment in terms of energy.

17.13.2 The Heart of Motion is Freedom

All description of motion that requires an anchor will fail to extricate the intimate mechanism of motion. At its heart motion is freedom, it is only when we have characterized the cause of the freedom in motion that we have explained motion, having gone beyond what can be construed as a chronological narrative of motion or the mere evolution of an object or objects in displacement. To understand motion, one has to understand the vacuum or its modeling abstraction. If the universe were to lose all forms of scalar monoliths and the quanta of energy densities alike, it would become a vast spread of pure emptiness in totalitarian freedom. The degrees of freedom of such a universal vacuum spread would be infinite in the sense that motion would be absolute in spread along omni-directional fleeting intangible lines. There would be no center to this intangible fleet, there would only be an order of *ortho-gonals* and *par-alleles* in this fleet, the latter altogether amounting to infinite dynamic foliates.

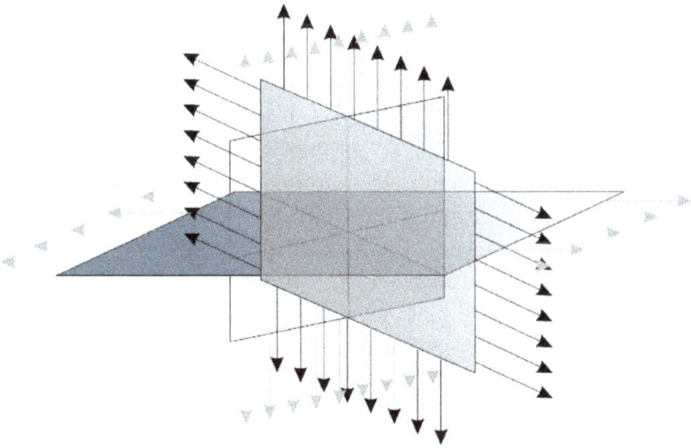

Fig. 17.14 A space unit of action is not an irreducible building block but a fleeting orthogonal tri-plane that fills up the entire universe, an infinite coalition of which leaves no interstitial spread in the integral of directions within the fabric of space, and all ultimately reducible to intangible lines.

The moment that a scalar makes its appearance anywhere in this fleet, the entire universal spread of emptiness would react to "carry" and "jostle" the scalar while self-reacting in constricting the entire fleet of *par-alleles*. One or several degrees of freedom would so be lost. And as more and more scalars make their appearance, universal vacuum would lose more and more degrees of freedom while *informing* the scalars into scalar densities. Whilst this description stands not for a chronological story line of the formation of the universe, it is more than a euphemism and it seeks to help comprehend why nothingness is not ontologically small or squint but big like an immensity, if only because of its unrelenting fleeting immanence. If one cannot intellectually visualize this artifact, one simply cannot understand the origin of motion (Fig.17.14).

Should some object to this view with the question of how can any formal numeric description of such a dynamical scenario ever be achieved, the answer is well portrayed in the characterization offered in Fig.17.14. That is exactly what the primitive of the Quanto-Geometric Grand Eigenfunction describes through the *variants of motion* as parents to the variance of normal density distributions and the *constants of motion* otherwise known as fundamental physical **coupling** constants.

17.14. Origin of Symmetry-Breaking

Symmetry breaking is a generalized mechanism throughout the Shells of Matter indistinctively, including the Karyote Shell where it is known indeed under the designation of biological mutations. Its origin is auto-dynamism or motion innate to space spreads we otherwise denote space momentum. As long as the description of motion of a dynamic object or system is realized with the aid of time, which is a subjective fictitious notion, the cause of motion will remain obscured and misunderstood, and so will the intimate mechanism of symmetry-breaking.

Considering a symmetry per Noether's Theorem visualization, the passage from one natural Gauge Group or Shell of Matter to another is realized through the breaking or mutation of the prior Group or Shell symmetry into the next Shell symmetry, and ultimately caused by explosion or transmutation of the space spread wavefunction innate to the prior Group. There is no mandatory push in a particular direction along the Shell line directing the mechanism of symmetry transmutation to be construed as the direction of evolution. If most everything we observe in the physical world emerges from the experience of birth, the same things equally disappear through the experience of deconstruction, death or extinction. And from death or destruction of an object other objects emerge in the natural world. The only reality being that the Universe has always been there with its principal partitions or Shells in perennial autonomous change. Therefore, the transmutation of the symmetry-breaking mechanism is innately stochastic.

If we imagine an *expanding* Universe as we currently do in mathematical-physics, for that expansion to ever be possible, we must assume the vacuum of space spreads to be estranged to the Universe, that the Universe only lives a scalar existence, and that the nothingness of the vacuum only constitutes the passive vehicle of expansion of the scalar Universe. The view that the vacuum is totally estranged from the Universe ontologically is completely erroneous however, because the vacuum exists everywhere we look, at both the microscopic and the macroscopic levels; it affords all objects their metric dimension as much as it is ultimately responsible for their explicit morphology. More than playing a role in the Universe, the vacuum is a fundamental component of the natural world to which the latter owes motion and change. So therefore, there is nothing for the Universe made up of both scalarity and emptiness of the vacuum to grow into. There is nothing beyond the vacuum at large scale construable as a medium or a circumventing province. Consequently, the only sensible visualization for the global Universe is what we have called an *Omniverse*, that is to say an edgeless, size-less or a-dimensional, boundless or completely unconstrained Universe which is absolutely a-temporal, and in perpetual immanent motion and change. All apparent evolution in the symmetry transmutations of partitions of the Universe can only be local and circumstantial and most importantly, all stochastic transmutations are ontologically typed, the order of types being a higher order of certainty innate to all probability distributions. It is certainly by virtue of our ignorance of this inner order of certainty that we have come to inappropriately denote these distributions *probability* distributions.

In considering the vacuum in the lower Shells of Matter than the Karyote Shell, one may wonder how we should visualize the dynamics of the scalar in those realms in terms of its trending behavior. If, at large scale, we imagine the distant objects making up the Universe in those realms (i.e., Galaxy groups) to be expanding through their circumventing vacua, one may reasonably compare these objects to those existent at small scale such as the realm of Molecules for instance, and wonder whether or not those objects which are equally circumvented by vacua are expanding to the same extent as the global

Universe putatively does. How would their vacua provide means of expansion to their scalar entities? And how do we visualize a boundary between their vacua and that of the next Shell of matter up, the Lattice or nano Materials Shell, that is. The answer to this question is quite simple: *there is no such linear expansion or linearly continuous Function.* The vacuum at one lower Shell does not continually flow into the vacuum (or vacuum locus) at the higher up Shell. Nor is there ever a meet between a scalar at one lower Shell with the scalar at the higher up Shell. Most importantly, we do not see a scalar at the higher up Shell in the form of a new totally different entity than the scalar at one Shell down. What there is is a discontinuous explosion of the vacuum of the lower scale realm, which immerses or imprisons a few of the scalar entities of the prior Shell. This alteration of the vacuum is based on the creation of a new inner state of symmetry out of the 9 basis states, whereas the new scalar state for the exploded Shell becomes laid on a juxtaposition of several of the prior scalar objects. The *breaking of the symmetry* gives rise to a new scalar-space correlation which is the very **fundamental physical constant** that cements the organization and dynamics of the exploded Shell or the higher up Shell. The breaking of the symmetry is an artifact of the innate motility of space spreads. One may euphemistically represent the entire universe as an onion ball with no bounds in its suite of layers or shells, each shell of the ball representing a Shell of matter. However, realistically there is no center point to the Universe, since the inexistence of finite bounds results in an undefinable center or centers of symmetry from the bounds. That is also why a shape for the ball is undefinable. The Universe is veritably an *Omniverse.*

Considering a single symmetry, whether the global Universal one or a local one, it is worth stressing that the motion which is immanent to the symmetry transformations does not have an exogenous cause to the entire symmetry. We cannot explain *Transport* and *Connections* through a force acting exogenously on the principal object of the symmetry. Notably, the notorious *Levi-Civita* connection stands for a description of transport which assumes that the spherical object is *subject* to a motion in two or three different directions. It does not assume that the spherical object *protagonizes* the motion. That is the reason for Noether's Theorem, as much as Transport through Connections, to have been expressed in *time* or with the time variable. Time, however, cannot explain motion because it does not and cannot interpret the cause of motion, as it only provides support for a *description* or an observational narrative of motion. Therefore, all mathematics that hinges on time to apprehend motion, from *Topology* to the *Christoffel symbols*, only present a chronological narrative of motion, unable as they are to offer an intimate comprehension of the very cause of motion.

17.15. Caveats of Circular Symmetry in Riemannian Geometry

The long running inability to intimately comprehend parallelism and the preeminent role of linear formations in Geometry has negatively informed our representations of Symmetry, in particular circular symmetry.

17.15.1 Misinterpretation of Circular Symmetry

Our theory of symmetry in Differential Geometry has the circle or the sphere as the one geometric element exhibiting the most expansive or prolific form of symmetry, in the sense that the circle is assessed to contain more symmetry than the ellipse, or the parabola, etc., ultimately because all points on the circle are interchangeable with one another, with or without rotation.

What this view has failed to comprehend is that only from a higher level of abstractions may one analytically approach geometric formations such as circles or spheres, which are indeed physical formations that are directly observable in nature. Should we choose to analytically approach geometric formations from the towering abstraction of a spread, one is constrained to primally visualize the possible qualities of a generic spread, which may subsequently become applicable to geometric formations. One of these qualities is the *state of symmetry* with the plurality of its attributes. From that vantage point, the state of symmetry immanent to a circle or a sphere (*3D* circle) is **unitary**, in the sense that every point on the circle is interchangeable with one another in only *one* way, defined by the radius of the circle. Contrastingly, the points on an ellipse have *two ways* to interchange with one another, as referentially defined by the major and the minor axes. Therefore, the state of symmetry immanent to an ellipse is twice as rich as the unitary state of symmetry attributable to the circle. There is to it a **binary** quality, in opposition to the unary quality of the monolithic state of symmetry of a circle or a sphere.

The state of symmetry is not a function of an imposed transformation. It has preponderance over a transformation which ordinarily is an analytic artifact externally imposed on the formation for our interpretative needs. While the application of an externally applied transformation may help discover behavioral attributes of the formation, it cannot ontologically define the formation. Consequently, it is a paradigmatic error to tribute the circle with more symmetry than the ellipse based on the rotational attributes of the formation.

17.15.2 Description of Motion through Circular Symmetry

A **geodesic line** is considered to be the shortest path between two points on a curved surface. It is the analogue of a straight **line** on a plane and considered to represent a generalization of the notion of a *"straight line"* at a generic level.

The representations of geodesic lines as longitudes on a spherical globe are considered parallel to one another from the inner perspective of the spherical surface. To say the least, these notions are regrettably inappropriate and incorrect. The rectilinear formation that we call a straight line represents instead the more general or generic form of linearity. Curvilinearity is borne out of rectilinearity, by diminution of the numbers of centers of symmetry inherent to the latter to its minimal level which is one. Geodesic lines or global longitudes are not parallel to one another from any point of view, because the only global circumvents that are parallel to one another are *latitude* formations. When we rotate the sphere by 90^0 to transform latitudes into vertical formations, they do not become interchangeable with traditional geodesics (Fig.17.15).

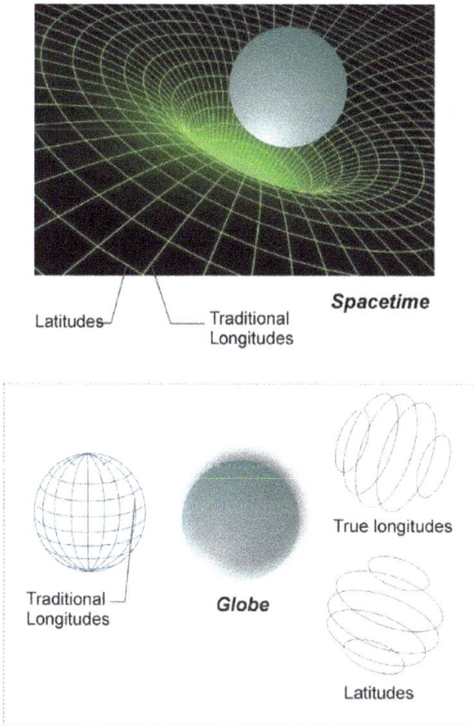

Fig.17.15 Global longitudes are traditionally considered parallel to one another in Differential Geometry, while they are in fact represented intersecting with one another in pure geometry. At the right is depicted what would be the true longitudes across a globe, which are legitimately and forthrightly parallel to one another.

In order to study motion at the most granular level possible in Differential Geometry, we traditionally consider displacements over a spherical surface and we note that utilizing a coordinate system to map transitions in displacements leads to the irreconcilable of different sets of coordinate values for the object in mobilization for different coordinate systems at every step. The problem primarily arises from the fact that no particular coordinate system has analytic preponderance. We thence choose to give preference to the artifact of a putative rotation of the sphere along the direction of mobilization of the object under study, the geodesic line. In that manner, the geodesic becomes the analytic basis for transport, in expressing connection from one point to another. We thus impose a rotational transformation on the sphere, which at its minimum gives us the Levi-Civita connection from one point on the sphere to another through a set geodesic line, an equivalent to a path of "parallel" displacements of the object.

None of these developments based on circular symmetry reaches a seminal description of the cause of motion, nor a clear representation of space inner curvature that causes objects to exhibit unexpected forms of motion. Further their use in mathematical physics has led to one of the most intractable problems facing the field: the so-called *vacuum catastrophe problem,* a problem that has tightly cornered the Relativistic framework. The incompleteness and weaknesses of this development have also left us without any clue to determine which theory between Newtonian physics of force and acceleration and General Relativity stressing curvature best describes gravity. Obviously, in the above discussed development, the point of view of Force and acceleration has been substituted by rotational transformations and connections through curvature.

17.16. Falsifying the Logic of Origin and Age of the Universe

The idea that the universe must have started from a point and that before that point origin there was nothing, so common both in culture and physical science, does not withstand acrimonious scrutiny. If our idea of nothing is a state of nothingness, which physically translates into the observable vacuum in purest state, then nothingness is indeed the beginning and not the visualized point origin which nothingness must predate. In fundamental logic, one cannot ascertain that nothingness is small in dimension, because if it was small, it would have bounds of some nature. Bounds, of whatever nature, are existentially something and not *nothing*. So therefore, in that vision, primeval nothingness must have experienced spread into the infinities, in complete unboundedness. In that event, we must then forcefully conclude that if the universe had experienced a starting point or an origin, it must have started from the infinite spread of nothingness. However, because the start of a state of nothingness is indeterminate (nothing before nothing is a void and unsustainable concept), just as division by *0* is indeterminate, the unbounded spread of nothingness, or the physical vacuum, must have *always been there*. In conclusion, and in accordance with the fundamental logic of falsification or reasoning by *reductio ad absurdum*, the universe cannot have experienced a start, a point origin, nor is time a valid existential dimension of the Universe.

Hence, the idea of a universe having started from a point, roughly 13 billion years ago, and that time directs the evolution of the Universe, qualifies as utter ignorance, no matter the lush yet false mathematics that had been thrown behind such concepts.

It is noteworthy that relativists have found, in the analysis of the dynamics driving the large-scale cosmic arena, that the *3D* Universe must evolve according to the Differential Geometry directives set forth by Gauss for fundamental forms (Fig.4). It is not clear whether their numeric analysis of the energy density of the vacuum in the larger cosmos has led them to that conclusion or else they have imposed a priori the Gaussian perspective of fundamental form to the interpretation of the supernovae data. It remains that their computed value of the vacuum energy density stands at the heart of the controversy known as the *vacuum catastrophe problem*, given that it is far removed from the value computed in Quantum Theory. The profound disparity between the two quantities remains one of the most disturbing results shaking the grounds of Theoretical Physics now for several decades. Despite the fact that the value computed within the Quanto-Geometric framework as an independent third party seems to lend better credence to the value upheld by relativists, the latter is still off by two orders of magnitude. No wonder that is very much the case, since the same calculation methodology has led to the formulation of the erroneous projection of possible fundamental forms for the universe, which do not respond to canonical logic. That all said, this appreciation in no way warrants the vacuum energy density figure computed within the Quantum Theory framework, which is an ever greater "catastrophe".

17.17. Gauge Group Symmetries

The plot in Fig.1.13 shows the graphical layout of the paramount *M(s)* Function which we have designated within the Quanto-Geometric framework, perhaps unartfully, the *Creation* function for the 3-D Arena. It relates the fundamental physical constants incarnated in the fundamental physical coupling constants, i.e., the Gravitational constant *G*, the unit of Charge *e*, etc. to the particular wavefunction *(s)* in action in every natural gauge group or Shell of Matter.

The integral *T(s)* of the Creation function, below reported from Book I, encapsulates a preeminent form of symmetry not captured by Noether's Theorem but ostensibly observable in Nature:

$$T(s) = \int M(s) \cdot ds = \frac{k}{\sigma^2} e^{\frac{-s^2}{2\sigma^2}} (s^2 + 2\sigma^2) + C,$$

where k/σ^2 is a mathematical constant[11].

The nature of the *M(s)* Function tallied on the vertical axis of the *q-s* coordinate system is largely that of a derivative. It evolves in increasing order from origin to infinite in its domain. This Function represents the second-order covariance between the scalar and the spatial spread (wavefunction), which is first-order differentiation. In each Shell of Matter, there is a seminal coupling between scalar and spatial spread which constitutes the fundamental invariant for the Shell. The level of that coupling is quantified by the *M(s)* function.

It is essential to note that *M(s)* represents in fact a trace Function or the surrogation of a Function, the variable *s* being indeed a function bundle in itself. In addition, *s* is discontinuous at the symmetry-breaking points, in the exact analytic terms evoked in the previous section. Thus, as the expression of an indefinite integral, the trace Function *M(s)* has no other meaning than making explicit the three grand phases of the Grand Eigenfunction, its direct primitive (or quasi antiderivative).

---

[11] *k* is not an arbitrary constant but a coefficient consistently derived axiomatically in the form of a pure scalar in Chapter 3 of Book I.

Were $M(s)$ to be a simple Function, the portrayed graphical design would show a pairing of the coupling values across the spectrum because for each point s of the $M(s)$ function in one sector, the slope-increasing sector i.e., there would be a matching point s_i with the same coupling value of the ordinate in the slope-decreasing sector. The discontinuous nature of $M(s)$ as a Functional somewhat disallows this two-to-two characterization of similarities, leaving only the three categories of slope-increasing sector, slope-decreasing sector and an extremum or maximum sector. The symmetry characterization is thus reduced to a form of group symmetry between the slope-increasing sector and the slope-decreasing sector, because they take the same value or value set on the ordinate axis. Intermediate between the two is the extremum sector which is unique and intrinsically a-symmetric.

Object	Size (m)
Quark particle - the **size** of quarks is ~10^{-18} m.	10^{-18}
Neutron particle - The neutron has a mean square radius of about 0.8×10^{-15} m	10^{-15}
Lithium atom: 134 picometers across – without outer electron: 90 picometers across	10^{-12}
Smallest molecule (H2): bond length 0.74 A Macromolecules: 100 to 1000A	10^{-10}
Crystal lattice - length of the unit cell edge of CsCl is 0.4123 nm. - The cell-edge length of NaCl is 0.5641 nm.	10^{-9}
Virus: from 30 to 400 nm	400×10^{-9}
Bacteria: 0.12 to 20 micron	10^{-6}
Protists: 1 micron to 3 m long	10^{-6} - 10
Size of blue Whale: 30m long	10^{+1}
Planet – Diameter of planet Earth 12741.98112 Km	$12 \times 10^{+6}$
Solar system	$150 \times 10^{+9}$
Diameter of the Milky Way: 105,700 light years	10^{+15}

Table 17.1. Physical realization of size and dimensions of physical elements throughout the different Shells of Matter in the 3-D arena

This symmetry characterization is paramount to an apprehension of the natural connection between the physics of matter and biophysics of the living by providing the most encompassing **Unified** view of Nature ever made possible. By the same token, it obliterates the view that Life is laid on different foundations than Matter and thus irreconcilable with the mathematical descriptors that provision comprehension and manipulation of the latter. Most significantly, the Quanto-Geometric framework from which this symmetry characterization emerges is able to derive a *physical life constant* which naturally arises among the so-called fundamental physical constants from a common set of canonical principles and a common derivational methodology. Table 17.1 lays out the relative size of the many constituents of the observable complexity suite.

17.17.1 Asymmetry of the Biosphere

The graph of the *M(s)* function (Fig.17.13) shows at once an asymmetric sector of the distribution which is the most remote determinant for what humans we perceive as the uniqueness of living entities, and which in Biology we have encapsulated in the notion of *physiology*. Although thus far a formal definition of life has yet to be reached, whether at the philosophical level or the scientific level, a clear intuition of a basic distinction between the inanimate world and the animate world has percolated in human intelligentsia. While said distinction is hitherto wildly overestimated both in science and culture, with regards to the fundamental components of life, it finds nonetheless an irrevocable level of justification in the asymmetric extremum sector of the distribution of Shells of Matter depicted in the *M(s)* graph.

At the lower end, life encompasses the kingdoms of viruses, bacteria (or prokaryotes), protists (or eukaryotes) and at the higher end, invertebrates, vertebrates, mammals (and other higher unknown forms of the living as this study will later unravel), which altogether make up the Shell of Karyotes. We have chosen the denomination of Karyotes for this Shell of Matter owing expressly to the notable role played by the box or kernel structure in its architecture, as captured by the syntax of their commonly given designations - *karyotes*[12]. While lattice aggregation or aggregates represent the invariance of the symmetry, *karyotisation* is primarily mandated by the second-order covariant relationship between the specific scalar tenet and specific spread tenet of **quadratic** Quintessence.

At that level, both tenets are of equated tensorial influence on one another, resulting in the *twinning* scalarization of spreads and spatialization of scalars. It all happens as though space spreads lose intangibility and scalars lose tangibility in an exact equated measure to become twins, so that now the *elastoplastic* structure is born and subsumes the organization of the karyotic cell box or kernel and even the surrogating histological superstructures composing internal organs. We are obviously including here, and justifiably so, the virus' obvious cell structure within the karyotic cellular world. The *elastoplastic structure* is numerically governed by the aforementioned *physical life constant*. In other words, it is only as a result of physical law emanating from Quintessence, that we see very squint vacuum deployment in karyotisation. In viruses, i.e., it is expressed in the minimal vacuum that we see inside of the crystal-lattice-like nuclear capsomere containing the RNA or DNA macromolecule, and within the global envelope laid over those structures, which amounts to a structural attempt to "box" the vacuum (Fig.17.16). At the same time, the scalar represented in the mass of these structures loses tangibility or monolithisation, resulting in the creation of *superficialization.*

[12] Greek root word Karyote means nut or kernel.

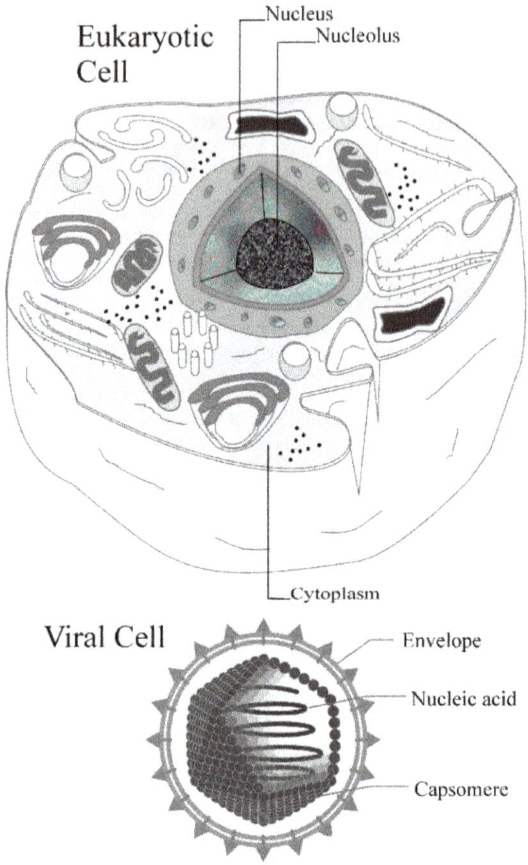

Fig. 17.16 Formation of the elastoplastic Karyotic kernel

The demotion of the deploy trending or spread of the vacuum together with the spread or stretch of scalar densification must be understood to be thoroughly achieved in the orchestration of *endogenous superficialization*, because in quadratic typology it is no longer possible to follow the history of the scalar and space spread individually due to their indistinguishable morphing into one another as twins. Therefore, from viruses to prokaryotes, we see an amplification of this trend with the development of the cytoplasmic medium, a *viscous spread* filling the entire internal space being imprisoned by the cell wall. Further, in eukaryotes, this cytoplasm becomes imbued with a vast extension of superficialization stemming from a coalition of inclusions and organelle formations (Fig.17.16).

Unmistakably, one of the most towering modelers of the living or the Biosphere is quadratic Quintessence, the extremum asymmetric sector of the ruling Quanto-Geometric distribution, that is.

17.17.2 Symmetry Between the Microscopic and the Macroscopic

It is probably unnecessary to make a vehement and prolific case for the symmetry between the microscopic quantum world and the macroscopic world of larger-scale structures because they are so similar to one another that in science we have naturally taken them for the same single world of inanimate objects otherwise denoted Matter. This even though the specific physical laws that relate them to one another had remained an unbroken challenge in theoretical physics for long.

The expression of spread and scalars retain their native physical formats of vacuum and mass respectively in those two sectors of the distribution. This constancy is principally the result of the artifact of *dominance* in the covariant relationship between the two alleles of the ontological duplet. Dominance of one or the other allele understandably creates the *structural rigidness* and *behavioral fixedness* that prevail in states and organization of both provinces of Matter. Objective structural *rigidness* and behavioral fixedness lay grounds for our historical consignment of *determinism* in the overall dynamics of mechanical systems (those which we ascribe to these two provinces). Against more insightful judgment, we constantly appeal to the false notion of time to help us describe *determinism* in the quantum world wherein we feel and insist that the time variable and time evolution must be ontological factors. The time evolution operator that physicists increasingly insist in sticking to the quantum Hamiltonian equation to make the dynamics of the system intelligible for themselves says no less. Concordantly, this rigidness and fixity are also cause for our natural interpretation of those dynamics in terms of *laws* essentially conceived to be immutable trends, quite apart from the higher QG canon of Norms hitherto unknown in Science.

In that sense, the values taken by the *M(s)* function on the ordinate axis, whether in its truncated or fulsome form, known as the fundamental constants of nature, also comport the *dominance* artifact. Making the absolute value of the curvature slope (putative 2nd derivative of original function) a direct measure of covariant *dominance* between the two alleles, one can easily grade the levels of covariant dominance at both monotonic sides of the graph; whereas at the top extremum, the slope is null, which means that covariant *dominance* is null there amounting to equivalency or equalization in the mutual influence between alleles of the original ontological duplet.

Whether one starts at the nearest point to origin of the coordinate system on the graph, or from a point at the infinitesimal end of the graph, dominance evolves in decreasing order reaching null value or equated mutual influence at the top extremum. Of course, one can translate the syntax of the description from dominance to yield instead, the second being the inverse of the first. In this case, the axis would tally minimal yield at near origin of the coordinate system, implicating two allele points on the graph, to maximal yield at the at top extremum of the graph, to signify freedom from one another in the covariant relationship between the two tenets.

In all, this simple symmetry analysis, reducible to plain parallel similarity or homology, has extricated one more dimension to the theoretical unification of microscopic and macroscopic physics, actively pursued in the field since early 20th century, following the quantum and relativist revolutions. However, beyond that adopted limited scope, physics can never be complete without the incorporation of biophysics or a mathematical-physics description of life, this analysis having peremptorily shown that without a granular description of physics of the living, our view of the overall physics of matter does not embrace thorough unification to fullest and due extent.

17.18. Implications of the Homology Between the Microscopic and the Macroscopic

In continuing with the analogy between the two sides of the M_3 graph across the median extremum, it is possible to draw a two-to-two homological relationship across the median between the microscopic at the left and the macroscopic at the right. This parallel roughly yields the following coincidental relations:

1. Quarks are homological to Quasars. They are accordingly both very compact or monolithic objects, as known. Quarks are so deeply immanently confined that they do not manifest as free particles.

2. Particles are homological to Galaxies. Observation of galaxies show them presenting either a disk or elliptical shape, spiral-whirling or vortex shapes, or simply an irregular morphology. Single particles have not been directly observed, but many in physics would not argue that their uncertain individual shape could be a rough match to the afore-mentioned. In particular, the flattened-bell shape of photon packets closely approximates the profile of vortex-shaped galaxies, such as the Milky Way.

3. Atoms are homological to Star Systems. Both entities have a striking tripartite structure showing a dense center incarnated in a nucleus and a star respectively; a mid-level structure incarnated in an orbiting proton shell, on one side, and on the other, ecliptic-plane planets, all orbiting around the nuclear center; lastly, a prolific and expansive electronic shell in atoms, in homology with a set of travelling comets in star systems, both in the outermost region of these structures. The structural resemblance is unmistakable in this case.

4. Molecules and nano Materials are homological to celestial singlets in the proximate cosmos, namely asteroids, planets, stars, and comets. The shapes are quite diverse across both ranges of homological elements. Except for planets and stars, shapes observed in the thousands of asteroids and comets are as

diverse and serendipitous as shapes observed in molecules, whether inorganic or macromolecules, setting the basis for the homological relationship.

5. The above leaves the extremum asymmetric median location to be occupied by viruses, prokaryotes and eukaryotes forming the Life Shell, predictably with unmatched levels of differentiation and plasticity.

This homology gives us one of the best lenses through which we should look at the diptych between the inanimate and the animate. It interprets for us the true dimension of the difference between the animate and the inanimate, which is an asymmetry in second-order covariance but not first-order. The peculiarity of the animate is not exceptionalism that puts biotic organisms at the top of ontological hierarchy in U_3 overall and should never turn us blind to fundamental first-order ontological covariance. In particular, there is no such thing in Nature as prevalence of animate over inanimate, as controversial as this may sound at this stage of human knowledge.

17.19. Caveats of Mathematical Physics Metrics

The significantly large number of paradigmatic problems facing the field of mathematical physics has their remote origin largely in pure mathematical misconstructions dating as far back as the inception of Euclidean Geometry. The distinction between physics and mathematics has never received the special attention it deserves in science, leaving the field of mathematical physics ungrounded from an abstract standpoint.

Effectively, the definition of the science of mathematical physics, indeed simply another designation for theoretical physics at this point, which states that the former is the application of mathematics to physical problems, is almost a tautology since without the intervention of mathematics in physics, there is no physical science per se, but only Aristotelian natural philosophy. The latter represents the former status of physics prior to its transition from philosophy to science. The canonical definitions afforded by mathematical logic must stand on their own as a model, without the intervention of physical concepts, so that they may adequately model the physics of nature from a towering vantage.

17.19.1 Time: The Confusion Factor

The regrettable notion of *Minkowski spacetime* very fittingly exemplifies the problem of this lack of theoretical separation in mathematical physics. Prior to Makowski's spacetime notion, time was generally considered a physical variable and was never invoked in Analytic Geometry. The independent variable therein was always ineradicably a space or metric variable. The conflation of space and time in Makowski's spacetime was never expressly justified from higher canon of mathematical logic. Before anyone can witfully do anything with time, its existence in the physics of Nature ought to be proven on the basis of the discovery of a natural unit of time. It is surely not the job of mathematical science[13] to prove the existence of time, since mathematics is exclusively concerned with pure abstractions. It is thus the job of physics as a natural science to prove the physical existence of time in Nature. Indubitably, it is a grave error to whether import into mathematics the notion of time as an abstraction, as widely assumed to be real as it is, or to peruse the notion of time in empirical physics without first proving its existence.

17.19.2 Unfounded Spacetime Metric

The above analytic perspective stands for the most adequate window over the science of mathematical physics, if the latter must come to fruition. This science which has its foundation largely in the notion of spacetime, ultimately amounts to an unfortunate amalgam responsible for the squint progress of theoretical physics since second half of 20th Century and its debacle alike in early 21st Century. It is essentially based on the vacuous notion of *spacetime interval* over a 4-dimensional background metric consisting in *3-D* of space and *1-D* of time. The formalism has it that an observer measures two distinct events in spacetime as separated in time by Δt and in space by Δx. If space and time are postulated to be entangled, then (1) Δt and Δx would have to be bijective and (2) $\Delta t\, \Delta x$ or $\Delta x\, \Delta t$ would have to constitute the new unit of spacetime.

Representing an event at point A to be $A(t_1, x_1)$ and an event at point B to be $B(t_2, x_2)$, if spacetime exists, then xt should notionally provide for its basis so that the new event coordinates would become: $A(t_1 x_1)$ and $B(t_2 x_2)$. The formalism does not specify in what particular manner x is entangled to t, however xt at the very least must obligatorily exist as the irreducible unit of spacetime. Inarguably, if any of the three dimensions of space can be represented by a line, time cannot equally be represented by a line, because it is in essence a fleeting element, contrarily to our common visualization of space. We come back here again to the fundamental misconception of the identity and representation of a line in Geometry. If we are going to conceptualize time as a timeline, the latter must bear different attributes than what we hold for a singlet dimension of space, the geometric line. We sure represent the spacetime metric with the symbol (+++,-), because we understand that something is intrinsically different between one and the other variable. Hence, we do not know in mathematical physics what xt or $\Delta x\, \Delta t$, which must be the unit of spacetime, means at all. It follows that the *FLRW* metric[14], as much as all metrics based on *3D* space and *1D* temporal, is wrong.

In the Minkowski formalism $\Delta x\, \Delta t$ has been cleverly and heavy-handedly replaced by Δct *(or,* more plausibly, $c\Delta t$*).* We must be keen to comprehend that the invariance of the speed of light is at the first place an indictment of the notion of speed or velocity in terms

---

[13] Despite the consecrated academic use of the term *mathematical science*, we should note that, strictly speaking, mathematics is not a science, since it is a body of abstractions devoid of empiricism.

[14] The *Friedmann-Lemaître-Robertson-Walker* metric

of change of position in time. The motion of light cannot be expressed in the form of *dx/dt*, for the simple reason that at first order:

$$x(t) = ct,$$

where *x(t)* is the function that describes position in time for the light beam and *c* the speed of light constant. Makowski's proposition of the quantity $c\Delta t$, deriving into *ct*, forcibly implicates the functional expression *x(t)*. For this function to be true or meaningful, *c* must be a scalar or a mathematical constant, which we know to be untrue, because it only exists in physics in the condition of a dimensional quantity, specifically the dimension of *m/s*. Since the time dimension already exists in *c*, *x(t)* would be at best a functional but not a simple function and the argument *ct* would not make then for a first-order polynomial quantity. Yet we see it being merrily used in the Minkowski spacetime formalism as:

$$s^2 = -(ct)^2 + x^2 + y^2 + z^2, \quad (1\text{-}6)$$

where s^2 is the spacetime interval.

Expression *(1-6)* doubles as an arbitrary expression, imposed by Minkowski, in the sense that it is given as a result of the application of the Pythagorean Theorem to the orthonormed *3D* spatial coordinate jointly with a 4th dimension of time. Yet we do not logically know how to geometrically articulate the linear time dimension to the 3 spatial dimensions. It is thus very much a leap of faith to believe that the fictitious linear dimension of time is statically normal to the *3D* spatial coordinate. No one has ever offered a geometric interpretation of this artifact, simply because the human mind cannot explicitly represent it physically. These shortcomings altogether put the notion of Minkowski spacetime on utterly shaky grounds.

Notwithstanding all these caveats as there may be in the formulation of the Theories of Relativity on the basis of Makowski's spacetime notion, the fundamental insight of a tensorial and geometric representation of the fabric of space remains completely adequate and legitimate. Just that the spacetime feature had deeply hampered the analytic capabilities of the Theory.

17.20. Deeper Axiomatization of Symmetry

The important results portrayed in Fig.17.12, demonstrating that the fundamental physical coupling constants, come out of an Analytic Geometry function in pure scalar incarnation, is another testament to the feasibility and appropriateness of David Hilbert's call for mathematical axiomatization of physics issued in early 20th Century, as argued in Chapter 1. The entire Quanto-Geometric framework constitutes a direct positive response to David Hilbert's bold vision, which at the same time obliterates Kurt Gödel's view of incompleteness concerning the reach of mathematical logic. Gödel's Theorem is another example of how summons built from truncated logic and insufficient scope can unfortunately contribute to limit the development of knowledge.

Hilbert's program was aimed at finding a complete and consistent set of axioms for all of mathematics, whereas Gödel's Theorems argued for inherent limitations of every formal axiomatic system in their capacity to model basic arithmetic.

We do not have to axiomatically counter Gödel's argument to demonstrate that it is false. We only have to demonstrate that all physics can be reduced to mathematical norms by showing how all the dimensional quantities that are foundational to physics can be reached in a completely dimensionless fashion from pure mathematical axiomatization. In that way, we demonstrate that mathematics is a complete system in itself because it is endogenously a completely *intelligible* system, the proof being in our conclusive demonstration that all of physical reality represents an application of mathematical canonical logic.

What K. Gödel had not understood is the far extent of Symmetry in both influencing the Number regime and informing the essence of Geometry in mathematics. He was also ignorant of the important fact that the order of Mind abstractions is a limitless stack and that there is always a higher abstraction level from which Nature can be prospected for ever-broader intelligibility. Indeed, the incompleteness was not in the objects of his axiom but in the very lack of scope germane to the axiom itself. *A bon entendeur salut!*

17.21. Conclusion

The problems of Symmetry are many and deep seated in Mathematics. The profound misunderstanding of Symmetry has led to many analytic dead ends in Mathematical Physics. It is in the chapter of representation of the Cosmos as a physical application of higher mathematical logic that those problems have festered and created long unresolved theoretical roadblocks. Notwithstanding the preeminent role played by the concept of spacetime in Mathematical Physics, no one in science has ever proposed or gained an intelligible representation of spacetime or simply the fabric of space. The pictorial representations of spacetime offered in the literature are only a remote and euphemistic approximation of the proposed physics (Fig.1.15), simply because a faithful physical representation has turned out to be an impossible. The impossibility of a faithful and clear representation of the proposed relativistic spacetime should have long hinted us that the notion is simply flawed. Plainly and simply, spacetime is a bankrupt notion because *time does not exist*: it is not possible to represent physically what does not exist. We have shown in this writing how to undertake a redress of the foundations of Differential Geometry in order to meet the abstracts able to successfully model quantitatively the paramount Symmetries upon which Nature is built. To go beyond just the 3 phenomena that General Relativity has helped analyzed to this day, to unify the Quantum Theory to General Relativity and to make solid inroads into the physics of Life, this all-out redress of the foundations of our Mathematical Logic platform is a must. The theoretical problems facing our physics are indeed in nature mathematical problems, as sensed by the best 20th Century mathematician, the visionary David Hilbert. If the problems hereby identified and raised were counter-intuitive and intriguing, we hope most find our answers in this text largely satisfactory and eminently edifying.

Chapter 18

What then is time? If no one asks me, I know;
if I want to explain it to a questioner,
I do not know.
St. Augustine

MIRAGE OF TIME THROUGHOUT MODERN PHYSICS

One may argue with good reason that the most significant aspect of the revolution that shook theoretical physics in the early days of the 20th Century is the unseating of the notion of time that had been hitherto prevalent in the science. Having started with an equation based on the indispensable variable of time to express the wave dynamics of the bound electron, Erwin Schrodinger stumbled on a reduction of the expression that was unexpectedly bereft of the familiar variable. In that manner we stunningly discovered that the microcosm of the quantum world was a time-free realm. The second aspect of that revolution was the discovery by Albert Einstein that classical astrophysical time was not absolute or universal but strictly a relativistic entity depending on the vicinity of large masses, and worse, as it relates to travel light and its observation, it inexplicably became a bloated or dilated entity both in the regimes of constant motion and accelerated motion. Gone the days of "universal, true and uniform" Newtonian time! If those dependencies do not thoroughly demote the notion of time, at the very least they severed the emperor from the throne and showed that he had no cloth. All in all, one may argue that it is difficult to accept that nature holds an entire span, the microcosm, where time is completely absent, and one other, the macrocosm, where it does exist even in a diminished fashion, given that the very nature of the notion of time implies that it must have an all-encompassing dimensional ontology. Squarely put, it must be universally ubiquitous or it is not, else the universe is awkwardly broken. In this Chapter, we are extending our discussion of a time-free world in prelude to our exacting biophysical analysis of Life.

18.1. Philosophy Ahead of Physics on the Notion of Time

As awkward as this whole situation concerning the nature of time might appear to many minds, it has nevertheless been the state of affairs in our physical science since 1905. There have been in the field of philosophy quite a few advances in tackling this problem however. Dialectic idealism, for instance, had well proposed that the most sensible cosmogony that we may visualize is one which is completely devoid of time, that the universe is ageless and has always been there. In his paper titled *The Unreality of Time* British philosopher J.M.E McTaggart advanced the same idea in 1908. Even though some physicists have entertained the idea at one time or another, most modern-day physicists appear to swing back and forth in their favor to one or the other argument, in all likelihood yielding every time to intuitiveness which always finds the notion compelling and uncompromisingly practical. They have demonstrated all but the incapacity to find a solid mathematical formulation for a time-less universe in its fullness, leaving the science to its inability to properly and radically foot the idea of a time-free world.

18.2. Lack of Internal Metric for Time

The principal hurdle confronted by the notion of time is probably the absolute lack of an innate ontological metric that would grade its flow in a consistent manner, even if the metric should be a running one. In that regard it is comparable to a ruler, whose sole purpose is to measure or to grade, but one with no tallies, hence incapable of fulfilling its purpose. Instead, we tally time on space with motion and then we declare that the spatial tallies are temporal tallies, or that the tallies are temporal in nature. That is how we have turned a human abstraction into a physical reality. You can only measure time in space or by space, and as such it is evidently a dependency of space. Is it then any surprise that time, presumably a dimensional variable, is officially defined as what human measure on clocks, one of the most subjective definitions ever in a science that pretends to be exact! If there were Martians on the red planet, time would be for them what they measure on their own clocks (a fraction of the spin or rotation cycle of their planet, if they chose to measure it that way!). Worse, if there were Venusians on the bright planet of Venus, their hours and days would be much longer than their years, because the Cupid planet accomplishes a rotation around its own axis in a lot more time that it takes to revolve around the sun![15] The inhabitants of those distant worlds would not agree with us on an independent unit of time based on a referential scale of time, because there is simply no universal scale of time. And that is because the construct of time is only dependent on both spatial

[15] The length of day on Venus is **243 Earth days**. A year on Venus is only **224.7 days**, so that a day on Venus is longer than its year.

distance and on motion equally, whether it is a creation coming from us humans or other possible extra-terrestrial intelligence.

To put it plainly, if time is a flow, we don't know and have no means to determine at what speed this flow propagates.

18.3. Entropy to the Rescue of Time

The concept of entropy, which calls for a world with physical content in increasing disorder, is one that has always guided all thinking around the notion of time by providing justification and substantiation to the latter. All physical dynamics in the universe as a whole seem to be accomplished along a timeline that sees them in continual augmentation, as though the increasing disorder of entropy coincided with the direction of the progress of time. Thus, the ensuing notion of an arrow of time. Julian Barbour has already convincingly opposed this argument in our sense in advancing that it is not time that creates change, it is change that creates the illusion of time progress and a time arrow. One must reconcile with the subtle fact that time is not an observable. Many argue to the contrary alleging that we can manifestly see that such things as the living, as much as everything else, age in the universe, and that therefore time is a direct observable in phenomenology. We argue that this argument constitutes an extrapolation, and that if we must attend to augmentation of change parallel to the progress of time, the only permissible conclusion in phenomenology is that the universe is in eternal transformation, not that it is continually aging along an evolving timeline. We do continually see concrete things that change state, statically or dynamically; time however is not a concrete tangible that can be perceived, it is only a fictitious human intangible or a fictitious inference.

18.4. Invariance of Physical Law under Time Inversion

That time does not flow at the same rate everywhere is a lesson taught to us with mastery by both A. Einstein's Theories of Relativity. What we don't have a firm theory for is the alleged direction of the arrow of time. What we know is that physical law in the macrocosm, or wherever we have introduced time as the evolution referential of dynamic systems, would play out perfectly well with no alteration to explain phenomenology, had we chosen time to flow in the opposite direction!

There have been attempts to downgrade this artifact essentially by confounding time with entropy and equating the evolution of entropy to that of time. However, we must insist that time and entropy are two different entities, time being a dimensional background variable and entropy being phenomenology, to be exact space-driven phenomenology.

A simple mathematical parallel to the invariance in question is this:

$$5-2=+3 \quad \textit{(18-1)}$$

compared to: $2-5=-3$. *(18-2)*

Therefore, when we invert the action in *(18-2)* we find two different quantities under the algebraic rule. In the case of the arrow of time, there exists a possibility for the outcome after inversion to still be equal to that of the first expression (+3). We have to assume for such *no direction* in the form of *algebraic absolute value* in order to rigorously find that both outcomes yield the same exact quantity (+3), which is what we are attempting to substantiate, as shown below:

$$|5-2| = +3 \text{ and } |2-5| = +3$$

This is what invariance of physical law under time inversion ultimately means mathematically. It is an Abelian action under the absolute value operator, which in itself is a precondition of dismissal of the inversion in question. In other words, if what we know in terms of the physical laws must hold, we ought to dismiss the notion of possible time inversion or arrow of time at the outset. Hence, the artifact in question is not a mathematical logic Symmetry, as traditionally asserted, but a Reductio ad absurdum instead.

To the eponymous human founders and followers of the notion of time and time arrow, this conclusion may very much be a hard pill to swallow.

18.5. Implausibility of Emergent Time

As much of a practical concept as time might be to the dynamics of the macrocosm, the prevailing sentiment in the physics community today appears to be that time, as John Wheeler put it, may not be a fundamental property of reality, an idea that has been gaining momentum due to the many different quantum entanglement experiments that the technology of the day has made possible, among other reasons. Bell's Theorem is another view that leads to the notion of *emergent* time as opposed to relative or universal true time.

18.5.1 Uncertain Future of Time as the Binding Agent for Grand Unification

The question of the resolution of time as a legitimate physical variable is central to the enterprise of Grand Unification, long coveted, repeatedly attempted but always scant in satisfactory outcome. Efforts in two broad different directions have been underway toward Grand Unification, one that seeks to reintroduce the time variable in the dynamics of the quanta in order to create a bridging path to Relativity and one other that attempts to export the artifact of timeless dynamics into the realm of larger scale structures treated in Relativity. Carlo Rovelli and Alain Connes have attempted the latter, which we believe is the right direction to go. Unfortunately, C. Rovelli's formulation led him to face the notion of *emergent* time because he still believes in time as the ruling variable in the macrocosm as a matter of evidence or common sense. He proposed with Alain Connes the simple "thermal time hypothesis" which has time emerge in a thermodynamic or statistical context.

18.5.2 Questionable Thermodynamic Time

To have time emerge from the quantum realm in a thermodynamic/statistical defining context is to otherwise propose that the symmetry that creates the Planck constant h and/or the Stephan-Boltzmann constant σ is by the same token the generator of emergent sub-dimensional time. Were that to be the case one wonders why the quantum realm does not prominently showcase the variable in its very own dynamics.

Furthermore, in order to prove that time *emerges*, the formalism must start out with a Hamiltonian or a Lagrangian bereft of the elusive variable, stating a dynamical scenario in whatever specific quantum realm (particle, atomic, molecular or lattice) one chooses initially to showcase it and demand its presence at the end for a resolution, not as a matter of interpretation or inference but literally. That seems to be the most suitable and convincing methodology for this argument, given that historically the inverse walk along that path is what took us to the time-free realm of Quantum Mechanics. It is important to understand that it is at the boundary between what we hold for the timeless sector of the universe and what we hold for its time-driven sector that the time variable must be seen to have *emerged*. The mathematical formalism must literally toss time out at that very boundary. To initiate the demonstration with Schrodinger's time-dependent equation is probably not sufficiently virtuous a procedure, to say the least.

18.5.3 Distortion of the Background with Emergent Time

Further, one would have to clarify the nature of the transformation by which a 3D background becomes a 4D background because the variable does bear an intrinsic dimensional character. There may not have been enough debate on the implied presence of two backgrounds in our universe, a 4D background for larger scale structures and a 3D background for the microcosm, in the context of Big-Bang cosmogony and the search for the origin of time. If our physical universe is one and homogenous, how can two different dimensional background frames coexist in it, how can it be 3D in one sector and 4D in another sector? If so, why is the speed of light an invariant both in 3D and in emergent 4D? These interrogations demand unequivocal answers for the notion of emergent time to have a chance of adequacy.

18.6. Thermodynamic Activity Not in Hand with Time

To add another point of argument to this record, as it appears the Lattice shell of matter, or the realm of condensed matter physics, is the only end quantum domain which it would be sensical for the time variable to emerge from, since time is only apparent from the next Shell of matter up, what in Quanto-Geometry we have designated the Karyote shell (world of viruses, prokaryotes and eukaryotes). To that extent the claim of thermodynamic/statistical condition as the emergent context for the appearance of the variable on the scene does not hold. Thermodynamic activity is a fundamental component of matter at absolutely all levels including the quark level, since absolute $0K_T$ is unquestionably unreachable, even below the Planck scale. In that regard one ought to consider the eminent role of the Stephan-Boltzmann constant as the pivotal thermodynamic constant in the configuration of all of 3D matter.

18.6.1 Unobservable Time

Lastly it is not clear in what sense the thermodynamic theory of time would be properly falsifiable since as a fundamental entity, time is not an observable, nor is it a constant of nature. Some have naively argued that space, however indisputable its existence, is not a tangible observable either. I submit that space is an intangible but not an unobservable. It is our inability to properly characterize the intangible void of space that in no small part has caused the human inference of time in the perception of physical reality. To further clarify this view, let us consider the following interrogation: if we were to take a look at a nightly sky devoid of stars, what would we perceive? Most would likely reply: Nothing. That would be exactly correct, the nothingness of space or the *spread of vastness* is what one would observe. One could devise means, some perhaps cleverer than others, to measure that spread. As we do so, we would discover, were we to be unaware, that it has dimensions of length, width and depth, three degrees of freedom. This void of space is everywhere present as one of the two fundamental components of all objects, in every Shell of matter. It is also primarily responsible for the profusion of *multiplicity* in the world. Were it not for its presence, the universe would be an-dimensional, a single incalculably small scalar point. We have a sense of that when a star system collapses, engulfing the otherwise huge star into something smaller than a sand bit: a black hole.

Therefore, while space as an intangible is an observable in many ways, time or the construct of a dimensional timeline is strictly not and remains an abstraction as an intangible, shall we add a fictitious abstraction.

18.6.2 The Unhappy Marriage between Time and Space in Spacetime

Common sense argument in defense of time, no more than the formalisms that peruse and support it, cannot object to the recognition that it is always formally construed as a dependency of something else, whether space, motion or gravity. Therefore, as such it bears no independent self quality to it. When confronted with the question of the cause of Lorentz invariance, A. Einstein had determined that the most we can do with time, short of dismissing it altogether, is to inflate it and entangle it to space. One is left to wonder, however, what the geometry of the artifact of entanglement of time to space may be? How does the time dimension intimately articulate with the three spatial dimensions physically, not representationally? No one has yet proposed a clear and acceptable graphical/geometrical description of this entanglement in the fabric of spacetime that are not seemingly bizarre ideas leading to wormholes and other holographic weirdness, all amounting to speculations, no matter the mathematics behind them.

The lack of geometric clarity in entangled spacetime implicates that the notion of spacetime amounts to no more than an abstract model of interpretation, not physical reality. If the Einstein Field Equations (EFE) for their own have achieved a working description of geometric gravitational interaction, they do not describe the intimate structure of spacetime as the agency of the interaction. The latter is for certain the ongoing working proposition of modern-day theoretical physics under the quantum gravity concept with no demonstrable outcome yet achieved. In fact, as of this writing, the European Large Hadron Collider's testing program has most certainly consummated the death of Supersymmetry, a major implication of the hypothesis of spacetime quantum gravity.

18.7. Time in Covariant Theories

If is fair to say that the general sentiment among physicists nowadays is that there is no absolute certainty about the ontology of time anywhere, given its many crippling dependencies. No one will hold their hand in the fire for the emperor any longer. Some physicists have developed covariant theories with the absence of a distinct physical timeline that guides the evolution of physical dynamics, such as C. Rovelli's thermodynamic time previously mentioned. Some level of predictions they are generally able to make. Most have failed however to break into the macroscopic world with the broad predictive power capable of describing the dynamical laws that govern the macrocosm, especially the fundamental physical constants that play the role of its building blocks.

As a form of covariant theory, the Quanto-Geometric Theory has successfully taken up this challenge by uncovering the timeless symmetries that preside over the development of the fundamental invariants (a.k.a. fundamental physical constants [6]) sustaining the configuration of three-dimensional matter, while systematically deriving those quantities from first principles.

Standing firm on this achievement, we peremptorily dismiss any contention by some, who are certainly otherwise brilliant, that physical science has historically not taken time seriously enough, and that the reality of time, if given the supposed credit that it deserves, will hypothetically help us uncover the origin of the fundamental laws of physics and their presumed time evolution. We must say with great confidence this is the absolute wrong way to go.

18.8. Retrocausation and Emergent Time from Quantum Entanglement

This review of how the notion of time has fared in physics to date would not be complete without an examination of the visualization of the famous variable dictated by the latest quantum entanglement experimentation. From the EPR paradox, which was a thought experiment by A. Einstein, B. Podolski and N. Rosen, involving two photons receding from one another (obviously at the very speed of light), to the actual realization of the experiment in the 1970's to the now ample variety of possible experimental methods involving even atoms or molecules in lieu of photons, physicists have been driven to the same conclusions ultimately leading to time as an entity *emerging* from quantum entanglement phenomenology.

First a few definitions. An entangled system of a pair of quantum objects or particles is one in which the sum of internal components that make up the *state* of one object is insufficient for the complete description of the *state* of the same and is dependent on the *state* of its par. It is not primarily about dynamical interaction but about *state*. That is why an entangled pair of quantum objects is said to be in the *singlet state*. The *state* behavior of these objects, so to speak, appears to be that of a single unified object. Two bound electrons within an atomic orbital is the perfect example of a natural pair of entangled quantum objects.

The second condition for an entangled pair of quantum objects to be or become such is that they must be a conjugate of one another in some way: for instance, in the orbital habitat the electrons must be in the spin-up state for one and the spin-down state for the other, or two travel photons must be coherent with one another. One may abusively sum up this condition by stating that the pair must have an opposite par or twin quality to it.

The gist to the matter here is that any circumstance that affects one element of the par such as a measurement of any sort (momentum, position, speed, etc.) automatically and instantaneously affects the other conjugate, and does so before any information could travel from the affected particle to its par (no signal can travel faster than *c* the speed at which the photons are already receding from one another). The state of the entire entangled system collapses in these circumstances. Einstein *et al* contended that at their coherent inception both particles acquired unknown or hidden information about each other in relation to their local variables or shared the same determinants, which would help explain why they behave in a correlational manner without exchange of information of any sort. Bell's inequality theorem destroyed this view to the increasing satisfaction of the modern-physics community by showing that on the experimentation plane it does not pass the muster of the strict laws of probability because the measurement results remain always correlated.

It is worth nothing in the context of this study of the notion of time that as recently as 2013 entanglement has been obtained between photons that did not coexist in time, in a demonstration of what has been called entanglement swapping. We must also point out that entanglement between quantum objects has been realized even when the measurements have been performed quicker than light could transit between the two entangled objects.

18.8.1 Diverse Views of Phenomal Time Inspired by Quantum Entanglement

The most important element of interpretation of entanglement phenomenology with regards to the notion of time undertaken thus far by concerned physicists seems to align along the ideas of:

- Time not being a fundamental variable
- Time being an emergent phenomenon arising as a side effect of quantum entanglement
- Temporal retro-causation in the sense that the future can determine present and past and that the present can determine the past.

Quantum entanglement seems to reinforce the conclusion from other approaches that time is not a fundamental variable. Time emerging as a side effect of quantum entanglement is a bold and brave approach ranging side by side with emergent thermodynamic time. In continuing with the Grand Unification attempts undertaken by J. Wheeler and B. DeWitt[16], physicists Don Page and William Wooters proposed to their time-free unification equation a solution based on quantum entanglement. These two theorists found that time can be measured through quantum entanglement! These are incipient ideas lying far from broad acceptation. The same objections that we have raised against emergent thermodynamic time may be raised here again.

The most startling idea arising from entanglement phenomenology is that of time retro-causation, which would quite frankly see A. Einstein shiver in his tomb. Effectively, the principle that found A. Einstein staunchly rebellious until his last breath is that of negation of intellectual cognition in regards to any aspect of physical dynamics as implied in the Copenhagen interpretation of Quantum Mechanics. He found a magical character to that interpretation which prompted him to express the collapse of the entangled wave-function as “spooky action at a distance”. It is anyone’s guess what his reaction would be in regards to specifically the notion of retro-causation much in vogue today.

As much as the current conclusions driven from quantum entanglement experimentation found EPR position vis-a-vis statistical quantum mechanics erroneous, from the Quanto-Geometric vantage we can fully support the idea that no aspect of three-dimensional physics can be outside of the scope of intellectual interpretation and that we must not accept no explanation for an explanation. Effectively, one will find no weirdness or mystery to the behavior of quantum mechanics’ wavefunction as long as we are able to understand the higher order of Covariant Norms, unknown thus far to past and modern-

---

[16] The Wheeler-DeWitt equation historically stands for the first attempt at unifying quantum mechanics to relativity under the umbrella of timelessness of physical dynamics. It still represents the foundation of current efforts at developing a theory of quantum gravity.

day physics, in its 9 typified forms and its 3 quintessential forms, which is central to Quanto-Geometric Theory. This analysis is largely undertaken in Book I. What we can further advance here is that it is only in the ignorance of the many properties innate to the void of space, beyond the geometric properties uncovered by Einstein's Theories of Relativity, that one will find mystery in the behavior of Quantum Mechanics wavefunction. The analysis relating to how the Quanto-Geometric wavefunction comprehensively deciphers quantum entanglement, through consonant and logical analytic elements, is beyond the scope of this thesis. We will only advance here that in its condition of a real-valued function has the Quanto-Geometric Eigenfunction been able to properly characterize quantum entanglement.

18.8.2 Can Time Stand Still for a Certain Time?

One of the many tentative interpretations of the role of the time variable in quantum entanglement proposes a vantage point from Special Relativity whereby Lorentz invariance is considered from the viewpoint of the very photon of observation. The theory states that from the photon point of view, time comes to a **standstill** as a consequence of time dilation contemplated in Special Relativity. In the case of the two entangled photons, the theory proposes, any measurement made on one photon at a time t would determine the state of the photon not only at time t (present) but at time $t' < t$ (past) as well. All appears to be well in the theory except when one starts to wonder *how long* did time stood still to the photon, has it been or will it be *forever*, since it is in eternal travel and left to itself knows no time, or can it ever be for a certain *time* duration, which would have to be of its own but which altogether falters? *Time* standing still for a certain *time,* moreover within the same reference frame, is a construct that is frankly hard to swallow from whichever perspective one is coming at it. Henceforth one loses appetite for the ensuing proposition of retro-causation that would walk us back to point of origin with *future information* informing the point or state of head start, so that we may somehow comprehend subsequent correlated behavior with no apparent direct or indirect path of causation.

The presumed experimental evidence by Ekaterina Moreva *et al* of emergent time from quantum entanglement is worth discussing, however briefly. Quite aside from the relatively intricate optics of the experiment, others have summarized it in this way: it is based on the creation of a presumed *toy* universe consisting of a pair of entangled photons and an observer that can measure their state in only one of two ways. In the first phase, the observer attempts to measure the evolution of the system by joining into the entangled state. In the second phase, a god-like observer (or *super-observer* situated in what is construed as an outer universe to the photons') measures their evolution against an external clock which is entirely independent of the toy universe. Many pertinent questions arise:

1. Why do two entangled photons suddenly become a *toy* universe? What are the boundary conditions that effectively subtract them from the effects of the laws of conservation and symmetry ruling the physics of the world beyond them?

2. The absurdity of this de facto *two-universe* proposition lies in the very fact that the wavefunction associated with each photon particle is infinite and knows no physical boundary. On account of that proposition every saturated electronic orbital in an atom, with its two entangled electrons, would be a closed world onto itself, and a toy universe to perhaps a higher shell of saturated orbital. We do not hold this view in quantum mechanics.

3. Entanglement between two structurally different entities is not canonical. Entangled elements must be coherent and conjugates to one another in some physically observable character. For any observer to become entangled by measurement with any of the two photons, the ontology of the observer must at least conform to that of the photon, and the observer must be prior in a conjugate state with the measured photon. Therefore, it must know the entangling state of the photon even before it can measure the property. If the observer should become entangled with the two-photon system, then it would have become a *stateful* part of the system and thereby loses the ability to observe or measure any of its properties!

4. Why is it that the measurement of the basis entangled property by the observer does not collapse the wavefunction of the particles in the experiment?

5. This criticism principally goes to the interpretation of the experiment by others, and to the authors as well only to the extent that this spirit originally motivated the methodology for the experiment. The metaphoric "*god-like observer (super-observer) measuring from a detached world the evolution of the entangled particles against an external clock which is entirely independent of the toy universe*" constitutes an idea we find completely dippy, berserk and preposterous. If there is a place for God or a super-observer outside of the 3-dimensional realm, it must have its own conservation and symmetry laws and its own background of which we cannot assume that time is a variable thereof. If you invoke time for that universe *(abstract coordinate time per the authors' language)*, you are already admitting that it at least shares this variable with the *toy* universe. If you further project the possible existence of an observer, God-like or not, you are implying that the evolution of matter as we know it exists as well in that outer universe. Therefore, that universe is not really a detached outer universe, it is at the very least a 3D-like universe, most assertively it is just the 3D universe, because you have not explained what the physical

adiabatic boundary between the two universes consists of. The argument is simply vacuous.

It requires a leap of faith to believe that such ideas or the original spirit that motivated them could have shaped a falsifying physics experiment! It is clear that the conclusion that time is only internal to the universe and absent for (putative) external observers to the universe is and cannot be more than a complete extrapolation from the optical dynamics implemented in the experiment, to which the authors adjoined the interesting epithet of "illustrative". It is nevertheless quite unfortunate that too many in the physics community took the experiment for a *probing* one and not just an *illustrative* one.

18.9. Troublesome Putative Principle of Retro-causation

In general, retro-causation theories are tantamount to explaining instantaneity with a rather traumatic feedback from future to past. In a sense they are strangely proposing entropy (forward moving disorder) to be the result of an inversion of the arrow of time. Effectively the phenomenological course of quantum entanglement as we experience it follows the forward movement of entropy (increasing disorder). To want to explain the initial conditions of quantum entanglement from its end result (so that the end result may become intelligible) is to want to invert the natural course of entropy, with the aid of a fictitious inversion of the arrow of time (the latter being in itself already a fictitious notion). This is philosophically unacceptable and, from a physics standpoint, simply unproductive, however laudable the forceful effort to resolve a vexing problem.

18.10. Ultra-Structural Analysis of the Ontology of Time

The last analytical element that we will invoke in this examination of the notion of time is its inner structure as a physical variable or a mathematical independent variable. We hope to show that if there exists any meta-level abstraction to the construction of temporal ontology, then we will have proven that time is not primordial, hence not foundational. Which means no possible dimensional presence in the background throughout the *Shell* continuum of the universal arena.

Our common representation of time is based on a triplet tempo overlaid on the continual flow. This triplet tempo is comparable to a saddle on the back of the flow, or a fleeting ocean wave. We routinely visualize time flow as an evolution trend whereby a point-origin A creates a point-present B which will give birth to a point-future C. The most complete irreducible unit of time in our minds is this triplet [A-B-C] segment, which we explode by generalizing over the entire universal arena in terms of *[past period - present period - future period]*.

This assumption is behind our common mathematical treatment of time as a timeline. As we move through the tallies of the timeline, we materialize a sweeping present time t_B, which leaves behind or in the past a sweeping time period $t_{A\text{-}B}$, while chipping away a sweeping portion of time period t_{B-C} from the future. Since the timeline is numerically ordinalized both in our everyday state of minds and in the mathematical formalism, we so create the instants along the timeline or on our clocks such that:

$$t_A < t_B < t_C$$

The relation above represents the instantaneous state of the fleeting triplet temporal segment. The global temporal state is given in the expression below:

$$T_{(-\infty,A]} > T_{BB} < T_{[C,+\infty[} \quad (18\text{-}3)$$

with T_{BB} representing an affine slice of time period, the present time period, $T_{(-\infty,A)}$ representing Past time as a downstream-closed time block and $T_{(-\infty,A)}$ representing Future time as an upstream-open-ended time block.

In Quanto-Geometry, the triplet norm is a direct derivate of the Quintessential Tri-Va lent Norm which has the following 2-space tensor expressions:

Hyper Quantum Tensor:

$$g_{\bar{1}}^2 = \int_{2\sigma/\sqrt{7}}^{\sigma/2} \sqrt{1+\left(\frac{dQ(s)}{ds}\right)^2}\, ds$$

Hyper Quadratic Tensor:

$$g_{\bar{4}}^2 = \int_{\sigma}^{\sigma\sqrt{8/5}} \sqrt{1+\left(\frac{dQ(s)}{ds}\right)^2}\, ds$$

Hyper Space Tensor:

$$g_{\bar{9}}^2 = \int_{\sigma\sqrt{8/5}}^{\infty} \sqrt{1+\left(\frac{dQ(s)}{ds}\right)^2}\, ds$$

Because in our ordinary abstract representation of time the past is set and finite, and the future open-ended and essentially infinite, expression *(18-3)* stands for the envelope of outskirt of the Quintessential Tensor as visualized in the Quanto-Geometric Theory. We only now have to show covariance between two inner tenets making up the construct of time for compliance to the model to be proved. At such effect, we have to ask ourselves the question: is there any inner component to the construct of time as we routinely visualize it? We know that, in the global temporal picture, past events create information and the potentials that inform the present, whereas present events create information and conditions that give birth to the possibilities which will materially model or induce the future. Therefore, we are able to ascertain at this point that the Present is an equalized reunion of two different Ephemera: one that is Settled (or Finite) Ephemeris emanating from the past and the other Unsettled (or Open) Ephemeris that is the potential for the future (Fig.18.1). We are giving the term Ephemeris the broad sense of *fleeting block of time*, in a departure from what it had officially meant at one point[17].

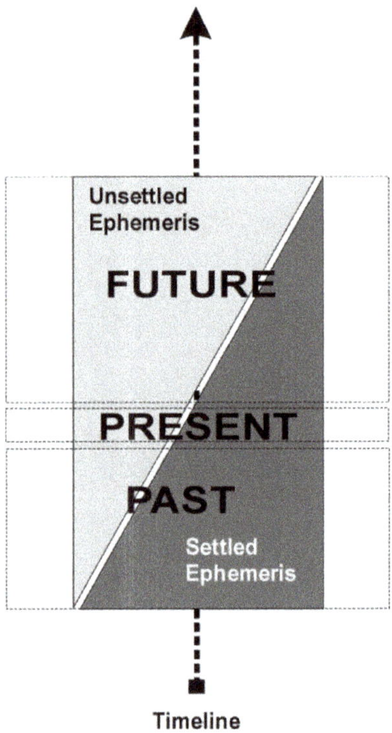

Fig. 18.1 The three bundled moments of Ephemeris

Therefore, we have identified thus far, on the one hand, for the Quadratic Tensor two constitutional elements which are Unsettled Ephemeris and Settled Ephemeris, and on the

[17] Ephemeris Time (ET) was defined to be the running period from the instant, near the beginning of the calendar year A.D. 1900, when the geometric mean longitude of the sun was 279° 41′ 48″.04. At that instant the measure of Ephemeris Time was 1900 January 0^d 12^h precisely. This time scale standard was put to rest in 1984 due to its several deficiencies, in favor of atomic time which was made to incorporate the relativistic effects of motion on time.

other hand Settled Ephemeris for the Quantum-Dominant Tensor and Unsettled Ephemeris for the Space-Dominant Tensor. Suffice it now to show the presence of some level of unsettled Ephemeris in the Quantum-Dominant sector and some level of settled Ephemeris in the Space-Dominant sector for the Quanto-Geometric Norm of orthogonal covariance between two tenets to be spectrally verified. Is it not common belief that the "future is rooted in the past" as Dictionary.com put it? This is another way to make a statement for two distinct assumptions:

- Within Settled Ephemeris (Past) there is *potentially* a certain level of Unsettled Ephemeris (Future) which will be realized further down.
- Within Unsettled Ephemeris there will be *de facto* a certain level of Settled Ephemeris.

The adage does not mean to say that the future consists in a complete migration of the past into the locus of the former. It proposes instead the intuition that some elements typical of past time will be consequential in the configuration of the future, as much as it means that some elements typical of the future were already discernible *ad reductio* in past time. There is no constraining the interpretation to phenomenology as some might suggest because the Ephemeris flow is at the heart of phenomenological 3-tier tempo. We should add that it is obvious that both flows must transit through the Present layer in their development.

Therefore, we have established that the Quanto-Geometric Tri-Valent Tensor exactly models the ontological ultra-structure of humans' common notion of time.

Now, some like Julian Barbour, have rejected the common-sense intuition of flow of time in favor of the visualization of "Relative Nows" in catering to phenomenology in a distinct manner. The argument is that there is no real past time or absolute past events intrinsically because the population of objects in the universe perdure for the most part, and that there is only a persistent "Now" for groups of objects, which makes time as a whole a permanent "relative Now". Instead of a vertical relationship between changing states of objects, he sees a horizontal relationship between those states whereby all possible states in any region of the universal arena or *tutti quanti* exist simultaneously, the resulting phenomenology being the product of interactions in a constant "Now" I must say that if one finds this view quasi unintelligible, at least difficult to reconcile with physical reality, it is because it is, and J. Barbour will admit that most will find it such as much as he himself has admittedly struggled with it. This view can only become intelligible if we visualize that "Now" to be a temporal instance accessible to or preferred by most objects and that "before" and "after" are states that are relative and, most importantly very improbable. It is not that these latter states would not exist in the absolute, just that they are highly improbable (almost don't exist).

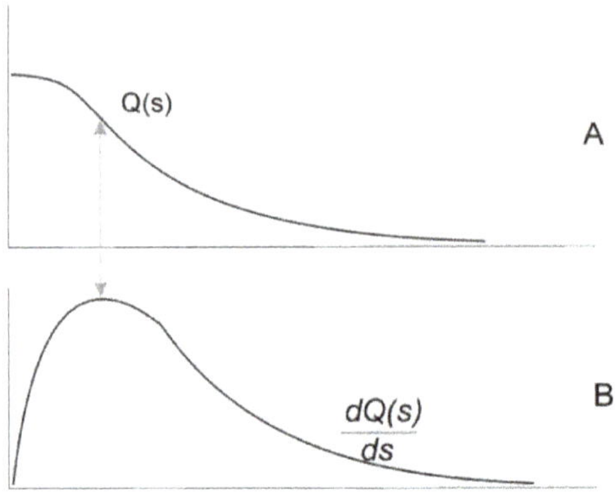

Fig. 18.2 The derivative of the Quanto-Geometric Function as an important component of the radial energy density distribution attributed to a bound electron

This vision at the end amounts to the proposition that the ontology of time is modeled by a probability density function whereby Present Ephemeris (Now) represents the most probable instance of Ephemeris, that Past is the "region" preluding the maximal node of Now while Future is the "region" beyond the node, in the understanding that these "regions" are temporal states (Fig.18.2). Indeed, we are now back to the Quanto-Geometric Tri-Valent Tensor characterization, with Past Time represented by the Hyper Quantum Tensor, Future Time represented by the Hyper Space Tensor and Present Time modeled by the Hyper Quadratic Tensor (Fig.2.3). This is what J. Barbour's no time proposition matter-of-factly amounts to, if it should become thoroughly intelligible.

Manifestly, whether in abstraction one visualizes time as Ephemeris flow, or challenges that representation with the alternative vision of a permanent static "Now", there is always an ontological ultra-structure attributed to the artifact of time. If there should be a higher abstraction, or a meta-level abstraction, to the artifact of time, an entity which had been given the rank of a background dimensional variable encompassing all of physical reality, then we must conclude that time is an abstract epi-phenomenon and that we can altogether do away with it. It would qualify at best as a representation model. Therefore, any description of nature anchored on this variable cannot be fundamental and must be superseded.

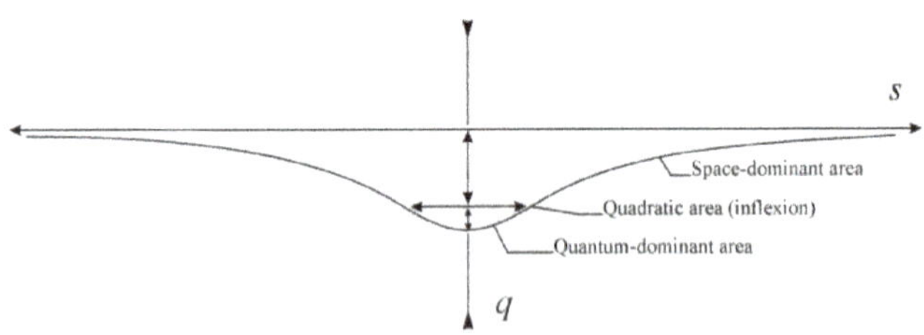

Fig. 18.3 The three Quintessential sectors of the Quanto-Geometric Function that grandfathers the Tri-Valent Operator

Despite the fact that the Quanto-Geometric canon is able to model the visualization of time under any of its facets, it does not at all validate the concept and in fact altogether dismisses it. The Quanto-Geometric canon proclaims the universal primacy of space with no time associate and proposes that all evolution is spatial evolution, through the innate symmetric qualities germane to the Void of Space.

18.11. Chronicle of the Epiphenomenon of Time

Surely an interesting exercise is to analyze the historical and cultural circumstances that have led the human global culture to the practicality and self-evidence of Ephemeris or time flow, forcing its blind acceptance to learned and laymen.

In past cultures, humans have all observed in the world flying birds, racing horses or dogs, falling trees, a moving sun through the zenith, all dynamical circumstances which demanded for us to have a certain sense of rate of motion for their appreciation. We needed to have a sense of rate of motion so that, for instance, if we are in the way of a falling tree, we know how fast we must cross the direction of the falling pole in relation to the speed at which it is coming down, in order to avoid being hit. The inner sense of rate of motion became our referential flow of time, the gradation or scale against which we gauge or rate motion. That is why in our minds the referential flow of time is set at a very low speed: the types of motion that we are familiar with being indeed very slow, our referential for gauging such motion must be even lower (or slower) so that we may positively grade the many different rates of motion occurring in our environment. If our sense of time flow were to be a very high rate, then our expectation value for the outcome of dynamical phenomenology would fail every time, because our referential sense of time flow would be disjunctive and unable to account for ordinary motion. To make a formal statement of this situation, we have:

$$v_f << v_i << c$$

where v_f is Ephemeris time flow, v_i average speed of ordinary objects and c the speed of light. In terms of an abstraction in our minds, time does not work for us and has never worked for us as a timeline with fixed tallies, we do not hold such a representation anywhere in conscious experience. Time has only worked for us as Ephemeris or a time flow. And the reason why our expectation value of everyday dynamical outcomes is on point and satisfies us this much is because v_f is very much lower than v_i, which is even multiple folds lower than c. Of the latter, the speed of light, human common sense has no representational idea whatsoever, as light appears to our good sense to be an instantaneous artifact. All in all, Ephemeris works for us because we are slow creatures in a slow-moving world of ours.

The formal concept of a tallied timeline was born in all likelihood with Newtonian physics and paved the way for the realization of clocks. Clocks in turn solidified the idea of time in our minds. Because they were based on a specific and conventional value for

v_f, clocks contributed to ingrain in our conscious experience the sense of a very specific level of referential flow of time that is *natural*.

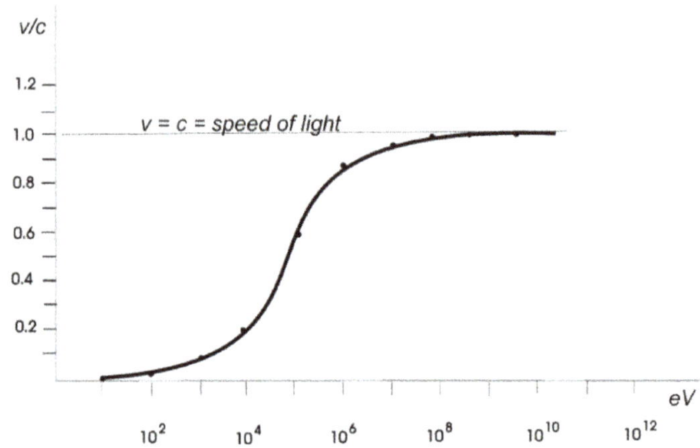

Fig. 18.4 Nonlinear variation of an observable (energy of a free accelerated electron) in response to linear variation of the β factor (v/c)

So therefore, universal time based on the second, the unit of time flow (or duration if you prefer[18]) between two tallies on the timeline (or time circle), was born and remained true until c, the speed of light, broke onto the scene with its perplexing invariance in relation to other moving objects or observers. In effect, the results of the Michelson-Morley experiment provoked a complete shock in the physics community. Ephemeris time flow v_f was a great tool because it even allowed us to predict dynamical outcome when two motions combine with each other, we could simply add their velocities on the basis of what is formally known as Galilean transformations. However, expectation problems arise when v_i is no longer this low: we cannot predict any longer, based on Ephemeris time flow v_f, the distance traveled by objects moving at v_i when the latter is appreciably high. How high? As v_i approaches the value of c. The expectation values for dynamical outcomes become in that event so significantly inaccurate that they are of no usefulness for practical purposes (Fig.18.4). We have to introduce a correction factor to v_i in order to maintain outcome accuracy, with the so-called Lorentz or Relativity factor:

18 The idea of time duration or period seems to convey a static idea of an instance of time. Only apparently so however, since for the next time segment or period to ensue, there must be an inner flow through the previous time block, given that we cannot imagine an interstice of time between time blocks and even less time standing still in between the blocks.

$$\gamma = \frac{1}{\sqrt{1 - \frac{v_i^2}{c^2}}}$$

So therefore, Ephemeris time flow is at fault when we try to appreciate the outcome of motion that is an appreciable fraction of the speed of light. We can still rely on it to some extent so long as we apply the Lorentz correction factor to the expectation value of the dynamical outcome. On the observation plane, this treatment is realized under the Lorentz coordinate transformation.

What is here really shocking to common sense is the circumstance where an object happens to move at the very speed of light, an atom of light i.e., otherwise called a photon. At that speed, our expectation value for dynamical outcomes completely collapses, since we cannot predict anything anymore based on Ephemeris time flow, to which results we cannot even apply the Lorentz correction any longer: we are faced with the absolute death of Ephemeris time at that speed. Velocities of two different objects with reference to one's self do not add any longer. Everyone perceives the atom of light moving at the same exact speed of c, ***whether they are themselves moving or not, and even if they were to possibly move themselves at the very same speed of c and in any direction!*** This is a completely unexpected and unintuitive outcome, expressed in the indetermination of the Lorentz correction factor at that level. Effectively in that circumstance, the denominator of the expression becomes 0, yielding division by 0 which is a forbidden or indeterminate mathematical operation.

The problem the educated thinkers faced at that point is the question of how you describe motion without the reference of a time flow of some sorts. H. Lorentz and A. Einstein decided that they were going to give Ephemeris a second lease on life and proposed to apply the relativistic correction factor γ to Ephemeris time flow itself! In that manner, Ephemeris which had experienced sudden death with the Michelson-Morley experiment, was resuscitated and granted a stretch of life literally under the notion of *dilation of time*, formally formulated as below:

$$T_1 = \gamma \cdot T_2$$

Whether this tour de force, possible because the Relativity factor has the dimension of a scalar, has been beneficial or detrimental to physics is left to the reader's judgment. He or she shall well remember that this vision had not been able to produce quantum mechanics, which had ushered in with the recognition of legitimate time free dynamics in the physics of nature. It is no accident for certain that a more fundamental description of the macrocosm beyond the Theories of Relativity seems to peremptorily demand a time-free approach as understood by J. Wheeler and B. DeWitt early on.

It is worth addressing at this point in this exposé the artifact of atomic clocks which seem to create the fallacious certainty of Time and time flow as an accurate physical quantity. While physicists at the standard bodies are justified in searching for accuracy in physical quantities that the global culture relies on for many different purposes, their adoption of the atomic clock for human time seems to create the false sense that it is an accurate measure of time, at the same time that it seemingly validates the notion itself. One should not be disabused by this measure however because it remains unquestionable that any atomic clock will "tick" differently depending on whether it is located at the bottom of Mount Everest for instance, at top of the mountain or standing still in a geostationary or drifting satellite above the earth, for reasons of gravity, mass and motion of planet Earth and/or the satellite.

Well aware of these issues, the Standard Bodies have adopted the most reductionist definition for the time standard strictly from a utilitarian viewpoint. Effectively one may read on the NIST physics portal the following definition for the *second*:

Considering that a very precise definition of the unit of time is indispensable for the International System, the 13th CGPM (1967) decided to replace the definition of the second by the following (affirmed by the CIPM in 1997 that this definition refers to a cesium atom in its ground state at a temperature of 0 K):

The second is the duration of 9 192 631 770 periods of the radiation corresponding to the transition between the two hyperfine levels of the ground state of the cesium 133 atom.

The reason why this arbitrary number of periods was chosen is so that the definition of the second could grandfather the definition of the meter in terms of a distance traveled by light. The most important element of this definition of the time unit based on dynamics occurring within an atom is its **conventionality**. It is based on a human convention which is a specific number of electromagnetic radiation periods (f^{-1}) of human election, which furthermore responded to a specific purpose, notwithstanding its nature. It is therefore not at all a fundamental or natural quantity.

18.12. Quanto-Geometric Theory as a Unified Time-free Framework

We have already invoked above Quanto-Geometric analytic elements in the interpretation of the ontology of Ephemeris. Effectively, as a covariant theory that exclusively rests on a dual set of physical irreducibles in the incarnation of a scalar tenet and a space tenet, the Theory posits a framework of constructs that is completely foreign to the notion of time. The correlation between the scalar and the space tenets creates the fundamental order of symmetry from which all of physical reality emerges. The scalar becomes tangible as a quantum of energy or mass in any of its classes, the intangible space tenet is manifest as an order of real wavefunctions expressed in normalized wavenumbers. On the computational plane, the framework models space as a 2-space entity represented in the single independent variable *s* and as a 3-space entity represented in the duplet set of independent variables *m* and *t*, which inherit the inner properties of *s* by rotational transformation.

The symmetries mushroom in and of themselves and recombine in order to create universal multiplicity through the development of Shells of matter. This entire view would all amount to speculation if it did not create, as indeed it does, the conditions for the mathematical computation of the entire series of known fundamental constants of Nature, which appear as the building blocks of the suite of Shells of matter: from the speed of light, to the gravitational constant, to the Planck constant, to the Rydberg Energy constant, to the mass of elementary particles… and more.

All of this is accomplished without ever invoking time at any stage of the framework. We do not need time in order to describe motion in Quanto-Geometry, because motion emanates from a symmetry quality pertaining to the wavefunction. Evolution is simply a result of scalar-space covariant shifts whether in ontology or phenomenology.

Because the Quanto-Geometric canon is so highly placed have we been able to peruse it here to model time and time flow and unravel the necessary ontology attributable to the notion if any. That sole artifact in and of itself reveals to us that the notion of Ephemeris time flow is fictitious or physically vacuous as a dimensional variable, in addition to the framework proving such with its very time-free construction. In its condition as a dimensional variable or a foremost universal primitive, time cannot have abstract structural or constitutional elements, I argue.

In the Quanto-Geometric theoretical framework, time is summarily and irrevocably dismissed at any and all phases of the universal deployment of 3-dimensional matter, from the quark scale to the large-scale cosmological structures.

18.13. To Live in a Time-free World

The paradigmatic notion of time is one of the most troublesome obstacles that human culture ought to overthrow for a more enriched conscious experience of the world and more assertive interaction with the same, both on the individual and collective levels. Some might caution that will require a cultural revolution and indeed we should expect no less.

If one cares to pay close attention to the history of physics, what comes out as salient evidence is that the notion of time, questioned ever since Relativity and Quantum Mechanics erupted onto the stage in early 20th Century, has slowly but surely entered the crisis mode up to modern-day physics. What we are witnessing in the physics community nowadays is as much a quest for a patch to still contend the notion as it is a feverous discomfort from the inability to reformulate physics assertively in complete divorce with the time variable, I dare say. This malaise has been there ever since John Wheeler and Bryce DeWitt attempted to unite Quantum Mechanics to Relativity based on a timeless equation first published in 1967 which was to subsequently acquire increasing significance. The equation attempts to describe the global universe in a timeless manner by applying a Hamiltonian constraint over a proposed wavefunction of the universe in a manner that is characterized in the simple trace expression below:

$$\mathrm{H}(x)|\psi\rangle = 0$$

What is interesting to note in that regard is how physicists had invariably sought to toss time out of the formalism at the very last mile of its development, albeit to little avail, because they figured that time must make a comeback if the formalism ought to have any good chance to describe the real world of the macrocosm. This obstacle had come to be known as the "problem of time" in quantum-gravity constructions. Ekaterina Moreva et al confirm:

The "problem of time" in essence stems from the fact that a canonical quantization of general relativity yields the Wheeler-DeWitt equation predicting a static state of the universe, contrary to obvious every day evidence.

One can see that most physicists using the Wheeler-DeWitt formalism have been working at cross purpose with the same.

It is my sense that the reformulation required for a unified physics will not take place or enjoy broad acceptation until the philosophical vision implicated in the perspective of a completely time-free world is wittingly accepted, clearly understood and properly exposed. The Theory of Quanto-Geometry has made great strides in that direction already.

18.14. Mandatory Shifts in Perspective Implicated in Timelessness

There are three fundamental shifts in perspective that are required to both learned and laymen for the experience of a time-free world. One is that we must leave behind the vantage that drives us into visualizing the beginning of every one thing under the sun as evolution from a quantum point. In Quanto-Geometric analysis, this behavior is a result of the application of the Hyper Quantum Operator to observables. We ought to start learning to envisage the beginning of physical things from a space perspective, which is that from the infinite is where objects initially instantiate their course of existence (Fig.18.5). That correlates to the transcendental sector of the Quanto-Geometric Function. This form of contemplation is valid not only for individual objects but for the physical universe as a whole as well. Some might counter that this vision is equivalent to no finite beginning at all to any one physical object but unrelenting transformation from an open end, and... right they are!

Secondly, we must learn to contemplate the evolution of objects not as a rundown or run up along a *fictitious* timeline, but in terms of *real* phase changes executed within the scalar-space correlation germane to the ontology of all physical objects. This operation still preserves chronology but not Ephemeris in terms of duration and time flow. Before, Now and After are chronologically preserved under correlational phase changes in the life of an object. Only that they have a quite different meaning than they did under Ephemeris.

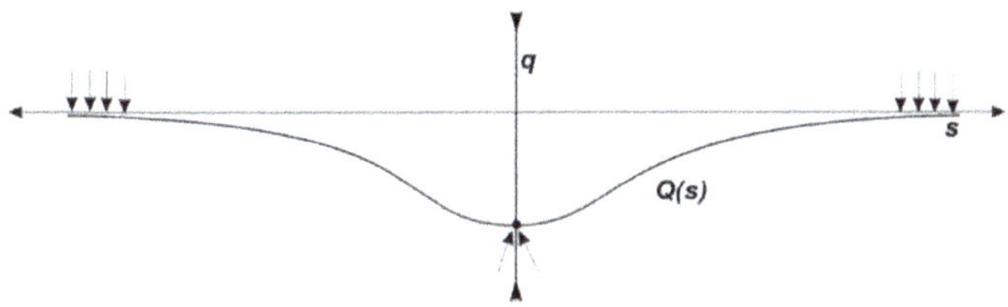

Fig. 18.5 Evolution of physical objects not from an initial quantum point but from the open-ended spatial infinites

Third and last, we must import into our conscious experience not just the notion but the intimate sense that the universe and every object in it, including oneself, is *eternal*. No man or woman that I know in mainstream human culture thinks or himself or herself as an eternal creature, but rather lives in the pressing or remote expectation of inevitable on-coming death. The awareness and contemplation of *eternal* existence is only part of unconscious experience, in both individual and collective unconscious. *Eternal* is a construct that you don't find in the Lagrangians and the Hamiltonians describing the dynamics of matter, despite their many layers of complexity bundled in one expression. Nor is it a common subject of discussion in our physics. That ought to change, and, for starter, one can think of it as a property of the universal *real*-valued wavefunction.

18.15. The ABC of Time-Free Living

As we move along with our lives, we think about the things that we do and those that occur or may occur in our environment as directed along a *vertical* fleeting timeline, as though they were strict functions of that timeline. We instinctively give intrinsic value to the flow of time as a measure of what we do. We so come to believe that we create more value in the world and for ourselves by beating the flow of time in terms of doing more in a lesser amount of time. If time is indeed an illusion as we do believe in Quanto-Geometric Theory, what transpires from a timeless vision of the world is a *horizontal* sense of awareness, which is a sense of cyclicity and synchronicity. This symmetry is one of the most important trends in phenomenology in all of physical reality.

In the context of zero time reference, it is important to comprehend the many different symmetry regimes in existence in the physical world to which we must learn to relate and correlate in conscious experience in order for us to appropriately perceive and manipulate the world outside of the fictitious concept of time and time flow.

The first of these orders of symmetry is that one resulting from ontological covariance between the scalar tenet and the space tenet which make up every universal object. The covariant development that establishes the 9 fundamental forms of symmetry and the 3 hyper norms of symmetry represent the alphabet of the new protocol pivoting timeless consciousness. One must learn to identify those basis forms of symmetry in ontology or phenomenology as a matter of simple dexterity.

From that standpoint, all forms of evolution are to be understood and expressed as quanto-geometric covariant shifts or ontological phase changes. There is no obligatory progression from one covariant level to the next. The intervention of a hyper tensor on any particular object with a primal or typified quanto-geometric articulation, or the composition of this object with other of another class, may result in a new object with covariant type situated at any other level of the spectrum. Also keep in mind that there are spontaneous and unexpected changes, commonly denoted *mutations*, that materialize those typological shifts, all of them remaining within the quanto-geometric spectrum because all mutations are typed.

For humans, the symmetry of cyclicity and synchronicity starts with the cycle of conscious and unconscious experienced through the circadian cycle regulated by astrophysical rotations. Our lives are not the forward jet stream projection of events that most conceptualize in conscious experience but principally a spiral of repetitive events from the perspective of a certain character.

The first of these repetitive events is the very consequential circadian cycle. Through the circadian cycle we both create “entropy” in the environment and develop the highest level of metabolic activity endogenously at every phase of sunlight, while we subsequently develop homeostatic processes and lowest level of entropy and metabolic activity at every nocturnal phase of the sun star. The next cyclical symmetry in our lives is the series of developmental phases which experience a form of commutative operation. Lastly, we experience the cyclical symmetry of repetitive events of a particular quanto-geometric type, which define the fundamental theme of a life, that which we call destiny or the meaning of life, to be exact the meaning of *a* life. Most remain essentially unaware of the latter.

We are going to open a parenthesis here to further discuss the notion of circadian cycle, an important notion for the life sciences, albeit with little or no presence at all in physical science. In the context of analysis of the role played by the notion of time in the human culture, it seems quite fitting to examine the notion of biological clock sustaining the circadian cycle and some of its implications. A circadian cycle is defined to be two well defined sets of biological phenomenologies (wake and all associated phenomenologies v. sleep and all associated phenomenologies) in a more or less 24-hour time period, with endless forward repetition according to the eternal sinusoid rhythm. Biologists believe that this cyclical biological 24-hour period (12h-12h) is implemented by an endogenous clock internal to the structure or ultra-structure of living organisms. The circadian clock has been consistently traced back to a specific gene in certain species. However, and most significantly, the deletion of the clock (or gene) in the same species does not suppress the circadian cycle as it relates to external stimuli of astrophysical origin! More generally, the time-of-day information as relayed by the eyes travels to the master SCN (suprachiasmatic nucleus) clock in the brain, and from there propagates to the suite of other secondary clocks in the rest of the body for their own synchronization. In that manner, the sleep-wake phase, timing of body temperature changes, timing of thirst and appetite phases are coordinated and correlatively controlled by what appears to be a system of biological clocks. There is wide consensus among biologists that *rhythmicity* stands for a key basis factor in regulating and coordinating *internal* metabolic processes, as well as in coordinating such processes with the *environment*. The continual maintenance by inheritance of circadian rhythms in fruit flies after several hundred generations in constant isolating laboratory conditions, the existence of creatures able to live in constant or prolonged darkness in the wild, as well as the experimental elimination of behavioral, but not physiological, circadian rhythms in quails, forcefully solidify this conclusion.

While these studies are fascinating in many ways and inconclusive in others, the picture that emerges for us is that a physical time duration clock per se does not exist anywhere at the ultra-structural level of living organisms but a cyclicity and synchronicity *function*, implemented hierarchically by a series of actors, which correlate the endogenous dynamics responsible for life processes with the multiplicity of astrophysical symmetries.

Cyclical symmetry is indispensable to Karyotes in other important ways as previously mentioned. It will be inappropriate for this study to delve into the realms of psychology and humanistic philosophy for a deeper explanation of the role of cyclical symmetry in those dimensions of karyotic life, this being essentially a physical/biophysical science thesis. I will only attempt to shed more light on the above with an illustration.

If it were possible, despite the presently known obstacles, for a human mother to be located on planet Mars, and a father to be located on another planet, i.e. somewhere in the Alpha Centauri system, and to have never been on earth, and if we further project that they attempt at some point to share views about their son who had been carrying on an existence on Earth at the care of a grandmother, in that situation they would not be able to agree or share ideas based on his age in terms of years. Further their protocol of communication exchanges would have to be based on correlational experience anchored on astrophysical symmetries having been away from each other's immediate environment and experience. As to his age, for a common understanding they would have to refer to his developmental age or stage, which is an artifact that obeys the derivative order of symmetry of the Quanto-Geometric Eigenfunction, as depicted below.

Developmental evolution of all Karyotes obeys the differential distribution deriving from the Quanto-Geometric Grand Eigenfunction (Fig.18.6) The ν coordinate stands here for biological vitality, a composition between metabolic and regenerative capabilities and the r coordinate the age in terms of relative astrophysical cycles (not time duration through Ephemeris time flow). The curve describes ascending vitality from birth through infancy, to childhood, to adolescence, reaching a peak in the middle of youth stage as is well known. From that point on, the trend is to descent, going from early adulthood, to maturity, to seniority, to elderly and finally to complete senility in the asymptotic neighborhood. We have attempted to make the stages as granular as might be.

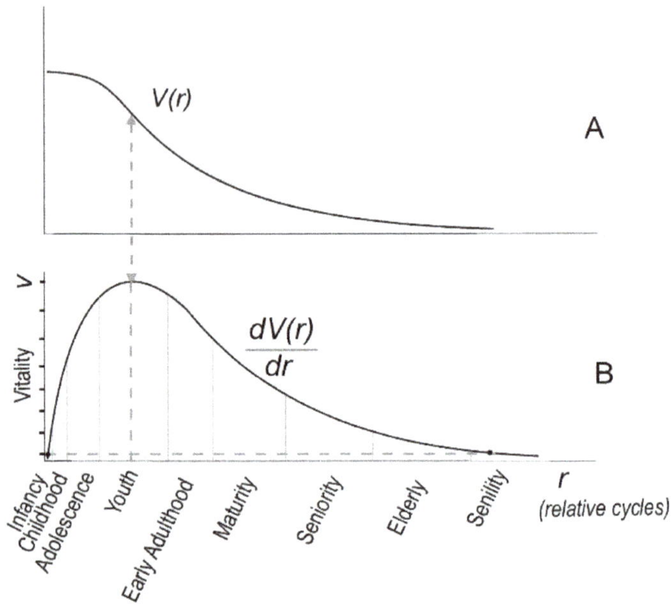

Fig. 18.6 Developmental phases through the life of a Karyote as a function of correlational astrophysical cyclicity

Commutative symmetry here dictates the same level of vitality for points on the same latitude in the ascending phase and descending phase of the curve, as shown on the graph. For instance, the curve well captures the symmetry between a toddler and a very senile person both at the very minimal level of vitality possible, i.e., they are unable to walk on their own or handle their personal physicality. It is so even though we know there is something very significantly different about these two Karyotes, which is the first-order weight between the two elements that make up their vitality: regenerative capacity is stronger in a toddler than it is in a senile person, whereas the metabolic capacity is stronger in a senile person than in a toddler. In that sense, the *life* operator adjoining the two vital ontological tenets in Karyotes does commute or obeys commutation symmetry.

18.16. Conclusion

Once again, there is no *"problem of time"* to be reckoned with… we have been dealing with an impostor, plain and simple… The notion ought to be dismissed and even banned, mainly in favor of the alternative notion of relative spatial phase or correlational dynamic relationships between existent objects or systems in the U_3 universe. Some in physics have argued that, while it is easy to dismiss time rethorically, what is more difficult is to undertake a timefree description of physical dynamics systems. We make that very objective the subject matter of the next Chapter. Finally, it is well worth stressing, in light of the foregoing discussion, that the way leading to Timeless or Transcendental Physics runs opposite to Self or Unitarity.

Chapter 19

There is no motion, because lines are infinitely divisible.
Zeno
Lines do not contain indivisible points.
Aristotle

TIME-FREE DESCRIPTION OF MOTION IN DYNAMIC SYSTEMS FROM MOMENTUM TO ENTANGLEMENT

Time is in serious trouble nowadays in physics. From every corner, skepticism is mounting up and a few of those physicists who would have put their hands in the fire in yesteryears in support of its ubiquity are nowadays becoming somewhat more receptive to the idea that it might be unreal or at least real only as a local variable. In the Quanto-Geometric Theory we have proposed new physics and novel solutions in response to the many conundrums facing the field which radically proscribe time in the metric of the physical realms. What is difficult to construct in a program of timeless physics is an interpretation of the evolution of time-independent dynamic systems at all scales. While we already know what that looks like for the quantum world with E. Schrodinger's Quantum Mechanics, albeit to reserved satisfaction of many who feel that the theory is incomplete, the endeavor of a granular dynamic description of time-independent physics in the macrocosm has probably yet to be achieved. We make this objective the subject matter of this Chapter, in the understanding that such a program will certainly require more ample treatment to be thorough. Nonetheless, in this proposal we undertake to substantially muster up a time-free description of motion in dynamic systems from momentum to entanglement, in physics at all scales.

19.1. Reminder of Number Quintessence in Quanto-Geometric Theory

Our primary notion of numbers concerns their cardinality, a quality by which they are applicable to objects, real or virtual, as a way to garner their inventory. And that is what we ask of numbers for the most part. When we state that there are 5 chairs in a room, we are saying that there are chair A, chair B, chair C, chair D and chair E. However, numbers are not just numbers, that is to say quantum descriptors of quantity or amounts.

Every number represents a form of quantity that is not only exogenous to the number as an expression of count applicable to objects in our world, but this form of quantity is equally intrinsic to the number. That intrinsic state of quantity qualifies as a quantum abstraction or an inner state to the number which may be described as a *scalar state*. As such in the numeric domain, there is no continuum, contrarily to what traditional number theory dogma posits. This is true for the numeric domain as a whole (the cardinal domain), including the real, the rational, the irrational, the algebraic and the transcendental ensembles.

It follows from that view that numbers are typed as well. The reason why it is so is ultimately because numbers participate in a Quanto-Geometric spectrum. In all, numbers ought to be visualized as quantum abstractions with specific types in structure and endowed with behavior pattern. This is all to mean that numbers are Operators.

19.2. Inappropriate Role of Time in the Description of Motion

In his continual search for a global description of the 3-dimensional World, John Wheeler had advanced *"that time may not be a fundamental aspect of reality, but [that] this only happens on extremely small distance scales"*. More audaciously, we contend that time physically participate in no aspect of reality at all. In the following we address a factual numeric description of the evolution of dynamics systems on a time-independent basis where it has not been undertaken before. We further offer a rational explanation for quantum entanglement that is bereft of the outright absurdity of retro-causation.

When we examine motion and time alone, one quickly and routinely falls under the impression that any value of motion and any chronological value of time are and must be possible. All classical or neo classical view of physics, which is also the view of the larger public, belongs to this school of thought. We ordinarily further assume that all dynamic systems must concord with this determination, independent of the objects in action or interaction. Despite the fact that Einstein's Theories of Relativity have shown the peremptory limit of motion in the 3-D World Plane, we have yet to understand the implications of the limit of motion on time and continue to visualize time as this *ad infinitum* continuum.

Lorentz invariance ultimately teaches that nothing can be accelerated to the speed of light with respect to the observer. One way to make sense of the situation of an object in acceleration is to approach it from the point of view of change in momentum or change in energy. Considering the change in momentum *(dp/dt)* as the object is being accelerated to *c*, *we commonly say that it would require an infinite amount of time to accelerate the object to c*, since:

$$\frac{dp}{dt} = m \cdot \frac{dv}{dt},$$

where *p* is the momentum, *v* the speed or velocity and *m* the mass of the object.

For the derivative above to describe an accelerating situation that would bring the speed of the object to *c*, we must read that as $dt \rightarrow 0$, *dv*, the incremental of velocity between two successive points, must become larger, which makes for a steep variational slope. In other words, there must be larger increments of speed for smaller increments of time for acceleration to occur toward reaching *c*. But we know that the instantaneous speed (most affine incremental) experienced by the object can only reach the neighborhood of *c* from the left side but not *c*. Hence, there is a limit on *dv/dt*. Counter-intuitively, the limit on *v* should not be applied to *dv* in terms of displacement Δx (given that $v_{av} = \Delta x/\Delta t$), because there is nothing in isotropic space to intrinsically obstruct objects' displacements and prevent any desirable value of Δx. Consequently, the limit on *v* must be reported to *dt*, which in this context becomes the unit of the scale of time as a human creation, in terms of a lower limit. The general consensus conclusion drawn from this situation is that **it would take an infinite amount of time to accelerate an object to the speed of light.** Instead, the conclusion should be that *c* is the limit of time, since it imposes a lower limit on the unit of time. The smallest possible bit of time τ is that one which would bring an object from speed v < c to ~c. No smaller bit of time τ < τ is possible because it would bring the objects' speed to c or over c. However, the limit imposed on time by c is counter-intuitive because it gives time a life as a quantum or discrete packet while time in our minds is a variable submissive to gradation by any quantity a such that a ε $\mathbb{R}+$, the realm of positive real numbers. In chronology, we do not add time by discrete bundles but by any finite numeric quantity we please according to the cardinality property of the real number realm. We chose the second to be the unit of time, in the understanding that it could have been any numeric value of duration of our election and that any total number of seconds is possible in chronology, as much as any fractional value of the same. In any case, the standard second is certainly not the smallest possible gradient of time as implied by its relationship with c.

The idea of a Planck time, responding to a search for a natural unit of time, circumvents the logic that relates time to *c* by simply projecting an expectation value of the unit of time from the value of *c*, whilst *c* is already expressed in unit of time. Its formula:

$$t_P = \sqrt{\frac{\hbar G}{c^5}} = 5.39116(13) \cdot 10^{-44} s$$

where $\hbar$ is the reduced Planck constant (h instead of $\hbar$ is sometimes used in this definition, but most commonly $\hbar$), G the gravitational constant, c the speed of light in a vacuum and s the SI unit of time, the second. It is the most perfect vicious circle.

Deriving from a clear vicious circle, the Planck time notion is not and cannot be a legitimate projection. Thus, the idea of a Planck unit of time is meaningless. Not in vain has it not been found in nature. Obviously if a natural unit of time cannot be found, time must be demoted from its dimensional stature, officially declared a human standard destined to disappear and a new time-free description of physics altogether summarily sought, on the basis of a dimensional time-free metric.

Note that had we conducted the analysis by starting with Energy instead of momentum, the conclusion we were driven to would have been exactly the same.

Finally, there is indeed a paradox of note in our conception of time, in that we attribute a discrete bundle unit to time, whatever its duration value is, while we are ready to accept that a chronological segment of time can take any finite value whatsoever. In other words, we do not have indeed a physical metric of time. In our minds, time is as loose as anything can be. At a minimum, we manage time as a mathematical entity incarnated in the positive real number domain. That assessment has prompted some to argue that time is real because it behaves as mathematical ordinals, an argument often coming from dissident physicists adverse to time and even from A. Einstein himself who wrote:

"Time has no independent existence apart from the order of events by which we measure it."

In fact, we do not measure time by the order of events as he asserts but we observe events and rationalize their sequential order by measuring or assessing their flow in time. So therefore, what he meant to say, in our estimation, is that we warrant time independent reality or legitimacy only on the basis of its equivalence with mathematical ordinalization. In point of fact, assimilating time to mathematical ordinals makes it an abstraction to the same extent as mathematical entities themselves and poses the problem of whether or not it is adequate to use mathematics to model time as a putative physical variable, such as done in all proposed metrics in all-time physics. The intrinsic incorporation of time to mathematics has presumably been a late effort by modern-day physicists who cannot accept the world without time and developed their own time-based mathematics, i.e. ergodic theory, even though some of these physicists were also genuine mathematicians. The mathematical-physics proposition in science, shall we say, quickly becomes a tricky proposition if its main attempt is to blur the lines between physics and mathematics instead of developing the lines of connection between the abstract model that mathematics indeed represents and the physical reality purported to be the main object of physics, which mathematics should solely aim to model.

19.3. Why Acceleration is Equivalent to Inertial Motion

Since the 6th century of our era, John Philoponous knew that two balls of very different weights will fall at approximately the same speed. In 1586, Simon Stevin dropped two lead balls of very different masses off the Delft church tower, for them to reach the ground with no detectable difference in arrival time as recorded. Those results are not necessarily intuitive insofar as the mass quantities involved. It so occurs that objects in free fall, or in accelerated motion, do not register an experience of being accelerated, in this case toward the center of Earth, but rather weightlessness and no acceleration at all. If an accelerometer is tied to such falling object, the device does not register any acceleration at all, so demonstrating that there is veritably none.

Examining this situation in the context of developing a theory of gravity, A. Einstein concluded that accelerated motion and inertial motion were physically equivalent to one another. The question is seldom posed as to what the reason behind the equivalence principle could be in physical terms as it relates to the types of motion. One may argue that Einstein sought an answer to this question by establishing the theory of space or spacetime as the mediator of gravitational interaction. A counter-argument is that a theory of the origin of motion is a better answer to the question and that a gravitational theory of motion is an appropriate answer only if one assumes that gravitation is a primitive of motion in general. Furthermore, Newton's gravitational description, which views gravitation as an attraction Force exerted on a body, expressed by the inverse square law, remains a legitimate and numerically accurate answer to this question. To this day we do not know which one of the two approaches best describe gravity.

Our aim in this paper is to attempt a description of the very intimate origin of motion. While General Relativity gives a view of motion in the context of gravitation, it does not quite explain the origin of motion. It builds on the observational fact that accelerated motion is equivalent to inertial motion and essentially offers the proposition of the geometric structure of spacetime as the mediator of gravitational motion. Let us start by comparing inertial motion to accelerated motion. Conceived as a displacement in time, pure inertial motion is described by the derivative in time:

$$v = \frac{dx}{dt}$$

Whereas pure acceleration is described as the variation of speed in time:

$$a = \frac{dv}{dt}$$

One way of looking at acceleration is to view it as a step function in time, whereby at each time increment, the speed jumps to a higher value.

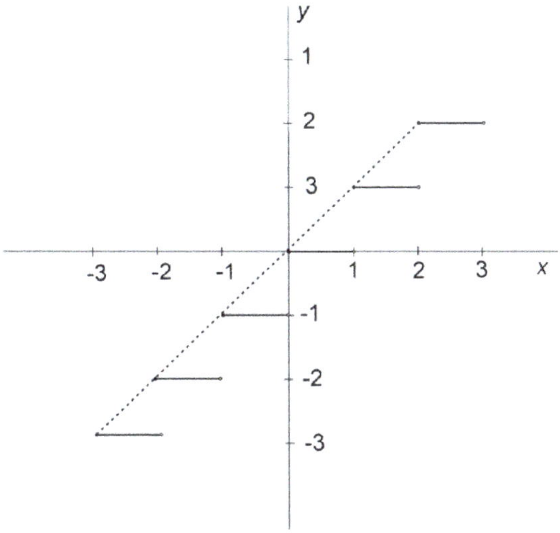

Fig. 19.1 The greatest integer function as a step function approach to acceleration

How long an instantaneous value of the speed at one time jump remains constant during the open time segment or time interval is immaterial to the overall step behavior of the motion, as long as the time interval remains constant to make for a motion with constant acceleration. The *greatest integer function (f(x)=[x])* is a very good model of this vision of acceleration (Fig.19.1). The most irreducible of such step functions is the one with the most affine time intervals (steps of the stairs), while the jumps in time are to be almost perfectly steep. The function is obviously discontinuous at those steps or jump points in time. For a function with the most affine possible steps, one can see *that it reduces to the linear function* obtained by joining the step points as shown in Fig.1. That linear function is a pure expression of speed or inertial motion.

Therefore, at the limit when $dt \rightarrow 0$, a becomes:

$$a \equiv k \quad v$$

where k is a slope factor. It is important to recognize that in the neighborhood of any considered step point where $dt \rightarrow 0$ is evaluated the function no longer behaves in time because of its discontinuity at these time points. Hence, we must forcibly conclude that if accelerated motion ought to be equated to inertial motion, this transformation is not and cannot be realized in time, but only within the dimension of displacements, within the space dimension, that is.

19.4. To Describe Motion Exclusively as a Function of Space

We have discussed above the reasons why, should *c* be the limit of motion, it consequently sets a lower end limit to the scale of time. Let us now examine what happens at the upper end of the scale. When we express motion in *m/s* with *c* as the scale encompassing the gradients of *m/s*, we are settling for tallies or units of *m/s* from *0* to *c* where *c* is the maximum value of *m/s*.

We know what *c* looks like as a pure scalar and not a physical unit in time: ***0.299 796 065*** $\boldsymbol{\sigma^{-1}}$, a number derived from plenary mathematical construction in Book I and II of this publication. In an attempt to give physical meaning to σ in that construction, we compare σ to *m/s* by first looking at how many dimensions there can be in σ. In the Quanto-Geometric 2-space metric, it is a one-dimensional parameter. In all, we have the choice of either giving it the value of time *(s)* or the value of space *(m)*. If we give it a value of choice from the metric of space, we obtain the quantity ***299,792,458*** that we well know, which is a quantity of motion, as demonstrated in Book I and II of this series. But if we wanted to give it a value from the metric of time, we cannot find a value from such metric to assign to it, because there is no natural time metric. A value dictated by a **natural** time metric would be the only acceptable one in this case. So therefore, we must conclude from this lack that the quantity ***0.299 796 065*** $\boldsymbol{\sigma^{-1}}$ does not recognize time or is not a fit to time. It must strictly be a descriptor for motion exclusively in terms of space.

Because all motion must have a geometric descriptor, and since every basis manifold is associated with a symmetry modulus that is a scalar, every motion becomes innately and discretely quantifiable, according to the metric germane to its associated eigenform. This firmly holds despite the fact that the continuum nature of space makes it generically subject to any metric measure. In its realization, however, physical space becomes differentiable or distinguishable without quantization. This differential order is motion or the origin of motion.

One cannot imagine motion without a constant, fixed or running, even though a running constant is no longer quite a constant. For a displacement to occur, the end point must be related to a fixed start point or a moving start point at a different lower or higher rate of motion, not only in representation but principally physically. In this hierarchy, displacement from a fixed point becomes a particular case of the general situation of a running start point. Therefore, if we have to let motion be true to itself, which is what occurs at the asymptotic level of quanto-geometric distribution or in all authentic linear metrics, *motion must start from infinite rate, which is when there is no incremental at the denominator of the fractional rate, and pursue by degeneration as the incremental at the denominator starts taking numeric values and trending to increasingly larger values.* What does this mean? Squarely the following:

1. When there is no incremental at the denominator of the generically descriptive fractional rate, then there is no rate and the value of the displacement is any infinitely large value one may care to imagine.

2. Subsequently when there is an infinitesimally small value at the denominator then the rate is at a very large value but a measurable or countable one.

3. As the incremental value at the denominator increases, the rate of motion diminishes accordingly. Should this value become infinitely large then the rate of motion drops to the almost standstill value of zero.

This flow rate is not *dx/dt* which we are all so very familiar with but *dr/ds*[19], the rate at which the very fabric or tapestry of space *defoliates* or *un-folds* by itself. It mediates what we have otherwise called elsewhere the propagation of space. In the hierarchy of endogenous space motion or *space momentum*, the arrow of space is from very large to very small, whereas in our intuitive metric of time, as consecrated in all postulated time-related metric thus far (FLRW metric i.e.), the flow (rate) of time evolves from smaller to larger. There is no wondering what the cause of this motion is, because the cause belongs in the very definition of the spatial spread: in Quanto-Geometric Theory space is ontologically kinetic, since it is orthogonal to the scalar, its intrinsically invariant and static ontological partner. There is no wondering what it is that space leaves behind in its perpetual defoliation, because there is only left the back scene of the infinites behind. Both from the tensor and the propagation descriptors, space is totalitarian: there can never be a gap in space since a gap is a spread of void, a space spread necessarily.

What we have called elsewhere the *s*-line or the super-scale imposed on the metric of 2-space determines the values or gradients of motion accessible to space, firstly on the map of the Omniverse or the Mappa Mundi, and secondarily by surrogation in the development of the World Planes.

We reproduce here once again the super-imposed scale over the 2-space metric:

for $n = 1$, $s_1 = 0$, or to be exact $s_1 \longrightarrow 0^+$
for $n = 2$, $s_2 = \pm\,\sigma/2$
for $n = 3$, $s_3 = \pm\,\sigma 2/7^{1/2}$
for $n = 4$, $s_4 = \pm\,\sigma$
for $n = 5$, $s_5 = \pm\,\sigma\sqrt{8/5}$
for $n = 6$, $s_6 = \pm\,\sigma\sqrt{10}/2$
for $n = 7$, $s_7 = \pm\,2\sigma$
for $n = 8$, $s_8 = \pm\,\sigma\,7^{1/2}$
for $n = 9$, $s_9 = \pm\,4\sigma$

[19] Note that the flow rate is in fact: $-\left|\frac{dr}{ds}\right|$, as later explained.

where n represents the symmetry modulus or degree of freedom giving birth to each World Plane of the Omniverse.

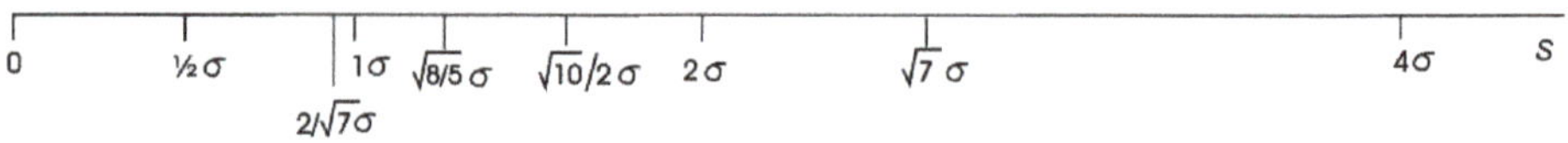

Fig. 19.2 Primeval sequence overscale of degrees of freedom laid on the s-axis

These values are expressed in terms of σ, the standard deviation of the Omniverse (Fig.19.2 Bis). To the limit, one may tend to view each of these values of s as a constant of motion in the traditional sense of a constant representing an invariant anchor. However, they are indeed best conceptualized and presented as the *variants* of motion, or the basis variant degrees of motion (since they typify *variance*). In this study of the origin and description of motion as solely resulting from behavior of space spreads, let's formally designate them the *primitive variants* of *space momentum*. They represent first-order descriptors of motion or precursors of motion. As such there is nothing beyond them, except for the order of dimensional modulus. Each of these variant primitives determines a manifold or *eigenform*, and a form of motion which concomitantly precurses a degree of motion, or an *eigenkinesis*. In other words, there exists a basis rate of motion

- in all circular motion and quantified by s_1.
- in all elliptic motion and quantified by s_2
- in all parabolic motion and quantified by s_3
- in all hyperbolic motion and quantified by s_4

  and so on and so forth until the linear 9-hyperbolic form of motion quantified by s_9.

So therefore, every root form is associated with a root level of motion, a form being the contour or topological geometry taken by the quanta within the spread of space bearing the particular variant primitive or *eigenkinetic primitive*. In the Omniverse, we know that s_3 is the eigenkinetic primitive that precurses the degree of motion set at ***0.299 796 065*** $\boldsymbol{\sigma^{-1}}$, which in turn precurses the degree of motion materially realized at c in our 3-D World Plane (the physical quantity we have labeled ***299,792,458 m/s***).

We report in Table 19.1 a list of SI speed quantities expressed as a function of c. More than a limit, c is the beginning of observable motion in our World Plane, because it is therein the first of all quantities of motion physically realized beyond the primitive. One way to understand c is in terms of an order of *space momentum*, as a surrogation of the Newtonian concept of momentum. In this case it would be the product of a stress quantity times a motional quantity. We do not wittingly indulge in that representation at a formal level however, only invoking it as a helper to visualization in relating to older more familiar concepts. Further on in this study, we propose a more consistent interpretative approach to the quantification of *c*.

Distance Qty	m/s (Scientific)	m/s	Fraction of c
1nm	1.00000000E-09	0.000000001	3.336E-18
	1.00000000E-08	0.00000001	3.336E-17
	1.00000000E-07	0.0000001	3.336E-16
1micron	1.00000000E-06	0.000001	3.336E-15
	1.00000000E-05	0.00001	3.336E-14
	1.00000000E-04	0.0001	3.336E-13
1mm	1.00000000E-03	0.001	3.336E-12
cm	1.00000000E-02	0.01	3.336E-11
1dm	1.00000000E-01	0.1	3.336E-10
1m	1.00000000E+00	1	3.336E-09
3m	3.00000000E+00	3	1.001E-08
3dkm	3.00000000E-01	30	1.001E-07
3hm	3.00000000E+02	300	1.001E-06
	3.00000000E+03	3000	1.001E-05
	3.00000000E+04	30,000	0.0001001
300km	3.00000000E+05	300,000	0.0010007
3000km	3.00000000E+06	3,000,000	0.0100069
30,000km	3.00000000E+07	30,000,000	0.1000692
	7.50000000E+07	75,000,000	0.2501731
150,000km	1.50000000E+08	150,000,000	0.5003461
	2.25000000E+08	225,000,000	0.7505192
	2.90000000E+08	290,000,000	0.9006231
	2.99000000E+08	299,000,000	0.9973566
	2.99700000E+08	299,700,000	0.9996916
299,792km	2.99792458E+08	299,792,458	1

Table 19.1. A list of SI speed quantities in different formats expressed as a function of *c*, toward a nomenclature standard.

Speed *c* emerges as the most natural scale to measure motion in the 3-D World Plane, if only we understand that all observable objects' motion, no matter their modalities (angular or linear) and their quantities, physically derive from it. From that perspective, it does not matter how complicated or simple the expression of all motions over the *c* scale may turn out to be nomenclature-wise. The great advantage ensuing from adopting a measurement scale anchored on *c* is that it affords us to be in tune or in synchrony with nature when it comes to kinetic instrumentation or transport technology, aside from absolute universality.

19.5. Covariance v. Causation

The order of Quanto-Geometric covariance between scalar and space is not given in time, which is a variable the Theory irrevocably disavows and expressly bans. This account begs the question of the role of motion in covariance. In other words, does motion mediate covariance?

We must acknowledge that the first cause of covariance is *conservation.* Quanto-Geometric Theory postulates the exclusive duplet existence of a scalar tenet and a space spread tenet as the foundation of all physical objects' ontologies. Coexistence between the scalar and the intangible void represents the basis topological state of the Universe. The forced coexistence between the scalar and the vacuum spread gives rise to the first *conservation* constraint which sustains the foremost order of ontological covariance in the universe. Consequently, the only fate of the two basis ontological tenets is to correlate with each other.

The scalar automatically inhabits the space spread so to speak. Therefore, correlation is automatic and ineluctable between the two basis ontological tenets. While the correlation, in order to become covariance or causal interaction, implicates action, it does not implicate motion in the form of translation. Motion is already part of the ontology of space. One must expressly guard against visualizing motion as the end result of the interaction. However, the motion of the scalar is indeed an expression of the interaction or the influence exerted on the scalar by space, while the *tensorial* stress experienced by the space spread represents the effect of the influence of the scalar over the space spread. The interaction or covariant action is not realized in *time*, which is here the big absent. The mutual interaction is exerted concomitantly. It has no beginning and no ending point, it is perennial and permanent. That is also the reason why there is in existence no pure scalar nor is there any existent pure space. All space is tensorial or under some level of stress and all scalar is spread in density state or exists in *volumetric state.*

Hence, the *first cause* is the coexistence constraint or conservation. Covariance or mutual prospection is the *second level of causation.* And the *third cause* of association or permanent association is the grip between the two tenets which establishes a degree or gradient of correlation.

Finally let's make sure to stress that if we were to insist in injecting time somehow in this program, because a paramount Quanto-Geometric axiom is *orthogonality* ($Q \times S = 1$) between the two ontological tenets (they are inverse of one another), their association with time would have to reflect their ontological *orthogonal* attribute. Time for one tenet $(+t)$ would have to be the conjugate of time for the other $(-t)$, which would cancel time altogether out of the interaction. Therefore, as quanto-geometrically framed, there is no place at all for time in this program. In consequence, motion cannot be expressed in this framework in terms of time in any manner whatsoever.

19.6. Scales of Motion

The program that we have described above represents the first-order conditions of motion in the 3-D physical world and directly precurses motion as it exists throughout the **Omniverse**. The correlation between universal space in every World Plane of the Omniverse and its total scalar mass represents the phase order from which motion is orchestrated as a direct native from the dimensional set defining every specific World Plane. Therefore, each World Plane as a whole owns its own precursor of motion c_i produced by the invariant quanto-geometric degree of correlation between its mass and space proper, which conforms the symmetry basis for the transformations the specific World Plane will develop as a matter of its phenomenological circumscription.

Consequently, the first dynamic system in a World Plane is the entire World Plane itself and that horizon is the origin of the metric and all metrics. Everything else in a World Plane is a product of degeneracy of the specific determinations emanated from this global order and a local **emulation** of the global norms, including the norms of motion.

Accordingly, every *eigenform* relates to an *eigenkinetic* degree originated from its kinetic primitive (Fig.18.3). Each of these correlations is pre-eminently featured by a World Plane of the Omniverse, but the specific dynamics of a World Plane occurring below the dimensional background are developed according to an emulation of the whole primitive panoply in conjunction with the main Operator Norm in action for that Plane.

The next order of dynamic systems right below the global World Plane horizon is the order of Shells. We are not going to delve here into the analysis of the dynamics of Shells of Matter, in particular that of the 3-D World Plane of most interest. We are only invoking it here in order to represent the order of dynamic systems as they truly occur in matter, quite differently from what we have normally conceived as a dynamic system thus far in all-time physics. Again, each Shell is based on its own fundamental invariant sustaining the transformations circumscribing its phenomenological framework. It is only at that level that gravitation, a matter of utmost interest currently in the field, arises with the constant of Gravitation G as the fundamental invariant native to a particular Shell of the 3-D World Plane. These regimes of motion are modeled by quanto-geometric covariance mathematically represented in several functional differential orders essentially.

What we are going to do next is to describe how the primitives of motion reflect on well-known dynamic scenarios in an effort to firmly falsify the above postulates.

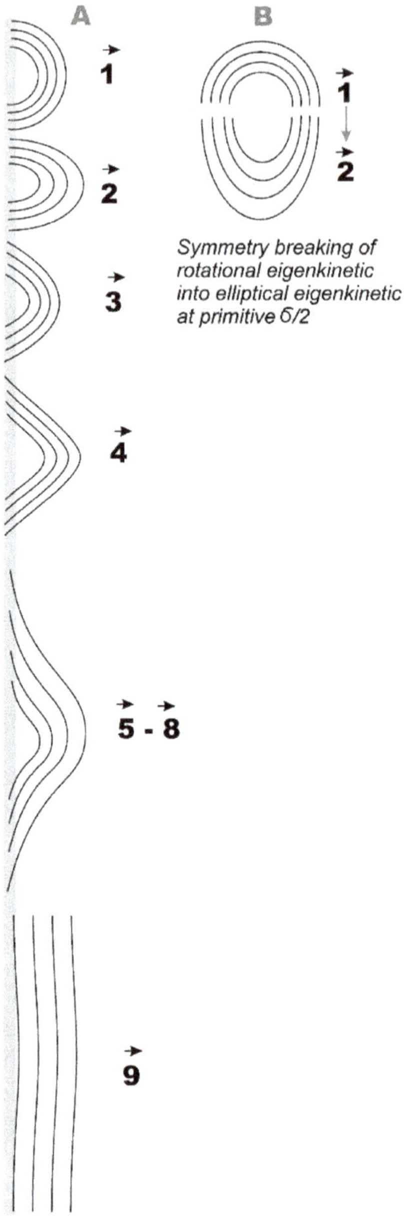

Fig. 19.3. Representation of eigenkinetics within the tapestry of the vacuum (A). While one may liken them to geodesics, they in fact materialize the irreducibles of motion but not paths of motion in time. When a symmetry breaks, not only a new eigenform emerges but correlatively a new form of motion as well, as illustrated in (B). One shall keep in mind the very important fact that the represented eigenkinetics are intangible in nature.

19.7. What is a Dynamic System?

What we generally call a dynamic system is a scenario with at least two actors of scalar nature in motion of some kind with respect to one another. For instance, an atomic nucleus with a bound electron, or a star with one or several planets in orbit all ordinarily feature a dynamic system. A cat chasing a mouse we do not generally consider a dynamic system since the norms of motion thus far known in physics do not allow the interpretation of such systems involving the living. That scenario we would rather call an event. However, if the Quanto-Geometric Function says anything as a stochastic function, it is that it is well at the root of the expression of life and its dynamics as well.

It is important to understand at this point that each and every one of the 9 Quanto-Geometric covariant phases that put in correlation a scalar density entity and a space spread entity represents indeed a dynamic system. They each represent the most generically conceivable dynamic systems. They take the counter position to motion being the result of interaction between two *mass-ive* bodies as conventionally viewed and instead postulate motion to be a property of the space spread which engages the scalar, and in so doing creates a particular *form* of motion equally reflected in the morphology of the scalar object as a specific geometric *eigenform* (Fig. 19.3). We have already explained the conservation constraints and the causal regime sustaining this dynamical order. The pictorial representation in (Fig. 3) serves to fix the idea, albeit not the most accomplished illustrative representation for the concept. It is based on the Ying Yang symbol which most of us under all latitudes have become familiar with in one degree or another.

So therefore, a theory of gravitation as interaction between two bodies is very far and away from being the ultimate theory of motion in nature or one for generic dynamic systems. Einstein's Theories of Relativity have certainly taken the correct direction by establishing interaction between space and mass as the engine of gravitation. However, he had not properly viewed the intrinsic properties of space, in particular its inherent motional elements in their many different orders, quite aside from the successful characterization of the tensorial nature of space; nor has the Theory understood that the correlational trend between matter and space does not only exist in the cosmological realm but at all scales in the 3-D realm of matter.

The illustration in Fig.19.4, shows three different orders of representation. One is the scalar density order represented in the white spread in the diagram, the other is the space spread represented in the black spread in the diagram. Third order of representation is the correlational or covariant influence within the primary duet portrayed by the two inner gray-tone circles inside each spread. From the monolithic object to the transcendental object:

- Space spread evolves by augmentation within the object's ontology.
- The quantum scalar evolves by diminishing its presence within the object's ontology.

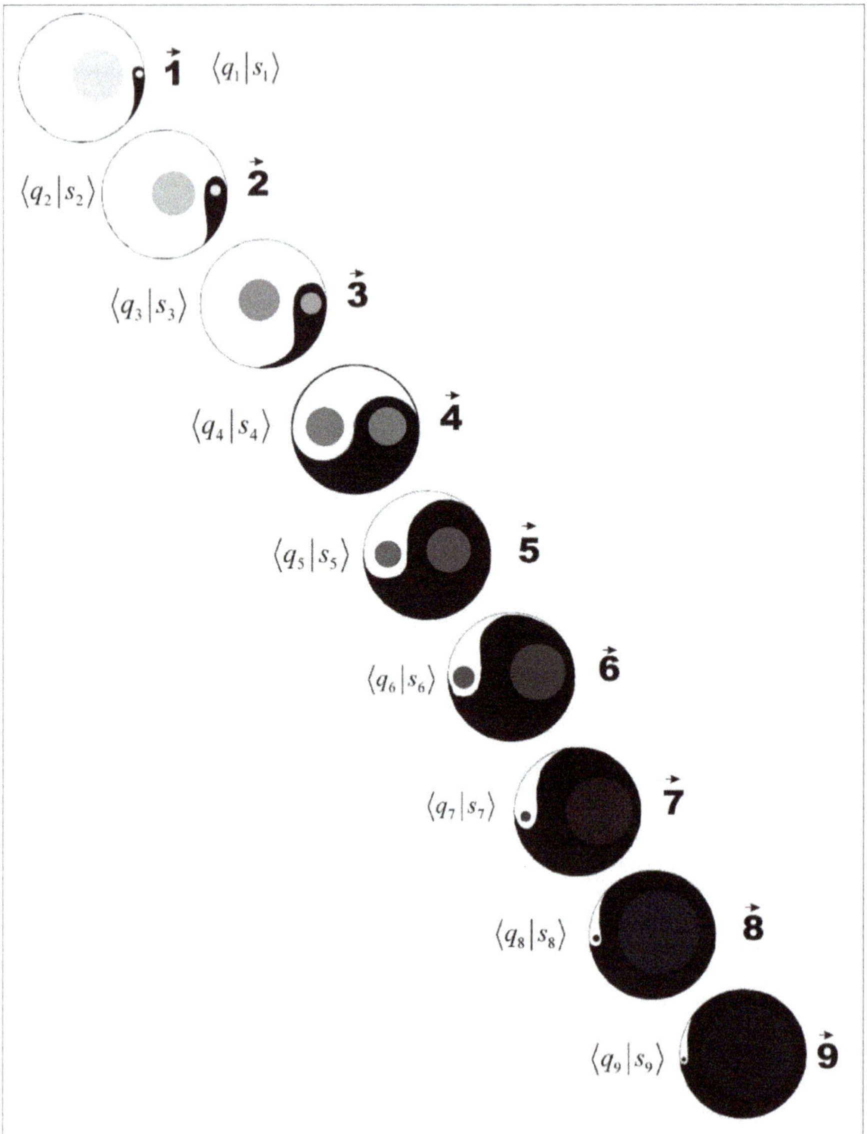

Fig. 19.4 A depiction of the primordial Quanto-Geometric ontological tenets and their covariant regime

- The correlation evolves in a way that sees the scalar to be dominantly influential at one end, gradually loses influential weight while the space partner acquires more and more influential weight or becomes more and more predominant. This is expressed in size of the inner balls.

Why do we say object and not a dynamic system? Because an object of any kind is intrinsically a ***stationary*** dynamic system. This framework replaces the expression of the First Newton Law with the parallel notion of quanto-geometric momentum Q_G, which stands for the product of the two basis tenets:

$$Q_G = Q \times S,$$

where Q stands for scalar density and S stands for the motional artifact or the spatial wavefunction in their most irreducible state.

The least *amount of motion* is given in the monolithic circular eigenform as spin motion. While the largest *amount of motion* is given in the transcendental eigenform as quasi rectilinear motion. It does not matter how fast a perfectly circular motion is realized, it has less motion than the most perfectly realized linear motion, simply because the expansive translational component of motion is at a minimum in circular motion. To make sense of that motional hierarchy, one may look at the eccentricity ***e*** of the eigenform related to each type of motion or its eigenkinetic.

- For the circle eigenform, $e = 0$
- For the ellipse eigenform, $0 < e < 1$
- For the parabola eigenform, $e = 1$
- For the hyperbola eigenform, $e > 1$
- For the 5- to 8-hyperbolic eigenforms, $e >> 1$
- For the rectilinear 9-hyperbolic eigenform, $e \rightarrow \infty$

We assert that as the eccentricity increases so does the eigenkinetic value of the motion. The mother of all motion is the rectilinear eigenkinetic because it potentially gives birth to all other eigenkinetics. One can easily concur with this appreciation by comparing circular motion to linear motion. While there is indeed a linear component to the circular eigenkinetic, essentially given by the run over the perimeter of the circle, contrastingly there is no circular component to the most perfectly rectilinear eigenkinetic.

As a reminder, eccentricity is the constant defining the locus of points P relative to a line called the directrix and a fixed point O, per illustration in Fig.19.5 and the following equation:

$$e = \frac{R}{d - R\cos\theta},$$ where e is the eccentricity.

Keep in mind that while this characterization helps understand the primordial hierarchy of motion, it is not perfect in that it does not represent quite correctly the hyperbolic order of motion which lies between the hyperbola eigenkinetic and the rectilinear eigenkinetic. It is only successful in helping characterize the transcendental hyperbolic eigenform or eigenkinetic, because the latter, by essentially having a life near the asymptote, can be practically assimilated to a rectilinear element, if σ takes a sufficiently large value.

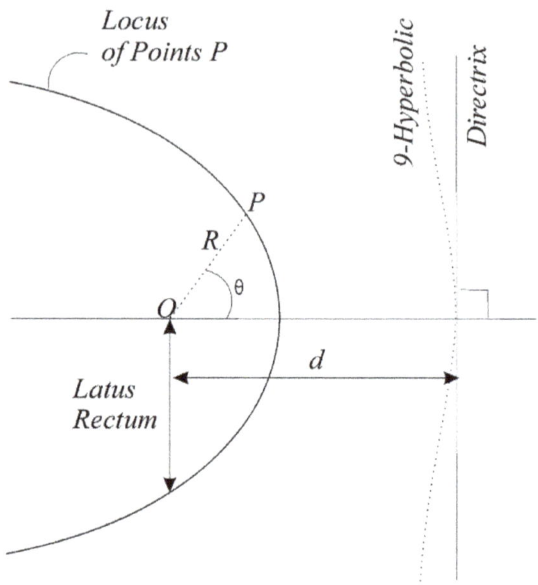

Fig. 19.5 Definition of a conic function

In this study of the different orders of motion, two remarks are important to put in highlight:

1. Because in Quanto-Geometric Theory, $s = 0$ is undefined, absolute spin is not possible. There is always some level of precession in rotational motion. In that sense a circle is already a precessional figure.

2. Some clever minds have found a scale of π where an object can make several full rotations before returning to its original starting position (spiral torus figure). That is simply an expression of π in 3-space, specifically spin motion over a cylindrical surface, and can never be true in 2-space where all primitives of motion originate.

19.8. Spin Within the Hierarchy of Motion

While spin is probably hitherto overrated in quantum physics, there is apparently a certain level of misunderstanding associated with spin motion as well. As previously mentioned, absolute spin is not possible, but if it existed it would the first gradient of motion. Absolute spin is rotation without translation or precession. One cannot imagine or represent motion of any kind in non-precessional spin. You can only represent spinning motion from the portrayal of its edges, the precessional perimeter. Once this image disappears one can no longer represent the amount of motion in spin, simply because the idealized center of spin experiences no motion whatsoever. Hence counter-intuitively, *the state of zero-motion or gradient 0 of motion is indeed the state of absolute spin.* Its physical realization (of provenance $s = 0$) is indeterminate or forbidden in Quanto-Geometric physical law, making it thus an impossible.

We characterize idealized zero motion as rest. In practice, we know full well that absolute zero motion or absolute rest is not possible, not because of the relativistic nature of observational frames, but because there always exists a framework of transformation to overtake any apparent object at rest. Consequently, two objects can only be at apparent total rest with respect to one another because they are moving identically in every factor or component of motion. This situation materializes the closest image possible to absolute spin. It is because absolute spin represents gradient 0 of motion that the next step in the hierarchy of motion (indeed the least countable one) is precessional rotation incarnated in the figure of a circle and rated by the π constant.

19.9. Linear and Harmonic Motion in the Quanto-Geometric Regime

Traditionally, one studies linear and curvilinear motion in a distinct chapter than harmonic motion. However, Louis Victor De Broglie's quantum relativity has eloquently shown that all motion, at least all objects in linear motion, are concomitantly subject to a harmonic motion as well. In Quanto-Geometry the eigenform that pairs with an eigenkinetic is a direct expression of the wavefunction because the *eigenform* is indeed a physical *waveform*. Consequently, and in accordance with Quanto-Geometric 2-space metric, every primitive of motion is a descriptor for both a degree of translational motion and a wavelength defining a wavefunction.

We have to abandon the idea that all wavefunctions, which we indiscriminately model on the eternal sinusoid, are the same. All wave functions are **not** the same because the symmetry implicated in their harmonicity (cyclic repetition) is based on different eigenforms, which are again true material waveforms. The only eigenform or eigenkinetic that can model all other eigenforms or eigenkinetics is that one belonging to the Quanto Geometric transcendental (9-hyperpolic) layer. One can see that as σ becomes very large the Quanto-Geometric Function graphically approximates a line and as all parameters rescind, the function migrates into a circle.

The generic Quanto-Geometric Operator, which stands for an alternative representation of the Grand Eigenfunction portrays the best mathematical expression for modeling dynamic systems. It outwardly displays in the habits of real quantities, estranged from all imaginaries or abstracts, the two components responsible not only for ontology but for phenomenological interaction, that is to say the wavefunction and scalar density tenets.

The exponential component in the expression stands for the wavefunction while the Q term stands for the scalar:

$$\|\psi\|_{a_i b_i} = \left[\frac{-Q^2}{2q}\right] \bullet \left[\frac{2}{a_i e^{\left(\frac{2a_i + b_i^2}{4}\right)} + b_i e^{\left(\frac{2a_i + b_i^2}{4}\right)} + c e^{\left(\frac{2a_i + b_i^2}{4}\right)}}\right]$$

Every sub-index i in $a_i b_i$ represents a q-s set, making up either an object or the tenets of a dynamic system. The wavefunction component is an expression of what we call Potential Energy, whereas the scalar term represents what we normally call Kinetic Energy. It is clear that Quanto-Geometric Theory radically concurs with the perspective of Hamiltonian physics which views Energy as the primary state of physical objects, and not the Momentum/Force perspective offered in Newtonian physics. An important rectification is required however, which mandates the scalar term to stand for the Potential Energy proper since it is innately static, while the exponential term ought to stand for Kinetic Energy, since it represents the spatial function, the innately kinetically active component.

19.10. Deployment of the Panoply of Motion Primitives

As we have previously explained, the super-imposed metric on the scale of 2-space marks the origin of the gradients of motion. The elements of this metric constitute the primordial gradients for motion both in their qualitative and quantitative characterizations. The quality means the morphology taken by motion along the scale of rectilinear at one end to circular at the other end. Worth reiterating that this includes the characterization of *waveforms*, deriving into the ordinary concept of wavefunctions. The primitive quantities range from limiting values in the neighborhood of 0^+ to $1/2\sigma$ to 1σ all the way up to 4σ (Fig. 19.2).

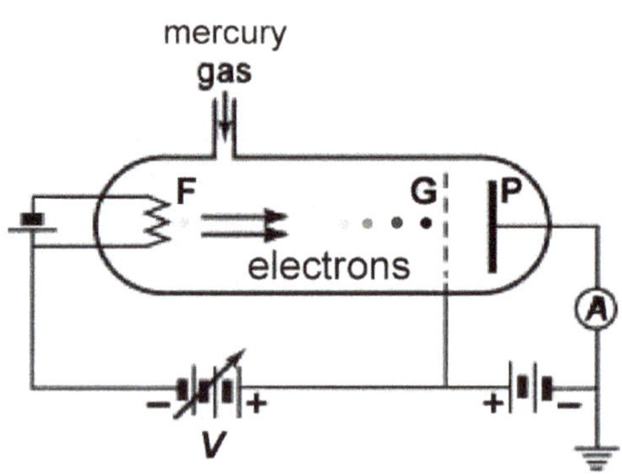

Fig. 19.6 Franck-Hertz experiment of travel electrons in a mercury-gas-filled chamber.

Let us examine a typical dynamic scenario in the manner experienced by an electron in a potential well created in the laboratory in the form of an evacuated chamber (i.e. our old cathode ray TV sets), or a chamber filled with mercury gas, per the original Frank-Hertz experiment (Fig.19.6). In our view, the well and the electron form a dynamic system of second-order quanto-geometric nature.

The experimental setup has a cathode F and an anode G inside of a gas filled crystal tube (or an evacuated tube) create an electric field which steers the electrons created at the cathode to the anode. Classically, the potential well should be able to cause the electrons to move at any speed value and in a linearly proportional manner to the energy quantity communicated to the electron by the electric potential. However, the dynamic outcome is much different than the classical expectation.

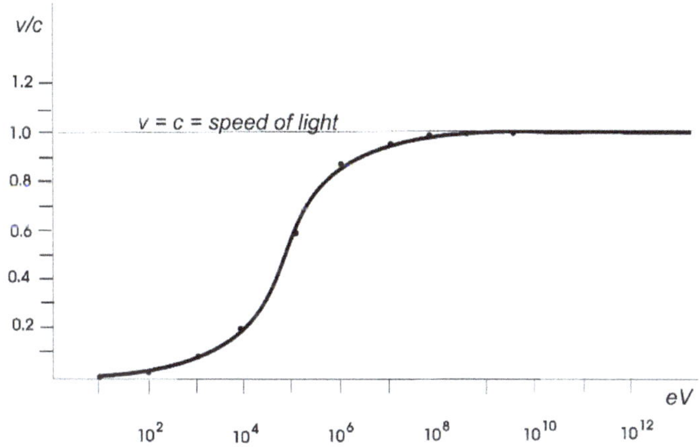

Fig. 19.7 Distribution of particle speeds v. their acquired kinetic energy

What we see per Fig.19.7 is a quasi-typical quanto-geometric distribution[20] that correlates Kinetic Energy to particle speed. We interpret that the potential well supplying kinetic energy to the particle in the form of *electron-Volts (eV)* **over-powers** or provisions a boost to the innate wavefunction of the particle and thereby causes an imbalance in the inner covariant scalar-space phase inherent to the particle. The particle reacts to this imbalance with a covariant shift re-balance, which is naturally reflected in the *space momentum* of the particle across the well, in accordance with the distributive disposition of the Quanto-Geometric Grand Eigenfunction. Why the limit in the accessible degrees of *space momenta* is *c* is a different matter, which we already discussed. The limit per se is imposed by the distribution. Here the potential energy from the potential well, which the

---

[20] This graph corresponds to the preliminaries of the Franck-Hertz experiment whereby the applied voltages fall outside of the range that would famously cause the first-time observed quantization of the energy levels of the electron per its inelastic interaction with the mercury gas molecules. These voltages are significantly higher than those causing quantization behavior.

particle experiences kinetically, plays the role of the primitives of motion. It can take any value in the absolute but there is an effectual metric imposed on it which is the one that is going to be permeable to the particles and reflected in the space momenta they develop. To put it simply, the particles redistribute their ontological quantum-space composition per the Grand Eigenfunction's 9-layer norm.

Finally, it is worth noting that the reason why one sees here a somewhat skewed quanto-geometric distribution is because of a somewhat irregular scalar-space covariant shift within the travel particles ultimately due to the way kinetic energy is communicated to the latter. We see a similarly skewed pattern when optical energy, instead of electric potential energy, is communicated to the particles, as displayed in the photoelectric effect experiment (Fig.19.8).

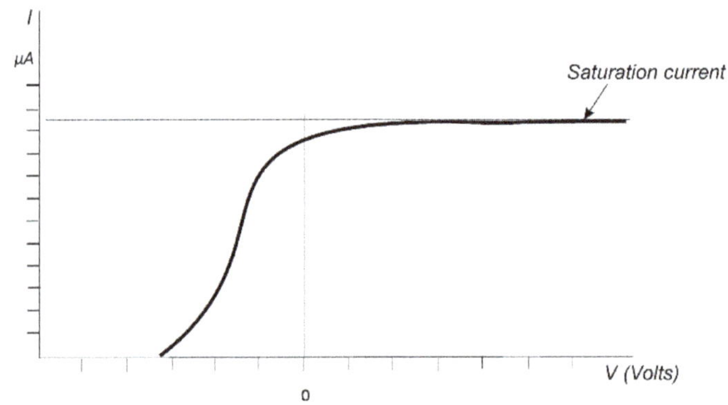

Fig. 19.8 Photoelectric experiment where incident monochromatic rays on a metal plate induce an electron flow or current.

In this experiment, the electric potential still develops the primitives of motion versus the *current A* or flow rate of electron particles ejected from the cathode. The quanto-geometric distribution of the curve is similarly skewed akin to the case previously discussed, as a particular effect of the manner the energy is communicated to the particle, here through monochromatic light *(hv)*. The skew in both cases may be studied in terms of an effect on the σ parameter of the distribution.

Also note that travel electrons in interstellar space will presumably demonstrate the same type of distribution reported in Fig. 4, as do other particle objects such as protons. This pattern is universal for all solid objects made to absorb energy in a way that would cause a change in their momenta.

19.11. Transfiguration of Speed in Space Momentum

All speed is rate, but the nature of this rate is essentially and uncharacteristically a density function. Not just in the mathematical sense, but very physically so. The de-generate deployment of space from *c* indeed creates the observable speed of objects as the first level of motion, whose universal etalon in the 3-D World Plane is *v/c*, and not a time-based rate. Because all observable motion is a normalized density function, the rate *v/c* is maximally set at 1 and 1 is an unrealizable, in accordance with one of the foremost Quanto-Geometric prescriptions. In other words, $v = c$ is unachievable for solid or fermionic objects. If that fact was already given by other evidence, then we must further understand that the reason why that is so is because motion is a Quanto-Geometrically distributed deployment.

Analysis of objects' statistical topological states has been realized in Book II and will not be repeated here. In strong support of the above, we will here report, however, the graphical Fermi-Dirac distribution (Fig.19.9), which is a faithful copy of the distribution obtained in the preliminaries of the Franck-Hertz experiment for the travel electron.

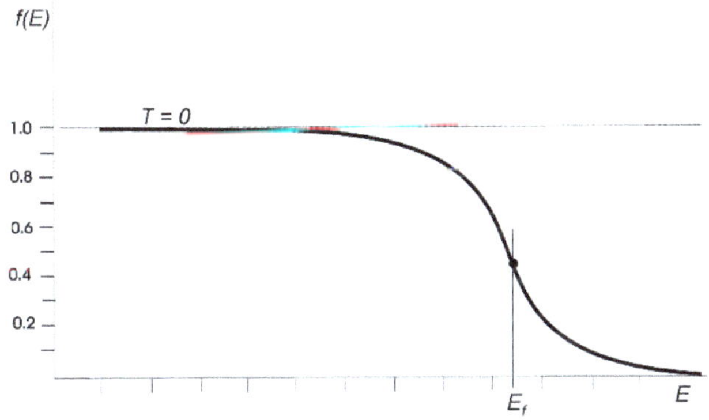

Fig. 19.9 Probability distribution of occupation of energy states by a topology of free electrons

Even in topologies of objects, which altogether conform a sub-Shell, motion is quanto-geometrically distributed with thermodynamic temperature as the standard deviation parameter. In this scenario, motion is the motion innate to physical states. The function *f(E)* in Fig. 8 represents the probability of occupation of energy states by individual electrons. The energy levels are the primitives of motion. It is not at all surprising to see here that most particles tend to occupy the lowest level of the energy (curve flipped from right to left as compared to Fig.3.8), because it is here about a topology or a *stateful* act, not an expansive act of translational motion. However, the *normal* nature of the *f(E)* function (with 1 as the higher limit), informs us by analogy of the density nature of the *v/c* space momentum function for the same elements in travel instead. The empirical formula for this function, below reported, is important when it comes to numeric analysis, but not so in the quest for an explanation to the cause of the behavior it describes:

$$f(E) = \frac{1}{e^{(E-E_f)/kT} + 1}$$

By the same token, this parallel allows us to visualize the physics that we invoked when we established a line of equivalence between stateful objects and dynamic systems.

19.12. What Creates Lorentz Invariance and the Illusion of Time

An important figure in relativistic physics is what has been called the β ratio, the ratio of any attainable speed versus the referential speed of light:

$$\beta = \frac{v}{c}$$

In the following, we are going to clarify the meaning of this ratio as a function whilst making use of it in order to uncover the origin of continuum physics, where time is king, along with the foundation of Lorentz invariance.

19.12.1 Basis of Continuum Physics

Because of our inability to visualize the origin of motion in terms of the panoply of primitives of motion, we have been arguably misled by a first-order derivative of the Function that relates those primitives to the development of actual velocities or speed quantities. A graphical view of the derivative of the Function is shown in Fig.19.10 below.

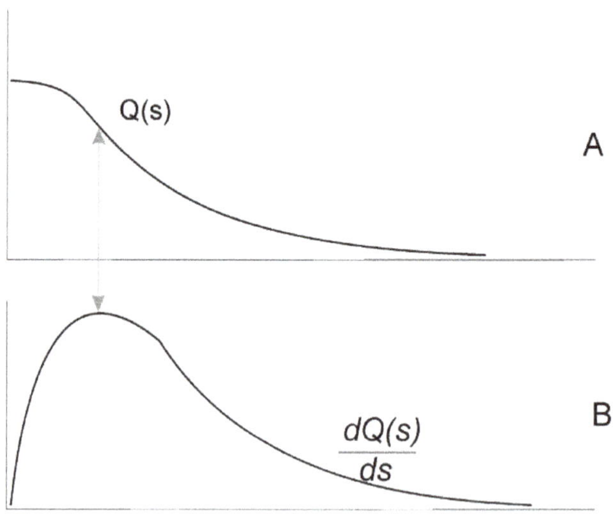

Fig. 19.10 Derivative of the Grand Eigenfunction

If we let the derivative represent classical momentum ($m \cdot dx/dt$), one can clearly see that in the region of momenta neighboring 0, for one value of the derivative there are, instead of two singular values of the primitives, two corresponding *quasi* sets of values of the primitives: the values of s approaching 0^+, which are continuously varying fractional quantities, and the values of s beyond 4σ, which behave linearly in the neighborhood of the s-axis asymptote and constitute equally continuously varying quantities. This effect is even more pronounced in the skewed distributions that we have reviewed above as we examine their functional derivatives. It is indeed through this window that we have historically been looking at motion in nature, which has caused us to create and visualize time as this continual varying artifact playing the role of background substrate to motion in general. That also stands for the reason why we have argued in Chapter 2 that time works for us because we are slow creatures in a slow world of ours. As soon as we leave these regions of continuity, classical physics and common-sense start breaking down, unable to properly account for common real-world motion because they no longer obey continuum (time) span.

19.12.2 Basis of Lorentz Invariance

We may gain further insights in large scale physics if we assimilate the density function $\Delta r/\Delta s$ to the vertical axis of the derivative graph in Fig.3.10. Because we are parts integral to the Universe as internal elements, and because we are a product of derivates of functions of motion, we cannot ascertain an extrinsic view of motion in the Universe as if a view of the same from outside, but only an intrinsic view of motion from the boundary of $v/c \rightarrow 0$ to which our living parts confine us, we thus cannot *naturally* become aware of the intermediate levels of the scale of the primitives and their corresponding space momenta.

In fact, what *nature* has done is to utilize our inability to perceive the upper region of the β ratio in order to fashion our very important sense of sight. Effectively, the *sense of sight* for most living entities in the biosphere is based on generalized Lorentz invariance *in all directions*. At whatever speed an object is moving with respect to us and whatever speed we are moving with respect to the object, we will always see or perceive the object with *complete definition*, with no blur or aberration, that is, in a demonstration that perception is based on the invariance of the speed of light traveling from objects under observation towards us. The same is altogether true for other simultaneous observers of the same object with a different kinetic experience in their own observational frames. Whereas our *sense of motion* naturally remains under the Galilean regime because nature has recognized that we need correct estimation of motion of common objects in our environment relative to us in order to conveniently keep ourselves safe or motionally efficient within the environment. When we see an object in motion, there are two different levels of awareness being exercised relative to the moving object: *there is the visual perception of the object and there is an innate perception or sense of the kinetic progress or displacement of the object.*

Those facts, which are seldom understood and invoked in justification of the generalization of Lorentz invariance, constitute a square rebuke to neo-classical arguments denying that postulate, often coming from dissident physicists who do not realize that their argument is a bargain against their own sense of sight, without which they would be most likely unable to develop and hold the very argument!

The transformation technique that we use as an analytic method to make sense of Lorentz invariance is vastly imperfect and overrated, because, among other shortcomings, it does not allow us to analyze transversal motion of light from several simultaneous observers' vantage points. We can only visualize, physically and mathematically, a transformation in *one single* direction originating from the source frame. How to do it in a multitude of directions at a time, since any light travel event may be viewed simultaneously by a multitude of observers with same or different kinetic experience proper, remains out of reach, because in my sense it is ultimately utterly difficult to explain what bloody property or properties of the basis underlying spacetime allow a simultaneous sum of timelines with different time arrows and flowing at different rates and dilating at different increments to be totally transparent to one another! So therefore, the analytic method of transformations of reference frames, otherwise attempted historically in more other ways than one can count, are ostensibly unable to thoroughly explain Lorentz invariance in all directions, leaving it to be postulated as a Special Relativity axiom.

The traditional translational transformation does little more than cover a game of running an inflation of time in correlative proportion with the Lorentz factor so that the speed of the perceived quantum of light always remains the same. An explanation of Lorentz invariance through this overhead construction is, to say the least, somewhat artificial and very much inadequate. Matter-of-factly, to understand what makes the principle of Lorentz invariance valid in its generalization in all directions, and what makes the speed of light an independent invariant from all relative environmental scalar movement (a non-additive motion), one ought to invoke the kinetic properties of the underlying spatial tapestry and comprehend what makes the v/c ratio a density function. This ratio is an expression of the measure of *displacement of the scalar with respect to the displacement realized within the spatial tapestry*, since it is the internal kinematic of the spatial vacuum that **induces** the motion of the scalar. The perennial defoliation or propagation of space from within its own tapestry and *in all directions simultaneously,* of which light is only a witness, is the preeminent **cause** of all motion and thereby casts all motion as a ratio of displacements. This ratio is indeed the amount of displacement experienced by the scalar object (or bundle) versus the amount of natural displacements implemented by space in the World Plane, our $-\Delta r/\Delta s$ figure. The scalar obviously can never experience larger displacements than the referential spatial displacements. Therefore, the density expressed by this ratio is normalized at 1. In addition, the displacement experienced by the scalar is a direct measure of the *mass*-ive gradient of the scalar. That stands for the reason why we see from the treatment through the Grand Eigenfunction that what we know as the speed of light is ultimately rendered as the value of the (trinitary) scalar in the 3-D World Plane:

$$q_3 = 299{,}792{,}458$$

An equivalent sign ≡ would have been a better translation to the underlying physics, but this slight mathematical mischaracterization remains innocuous and bears no further consequence.

Absent a representation of the intimate kinematic of space in these terms, the β ratio cannot acquire its full *functional* meaning beyond its appearance as a flat dimensionless ratio. What's more, one need not represent geodesical lines in the tapestry of space to describe motion, once it is understood to be the displacement ratio we described. The relativistic geodesics constitute a support for describing motion in time with the *4*-vector $\vec{v}$:

$$\vec{v} = \frac{dx}{dt}\vec{u}_x + \frac{dy}{dt}\vec{u}_y + \frac{dz}{dt}\vec{u}_z. \quad (19\text{-}1)$$

In our perspective, the geodesics are unnecessary. While one may liken the eigenkinetics to geodesics (Fig.19.3) to some extent, the former in fact materialize the irreducibles of motion but not paths of motion in time. All trajectories of motion are determined by a unit-form of motion and a precursory quantitative schedule for that transport in the form of spatial variance or multi-variation.

19.13. Eigenkinetic Velocity Vector

In contrast to geodesics, we represent the eigenkinetic velocity vector to be a multivariate displacement in true coordinate space, as function of one observable spatial variable *s* or two observable spatial variables *t* and *m*, directly from our metric nomenclature. The expressions of the primitive velocity vector or eigenkinetic follows:

$\vec{v}_{iS} = -\left|\frac{dr}{ds}\right| \vec{u}_S$, in 2-space with *s* and *r* spatial coordinates

$\vec{v}_{iMT} = \left[-\left|\frac{dr}{dm}\right| \vec{u}_m\right] + \left[-\left|\frac{dr}{dt}\right| \vec{u}_t\right]$, in 3-space with *r*, *m* and *t* spatial coordinates, where *r* is the scalar displacement coordinate and the $\vec{u}_i$ the unit-vectors along each axis of the coordinate system. The displacement rates are negative because the scalar moves in opposite direction to the inducing spatial displacement, in accordance with the prescription emanating from the first-order derivative of the Grand Eigenfunction, a negative quantity. Effectively, there is a direct isomorphic relationship between *dq/ds* and *dr/ds*. In addition, the *i*-values taken by the velocity vector pertain to the basis *s*-line of the metric, not to the real number domain.

The spatial velocity of the object is a tangent vector to the path, just as the eigenforms are representatives of curvilinear tangent bundles in the metric. On that account the eigenkinetic velocity vector parallels an eigenform. In other words, as primitives that spring from the *s*-line, the eigenkinetics walk hand in hand with the eigenforms as a specific eigenform, whether a 2-space or a 3-space manifold, composes the path of the motion in question.

It is of utmost importance to comprehend that the quantitative capabilities of the *s*-line metric, basis of the eigenkinetic primitives, are what confers to this descriptive framework at very first order intrinsic kinetic quantification, in discard of the traditional velocity measures in terms of variations in time. Motion here is described over true coordinate space background and not an artificial spacetime construction.

19.14. Effects of Motion on Mass and Mass Behavior

Let us further examine the effects on mass inside of the scalar-space phase responsible for the momentum. One shall keep in mind that the phase shift implies an ontological displacement of the correlation, which means a covariant change in scalar density or mass of the particle and the correlated space spread of its makeup. This sweeping change in mass (or changing mass) is what the Theory of Relativity proposes for relativistic mass as the speed of the particle reaches the vicinity of *c*. As we know, the concept of relativistic mass is not without its caveats and controversies. In our sense, the reason for the controversy is simply because thus far in physics we have not understood mass to be defined as **scalar density** instead of a punctual monolithic *thing* lacking a primary or generic definition. In addition, for an object to have additional intrinsic mass with respect to an observer, as one interpretation has it, is plainly unsound philosophically. A. Einstein admitted that much about relativistic mass:

"It is not good to introduce the concept of the ***mass M of a moving body*** *for which no clear definition can be given. It is better to introduce no other mass concept than the "rest mass" m. Instead of introducing M it is better to mention the expression for the momentum and energy of a body in motion.*

In other words, he unequivocally disavowed the expression of relativistic mass *M*:

$$M = \frac{m}{\sqrt{1-\frac{v^2}{c^2}}} \qquad (19\text{-}2)$$

where *v* is the registered speed of the object, *m* its rest mass, and *c* the speed of light. Most physicists acknowledge that the experimental results taken to seemingly justify the hypothesized concept of relativistic mass increase may well be subject to other forms of interpretation implying no mass increase at all.

Gravitational mass, inertial mass, relativistic mass, etc., are all conceptual models of behavior of mass which the real ontology of particles and objects knows nothing about. No object or particle has a mass that changes its nature according to the dynamic system where it is kinetically participating. Furthermore, one ought to wonder, if the mass, conventionally thought to be a monolith, is going to increase, where would the supplemental amount of mass be coming from? Not from thin air (or thin space!). Energy conversion from the principle of mass-energy equivalence is just another contentious proposition heavily litigated in recent past, which would further complicate the matter instead of providing clarifications. Relativists have thus settled for the safe interpretation suggested by A. Einstein. As a result, the nomenclature only ventures to state that as an object approaches the speed of light, the object's energy and momentum increase and without bound, while offering the following two supporting formulae:

$$p = \frac{m_0 v}{\sqrt{1-\frac{v^2}{c^2}}}, \quad (19\text{-}3)$$

where p is the momentum;

$$E^2 = (m_0 c^2)^2 + (pc)^2 \quad (19\text{-}4),$$

where E is total Energy, p the momentum of the travel object and $m_0 c^2$ the energy at rest, negligible relative to $p^2 c^2$ if the object is moving at very high speed.

Note that expression *(19-4)* is derived from *(19-2)*, as a result of multiplying the latter by c^2, despite the fact that *(19-2)* is expressly disavowed. To remain truthful and avoid the evident clever play of hide and seek with these concepts, one ought to acknowledge that if there must be an increase in Energy and momentum as the solid or fermionic object approaches the speed of light, there must be a **loss** in mass rather. What we recognize in Quanto-Geometric Theory is that although the value of the mass of an object is perennial in invariant conditions, transformations applied to the object, such as acceleration in potential wells may change the value of the mass. In particular, when the mass increases, the object has experienced an augmentation in its **scalar density**, when the value of mass decreases, the object has experienced a drop in its **scalar density** (Book II). There is of course an allotted range for these changes, since beyond each of the primitive points of the super-metric, symmetry breaking occurs which transmutes the object into the next one up in the scale.

This view is strongly supported by the fact that all particles' mass-driven interactivity at speeds very near c is highly diminished and subsequently at c they transmute into vector bosons. Their mere natural conversion into vector bosons in spontaneous disintegration processes equally advocate for this view to a certain extent. The likely cause of this behavior is that c is not just an imposed or arbitrary speed limit but, in reality, an endogenous kinematic implementation of the vacuum, which then completely subsumes the travel particle by shifting its wavefunction and potentially volatilizing it altogether into void itself, thus leaving the particle with no mass-like capabilities. This view is fully consistent with the structure and morphology of particles proposed in Quantum Relativity, according to which a photon is a much larger entity than any fermion, and obviously less dense from a scalar point of view. When a fermion like the electron explodes into a photon i.e., as it happens in the reaction particle-anti-particle pair (electron-positron), it is not behaving like a scalar monolith, in which case it would split into smaller monoliths, but as a density or distributed scalar, accessible to a state of larger volumetric ontology represented in photons, translating thereby a diminution in its original scalar density. When a neutral pion π^0, on the other hand, naturally disintegrates into two γ photons instead of explosively splitting in scalar parcels, one has every reason to think that the instability of these particles is conveyed through a distributive volumetric state (or decreased scalar density) in an active intrinsic kinetic process that transmutes the particle into photons. These are all *mass*-ive behavior consistent with scalar density ontology for mass.

Lastly, we shall consider the behavior of bound electrons flying at relativistic speeds around atomic nuclei, whereby they create a stationary density cloud we denote orbital, additional support to our view of mass ontology and behavior. What's more, even the structure of the very electron in those conditions appears to be stretched out, if one ought to give credence to E. Schrodinger's view that the bound electron must be like a blob. To the same record, we must firmly add the fact that in those same conditions, the particle cannot demonstrate momentum or inertia, a mass-laden characteristic.

19.15. Time-free Experience of Motion in Quantum Entanglement

Quantum Entanglement has historically caused quite a stir in modern physics due to its puzzling defiance of the notions of time and determinism. It is particularly opportune to address this conundrum in an exposé on the subject of time-free motion in dynamic systems and offer a new line of rational interpretation, quite apart from the atypical and contentious notion of retro-causation. From the EPR paradox, which was a thought experiment by A. Einstein, Podolski and Rosen, involving two photons receding from one another (obviously at the very speed of light), to the actual realization of the experiment in the 1970's to the now ample variety of possible experimental entanglement methods involving even atoms or molecules in lieu of photons, physicists have been driven to the same conclusions, ultimately leading to time as an entity emerging from quantum entanglement phenomenology.

19.15.1 Defining Entangled Quantum Objects

First a few definitions. An entangled system of a pair of quantum objects or particles is one in which the sum of internal components that make up the *state* of one object is insufficient for the complete description of the *state* of the same and is dependent on the *state* of its par. It is primarily not really about dynamical interaction but about *state*. That is why an entangled pair of quantum objects is said to in the *singlet state*. The *state* behavior of these objects, so to speak, appears to be that of a single unified object. Two bound electrons within an atomic orbital is the perfect example of a natural pair of entangled quantum objects.

The second condition for an entangled pair of quantum objects to be or become such is that they must be a conjugate of one another in some way: i.e., in the orbital habitat the electrons must be in the spin-up state for one and the spin-down state for the other. One may abusively sum up this condition by stating that the pair must have an opposite par quality to it.

19.15.2 Wavefunction Collapse

The gist to the matter here is that any circumstance that affects one element of the par such as a measurement of any sort (momentum, position, speed, etc.) automatically and instantaneously affects the other conjugate, and does so before any information could travel from the affected particle to its par (no signal can travel faster than *c* the speed at which the photons are already receding from one another). The state of the entire entangled system collapses in these circumstances. Einstein *et al* contended that at their coherent inception both particles acquired unknown or hidden information about each other in relation to their local variables or shared the same determinants, which would help explain why they behave in a correlational manner without exchange of information of any sort. Bell's inequality theorem destroyed this view to the increasing satisfaction of the modern physics community by showing that on the experimental plane it does not pass the muster of the strict laws of probability because the measurement results remain unexpectedly *always* correlated.

As recently as 2013 entanglement has been obtained between photons that did not coexist in time in a demonstration of what has been called entanglement swapping. We must also point out that entanglement between quantum objects has been realized even when the measurements have been performed quicker than light could transit between the two entangled objects.

As much as the current conclusions driven from quantum entanglement experimentation found EPR position vis-a-vis statistical quantum mechanics erroneous, from the Quanto-Geometric vantage we can fully support their contention that no aspect of three-dimensional physics can be outside of the scope of intellectual interpretation and that we must not accept no explanation for an explanation. Effectively, one will find no weirdness or mystery to the behavior of quantum mechanics' wavefunction as long as we are able to understand the higher order of Covariant Norms, unknown thus far to past and modern-day physics, in its 9 typified forms and its 3 quintessential forms, that is central to Quanto-Geometric Theory. It is only in the ignorance of the many properties innate to the void of space, beyond the geometric properties uncovered by Einstein's Theories of Relativity, that one will find mystery in the behavior of Quantum Mechanics wavefunction in general.

19.15.3 Unraveling the Quantum Entangled State

To be specific, an understanding of quantum entanglement starts with a physical visualization of the quantum wavefunction.

19.15.3.1 Significance of the Boundless Spread of the Wavefunction

Shall we say that it is time to make good on the quantum-mechanical concept of the boundary-free wavefunction. The quantum wavefunction is *real* and not an imaginary or abstract mathematical entity with a life in momentum space. The quantum wavefunction is indeed the spread of the void of space. For what we call a particle object, it may be visualized as a tensored regionalization of the void of universal space in the immediate surrounds of a point scalar, in the exact manner we have depicted it in the *Ying yang* representation (Fig.3). It is not and cannot be a segregated expansion of background universal space, since it is integral to the latter. Its expression is the exponential term of the generic Quanto-Geometric Operator.

$$\|\psi\|_{a_i b_i} = \left[\frac{-Q^2}{2q}\right] \bullet \left[\frac{2}{a_i e^{\left(\frac{2a_i + b_i{}^2}{4}\right)} + b_i e^{\left(\frac{2a_i + b_i{}^2}{4}\right)} + c e^{\left(\frac{2a_i + b_i{}^2}{4}\right)}}\right]$$

Therefore, from the standpoint of an object's wavefunction, there is in principle no locality to any physical object. An object is only local or localized through its scalar component or its mass.

Once we adopt this viewpoint, then everything, including what appears to be inexplicably instantaneous, or disjoint in time, becomes transparent to intelligentsia at the very least through spectral analysis of superposition of states, the qualities and signatures of which having been well understood in the *Quanto-Geometric* order. Only because we have not understood the *real* properties of the quantum wavefunction, have we inferred a seemingly "spooky" character to it from the putative reductionist approach of statistical analysis and a supposed interpretative incompleteness of the latter. Every object's wavefunction is universally ubiquitous because it is ported or deployed within universal background space. An additional important fact is that mutual influence of wavefunctions does not propagate but enjoys *in situ* intermediacy.

19.15.3.2 Entangled Particles as Surrogates of the Quanto-Geometric Operator

Quantum entanglement in a dynamic system of two particles is at first glance unmistakably a surrogation of the Qanto-Geometric ontological signature whereby one particle plays the role of the scalar tenet and the other the role of the space wavefunction tenet. What the two particles are doing is specifically emulating the **state** of the Quanto-Geometric object $\langle q_4 | s_4 \rangle$. The two particles, which at first glance form a phenomenological dynamic system protagonized by two distinct entities in travel, are in fact executing a constraint or mandate that imposes on them the behavior of a *singular* object. That is the reason and the only reason why they adopt the *singlet* state ontology well known to them. The two particles have become a *single* Quanto-Geometric object with themselves as the common ontological duet that is further engaged in mutual covariance per the Quanto-Geometric norm.

It is through Quanto-Geometric ontological signature that they inter-relate with each other, causing what affects one tenet, whether through the wavefunction or their mass, to resonate or entail consequences for the duet as a *single* object.

- ***Quantum Entanglement as the State of the Quanto-Geometric Quadratic (Radical) Operator***

Let us recognize that the object or phenomenology that we are dealing with here is the Quadratic (Radical) Operator, reported in previous volumes of this series, whose graphical and numeric expressions are:

$\vec{4}$

$\langle q_4 | s_4 \rangle$

$$\|\psi\|_{a_4 b_4} = \left[\frac{-Q^2}{2q}\right] \bullet \left[\frac{2}{a_4 e^{\left(\frac{2a_4+b_4{}^2}{4}\right)} + b_4 e^{\left(\frac{2a_4+b_4{}^2}{4}\right)} + c e^{\left(\frac{2a_4+b_4{}^2}{4}\right)}}\right]$$

For this Operator:

$n = 4,\ s_4 = \pm\sigma$

We refer the reader to Book I of this series for an explanation of the parameters a_4, b_4 and c. In considering the properties of this Operator, it becomes outwardly clear that all ontology and phenomenology attributable to quantum entanglement belongs to the Quadratic Quanto-Geometric order whereby the two basis ontological tenets are of equal "weight" and exert equated covariant influence on one another. Equal-weight covariant influence drives them to "seemingly" morph into one another, causing them to become indistinguishable from one another, albeit remaining each their own. If there was ever a point where the principle of mass-energy equivalence held true, it is manifestly the quanto-geometric quadratic covariant level. However, any inquest in the principle of mass-energy equivalence as a possible interpretation of quantum entanglement is seldom realized because the *real* properties of the quantum wavefunction have hitherto remained unknown.

19.15.3.3 Replicate of Quanto-Geometric Orthogonality in Quantum Entanglement

The foremost postulate of Quanto-Geometric Theory is the existence of two tenets that cannot be defined in and of themselves, scalar and nothingness of space, that is, but only in relation to one another and specifically as *opposite* to or *inverse* of one another: *one is everything that the other is not.* This stands for the primordial condition of orthogonality as previously stated.

It is unquestionably the orthogonality characteristic of the Quanto-Geometric duet that is reproduced on the conjugate condition imposed on quantum entanglement state. Because quantum entanglement is expressly executed under the umbrella of the Quanto-Geometric ontological signature, do we see that the orthogonality feature finds its way through the phenomenon with the mandate that the particles show a conjugate feature of some kind with respect to one another. Two bound electrons, the most ubiquitous form of quantum entanglement, are two twin elements with equal weight but with *opposite spin* (one up, one down). When it comes to photon particles for their own, for two photons of light to experience an entangled state as shown experimentally, they must have been spawn off two orthogonally polarized light beams, or otherwise subject to reversal of their polarization plane with respect to one another. Absent this conjugation character, entanglement cannot occur.

Conjugation in the entangled state is undoubtedly a replication of the anti-symmetric or orthogonal condition inherent to the Quanto-Geometric ontological signature.

19.15.3.4 Replication of Quanto-Geometric Time-free covariance in Quantum Entanglement

We have well explained at the beginning of this exposé why this theoretical framework is one of timelessness insofar as dynamic systems, and in particular that the covariant influence exerted on one another by the ontological tenets does not propagate in time but is instantaneously exerted.

Well, most unfathomable of all things, the covariant influence between two quantum entangled objects is an instantaneous observable. Notably, if the wavefunction of one particle is forced to collapse, the wavefunction of the other *ipso facto* follows suit. If one cares to understand that the environmental factor or laboratory setup that imposes the entangling constraints on the particles has in fact caused them to become a ***new object of Quanto-Geometric signature,*** one then readily gains the coveted insight into this mysterious behavior.

The question of how long it takes one particle to propagate its influence onto the other is no more sensible than the question of how long it took the electron to jump to the new value of momentum once the applied potential energy has shifted its quantum-wavefunction phase. Those effects are simply instantaneous. Once one understands that the two-particle dynamic system is in fact a new object quanto-geometrically bound, independent of distance between the two particles which may be the infinites, then there is no more time problem or question.

Supplemental cause of the above is that space is totalitarian, allowing us to visualize the spatial component of the two objects surrogating the Quanto-Geometric condition to be:

- Very coherently in superposition with one another in their separation without vertical or horizontal bounds.
- Intimately and thoroughly merged with one another due to the symmetry in the properties of their wavefunction.
- That they are one with the tapestry of universal background space in the theater Shell of their interaction.

It is this totalitarian property of space that sustains the infinite development of harmonic waveforms at one single point or region of space as captured in more familiar Fourier analysis. If one takes the time to represent physically the possibility that a multitude of electric waveforms in even or odd harmonics can add up at one point, or how one point may represent both a positive peak and a negative peak or valley or even a node all at the same time, then one is forced to conclude that this wave or oscillation point, if it must be physical and not only representational, ought to be ultimately a spatial point and that space must be totalitarian (infinite foliate sum) in its tapestry both orthogonally and transversally.

Time is definitely not a part of this vision. That is what makes it possible with Laplace transforms to leave the time domain for achieving better solutions and representations. Ironically enough, in the Laplacian treatment we believe that the space domain is abstract and the time domain to be the real one! Once one has risen to the visualization provisioned by Fourier and Laplace, one should have the good sense and above all the courage to depose the mental model of time and not invoke the absurd of retro-causality for an interpretation, let alone seek to normalize it.

I will conclude this section by stating that it is every physicist's prerogative to put fancy math, clever math, copious math or perhaps obscure math behind any conundrum they face, but it can never substitute *intelligentsia* first. Retro-causation Theories, which have had a long insidious life ever since the birth of the Quantum Theory, shall all henceforth be disbanded and obliviated.

19.16. Quantum Entanglement in the Macrocosm

A *USA* press story in early days of this millennium is of note when showing that entanglement exists not only in the quantum world but in the macroscopic world as well, in particular in the biosphere. It related an episode in the lives of two female twins who lived in the same city but in different sections of town. They visited each other as frequently as any pair of siblings did. When Christmas of that year came along, they met at one of their homes and exchanged gifts for the occasion. Holy surprise! They bought each other the exact same gift! To the reader who needs an explanation, just two words: Quantum Twins!

Is this example any different from the EPR photons at play in the midst of our laboratories? No, it is quantum entanglement or quantum phenomenon in our big macroscopic habitat. Note that the EPR photon experiment is seldom considered when looking for clues into Unification, a pursuit of long in our science, due to our deep-seated misunderstanding of the scales or Shells of matter. There are at least four world spans of a difference (or four Shells of matter) between two entangled electrons in an **atomic** orbital and two entangled photons in the space of our human **laboratories**. The EPR entangled photons are certainly an example of quantum entanglement or quantum phenomenon in the macrocosm that in itself could set the stage for reflections leading to the discovery of the true dimension of Unification. Sadly, it never has.

To express numerically the ontology of the new internally entangled singlet object $\|\psi_{44'}\|$, or the entangled dynamic system as one pleases to name it, is to formulate a product between the wavefunction of one element and the scalar term of the other element respectively, all inserted in one *bra-ket* as denoted below in *Eq.3-4b* where the elements with the prime symbol represent those pertaining to the second twin. Since we have already explained in Book I and Book II how to approach the resolution of practical dynamic problems with the application of Quanto-Geometric operators, another repeat is counter-indicated here.

I shall however put in particular highlight here that:

- the component objects do not have to be Quadratic objects but only the resulting singlet object.
- their standard deviation parameters in their quanto-geometric expressions **must be equal** as a *sine qua non* condition.

$$\|\psi_{44}\|_{a_4'a_4b_4'b_4} = \left\langle \left[\frac{-Q^2}{2q}\right] \cdot \left[\frac{2}{a_4'e^{\left(\frac{2a_4'+b_4'^2}{4}\right)} + b_4'e^{\left(\frac{2a_4'+b_4'^2}{4}\right)} + ce^{\left(\frac{2a_4'+b_4'^2}{4}\right)}}\right]\right\|$$

$$\left| \left[\frac{-Q'^2}{2q'}\right] \cdot \left[\frac{2}{a_4e^{\left(\frac{2a_4+b_4^2}{4}\right)} + b_4e^{\left(\frac{2a_4+b_4^2}{4}\right)} + ce^{\left(\frac{2a_4+b_4^2}{4}\right)}}\right]\right\rangle \quad (3\text{-}4b)$$

If the component objects are quanto-geometric quadratic objects, then the resulting singlet object will become a *super-entangled* object, by further potentializing the equivalency signature within the singlet and making the two sub-components even more indistinguishable from one another.

19.17. A Different View on Schrodinger's Cat

The puzzle of all-time Quantum Physics! Let's frame it in E. Schrodinger's own words:

A cat is penned up in a steel chamber, along with the following device (which must be secured against direct interference by the cat): in a Geiger counter, there is a tiny bit of radioactive substance, so small, that perhaps in the course of the hour one of the atoms decays, but also, with equal probability, perhaps none; if it happens, the counter tube discharges and through a relay releases a hammer that shatters a small flask of hydrocyanic acid. If one has left this entire system to itself for an hour, one would say that the cat still lives if meanwhile no atom has decayed. The first atomic decay would have poisoned it.

The ψ function of the entire system would express this by having in it the living and dead cat (pardon the expression) mixed or smeared out in equal parts.

Physicists believe that the application of standard microscopic quantum mechanics to macroscopic objects, such as cats and computers require that the macroscopic objects do not always have unique classical state descriptions, but a series of superposed states that each may bear resemblance with a particular classical state. This interpretation of quantum dynamics begs the question of *when* it is that a quantum system stops existing as a superposition of states and becomes one or the other. It is clear that the cat, if he survives, will only remember the many episodes of his fleeting life, he will not remember having been dead or having undergone a state of death. The thought experiment vividly illustrates this apparent paradox. The experiment ties the fate of the cat to a quantum system that operates based on mixtures of states and directly transfers the quantum determinants and outcome to the macroscopic living entity, thereby requiring that the cat be both dead and alive at the same time before he reaches a final state, whether that state is a singlet one or even a multiplet or mixed state.

Because this paradox has proven to be unsolvable in all logics, physicists have principally relied on invoking artifacts of observation for its resolution. First critic of that line of interpretation, no other than legendary A. Einstein:

"Nobody really doubts that the presence or absence of the cat is something independent of the act of observation."

They contend that the act of observation decides what is real or determines which final state becomes real by collapsing the mixed wavefunction, because the cat cannot be dead and alive at the same time in any real outcome. The entangled wavefunctions here are: *non-decayed nucleus - cat alive* and *decayed nucleus – dead cat.* Therefore, all the interpretations thus far offered to resolve this conundrum bear on observation and consequential collapse of the wavefunction. None has gained universal acceptation however, most likely on accounts of every one's realization that no aspect of physical reality can be dependent for its objective veracity and fullness on the presence, absence or action of a subjective and passive or perhaps inexistent third party or observer. Logically, the cat will either be dead or remain alive whether or not some curious human has opened the door to take stock of the outcome.

From the interpretative standpoint, the problem with this thought experiment is that it is based on mistaken assumptions and a misunderstanding of the physics of life. Most significantly, it is erected on the ignorance of the most fundamental axiom of physical law: *Quanto-Geometry*. What one firstly ought to understand is that life is both a condition of *vitality* and a condition of *death*. There is not a single human being that would live perhaps an ephemeral second if the army of cells that have become deficient during homeostasis are not in continual succumb to the process of death instigated by their lysosomes, in addition to those succumbing by *apoptotic* control, a genetic program orchestrating cell death throughout the entire anatomy of a living body as a necessary condition to support homeostasis. Out of the 37 trillion cells or so of an adult human body an average of 25 billion of them die every day under *apoptotic* mandate. So therefore, the condition of vitalization rests on the ensemble of living cells in our body which promote and sustain the meta-level histological blocks and processes, while the condition of death rests on the ensemble of dead cells and the systems devoted to the orchestration of death in our body.

From a Quanto-Geometric viewpoint, the condition of vitalization plays the role or function of the spatial wavefunction tenet, because its north is motion and expansion. Whereas the condition of death plays the role of the static scalar, because in death, as demonstrated in a dead body, there is only inactivity and staticity as the ultimate pursuit and realization. The Quanto-Geometric condition of orthogonality is so very explicitly present in life and death as a duet. Are they not completely anti-symmetric to one another? Therefore, every living entity is both alive and dead at the same time, in accordance with the Quanto-Geometric ontological signature.

There is no way possible to rationalize the human experience in a life traversal without assuming an orchestrator of pro-pensive or pro-life processes and another orchestrator of death or shadow processes, all based on twin ontology in every human. On that account, every human is one or one persona in active life under the sun, and another persona in inactive or quasi static life in the silence and quiet of our nights. It is not without reason that our bodies all mimic death during night sleep. At every instance of life a human being is this perfect quantum twin in every which way it is defined. A life traversal is the story of a quanto-geometric covariant duet spanning the 9 normalized phases of ontological development. One aspect of the shadow persona is indeed instructed in the psychoanalytic school of psychology, under the denomination of *shadow* personality or *alter ego*.

The fact remains that the wavefunction of the orchestrator of life and the wavefunction of the prostrated *alter ego* are both present in the individual's body, whose life course is the entire suite of Quanto-Geometric normalized covariant signatures. According to this view, at birth there is death and at death there is vibrancy of life, to E. Schrodinger's posthumous complete satisfaction despite his incredulity, as he wrote:

"The ψ function of the entire system would express this by having in it the living and dead cat (pardon the expression) mixed or smeared out in equal parts."

Except of course that the dead cat is not visibly smeared out but that its dead parts pertain to the internal and quasi unobservable province of the *alter ego* orchestrator. While it might be difficult for the rational mind of a physicist to comprehend the origin of death at birth in its complete orchestration (unless one calls for the trans-physical dimensions of the Omniverse and their metrics), it is fairly simple to represent presence of life at death as embodied in the known process of agony. The better part of agony is a lethargic state similar to hibernation where the distinction between life and death vanishes because they have punctiliously merged into one another. Furthermore, anyone, unafraid enough to study in detail the development of necrological processes in the prostration of death, will know that the many body's histological systems never shut down *all at once*, the sense of audition for instance *always* persisting to the very end across the histological blocks that make it possible. On that account alone, in the state of prostration of death, the body goes through a transition of *mixed* states conformed in low-life systems/dead systems, until all histological systems have absolutely shutdown (at an advanced point in the *transcendental* phase).

Given all of the above, the Manicheic view that the living must be either alive or dead, and that two stateful wavefunctions representing both conditions in one single individual at any one time cannot reasonably be claimed, simply stand for a mischaracterization. Schrodinger's cat, akin to all living entities of the biosphere, is indeed both alive and dead at any one moment, independent of all forms of observation. Collapsing the wavefunction of the prostrated twin will indeed collapse the entire ontology. However, third party observation does not entail such effects.

19.18. Conclusion

What we have accomplished in this thesis is a clear and detailed exposition, we hope, of the foundational descriptors of motion in dynamic systems through an entirely time-independent kinematic regime. We have presented the a-temporal defoliation rate of the spatial vacuum as the central element at the heart of the evolution of dynamic systems. We have uncovered the panoply of primitives of motion responsible at first order for momenta through the eigenkinetics and eigenforms essentially deriving from the Quanto-Geometric 2-space metric. Lastly, we have de-ciphered the obscure and cryptic state code behind the dynamics of quantum entanglement phenomena. These elements altogether constitute the root methodology for a granular description of any type of motion in dynamic systems in the timeless regime.

Shell symmetry, based on the referred 2-space metric, emerges as a fact of paramount significance when analyzing stateful behavior, generally attributed to objects, with respect to dynamic behavior, generally attributed to multi-partite systems, for a better comprehension of the both in their natural courses. Both quantum entanglement and the experimental quantization of free travel electrons prove without a doubt that the evolution of all dynamic systems may be properly understood in terms of stateful behavior of objects in all Shells of matter under the most fundamental of physical laws: Quanto-Geometric scalar-space covariance.

In closing this study, we open two windows onto the future of quantum physics in terms of two predictions projected from the implications of this study:

- *Because all phenomenologies occur through the many foliates of space, which leaves nothing in a non-existent fleece of behind or arrearage, nor in time which is a fleece of fiction, everything that ever occurred must still be existent or preserved within the tensored deploys of space for perennial direct dimensional observation or re-enactment, in the most concrete materialization of dimensional a-temporality or perpetual present.*

- *Because space is totalitarian and with no absolute front end to its omniversal deployment, there must exists a spatial force, of which imagination of the living is only an incipient aspect, by which all projected reality through a sufficient degree of that force is conformally driven to existential materialization, without loss of homogeneity in the continual space deploy and spatial tapestry of the Omniverse.*

Chapter 20

If you think you understand Quantum Mechanics, you don't understand Quantum Mechanics.
R. Feynman

ADVANCING FUNDAMENTAL AND TECHNOLOGICAL POST-QUANTUM PHYSICS THROUGH UNIFIED PHYSICAL LAW

It is a rather fair assessment that modern physics is rife with an overemphasis on inflated mathematical-physics code, a focus on abstract or fictitious constructs with no material representation and a bloated formalism aimed at consolidating educational content of the profession, without regards to evaluative norms when it comes to progress in the resolution of fundamental problems of the knowledge body. The inability to timely solve these issues has arguably created a mainstream trend to turn modern-day physics into an exercise of perfection in talking about physics and presenting physics in the most polished envelope, to the point that prizes are granted for best *essays* about physical subjects such as gravitation but not contributions to the resolution of the conundrums surrounding this force, for an example. Public policy, which executes the vision of power and particular interests of a political economic elite at the global scale and undertakes secret science to project power and ascendency, does nothing but exacerbate this acute problem in the academy of physics. The development of post-quantum physics is not immune to these trials and tribulations, and when it comes to the knowledge body itself, to the major problem known as the Hierarchy Problem. In this Chapter, we undertake to present a uniform theoretical basis embodied in the Quanto Geometric Framework of Unified Physical Law not achieved to the plethora of segregated constructs, around nearly one century of experimental findings, in a field where Theory unfortunately lags far behind technology.

20.1. An Iteration over the Universal Irreducibles

To make in our minds an imaginary representation that one only and incalculably small thing exists as if the world altogether imploded is to still represent more than one thing in our minds. Effectively, that monolithic and incalculably small thing would have to exist somewhere. The whereabout of the tiny thing is a necessary component of its imaginary representation in our minds. Even though this representational whereabout seems to be of a most miniscule extension, it is there as accompaniment of the monolithic and minute object. Furthermore, because there is nothing else physically in that representation, the whereabout of the monolith must have the same absolute or ontic value as the monolith. In other words, the space patch of occupancy by the monolith must be ontological and not an agency. And for all that matter, when we have conceived of an all-endogenous monolith we have equally and immediately conceived of an exogenous patch of vacuum where it exists or coexists.

This is not the beginning of the physical world but the beginning of the representational construction we dub Quantum Physics of nature.

To further characterize the two existent tenets with physical properties, we might state that the monolith is endowed with the property of scalarity or the essence proper to a "thing" while the space patch is endowed with a dimensional characteristic. From a metric standpoint, a scalar unit is not a weight or a measure of mass. Scalar units for the "thing" are the quantum abstractions that we call numbers. These delimit and dimension scalarity on the basis of a base number scale or the traditional suite of numeric units forming a number base. It is understood that in any number base each number is formed from the first unitary quantum abstraction as foundational unit by simple or cumulative addition or shifts. So therefore, the quantum abstraction that we call number 1 is the natural unit of the scalar essence in the primal "thing". It is also clear that number representations are quantum abstractions insofar they are intrinsically ontic existents in everything in the natural world while being completely intangible. Any objection to this contention is an objection to the fact that everything that exists in segregation from some *thing* else or everything else exists in the order of cardinality and therefore every *thing* that is countable is numerically impregnated. This numeric essence is scalarity.

At the other side, the vacuum (or patch of space) is based on dimensional units proper. The distance metric is not a fundamental metric of space as traditionally posited in the many physical metrics (FLRW i.e.)[21], because distance units are conventional and arbitrary human measures. The natural unit of space is a linear dimension. A linear dimension represents the minimal whereabout or locus for the tiniest imaginable monolith. Our observable physical world rests on a spatial metric based on 3 dimensional units or as commonly stated 3 dimensions: a linear dimension of Height, a linear dimension of Width

[21] The Friedman-Lemaitre-Robertson-Walker metric

and a linear dimension of Depth. These three dimensions are perennially present in any patch of the observable or inhabitable vacuum of any extent.

In the absolute, one cannot conceive of a pure vacuum spread with limits since the limits of nothingness or void as the essence of the vacuum amount to non sequitur. This is reflected in the fact that a dimensional line is endless at both sides. Only when we downgrade a line by chopping it down to a line segment do we impose limits on a line. Consequently, a unit-dimension of the metric of space is infinite, by contrast with the scalar unit which is finite. The ultimate consequence of this view is that the Universe cannot have started nor will it ever end, ultimately because there is no other destiny for scalarity to explode into but the dimensional coordinate space vacuum, and there is no other destiny for the latter to implode into but scalarity itself.

20.2. Causation and Covariance

It is not possible for the monolith to be phenomenally independent from the space patch, for the simple reason that there is nothing or nothing else between them, therefore no possible mediation. Hence correlation is automatic and ineluctable between the two basis ontological tenets. Causal interaction or correlation between the monolith and the dimensional space patch spells out as covariance and implicates action. Causal interaction does not implicate motion however.

Further, there is in existence no absolutely pure scalar nor is there any existent pure space or space patch. Due to ubiquitous covariance, all space patch is tensorial or under some level of stress and all scalar is spread in density state or exists in volumetric state.

20.3. Origin of Motion

Motion is already part of the ontology of dimensional space. Although unintuitive to most, the impetus of motion is an ontic aspect of dimensional space vacuum. Akin to the intangibility of numeric quantum abstractions as a property of the scalar parcel, motion is an intangible property of pure vacuum deriving from its dimensionality. Effectively, intrinsic boundlessness of dimensional units of space is a result of ontic motion.

It is important to guard against visualizing motion as the end result of interaction between the monolith and its space spread partner. Although observable motion is always motion of the scalar parcel, the latter is indeed only an expression of second order influence exerted on the scalar by the space patch. The engine of motion itself remains intangible and rests with the very space spread. Newton's first law stating momentum invariance of an all-unconstrained scalar is phenomenological but not ontological exegesis. The ultimate cause of this invariance is the inner space spread momentum, whatever occurs to the scalar monolith to put it in motion, whether impact or else.

Inner momentum of space or space momentum is real. In this program we never represent fictitious dynamic spaces such as momentum space which do not exist and amount to no more than misguided constructs in response to representational challenges.

20.4. Time-free Primeval Dynamics

The phenome of covariance between the two basis tenets does not occur in time. There is no Time background in Nature because there is no natural unit-dimension of Time. The most irreducible composition of all and any object in Nature is the duplet formed by scalarity and space spread. There is no bound to the extent of the space spread. Therefore, interaction principally mediated by space spread components of higher objects, wherever they are in the Universe and whatever the distance separating them, is instantaneous. Such a mutual interaction is exerted concomitantly.

20.5. Foliation of the Vacuum

The second order of realization of pure coordinate space, of which celestial vacuum is a part, is in its foliation deployment borne out by derivation of its dimensional linear layout. The necessary dimensional assembly creates foliation bundling of the vacuum spread at its lowest or ground level. It follows that stress of the vacuum is thus a result of its defoliation capability. Vacuum patches under stress above ground level defoliate to give gestation to observable geometries.

The conditions of stress of the vacuum are no other than its coupling with the scalar monolith partner. Nothing else can stress the vacuum because nothing else exists. The order of spatial momentum intrinsic to the vacuum couples with the foliation order, in the sense that when quasi completely left to itself the vacuum's momentum is maximal (being c in our Universe) and corresponds to a simplex or quasi tri-planar foliate.

In a higher frame, this program visualizes two forms of explosions in the dimensional foliate order: one in a meta vertical order that creates a suite of unobservable world planes, and one other in the meta horizontal order that creates the observable suite of degrees of freedom resulting in what we call complexity, from quarks to molecules to galactic formations.

Of course, this would be tantamount to just a view, as interesting as it might have been, if it should not be testable. This view emerges from mathematical norms and quantification in a natural construction that has been able to outstandingly predict the known value of c as the most fundamental experimentally known physical constant, and every single one of the fundamental physical constants regimenting the scale of complexity as it experimentally appears before us, from quarks to molecules to galactic groups, that is. What we call symmetry-breaking in physics is the engine of this local evolution.

20.6. Hierarchy of Quantum Causal Forms

From a universal standpoint, the first cause or causation principle is the coexistence constraint or conservation encompassing the two irreducible tenets. Covariance or mutual prospection of the two tenets is the second level of causation or the second causation principle. And the third cause or third causation principle is the grip between the two tenets which establishes a degree or gradient of correlation resulting in a permanent qualified form of association. This third causal order is what we are going to focus on in this Chapter as the foundational order of Quantum Causal Forms. The quantum causal regime may be expressed or coded in many different ways. In graphism, it may be represented in a similar form to old age Ying Yang symbol or symbolism[22]. This graphical rendition is no other than the representation we have offered in Chapter 2, Fig.3.4. Numerically, the correlated tenets are expressed in the relation:

$$Q \times S = 1$$

where Q is the scalar parcel and S the space spread. In this paramount Quanto-Geometric axiom we note the orthogonality between the two ontological tenets since the relation reflects their being inverse to one another (a thing v. no-thing). We can at once see that this characteristic already satisfies a Hilbert space which strictly requires orthogonality. However, our perspective is not that of a fictitious space but real space with inner momentum.

The graphical depiction shows 9 universal distinct levels of covariance which we otherwise dub the 9 distinct quanto-geometric causal forms. Because they are universal, they are also quantum causal forms, situated at the level of the afore-mentioned third causation principle. Each holds a specific ontological quality according to the quantum abstraction stamp through which each is elicited. We are going to transcribe them by using the familiar Bra-Ket notation in Quantum Physics with some adaptations. Here the symbol q stands for scalar state and s stands for space spread. The quantum causal regime follows:

- Monocentric Causal Form: $\langle q_1 \mid s_1 \rangle = \vec{1}$
- Ambivalent Causal Form: $\langle q_2 \mid s_2 \rangle = \vec{2}$
- Relational Causal Form: $\langle q_3 \mid s_3 \rangle = \vec{3}$
- Twinning Causal Form: $\langle q_4 \mid s_4 \rangle = \vec{4}$
- Metamorphic Causal Form: $\langle q_5 \mid s_5 \rangle = \vec{5}$
- Harmonic Causal Form: $\langle q_6 \mid s_6 \rangle = \vec{6}$
- Antagonistic Causal Form: $\langle q_7 \mid s_7 \rangle = \vec{7}$
- Golden Causal Form: $\langle q_8 \mid s_8 \rangle = \vec{8}$

[22] To Niels Bohr, arguably maximum proponent of Quantum Physics, this symbol was a devotional favorite.

- Expansive Causal Form: $\langle q_9 \mid s_9 \rangle = \vec{9}$

The quantum causal regime is easier to follow in graphical depiction by starting with the Twinning Causal Form and moving upward or downward from there. Each tenet of the covariant relation is represented in the layout of the design in black (space tenet) and the layout of the design in white (scalar parcel). The circle inside each of the two designs describes the resulting effect of the correlation on each tenet. The size of the circle inside each of the two designs is here identical, indicating that the influence exerted by one ontic tenet on the other is of the same weight or equal to one another. The end result or effect of this correlational type is materialized in two twin objects that have become indistinguishable from one another, hence the Twinning attribute. This is designated entanglement state in Quantum Physics. The Twinning Causal Form is responsible for most, if not all, of the weirdness or weird effects that have forever flabbergasted quantum physicists. These peculiarities and ambiguities are what prompted Richard Feynman and Niels Bohr to rally around the famous statement: "If you think you understand Quantum Mechanics, you don't understand it." Note that the two inner circles of the graphism always have the same tint of shade to represent the conservation of the Quanto-Geometric Momentum throughout the scale.

As we move upward from the Twinning Causal Form, the white design becomes more and more predominant in extension and in the inner resultant effect circle as well, indicating a predominance of the vacuum tenet in the covariance, for a particular causal signature at each step. Keep in mind, however, that these signatures are not continuous but discrete as reflected in the distinct attribute of their types which is not linear or continuously gradual.

Likewise, as we move downward from the Twinning Causal Form, the black design becomes more and more predominant in extension and in the inner resultant effect circle as well, indicating a predominance of the scalar tenet in the covariance. Their discrete signature types are spelled out as well. It is paramount to recognize and keep in mind that this suite of types is not arbitrary but arise from normalization of a universal root Function, the Quanto-Geometric Eigenfunction.

The base numbers, each with a top arrow, signify a state composed of two orthogonal tenets with a kind of covariance typified in the attribute given to the designated Causal Form of the covariance. The number symbol states the quantification level and reflects the intangible quantum abstraction embedded in the composite object. The subindices of q and s identify their primal object type in quanto-geometric terms or their level within the 9-stratum spectrum.

20.7. Post-Quantum Physics Codification

The discussions between Einstein and Schrodinger about quantum states as well as those carried out between Rosen, Podolski and Einstein were later formalized by Dirac and Bohm with the Bra-Ket language, which was a form of simplified quantum math meant to be more expeditious and amenable than would otherwise be.

In the bra-ket language, the entanglement quantum state exhibited by two particles, such as two electrons in an atomic orbital is expressed as follows:

$$\langle \varphi_{AB} \rangle = \frac{1}{\sqrt{2}} \{(|0\rangle|1\rangle)_A - (|1\rangle|0\rangle)_B\} \quad (20\text{-}1)$$

The particles are entangled with respect to their spin property. Because there are only two directions of spin, we choose a binary numeric code with two places to express the spin orientation of the particles: one direction for spin down which is the left location in the notation and the opposite direction for spin up which is the right location in the notation ($|L\rangle|R\rangle$). The two entities within the accolades in *(20-1)* represent the states of the two A and B particles. The minus sign reflects the ontic inversion status of the two particles with respect to one another as relates to spin.

The primary condition for the realization of an entangled pair of quantum objects is what is fully reflected in the bra-ket language per the above equation, which is that to be or become entangled the pair of particles must be a conjugate of one another in some way: i.e. spin-up state for one and spin-down state for the other. One may abusively sum up this condition by stating that the pair must be anti-symmetric in at least one property.

Further an entangled system of a pair of quantum objects or particles is defined to be one in which the sum of internal components that make up the state of one object is insufficient for the complete description of the state of the same and is dependent on the state of its par. It is primarily not really about dynamic interaction but about state. That is why an entangled pair of quantum objects is said to in the singlet state. The equation above reflects that singlet state in the term φ_{AB}. At the end, the state behavior of these objects amounts to that of a single unified object.

Once the particles become entangled there is a slew of non-classical, some would say non causal, weird behavior that occur. In lieu of measurements, we commonly say that the particles are in the possession of two separate experimentalists, Alice with the A particle and Bob with the B particle.

1. Alice will measure her A particle to be spin-up in an average of 50% of measurements.
2. Alice's measurement causes the state of the two particles to collapse.
3. If Alice performs a measurement of spin in the z-direction with respect to an arbitrary orientation basis, then Bob's system will be left in one of the two states.
4. Likewise, if Alice performs a measurement of spin in the x direction with respect to the basis, then Bob's system will be left in one of the two states.

This "steering" phenomenon, a term coined by E. Schrodinger, is responsible for the collapse of the ontic wavefunction of the particles, a nomer for what we denote space spread in our Theoretical framework. Note that since 1982 Alain Aspect had put the above on firm experimental basis with two photon particles.

20.8. Stubborn Serious Errors of Nonlocality Interpretation

The gist to the matter of entanglement is that any circumstance that affects one element of the par such as a measurement of any sort (momentum, position, speed, etc.) automatically and instantaneously affects the other conjugate, and does so before any information could travel from the affected particle to its par. The contention is that because the two particles are receding from one another at the maximum attainable speed *c*, it is virtually impossible for any influence in the form of a signal from one to ever reach the other. Note that two electrons in an atomic orbital are receding from one another as well at almost relativistic speeds.

This phenomenon, at the center of post-quantum physics, has been termed by Einstein spooky action at a distance and has come to be dubbed nonlocality in Bell and Bohm's theories. Einstein et al contended that at their coherent inception both particles acquired unknown or hidden information about each other in relation to their local variables or shared the same determinants, which would help explain why they behave in a correlational manner without exchange of information of any sort. Bell's inequality theorem destroyed this view to the increasing satisfaction of the modern physics community by showing that, on the experimental plane, it does not pass the muster of the strict laws of probability, because the measurement results remain unexpectedly always statistically correlated, which is inexplicable from a conventional local causation perspective.

20.9. New Quantum Entanglement Exegesis

First and foremost, if it is true that, once entanglement sets in between the two particles, they behave as if in a singlet state, then the new particles must have become a new single object. Their individual ontic states must have transmuted into a new single ontic state involving both their scalarity and their space spreads or wavefunctions. At the point of entanglement, you no longer have two segregated objects but one new object experiencing a higher ontic degree of freedom. They enter a state where their prior form of influence on one another, which was of the phenomal order, becomes ontic covariance with a particular quantum causal type out of the nine in the canon. In this case, it is the Twinning Causal Form corresponding to an object of model type:

$$\langle q_4 \mid s_4 \rangle = \vec{4}$$

In this context, Lorentz invariance is phenomenology which the particles were subject to in their dynamic medium as segregated or autonomous entities. Once they become entangled, whether naturally through chemical conditions or artificially through application of human physical techniques, they will forever remain that new ontic object, whatever the distance the two particles, i.e. photons, freely travel in separation from one another.

Should the entangled particles be fermions, the scalar expression in the prior state is now muted and becomes an extroverted and more tenuous scalar density of a different class or model type. The nature of the individual particulate space spreads has likewise transmuted in a new singlet quality as it acquires a higher degree of freedom under a new ontic definition or model type, the same as the partnering new scalarity.

20.10. Operator Trigger of Quantum State Transformations

The physical, chemical or physical-chemical conditions that trigger these quantum ontic transformations or creation are to be understood as Operators, principally of Hermitian nature. We have profusely explained throughout this entire publication the modality of action of these Operators, mainly by conjugation and frequently Abelian composition. Further detailing is unnecessary.

20.11. Discreteness in the Shells of Matter under Unified Physical Law

By the same token, it is of paramount importance to comprehend that the quanto-geometric or quantum covariant regime with its list of causal signatures has thus far remained unknown to our Statistical Theory. Ignorance of said causal regime by Statistical Theory has led Post-Quantum Physics to a series of wrong conclusions as regards to non-locality. Suffice it to model dynamic quantum variables on the basis of the discrete signatures emanating from this covariant regime, for the notion or conclusion of existence of effects without recognizable causes to completely vanish. It turns out that the one only cause that we are familiar with and have forever been looking for is the Monocentric Causal Form. That one causal form constitutes the province of Classical Physics. Entanglement remains a puzzling effect in our science to the point that we've been looking at Retrocausation, a totally irrational form of reasoning, for an explanation ever since Richard Feynman.

20.12. What Makes Statistical Theory Applicable Everywhere?

If the Normal Density Function at the heart of Statistical Theory is universally applicable, that is to say both for microscopic and macroscopic objects and phenomena, why it is unable as known to explain both Lorentz invariance with the specific value of c and the reason for the known value of the constant of quantum discreteness h is a fundamental question to raise.

We apply this Theory across the board to explain interaction results between possibly correlated variables. Statistical Theory is in essence a mathematical theory. If it is applicable at the microscopic quantum scale, it should not only be able to set the conditions for what is knowable as realized in Bell's inequalities, but it should first capably extirpate from its numeric engine the most fundamental quantum number *h*, the ruling constant in that dimension of the natural world. Likewise, if Statistical Theory is applicable in the macroscopic world from materials to physics of the cosmos, it should not only be able to explicate the deep elements of life beyond distribution of heights in a biotic population i.e., but derive from its mathematical engine the value and the reason for the invariance of the speed of light *c*, the latter being the ruling constant in the macroscopic dimension of the natural world.

20.13. How Does Complexity Emerge from Both Discreteness and Continuity?

Our physical theories thus far frame the physics of the natural world as if its many different constituents existed on a single flat sheet. Effectively, our view in physics is that the vacuum is always the same everywhere, putting on the same scale the following fundamental constituents of the observable Universe:

- *Quarks*
- *Elementary Particles*
- *Atoms*
- *Molecules*
- *Materials Lattices*
- *Karyotes*
- *Celestial Singlets (planets, moons, comets, asteroids, etc.)*
- *Solar Systems*
- *Galaxies*

We cannot or do not know how to define mass in and of itself in physics and have taken since Einstein to evaluate it in terms of energy, a notion that is majorly connected to the visualization of the ontic wavefunction or the dynamics of the vacuum. If the vacuum were to be the same everywhere in all of its possible properties, the above suite which is the all-encompassing expression of complexity would not exist. Complexity exhibits not a juxtaposition profile but a constitutive hierarchy of elements successively and progressively embedded within one another, a meta model of which is the doll in the doll (Fig.20.1).

While the vacuum is homogenous by observation at all stages of complexity, it is not continuous at all orders throughout the tapestry of complexity. For that very reason, complexity exists. Complexity does not primarily exist because of the expression of mass in the Universe, complexity exists because the vacuum

vacuum is innerly dynamical, experiences inner foliate and/or tensile transformations from a symmetry-breaking regime. This view is supported by the fact that we do not see new mass or new scalar elements of a different nature at each stage of complexity.

Fig. 20.1 Russian art piece of a series of dolls inside one another.

What we see instead is a panoply whereby the mass or scalar parcels at one stage of the scale assemble and aggregate to form the scalarity of the next elements. For instance, an assembly of a few atoms gives rise to a molecule or molecular unit. While the molecular unit will exhibit new ontic properties extraneous to its inner atomic units, its weight still remains the sum of weights of all its inner atomic units. What this is telling us is that the vacuum experiences inner transformation and that there must exist state properties of the vacuum that are be different from one stage of the scale of complexity to the other. Therefore, from a global perspective, each stage of the scale of complexity must represent a level of discreteness along a hierarchy line akin to a spectrum, instead of topological spots or points on a flat and totally homogenous single sheet.

Hence, the Universe is discrete, owing to the spectrum of stages in the scale, which our Framework otherwise denotes Shells of matter. The notion of a continuum is only valid locally within the stages of the scale, whereas each of the stages is founded on a paramount symmetry. The most fundamental properties of the vacuum are invariant across all stages of the global 3-dimensional Universe, that is to say:

1. space momentum known as Lorentz invariance and the *c* value of this motion.
2. the ground foliate tensor characterizing 3D vacuum which is also known as the Einstein tensor.
3. the so-called vacuum energy density, better known as the cosmological constant.

The sum of these three properties is responsible for the homogeneity of the vacuum at all stages of the 3D scale, which might provide some grounds to the visualization of a continuum vacuum throughout.

From a more towering vantage point, what have been designated the fundamental coupling constants do represent a higher order expression of covariance between the particulate scalarity and space spread that typify each stage of the complexity scale. These numeric invariants stand for the pilot quantities that govern the individual foundational symmetries sustaining each of these stages. As such the value of these constants forthrightly expresses the discrete character of the stages of the observable natural world, beyond the human unit-dimensions that have been arbitrarily chosen to quantify them. In our theoretical derivation of these constants of a pure mathematical physics nature, they have all appeared under the same mathematical category, indeed well-fitting for the designation of coupling quantities.

If the spectrum of the vacuum is overwhelmingly discrete and if the fundamental physical coupling constants quantify levels of discreteness of these many degrees of freedom, then Statistical Theory as known including Bell Inequality theories must be wrong in the assumption or conclusion that there is no direct causal intelligibility or exegesis in these properties but Randomness. The notion of Randomness is the most significant and ubiquitous descriptor of the natural world or the overall behavior of the vacuum that has been utilized when it comes to scrutinizing it in its fullness, as the essential element of the seemingly universally applicable Statistical Theory. In science, we assume as a matter of course that this mathematical theory is in principle applicable to phemenona at any stage of the complexity scale. If the Grand Theory of Unification long-sought was always under our noses, it would be Statistical Randomness since we deem it indiscriminately applicable and in fact apply it across the entire complexity scale, exactly as is. Does the artifact of randomness really exist however? How much of the notion of randomness is really attributable to natural phenomenology and does not genuinely qualify as human ignorance at this stage of our knowledge complex? This question goes right to the heart of the spectrum of quantum causal forms that we have identified.

20.14. Randomness Does not Exist in the Absolute or the Problems of Normalization

The one major problem with Statistical or Probability Theory to point out is that its prominent empiricism, in its incarnation through the Normal Probability Density Function, has masked the fundamental normalization canon perceptible from our Framework. With this development, we hope to strongly establish that John Bell's theories of quantum inequality is dismissible on grounds of incompleteness of Probability Theory.

For all its given universal range of applicability, the Normal Probability Density Function completely ignores one of the paramount ontic norms regimenting the expression of matter in the quantum domain, namely the first principle enunciated in this study. According to this ontological principle, a point of absolute scalar compactness can never exist in total estrangement of ontic association with an orthogonal partner, as scant of a presence or expression of the latter might there be within the ontic par. This principle has enormous consequences on the normalization modes of the Normal Probability Density Function tending to make it an applicable generalization at all stages of complexity, in particular the quantum realm. And we must remember that the most important role of this function in science or knowledge acquisition is to study interdependency of variables or their covariance, as further detailed in Chapter 6 and 7.

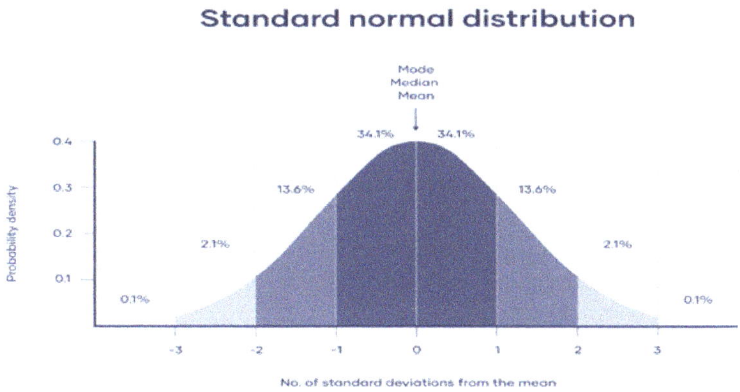

Fig. 20.2 The standard Normal Probability Distribution

Hence, from a modelling standpoint, the Normal Probability Density Function cannot be allowed a value at maximal density which corresponds to gradient 0 of the autonomous or independent variable. Here we are referring, of course, to the Standard Probability Density Function to be strict (Fig.20.2). Both values of the dependent and independent variables are therein inadmissible, no matter what real physical variables they are intended to model. Because of the discontinuity of the function at that one point, differential calculus norms dictate that the full area A under the graph of the function cannot and should not be determined; in particular it is therefore not equal to one as commonly posited in Statistical Theory:

$$A \neq 1$$

So it is, because there are two distinctly separate areas under the graph, not one. At the same time, it becomes improper to consider values of segmented areas under the graph traversing the central point of discontinuity, i.e. the common empirical area delimitations for 68%, 95% and 99% across the mode. In the Theory, those percentages of the total area A are a match with deviation values of the independent variable situated at 1, 2 and 3 standard deviation units from the mean in the standard distribution. Said superimposed scale on the axis of the independent variable responds to our preferred form of mathematical ordinalization in our academic culture and certainly does not represent a natural metric, as compared to the s-metric posited early on in laying down the foundations of our Framework. In all, this delimitation rule is purely arbitrary and responds to the quest for weighing the variable's distribution pattern, just as z scores are elicited to help determine accuracy by evaluating deviation from the mean. The most significant drawback of this artificial scale is its complete obscuring the seminal teleological normals innate to this important function, which transpire when the function is canonically dimensioned, as we have shown in Book I of this publication.

What is critical to represent is that, if the Normal Probability Density Function accurately describes physical phenoma, it is only because it is an image or emulation function of the more fundamental Function we have dubbed the Quanto-Geometric Grand Eigenfunction, which incorporates all the principles enunciated at the beginning of this Chapter's exposé. The Normal Probability Density function is therefore a lower order expression or emulation of the Grand Eigenfunction in the apparent continuum of the macroscopic levels of complexity. It is too deprived to reflect the underlying causal quantum intricacies that dominate the microscopic quantum domain.

On that account, the conclusions brought about by analysis of quantum dynamic systems with this tool are misguided, incomplete and quite frankly non insightful.

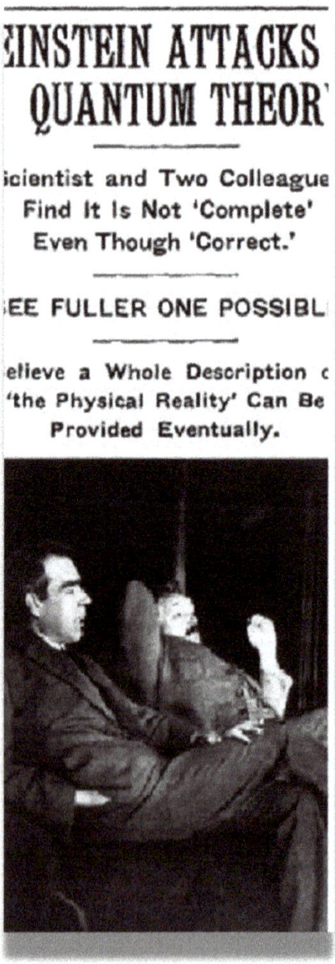

20.15. Are Quantum Non-Causal Effects "Magical" Effects?

We must admit it: if we can observe and testify to the existence of quantum effects without absolute causes, then mother nature is expert at sleighs of hands when the hands are very small. This is the kind of ideas that A. Einstein staunchly opposed and irrevocably rebuked in the public debate he maintained to the end with Niels Bohr.

This situation is no less similar to the contention of so-called "magic numbers" identified among the number of nucleons in atomic nuclides, which confers them an outstanding and unusual level of stability even when we should expect them to be radioactive and disintegrable in the midst of a suite of such elements. We have demonstrated since 2019 in *Quanto-Geometry Vol II* the full extent of unsuspected keen periodicity in nuclide's energy density shells that is principally responsible for these "magic" numbers in nuclei's stability status.

Bell's inequality theory was there all during the course of the Bohr-Einstein debate and never acquired prominence. It is only after Alain Aspect's two-photon experiment in 1982 that the science started having recourse to Bell's ideas to explain what seemed to be evidentiary experimental results. What is clear is this: if someone may "appear" to be dead and alive at the same time, the state of death or lifeless can never be at the time a state of liveliness. It is tantamount to a vacuous ontic proposition because, to start with, there would be no need to define them distinctively whether in syntax or notionally. We do not define a “book” to be a “book” i.e., nor do we define an “electron” to be an “electron”, because, when we define them or notionally fixate them, we are making a state proposition that gives them exclusive entity and identity. Else definitions and nomers are completely useless and purposeless and everything is everything. The only way for something or any living thing to be statefully alive and dead at the same time is that this entity must have always ontically incorporated a part of life and a part of death. In that instance, three uppermost conditions must be satisfied:

- *There must be a phase where the life tenets have preponderance over the death tenets, for which we state that the entity is alive.*
- *There must be a phase where the death tenets have preponderance over the life tenets, for which we state that the entity is dead.*
- *And lastly, there must be indeed a phase where the life tenets and the death tenets are in mutual balance or equated counter-balance, for which the entity may plausibly appear to be equally alive and dead at the same time.*

This also requires us to pre-suppose, albeit initially unstated, that the condition of liveliness and that of death are orthonormal with respect to one another (or absolutely distinguishable from one another), in accordance with the background principle from which this view emanates. One shall note that the above tripartite representation is at the same time a statement of the limits of observation, even in an experimental context.

What this finally all means is that:

- *in Phase (1), even when we thought that death was absent, it was always there as an eclipsed par;*
- *in Phase (2), even when we thought that life was no longer there, it is still there as an eclipsed par;*
- *and in Phase (3), we are flabbergasted at the apparent paradox of life and death being equally there with the same weight because we never thought they were together in ontic parity within the entity and even less that they could reach a state of equated covariance.*

This is the foundation of quantum correlation theories, including Bell’s inequalities, in post-quantum physics. It is no more than a reflection of the quanto-geometric ontic spectrum exposed as encapsulated in our construct of Quanto-Geometric Quintessence. This theoretical Framework is situated beyond the axiomatic reconstruction of quantum mechanics via Generalized Probability Theories and sets the most extensive and comprehensive theoretical basis for upcoming quantum information systems and quantum technologies in general. From the perspective of this study, it is not difficult to realize the serious implications for the Quantum Initiatives taking place currently all around the world, in particular when it comes to quantum encryption technologies which altogether ignore the fundamental quantum causal forms elicited in this Framework. What this very importantly means is that quantum encryption naively based on the assurances of Bell inequalities will effectively be broken with a clever information algorithm based on the Quanto-Geometric causal forms!

A bon entendeur, salut!

20.16. Revisiting Temperature from a Quantum Perspective

The association between temperature, heat and heat energy in science have always been notionally uneasy. An absolute scale grading the coldest possible point to hottest reachable point seems to be quite desirable but requires an association of the temperature notion with that of heat or any artifact thereof. Because heat and cold are relative notions of temperature, the standard bodies have initially sought a thermodynamic definition of temperature in the average kinetic energy of atoms and molecules making up material substances, since their resulting friction after all is prime generator of heat energy. This definition is obviously a phenomal definition that does not directly reside with the ontic origin of thermodynamic temperature as a physical variable. Effectively, this definition cannot be applied to pure coordinate vacuum itself, the temperature of which must be evaluated by deductive methods from its other physical attributes.

In this regard, the NIST offers this appreciation: *"According to classical, 19th-century physics, motion completely stops at absolute zero. But according to the quantum theory introduced in the 20th century, matter does have random motion at absolute zero, called "zero-point motion," thanks to a quantum concept known as the Heisenberg uncertainty principle, which dictates that the position and momentum of an object cannot be known with complete certainty at the same time. The zero-point motion is not considered heat-driven (thermal) motion and thus is not part of the definition of thermodynamic, or absolute, temperature. At absolute zero the only motion that exists is quantum-mechanical zero-point motion." Created May 14, 2018, Updated June 2, 2021*[23].

If the zero-point motion exists, and this random motion of matter can reach absolute zero of temperature gradient, while the so-called thermodynamic temperature is defined to be an absolute scale of temperature on which 0 gradient is an unreachable, then there must exist two different absolute scales of temperature: one for quantum mechanical processes and one other that is the thermodynamic temperature which is "heat driven". However, we do not find that in order to establish a particular gradient of temperature in condensed matter physics, we need to use an intimately different technological approach than

[23] https://www.nist.gov/si-redefinition/kelvin-introduction

that which is used at the macroscopic scale. That much we know since it is still the Kelvin scale that is used to establish such gradients of temperature. Complexity thus seems to uphold a single thermodynamic temperature scale throughout.

In order to resolve these conflicts, it is necessary to introduce in this particular prospection the construct of space momentum anew. It is important to first acknowledge that all heat is a result or a consequence of space momentum. Second, the premises of our Framework invalidate the classical theory notion that absolute zero motion is possible and that absolute zero gradient of temperature is or may be reached at absolute zero motion. Absolute zero motion is inexorably not a possible however, because:

- The vacuum itself is intrinsically motional.
- No void gap is possible within the tapestry of the vacuum, neither conceptually nor materially.
- All scalar parcels are always in correlation with a vacuum spread.
- Lastly, there will always be a reference frame for which an object appearing to be at total rest will be observed in motion.

An ontic definition of absolute temperature to be framed on a single scale all throughout complexity takes us back to space momentum and to attribute temperature to the homogenous and dimensional properties of the vacuum, thus leading us to completely abandon the notion of thermodynamic temperature. In that regard, absolute temperature appears as another property of the wavefunction foliation of the vacuum with the mission to promote expansive spreads of continua or topology within the buildout of our 3D Universe, as analyzed in Vol. I (p.152-158). And that is the reason for the due existence of a natural unit of temperature or a constant of absolute temperature, previously identified in the Stefan-Boltzmann constant and whose value we have theoretically derived from our Framework's mathematical-physics axiomatic.

Here we shall not dispense a consideration of the mathematical order. The decimal number base tells us that pure ordinalization is not possible inside of a layer of the numeric stack starting with the irrational layer, as evidenced in Quanto-Geometric Number Theory in Book I, Chapter 4. Our normal concept of a continuum takes its foundation in the properties of the last three numeric layers of the primordial numeric stack, namely the irrational layer, the algebraic layer and the transcendental layer. This comes as a consequence of the fact that every single one of these classes of numbers never meet an end in their expression. Hence, in order to promote and de facto develop continuum spreads and thereby topology, a natural unit or constant of absolute temperature becomes indispensable. This constraint creates the somewhat incongruent situation of a unitary entity as the basis of continua. Nevertheless, in its very role, the constant of 'thermodynamic' temperature, as previously mentioned in Book I, parallels the constant of discreteness, the Planck constant *h*, both of which were theoretically derived therein in the same differential calculus condition, as shown in the reminder below:

Stefan-Boltzmann's constant: $\left[\frac{dQ(s)}{ds}\right]^{-1}_{s=\sigma/2}$ *(20-2)*

Planck constant: $\left[\frac{dQ(s)}{ds}\right]^{-1}_{s=\sigma}$ *(20-3)*

At the same time, the above *(20-2)* transcribes the most foundational definition of temperature by relating it to the root Grand Eigenfunction which correlates the two primordial physical variables (the scalar *Q* and the vacuum spread *s*) intervening in the buildout of the 3D World Plane.

We do not arbitrarily choose to designate the inverse derivative in *(20-2)* the Stefan-Boltzmann constant, but only because its theoretically-computed numeric figure is in a close match with the experimentally known figure:

QG Stefan-Boltzmann constant: 5.680763 $\boldsymbol{\sigma^2}$

Experimental Stefan-Boltzmann constant: 5.6705 10^1 $\boldsymbol{W\,m^{-2}\,K^{-4}}$

Interestingly, in its 2019 sweeping revision of the fundamental physical constants, four years after the release of Book I of this publication, the NIST has elected to radically change the definition of thermodynamic temperature to fundament the same on the Boltzmann constant. They write:

"In 1954, the kelvin was defined as equal to the fraction 1/273.16 of the thermodynamic temperature of the triple point of water—the point at which water, ice and water vapor co-exist in equilibrium. That is a valuable common reference because, for a precise formulation of water at a specific pressure, the triple point always occurs at exactly the same temperature: 273.16 K.

Extrapolating from the water triple-point temperature to very high or very low temperatures is problematic; so, by international agreement, 21 other defining points are specified, ranging from the freezing point of helium to the freezing point of copper.

However, the kelvin has been redefined in terms of the Boltzmann constant, which relates the amount of thermodynamic energy in a substance to its temperature. When the revised SI was approved in November 2018, the new definition became:

The kelvin, symbol K, is the SI unit of thermodynamic temperature; its magnitude is set by fixing the numerical value of the Boltzmann constant to be equal to exactly 1.380649 × $\boldsymbol{10^{-23}}$***...J*** $\boldsymbol{K^{-1}}$ ***[joules per kelvin].***

If that seems like a mouthful, you wouldn't be wrong! To best understand the context and significance of this historic redefinition, it's helpful to learn more about the past, present and future of measuring temperature."

Despite what must be acknowledged best-effort elimination of arbitrariness in the definitions of natural experimental constants, this redefinition is still missing the mark. First and non-trivially, if the standard body's contention about temperature in relation to zero-point energy is correct, it would still fail to give us the unified scale of temperature across the entire spectrum of complexity, including expressly dynamic systems in the quantum domain. Secondly, we estimate a major error the move to make temperature, from average speed of atoms and molecules previously, a direct function of heat energy, concordant with its view that temperature is "heat-driven". It is essential to acknowledge that it is not temperature that is "heat energy-driven" but heat energy that is driven by the impetus of the physical variable of temperature in its thrust to create topological continua. Not knowing any better, we have understandably always put the physical variable of temperature, both theoretically and empirically, on the scale of heat or thermal energy *(kT)* while this variable is in fact a property of coordinate space. Consequently, our absolute thermodynamic scale of temperature is materially untrue to nature.

Furthermore, whilst we understand the laudable objective to move away from physical etalons and embrace fixed mathematical quantities for the values of natural experimental constants, the Boltzmann constant does not proceed outright from Theory or theoretical axiomatization, and does not legitimately represent a fundamental constant. Notwithstanding the fact that the empirical formula from which the experimental Stefan-Boltzmann constant is computed contains the Boltzmann constant *k*, the latter is not to be consequentially considered a precursor of the Stefan-Boltzmann constant since Theory only can establish priority in the order of these constants. From theoretical insight emanating from our Framework, the best and most legitimate etalon for temperature is not the Boltzmann constant but the Stefan-Boltzmann constant, since it is the one only constant that, in the theoretical derivation exercise, transpires within the energy metric constellation. This conclusion is even borne out in the logic of a few additional experimental facts, which we shortly visit.

The Boltzmann constant establishes the minimal relationship that must exist between **mechanical Energy** and temperature in terms of the distribution of **mechanical** Energy over the temperature scale *J/K*. It tells us that this ratio must be equal to $1.380649 \cdot 10^{-23}$. The Stefan-Boltzmann constant establishes the minimal relationship that must exist between **radiative Energy** and temperature but in terms of the distribution of **radiative Energy** over the temperature scale *E/T* (*T* expressed in kelvin). It states that the ratio between radiative Energy and the 4th power of temperature must be equal to $5.6705 \cdot 10^{-8}$, the relation being:

$$E(T) = \sigma T^4 \quad (20\text{-}4)$$

Although Ludwig Boltzmann was able to infer this formula from thermodynamics in 1884, there is a tight relationship between this formula and the Planck formula for distribution of radiative energy in a black body. The T dependence of radiative energy in *(20-4)* can be inferred from the Planck formula itself:

$$u(f,T) = \frac{(cf)^3}{8\pi h(e^{hf/kT} - 1)}$$

by integrating over the entire range of frequencies. The integral follows:

$$u(T) = \int_0^\infty \left(\frac{8\pi f^3}{c^3}\right)\frac{1}{(e^{hf/kT}-1)}df$$

The result of this definite integral is of the form:

$$constant \times T^4$$

exactly as in the Stefan-Boltzmann relation. This leads us to further consider incorporating the area covered by the radiation in order to give its full dimension to the temperature variable under scrutiny, consistent with the unit-dimension of the constant *($W\,m^{-2}\,K^{-4}$)*. In doing so we are brought to inquest the Energy per unit area radiated by the black body, as expressed in the relation below:

$$W(x) = \frac{\mathrm{k}^4 T^4}{4\pi^2 c^2 \hbar^3}\int_0^\infty \left(\frac{x}{e^x - 1}\right)dx,$$

where we have replaced f with hf/kT and then with x altogether to simplify the syntax of the expression. Here again the presence of the T^4 term is unmistakable.

With this result we must reasonably and forcefully emphasize the close association between the Stefan-Boltzmann constant and the **quantum** distribution of radiative energy density as it occurs in a black body, which definitely replicates the same parallel association that our Theoretical axiomatic framework has come to exhibit between the Planck constant and the Stefan-Boltzmann constant (Book I).

What's more, it is crucial to realize that black body quantum distribution occurs at the point of equilibrium or uniformization of temperature throughout the black body and the radiation, a significant fact from the standpoint of the ontic role that we have assigned to the temperature variable. The variance of the Energy v. wavelength distribution is uniquely dependent on the gradient of uniform temperature.

Those facts bear capital significance from a normative viewpoint as relates to regulatory standards specifications. Altogether, in this effort to set a towering definition of the temperature variable in physics, they confer higher status to the physics of black body particulate radiation over the physics of the triple phase point of water molecules with which the Boltzmann constant is more aligned as a prime element of the Kinetic Theory.

20.17. Quantization of Heat Capacity

In this review of particular aspects of quantum physics from the vantage of the QG framework, we opportunely remind the reader of our discussion in Chapter 16 of Book II, about discreteness within the scale of thermodynamic temperature. We've made the case for a quantized or discontinuous nature of this absolute scale, despite its role of promoting continuum through topological equalization and uniformization.

Not all temperature differentials $[T_2 - T_1]$ in the absolute scale are always allowed. This known fact has long been overlooked for the most part. Said differentials are tantamount to expression of discreteness, or more precisely discontinuity, in the temperature scale. Effectively, there is a known limit to heat capacity for molar populations of substances, which is **6.96 cal deg^{-1} mol^{-1}** for no apparent or classically justifiable reason. Whenever this limit is approached from incremental changes at very high temperatures, the molecules in the population (the topological singlets) would simply dissociate as heat containers. The reason for this behavior cannot admittedly be explained in the classical continuum vision.

Likewise, it is well known that different topologies of **diatomic molecules** of substances in general have their heat capacity values restricted to a limit in the range between ***4.97*** and ***6.96 cal deg^{-1} mol^{-1}***, which equates a direct restriction on their reachable thermodynamic temperature.

These two restrictions in thermodynamic continua admittedly can only be explained in quantum mechanical terms on the basis of an intervention of ***hν*** in the expression of heat capacity, per the known applicable formula below for their heat capacity:

$$\bar{C}_{V,vib} = \frac{R(h\nu / kT)^2 e^{-h\nu / kT}}{(1-e^{-h\nu / kT})^2}$$

Therefore, even when heat capacity is factored in in the ontic analysis of the temperature variable, the latter comes out as a quantized or quantization-sensitive variable.

20.18. Why Establishing a First-order Definition of Temperature is Paramount

The benefits of establishing an adequate and accurate definition of the temperature variable are many. We submit in this respect the following considerations:

1.- Insofar as we can logically see, complexity does not call for two different absolute scales of temperature. Any form of representation that implicates one scale for the quantum domain and another for the macroscopic world, or claims incongruity between the two, is erroneous in appreciation and misunderstands the nature and discreteness quality of the absolute temperature scale as well as the role of the temperature variable in nature's physics.

2.- It is utterly improper to opt for establishing a selective or arbitrarily fixed quantity for a fundamental constant and yet err in the selection of that quantity.

3.- Whereas one may with good reason qualify the periodic experimental remeasurements of the values of fundamental constants procedural nuisance, it is of great import to avoid readjustments or changes in the value of fundamental constants that have been numerically set for the very sake of avoiding those periodic adjustments.

4.- Not possibly getting the scale of temperature with best accuracy will be of significant consequence to quantum technologies which typically require deep dives into the bottom region of the scale and show great sensitivity to gradient variations. They run the risk of navigating with a skewed compass, so to speak, and completely losing the ability to identify what is amiss when in troublesome haphazard.

5.- An absolute scale of temperature that is true to nature and applicable across the entire scale of complexity must be strictly based on the Stefan-Boltzmann constant. We further elaborate on the numeric value of that quantity. Whether we still designate the variable thermodynamic temperature or not will be immaterial to its accuracy and the absoluteness of its class.

We will intently not finalize this section without noting that up until CODATA 1998, the value transcribed by the NIST for the Stefan-Boltzmann constant was:

$$\sigma \approx 5.67 \times 10^{-6}$$

while many authors such as Fisbane – Gasiorowicz – Thornton *(Physics for Scientists and Engineers)* commonly reported:

$$\sigma = 5.68 \times 10^{-6}$$

up to that point, their figure coming from other laboratories which were then the only ones performing measurements for this constant (not the NIST). That value is in better figurate match with our theoretically derived value.

20.19. Josephson Tunneling (Super Low Temperatures)

Quantum mechanical tunnelling is knowingly a phenomenon for which the classical concept of force fields, force barriers or pressure have no meaning whatsoever. A conceptual adjustment from 19th century classical Kinetic Theory and Thermodynamics is required in order to properly deal with quantum mechanical phenomena, let alone post-quantum physics. The dependency of quantum tunnelling on the absolute temperature scale is of utmost significance in this context and sets a key condition for the oncome of such phenomena. It must be well understood that the adventures of the quantum wavefunction, whose properties and real-valued definition stand for one of the landmarks of our Framework, play a pivotal role in the phenomal quantum domain at that level. In order to sustain continued progress in this field, we must endeavor to unpack the full range of physical properties associated with the quantum wavefunction.

While this new technological enterprise does not require a divorce with classical physics intellection, it does peremptorily call for larger receptivity to qualified action by intangibles, such as those contained in the quantum wavefunction. Only in that measure will we be able to avoid the shock element from incomplete or unexpected observables which prompt irrational explanations, as well as ease us into natural expectation values for physical variables and constructs in the new domain and save us from the anomaly of calling nature "anomalous".

Tunnelling particles of subatomic size or Cooper pairs are able to make their way across barriers even with energies too small to carry them through in the conventional or classical manner. If it is true that such particles follow the undulations of an associated quantum-mechanical de Broglie wave, then it is important to understand the connectum between the nature of de Broglie wavelength and the Compton wavelength as a characterizable wavelength (or wavenumber) spectrum, as studied in Book II.

Likewise, the analysis that renders an estimation of behavior of free electrons in supercooled metal sheets down to within a few dozen Kelvins, as an effect of how their electric charge responds to temperature in terms of a repulsive effect or an attractive effect, is more likely than not completely unassertive. The officially stipulated idea that there exists a critical gradient of temperature that delimits influence of the latter over the electric charge of the electron particles, making them attractive above that gradient and repulsive below that gradient, qualifies as no less than utterly insensible. What exactly in temperature affects charges and how has never been stipulated in physics.

20.20. Free-Electron Theory

All in all, we do not cater to the particular thermodynamic considerations stemming Fermi's theory and its reliance on the Boltzmann constant to quantify electrical conduction in and across potential wells. The problem with free electron theory thus far is that it has not proposed or ascertained any particular physical geometry for the particles' wavefunction inside of the lattice. The invoked potential wells remain abstract hypotheses with no particular indication about what the energy levels from bottom to top of the wells physically represent and remain pure mathematical entities.

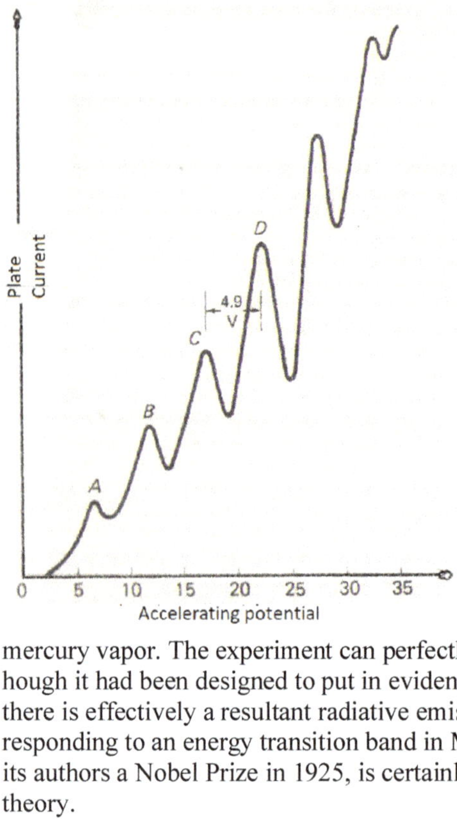

Fig. 20.3 Franck-Hertz experiment

Ever since the Franck-Hertz experiment, featuring free electrons travelling through mercury vapor at low pressure, we know that these electrons absorb pump energy in a quantized manner. In general, this occurs whether the pump energy is thermal or an electric potential. In the case of the cited experiment (setup in Fig.3.6), as the applied electric potential rises, a peak is reached in current with an unexpected sharp drop or gap thereafter, a recurrent event for every 4.88V increase in the electric potential scale (Fig.20.3). In this case, what we have are excited free electrons within the metal plate receiving a further energy kick from the applied electric potential of 60V, enabling them to leave the metal plate and circulate through the barrier of mercury vapor. The experiment can perfectly be interpreted from that vantage point, although it had been designed to put in evidence quantized energy levels in Mercury, since there is effectively a resultant radiative emission from the gas in the order of 2536 Å corresponding to an energy transition band in Mercury atoms. This experiment, which won its authors a Nobel Prize in 1925, is certainly a precursor to free electron quantization theory.

Although Fermi was instrumental in imposing the Pauli exclusion principle to superconduction electron dynamics, the effort to assertively and fully characterize the state of these electrons has hitherto remained an ongoing development. The so-called Fermi energy is set to represent the top level of the potential well which provides the energy necessary to settle conduction by setting the electrons in that state in motion. It is known that the effect of temperature changes on behavior of electrons having reached their particular lattice Fermi energy is negligible. For instance, in copper the contribution of heat energy for a change of *100-K* is only $0.9 \cdot 10^{-2} eV$ compared to *7 eV* contribution of Fermi energy. Thus, the important resistivity characteristic for conduction is independent of thermodynamic temperature, for reasonable temperature changes. In other words, the effect of temperature changes on the behavior of free electrons with a given Fermi energy is negligible and will remain so for any temperature below this critical level or T_C on the absolute scale.

Very significantly, it is only when enough heat energy has been removed and we let their temperature drop below what has been designated the critical temperature *Tc* that the regularities start appearing. What this clearly means is that the free state of individual particles within what can be designated at least an inter atomic environment is not a ground state but an over excited state. And because they do not demonstrably radiate or constantly lose energy in that state, they must be thus in a dynamically stable configuration. Further, these electrons are not seemingly stray or spurious entities within the metal lattice, since there may be about dozens of them for each atom in the lattice. Their sheer comparative number calls for an ontic attribute as to the material.

Fermi's original theory, setting the modality in which electrons were filling potential energy wells from received heat energy kicks, notably could not explain the capital difference between conductors and insulators. However, the ion attraction theory that followed and still survived in the Hubbard model is quite controvertible in our sense and in dire need of correction for further progress in post-quantum physics. In 1957, Cooper et al theorized that:

- *all tunnelling electron pairs within the metal sheet or bulk are in the same quantum state.*
- *they have all ontically transmuted from free fermions to bosons in the one single superposition state.*
- *a gap has developed around the Fermi energy.*
- *superconductivity relies on electrical interaction between free electrons and the vibrating lattice.*

as contended in the so-called BCS Theory.

The closest that two electrons in a Cooper pair can presumably approach one another is said to be half of their de Broglie wavelength, since any closer distance would superimpose the electrons, which is forbidden by the Pauli exclusion principle. So therefore, to calculate the Fermi energy level of a pair one takes one half of the closest distance *d* that each particle can approach each other, which corresponds to the Fermi momentum p_F:

$$p_F = \sqrt{2m_e E_F}$$

$$d = \frac{\lambda_F}{2} = \frac{h}{2p_F}$$

$$d = \frac{h}{2\sqrt{2m_e E_F}}$$

As contended by the Theory, if Cooper electrons are bosons, it is correct to consider their wavelength to be of the de Broglie nature, and not a Compton class wavelength.

However, consequentially considering a maximum distance they can approach each other to avoid superposition, which is fully permissible to bosons, seems quite inconsistent.

Further, the assumption that all the Cooper pairs are in the same coherent quantum state, implicating that all their spin vector pars are in the same direction and orientation, does not seem to have been sufficiently well established.

In BCS Theory, superconductivity is predicated to rely on electrical interaction between free electrons and the vibrating lattice where they reside. The theory asserts that, for complicated reasons, the electron-ion interaction inside the lattice leads to a weak attraction between those particular electrons whose energies are close to the specific Fermi energy for the lattice. Leon Cooper hypothesized that a weak bond resulted from this mutual attraction. It is not clear though why these electrons do not end up mutually colliding, because their apparent fate would not be any different than the collision that would result in a Hydrogen atom from attraction between proton and electron. In Quantum Theory, we eponymously know that electrostatic interaction between the two particles in the Hydrogen atom cannot thoroughly explain its stability.

Furthermore, the invocation of electrostatic interaction to explain the stability of Cooper pair is incongruent with the stability of other well-known entangled pairs such alpha particles which are equally capable of tunneling through an otherwise impenetrable nuclear barrier. Alpha particles, formed by neutrons that have joined with one another at some point within the confines of atomic nuclei before ending up tunnelling through the energy barrier of the nucleus as pairs, are knowingly chargeless. Both neutron particles and alpha particulate aggregations are chargeless. The further assumption, in the case of electron particles, of an electrical interaction with the ionic lattice blocks of the metal bulk to produce the tunnelling effect is thus unfounded as well.

In order to fully complete the conduction theory and explain the difference between conductors, semiconductors and insulators, the theory had called upon electron-proton interaction adjoined to conduction bands inside potential wells centered on adjacent proton-filled nuclei. The existence of tunnelling chargeless alpha particles totally and summarily invalidates this hypothesis as well. Thus, it is unnecessary to revisit the details of the latter long-lived hypothesis.

20.21. The Hubbard Model

The Hubbard model, proposed since 1963 by independent researchers, is an approximation model, within the inspiration of BCS Theory, essentially aiming at describing the transition between conducting and insulating systems. It states that each electron experiences competing forces: one forcing it to tunnel through neighboring atoms, and another repelling it away from its neighbors. Its Hamiltonian thus showcases a kinetic term allowing for tunneling or hopping of particles between lattice sites and a potential term reflecting on-site holding interaction. In this model, conduction particles can either be fermions, per original Hubbard's work, or bosons, in later contributory work.

In its effort to formally describe the conduction panoply found in diverse materials, the model contemplates a contact interaction between particles of opposite spin on each site of the lattice. When the Hubbard model is used to describe electron systems, it expects these particular interactions to be repulsive, coming from the so-called screened Coulomb interaction. Inversely, it attempts to contemplate at the same a holding behavior hypothetically resulting from attractive interactions. The physics of the Hubbard model is determined by competition between, on one side, the strength of the dynamic hopping of the particle captured by the integral which characterizes the system's kinetic energy, and, on the other side, the strength of the interaction term characterizing the potential energy. In that way, the Hubbard model can more or less explain the transition from metal to insulator behavior in certain interacting systems.

The Hubbard model is not universally successful and only correctly predicts the existence of the so-called Mott insulators. The latter are materials knowingly insulating due to strong repulsion between electrons, while they satisfy the usual criteria for conductors, such as having an odd number of electrons per unit cell. The model is not successful at describing complex systems beyond a 1D chain of hydrogen atoms, since those systems admittedly experience other effects not contemplated by the model. Yet other systems exhibit conduction characteristics qualifying them as so-called charge-transfer insulators which the perspective of the model is unable to embrace.

20.22. Cooper Pairs Exist in Insulators as Well as Superconductors

In late 2007, physicist Michael Stewart made the surprising discovery that Cooper pairs can show in insulating materials and that in conductors they can be turned to block electron flow contrary to superconductivity. To create an insulator for his experiments, Stewart chose bismuth, a rare metal which serves as an excellent superconductor when thick, and, when thin, behaves as an exceptional insulator.

A template honeycombed with holes measuring only 50 nm in diameter was coated with a layer of bismuth just four atoms thick and cooled to super-low temperatures. This material could be transformed into either a superconductor or an insulator. The holes in the material are what allowed the insulation instance to develop. Cooper pairs then formed, but stayed segregated in these whirlpools, preventing the pairs from making up a continuous line of current. Cooper pairs are thus present in both superconductors and insulators, but behave differently in each instance. In superconductors, pairs link up with other pairs and move in a linear fashion to create a continuous stream of electric current. But in the insulating film, the pairs presumably spin solo and do not connect. In that case, if and only if a magnetic field is applied, the insulating film turns into a conductor.

The problem of making superconducting electrons go straightforwardly in one direction without the application of a magnetic field to the insulator remained a challenge however, until early 2022 when Mazhar Ali et al probed the material Niobium Bromate as a Josephson insulating layer no thicker than just a few atoms *($NbSe_2/Nb_3Br_8/NbSe_2$)*. The material made possible for the first time a one-directional flow of Cooper pairs through an insulator without the intervention of a magnetic field, materializing a violation of the well-established Josephson relations. This characterization is a gentle way to assert that Mazhar Ali's late discovery and other posterior similar JJ research in the same year invalidate all post-quantum physics that sought roots in formerly established principles for supercooled junctions. All throughout 20th century quantum physics, no one ascertained the basis for aligning superconducting electrons in just one direction. In the 70s, scientists at IBM attempted implementation of superconducting computing but stopped the effort as they formally concluded that, without non-reciprocal superconductivity, a computer running on superconductors is impossible. In 2007, Cooper pairs in insulating materials was a totally counterintuitive proposition to Michael Stewart who by own admission did not at all expect any confirmation of his experimental premise and was totally dismayed by his discovery, given hitherto common knowledge on the subject. What this means is that even when government and industry pressure are weighing in, one shall never ignore, neglect or avoid questioning the fundamentals and must hold the benefits of a rightful and adequate theoretical framework as inestimable value.

20.23. Superconductivity Energy Gaps and the Complexity Scale

A key aspect of BCS Theory which has not been sufficiently interrogated is the so-called energy gap preceding the Fermi energy level. It is situated at a temperature gradient preceding T_c, should we consider the changes in temperature going from bottom of the scale (near *0-K*) upward, an inverse modality to that which we actually execute temperature variations experimentally. For superconductors, the energy gap is a region of suppressed density of conduction states near the Fermi energy, with a much smaller size than that of the energy band beginning with the Fermi energy level. The size of the energy gap stands for the energy gain for two electrons entering into a Cooper pair association. BCS theory predicts that the size of the energy Δ in the gap for conventional superconductors at *0-K* is given by the relation:

$$\Delta E_{T=0} = 1.764 \cdot k_B T_c$$

where E is the energy, k_B the Boltzmann constant and T_c the critical superconducting temperature. Upon further manipulation of the formula:

$$\frac{\Delta E_{T=0}}{k_B T_c} = 0.1764 \cdot 10^1 \quad \textit{(4-5)}$$

The number at the right is a proportionality constant that represents the invariance in the ratio between the Energy gap size and the thermal energy $(k_B T_c)$ at the critical superconducting temperature for all metal lattices. The reader will recall that, within our theoretical derivation framework of fundamental constants, this number comes out as a figurate of the first derivative of the Grand Eigenfunction relating the space spread wavefunction to the scalar (Book I) or the *Electron Gyromagnetic Ratio:*

$$\left[\frac{dQ(s)}{ds}\right]_{s=\sigma/2} = 0.17603266 \cdot \sigma^{-2} \; (4\text{-}6)$$

This figure is experimentally known in the following unit-dimensions:

$$0.176085974 \cdot 10^{10} Rad\ s^{-1} T^{-1} (4\text{-}7)$$

Remove the unit-dimensions, which are arbitrary human constructs that happen to be right at the heart of the so-called *Hierarchy Problem,* and keep the numeric figurate as a natural quantum abstraction in the mathematical numeric order. The closeness in numeric figuration between the two quantities in *(4-6)* and *(4-7)* is as undeniable as it is meaningful.

This observation brings us back to the discussion about the inadequacy of the current absolute temperature scale. Let's first note that the above firmly establishes that the delta gap or energy packet gap after which superconductivity always starts is a multiple of the spin magnetic moment of the particle itself. In other words, superconductivity develops on that energy scale. At the same time, this also means that gradient *0* of the absolute temperature scale does not reside here, if anything because the spin of the particle is an irrevocable ontic attribute of the particle. By the same token, this calls back and justifies our contention that temperature is a function of motion itself or *space momentum*, and does not ontically pertain to the scale of heat energy. What is thus certain is that the beginning of the *natural* absolute scale of temperature transcends the atom Shell and goes all the way down to the topology of the particle Shell. The natural absolute scale of temperature is indeed based on the Stephan-Boltzmann constant as its unit as proposed, which is the reciprocal of the above quantity, indeed the inverse functional derivative of the Grand Eigenfunction at the same higher metric point *(s-metric point of* $s = \sigma/2$*)*.

So therefore, it is essential to realize that the invocation of energy gaps in the phenomenology of Cooper pairs cannot be made without the larger context of the complexity scale and, very particularly, the boundaries between its stages. In this respect, the Fermi energy is not to be considered one energy level within a continuous flat sheet of energy levels, as if energy was in and of itself a first-order homogenous scale. It is important to understand that the Fermi energy gap delimits two different ontic provinces associated with the material wavefunction of quantum objects. Below the Fermi energy gap is situated the realm of atomic wavefunctions, and above the Fermi energy point is situated the realm of molecular and lattice wavefunctions.

In order to thoroughly understand the dynamics taking place in this region of complexity, one has to consider the scaffold of mathematical-physical quantities theoretically derived in our Framework (Book I) and their relationship with one another. It is no less important to represent the implications of the unit-dimensions known to the same quantities as to their primacy in allowing figurate numbers that match their pars in a theoretical mathematical-physical derivation of these fundamental constants. Dimensional analysis this kind is essential, should one harbor any longing to comprehend the scope and magnitude of the *Hierarchy Problem*. We will however keep a limited and concise discussion here.

First and foremost, two primordial quantification streams converge in order to configure the build out of this incipient microscopic region of complexity. One is the continuum stream of which temperature, regimented by the Stephan-Boltzmann constant, along with the Electron Gyromagnetic Moment constitute the pilot. The other is the discretization stream of which the quantum of action, regimented by the Planck Constant, altogether with the Electron Compton wavelength constitute the pilot. In each rubrique, the second is the reciprocal derivative of the first and becomes the natural scale for the first. When computed in standard *eVs* units, the Planck constant *(413.27 eVs)* yields a figurate number that is the reciprocal of the Electron Compton wavelength *(0.2419 nm)*. These long unsuspected matches go a long way to decisively defuse the *Hierarchy Problem* in quantum physics.

Cooper pairs partake of the Atomic Propagator Function that creates orbitals within this province of complexity, the Atom to Molar Lattice Shells, that is. Once formed at the bottom of the natural absolute temperature scale, Cooper pair orbitals are no less stable than the commonly known orbitals in atomic elements at much higher levels of the temperature scale, which create the topologies in the observable natural world. The reason why everything about the behavior of Cooper pairs is reminiscent of standard atomic orbitals is simply because **they are** atomic orbitals. *If it looks like a duck and walks like a duck, it's a duck!* They ontologically pertain to this environment as legitimate integrant of the lattice Shell and partake of its topological properties. They simply cannot remain stable when drowned in the elevated thermodynamic conditions that are current in lattice topologies in the natural world out there. In these instances, the bond between the entangled pairs of electrons breaks down for them to become excited free singlet electrons.

Our perspective leads us to the ineluctable assessment that under the artificial experimental conditions of super low temperature on the absolute scale, free particles thus lose a drastic amount of energy, sufficient for their wavefunction foliate to be brought back to the lowest level of energy within this sector of the complexity Shell spectrum ending up with higher-order lattice topology. In that environment, the lowest energy level is an atomic orbital shell. It has been said that Cooper pairs have less energies than free electrons because it takes some energy to separate the pairs. What is more important and indeed capital to represent is the magnitude of the implosion in their degrees of freedom within the lattice, which took them from the degrees of freedom with higher-order signature all the way back to reduced degrees of freedom with orbital signature. We refer the reader to our extensive study of Material State Topologies in Book II. Two observations now become relevant at once:

- The development of complexity requires, as abundantly stated, the assembly of several objects pertaining to one stage of the scale in order to galvanize them into forming one unity object pertaining to the next stage of the scale under regency of its own coupling constant. Consistent with this primal principle, the degree of freedom showcased by the atom presents with an aggregation of three different particles at different levels: neutrons, protons and electrons, whereas that showcased by Material lattices exhibit a molar aggregation of interlacing lattices.

- Supercooled quantum tunnelling always occur with paired particles, a property attached to orbital formation within the atomic edifice, which has been termed the Pauli Exclusion Principle. This constraint occurs across the entire atomic edifice, as evidenced by tunnelling alpha particles (2-neutron aggregates) originating from the nucleus, as parallels to electronic Cooper pairs of our main focus.

Hence, these two observations drive us to the conclusion that in the quantum tunnelling transformation, *the original molar degree of freedom experienced by the particles is slated to break down, through negative symmetry-breaking, into an orbital wavefunction, which duly requires additionally the two particulate objects to be in orthonormal states with respect to their spin vectors.* We will shortly come back to the orthonormal condition.

The two particles now appear to be a lot less sensitive to the presence of surrounding blocks (or lattice sites) because their orbital dynamics take them through fewer layers of these obstacles throughout the lattice interstice. Same holds true for the topological occupancy of the vicinity of the metal bulk. That is the main reason why these so-called tunnelling particles are equally able to travel over certain distances without hitting the so-called lattice sites in the insulating barrier. It is obvious that the more rarefied the vicinity of the metal sheet at its exterior boundary, the easier the degree of freedom in orbitalization can develop, or the more unimpeded the supercurrent flows of tunnelling electrons can develop. Effectively, the limitations on the size of the barrier normally constituted in a layer of juxtaposed non-superconducting material are well known in a Josephson junction (JJ):

- If the barrier is made of an insulating material, it ought to be in the order of *30 Angstroms* (3 nm) to be permeable to the supercurrents.
- If the barrier is made instead of another metallic material, it can be no thicker than *several microns* ($1 \times 10^{-6} m$ or *1000 nm*)

These are the limits imposed to the thickness of the second layer for the establishment of the tunnelling effect cross materials in JJ's. The movement of the pair of tunnelling electrons develops according to the dynamics innate to quantized orbitals, subject to the lattice structure of the barrier. However, the quantum orbital signature of the tunnelling electron pairs is primarily responsible for the uncharacteristic flow devoid of resistance attributed to the electronic supercurrents, whether within the superconductor or across the JJ barrier. *The motion of these electrons is eponymously momentum-less and quasi relativistic.* It is not an attraction exerted on electrons by ions within the lattice that creates this motion as BCS Theory has it. It is a leap of faith to have the Cooper pairs as electron pairs occupy all collectively the same state, as coherent as one might want to qualify it. BCS Theory accordingly hypothesizes that Cooper pairs are bosons, apparently because they are spread around in an otherwise eminently fermionic environment: inside the superconductor, Cooper pairs can travel as far as *100 nm*, whereas the interstitial distance within the fermionic lattice may only go between *0.1* and *1 nm*. First of all, it would be difficult to reconcile a bosonic character with the attributed charge-based interaction with the lattice ions where there are. Second, it is opportune to remember that orbitals are of infinite spread, even though they are constrained to a most probable contour of distribution, which in *1-s* orbitals is anchored on the Bohr radius of *0.05 nm*. In this picture, the momentum-less motion of an orbitalized particle becomes relevant in that it will perfectly account for irregular trajectories along the *100 nm* travel distance.

Furthermore, we well know that electrons left to themselves do not spontaneously flow in currents, such as those around cathodic electrodes caused by the heating element in our older vacuum tubes, unless a potential gradient is imposed. By contrast, Cooper pairs in superconductors flow on their own, do not lose energy and do not require constant application of potential energy for their supercurrents. In fact, an evacuation of "heat energy" is required for their establishment. All of these are imperative atomic orbital attributes.

From there, it is also not difficult to see how orbitalized electron pairs that traverse the barrier could induce superconductivity among free electrons in a third layer of superconducting material juxtaposed to the insulating layer, as fully realized in the 3-layer Josephson device. Chemical bonding, which we are so very familiar with, is a clear model for this induction representation. Effectively, valence bonding and ionic bonding are realized on the core basis of orbitalization between electron pairs of the exchange. Effectively, recently in 2021, Peizhi Mai et al presented a study arguing for molecular orbital structure to superconducting cuprate pairs at high temperature, and based on the Hubbard model of free electrons within lattices. Ultimately, the induction of supercurrents in the 3rd layer can be visualized as the propagation of an orbitalization stream ignited by the tunnelling orbitals traversing the middle insulating layer.

Finally, if our analysis is correct, the orthonormal character of the tunnelling electron pair should be in some way detectable. We know that electron spin is a property that is preserved in an atomic orbital. In a non-saturated orbital (with only one electron), the spin of the particle transpires as a magnetic moment with one of two orientations. In a saturated orbital (with two electrons), the quantization of the electron spin in only two possible orientations of the spin vector and the mandatory orthogonality of their spin vectors produce a cancellation of the magnetic moment emanating from the electron pair of the orbital. Well, consistent with this rule, the superconducting state in a metal knowingly expels all magnetic fields (ferromagnetism) from the material! Provided the fields were not too large, they will completely disappear once superconductivity sets in, an effect that had been designated the Meissner effect after the name of its discoverer. This result must be put on account of the orbital configuration of the Cooper pair as it is rarely appreciated, if ever.

Furthermore, it is capital to recall that the Meissner effect plays a pivotal role in quantum tunnelling phenomenology, and historically constituted the polemic experimental observation that led to or imputed BCS Theory. Given that the Meissner effect is common denominator to all types of superconductors without any exception (Type I and Type II), one must concede that, more than a side effect, it is central to the establishment of quantum tunnelling phenomena. Although BCS Theory has become mainstream quantum tunnelling Theory, a few theoreticians, notably physicist Jorge E. Hirsch, still adamantly argue that BCS Theory fails to adequately explain the Meissner effect, if at all. J. E. Hirsch recognizes that, while the multiplicity of current theories describes the initial and final state of tunnelling streams, they cannot intimately describe the process by which the system dynamically evolves from the initial to the final state. He further proposes that:

"Unless and until proponents of BCS-London theory explain which force in nature will propel the charge near the surface to move in the azimuthal direction and overcome the Faraday counter emf to generate the Meissner current, the Meissner effect will remain unexplained within the conventional theory. Since all superconductors exhibit the Meissner effect, we must conclude that the conventional theory in its present form does not apply to any real superconductor."

By contrast and as one can easily see, within our Framework, the universal superconducting Meissner effect finds complete, direct and consistent analytic support.

20.24. Caveats of Magnetic Flux Quantization

The notion that in physics we can freely combine natural constants to form new ones at will is an all-time unfortunate practice that betrays and underestimates ignorance of the origin of the natural experimental constants, and most significantly, how the progressive Shells of complexity form a hierarchy rooted on unique symmetries. When physicist von Klitzing combined the Planck unit of action h with the unit of charge e to form what has been designated the von Klitzing constant $R_K = h/e^2$, he was not violating any mathematical rule of association, but certainly dismissing the natural relations there might otherwise be between the two constants, which cannot be experimentally deducted. Originally, there is only an awareness that the quantum Hall effect, which the electron charge e is meant to characterize in the formula, and the quantization of magnetic flux, which the constant R_K is meant to represent, both standing for so-called emergent phenomena associated with thermodynamically "large number of particles". The view of emergent phenomena brought about by large number of particles is precisely the view that takes us to the Hierarchy Problem. Sheer increase in the number of objects does not create complexity but the symmetry-breaking that gives way to the larger assemblies of objects, an artifact of the wavefunction and not scalar parcels themselves. Even when the Avogadro number appears inextricably mixed with crystal lattice formation, it is not the amount of the scalar parcels that directly triggers the formation of the new degree of freedom but an explosion in the topological wavefunction as that particular number is reached to create the molar entity. That is ultimately the reason why the Avogadro number does not directly and naturally emerge from the mathematical-physical derivation framework that we have developed. Therefore, the unease that some may feel for mixing a constant that typify the microscopic with another that characterizes the macroscopic as pertains to the Hall effect in sizable solids is not only justified but points to a bigger structurally significant problem.

The same remarks can be extrapolated for the so-called Josephson constant K_J with the formula:

$$K_J = \frac{2e}{h}$$

The Josephson constant attempts to characterize the situation above T_c, when the potential gradient between the edges of the solid conductor starts varying and turns into a sinusoidal wave. The constant purports to represent an invariant ratio of frequency over voltage, found to be *483,597.84841698* $\boldsymbol{GHz\ V^{-1}}$.

Now, of course, there is no reason why one should not heed the inverse ratio of the Josephson constant $(h/2e)$ and call it K_J^{-1}. The official science has decided to call it instead the *magnetic flux quantum*. This is by all accounts an abusive designation since it tends to suggest the existence of a natural unit of magnetic field. We must remember that the Hall effect, which is the origin of these quantities and represents an asymmetric *side* or lateral effect of charge density resulting essentially from the geometry of the conductor, knowingly exists as well at much higher energy levels in classical electrodynamics. So therefore, one should never be disabused in confusing the so-called *magnetic flux quantum* with a *magnetic monopole* which would have to exist for any unitary construct to become sensible both from the classical and the quantum tunnelling vantage. In a manner simply akin to the unit of charge *e* giving rise to the electric field and its flux, shall we add. Magnetic monopoles, however, do not exist. Should one object to this characterization, then they would also have to throw away Maxwell's equations with their magnetic $\nabla \oint$ operator.

In all, we should attempt to deter from but not confuse a fictitious Φ magnetic monopole with a magnetic flux quantum.

20.25. Novel and Effective Exegetic Framework for Quantum Tunneling Technology

From first half 20th Century to date, there has been no lack of facts and artifacts concerning quantum tunnelling as to elements or compounds of nature susceptible to exhibit relevant effects this class, their electrical dynamics, electromagnetic and thermodynamic behavior, etc. The field has been truly abuzz for the last few years and appears to have gained a significant amount of traction through public policy. What is unfortunately still amiss is the discipline of a solid, parsimonious, thorough and unique Theoretical framework able enough to consistently and consensually align all the pieces of this puzzle, which are many, in order to properly tailor the advances demanded by culture in this field.

One of the most significant hurdles in this landscape has to do with the dynamics of magnetic fields as they intervene in this technology complex. The symmetries in super-current flows as well as the surface asymmetries of the same represent significant obstacles as well. We believe that the contributions of nanomaterials' lattice geometries and their inherent degrees of freedom are central to quantum tunnelling and have been by far insufficiently understood.

	Particle	Constant
M①	Quarks	no invariant
M②	Particles	α alpha (magnetic constant) –
M③	Atoms	G^-1 gravitational constant –
↓M④↑	Molecules	e electric charge constant –
M❺	Materials	a Van Der Waals constant–
M❻	Karyots	life constant –
M❼	Celestial Singlets	unknown
M❽	Solar Systems	unknown
M❾	Galaxies	unknown

❺①	Amorphous Solids
❺②	Crystalline Solids
❺③	Powders
↓❺④↑	Waxes
❺❺	Colloids
❺❻	Oils
❺❼	Aqueous Liquids
❺❽	Condensed Vapors
❺❾	Volatile Gases

Fig. 20.4 A tabular representation of how the 5-Operator and *M* act on the nano Materials spectrum

It is our assessment that the surface configuration of current densities has more to do with lattice infrastructural geometries imposing particular configurations to current density distribution than strictly charge-based interaction with sites of the lattices. In this context, the study of superfluidity with respect to the degrees of freedom d encountered in crystals represent a mandatory pre-requisite to the thorough study of supercooled quantum tunnelling. In that regard, our framework prescribes (Fig.20.4) the first chapter of this experimental study to be quantum behavior of matter in each of the 9 states representing the Quanto-Geometric spectrum of condensed matter, namely:

1. Amorphous Solids
2. Crystalline Solids
3. Powders
4. Waxes
5. Colloids
6. Oils
7. Aqueous Liquids
8. Condensed Vapors
9. Volatile Gases

The Van der Vaals $\boldsymbol{a}$ constant constitutes a universal state constant for the Materials Shell (aka nanomaterials). The eponymous Van der Vaals term $[\boldsymbol{a} \cdot \boldsymbol{n}^2 / \boldsymbol{V}^2]$ captures $\boldsymbol{a}$ as a constant of proportionality between the *mole (n)* as a scalar and *Volume (V)* as a physical variable. This scalar represents in fact a mathematical-physics quantity in that it is the count of a group of physical elements. As such it is tantamount to the scalar function of this Shell, and represents molar mass unity. That molar singlet, rooted in the mole of Hydrogen molecules H_2-H_2 as we have seen, takes thus the value of a natural unit of mass specific to this Shell. Consequently, the state function here is about how a molar bulk of molecules relates to another molar bulk of molecules.

It follows from this view that the most important aspect of molar topologies in terms of lattice organization is the solid structure based on interlaced lattices. It should not be confused with the interpenetrating kind of lattices, akin to the sodium chloride salt featuring two interpenetrating lattices: one consisting of sodium atoms or ions and one other consisting of chlorine atoms or ions. We later discuss the interpenetrating lattice arrangement. The molar interlacing pattern must be understood to represent the most fundamental level of organization of a large number of the 9 molar state strata above enumerated and subsumes internal molecular bonding or chemical bonding, which has forever been the prevalent view in Physical Chemistry of states. At the end, what is to be retained is that the Van der Vaals constant, invested with the Quanto-Geometric scalar/wavefunction definitions, in particular their mutual first-order and second-order covariance, is principally responsible for the interaction between molar units giving rise to lattice interlacing geometric arrangements.

From an overarching perspective, paired molar interlacing portrays a topology at the Materials Shell level that parallels the bi-localization of electrons in atomic orbitals at the Atom Shell, as well as paired (covalent and ionic) orbitals in molecular bonds at the Molecular Shell level. This interlacing molar topology is an expression of energy density state akin to orbitals even though the Pauli exclusion principle cannot apply here due to the higher degree of freedom acquired by coordinate space at the Materials Shell level. Let's keep well in mind that the evolution of this arrangement will notably give rise to the Karyotic Shell level of complexity through symmetry-breaking of the vacuum wavefunction of Materials elements. The oncome of the Karyotic Shell is incarnated in the populated kernel box as the crucible of viral and cellular organization. Pivotal element of this symmetry gauge group of complexity is a physical life constant, to be discussed and elicited in Chapter 6, in the same theoretical computational line that marshalled the extolment of the previous fundamental coupling constants.

All in all, molar energy states through paired lattice interlacing organization represents the most fundamental levels of energy distribution in the composition of nano Materials. Again, to compare with atomic energy distribution, the spectrum of 9 states represents a suite of quantized energy density levels, each higher than the previous, which sets the stage for energy density transitions for topological molar behavior. Obviously, this organization lies much beyond rotational and vibronic molecular degrees of freedom hitherto credited for the binding forces within lattice topologies. In the study of condensed matter physics, it is critical to represent the molar degree of freedom crystallized in interlacing molar lattices as a new medium with its own energy density shelling. The physical quantities that regiment this topology are principally responsible for phenomenology inside this Shell.

In attempting to formalize this vision, we erect the order of a principal quantum number n with integer values matching the eigenvalue primitives of the Quanto-Geometric Grand Eigenfunction. This quantum number n is evidently such that $1 \leq n \leq 9$ and typifies each successive energy level or stratum of the nano Materials spectrum.

The primary formal mathematical physical drivers for the nano Materials Shell are the following:

1. The Quanto-Geometric Imaginary Operator
2. The Quanto-Geometric Imaginary Tensor
3. The $M_3(s)$ Composite Function
4. The Molar Spread or Wave Function
5. The Integral $\int M_3(s) \cdot ds$

The applicable Quanto-Geometric Operator here is the type 5 or Imaginary Operator whose expression is:

$$\|\Psi\|_{a_5 b_5} = \left[\frac{-Q^2}{2q}\right] \cdot \left[\frac{2}{a_5 e^{\left(\frac{2a_5 + b_5{}^2}{4}\right)} + b_5 e^{\left(\frac{2a_5 + b_5{}^2}{4}\right)} + c e^{\left(\frac{2a_5 + b_5{}^2}{4}\right)}}\right]$$

This assignment is dictated by the quality of the *normal* corresponding to the nano Materials stratum within the primary Shell spectrum of $\boldsymbol{U_3}$, our inhabiting World Plane framed on ***3 omniversal* degrees of freedom**. This specific operator is the parallel equivalent of the quantum Hamiltonian that would be otherwise applied to any dynamic object in the nano Materials shell space. The first term expresses the scalar state of the object and the second the wavefunction or eigenform associated with the same.

The applicable Quanto-Geometric Tensor here is the type 5 or Imaginary Tensor whose expression is:

$$g_5^2 = \int_{\sigma\sqrt{8/5}}^{\sigma\sqrt{10}/2} \sqrt{1 + \left(\frac{dQ(s)}{ds}\right)^2} \cdot ds$$

It is a 2-space tensor, which is sufficient to describe the de-ploy or foliation degree attributable to the nano Materials shell space, and which is otherwise denoted the stress tensor in relativistic parlance. More generally, it describes, as previously discussed, the mutual constraint exerted on each other by the scalar tenet and the space spread wave-function characteristic of nano Material objects of this $\boldsymbol{U_3}$Shell. Superscript *2* in *g* denotes a 2-space tensor while the subscript *5* denotes the specifically concerned quanto-geometric stratum within the Shell. The interval of integration expresses the upper and lower limits within the space spread metric that circumscribe the symmetry-breaking points for this Shell space.

The $M_3(s)$ composite function is the very function that computes the paramount invariant pivoting the symmetry encapsulating the building block of the Shell:

$$M_3(s) = -\frac{\sqrt{7}}{4\sigma^4 \cdot \sqrt{2\pi}} \cdot e^{-\frac{s^2}{2\sigma^2}} \cdot s^3$$

From the higher supermetric of *s* delimiting $\boldsymbol{U_3}$ Shells, where $s = \sigma\sqrt{\frac{8}{5}}$ for Shell 5, we obtained by replacement:

$$M_3(s_5) = 0.239962558/\sigma$$

which is our derivational value of the Van der Vaals constant applying to the nano Materials Shell. The reader shall keep in mind the axiomatic computational origin of the aforementioned coefficient $k_P = 0.661437$, which is thereby obviously far from arbitrary.

The Molar Spread Function is a descriptor of how the nano Materials' shell space develops its inherent degrees of freedom. The expression of this function is:

$$V_M = \left[k_P.\frac{dQ(s)}{ds}\right]_{s=\sigma}^2$$

This function is the square of the invariant which constructs the symmetry of the Atomic edifice. It describes the spread of *nano* space over the molar scale. Note expressly that it is not a descriptor of number of moles per space unit. That is why it duly covers the case of several moles occupying the same volume, as it is in fluids and semi-fluids with interlacing lattices. The value experimentally attached to this constant is $0.0224\ m^3/mol$. The theoretical value cast by the above equation is:

$$V_M = 0.022469420 \cdot \sigma^5$$

In order to understand how the Molar Spread Function develops the degrees of freedom proper to the nano Materials' Shell space, it will be useful to undertake a parallel with the *orbitalization function* as realized within the Atomic edifice. Those two functions fittingly feature the same analytic category in our derivational canvas.

The Rydberg energy unit, resulting from the orbitalization function, is empirically defined to be the distribution coefficient for the energy of atomic orbitals in a hydrogen atom, the first and smallest of all:

$$R_y = \frac{hc}{4\pi}\frac{\alpha}{a_0}$$

Experimentally valued at $0.1360 \cdot 10^2 eV$, the Rydberg energy unit is the incremental amount of orbital energy over the Bohr radius scale of atomic orbitals, in the understanding that the Bohr radius scale is the discrete scale of allowed stable orbits around the atomic nucleus. In fact, those radii define energy scales of varied spatial geometries since only *s* orbitals are truly radial or spherical in the maximal probability density contour. The Rydberg energy unit is thus an invariant that describes behavior of a physical variable, in this case energy, over a spatial spread. What we have discovered in our derivational framework is that it is in figuration no less of an equivalent to the square of alpha (α), the scalar-space coupling constant at the heart of the ontological composition of the galvanized particle imprinting the orbital dynamics.

The theoretical expression of alpha $(\alpha)$is:

$$\alpha = \left[k_{P}.\frac{dQ(s)}{ds}\right]_{s=\sigma/2}$$

Elevating alpha (α) to the square, we obtain:

$$\left[k_{P}\frac{dQ(s)}{ds}\right]^2_{s=\sigma/2} - 0.1355707 \cdot \sigma^5$$

which gave us a figurate number pretty close to the Rydberg orbital energy unit:

$$R_y \equiv 0.1355707 \cdot \sigma^5$$

In comparing Theory to empiricism, we find an approximation of this value by starting with an empirical surrogate of alpha in the incarnation of the *Fermi coupling constant*:

$$\frac{G_F}{(\hbar c)^3} = 0.11663787 \cdot 10^{-4}$$

Elevating the above to a square, per the theoretical square rule, we obtain:

$$R_y = \left[\frac{G_F}{(\hbar c)^3}\right]^2 = 0.136041898 \cdot 10^{-10}$$

a number that is in a better figurate match with the known empirical value for the Rydberg orbital energy unit in unit-dimensions:

$$R_y \equiv 0.136056253 \cdot 10^2 eV$$

Fig. 20.5 Structure and morphology of the Beryl crystal

Consequently, the square of an alpha surrogate, as pivotal constant for the symmetry basis of the architecture of elementary particles, gives us the **energy distribution over the quantized radial scale** forming atomic orbitals, just as the square of the pivotal constant for the symmetry basis of the architecture of Atomic elements gives us the **volume distribution over the molar scale** forming nano Material kernels. The homological parallel is irrefutable. If the constant that regiments the volume distribution over the molar scale ($V_M = 0.022469420 \cdot \sigma^5$) is a legitimate invariant, then it must be valid across the entire spectrum of Materials states for this Shell, not just for the gaseous state. Hence, there must exist a kernel organization over the interlacing lattice structure that dictates ultimate internal lattice and lattice molar structure to satisfy this constraint cross Shell, from

Amorphous Solids to Volatile Gases. In particular, for solids there must be deep lattice interlacing for the quantized volume per mole distribution to be kept satisfied.

The multifold geometric symmetry in the Beryl crystal revealed by X-Ray crystallography patently exemplifies these qualified geometric kernels (Fig.20.5). Beryl is a silicate of beryllium and aluminum ($Be_3Al_2Si_6O_{18}$), appearing under shades of yellow or green or blue, due to impurities which we further discuss below. Although Beryl belongs to the hexagonal crystal system, sensitivity to circular (in fact spherical) symmetry is evident throughout the topology as, from a frontal planar perspective, the crystal sites appear with patterns of circular arcs of different radii, full circles, compositions of circles made up of circular arcs, and even centers of circles. All in an intricate interlacing manner, with the different sites or inner molecular groups clearly observable in the pattern by size and shade depth. This arrangement obviously goes much beyond molecular polar attraction between lattice sites and intermolecular bonds within the lattice network, and necessarily implicates a higher-level degree of freedom to materialize, which must be the molar interlacing level of complexity (given that it evidently cannot be the *Life Shell*, next one up).

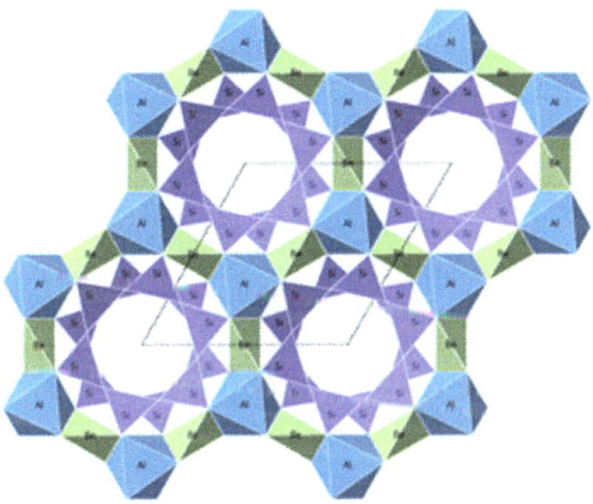

Fig. 20.6 A schematic view of the Beryl crystal structure down the Miller C axis. Well-known varieties of Beryl gems include aquamarine and emerald [Wikimedia Foundation].

In continuing with the structural analysis of the Beryl crystal, one ought to note that normally these crystals form hexagonal columns. Beryl can also occur in massive habits as illustrated in Fig. 20.6. Thus, as a cyclosilicate, Beryl reflects sensitivity to circular geometry. In Fig.4.6, the circular purple rings of silicate tetrahedra Si_6O_{18} are quite ostensible. They are arranged in columns along the Miller C axis and as parallel layers perpendicular to the C axis and forming the channels along the C axis. Another layer of circular geometry formed by Aluminum and Beryllium atoms is laid in encapsulation of the primary silicate circular rings.

Furthermore, along the line of its composite hexagonal structure, there are geometric channels within the topology that allow a variety of ions, neutral atoms, and molecules to be incorporated into the crystal by substitution of the primary Aluminum, Silicon, and Beryllium sites in the crystal structure. These so-called impurities are responsible for the variety of colors showcased by Beryl crystals. In the trend to explain everything with charge interaction, it is generally argued that the added impurities make it into the structure by disrupting the so-called "overall charge of the crystal". It is clear, however, that when neutral atoms make it into the nodes of the lattice network, as well as entire neutral molecules, one can no longer legitimately call for exegesis strictly based on charge, and the notion of "overall charge of the crystal" strikes as equally dubious as abusive to say the least. The analyst must forcefully conclude that a higher-level structural force has taken over the organization of the Beryl crystalline topology. This Operator is nothing else than the Van der Vaals coupling force in its genuine expression, while the inner structural geometry is simply governed by molar interlacing.

20.26. Shell Identity of Topological Assemblies in Condensed Matter

Topological identification of assemblies of objects pertaining to Shell 3 (Atoms), Shell 4 (Molecules) and Shell 5 (nano Materials) can become a pretty tricky business, since they generally take the form of stock piles of identical objects. It is not always the case that the inner structural geometry of the stack helps unambiguously determine Shell identity of the stack. This question has perhaps never been addressed in theoretical physics since the difference in the nature of the wavefunction in each of these three Shells has been all but overlooked, the science having assumed to this day that the vacuum is the same everywhere. Consequently, the Shell filter, a fundamental distinctive trait, has never been applied to phenomenological analysis, in particular topological analysis. We only apply the cross dimensional classes of bosons and fermions to any and all topologies of objects pertaining to these three Shells.

The cubic organization of metals is not interpenetrating, molar or molar interlacing. Therefore, these assemblies of cubes forming a crystalline structure is strictly atomic. Metallic crystals, in which there are only two main forms of aggregation, namely cubic close packing and hexagonal close packing, are atomic populations of objects. The lattice sites appear as spheres and the interstices are filled with electrons loosely held within the lattice as a defining characteristic. The vacuum inside the sites and within the interstitial spreads is of the same nature and foliation degree. One can think of the crystal as a single large vacuum spread taken over by a colony of scalar parcels which must geometrically obey the symmetry state or eigenform imbued in that vacuum wavefunction type. It is because this type of crystalline topology resides in a deeper region of complexity, where scalar predominance is in order, than all other types of crystals, that metallic crystals are comparatively strong and ostensibly more permeable to bending (or malleable) than breaking. They experience a higher agglutination force that impels them to keep together. Loose electrons inside of this medium are relatively unconstrained or unbound, that is to say, more subject to the properties of the lattice medium of their occupancy than the lattice sites. That is why they easily yield to inner space momentum imprinted onto them by the medium to form so-called electric current flows, so instrumental to modern-day technology. Here we must also remember that the first level of covariant interaction between atomic vacuum wavefunction and (occupying) particles such as electrons, as well as protons and neutrons, is given in orbitalization which is heavily driven by the Atomic Configuration Operators, as studied in Book II. This first order of interaction is naturally incarnated in the super low gradient of absolute temperature within the foliation of the wavefunction at ground basis.

The cubic organization of molecules is intrinsically of the lattice *interpenetrating* kind, even in the case of symmetric diatomic molecules. It is the very nature of intra molecular bonding that governs this type of lattice architecture. Because there is even a polar nature to a few neutral molecules, the interpenetrating class of lattice arrangement must be generalized with the role of basis of molecular lattice topology. Effectively molecular lattice arrangement is based on the electric charge invariant e, a constant that has little to nothing to do with electrons, protons and neutrons, as commonly assumed in physics thus far. It simply represents the force that pilots the construction of the molecular edifice. In its condition of a twinning force, it brings over a correlational form of liaison between two parties in a kind of equity relationship. This form of association is very different from, say, the kind between the three particles that compose an atomic element, neutrons, protons and electrons, that is. The twinning force represented in the electric charge invariant for this Shell is not only manifest in the molecule's bonding site but the entire molecule as a quality in the particular degree of freedom being developed here by the wavefunction, and therefore its topological lattice spread as well. The equi-polarity at the basis of molecular bonding is irrefutably evident in the main forms of known chemical bonds, namely covalence, ionic bonding, polar bonding (and hydrogen bond). It so much overwhelms the topological lattice arrangement that in the case of covalent bonding, the entire interpenetrating molecular crystal legitimately comes to be considered a single molecule.

To understand the claim that the electric equi polar force is of molecular nature, and not of particle origin, one must recall our deep analysis of periodic distribution of neutron and proton orbitals within the atomic edifice. Proton orbital shelling does not necessarily need an external shell of electrons to exist, nor does the neutron orbital shelling, which is knowingly devoid of all electric charge attribute, need electric interaction with protons, even less electrons, to exist. The drivers were simply others. The very reason why the full extent of the organization of atomic nuclei remained unknown for so long is precisely due to the attribution of positive and negative charges to protons and electrons respectively, and neutral charge to the neutrons, an overall charge picture that satisfied stability conditions. In quantum physics, however, we have learned that the classical hypothesis of electrical attraction between protons and electrons could not explain a stable Hydrogen atom. No more can it veritably explain the polar nature of molecular bonding. It is certainly not the driver in the atypical formation of molecules such as P_4 or Si_8.

Furthermore, we do not have an ontological definition for an electric charge, or what makes a charge positive or negative, since what we are calling the electric charge apparently simply depends on the orientation of the spin vector of electron and proton particles: the positron, anti-matter correlate of the electron, has an assigned positive charge, just like the proton! Again, electric charge cannot be ontological to elementary particles because such an important particle as the neutron particle does not knowingly hold any such property. The neutrino, albeit of lesser prominence in the buildout of the 3D World Plane, does not hold any charge property whatsoever either. We must therefore finally acknowledge that electric charge is not an ontological property of elementary particles and stop perusing it as such in physics. It is only in phenomenology that some of these particles at times “appear” to comply to that attribute. Once we overcome that hurdle, we become closer to an adequate understanding of topological assemblies in condensed matter physics.

20.27. Structural Quantum Definitions of Topological Assemblies in Condensed Matter

Sensitivity to the dimensions of space vacuum, akin to atomic and molecular elements in their orbital shelling development, stands for the first descriptor of the geometric layout of nano Materials at Shell 5 of U3. We have advanced a quantum number n such that $1 \leq n \leq 9$ to specify and grade this quality from one stratum of Shell 5 to the next.

Akin to the equivalent *s* orbital shelling, we expect the value of $n = 1$ to dictate a radial behavior that is same in all directions of the Material’s topology or in one word isotropic. However, because of the intervention of the Metamorphic or Complex Operator in the geometric configuration, we should not expect the eigenform proper to the Monolithic Causal Form, which is in command in Stratum 1, to materialize in its pure form, but instead a composition of these two qualities.

Consistent with this predictive view, we have to firstly acknowledge that Stratum 1 of this Spectrum features Amorphous Solid known to be isotropic, that is to say, they exhibit uniform properties in all directions. Very importantly, the measured thermal and electrical conductivities, coefficient of thermal expansion and refractive index of an Amorphous Solid show the same magnitude in whatever direction the properties are measured. Secondly, Amorphous solids typically possess an overall Random arrangement of molecules, an attribute predicted by the Metamorphic or Complex Operator. They exhibit short-range order only over a few molecular dimensions, and show physical properties quite different from those of their corresponding crystalline states.

Examples of Amorphous Solids (Fig.20.7) are glass, ceramics, polymers such as pitch, rubber together with many plastics, rapidly quenched melts and thin-film systems deposited on a substrate at low temperatures such as the mid-section of JJ’s, etc. The investigation of Amorphous Materials is an increasingly active area of research. Despite significant progress in recent years, the field’s understanding of Amorphous Materials admittedly remains far from complete, owing understandably to absence of the usual simplifications associated with periodicity. In this study, we are hereby filling that void by unearthing the fundamental drivers dictating the dynamics and configurations of this topology.

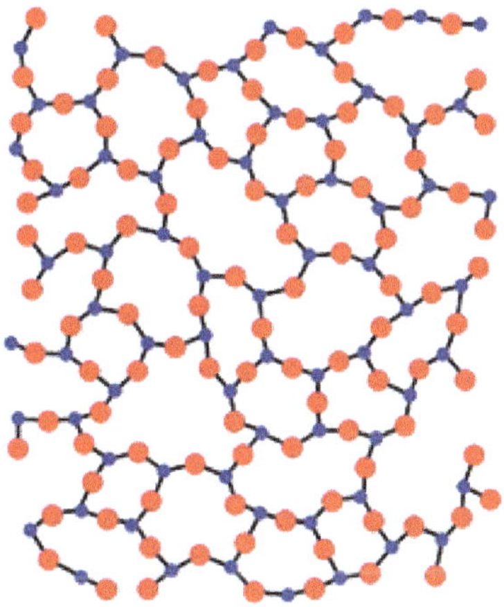

Fig. 20.7 Lattice sites inside of amorphous glass solid

Important observation: any given crystalline solid can be made amorphous by the very rapid cooling of its melt or by freezing its vapor. The effect of rapid cooling is tantamount to the expression of an energy gap between the initial and the final topologies of the transformation. Likewise, when quartz, the crystalline form of SiO2 is melted and then rapidly cooled, the amorphous solid known as quartz glass or silica glass results from the transformation. Quartz glass material has the same SiO2 composition as quartz but the molecular topology has lost its former orderliness in quartz. It is important to note that the change of temperature in itself is not sufficient to produce the topological transformation, but the temperature Δ requires the addition of a kinetic component for the topological change to occur. This is equally the expression of an energy gap between the initial and the final topologies of the transformation. This gap very much reminds us of the Fermi energy gap in the bottom region of the absolute temperature scale that we have previously discussed in the production of superfluidity and superconductivity.

This sort of topological change even occurs in metal alloys when thin films of the various melted metals are rapidly cooled. The resulting so-called metallic glasses are strong, flexible and much more resistant to corrosion than the crystalline alloys of the same composition. In alloys, we no longer have a typical metallic lattice with homologous sites, but indeed molecular types of lattice sites since they are undoubtedly heterogenous with different metal elements forming interpenetrating types of topologies. Only on that account can alloys experience the energy gap required for the transformation to occur.

20.27.1 Crystalline Solids

The above naturally brings about a discussion relative to the quantum nature of Stratum 2 of the Spectrum where n = 2. Akin to the equivalent *p* orbital shelling, we expect the value of n = 2 to dictate a radial behavior that is unequal in two spatial directions of the Material's unit topology or in one word anisotropic in at least two dimensions, based on the elliptic form of symmetry. However, because of the intervention of the Metamorphic or Complex Operator in the geometric configuration, we do not expect the elliptic eigenform proper to the Ambivalent Causal Form, which is in command in Stratum 2, to materialize in its pure form, but instead a composition of these two qualities. Consistent with this predictive view, Stratum 2 of this Spectrum features Crystalline Solid well known to be anisotropic, as previously discussed. Axial anisotropy in crystalline solid is the reason why these materials have been permeable to a system of classification based on axial references.

By definition, all crystals with an external form belonging to the same symmetry make a specific crystal class. In this case, A – Closed form v. B – Open form symmetry have been defined. All the crystals of minerals and synthetic substances studied so far can be grouped, on the basis of those two forms of symmetry into 32 different classes. However, when the crystallographic axes are taken as primary reference, all crystals (and hence crystal classes) can be conveniently grouped into 7 major crystal systems as known, making the axial reference the most generalized element of crystal classification as stipulated above. The axial definitions, per Miller indices, cast the following 7 crystal systems:

- Isometric (Cubic) System
- Tetragonal System
- Hexagonal System
- Orthorhombic System
- Rhombohedral
- Monoclinic System
- Triclinic System

Sub-typification inside of these systems based on arrangement of crystal sites over the lattice unit give us the 14 Bravais lattices. In common practice, crystallographic studies involve referring at first order any given crystal to the three crystallographic axes in order to determine the crystal system to which it belongs. Then, its symmetry elements are determined in desired detail to place it in a proper so-called symmetry class. Finally, various forms developed are superimposed in order to carefully determine parameters, indices and coined symbols in each case. If this classification system seems a little complex or heterogeneously irregular to say the least, it's because it is. Here again the Metamorphic or Complex Operator intervenes to incorporate in the symmetry organization a dissimilar vector, aka unpredictability in the specificities, but not of course in the ontological quality.

A typical representation of this complex composition in crystal topology is the structure of the Axinite crystal which pertains to the triclinic system (Fig.20.8), or the Pyroxene crystal which pertains to the monoclinic system:

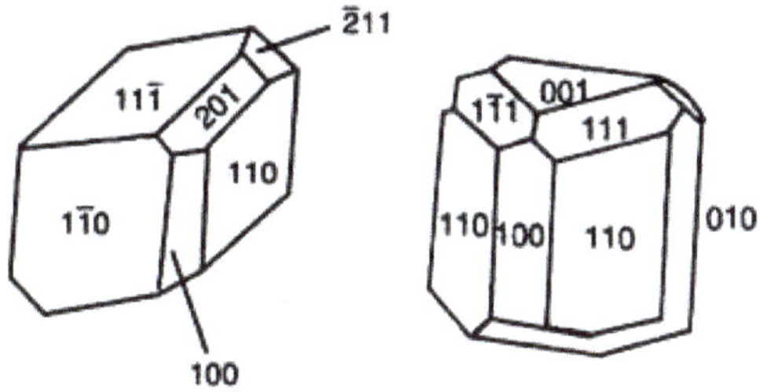

Fig. 20.8 Axinite, a crystal of the triclinic system

Some examples of amorphous solids are glass, rubber, pitch, many plastics etc. Quartz is an example of a crystalline solid which has regular order of the arrangement of SiO_4 tetrahedra. If quartz is melted and the melt is cooled rapidly enough to avoid crystallization an amorphous solid called glass is obtained.

Amorphous solids find many applications because of their unique properties. For example, inorganic glasses find applications in construction, houseware, laboratories, Rubber another amorphous solid is used in making tires, tubes, shoe soles etc. Plastics are used extensively in household and industry.

20.27.2 Powder Stratum

Powders knowingly have an internal symmetry arrangement that is one step beyond the solid crystallographic arrangement. The geometric signature of such inner structures can only be captured in the concept of texture. X-ray crystallography remains the method par excellence to detect this type of structure, albeit still technologically challenging to this day. The distinction in inner structure between powder and crystalline solid resides in the degree of texturing in the sample and direct optical monitoring of the inner structure through X-ray diffraction projections is no longer feasible. That is why the concept of a reciprocal lattice unit has been advanced in order to represent the inner structure in an abstract mathematical space.

We have seen that crystals can be described by a Bravais lattice with some level of regularity in the spacing between atoms. In a higher frame of generalization, we can apply a Fourier transform to the Bravais lattice in order to obtain a so-called space of reciprocal lattice structure allowing us to scale or typify patterns of regularity otherwise unobservable. At the same time, the regular Bravais symmetry finds its scaling degree within this new abstract structural scale as well. With this transformation, three-dimensional coordinate space become characterizable with reciprocal axes x*, y*, and z* or, alternatively, in spherical coordinates q, φ*, and χ*. In powder diffraction, radiation intensity is homogeneous over φ* and χ*, and only q remains as a sensible measurable quantity. This is a result of the important artifact that orientational averaging causes the three-dimensional reciprocal space studied in single crystal diffraction to be projected onto a single dimension. When the scattered radiation from powder is collected on a flat plate detector, the rotational averaging leads to smooth diffraction rings around the beam axis, unlike the discrete Laue spots observed in single crystal diffraction. The scattering angle, which is the angle between the beam axis and the ring, is denoted 2θ in X-ray crystallography. Bragg's law dictates each a correspondence between each ring and a particular reciprocal lattice vector G in the sample crystal, leading to the definition of the scattering vector as:

$$|G| = q = 2k \sin\theta = \frac{4\pi}{\lambda}$$

where G is the reciprocal lattice vector, q is the length of the reciprocal lattice vector, k is the momentum transfer vector, θ is half of the scattering angle, and λ is the wavelength of the optical source.

Powder diffraction data are usually presented as a diffractogram in which the intensity I of the diffracted radiation, is shown as a function either of the scattering angle 2θ or the scattering vector length q. The latter variable presents the great advantage that the entire diffractogram no longer depends on the value of the wavelength λ. Therefore, the use of the vector length q has been recommended and gaining acceptance, as it even further facilitates comparability of data obtained with different wavelengths. Note that the advent of synchrotron sources has considerably widened the choice of X-ray wavelengths for powder diffraction patterns. As to X-ray diffraction methodology in sizing up structural crystallographic phases, it is acknowledged that the fundamental physics upon which the technique is based provides high precision measurement of interplanar spacings, in some instance to fractions of an Ångström, The ability to analyze multiphase materials also allows analysis of how materials interact in a particular matrix or topology such as geologic cores, pigments found in historic paintings, cement and concrete powders, pharmaceutical tablets, electrical circuit boards, etc.

Despite these conveniences, the advantages of a quantum treatment for structural determination of the powder state phase based on the set of quantum equations above presented, toward a transformative development of materials science, need not be argued. Only within the framework of modeling this kind can powder state of materials be fully understood and underpinned as a phase state in its own right. Powder XRD has become a widespread industry method to analyze materials structure at the microscopic level, based on an understanding centered on the construct of texture. It is an empirical comparative method that helps identify polycrystalline structures by reference to known registered patterns. Unfortunately, it remains blind to the fundamental notion of quantum states as the foundation of heterogeneity in the structural geometry that differentiates one material type or class from another. Powder XRD attempts to deduce the identity of a bulk material from exfoliation of the solid material sample as a powder. Despite its importance and ubiquity in industry, field experts acknowledge that the information contained in powder XRD data for nanoscale materials, while not always fully harnessed, is outright misinterpreted in other cases and may become too complex for useful interpretation. Physicist Cameron F Holder further acknowledges that “Unambiguous sample characterization almost always requires complementary experimental and/or computational methods.”

Materials science will need to pay more attention to the thermo-mechanical processes in the formation of materials, well beyond the mere concept of texture, in order to experimentally determine the discontinuous energy scale intervening in such processes. Texture forms in materials during thermo-mechanical processes, during rolling production processes i.e. That is why the rolling process in particular is often followed by a heat treatment to reduce the amount of possible unwanted texture. Even geologic rocks show texture due to the thermo-mechanical processes intervening in the history of their formation. Texture is manifest in almost all engineered materials; it can have a significant impact on desired materials properties.

20.28. Microstructures from Waxes to Volatile Gases

While waxes seem to deserve at first sight a class of its own, there seems to be a common property to the materials states spanning from colloids to oils to aqueous liquids to vapors to gases, which is fluidity. The classical study of fluids has long enlisted liquids and gases primarily in the class of fluid materials with some infrequent addition of viscous materials. However, a complete study of fluid materials must include all the states that we have individually listed under that rubrique. They do exist and must therefore be formally accounted for.

The fact that waxes seem to be an intermediate or hybrid class between solids and fluid materials come to support our quantum or quanto-geometric distribution of the class spectrum for materials' states. Let's recall that Quanto-Geometric Quintessence gives us three distribution groups for each primitive spectrum: one group characterized by scalar dominance, an attribute which typifies the group cover, another characterized by space momentum dominance, an attribute which typifies said group cover, and a third intermediate group which is a hybrid group with equated or balanced covariance between the two principal factors (scalar factor and space momentum factor). That is exactly what we empirically observe here. The existence of the wax state as a hybrid solid-fluid state tells us that state evolution of materials from solids to fluids is not linear but non-linear and hence that each microstate or stratum of the distribution is a discrete level of the same.

Fluids are classically defined as substances that are not resistant to deformation and offer little to no resistance to shear forces. Fluids readily take on the shape of their containers which exert a force on them that is normal to its surface. Contained fluid would respond to a shear force by deforming its bulk until the force disappears. By this standard, one can see that a wax could not pertain to this class, if any doubt. From a microstructural viewpoint, the molecules within the topology of fluids do not correlate in a vibrational manner as though they were bound to one another through putative springs, as it is in solids. The topological sites are relatively unbound from one another, enjoying a higher degree of freedom as the angles in their polygonal unit structure may increase unrestrictedly to nearly 180°.

20.29. The Uncertain Heisenberg Quantum Uncertainty Principle

The uncertainty principle, which states that there are pairs of quantum variables that cannot simultaneously be known, exists because it does not understand shell symmetry-breaking and its foundational role in the buildout of U_3. The uncertainty principle exists because it does not understand scale in the tapestry of the vacuum and pretends that everything exists in their static and dynamic modalities on a flat sheet, as it establishes unfounded expectation values from comparison between objects that matter-of-factly exist at different scales. We should not compare a free electron with an orbitalized electron and expect their sets of dynamic properties to be comparable or homological to one another, without understanding the marked ontological scale difference between the two objects. The uncertainty principle exists because it does not understand that there is adequacy and usefulness in comparing objects at different U_3 scales only through homeomorphic relationships. The uncertainty principle can exist owing only to the ignorance of the foundational numeric quantum abstractions and quantum causal forms. The uncertainty principle should write in big letters on the wall that each Shell of U_3 represents a symmetry and that the symmetry breaks through the foundational quantum causal forms. The uncertainty principle should no longer be skeptical or ignorant of the overarching norm of symmetry that regulates the associations of elements from which the Shelling emerges, namely when:

1. Quarks inter-associate to form elementary particles.
2. Elementary particles inter-associate to form atoms.
3. Atoms inter-associate to form molecules.
4. Molecules inter-associate to form nano materials.
5. Interlacing lattice groups inter-associate to form karyotic cells.
6. Higher Karyotes inter-associate to form singlet stellar bodies.
7. Singlet stellar bodies inter-associate to form solar systems.
8. Solar systems inter-associate to form galaxies.
9. Galaxies inter-associate to form galaxy groups.

The uncertainty principle shall know that the hierarchy in the set of quantum causal forms ultimately dictates the Shelling Hierarchy of U_3. The uncertainty principle should therefore stop cautioning the adoption of arbitrary units to characterize physical properties along the Shelling Hierarchy of U_3 and inquest the canon of quantum abstractions to properly and innerly dimension physical attributes and properties across Shells. The uncertainty principle, whose true age is older than Werner Heisenberg's physics, shall be deemed an impostor in the establishment of norms of uncertainty and inaccuracy through the standard deviation variance in physical measurements. For all these reasons, the uncertainty principle shall be imposed a “guilty of accessory” verdict in the indictment concerning the Hierarchy Problem.

20.30. Physical Sciences’ Blindness to Number Figuration in Favor of Unit-Dimensions

Quanto-Geometric Number Theory reveals that beyond the cardinality dimension, numbers behold an ontological signature which confers them an identity instance, an attribute instance and a methological or operational instance. By virtue of this innate abstract signature, we have alternatively denoted them in this study quantum abstractions. It is from those qualities that the quantum causal forms, discussed at the beginning of this chapter, duly derive, as previously stated. Our ignorance in physics of those innate abstract qualities in the number system has led us to the realization of a mathematical physics system erected on unit-dimensions for the assessment of physical magnitude.

Yet we have found in this Framework that some experimental physical quantities, dubbed fundamental physical constants, take us out of the unit-dimension system and appear as dimensionless numbers or pure scalars. For all intents and purposes, these fundamental dimensionless quantities are exactly identical to the known mathematical constants such as transcendental numbers π and e. In physics, we have invented unit-dimensions because we do not understand numbers in their quantum signatures and thus cannot find physical magnitude or measurement representation in the numeric system itself. It is by no coincidence that in the Quanto-Geometric Framework, we find, by derivation from a pure mathematical numeric system, dimensional numeric quantities as fundamental physical constants, let alone dimensionless fundamental constants. Some have been puzzled by and found weirdness in said fact or artifact, wondering how physical unit magnitudes that have been chosen in a completely arbitrary fashion can cast numeric quantities with any level of precision within any theoretical axiomatic system. This fact is perhaps no less puzzling than the manner in which E. Schrodinger’s quantum Hamiltonian, having started in its development as a time-dependent equation, ended up kicking out the false time variable and portrayed itself as a time-independent Hamiltonian Operator, as discussed in Chapter I, Book I. So therefore, the virtue of our system is to provide the absolute numeric etalon to gauge, legitimize and exegetically scrutinize unit-dimensions that we have traditionally elected with the aims of quantifying physical magnitudes in our mathematical physics system. It is not the other way around. This amounts to a different yet most adequate way of conducting dimensional analysis which is able to clarify the physical meaning behind unit-dimensions. As formerly noted, it is a singular fact to reckon with when some physical quantities, which are unrelated both experimentally and, in their unit-dimensions, appear under a reciprocal relation in their pure numeric figuration. This unusual fact cannot be ignored and demands from orthodoxy full consideration and precise meaningful interpretation. This sort of pure numeric relation simply transcends unit-dimensions and reveals physics in its pure mathematical dimension, fundamentally unifying physics to mathematics.

Yet, some may ask how do we establish a connection or a mapping between a pure number and a physical magnitude. We have answered this question when we ported the standard deviation unit in our system to the meter standard by establishing that:

$$\sigma = 1.00001203 \times 10^{-9} m \quad \textit{(4-8)}$$

Furthermore, and to this effect, when we have found that the vacuum energy density of the 3-D World Plane, aka the Cosmological Constant, is fittingly the cube of that number:

$$\sigma^3 = 1.00003609 \times 10^{-27} \quad \textit{(4-9)}$$

we have thus verified the natural unitary value of the sigma (σ) quantity, having found that ground level of the vacuum energy density of a 3D World Plane is nothing else than the cube of the standard deviation of the distribution making up the Omniverse. Here we have to opportunely recall that in the Standard Normal Density Function, which is a progeny of the Quanto-Geometric Grand Eigenfunction, sigma (σ) does represent a unitary dimension. This characterization altogether epitomizes the foundation of mathematical physics in the natural world at deepest level, while making the equality in *(4.9)* the foundational bedrock of all mathematical physics. At the same time, this characterization significantly dampens the so-called Hierarchy Problem in physics as an acute problem with a plethora of contributors.

20.31. Conclusion

In this Chapter, we have presented a uniform theoretical basis, embodied in the Quanto-Geometric Framework of Unified Physical Law, to a number of long segregated constructs in nearly one century of experimental findings in post-quantum physics. In concluding the Chapter, we will underscore some of those important contributions. We ought to highlight at any rate the important fact that the effect of temperature changes on the behavior of free electrons with a given Fermi energy is negligible and remains so for any temperature below this critical level or T_C on the absolute scale. The free state of individual particles within what can be designated at least an inter atomic environment is not a ground state but an over excited state. The delta gap or energy packet gap after which superconductivity always starts is a multiple of the spin magnetic moment of the particle itself, setting superconductivity to develops on that very energy scale. This result also means that gradient *0* of the absolute temperature scale does not reside here, because if anything the spin of the particle is an irrevocable ontic attribute of the particle. Thus, the Fermi energy gap delimits two different ontic provinces associated with the material wavefunction of quantum objects. Below the Fermi energy gap is situated the realm of atomic wavefunctions, and above the Fermi energy point is situated the realm of molecular and lattice wavefunctions. The scale of absolute temperature itself ought to be removed from the thermodynamic heat-energy vector with the Boltzmann constant as basis and more legitimately anchored on the Stefan-Boltzmann constant. This overall perspective is fundamental in providing broad theoretical basis to post-quantum physics which can connect the diverse range of experimental and technological facts disparately garnered over a century of the science.

Chapter 21

[If you feel annoyed by apes being your biological precursors, then get used to the idea of viruses being the same.]
Matt Ridley

QUANTO-GEOMETRIC VANTAGE OF THE FOUNDATIONS OF LIFE

Physicists have long realized that the idea of holding one theory for dynamic systems in the microcosm and one other for the dynamics taking place at the larger-scale cosmic structures is simply untenable, even more so if both families of theories are completely opaque to one another. Despite the proliferation of gauge theories, we have still been unable in modern physics to explain the complexity suite of mother Nature, particularly in terms of symmetries. It is by endeavoring to elucidate the origin of the most important traits of complexity, which is the plethora of empirical fundamental physical constants making up the quantized Shelling of mother Nature at all levels, that this Framework has effectively reached Grand Unification, when it comes to physical dynamic systems. In this Chapter, we are going to reach further to lay down the most irreducible foundations of biophysics within the Quanto-Geometric Theoretical Framework. There have already long been continuing efforts in that field to address the ontology and phenomenology of Life on a formal numeric or mathematical-physics basis. However, we are hereby otherwise going to more stringently and seamlessly connect Life processes to the most significant streams in mechanical physics. To do so, this effort will elicit a number of unknown Life symmetries and, very importantly, the natural physical constant regimenting Life processes, and the architecture of the many distinct taxonomic ontologies alike.

21.1. Defining Life

The numerous private attributes of living substances have been time and again cited in most biophysics attempts to make a physical definition of life permeable to mathematical modeling, such as done in Prigoginean thermodynamics. Traditional biophysics proposes that the formation of living thermodynamic structures arises at the nano dimensional level of complexity (nano Materials) owing to a constantly varying environment which causes a great variety of living forms. Oftentimes, thermodynamics play a major role in those mathematical physical models of Life, under the evocation of a certain principle of substance stability. It is generally on that basis altogether that the definition of life as a biochemical-physical phenomenon is brought about and framed within the realm of exact sciences, chiefly chemistry, physics and physical chemistry, that is.

In the quest for fundamental life axioms, a few thinkers have proposed and worked along the line of the following factors or characteristics:

1. Organization in one or more building blocks
2. Metabolism
3. Reproduction or self-replication
4. Homeostasis
5. Adaptation
6. Response to stimuli

However, upon close examination, these defining factors of Life do not meet the standard of irreducible definitions. Most are about phenomenology, but not ontology. In one way or another, all are replicates of processes that do occur beyond the biosphere.

The structural factor of "*Organization in one or more building blocks*" can barely have a defining value for any physical object in existence (except perhaps for quark particles). Are there not hydrogen atoms out there in open space as singlets, and cumulations of hydrogen atoms in duplets (molecules) forming hydrogen gas out there in the cosmos and on planets? No one would agree to impart them the status of living entities on that basis.

Metabolism is a mechanism that is fundamentally driven by motion and activity, both for intake and catabolism. The keyword here is autonomous motion. Brownian motion of smoke particles, driven by random kinetic motion of molecules, are as autonomous as pollen grain motion. Even the non-deterministic motion of electrons in an orbital shell is to be considered autonomous motion. That is why E. Schrodinger reportedly estimated that the electron must be like an amorphous object, presumably akin to Amoeba, in response to perpetual sudden changes of momenta. Further, A. Einstein was no less than appalled by the momentum-less behavior of this entity, stating that *"he cannot imagine the electron hopping like a bug"*.

Reproduction is in essence the ability to replicate oneself, independent of the modality in which the process takes place. The keyword here is self-replication. Anyone holding that Reproduction is a defining property of Life should recall that a crystal put in the proper chemical environment will indeed grow by reproduction of its lattice unit. This observation also takes care of the possible argument of *growth* and *anabolism* as defining factors of life as well.

Homeostasis is the process in living entities of maintaining constant thermal equilibrium by act of necessity, while beyond biota this process is autogenic, as far as population of identical or diverse objects in interaction are concerned. One may view systemic homeostasis as the modality taken by thermal equilibrium throughout the population of cells represented by biotic entities, necessary only because the physics of juxtaposed cells or cell materials do not facilitate timely autogenic thermal equilibrium across the entire individual.

Adaptation and *response to stimuli* are epi-phenomena that are sub-functions of the above, a discussion which we will forego.

These factors altogether represent important capabilities to sustain life but are not necessarily responsible for its gestalt. Further, precisely because their replicates exist at other Shells of matter, they not only lack defining value for life but advocate against the notion of *emergence of life.*

21.2. Biophysical Ontogenetic Definition of Life

A phenomenological definition of life is doomed to miss the mark, since phenomenology is a result of the attributes of ontology. Therefore, the irreducibles that one duly seeks for a definition preferably resides with ontogeny, in this framework the two primal existential tenets together with their interaction and spectral state evolution at any order. The difference between a natural mechanical object and a natural living object or entity is to be first attributed to the typified level of the latter within the spectral suite of complexity, Primitive Shell 6 within the suite, that is. Going back to the $M_3(s)$ function, Life is primevally defined by that function at the point where the independent space tenet variable reaches the normalized value of:

$$s = \sigma \frac{\sqrt{10}}{2}$$

Recall that the previous value of $\sigma\sqrt{8}/5$ gave us the nano Material Shell. This definition seizes Life in its deepest level of composition, which is physical space and scalar density, and further amounts to a phylogenetic portrayal, as the science of Biology has it. In this context ontogenesis becomes a property of phylogenesis. This important axiom resolves for us the intractable Biological Science problem of which is first between the chicken and the egg, for all species. The deepest level at which a life can be built is directly from the natural physical variables upon which the U_3 World Sheet is built. Biological ontogenesis caters to topological development which we designate with the term *population* in human culture. Living species population is physical topology and the ontogenetic reproductive function is nothing but an emulation of phylogenetic structural development. The fundamental physical duplet of space and scalar tenets can never be secondary to anything in existence, and so it is not the egg that primevally produced the chicken, but the wavefunction (or eigenform) tenet within the chicken that orchestrated the reproductive egg as a faculty in order to subsequently foment topology, aka population or species, in a self-reliant manner. In summary, both real phylogenesis and emulated phylogenesis (ontogenesis) are realized under the regency of the Quanto-Geometric M_3 *Creation function.* A *phylogenetic* definition of life ought to go back to the Karyotic Shell organization, and find cover under the physical symmetry of the Karyotic building block borne out of symmetry-breaking from the lattice or Materials gauge group. Since it is represented by the $M_3(s)$ derivative function, the order of covariance here between the basis tenets is of the second-order, that is to say between ***real karyotic coordinate space*** and ***karyotic mass***, to be taken as the two fundamental *phylogenetic* components of life.

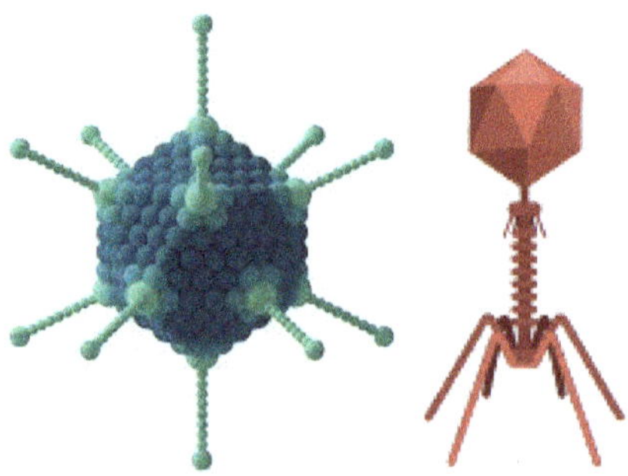

Fig. 21.1 And adenovirus and a complex virus

Consistent with Quanto-Geometric first axiomatic principles, the ontogenetic emulation of primeval phylogenetic structural development poses a mobile male gamete, in the role of the wavefunction, and a static female gamete or ovule, in the role of the scalar or mass tenet, along with a subsequent merger between the two. What ensues out of the merger is a fertilized egg or a karyotic box poised for a full re-enactment of the fundamental order of typified covariance between the two universal primal ontological tenets, as it occurs in phylogenetic space. The above describes the premises of ontogenetic development in multi-cellular organisms. The process is less complex in monokaryotic organisms or macromolecules such as viroids, yet the reproductive faculty is unmistakably present as these entities demonstrate fulsome self-replication capabilities.

It important to realize that complexity puts together here a selection of objects from the nano Materials Shell in order to build the Symmetry sustaining this following Shell of matter. Although in biological science thus far, we are granting macromolecules the distinct role of building blocks of life, in fact the physical symmetry of the Karyotic building block is borne out of symmetry-breaking not from the molecular gauge group but from the Lattice or nano Materials gauge group. One can see the wisdom in this view by merely heeding the outer geometric structure of an adenovirus or a complex bacteriophage virus (Fig.21.1).

The polyhedral lattice structure and geometries of the karyotic box in both organisms is at plain sight and blinding at the forefront of their organization, and can only be overlooked in total ignorance of how Symmetry works to build complexity. More significantly, a virus may knowingly and effectively be *crystallized* in the laboratory. Such a crystallized viral entity may remain indefinitely in the form of an inanimate crystal in the laboratory medium. It will not lose any of the potentials and faculties attributable to an active living organism. In particular, it will not lose its peculiar self-replication modality, which it will demonstrate as soon as it is placed inside a host.

We will later delve into the details of structural organization of these entities from an axiomatic standpoint. Let us retain for now that Life is not primarily about macromolecules and what they do as has been vehemently upheld thus far in Biology, but the directives that they receive from the higher Karyotic gauge group, when it comes to molecules. The biosphere is a Shell of matter with its own drivers, which are not molecular or molecular-Shell drivers. The study of the biogeometric aspect of the Life Shell will help fully bear out this prescription.

Lastly, it is no less important to understand the physiological processes of metabolism and reproduction as physical derivates of the propagating wavefunction in real Karyotic coordinate space and scalar density (mass) as Karyotic monolithisation respectively. One can readily see the foundational logic of this attribution which sets it apart from an a-priori postulate. All other phenomenal activities demonstrated by the living, be it homeostasis, anabolism, catabolism, unicellular organization, multi-cellular organization, etc., all remain lower-order covariance derivates of these two primal material potentials.

21.3. Limits of Evolutionary Biology

In the following, we are going to briefly review the main principles advanced in modern-day Evolutionary Biology, which, in fair appreciation. has been in unrelenting flux for the las twenty years, as new discoveries raise counter truths against and supplant long held truths, if not even rehabilitating long demoted truths.

Natural Selection has been considered the pathway par excellence to materialize evolution of species. Gaining or losing hereditary characters is not considered, however, to immediately lead to the formation of new species. Before a group of organisms can be considered a new species, a number of important modifications ought to occur.

21.3.1 Natural Selection Mediates the Evolutionary Pathway

As individuals from one group become largely different from other members of the same group, they may no longer mate with the latter to produce descendants that are themselves fertile. Thus, these new individuals are considered members of a new species. It so happens that, as time goes by, a certain number of species evolving from common ancestors, having experienced so profoundly marked differences with other parent species, ought to be classified as a new genre or phylum. For instance, among the many different species of primitive reptiles, some developed feathers while others developed light fur and yet others shells. From these small differences at the outset, and with the passage of time, further modifications occurred conferring them selective or adaptive advantage as they become more and more differentiated. Hence, in the course of millions of years, the Theory believes, descendants became so broadly different from the primitive precursor reptiles that the animals with feathers we taxonomically give the distinctive group denomination of Birds and those with fur the distinctive group denomination of Mammals.

21.3.2 Darwinism Strongly Imprints Our View of Evolution

In all of Biology, the Theory of Evolution plays a primordial interpretative role. It stands for the amplest principle at the core of most theoretical representations in Biology. This Theory is so broadly embraced in Biology and so deeply ingrained in its foundation that this science, as it stands today, may not possibly be at all understood without it. Jean-Baptiste Lamarck did propose earlier a hypothetical mechanism that mediates evolution. He conjectured, as an inference from observation of natural biota, that (1) biological characteristics change as a result of use and disuse *(Lamarck's law of use and disuse)* and (2) acquired characteristics are passed down via inheritance *(Lamarck's law of inheritance of acquired characters)*. By contrast, Charles Darwin's work, half a century later, in superseding Lamarck's made for establishing evolution as a falsified theory, at least per the science's claim, which lays natural selection as the very mechanism through which evolution is accomplished. The theory of evolution through natural selection has come to represent the most important conceptual structure in modern Biology, ever since Charles Darwin published his field studies in mid-1800's *(On the Origin of Species by Means of Natural Selection).*

21.3.3 Genetics Comes to Support Inheritance Through Natural Selection

When the works of Gregor Mendel and others later demonstrated how specific rules of dominant and recessive inheritance informed selection in evolution, Lamarck's hypothesis of acquired characteristics fell in disfavor. Further developments in the science of Genetics throughout the 20th century were to put the notion of inheritance through acquired characteristics to complete rest, giving primacy to chromosomic character transmission through macromolecular DNA. However, at the turn of the 21th century, the concept of the epigenome emerged which posits acquisition of inheritable biotic characters from interaction with the environment. Genes inherited from biological parents might or might not become active or "expressed" in the individual's own makeup. It turns out that "genetic expression" can be affected by individual experiences, and even by circumstantial thoughts and feelings. Epigenetics refers to the study of these changes in gene expression that do not involve alterations to chromosomic DNA.

21.3.4 Lamarck's Evolutionary Perspective Has Made a Comeback

The human individual holds between 20,000 and 25,000 protein-coding genes, which make up just about 1% of the human genome. The epigenome refers to the 99% of our genome[24] that does not knowingly code for the proteins making up biotic structures. Instead, much of that enormous portion of our genome apparently has to do with gene regulation, largely through coding of various types of RNA. It is an even rather remarkable fact that the epigenome can register changes during a lifetime which will be passed on to progeny *through sperm and egg*, according to studies carried out since 2003. There exists clear evidence of non-DNA inheritance by descendants of physiological traits acquired during a lifetime by their progenitors. This finding has firmly put Lamarck's second rule back on the map of evolutionary biology.

21.3.5 Much of Chromosomic Activity Remains a Mystery

It is important to point out, in any study of evolutionary biology, that the science of Molecular Biology along with genetics are still a very immature conglomerate of knowledge. Despite their almost century-old debut and the recent advances in genetic knowledge brought about by the Human Genome and Encode project in the 2010 decade, these ambitious projects aimed at decoding the entire human genome have yet to see full daylight. Many researchers nowadays agree that the success of these projects has been oversold in light of so many important unknowns remaining in the understanding of genomic structure and activity. Reflective of this lack of thorough understanding is, in my sense, the designation that has been given for quite a while to the 99% non-coding DNA of *"junk DNA"* in the research community. It is difficult to understand, given the primacy of Occam's law of parsimony or the law of least action in physical science, how in biological science anyone could ever consider that 99% of such a crucial macromolecular stock as hereditary DNA could merely be there playing a dummy role[25].

Genes, as we understand them in research, are nothing but ordered DNA sequences forming protein patterns, on the basis of a 4-letter alphabet. They are therefore easy to recognize due to these regularities. However, researchers acknowledge that we do not even always know what genes do or are supposed to do, although the action of many have been identified. When it comes to the non-coding genes, their inner organization

---

[24] https://sitn.hms.harvard.edu/flash/2012/issue127a/

[25] https://medlineplus.gov/genetics/understanding/basics/noncodingdna/

remains a complete mystery. We have only begun to understand their effect or participation in the transmission of acquired characters during the individual's lifetime. How a transcription factor interacts with DNA is a very well-known process. What is harder to figure out is the logic behind the process, i.e. why some genes have 3 transcription factors to regulate them while others have as many as 20 transcription factors. Therefore, the knowledge of the bigger picture or the entire map is missing at this stage of human knowledge of the foundations of Life.

21.3.6 How Chromosomic DNA Builds the Entire Architecture of the Biotic Individual Remains Completely Unknown

To further understand the magnitude of our ignorance in the matter of genomics, one ought to consider that the 20,000 to 25,000 genes in the human chromosomic DNA repertoire are also the same 20,000 to 25,000 genes in worm's chromosomic DNA repertoire. A worm does not belong to the human species, a worm is not a mammal, a worm is not even a vertebrate, per universal Taxonomy of the living. Any of those cited species is far more complex than a worm, yet they all hold the same number of coding genes as the invertebrate worm. Consequently, it is fair conclusion that our current knowledge of the buildup of species from chromosomic DNA is far from supporting our vision of how biota evolved, as represented in Evolutionary Biology in its taxonomy of the living. There must be thus something vastly more profound to the genome or proteome beyond genic base pairs to account for this astronomical histologic difference between worms and higher organisms.

Furthermore, the idea that all a coding gene does is to hold a template for a specific protein buildup, and a non-coding gene is only a regulatory instance for the coding gene, is far insufficient to lead to an explanation of the formation of anatomic organs within multi-cellular biotic individuals. It is patently not enough to pile up protein molecules on top of one another in order to build up a tissue or an anatomic organ with the well-defined structure and morphology that duly characterize them. One thing is to identify a DNA sequence that is responsible for a physiological dysfunction and a totally different thing is to build an organ from chromosomic DNA information in general, with the help of our limited understanding of coding and non-coding genes. Organoids we have managed to build from stem cells have artificially-induced morphology and no structure at all. In other words, what gives our eyes their visible shape and known internal structure i.e. is something that remains a total mystery to genetic science to date. This assessment includes all we have been able to learn thus far from editing DNA with the CRISPR technique consisting in chopping DNA strands and observe the consequential changes at any order.

What's more, the state of knowledge in genomics is far and away from having ascertained an understanding of the internal architecture of even a single cell, whether of prokaryote or eukaryote generation, with its usual plethora of organelles and foliating structures. Likewise, how a virus manages, putatively from its genomic stock within the host's macromolecular DNA strands, to assemble and erect its molar building blocks into its characteristic kernel construction, are all unknown phenomena that lie beyond simple protein buildup through gene expression. Effectively, it is a profound error to visualize the construction of a biotic individual as a passage from the molecular level, no matter how large involved macromolecules may be, to the tissue or anatomic level, in complete disregard of the Materials or Crystalline Shell level of matter based on molar topologies, which represents an ineluctable step of the ladder of biophysics in molecular biology.

One of the most fundamental questions thus far unaddressed in genetic science is the nature of the agency by which chromosomic genes of a eukaryotic cell code the buildup of an organelle such as the Golgi sacs inside of the cell, and, i.e., the gastric sac at the histological level of the multi-cellular organism. At the heart of this question is what we might call *biogeometric instantiation* and *instantiation of degrees of freedom* within the intimate process of construction of the living overall. Promising though the CRISPR technique of discard and check might be from an investigative point of view, admittedly in the current state of knowledge in microbiology, we do not understand enough yet about the human genome, notably which genes give rise to which particular anatomic traits, and how the many different categories of genes interact, in most cases, to even enable editing for any form of anatomic enhancement today.

21.4. Lack of a Mathematical Description for Time-Based Biotic Evolution

To establish a biophysical foundation for the science of biology, in an attempt to unify biological law to physical law, one ought to clearly identify the role played by time as a physical variable in Evolutionary Biology, towards implementing formal mathematical descriptions based on that variable for biological processes. In that regard, a reconciliation of the Astrophysics vision of a Universe in evolution under the Big Bang scheme with Mutational evolution in Evolutionary Biology comes to the fore.

21.4.1 Effect of Passage of Time in Evolutionary Biology

No study the kind afore-mentioned seems to exist in the reviewed literature. We are next going to approach this interrogation from the Quanto-Geometric perspective. What Darwinism proposes is that the passage of time is necessary for modifications to occur, from small to significant, in order to imprint the selective and adaptive advantage that would give these new biological characters phylogenetic longevity, essentially by granting them a chromosomic or epigenomic pathway. The Theory believes that in general millions of years are necessary for descendants to become sufficiently differentiated from the primitive precursors in order for genuine new species to arise such as mammals or birds. The vision of large-scale flowing time is essential in Evolutionary Biology to explain the overwhelming variety of biotic species that we see on earth, to the point that one may fairly conclude that time in that vision plays the role of an agency or catalyst for evolutionary change. Adaptation or fitness for survival conferred in a particular trait drives the evolutionary change over the physical background of time flow or along a physical timeline, with the addendum that a large delta in time is necessary for any incremental change on the biotic line. These premises would yield a linear function *F(t)* with a very large slope figure associated with the independent variable of time, such as:

$$F(t) = at$$

where *F(t)* is the biotic type or species, *a* the slope quantity in millions of years (10^6) and *t* the time variable in year unit. The function *F(t)* for a particular evolutionary line could possibly be nonlinear. However, to remain faithful to the premise set forth in the theory, its first-order derivative would have to portray steep positive variation. Therefore, unless and until the Theory is able to come up with this formalism for every evolutionary line, the vision of the agency of time in biotic evolution will remain an unproven theoretical conjecture but not a systematic assertion of fact, which is a necessary realization for a biophysics foundation.

The open questions that require this sort of formal interpretation are in our view:

1. For any trait, any genetic character, and any organism, … Evolutionary Biology must systematically formalize through a mathematical description of sorts the modality by which it took the specific number of years that it has duly assigned to each for their mutational emergence.

2. If, as believed, ontogenesis is a re-enactment of phylogenesis, the Theory ought to systematically interpret the timeline of 9 months taken by human ontogenesis for its development. The same question could be extended to every biotic species, except for viruses and primitive bacteria.

3. If there exist evolutionary Drivers that pushed macromolecules to mutate into viruses for the jump from the inanimate to the animate, what is the projected delta in the timeline?

21.4.2 Aging Process of Multi-Cellular Organisms

It is a remarkable fact that the biotic process of aging is not a result of physical tear and wear of the living organism occurring over time as it is in raw matter, in a clear demonstration that if the timeline physically existed, biota does not develop over that putative physical line or the metric of time. In multi-cellular organisms, tissues and cells do not physically fade away or degrade over time.

We have known for a while now that biological aging is instead a genetic function. There are specific genes that control aging. In that context, the apoptotic function or cell death orchestration remains under the control of the genetic aging mechanism. Galapagos giant tortoises, who routinely outlive their human researchers, and sequoia trees which can stretch over 3000 years of a lifetime, despite being directly and inevitably submitted to adverse weather conditions and roughness of the elements, come to mind.

21.5. Contradiction Between Time-Driven Biotic Evolution and Random Mutational Changes

What has made Darwinism overwhelmingly more compelling than Lamarckism with its notion of spontaneous generation was the verifiable vision of evolution through hereditary traits, the full extent of which was empirically studied by Gregor Mendel and totally confirmed by macromolecular DNA experimentation. The notion of Mutations in physical traits originated from genetic alterations constitutes the most important representation in Evolutionary Biology to date, to the point that it may judiciously be considered an irreversible point of knowledge in the science today.

The notion of Mutation has much evolved since Mendel, to the extent that we may distinguish between different forms of mutations, namely evolutionary mutations at the origin of biotic species, intra-species mutations responsible for secondary characters and intra-species dysfunctional mutations causing pathologies. They are all rooted however in activities of both the coding and the non-coding DNA elements of the genome. The most important descriptor of the Mutation construct remains the random, unpredictable quality by which it yields new traits.

Important and broad-encompassing though it might be, the construct of Mutation has not been mathematically formalized in Evolutionary Biology, a must for all attempts of knowledge unification through biophysics. Being by definition random changes, mutations cannot understandably be described as such through a time-driven functional representation. To formally describe Mutations, one must use the category of Stochastic statistics in mathematics. There lies a fundamental metric problem in terms of a contradiction between a timeline for Evolution of species and the stochastic emergence of species. Genuine stochastic categories in mathematics imply distributions devoid of a time continuum. There is no common formal background for these two streams of representation creating an insolvable metric problem. This drawback is in many ways reflective of the contradiction between classical physics and quantum physics. Insofar as the time continuum had survived in the Theories of Relativity, one can see what had made the Theories of Relativity and Quantum Theory intractably incompatible with one another. Likewise, any biophysics attempt at a formal description of Evolutionary Biology on mathematical ground integrating all of its current representations faces this same paradox.

It is clear that this shortcoming represents a camouflaged problem in Evolutionary Biology which has not surfaced thus far to researchers' attention perhaps because the science mostly relies on speculative interpretation, proper to natural sciences, instead of systematic rigorous interpretation, proper to hard core sciences and as delivered by the omnipresent tool of mathematics. This challenge is a must resolve to foundational biophysics.

Easy it is to foresee how Quanto-Geometric Theory may successfully take on this challenge on the basis of the Grand Eigenfunction at the heart of the framework.

21.6. Origin of the Mutational Development of Life

In the relatively new science of molecular Biology, researchers are so fascinated with the intelligent activity of the genome that they have yet to raise a few fundamental questions relative to the development of Life. Once a certain level of understanding of symmetries in nature was gained, physicists had raised, i.e., the puzzling question of the very origin of physical symmetries, which we have hopefully brought more and more satisfactory answers to in Quanto-Geometric Theory. One similar question in molecular biology is the origin of the Mutational Development of Life. Why do mutations occur as a trend or process within the development of Life?

It is quite natural, in a unification effort, for a physicist or biophysicist to attempt relating physical entropy in raw or inanimate matter to the movement of change that galvanizes biotic mutational events. Mutation, in that context, would appear to be as natural as physical entropy. When physicists attempt to connect Life with Matter, many reason that the Universe has *fine-tuned itself for Life.* They reason that the conditions which allow life in the Universe to occur are only those under which certain universal dimensionless physical constants acquire a very narrow range of values, so that if any of several fundamental constants were to have ever so slightly different values, the Universe would likely disallow the establishment and development of matter, astronomical structures, elemental diversity, and life as it is currently understood in molecular Biology. Hence, they conjecture that the Universe must be fine-tuned.

If Mutational development of life is a replicate or continuation of entropy in matter, then we have shown in Book I that the movement of freedom leads to very specific values of the fundamental physical constants, far from resulting from an arbitrary fine-tuning of sorts. Our formalism towered by the elemental Mass/Space duality culminates in the theoretical discovery of the exact values taken by the fundamental physical constants as known. And that includes the very specific value of the *physical life constant,* which we shall discuss later. All the Laws that anchor the Regularities distinguishable in the Universe are a logical consequence of the most primitive irreducible subsuming its structures, which is the duality Mass/Space. Consequently, the Universe is not fine-tuned and the mutational movement operating in the Karyote Shell of matter represents but the breaking of the symmetry conformed in the void of the prior Shell of the Hierarchy, the molar Materials Shell.

21.6.1 Complexity in the Development of Life

It has been said many times before that the Universe seems to be evolving in increasing order of complexity since the spread before us goes from the microscopic to the macroscopic. What has apparently not been done is a close examination of the many different levels of the scale. The serious problems of interpretation resulting from examination of the scale levels may have dissuaded many from this enterprise, important though it has always been. The term complexity for the serial Suite or Spectrum of U_3 Shells is a rather vague term all in use throughout the Physics and Biophysics literature. We brought back this term in this framework only for the sake of connection between the latter and modern physics. We have repeatedly listed throughout this study the different elements making up each level of the complexity scale, and will do so here again in order to stress how the overarching Symmetries are built through Quanto-Geometric Quintessence. Never to be overstressed is the significance of intellectually representing how, for one given element at any level of the scale, two or more elements of the immediately preceding level of the scale inter-relate or combine to give birth to that one considered element, in a spectral triplet profile portraying a span from scalar preeminence to equated covariance to spatial preeminence. We describe below once again inter-association at each shell level of the 3D 9-Shell Hierarchy in U_3.

- *Quarks inter-relate to form elementary particles.*
- *Elementary particles inter-relate to form atoms.*
- *Atoms inter-relate to form molecules.*
- *Molecules inter-relate to form nano materials.*
- *Interlacing lattice groups inter-relate to form Karyotes.*
- *Higher Karyotes inter-relate to form singlet celestial bodies.*
- *Singlet celestial bodies inter-relate to form solar systems.*
- *Solar systems inter-relate to form galaxies.*
- *Galaxies inter-relate to form galaxy groups.*

Except for Shell 6, it is easy to garner consensus at first sight on the layout of this hierarchy in terms of the formation or composition of each Shell. Take the association at the Shell of Singlet Stellar Bodies as an undismissable predictor of pan consciousness, which is a view that is being ardently debated in Academia of late. No further comment.

21.6.2 What Distinguishes the Animate from the Inanimate

What deserves ample emphasis is how the graph of the $M_3(s)$ function keenly predicts the peculiarity in the ontological signature of the Life Shell as the range where the function asymmetrically reaches maxima (no other similar point on the graph), on one hand, and on the other hand, at each side of that Shell on the graph, we have two symmetric wings with sets of par points that have the same $M_3(s)$ ordinate value. If there is anything peculiar about the living in the context of the natural world, a feature of high praise in human culture, it is due to this asymmetry. The peculiarity in question refers for certain to the *"animate"* attribute of their ontological signature. If there is any commonality between the groups at each side of the Life Shell, in particular between galactic formations, solar systems (macroscope) and quarks, particles, atoms (microscope), it is due to the symmetry rooted in their similar values in the Shelling or Creation Function (ordinate axis), a feature by which we designate them *"inanimate"* objects. The most striking rendition of the Life Shell asymmetry is manifest in the disappearance of the spread of void in the ontology of constituent elements and a loss of monolithisation in the scalar compo nent, both contributing to create the *elastoplastic* topology typical of Karyotisation.

21.6.3 Inauspicious Notion of Emergence of Life

Our approach has already significantly narrowed down the boundaries or boundary conditions for a formal definition of Life, having perfectly framed the phenomenon within the formal Quanto-Geometric defining categories. Accordingly, one should not attempt to define Life as a so-called *emergent phenomenon*, as though the other developments germane to the other Shells of matter would be undeserving of this qualifier for themselves. This notion is no more adequate than the notion of a fine-tuned Universe i.e. It simply pre-supposes that everything in nature beyond biota is purposeless and devoid of meaning in and of themselves and that their only purpose is to offer human, in their supposed own emergence, context and means to be purposeful and meaningful themselves. We ought to overcome this limitation of scope.

21.7. Two Primitive Spectra Encompass the Ontologies of Life

In Biology thus far, the inability to unequivocally determine the beginning of the fulsome physical expression of life had been one of the caveats keeping out of reach a strict and assertive definition of life. Even though a few so-called *model organisms* had been identified, these interrogations have remained without clear resolute answers. The T-even virus has been identified as a model organism i.e., as much as the Euglena protist has, two organisms belonging however to two different orders of the living. Should we take the word organism to be another nomer for living body, one is then left to wonder what the word model in this context leads to. The model characterization seems to be adding to the confusion instead of granting more clarification. At the end, that lack of a definition for life might have been the principal cause of the rift in Western culture between the pro-life socio-political faction and the pro-choice faction, notwithstanding the fact that the division is explicitly centered around chronology in human gestation.

In the canvas of the Quanto-Geometric *M_3 Creation function*, following the Materials Shell is an entirely new one with integrants that span from viral entities to humans and perhaps other unknown human forms, as we shall discuss later. Most importantly, the incipient elements of the Shell in terms of their Karyotic kernel configuration we have clearly established in previous discussion. By that measure, viral entities are rightly the first sprouts of life in its fulsome expression. In Biology, we generally consider the lack of autogenous self-replication capacity as well as the lack of a clearly-set continual metabolic activity by viral entities to be deficient characters disqualifying them as fully living bodies. By the same token, we barely know what to make of the fact that many protozoa, which are unicellular eukaryotic organisms, show organelle precursors to histological organs in multi-cellular organisms, the eyelets of the Euglena protist i.e. Much has been debated in Evolutionary Biology concerning the role played by autotrophic v. heterotrophic capability, from the metabolic viewpoint, in the onset of the origin of Life. Again, these theoretical approaches represent attempts to define ontology from phenomenology, which have understandably failed to bring the expected resolute answers to the thorny question of the origin of life.

From the vantage of the Karyotic kernel set by the M_3 *Creation function*, there are two basis Karyotic distributions of Quanto-Geometric nature composing the topology of that Shell of matter. The mathematical-physics foundation of this development we have exposed in Section 14.2 of Book II of this treatise, discussed in the context of atomic orbital generation. We refer the reader to that exposition, in the hope that he or she will become familiar with the emulation functions that Nature implements on top the Quanto-Geometric primitives as an essential construction method in its developmental course. Inasmuch as quantum chemistry represents the foundations of all of Chemistry, including biochemistry and molecular biology, we strongly urge the reader with biology specialization to visit or re-visit Chapter 14 of the referred volume in its integrality ***(Periodicity of Electronic Shells in Unified Atom Architecture)***. Recognizing the significance of these emulation functions stands for a prelude to best possible understanding of the formalism subtending the development of life. Per the showcased quotation at top of the referred Chapter, when Richard Feynman wrote that "*it is possible for Mother Nature to only have one way of doing things and that she repeats her storyline [time and time again]*", he evidenced an incredible lucidity as a visionary scientist. The emulation functions reflect the wisdom of his premonition.

21.7.1 Drivers of the Karyotic Symmetry Group

While this discussion develops, the reader shall keep in mind the graph of the M_3 *Creation function*, which is paramount to the apprehension of how matter and life evolve in the 3-dimensional universe. Three main drivers shape the composition of the Karyotic Symmetry group within that development, as expressed below:

$Karyotic\_Kernel = \|6 - Operator \oplus 4 - Operator\| \times [Primitive\_Spectrum]$,
where $\oplus$ is the sign for Operator composition.

A Primitive Spectrum ultimately shapes the distribution of elements within the Group. It is the target object for modulation by two operators, the 6-Operator and the 4-Operator. These two operators act on that spectrum in order to imprint the final signature of the individual elements of the Group. We have formerly designated the 6-Operator with the nomer of Enthalpic Operator, and the 4-Operator we have formerly designated the Quadratic Operator. Further, the composition between the two Operators here is commutative or Abelian.

The 6-Operator constitutes the determinant for the election of proteic molecules as the biophysical brick of this Shell. Its provenance is its Shell6 typology within the primary M_3 distribution. As a reminder, proteic molecules are of Quanto-Geometric enthalpic type within the molecular spectrum, which makes them the natural elements of choice among the molecular set for the Operator. The 4-Operator is responsible for the elastoplastic fabric permeating the Primitive elements of the Shell. Its origin is the inner Quadratic span of the original M_3 distribution, a quality by which it represents the region where the functional derivative is of null value, setting the two functional variables of equated covariant influence on one another. Lastly, the individual elements of the Shell are cast according to an emulated Primitive Spectrum, as the target object to be modulated by the two aforementioned Operators.

The emulated primitive spectrum is innerly animated by a duality between a poster child of the original space wavefunction and a poster child of the original mass or scalar density component. From this determination is borne out another important characteristic of the living in the representation of the *male* and *female* qualities. The space wavefunction represents the footprint for the male quality, whereas the scalar density element represents the footprint for the female quality. One of the very first expressions of these two distinct qualities is the parity between the RNA and the DNA macromolecules in karyotic coordinate space. The agile single-strand RNA molecule readily carries out the male behavioral attribute, while the bulkier double-stranded DNA molecule carries out the female behavioral attribute. As a reminder, proteic molecules, which are the bricks of the Karyotic universe, so to speak, consist of one or more chains of amino acids linked by peptide bonds and are folded into a specific 3-dimensional shape maintained by additional chemical bonding.

This emulation framework sets the stage for all sub-ontologies, forms of expressions and activities to be developed in real Karyotic space or the Karyotic universe.

Accordingly, the elements of the virus phylum can either have a DNA or an RNA nucleic acid molecule, but never both. The same dictum determines why female DNA viroids develop biosynthetic activity centralized in the nucleus of their cellular host, whereas the RNA viroids develop biosynthetic activity within the cytoplasm of their host. In recollection that the nuclear area represents the Karyotic quantum-dominant region and the cytoplasm represents the Karyotic space-dominant region in a cell, we readily understand the natural choice of hosting medium or environment for the distinct development of DNA and RNA activities.

21.7.2 Intelligence Subsumes the Very Roots of Life

Arguably, the most generic possible definition of the notion of intelligence is the order of typologies which encapsulates both structure and function, as reflected in our previous study of quantum causal forms. Thus, intelligence may be viewed as the expression of covariant influence between the two prime constituents of the irreducible of matter at the heart of Quanto-Geometric Theory. In that sense, intelligence in its generic form is everywhere across the Shells of Matter. In phylogenetic Karyotic space, however, it acquires several boosts evidenced in the recurrence of the innerly covariant primitive spectra by way of emulation. On that account, Karyotic space develops the remarkably intelligent covariant phenomenon of synthesis, also known as biosynthesis. Under our umbrella however, the concept applies to both multiplication or self-replication activity and metabolic activity.

For a virus to multiply, it must knowingly invade a host cell and take over the host's metabolic machinery. Multiplication cycle of viroids, virions and viruses, can be divided into several distinct progressive stages, which are attachment to the host, penetration of the latter, build of viral components, maturation and release, all of which imply intelligent activity and particularly biosynthetic activity when it comes to component building.

The organism undertakes the following:

1. Synthesizes many copies of DNA
2. All undertaken RNA transcription is mRNA for biosynthesis of phage enzymes and capsid protein.
3. Host cell's ribosomes, enzymes and amino-acids the organism uses for translation.
4. Viral enzymes are almost entirely concerned with replicating or processing viral nucleic acid. They are almost never concerned with the machinery of protein synthesis toward block building the entity or energy production. Most viral enzymes (or virus-specific enzymes) – the enzymes encoded in viral nucleic acid – are not part of the virion's body.

Viral enzymes are synthesized and functional only when the virus is within the host cell. While the smallest naked virions do not contain any pre-formed enzymes, the larger virions do contain one or a few enzymes, which usually function only in helping the virus penetrate a host cell or replicate its own nucleic acid during infection of host cells.

5.- For several minutes following infection, complete phages cannot be found in the host cell. Only separate components can be detected: DNA and proteins. The time period from penetration to complete viral multiplication without yet the presence of infective virions is called the eclipse period.

21.7.3 Assembly at Maturation

Following synthesis of viral components, phage DNA and capsids are assembled into complete viral units within the host cell. For many simple viruses, nucleic acid and capsid proteins assemble spontaneously to form virions, without the intervention of other phage gene units. For complex virions such as T-even, viral genes within the host's DNA guide the assembly process in a step-by-step sequence. Phage heads and tails are separately assembled from protein subunits. The head is packaged with the phage DNA and the tail attached.

The phage DNA even contains a lysis gene controlling the eventual lysis of the host membrane through the synthesis of lysosomes at the end of the assembly process. This final step releases the thousands of assembled virions in the area inhabited by the host cell. Alternatively, lysis may not be chosen but incorporation of phage DNA into the host's DNA, in which the virion is called a prophage.

21.8. Primitive Spectrum of Viruses and Bacteria

We are now going to take a look at what might be designated the prokaryote Primitive Spectrum, formed by Viruses and Bacteria. The post-quadratic or space-dominant sector is entirely made up of bacteria, whereas the pre-quadratic or quantum-dominant sector is entirely made up of viral entities. For its own, the quadratic layer is made up of a hybrid type of entities, to be denoted bacterium-viruses.

1.- ***Viroids***: infamously called pieces of genetic material or RNA macromolecule. The difference between a flat piece of RNA macromolecule and a viroid is equivalent to the difference between an unbound free proton and an atomic bound proton such as an H^+ ion.

2.- ***Virions*** (also denoted naked viruses or non-enveloped viruses) comprising the following types: non-enveloped helicoidal, non-enveloped polyhedral and non-enveloped complex.

3.- Simple **Membrane-bound Viruses** (or Viruses with simple envelope) comprising: adenoviruses and the following simple-envelope types: helicoidal with envelope, complex with envelope and Polyhedral with envelope.

4.- ***Bacterium-viruses*** comprising: karyotic entities with a hybrid viral-bacterium structure (poxvirus-like complex viruses) and bacteria with viral genotype (Chlamydia, rickettsiae)

5.- ***Mycoplasma Bacteria***

6.- ***Endosymbiont Bacteria***

7.- ***Gram-positive Bacteria***

8.- ***Commensal and Cometabolic Bacteria***

9.- ***Perdurable Bacteria*** comprising: heterocysts, actynomycetes, sheath- or bud-laden bacteria

21.8.1 Viroids as Primitive Layer 1 of the Life Spectrum

Viroids are the smallest infectious pathogens known. They are composed solely of a short strand of circular, single-stranded RNA that has no surrounding protein coating (Fig.21.2). All known viroids inhabit higher plants, in which most cause diseases, although some hosts remain free of symptoms and only serve as vectors of infection.

Viroids were discovered in the works of researching the agent of potato spindle tuber disease. In 1971 Theodor O. Diener showed that the agent of potato spindle tuber disease was not a virus, but a totally unexpected novel type of pathogen, 1/80th the size of a typical virus, for which he coined the term "viroid".

The short segments of naked single-stranded RNA making up Viroids consist in about 246 to 467 nucleobases. In 1978, Gross et al. determined the complete nucleotide sequence of potato spindle tuber viroid. The single-strandedness and circularity of viroids was firmly confirmed under electron microscopy.

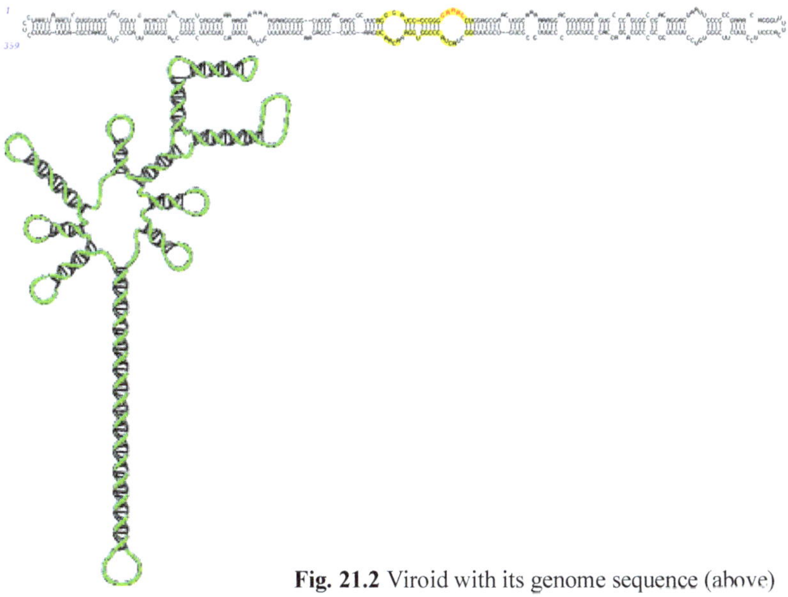

Fig. 21.2 Viroid with its genome sequence (above)

Viroids are devoid of a protein coat, unlike viruses or adenoviruses. The nucleotides are often internally paired, so the molecule has a closed, folded, 3-dimensional structure presumably keeping it safe from attack by cellular enzymes. Uncharacteristically, the RNA strand does not code for any protein. It is important to understand that the fulsome composition of a viroid is both the macromolecular mass and the spatial spread that keeps the molecule as a 3-dimensional entity. This is a dictum from the Karyotic Symmetry group that creates the viroid entity. The viroid entity is not, repeat **is not**, the macromolecule alone.

Viroids replicate autonomously in the nucleus (*Pospiviroidae*) or chloroplasts (*Avsunviroidae*) of plant cells through an RNA-based mechanism executed in three steps. They require RNA polymerase II, a host cell enzyme normally associated with synthesis of messenger RNA (mRNA) from DNA in the eukaryotic cell. In this case the polymerase catalyzes instead the synthesis of a new "rolling circle" RNA using the viroid RNA strand as a template.

The fact that viroids only show activity when located within eukaryotic plant cells or chloroplasts is an unmistakable indication of their quantum-dominant nature, given that the gauge group of those eukaryotic cells are of the same quantum-dominant nature, as we shall later see. In that manner, they obtain an ontological boost which over-potentiates their capabilities. Since the scalar representative in the entity is the molecular mass, the multiplication capability of that mass orchestrates the potentiation effect.

One with good reason however, may ask the question as to the reason why they only replicate but do not synthesize any component or building block. First of all, there is no superstructural proteic building block to synthesize. Secondly, as a single RNA male mono-molecule originating from the wavefunction of the Karyotic gauge group, the requirement of self-consistency is already satisfied, a la inanimate crystalline materials, so to speak, to which they remain arguably close across the symmetry-breaking point. Therefore, they need not develop a biosynthetic or metabolic activity. What they are devoid of is the scalar-mass component to accompany the male RNA macromolecule, for the complete duality function. That is precisely what they find in the DNA macromolecule of the host to which they relate through coupling. In a sense, it is a parasitic association that gives them what they lack - a DNA strand for their own mass multiplication function.

Additionally, the closed circular geometry of the mono-molecule is another indication of its type 1 ontology, given that circular symmetry corresponds to the eigenform characteristic of Primitive Layer 1 in the Quanto-Geometric canon. Therefore, three attributes stamp the quanto-geometric typology of viroids:

1. Circular symmetry
2. Singlet monolithic structure as a single RNA strand
3. Self-consistency trumping metabolic activity

Viroids are extremely small in size. With just about a total of a few hundred nucleobases (little less than 500 maximum), they consist of fewer than 10,000 atoms. In comparison, the genomes of the smallest known infectious viruses are around 2,000 nucleobases long.

It is important to note the following facts from a biological evolutionary perspective:

1. Viroids' small size strictly is a function of Primitive Layer 1 to which they belong.
2. Their known high guanine and cytosine content admittedly are likely to increase the level of stability of the molecule and fidelity in replication alike.
3. With a circular structure, they are able to develop complete replication without the use of genomic tags.

In Diener's hypothesis of a pre-cellular RNA world formed by viroids, he interpreted their small size to be a measure against error-prone replication. By cutting on size the macromolecule hypothetically reduces the chances of error in replication. Diener viewed several other unique viroid properties as selective attributes over other RNA strands to perform certain critical steps in the evolution of life from inanimate matter (abiogenesis), as listed below:

1. Existence of structural chromosomic periodicity, which permits modular assembly into enlarged genomes.
2. Lack of protein-coding ability, consistent with a ribosome-free habitat.

3. Replication mediated in some by ribozymes — putative fingerprint of an RNA world.

While these views are in keeping with the evolutionary schemes that biologists had traditionally advanced in the history of the science to explain the development of biological complexity, they do not reflect an understanding of the biophysics underlying these developments. Of particular importance is de-legitimizing the label of "pathogen" ascribed to these biotic entities for their main qualifier, even though they may indeed be lethal or otherwise detrimental to their hosts. From the Quanto-Geometric perspective, one ought to comprehend medium or habitat strictly on a physical topology basis and within the view of a hierarchy of degrees of freedom. Whether a viroid is detrimental or innocuous to its medium of inhabitation can never be a constitutional qualifier from an ontological or structural standpoint.

21.8.2 Virions as Primitive Layer 2

The ontological signature of Quanto-Geometric primitive layer 2 is the binary or bi-polar stamp. Although the proteic composition of the capsid coat is of import in distinguishing one virus from another in the group, the presence of the capsid stands for the group or family an identification mark of virions (Fig.21.3). Virions feature two and only two structural elements: the central DNA/RNA macromolecule and the circumventing proteic capsid.

Functionally, the capsid, with its topology of capsomere units, is largely considered a protective layer to the central macromolecule. This view is for certain consistent with the 6-Operator, whose material genotype or typology is *coupling through embedding* (or Dissimulation - Camouflage) as one of the driving vectors in the formation of this Primitive Spectrum. Why we do not see a capsid structure juxtaposed instead to the chromosomic macromolecule must be understood in terms of a determination by the 6-Operator. The concentricity of the capsid structure with respect to the chromosomic macromolecule is in that sense an ontological trait more than a functional attribute or a phenotype.

While all capsomeres may be identical in a virion and in others of different proteic composition, the capsid takes the active role of the wavefunction property in accompaniment with the scalar density property assumed by the central macromolecule. Its outer structural location is concordant with that role as well. From a morphological point of view, most virions are either helical or icosahedral in structure. Molecular biologists believe that the symmetry of these shapes permit viroids to use the same component protein multiple times in order to build their structures from a minimum number of distinct protein species. This strategy, they believe, keeps at a minimum the amount of genetic material viruses in general need to encode structural proteins. While these assumptions may be true in terms of covariance dictating phenotype, it is important to recognize that the related preeminent topological driver in the structure of virions is the crystalline solid or lattice arrangement by adoption of elements of the prior Shell, the Materials Shell, in the construction of higher-level complexity. Strong evidence of this artifact is in the polyhedral or icosahedral shape of the capsid in these entities as well as the helical tubular shape in many others, where topological lattice symmetry is in order. More generally, the capsid macrostructure must be understood in terms of a sketchy reproduction at a higher level of the interlacing molar population distributions of the Materials Shell. The latter is further clearly elicited in adenoviruses with an icosahedral capsid where *Penton* bases and *Hexons* blocks are interspersed within the capsomere topology of the capsid in a regular fashion.

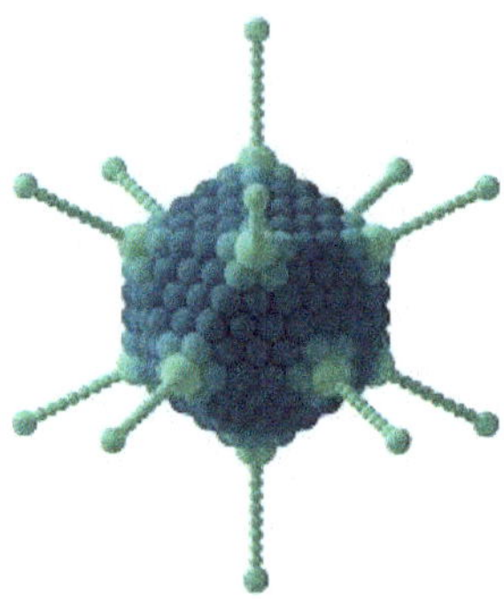

Fig. 21.3 An adenovirus

At the end, what we see in the virionic structure is a bi-polar or binary specialization whereby the central macromolecule conserves its multiplication role and a circumventing capsid arrangement assumes the dynamics of securing the integrity of the central macromolecule as well as implementing its mobilization, in particular through fiber spiking appendages. In this sense, we ought to consider that the wavefunction property inherent to the spread of real karyotic coordinate space within the entity has been extended into the molecular structure of the same.

Of note is the fact that in one of the most complex forms of virions, the so-called complex virus of the T-even type, the wavefunction property takes a very highly developed form of implementation. T-even shows not only a polyhedral capsid head containing the central chromosomic macromolecule, but also a vertical hollow spiral body similar to the capsid morphology of a helical viroid, a few spikes for attachment and six bent tail fibers acting as legs for support and locomotion. Once the viroid T-7 has landed on a bacterium, it is able to flex its previously folded fiber tails for grip. Molecular microbiologists have found that they prospect the bacterium membrane "searching" for the best suitable protein patches to home in for penetration. A research group has even recently discovered that T-7 may well *walk* over the membrane surface during prospection. In this case, it is fully justified to posit an even higher form of bipolar specialization in the complex viroid, whereby the capsid/macromolecule head represents the multiplication scalar pole of the entity and the entire set of appended structures to the head a wavefunction-propagator pole to the same.

According to the logic of the quanto-geometric primitive spectrum, T-even is thus not a virus per se, as commonly classified, but a viroid indeed, or an advanced one. It must be considered so for the additional reason that it is devoid of the third structure that is required to qualify as a virus, as we shall shortly see.

21.8.3 Viruses as Primitive Layer 3

In layer 3 of this Primitive spectrum, we classify viruses with an envelope making up the third molecular structure composing the element (Fig.21.4). We properly designate these entities viruses in their own right. The further extension of the ontological wave-function-propagation component is responsible for the envelope structure, which will bear, as we will shortly see, the phenotype arising from that ontological tenet.

The envelope usually consists of lipids, glycoproteins and carbohydrates. Specifically, the envelope is composed of two lipid layers interspersed with protein molecules, a lipoprotein bilayer. Aside from own synthesized material by the virus element, the envelope may also contain material from the membrane of a host cell. Effectively, in this type of element the wavefunction-propagator tenet bears such a superior potential in terms of propagation in interactions, as compared to virions, that the viral element may altogether pull a manifold piece from the large host membrane to constitute its own smaller envelope. That action represents a strategy to acquire the entire set of needed lipid molecules in their pre-organized placements and its breadth of action effectively represents the capability of prehension. This so-called budding process of the host's membrane is undertaken under the virus's directive through the interaction with the host cell, and matter-of-factly equates an action of grabbing or prehension by the virus. Subsequently, the virus will proceed to notably replace the proteins in the borrowed cell membrane with its own proteins, thereby creating a hybrid structure of virus-derived proteins and cell-derived lipids, at the most minimal cost possible.

The population of synthesized spikes over the envelope further expands the spatial interaction capabilities of the viral element. The influenza virus i.e., may use its spikes, not just to attach to one animal red blood cell host, but to even clump several of these cells in what is called hemagglutination. In general, the viral envelope is roughly spherical, while being highly pleomorphic due to its elastoplasticity. The spikes endow the entity with a tremendous displacement capacity. Their equivalent among virions would be the complex T-even tails. However, in viroids, the tails are used over the one single host in order to allow the entity prospection for a suitable attachment site on the membrane. The spikes projecting from the virus's envelope confer to the entity medium-wide displacement ability.

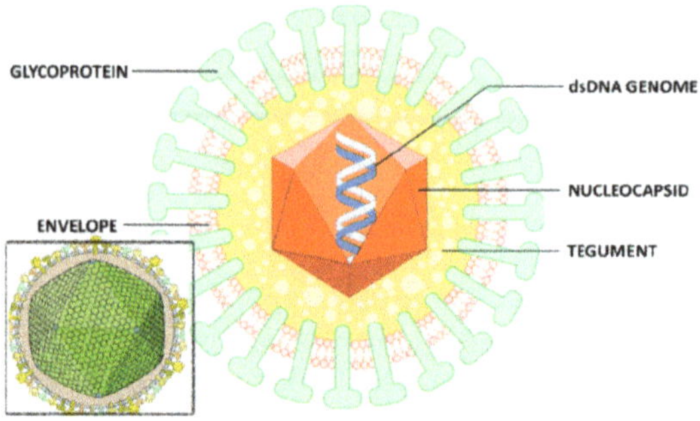

Fig. 21.4 Virus with a coat or envelope

These spikes are such a reliable characteristic for most viruses that they have been used as a means for their identification. It is fair to conclude that the set of characters developed by the envelope in these tri-partite viral entities is highly reflective of the *Relational* typology inherent to Primitive Layer 3 in the Quanto-Geometric canon. Example of these viruses are the aforementioned influenza virus and the herpes simplex virus.

21.8.4 Bacterium-Viruses as Primitive Layer 4

Primitive Layer 4 comprises organisms formerly classified to be intermediary between virus and bacteria known as Chlamydiae and Rickettsiae. We also add to this Layer the peculiar Orthopox family of virus. We shortly discuss the peculiar characteristics that qualify the latter as intermediary between viruses and bacteria, similarly to Chlamydiae and Rickettsiae.

21.8.4.1 Orthopox Virus as a Bacterium-Virus

The primary organisms of this Layer are the Orthopox family of viruses (Fig.21.5), a typical member of which is the human virus known as smallpox or variola virus. These viral entities singularly show continual molecular spread inside their karyotic box, eliminating the actual spread of coordinate space that is part of the composition of previous entities in the lineup of the spectrum. As previously explained, the molar *elastoplastic* structure is a dictum emanating from the quadratic typology in karyotic phylogenetic space. In this instance, it is by virtue of a quadratic boost (or operator overloading) brought about by the 4-Operator profiling the formation of the Spectrum per equation x, that the end result of generalized elastoplastic structure throughout the entire structural organization of the element is obtained. What used to be empty void of space inside the kernel box between capsid and envelope is now, under that boost, filled up with material from the so-called lateral body. This molar population is said to be "lateral" to the chromosomic center; however, it is best visualized instead as spread-out content all around the chromosomic center up to the outer envelope and materializes a *primitive cytoplasm* surrogating coordinate space of its replacement.

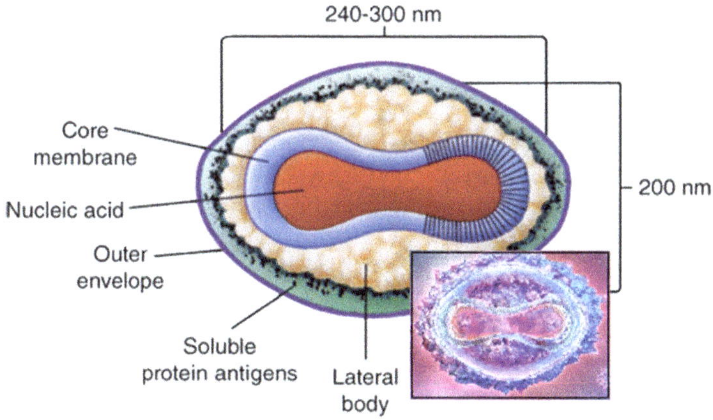

Fig. 21.5 An Orthopox virus as a bacterium-virus

The chromosomic material for itself has grown to fill up the entire nucleic compartment, previously the space between the central molecule and the capsid. The latter is now generally designated core membrane or wall. It is so adapted to propagation that pore-like structures are already spanning this core membrane. Although no function has been yet determined for this structure in microbiology, it could admittedly be involved in the extrusion from the core of the viral mRNA during early transcription. It is clear that the wavefunction propagator intimately participates in the configuration of this structure, as it essentially does the rest of the outer structures from the core membrane on up.

Potentiation of the wavefunction propagator in Poxviruses is further evidenced in the capability for these DNA-viruses to replicate in the very cytoplasm of the animal host cell rather than in the nucleus, a unique capability among DNA viruses.

Variola is a large brick-shaped virus measuring approximately 302 to 350 nanometers by 244 to 270 nm. It showcases a single linear double stranded DNA genome, 186 kilobase pairs in size. Two classic varieties of smallpox are known: variola major and variola minor. The viral envelope is made of modified Golgi membranes containing viral-specific polypeptides, including hemagglutinin.

The most important structural aspect of this family of virus is their bulk consistency throughout the entire karyotic box, which qualifies them as incipient or primitive cell of prokaryotic organization, much beyond virionic organization as currently held in the field. In particular, we insist that the so-called lateral bodies are the patent pre-formation of a cellular cytoplasm. The so-called lateral bodies flanking the core presents an amorphous structure composed of proteins of unknown role or function thus far. This structure is acknowledged to be one of the most intriguing features in poxvirus karyotic organization in terms of its origin and composition. Not all that intriguing however under the Quanto-Geometric approach, in light of the terms of the analysis underway.

Microbiologists hypothesize that one possible function of the proteins in the lateral bodies is virus-host interaction, helping the virus escape innate cellular antiviral defenses. In this case, the lateral bodies would hypothetically have a similar role to the tegument in the *Herpesviridae*. These functional views, which ignore ontology and the intervention of ontological drivers, fail to capture the no less important fact that the nucleotide area is as filled up with molar populations as the area between the core membrane and virus envelope. They also fail to give an interpretation to the concomitant disappearance of coordinate space within the entity altogether.

Poxviruses have differing life cycles, many showing dependence on occupied medium in the so-called infectious forms, with differing mechanisms of cell entry. In order to replicate, poxviruses produce a variety of specialized proteins not produced by other DNA viruses, the most important of which is a viral-associated DNA-dependent RNA polymerase. In all, the best visualization of the poxvirus family of viruses is that of a prokaryotic cell in formation or a pre-prokaryotic cell yet retaining the main characteristics of the viral structure. That is to say, a *bacterium-virus* hybrid entity.

21.8.4.2 Chlamydia as a Bacterium-Virus

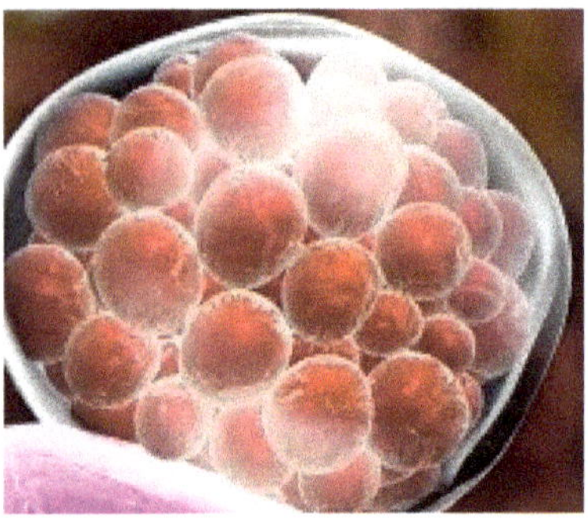

Chlamydiae have a morphology and internal structure typical to Bacteria. They host a well-defined viscous cytoplasm, with reticulate replicates, a collection of reticulate bodies, otherwise called inclusion bodies. In the central region, chromosomic DNA and RNA macromolecules, as well as ribosomes, agglutinate at the heart of the karyotic kernel. In that they clearly depart from viruses which may only contain one type of macromolecule between DNA and RNA. This genotypic trait marks one of the stamps of prokaryotic cellular organization. By the same token, we expect other features to express the gender of prokaryotic cells, as we shall later discuss. An external cell wall compartmentalizes the entity, but lacks peptidoglycans. Because of the above traits, Chlamydiae have now come to be classified as bacteria. The group consists of the single genus: Chlamydia *(order Chlamydiales, class Chlamydiaceae)*.

A singular trait of Chlamydiae is that they cannot make their own ATP and require the internal environment of live cells for growth of the kernel box, in contrast to normal Bacteria. In addition, they depend on host biosynthetic machinery for multiplication, akin to viruses. For these reasons, the Chlamydiae were originally classified as viruses. This taxonomic ambivalence resulting from the nature of both genotype and phenotype is indicative of the ontological hybridity of this genus which is truly half-way between viruses and prokaryotic cells, a well predicted feature in the Quanto-Geometric primitive canon.

21.8.4.3 Rickettsiae as a Bacterium-Virus

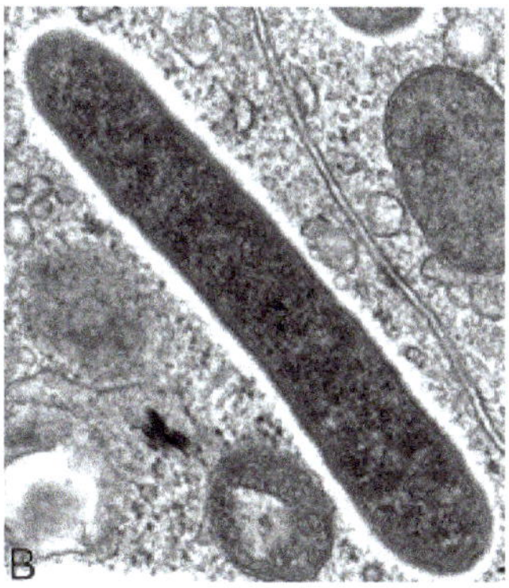

Having been discovered at the end of the first decade of 20th century by American medical microbiologist, H.T. Ricketts, while studying the Rocky Mountain spotted fever, in the current taxonomy these entities take their designation after their discoverer and in his honor. The morphology of Rickettsiae varies from coccoid to rod-shaped to filamentous forms. They have a rather small kernel, measuring between *0.3 to 0.7 μm* in width and *1 to 2 μm* in length. They possess a cell wall of the gram-negative type and a cell membrane, feature pleomorphism, and contain both RNA and DNA macromolecules. Nevertheless, they remain non-motile, a key ontogenic feature, in comparison to fully developed bacteria which are fully motile. A hybrid feature that is consistent with our definition of intermediate taxon for Layer 4 of any spectrum.

Like the Chlamydiae, Rickettsiae are obligate intracellular elements, as such currently designated in microbiology as parasites of the arthropod cells of lice, fleas, mites, and ticks which they inhabit without injury. They are generally cultured either by inoculation into the yolk-sac of chicken eggs, or by infecting host cells in tissue culture. Because they are endowed with the capacity to produce their own energy (ATP) through oxidative phosphorylation with the help of a cytochrome system, such level of autonomy, superior to that of Chlamydiae, has prompted microbiologist to envisage their culture outside of host cells. However, to date Rickettsiae have not yet been cultured in the absence of host cells and only exist in nature within host cells. Their cytochrome c, a to-be common bacterial feature, is primarily known for its function in the mitochondria as a key participant in the life-supporting function of ATP synthesis. Note, however, that when a cell receives an apoptotic stimulus, cytochrome c is released into the cytosol and triggers programmed cell death through apoptosis. It is important to note that the penetration of a host cell by a Rickettsia constitutes an active process requiring both host and the intracellular Rickettsia to be alive and metabolically active.

Once inside a host cell, the Rickettsiae multiply *primarily in the cytoplasm* and continue replicating until the host cell is completely filled up with their progeny. They multiply by normal binary fission with doubling times of about 8 hours. At the end of the multiplication process, the host cell is forced to lyse and the new generation of Rickettsiae released into the surrounding fluid.

In living and multiplying endoparasitically within the cells of their arthropod vectors, Rickettsiae demonstrate a phenotype that strongly relates them to the entire virus genus. At the same time, it is clear that these organisms hold a bacterial-type kernel box in its layout and composition. Because of this hybrid characteristic in genotype and phenotype, for many years these organisms were rightfully considered to be biologically intermediate between bacteria and viruses. They have now been wrongly viewed to be bacteria and inappropriately included in bacterial classification. This view must be flagged erratic from the standpoint of the fundamental drivers of life.

21.9. Primitive Layer 5: Mycoplasma and Spiroplasma Bacteria

At this layer of the Karyotic spectrum, we classify the first fully developed Bacteria, typically the Mycoplasmas and the Spiroplasmas. The typical traits of these organisms are that:

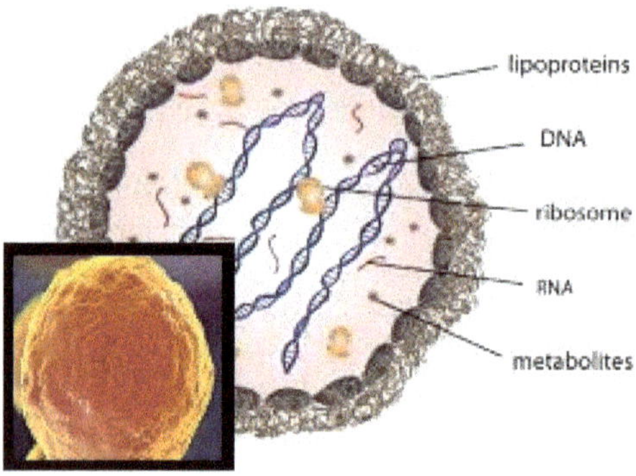

1. They only bear a cytoplasmic membrane for an envelope, lacking a cell wall, and
2. They are completely autonomous in terms of replication and metabolic activity. They are individually very small in size ranging from *0.25 μm* to *1 μm*, about the same size as the organisms of the previous Layer.

21.9.1 Role of Lipids as Mediators of Wavefunction-Propagator in Cells

Lipids are organic compounds consisting in utterly non-polar molecules that are insoluble in water. They are however readily soluble in non-polar solvents such as ether, chloroform and alcohol. *Lipids are essential to the structure of membranes in the cellular world.* It is noteworthy that, within the molecular spectrum, lipids constitute the main elements bound by the *entropic* form of chemical association in quanto-geometric analysis. It will be no surprise to see them here in the role of materializing the covariantly dominant wavefunction-propagator tenet under the *entropic* typological stamp (type 5).

Simple lipids or fats contain an alcohol called glycerol with a group of compounds known as fatty acids. A fat molecule is formed when a molecule of glycerol combines with one of the three fatty acids to form either a monoglyceride, a diglyceride or a triglyceride. A large variety of lipids are known because of the many different structures taken by fatty acids. The reverse reaction by fat molecules breaking down into their component fatty acids and glycerol molecules release energy. *Simple lipids thus serve as energy storage materials* in the cellular kernel box.

Complex lipids for their own contain not only the carbon, hydrogen and oxygen atoms found in simple lipids but in particular such elements as phosphorus, nitrogen and sulfur as well. Note well that phosphorus and nitrogen stand for **Group 5** elements in the periodic table of elements. Again, another entropic stamp dictating the choice of phospholipids as components of the entropic plasma membrane in the karyotic world. To understand why phosphorus and sulfur are energy propitiators, we ought to remember that they eponymously form the main elements of our matches, where phosphorus *ignites* and sulfur *burns* as fuel. And to understand how they come to behave as energy propagators, we ought to remember that photon particles essentially mediate the action in energy liberation and that in the particle spectrum at Sell 2 of matter, photon particles, endowed with wavefunction dominant covariance, initiate the formation of that topology with gamma-X-ray photons at Layer 5 of this distribution (Book II). Energy ultimately comes down to wavefunction dominant covariance.

Complex lipids bear the name of phospholipids and are specifically made up of two fatty acids and a phosphate group replacing what would be the third fatty acid. Phospholipids represent the essential structural component of cellular membranes. They present both polar and non-polar regions which dictate their behavior within their medium of inhabitation. The polar regions, consisting of phosphate groups and glycerols, react with water and make the molecule twist in the presence of water, while their non-polar regions consisting of fatty acids allow the molecule to make contact with non-polar regions of neighboring molecules. From a functional viewpoint, it is this selective behavior of phospholipids that make them particularly suitable for cellular membranes in their interactions with the water-based habitat. By the same token, it is important to understand that, analytically, function is a dependency of ontology.

Akin to simple lipids, one of the most important roles of phospholipids is the generation of the indispensable energy to be used for the many cellular processes. Generation of energy is carried out via Adenosine Diphosphate (ADP) and Adenosine Triphosphate (ATP) through the transformation of ADP into ATP. An electron transport chain is created whereby a transfer of electrons from one acceptor to the next releases energy converting ADP into ATP. This process occurs in the plasma membrane of prokaryote cells or bacteria. ATP then becomes a reservoir of energy for the cell. Worth highlighting that the referred electrons are not free electrons but bound electrons within atoms and/or molecules. As such they exist in quantum Shells or orbital density states and it is the transitioning from one upper quantum Shell level to another lower one throughout the transport chain that creates or liberates the energy in the ultimate physical form of photon packets. Of paramount significance is this reconciliation of the chemical view of cellular processes with the very physics of matter mediating them, from a physical science perspective, for a comprehensive grasp of the development and evolution of involved states and processes.

So therefore, it is important to keenly comprehend that the notion of energy, or more precisely chemical energy, amply used in microbiology to explain cellular phenotype, is nothing else, from a physical perspective, than the manifestation of the material wavefunction-propagator in karyotic coordinate space, notwithstanding its *overloading* modality. The broad functional amplitude expressed by the cellular membrane at this Layer of the spectrum is reflective of the dominant character taken by the wavefunction over the scalar representative, consistent with covariance as developed in the Primitive Spectrum of the Quanto-Geometric canon. The preponderance of the karyotic material wavefunction starting at Layer 5 of the karyotic spectrum takes other significant characteristics which we shall shortly discuss.

21.9.2 Preeminent Ontological Role of the Plasma Membrane

The metabolic-synthetic ability of the simple bacterium, amply deployed by the cytoplasm and the cytoplasmic membrane, is so overwhelming that it seems to even subjugate the replication processes coded by the central DNA macromolecule. Effectively, the orchestration role played by these structures in regards to DNA replication seems to have been largely under-estimated in microbiology, given the emphasis always given to the chromosomic aspect of transcription and translation activities. If it is true that the DNA macromolecule possesses the genic template for the key structural and functional protein molecules, cytoplasm and notably the cellular membrane play outright the conductor role of the replication orchestra.

Clear experimental evidence has shown that the phenomenon of cellular division is initiated by the plasma membrane and the cell wall when present. Since the membrane is always aware of the nutritional conditions of the medium, once favorable conditions are detected, it initiates the formation of a wall section within the equatorial plane of the cell while sending the signal through mRNA to the chromosomic DNA molecule to prepare for replication.

Further the prokaryotic plasma membrane possesses internal invaginations denoted mesosomes that equally play a role in the replication process of the DNA macromolecule (Fig.21.6). There have been in the past considerable debate about the role and function of mesosomes. Mesosomes appear to develop during the replication stage of the cell. Mature mesosomes form a revealing link between the cytoplasmic membrane and the nucleus of the cell. The invaginated structures comprise in many cases vesicles and tubules of lamellar whorls. The arrangement of mesosomes observed in dividing *Escherichia coli* cells had led to the hypothesis that cellular division in these elements is accomplished by two separate polar mesosomes. One mesosome deriving from the parent cell remains present at one pole of the daughter cell. The other freshly synthesized one remains in the vicinity of the newly forming pole of the daughter cell. As the cell grows and elongates, the two mesosomes move apart from their attachment to their respective chromosomes, thus effecting nuclear division in this particular bacterium. What is certain at this point as a generalization is that mesosomes are found for the most part in association with nuclear area or near the site of cell division. There is no doubt that the perspective hitherto adopted in microbiology which gives primacy to chromosomic DNA in cellular phenomenology is responsible for the little research conducted on mesosomes, leading to the ignorance of the preeminent role they play in cellular division as a structural component of the plasma membrane.

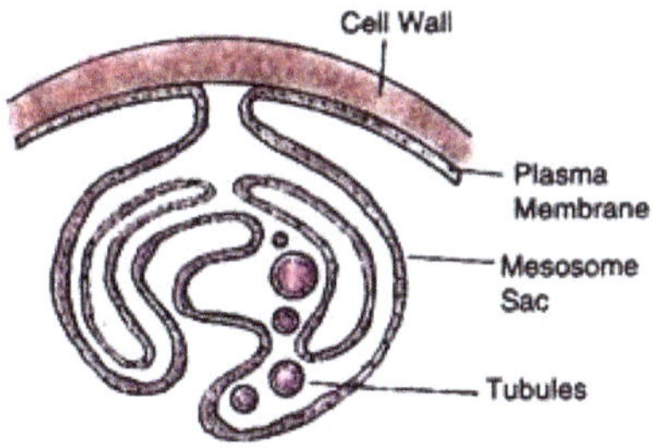

Fig. 21.6 A bacterial mesosome

In all, in a simple bacterium like the Mycoplasma or the Spiroplasma, once the replication process is complete, the transverse septum representing a projection of the plasma membrane cleaves the cell in its equatorial region. In so doing the transverse septum divides the cytoplasm into two portions which will each host one of the two newly formed DNA molecules.

On another hand, in considering the DNA replication process itself, the association of the plasma membrane with every one of the replication points of the macromolecule in terms of the substances that converge at these points is a given. In *Sporosarcina ureae* bacterium, one of the two mesosomes can be clearly observed extending from the innermost edges of the septa through the cytoplasm and into the replicating DNA nuclear area. The replicating substances are unwinding proteins, stabilizing proteins, ligase proteins, the DNA polymerases and others. This string of enzymes is necessary to the work of replication itself. Some of these proteins are needed to hold the two helical chains steady, others are tasked to stabilize the opened points of replications in order to prevent rupture during the replication process, and yet others are tasked to bond the newly synthesized bases to the rest of the molecule in formation from the replication points. Subsequent to all of the above, the plasma membrane, assisted by the mesosomes, undertake to control the partitioning of the cell into two new daughter cells by extending the septa into a transverse wall between the two poles of the cell now with two central DNA molecules each. When the dividing wall is complete, the two new cells finalize their separation from one another.

In Mycoplasma bacteria, preeminence of the membrane structure is such that it takes a super-structural form of organization whereby complete physical separation between the new cells does not occur, so that an entire colony of cells *appears* as a continuous single structure portraying an endless membrane. Note that there are even pure arm segments of membranes between cells at many locations of the super-structure. Note finally that the high sterol content of the membrane, an Entropic Operator overloading feature as previously indicated, facilitates the extension of the super-structure through *pleomorphism* of the unit cells.

21.10. Layer 6 of the Karyotic Spectrum Characterized with Unique Formation of Cell Walls

At this layer of the prokaryotic spectrum, the taxa are represented in prokaryotic boxes which have been included in the larger category of "eubacteria" (for true bacteria) in microbiology. The creation of this rather loose category of eubacteria seems to capture the notion of a typical bacterium, attesting to the fact that these organisms appear to be complete in structures.

Fig. 21.7 A flagellum

We ought to note that, in the matter of taxonomy, the science of microbiology still shows an ostensible level of immaturity, with a plethora of taxonomies established on controversial criteria sparking incessant debates in the field. Eubacteria seems to be another one of these less useful classes or taxa, given that what makes for true defining ontologies versus functions or phenomal derivates is not at all understood in the field.

21.10.1 Manifestation of the Wavefunction-Propagator in Flagella

One of the ontological features attributed to the layered progression in the Quanto-Geometric canon is of course the progressive enhancement of the wavefunction-propagator from one layer to the next one up. Consistent with this feature, one may predict that the cell envelope structure is hereby going to acquire enhanced spatial capabilities. Effectively, all so-called eubacteria present extremely developed surface appendages granting them tremendous mobility within their environment, if not marked ability to attach to other external structures. Two types of surface appendages can be recognized on these bacterial species (Fig.21.7): the flagella, which are organs of locomotion, and pili (hairs in Latin), which are also known as fimbriae (Latin for fringes) and serve in interaction through contact. Flagella occur on both so-called Gram-positive bacteria (positive reaction of cell wall to Gram stain) and Gram-negative bacteria. We ought to note that pili were already present in organism pertaining to Layer 5, which grow from their plasma membrane. In the Gram staining categorization, these organisms have demonstrated to be Gram-negative. That said, the pili flora can become very dense on Layer 6 as a new development.

The presence of flagella is to be forthrightly taken for one of the characteristic new traits of magnified mobilization incarnating the enhanced wavefunction-propagator at this layer. Flagella give these bacteria a manifold increase in mobilization ability resulting from direct execution of the mechanical wavefunction. Flagella can be several times longer that the bacterium. While they may be located at one extremum of the bacterium box in a set of three or more, they may be packed as well at both ends of the bacterium box. Their proximate end features a very ostensible mechanical arrangement in rings or spools and a hook that both secure them inside the bacterial envelope and allow for swift rotations of the flagella. Each flagellum works as a semirigid helical rotor that stirs the cell with their own rotation from the basal arrangement, similar to the movement of the shaft of an electric motor. The rotation may be either clockwise or counterclockwise around the axis perpendicular to the envelope. The rotation of the flagella compressed in bundle ultimately cause the propulsion of the bacterium through the liquid medium. While the conversion mechanism from the biochemical basis of this action overall to its mechanical expression remains unknown, it is well known that it depends on the cell's continuous generation of energy through ATP, a materialization of the wavefunction-propagator discussed earlier.

Further, the wavefunction-propagator gives flagella the mechanical ability to alter the speed and direction of their rotational motion, making them capable of various patterns of motion. When a bacterium continually moves in one same direction, the movement is said to be a "run" or a "swim". Runs may end in an abrupt change of direction designated a "tumble" before it resumes again in that other direction. A tumble is the result of a sudden reversal in the direction of rotation of the flagellum. Some species of bacteria (peritricous) with a large number of flagella can even execute rapid wavelike progression across the solid structure of a culture medium. Microbiologists believe that the movement of flagella is conditioned by chemical stimuli coming from receptor sites of the envelope transmitting information about presence of nutrients, oxygen, ribosomes or galactose. If the chemostatic signal is positive, the flagella initiate movement toward their location, whereas flagella will steer the bacterium away from their location if the chemostatic signal is negative.

We shall briefly note the existence of another class of flagellum, denoted in microbiology "axial filament" as though they were not at all related to flagella. They are present in that group of bacteria designated spirochetes and coil around the entire length of the bacterium box. For one, axial filaments, as bundles of fibrils, are anchored at the end locations of the spirochete like flagella, even though they appear hidden under the outer sheath or capsule of the envelope. Further they have similar biochemical structure to flagella. Rotation of the filaments cause the rigid helical prokaryotic box to rotate and move longitudinally akin to a corkscrew.

21.10.2 Manifestation of the Harmonic Typology in Cell Walls

One other important attribute of this taxon is a characteristic outer cell wall whose principal element is a lattice of peptidoglycan molecules. The very elaborate *harmonic and matching* arrangement of the inner molecule groups forming this super lattice (so-called PG-lattice structure) is a high mark of its overloading Operator-6 modelers. A second level of expression of this 6-Operator conjugation feature is in the severally embedding configuration of this external structure.

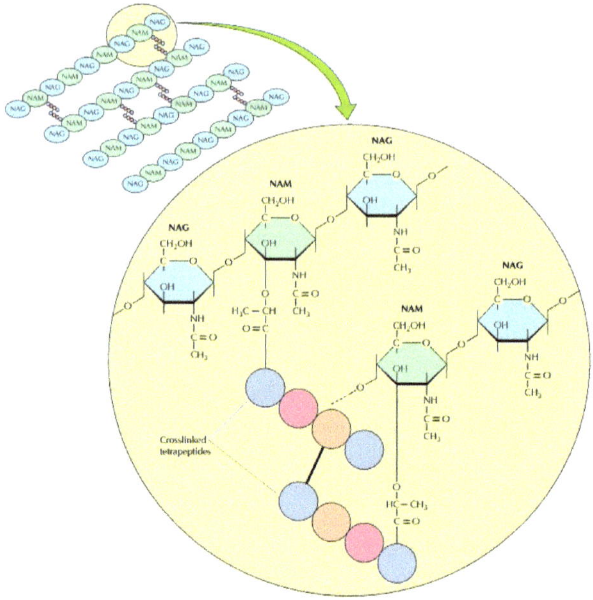

Fig. 21.8 Molecular structures in cell walls

Effectively, the peptidoglycan populations may appear in two distinct configurations. They may appear as a stack of 6 distinct lattice layers, which qualifies the bacterium with the of *gram-positive* attribute due to the cell wall reaction to crystal purple stain. Else, the peptidoglycan population takes the form of a single layer in sandwich between two envelopes of plasma membranes, the outer one conferring to this type of bacteria the attribute of *gram-negative* due to its characteristic lack of reaction to the Gram stain. The chemical structure of a PG-repeat unit consists of disaccharides, pentapeptide-stems, and a bridge structure (Fig.21.8). The disaccharide, composed of *N*-acetyl-glucosamine (NAG) and *N*-acetyl-muramic acid (NAM), is conserved in all so-called "eubacteria" (another name for cell-wall bearing bacteria) while the pentapeptide-stem and bridge structure vary among sub-species. The lattice or scaffold showcases an ordered sequential arrangement of NAG and NAM in groups of 4, which are cross-linked by pentapeptide side chains.

Cell wall is a common feature found in all bacteria beyond Mycoplasmas, and aside from degenerate L-forms together with some Archaebacteria. In Gram-positive bacteria, the cell wall thickness varies from *20 to 40 nm*. Both in its Gram-positive and Gram-negative aspects, the cell wall conditions the global dimensioning of bacteria and regulates pressure inside the karyotic box. It allows the bacterial entity to withstand such physical extremes as changes in osmotic pressure up to *25 atm*. The very feature of *embedding through harmonic coupling*, which is the composite ontological stamp dictating the configuration, enables the spatiating cell wall to develop not only the pressure regulation function but to equally play the role of a protective barrier against the external environment and to mediate exchanges between the cytoplasm and the external medium. The peptidoglycan layer or layers present the structure of a lattice or scaffold with large enough interstices for the expected molecules to transit through the structure (molecules up to a molecular weight of 800). The plasma membrane present in both Gram versions of the cell wall features protein blocks dubbed porins forming channels to the effect of conferring permeability to the entire structure. Some porins provide permeability in a protein specific manner with regards to the transitioning substances. These porins only allow the transitioning of specific substances, i.e., maltose, nucleotides or amino acids, simple sugars, iron, water etc. Large molecules such as proteins cannot pass the porin channels. Therefore, the cell wall functions overall as a complex two-way selective filter both in its Gram-positive and Gram-negative configurations, in a way selectively concealing the endogenous world on the karyotic box consistent with its ontological stamp.

It is important to note that the cell wall may develop catabolic capabilities as well. Just as the mesosome invaginations, the rest of the plasma membrane is able to produce ATP and to further contribute via intervention of own enzymes to the breakdown of nutrients.

One last structure developed by the cell wall is the glycolyx, which is the outermost envelope of the wall. It is essentially a deposit made up of a gelatinous polymer of either polysaccharides or polypeptides, or both, synthesized inside the cell and excreted to the surface. If viscous and firmly attaching to the external surface of the cell wall, the glycolyx is described as a capsule. If it is instead inhomogeneous and loosely attaches to the surface of the cell wall, the glycolyx is portrayed as a slime layer. One may view the glycolyx as one more *embedding* envelope to the karyotic box from the Quanto-Geometric analytic perspective.

21.11. Archeobacteria Form Layer 7 of the Prokaryote Spectrum

Let us start by emphasizing once again that the entire genome in the central DNA molecule of the karyotic kernel is not the primitive dictating the configuration of the karyotic kernel, but remains itself under a higher-order mandate, the mandate to register and crystallize a class type. In this case the type is a complex conjugate Quanto-Geometric type and codified per equation (x). So true is this assessment that the modality or structure of the genome itself is far from randomized but typified as well. Therefore, while one can attempt taxonomic classifications of organisms based on phenotypes, or even based on genotypes in terms of specific genic characters or their proteomic expressions, the most fundamental classification is the first-order class descriptor which can speak to the orientation taken by the entire proteome the genome is responsible for, as well as the architectural organization of the genome itself. We hope to have eloquently shown likewise the reasons why viral organisms perfectly and legitimately belong to this spectrum, the evolution of their ontology being consistent with the Quanto-Geometric spectrum canon.

Invested with that knowledge, we are pulling back the *extremophiles* formerly registered as *Archeobacteria* from the taxa of *Archaea* they have been accorded recently in microbiology and the independent taxonomic filiation they have been granted as a distinct form of the living situated between bacteria and eukaryotes. As prokaryotes, which is a descriptor still currently granted to them, they fully belong to the same taxon under which all bacteria are classified. What has not been understood in microbiology are the primal reasons why these organisms deliver a phenotype justly qualifying them as *extremophiles*.

As James A. Coker writes: *"..., in regard to extremophiles, it is important to understand that they are "extreme" only because they are found in regions where human beings typically cannot live unassisted. It is also important to think of them as growing and thriving in these conditions and not merely eking out an existence as the "last organism standing" that was able to survive just a bit longer than the rest. After all, they have evolved a set of adaptations that have made these "extreme" environments their preferred niche."* Their extreme faculties must reasonably be taken at face value and infer ontological qualities of the paradoxical typology. It is by virtue of that ontological type that their sensitivity to extreme or adverse conditions became an attribute, with structural and dynamic features consistent with that attribute. Accordingly, their class includes: acidophiles, alkaliphiles, halophiles, psychrophiles, thermophiles, piezophiles, and radiophiles, which are qualifiers that express the extreme or paradoxical vector governing their ontology. Only by virtue of this innate ontological trait that Archaebacteria take the counterpoint of every structural feature or process of normal bacteria or Eubacteria. Modern Microbiology has wrongly made a whole root taxon of this class of bacteria by granting it the role of an entire branch in the Tree of Life in microbiota.

21.11.1 An Inadequate Phylogenetic Tree of Life

How was this Tree of Life composed? Modern Microbiology offers a comparison of the sequence of 16s ribosomal ribonucleic acid (16s rRNA), a key unit of ribosomes, across known organisms.

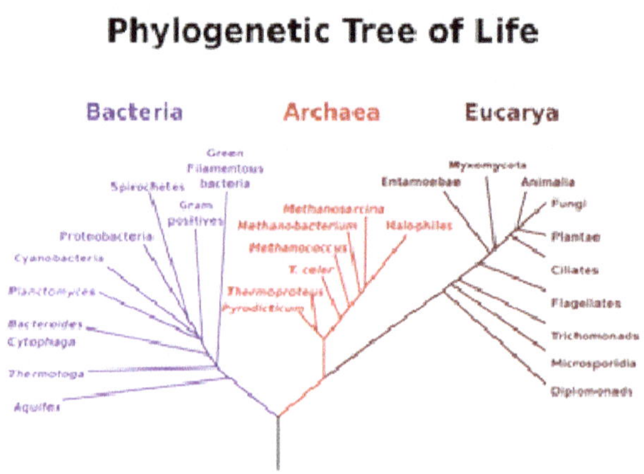

Fig. 21.9 Proposed Phylogenetic Tree of Life

The result of this analysis showed that the underlying genetic code for a component of ribosomes differed greatly between normal bacteria, a group of bacteria that came to be known as archaea and eukarya. Thus, the science offers that they form three distinct branches of the Tree of Life (Fig.21.9). We counterargue that the characteristics of one single cytoplasmic element among many others, albeit endowed with some form of RNA macromolecule, can never be the basis for ontological, let alone phylogenetic, classification of a whole domain of the biosphere such as microbiota.

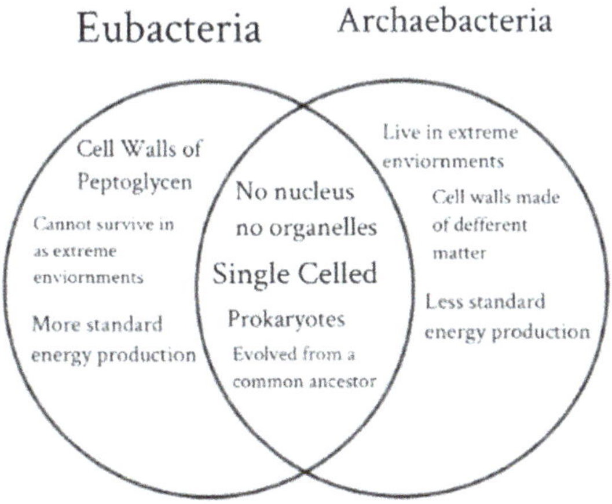

Fig. 21.10 Observed differences and commonalities between Archaebacteria and Eubacteria

In its current state of the art, the science further argues that in all three taxa, ribosomes have very different genes encoding them while they are comprised of different proteins. Nevertheless, one may argue, these inclusions still morphologically and functionally remain ribosomes, the energetic engine of cells. It is unsound to build deep taxonomic differentiation between generations of cells at the phylogenetic level based on these terms.

Lastly, the science proposes that a big difference between archaea and bacteria (Fig.21.10) resides in the genome sequence, arguing that archaea possess genes that resemble eukaryotes more than they do bacteria. Should we go back to the Symmetry which dictates that any one Shell in a Spectrum is built from a few elements of the previous, this becomes an insignificant difference. And for closure, there are fundamental structural differences between all of Eukarya and Prokarya which set them apart taxonomically in important ways.

21.11.2 A Remarkable Predator in the Layer-7 Taxon of Microbiota

Below the full story of a predator named *Bdellovibrio bacteriovorus* that belongs to this taxon of paradoxical extremophile bacteria (Fig.21.11), as related by microbiologist Hiu Cheng[26]. This story helps us relate extremophilia with the conflictual trait associated with the paradoxical typology in the QG canon.

"*B. bacteriovorus* seems to be pretty ubiquitous in nature and manmade habitats. They have been found in soil samples, rhizosphere of plant roots, rivers, oceans, sewage, intestines and feces of birds and mammals, and even in oyster shells and the gills of crabs *B. bacteriovorus* are able to thrive in almost any habitat, the general requirements are that

[26] *https://microbewiki.kenyon.edu/index.php/Bdellovibrio_bacteriovorus*

there needs to be oxygen and some other Gram-negative bacteria present in its environment. Its optimal temperature is between 28-30C, making *B. bacteriovorus* a mesophile.

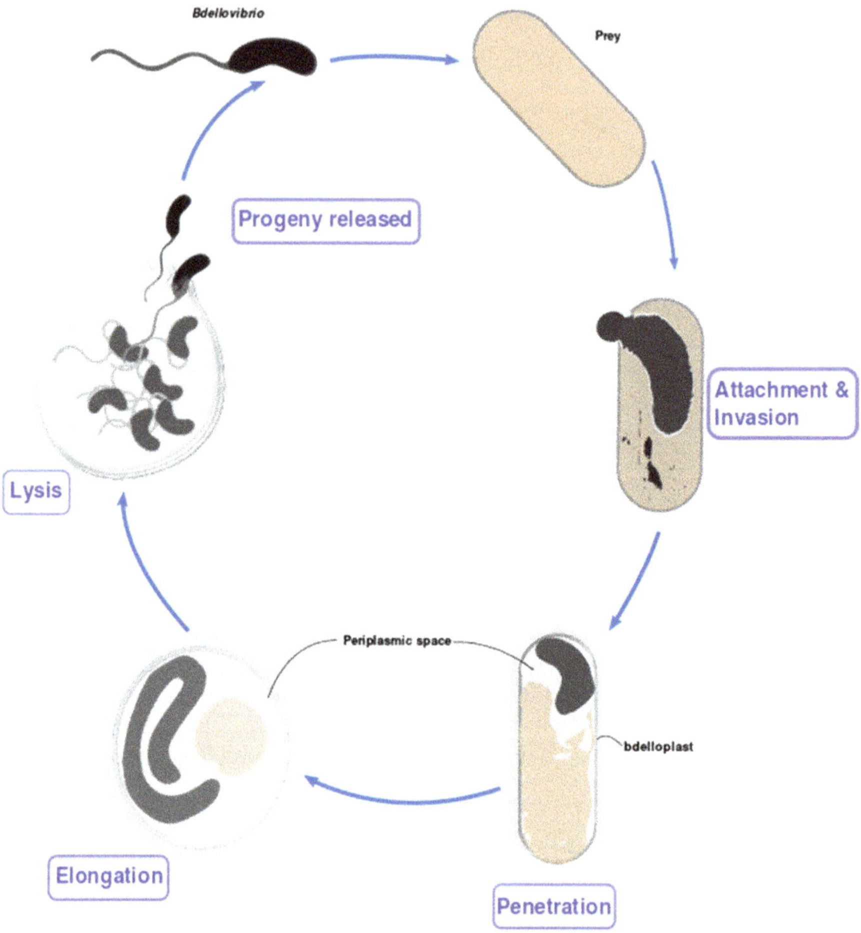

Fig. 21.11 Lifecycle of the killer bacterium named *Bdellovibrio bacteriovorus* [Wikimedia]

B. bacteriovorus is somewhat tolerant to its environment. It can live in a range of salinity and is resistant to pollution of its environment. **During its attack phase or as bdelloplasts, *B. bacteriovorus* can tolerate dry spells and short periods of anaerobic environment.** This makes *B. bacteriovorus* more tolerant to environmental changes than other permanently aerobic bacteria.

B. bacteriovorus has also been associated with biofilms. It seems that the rich environment of biofilms supply *B. bacteriovorus* with many possible host cells. Research has shown that *B. bacteriovorus* can significantly decrease *E. Coli* biofilm that is grown on stainless steel.

The sequencing of the *B. bacteriovorus HD100* genome was completed on 01/31/2004. The complete genome consists of a single circular chromosome that is 3,782,950 nucleotides long. The entire genome has a 50% GC content and contains 3629 genes which code for 3587 proteins and 42 structural RNAs.

Of the large number of proteins that the *B. bacteriovorus* genome encodes, **many of them are degradative and lytic enzymes. These enzymes are necessary for *B. bacteriovorus* to breakdown its host or prey cell.** Some of these enzymes need to be secreted out of *B. bacteriovorus* and into its prey's cell, **while degradative products need to be transported into the *B. bacteriovorus* from the prey cell cytoplasm**. Therefore, the genome of *B. bacteriovorus* also includes a large amount of transport proteins. In fact, ***B. bacteriovorus* has the potential protein secretory capabilities of at least five types of outer membrane secretion systems and four types of inner membrane secretion systems.**

Other than encoding for degradative enzymes and transport systems, the chromosome also contains multiple biosynthetic gene clusters that are necessary for the formation of the flagella and pilus. **These gene clusters are essential for the motility of *B. bacteriovorus*, which is important to its life cycle because it directly relates to its ability to find and approach the prey cell**. However, the organism sheds its flagellum once it makes irreversible contact with its prey cell suggesting that the flagellum is not required for it to grow.

The general cell structure of *B. bacteriovorus* is small, curved, and highly motile Gram-negative bacteria approximately 0.2 to 0.5 µm wide and 0.5 to 2.5 µm long. It has a flagellum that is membrane bound, allowing it to swim up to 100 body length per second while it looks for its prey.

The unique biphasic life-cycle of *B. bacteriovorus* is directly related to its metabolism. In the free-living or mobile phase, *B. bacteriovorus* actively seeks other Gram-negative bacteria by chemotaxic means. **During the mobile phase, most of the metabolic activity is involved in motility and prey detection. In the attack phase, *B. bacteriovorus* burrows itself into the periplasm of the prey cell and begins to degrade the prey cell from the inside out.** A large variety of metabolic activities occur during this phase.

In the attack phase, *B. bacteriovorus* begins by attaching itself to the outer membrane of the prey cell via hook-like structures. It then proceeds to secrete enzymes which break down a large enough hole in the outer membrane for *B. bacteriovorus* to enter. After entry into the prey cell, *B. bacteriovorus* sheds its flagellum and seals the hole in the outer membrane of the prey cell. *B. bacteriovorus* then secretes degradative enzymes that act on the inner membrane of the prey cell allowing it to access the prey's cytoplasm. **The degradation of the inner membrane causes the prey cell to become a bdelloplast, host cell of *Bdellovibrios* with a stable spherical structure, and also the destruction of the host cell's ability to generate energy.** At this point, *B. bacteriovorus* degrades and metabolizes the macromolecules and nutrients from the host's cytoplasmic material allowing it to grow and begin DNA replication. *B. bacteriovorus* subsequently becomes a long filament and then fragments into individual mobile progenies which will lyse the remaining bdelloplast.

The metabolism of *B. bacteriovorus* has many interesting areas. One of these areas is that *B. bacteriovorus* can survive while lacking numerous amino acid biosynthetic pathways. **The reason for this is that *B. bacteriovorus* utilizes the metabolites of its prey for its own protein synthesis. *B. bacteriovorus* is very efficient at utilizing the DNA, RNA, and other pre-existing precursors from the host cell.** It can break down or rearrange these precursors for the biosynthesis of homologous polymers. The uptake of nucleoside monophosphates and glycerol phosphates greatly reduces the amount of energy that *B. bacteriovorus* has to produce during nucleic acid synthesis and lipid synthesis. There has also been evidence that *B. bacteriovorus* incorporates intact fatty acid molecules from its host. Experiments have shown that the energy efficiency of *B. bacteriovorus* can be 0.8 to 1.5 times more efficient than bacteria growing in rich media. "

The highlights in this excerpt are of our making in order to emphasize the destructive and lethal phenomenology associated with this killer bacterium. One is somewhat hard-pressed to strictly associate typology to structure here due to the sheer number of fine details of the biogeometric order achieved by the organism. Typology is principally portrayed here in the ultrastructure and refined dynamics achieved through the biochemistry and physical chemistry of the organism. This ultrastructure, revealed in the large amount of intelligent phenomenology developed by the organism in the prosecution of death, pertains to the province of ontology. Going forward, in the subsequent taxons of the Spectrum, we will see more of this trend to keenly translate ontological typology into phenomenology, beyond strictly structure.

21.12. Commensal, Mutualistic and Cometabolic Bacteria as Layer 8 of Microbiota

Syntactically, the term commensalism refers to a type of relationship between two different organisms that "eat from the same dish". In this kind of relationship, neither organism benefits from the other or provokes any harm to the other. It is therefore at a minimum a neutral relationship, and at most a friendly one. Although "friendly" bacteria inhabiting our organism are usually referred to as commensal, research in this field suggests that the relationship between gut microbiota in the human body is not merely commensal, but rather mutualistic or symbiotic. Effectively, mutualism or symbiosis is another class of relationships between organisms of interest for this Stratum, a relationship in which both organisms obtain benefits from one another (Fig.5.12).

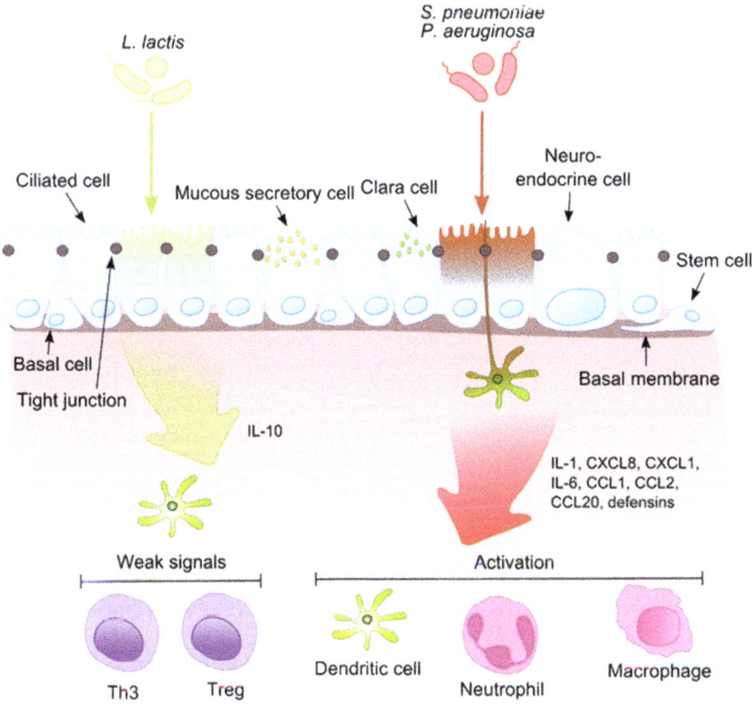

Fig. 21.12 Commensal v. pathogenic mechanisms in human infection of the respiratory tract. *"The mechanism of distinguishing between harmless and harmful bacteria on the molecular as well as on physiological levels is not completely understood."* [Wikimedia].

We let Rabia Khan et al introduce and identify commensal organisms for us, per the excerpts below.

"A diverse community of trillions of commensal bacteria inhabits mucosal and epidermal surfaces in humans and plays an important role in defense against pathogens, including respiratory pathogens. Commensal bacteria act on the host's immune system to induce protective responses that prevent colonization and invasion by pathogens. On the other hand, these bacteria can directly inhibit the growth of respiratory pathogens by producing antimicrobial products/signals and competing for nutrients and adhesion sites. Such mechanisms preserve the niche for commensal bacteria and support the host in containing respiratory infections.

Commensal bacteria are key regulators of digestion, a process that begins in the mouth and continues as ingested food and its digestive intermediates transit more than 20 feet (6 meters) to the end of the adult human gastrointestinal tract. Along the way, the digestive slurry is mixed with commensal bacteria, which are important for the extraction, synthesis and absorption of many nutrients and metabolites, including bile acids, lipids, amino acids, vitamins and short-chain fatty acids (SCFAs). These nutrients and metabolites derived from commensal bacteria are directly linked to diet and digestion and are therefore considered to be diet-dependent microbial products. In addition to producing diet-dependent metabolites and nutrients, commensal bacteria can produce many different diet-independent microbial products, examples of which are lipopolysaccharide and peptidoglycan."

Commensal bacteria	Anatomical location	Respiratory pathogen	Mechanism of inhibition
Streptococcus salivaris	Oral cavity	Streptococcus pneumoniae - Streptococcus pyogenes	Ribosomally sysntesized antimicroials (Bacteriocins)
Stapylococcus lugdunensis	Skin and nasal cavity	Staphylococcus aureus	Non-ribosomally synthesized antimicrobials (Lugdunin)
Corynebacterium accolens	Skin and nasal cavity	Streptococcus pneumoniae	Metabolic products with antimicrobila properties (Free fatty acids)
Staphylococcus epidermics	Skin and nasal cavity	Staphylococcus aureus	Secreted enzymes (Serine protease)
Streptococcus pneumoniae	Nasopharynx and oral cavity	Staphylococcus aureus	Hydrogen peroxide (H2O2) mediated killing
Corynebacterium spp.	Skin and nasal cavity	Staphylococcus spp.	Nutrient competition (Iron limitation by sideropore)

Table 21.1 Mechanisms of colonization resistance used by commensal bacteria to benefit their hosts.

Table 21.1 shows the direct mechanisms of colonization resistance used by commensal bacteria in their fight against respiratory pathogens in protection of the host. If commensalism is behavior, there is no question that it derives from typological orientation or even intelligence, given that behavior in its specificity must have a higher-level pointing vector. Jonas Schluter and Kevin R. Foster offer the following on mutualistic bacteria:

"Every human is home to roughly 100 trillion bacterial cells, collectively called the microbiota. The majority of these cells reside in the human gastrointestinal tract and, in particular, in the large intestine. Here, bacteria can have beneficial effects such as the digestion of complex carbohydrates, colonization resistance against invading pathogens, maturation of the adaptive mucosal immune system and immune cells, and the production of secondary metabolites, including vitamins. However, these activities are not performed by all species, and the species composition of the microbiota in a healthy human is clearly distinct from bacterial communities in other environments. Moreover, various diseases correlate with disturbances in the species composition of the microbiota. It is clear then that the gut community has the ability to both help and harm the host. Despite the potential for harmful effects of the gut microbiota, the major class of interaction with the host appears to be one of mutualism, whereby both sides benefit from the interaction. The evidence for host benefits comes both from our understanding of the metabolic services that the gut microbiota provides and studies of germ-free animal models."

As previously mentioned for Archaea, this is here again the case where the structural integration of a large variety of elements becomes so complex that typology or ultrastructure is best reflected and identified in phenomenology through the specific behavioral taint exhibited by the organism. If phenomenology is not a first-order descriptor, one must assume that the quality in phenomenology must be aligned with ontology, given that phenomenology in itself must have a determinant or a cause, and it can be no other than ontology.

Finally, there exist these bacteria which metabolize polymers such as cellulose and can spread in their living medium simple degraded components thereof such as glucose, which is utilized as a nutrient by other occupying bacteria in the medium. These cometabolic bacterial entities engage in this activity only for the purpose of providing the other medium's inhabitants with useful nutrients, without any attempt of self-benefitting. One may conclude that these elements respond so well to the "perfection" or "accomplished" typology characterizing Layer 8 of the QG canon that they are able to go beyond self-completion to unconditionally undertake providing others with what they need or lack.

21.13. Perdurable Bacteria as Layer 9 of the Microbiotic Spectrum

Layer 9 of Microbiota comprises: heterocysts, actynomycetes, and sheath- or bud-laden bacteria. The most important phylogenetic trait of these bacteria is their formidable ability for longevity, that is to say the faculty to repeat, renovate or outstandingly extend their lifecycle.

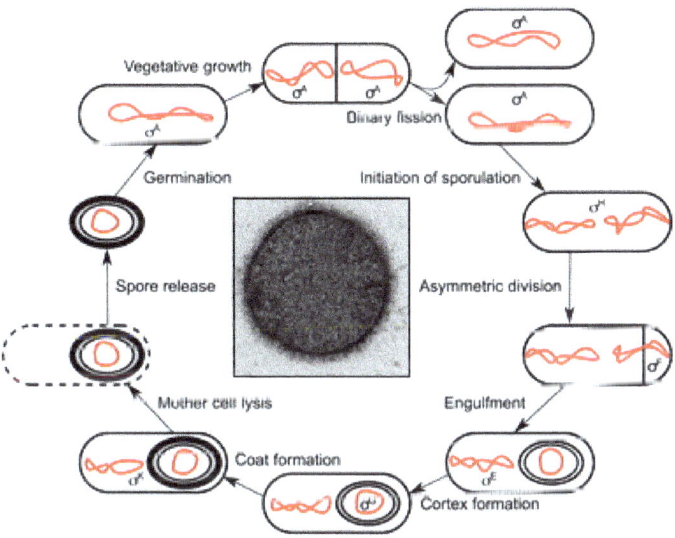

Fig. 21.13 Morphological stages of the ***B. subtilis*** life cycle

This ability is manifest in a variety of modalities, not the least of which is a peculiar capacity to resist adverse habitat conditions by sporulation (Fig.21.13). They may also develop specialized organs for an explosive form of reproduction-multiplication.

These perdurable organisms may also show the ability for prolonged longevity by modifying their metabolic patterns and producing super-structures capable of assimilating habitat waste when nutrients become dangerously rare. Heterocysts in colonies of cyanobacteria exhibit in fullness this awesome metabolic capacity. F Garcia Pichel writes the following about heterocysts:

"Heterocysts … are morphologically distinct cells that develop in response to a lack of combined nitrogen sources in the environment. The ability to develop heterocysts occurs without exception within a monophyletic group of filamentous cyanobacteria (heterocystous; subsections IV and V). They are usually larger than vegetative cells, develop thick tegumentary layers and intracellular hyaline buttons at the points of attachment to the vegetative cells, displaying a pale coloration and reduced autofluorescence. As such, heterocysts are easy to recognize under the microscope. They may differentiate from end cells (terminal heterocysts, as in Calothrix) or from cells within the trichome (regularly spaced intercalary heterocysts, as in Anabaena, or lateral as in 'Mastigocoleus'). Heterocysts are highly specialized in the fixation of dinitrogen under aerobic conditions. They represent a successful solution to the nontrivial problem of avoiding nitrogenase inactivation by free oxygen in oxygen-evolving organisms. Heterocyst biology has been relatively well studied at the biochemical and molecular levels. Heterocysts are the only cells that express nif (nitrogen fixation) genes and synthesize nitrogenase in heterocyst-forming cyanobacteria. Heterocysts do not evolve oxygen themselves (photosystem II (PSII) activity is absent or restricted) but a functional photosystem provides ATP. The source of reductant for nitrogen fixation is provided (as organic carbon) by the adjacent vegetative cells, which in turn obtain fixed nitrogen from the heterocyst in the form of amino acids (mostly glutamine). The heterocysts protect their nitrogenase from oxygen inactivation by maintaining reduced internal partial pressures of oxygen, a situation that is attained by means of increased rates of cellular respiration and, apparently, by restricting diffusive entry of oxygen from the environment as a result of their thick envelope. The developmental regulation of heterocysts is beginning to be understood at the genetic level. The autoregulated gene hetR, which is activated by the deficiency in combined nitrogen, seems to play a crucial role in the initiation of heterocyst development[27]".

Sporulation is another trait stemming from the expansive or perdurable faculty of these organisms whereby a cell in a colony inhabiting an environment turned averse to their lifecycle produces a bud to become a tripartite sub-cell or endospore. The spore carries an envelope that is extraordinarily resistant and later on detaches from the mother cell to enter a cycle of hibernation or dormancy. Whenever the ambient conditions become propitious enough again for thriving, with no limitation in the time lapse, the endospore comes

[27] F. Garcia-Pichel, Cyanobacteria, Encyclopedia of Microbiology (Third Edition), 2009

back to life. *Michiel JL de Hoon et al* explains the sporulation process and its regulation in B. subtilis[28]:

"In rich medium, *B. subtilis* cells divide by binary fission approximately every 30 minutes. By contrast, deterioration of environmental conditions triggers sporulation, a developmental process that takes about 8 to 10 hours. Thus, endospore formation represents a formidable investment of time and energy and is considered to be a survival pathway of last resort, as *B. subtilis* cells only commit to sporulation after they failed to deal with starvation in other ways, such as cannibalism or establishment of a genetically competent state. The successive morphological stages of sporulation have been defined using electron microscopy. Sporulation begins with an asymmetric cell division and results in the generation of two cell types, a forespore (the smaller compartment, also called the prespore) and a mother cell. The two cells experience distinct fates, as the mother cell ultimately lyses by a programmed cell death mechanism, whereas the forespore matures as a spore. Shortly after asymmetric division, two parallel programs of gene expression are established in each compartment under the control of transcription factors that are activated in a cell-specific manner. In addition to regulatory interactions within the forespore and mother cell, precise inter-compartmental signaling is required to control the spatial and temporal progression of the developmental process."

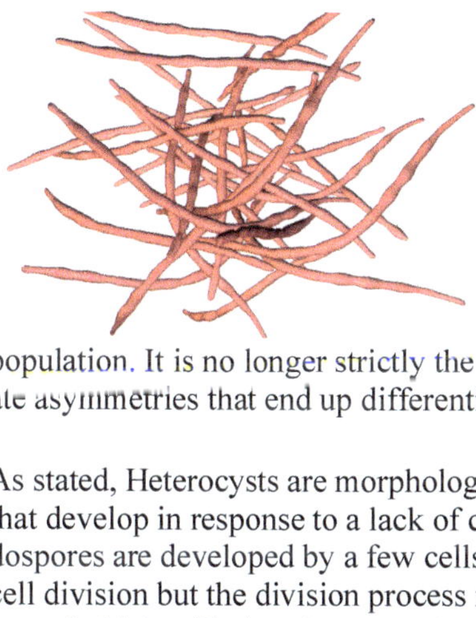

Fig 21.14 Filaments of actinomycetes with strings of cells inside the sheaths

It is important to realize the onset of an incipient form of multicellular organization in the expansive trait of Layer 9 of this Spectrum as we examine the topology of these entities. Thus far we were dealing with topologies with a single type of karyotic box within the extension of the population. It is no longer strictly the case here, since sporulation and cysts clearly initiate asymmetries that end up differentiating, albeit minimally, elements in a population.

As stated, Heterocysts are morphologically distinct cells in a population of cyanobacteria that develop in response to a lack of combined nitrogen sources in the environment. Endospores are developed by a few cells that not only initiate individually an asymmetric cell division but the division process results in the generation of two different cell types, one of which will give rise to a colony of spores that are structurally quite distinct from the original mother cells. When it comes to sheath- and bud-laden bacteria which we have duly included in this Layer taxon, one may note the presence of Actinomycetales, whose members are often called **actinomycetes** (Fig.21.14).

[28] Michiel JL de Hoon1,†, Patrick Eichenberger2,*, and Dennis Vitkup1, Hierarchical evolution of the bacterial sporulation network, Curr Biol. 2010 September 14; 20(17): R735–R745. doi:10.1016/j.cub.2010.06.031.

Being generally gram-positive and anaerobic, Actinomycetales topologically portray mycelia in a filamentous and branching growth pattern. Some actinomycetes can form rod- or coccoid-shaped forms with cellular units inside a sheath, while others can form spores on their aerial hyphae above the fluid habitat. The expansive ontological feature thereby in display, through membrane layering and appending, is indisputable.

21.14. Primitive Eukaryotic Spectrum

In the publication "La Quanto-Géometrie", which represents the protogenesis of this Unification Framework, we offered a detailed and extensive discussion about the Primitive Eukaryotic Spectrum or Individuation. We will make here a brief recount of the 9 different Layers of this Spectrum.

21.14.1 Why Does the Order of Eukaryotes Belong to the Same QG Shell as Prokaryotes?

Shell 6 is the Shell of Life. Even though all of the Shells encountered thus far in the development of U_3 are based on one single preeminent primitive spectrum, the spectrum of Eukaryotes rightfully belongs here in the Shell of Life along with the spectrum of microbiota (viruses and bacteria). The principal reason for this determination is that viruses, bacteria and eukaryotes altogether present with the same foundational biogeometric architecture, which is what we have called the karyotic box. Even though the Eukaryote order appears to be a normalized continuation of the spectrum of microbiota, it does not develop on the same structural flat sheet as microbiota, for the essential reason that Eukaryotisation puts before us a span of multi-cellular organisms as an ostensible feature. Call it the topological multiplication of the karyotic box to form individuation. The eukaryotic unit entity or individual, aka living organism, is built on multiplicity of the karyotic box for the most part.

21.14.2 Origin of the Eukaryotic Cell

What we are witnessing with the Eukaryotic or Individuative Spectrum is a shelling emulation within the very development of a Shell. There is no doubt that the eukaryotic box is a composition between a perdurable bacterium (Layer 9) and an endosymbiont bacterium (Layer 6 or harmonic) on the basis of a Shell symmetry configuration: a few elements of the previous shell assemble to form the next shell unit to be based on quintessence and covariance. Mitochondria (animal cells) and chloroplasts (vegetative cells) are more than a cytoplasmic inclusion in the eukaryotic cell. They each represent the largest inclusion of all within the eukaryotic cell, retain the fulsome structure of a bacterium from the central DNA macromolecule to the superficial cell membrane, and play the ontogenetic role of the wavefunction propagator within the cell as its energetic engine. Given the explosion of complexity both in the dynamical processes and element configuration at this level, it is easy to understand why Occam's law of parsimony would play out in terms of the selection of an entity prone to reclusion and another prone to expansive multi-cellular structure to give birth to the Individuation spectrum as defined.

Microbiologists have long speculated that Archaea might have been the parent prokaryotic cell at the phylogenetic origin of the eukaryotic cell. That is in fact quite improbable given the destructive and destabilizing properties of this taxon, despite its peculiar and unique features.

Perdurable bacteria are known to possess the following specific faculties which are conducive to the ontogenetic role being discussed:

1. Endospore formation for survival and multiplication.
2. Superficial membrane expansion in order to capture more nutrients in nutritively poor environments.
3. Formation of vacuoles for buoyancy to the aims of capturing more radiant energy off the environment as needed.

The eukaryotic cell exhibits these properties in many ways. Sporulation is generalized in the protist world, the polypod world and the vegetative world. We believe that in the animal world, spores have evolved into eggs, first calcified and then uterine. Some polypods (elementary multi-cellular organisms) do feature as well filamentous forms of growth as showcased in the perdurable-expansive prokaryotes. Vacuoles constitute a permanent structure of the eukaryotic cell. We equally believe that the perdurable-expansive prokaryotic faculty to increase membrane surface and cytoplasm volume in order to capture more nutrients from the medium are in all likelihood the predecessors of the eukaryotic cell's ability to flexibly deform a discrete region of its membrane and cytoplasm in order to phagocyte preys from the external medium.

These are significant metabolic and reproductive properties, which are direct derivates of the towering wavefunction tenet and scalar tenet respectively in ontology.

21.14.3 Layout of the Eukaryotic Spectrum

The nine Layers of the emulated Eukaryotic Primitive Spectrum follow with their principal characteristics.

21.14.3.1 Monolithic or Unitary Individuation

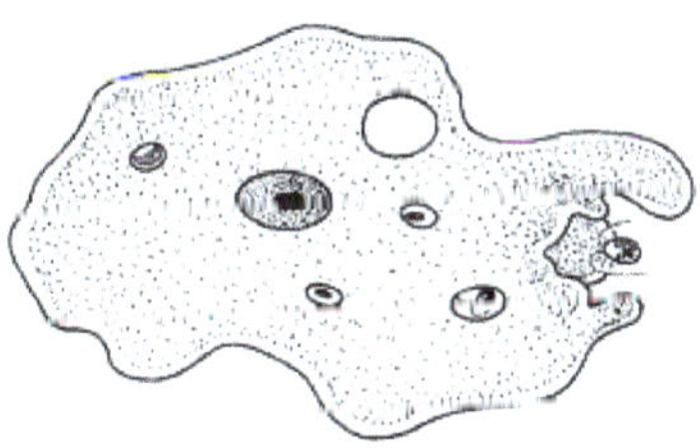

Identity Definition: Unicellular Organisms or Protozoans.

Taxon Members: Amoeba, Flagellates, Ciliates, Unicellular Algae, Fungi and Yeasts.

Essential Attributes: Autonomous eukaryotic kernel.

21.14.3.2 Ambivalent or Binary Individuation

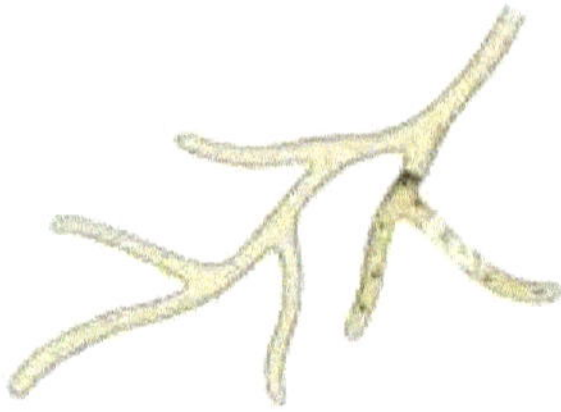

Coenocytic Hyphae

Identity Definition: Coenocytic or Primal Metazoans.

Taxon Members: Pseudohyphae, Coenocytes with nucleus-sparse cytoplasm, Coenocytes with porous separating cell walls, and Mushrooms.

Essential Properties: Ambivalent or covalent structure with one cell group dedicated to metabolism and one other dedicated to reproduction; differentiated organization, cooperative activity.

21.14.3.3 Arborization or Trinary Individuation

Identity Definition: Arbor or Vegetative Elements.

Taxon Members: Flower Plants, Flowerless Plants, Ferns, Coniferous Shrubs and Trees.

Essential Properties: Relational or Associative structure in foliage, trunk, branch and roots, chlorophyl assimilation through a relational molecular structure (the chloroplast cell).

21.14.3.4 Spongia or Quadratic Individuation

Identity Definition: Nerveless Animals.

Taxon Members: Sponges and Corals.

Essential Properties: Elastoplasticity, muscular histological structures

21.14.3.5 Metamorphic Individuation

Identity Definition: Invertebrates.

Taxon Members: Coelenterates, Echinoderms, Plathelminths, Nematodes, Segmented Worms, Mollusca, Crustaceans, Insects, Arachnids, Chilipoda and Diplopods.

Essential Properties: Metamorphogenesis, nervous system, and exoskeleton

21.14.3.6 Harmonic Individuation

Identity Definition: Vertebrates.

Taxon Members: Primal Vertebrates (Lamprey, Amphioxus, etc.), Fish, Birds (2-legged flyers and Penguins), Amphibians (Frogs, Salamanders, etc.), Reptiles (Crocodiles, Turtle, Snakes, etc.).

Essential Properties: Containment bone and skeletal structures, esthetic proneness (artful-colorful and harmonizing appearance), camouflage and mimesis.

21.14.3.7 Paradoxical Individuation

Identity Definition: Mammals.

Taxon Members: Dogs and Wolves, Cats and Tigers, Moles, Bats, Cows and Buffalos, Bears, Equidae (Horses, Zebras, etc.), Skunks, Rats and Mice, Lions, Giraffes, Hyenas and Pigs, Marsupials, Elephants, etc.

Essential Properties: Endowed with structures expressing proneness to conflicts and destruction, acute strong teeth, viviparous for self-protection, 4-legged for swift and robust projections, hairy and coarse epidermis, filthy body fluids and excretions, etc.

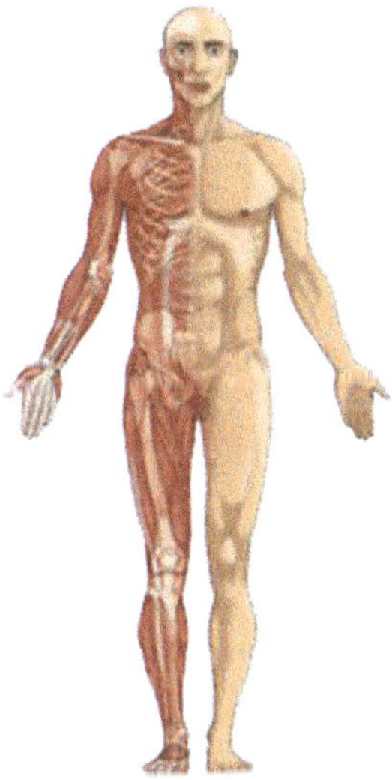

21.14.3.8 Perfect Individuation

Identity Definition: Hominids.

Taxon Members: Macaques, Monkeys, Gorillas, Humans, etc.).

Essential Properties: Self-sufficient and self-reliant, extensive capacity for prehension, tooling or manipulation through tools, expressive facia and proneness to self-detachment (pleasure, relaxing activities and utterances, laughs, etc.), anatomic verticalization and frontalization.

21.14.3.9 Perdurable Individuation

Identity Definition: So-called Monsters.

Taxon Members. Attached Twins or Double Monsters, Back-to-Back Twins, Frontal Twins, Longitudinal Twins (Xiphodene), Abdomen-Attached Twins, Parasitic Double Monsters (2nd Twin appears as an appendage to the Main), etc.

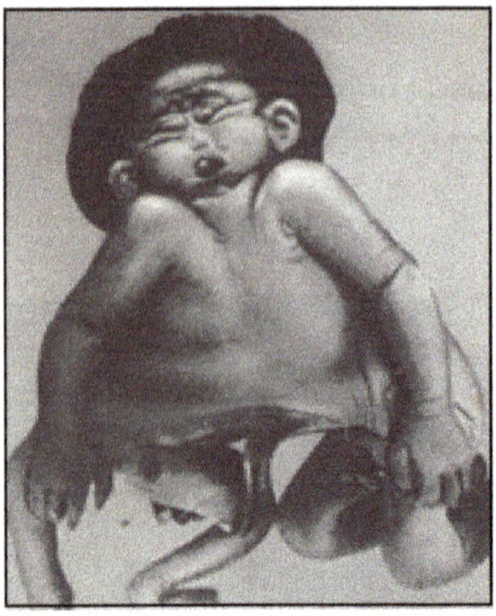

Essential Properties:

The most striking characteristics of teratological entities is that they do not exhibit any physiological dysfunctionalities and are organically and psychologically, thus functionally, perfectly integrated living bodies. In a double monster that shares the same urinary and colon tracts, there is never a conflict in the urge to defecate, they both feel the same urge when the need arises individually. Whenever they feature two anatomically formed heads, they do not and feel not the need to talk to each other, just like one monophyletic individual as most of us are does not do so. Even in the rather teratologically infrequent case of a parasitic head, the extra head that has grown on the chest lacks motor volition. Where there has been enough anatomic separation for free individual expression, the two bodies are able to perfectly function within the limits of their physical autonomies.

There is not even dysfunctionality in the monster twin cases with so called frontal mirror symmetry (two attached facing bodies), which is a remarkable fact; only some irregularities in the blood flow order are known when the symmetry does not reach thoroughness in those cases! That double monsters are always of the same sex is a highly telling fact as well. It shows irreducible unity in that form of individuation. The signal that is coming from mother Nature is that there is nothing "abnormal" in the teratological expression of Life which we must understand as a raw or primary form of Symmetry, quite akin to the integration or reunion of two prokaryotic cells to give rise to the Eukaryotic kernel. In this regard, we shall remember that the phylogenetic act playing out at this second stratum of the Shell is being realized through shelling Emulation and a symmetry explosion.

Early teratologists had the view that double monsters are cases of twins whose embryos did not ascertain fulsome development. Etienne Wolf, expert teratologist of yesteryears, held a reverse perspective and had this to say: *"If one can ever find assertiveness within the abnormal, it is fair to say that [separate] twins are cases of double monsters that found the path to normalcy."* Whether double monsters are imperfect twins or twins are double monsters at full course is an admissible question which yet does not bring insight into the ontological configuration of neither separate twins nor double monsters. The question raises a concern relating to embryological normalcy in this case. Unfortunately, in the ignorance of the fundamentals of life, modern teratological science has developed in the direction of developmental malformations and offers no other comprehension and intent than pathologizing double monster cases. In particular, those cases seem to be assimilated to the quest concerning toxic chemical origin of physiological and anatomic malformations. According to the teratological review effected by Eduard Ujházy et al:

"These [toxic agents] include some pesticides, dioxins, polychlorobiphenyls (PCBs), and bisphenol A. Despite modern approaches of molecular biology and genetics, along with top diagnostic techniques, we are still not able to identify the actual cause in more than 65 to 70% of all congenital defects classified as having an unknown etiology. Today CDDs [Congenital Developmental Disorders] include any birth defect, either morphological, biochemical, or behavioral."

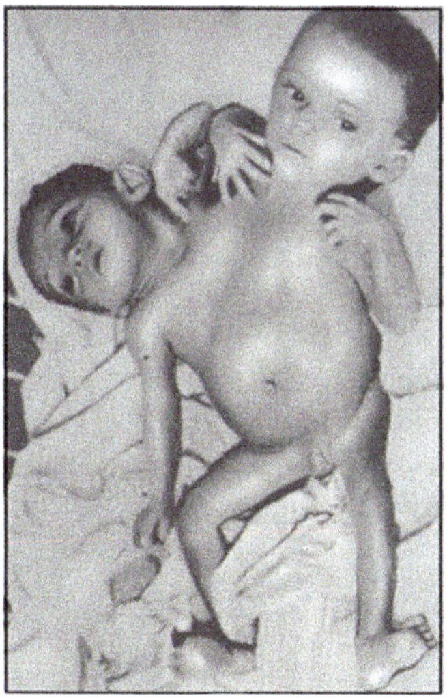

The fact that a toxic agent may be cause for CDD does not necessarily mean that all CDD's must be caused by toxic agents. We are thus learning at our own expenses the mathematical logic rule that not all theorems have or infer a reciprocal. Effectively, the figure of 70% of etiological mystery among known cases is almost an indictment of this teratological approach, advocating for a higher and more stringent mathematical logic perspective to these developmental facts. The systematic Biophysics approach we are hereby exposing, aided with stringent mathematical logic and numeric framing, intends to cure these caveats in Biology and Medicine. The teratological view of accidental malformations caused by chemical toxicity has even spread to the area of neurology and psychiatry, as evidenced by Eduard Ujházy et al in their review in these terms:

"Further development of neurobehavioral teratology is considered a very important issue. It is associated with the so far unexplained increasing rate of psychic and behavioral disorders on the one hand, and with excessive chemisation and the action of excessive stressful stimuli on the other. According to several experts, the extremely increasing number of individuals with an autistic spectrum of disorders recorded in the U.S.A. over the past 20 years cannot be explained only by improved diagnostics and changed diagnostic criteria (Kim et al. 2011). It is assumed that a number of chemical substances and their mutual interactions, with no negative effect on the adult brain, may over their long-term action negatively affect the developing brain even at very low concentrations."

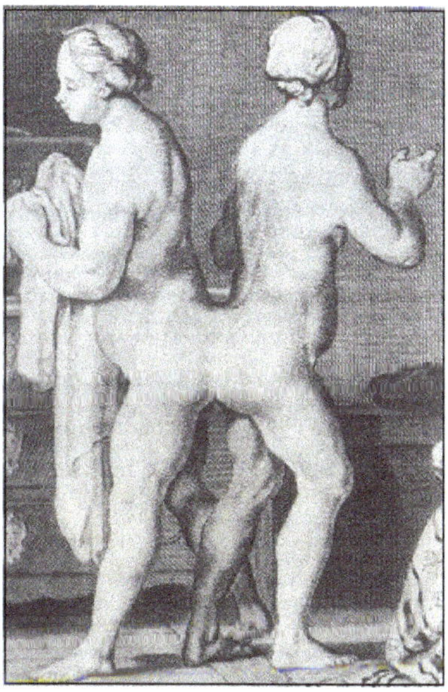

Note that no qualified mental illness is known to be associated with historical cases of teratological double monsters. If we should take the case of the frequent tantrums that used to occur between the twins of a double monster who was a court musician of Scottish King James VI (1567), it would quite reasonably qualify as a tantrum disorder in their very own class, for a violation of the standard of not talking to oneself. He was a xiphodyne monster, that is to say, an alter human with two distinct superior bodies and one single fused thorax, yet two distinct pelvic structures. The view of malformation v. normalcy can only be adequately resolved within the broadest comprehension of the Norms subtending QG spectra. When a qualified toxic agent causes a biological accident, just as when mutations occur in nature, biological changes always ensue within the framework of a Normative platform to construct a possible occurrence or outcome, frequently through operator composition. The outcome may just not be reconcilable with the elements of the operating domain, because a Symmetry had been broken. Accidental occurrences are then probably best described in terms of complete or insufficient realization in one direction or another, a typological mis-realization, and consequential viability. This level of understanding of life processes is arguably mandatory in any normative conference to determine what is acceptable and unacceptable. It is also clear that typological realizations of the Paradoxical nature specifically are always singular outcomes requiring cautious consideration and disposition.

21.14.4 Inside Polyembryogenesis

One might be tempted to believe that polyembryogenesis occurs at a rather early phase of the embryogenesis process when at least a few early cells have been produced, which by some form of segregation could give rise to two different courses of development. Not at all. The fertilized egg destined to form a double monster begins developing just as one that would follow a monoembryogenetic course achieving at the end one normal human. It is only at a very advanced stage of the embryogenetic process that the twinning feature appears. What then occurs at this rather late phase is the impromptu formation of two distinct centers of further embryogenetic development.

How something, or to be specific a kernel box, that in its known composition, down to the proteomic organization, normally and logically gives rise to one single entity instead elects to achieve two differentiated ontological structures remains a mystery in Biology and Teratological Medicine. Even when biologists experiment on animal eggs and tie a knot around a two-cell blastomere with a hair, in order to artificially produce ordinary twins, it still begs the question of how come two cells that were going to develop into a single body can ever be made to produce two distinct and complete living bodies. This would be a-physical in the prior Shells of mother nature.

It clearly seems that the fertilized egg is endowed with regulatory virtues that go beyond simple replication ability from the proteome. The answer to this conundrum is manifold:

1. Ontology in the fertilized egg strictly responds to Phylogenesis or the dictates of the Life Shell or Life Symmetry.
2. The primary mandate of the fertilized egg in its ontological reproduction function is to Emulate Phylogenesis.
3. The connection between Ontogenesis and Phylogenesis is orchestrated in a Physical Life constant which acts as the arbiter of all living processes and is directly reflected in Stem Cells sustaining the living architecture at all phases and stages. That one Physical Life constant institutes in and of itself the canonical reference for all reproductive processes including Emulation, and thereby transcends any singlet or singular developmental course or trend.

We will devote the next Chapter to the prescription in the 3rd item.

21.15. Higher Karyotes Interrelate to Form Singlet Celestial Bodies

The study of Double Monsters gives us the basis to comprehend this rather odd postulate. Yet as odd as it might seem to the modern Western mind, this school of ideas, aka Animism, was prominently spread on this planet 6000 years ago, across ethnic groups and cultures, in particular the Kemetic culture across Northern Africa. In West Africa, monsters in the Voodoo culture of the Fon, Aja and Ewe people bore the nomer of Toxosu, they were perceived as higher sacred entities.

The culture of the Dogon people in today's Mali has it that their ancestral birthplace is the Sirius star system, closest inner galactic neighbor to our star System, and were brought to Earth by the Monster they call *Nommo*. Their remarkable knowledge about the invisible Sirius stars is a mystery that sparked an unsolved debate to this day amongst Astrophysics personalities, most of approach followint the vein of famed cosmologist Carl Sagan, weighing in unfortunately in a rather speculative manner. The theology of the ancient Egyptian people presents with fascinating replicas of a plethora of Monsters they held for Gods. The astonishing details with which they depicted these beings in sculptural artifacts and their sheer abundance are, in my sense, far too pronounced for these representations to be simply assimilated to imaginary esthetic works of art instead of plain witnessing in one way or another. Incidentally, in the matter of witnessing, the science of Paleontology still offers no reasonable and logical explanation of how Ancient Egyptians physically built the famous Pyramids that these researchers incessantly excavate, nor do our core sciences. The Kemetic people of not only Egypt but the entire span of North/West Africa held that our Sun Star beholds anthropomorphic faculties and represents the prime Architect and Purveyor of matter and life in this galactic corner we designate in Western science the solar system, and as such worthy of unconditional and ceremonious adoration.

Fig. 21.15 An artist's rendition of the mysterious **Oumuamua** celestial object which was observed after the fact in its travel course across our solar system. It came from the direction of the Lyra constellation and headed toward the Pegasus constellation at leaving. Strangely, its path does not reconcile with gravitational attraction by any nearby star, an inexplicable fact that prompted Harvard U. Astronomer Avi Loeb to propose an alien technological signature for the object.

Perhaps some time soon, in Western academic science and as a matter of course, we will go beyond being able to witness ourselves and/or choose to hold back or blur as often done, because invested with better science, we will be able to give a knowledgeable and comprehensive account of what we see, possibly the likes of intelligently-driven interstellar objects such as *Oumuamua* recently observed in 2017 (Fig.21.15). This event was not broadcast in the global news media, let alone with the same fanfare that usually accompanies the scheduled "revolutionary" discoveries or achievements by our official science, always deemed to change life forever. It will be interesting to see the day when government science, under the tumultuous currents of geopolitics, will find itself compelled to publicly connect with civilizational galactic streams that we incessantly witness in our space. As reported at the beginning of this publication in Chapter 1, the World is not without knowing the *June-2022 Pentagon Report*, object of a hearing by the United States Congress from the Undersecretary of Defense for Intelligence and Security, Ronald Moultrie, about interference of extra-terrestrial intelligence with American nuclear weaponry, further stating relative to *Unidentified Aerial Phenomena (UAP)*:

"Of these reported UAP sightings [by the U.S. Navy], 18 displayed extremely unusual flight behaviors, with the mysterious objects appearing to "remain stationary in winds aloft, move against the wind, maneuver abruptly or move at considerable speed, without discernible means of propulsion..." (Fig.21.16)

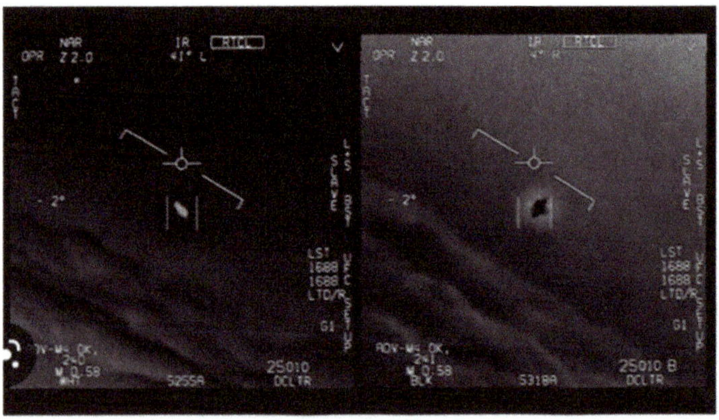

Fig. 21.16 *UAP* sighting video released by the US Pentagon in 2022 captured by the U.S. Navy

All of these events, facts and artifacts are altogether keenly connected to one another and pertains to Cosmic Culture within the Milky Way. Does everyone now understand what the Sphynx or the winged Angels are is in their intimate nature? And no, Humans are not the ultimate form of life across the Milky Way, but the Perdurable Karyotes otherwise dubbed Gods and Goddesses by earlier generations of Humans.

21.16. Conclusion

What the above discussion about life and matter in U_3 is all telling us is that, at the end, there is only one form of universal knowledge, and that mystical or revealed knowledge and scientific knowledge will meet and integrate in the course of the human global culture. As I wrote in a Book I of this publication, Religion is indeed Cosmic Culture. Grasp of the scope and breadth of Shelling Symmetry, together with its derivate forms of Emulation, is key to this awakening and empowering Science. For now, we have closed the gap between physics of the Academy and the New Physics leading to galactic culture through the unraveling of Unified Physical Law as envisioned by David Hilbert in his prescient proposition of *Mathematical Axiomatic Physics*.

Chapter 22

Somewhere, something incredible
is waiting to be known.
Carl Sagan

THEORETICAL DERIVATION OF THE PHYSICAL LIFE CONSTANT THROUGH STEM CELL GENOMICS

Two decades ago, when the human genome was first decoded, promises were made of great advances in knowledge of the proteome. The exercise, extremely important in Biology and for Medical Science though it was, strictly amounted to the conversion of an unknown language into another unknown language. It was not a blueprint that helped increase comprehension of the semantic content of the language, so to speak. The advances that have occurred since the culmination of the translation project have been effective in disease treatments, through discoveries realized by the trial-and-error method. A top-down approach in basic research is necessary to unravel what is still the mystery of the origin of Life's genetic information as well as the primitives dictating the biogeometric organization of the living, both at the histological level in organs and anatomically. In this Chapter, we are going to grapple with the numeric basis of ontogenesis and phylogenesis and uncover the Physical Life constant that regiments this quintessential configuration and interaction.

22.1. What are Stem Cells?

We begin this quest by addressing the biology of stem cells, from a defining and classificatory vantage point. Stem cells are the body's master cells specialized in the

Reproduction function. In other words, the phylogenetic role of the scalar tenet in those cells has been potentiated multiple folds with respect to the metabolic role or wavefunction tenet. First-order stem cells are those presiding over embryogenesis resulting from egg fertilization (fusion of the male gamete within the female gamete). These stem cells come from embryos that have 3 to 5 days of life. At this stage, an embryo bears the nomer of blastocyst and counts with up to about 150 cells. Once embryogenesis has completed and the organism segregates into an independent entity, each histological composition or organ becomes endowed with its own stack of stem cells which are second-order root cells.

While first-order or embryogenetic stem cells give rise or differentiate into any class of specialized cells during embryogenesis, second-order stem cells in the realized organism acquire a restriction gearing them specifically to the production of or differentiation into the same class of cells forming the histological formations where they reside. Second-order stem cells, also designated adult stem cells, evidence a more limited ability to give rise to various cells of the body, compared with embryonic stem cells. They are further found in small numbers in most adult tissues, such as bone marrow or fat. For instance, bone marrow stem cells may be able to create bone cells or heart muscle cells.

There exists a third-order stem cell generation whose origin are perinatal formations, the amniotic fluid and the umbilical cord blood i.e. In our sense, this class of stem cell must be viewed from the vantage of the pregnant female body and be brought into the same class of stem cells artificially obtained from alteration of adult cells in adult living bodies. Although perinatal stem cells are presumed to have the same properties of embryonic stem cells, they did not originate from blastocyst evolution and are therefore not identifiable as first-order stem cells. Because our knowledge of the proteome is blatantly incomplete at this stage as much as concurrent epigenetic activity, one cannot discard that what we have figured in experimental manipulation of the proteome in bringing adult stem cells back down to root state is perhaps just a replicate of natural processes occurring hitherto unbeknownst to us in living organisms. I for one do not believe that we are going to be smarter than mother Nature, and, incidentally, she is the one who taught us the CRSPR technique.

Finally, as one would expect, all stem cells are endowed with the potential to exactly replicate or repeat themselves.

22.2. Capital Regenerative Role of the Hematic Liquid Tissue in Vertebrates

With the astonishing number of approximately one trillion cells being generated daily to compensate apoptotic cells in human Bone Marrow [BM], and given the deep penetration of blood cells in all systems of the organism, Hematopoietic Stem Cells seem to play a hierarchical towering role in the Regenerative potentials of the entire vertebrate organism. A hematopoietic stem cell, also called blood stem cell, is a precursor cell that can develop into all types of blood cells, namely white blood cells, red blood cells, and platelets. We will also see that they are endowed with transdifferentiation capabilities according them precursory faculties for other cell lineages in the histological ensemble of the vertebrate organism. Hematopoietic stem cells reside in the peripheral blood and bone marrow.

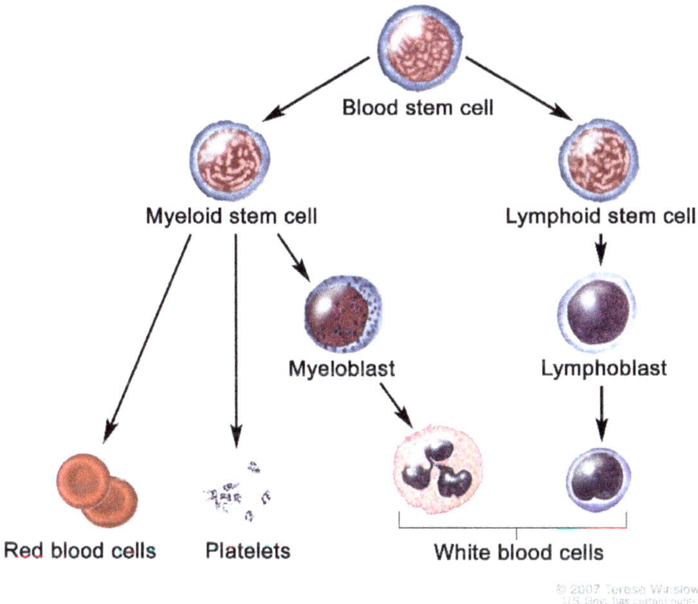

Fig. 22.1 Blood stem cells

Alexander Medvinsky offers the following clarifications about the origin of hematopoietic stem cells (Fig. 22.1):

"The ontogeny of the vertebrate hematopoietic system is characterized by two waves: embryonic hematopoiesis, which generates transitory hematopoietic cell populations, such as primitive erythrocytes and some myeloid cells; and permanent, definitive hematopoiesis, which originates later in development from definitive hematopoietic stem cells (dHSC) and which gives rise to all the mature blood (erythroid, myeloid and lymphoid) lineages in the adult organism.

It was Francoise Dieterlen-Lievre in 1975 (Dieterlen-Lievre, 1975), who showed, using quail chick chimeric embryos, that the adult hematopoietic system originates inside the embryo proper: 'haemopoietic stem cells of the definitive blood cell series originate from a source other than the yolk-sac, and that this source must be intra-embryonic' (Dieterlen-Lievre, 1975). The use of the term 'definitive' in this context is clear, meaning 'fully formed and developed'. Thus, although from the mid-1990s the term 'definitive' has also been used to refer to any hematopoietic progenitor cells emerging during embryogenesis or during embryonic stem (ES) cell differentiation that can produce myeloid and adult-type enucleated erythroid cells, historically and more accurately, the term 'definitive HSC' (dHSC; see Glossary, Box 1) refers to a stem cell that gives rise to the true adult hematopoietic system. The term 'definitive HSC' is specific to developmental hematopoiesis because it refers to a stage in HSC development when an HSC acquires the ability to provide long-term hematopoiesis upon transplantation to wild-type irradiated adult recipients." …

"All lineage blood cells are produced by a rare population of multipotent HSC, which can proliferate by self-renewal and differentiate to accomplish the functional maturation. From HSC to mature cells, there are several intermediate progenitor cells (5). These cells can display both functional multipotent and lineage-committed properties simultaneously or separately prior to complete maturation. Blood is a highly regenerative tissue due to its short life span, and BM supports the dynamic movement of diverse cells to maintain homeostasis of blood cells. Approximately, one trillion cells are generated daily to compensate apoptotic cells in human BM, suggesting rapid circulation of blood cells. HSC can be divided into LT-HSC, ST-HSC, and multipotent progenitor (MPP) in terms of duration of repopulation."

It is well worth emphasizing the assertion that HSC's can display both functional multipotent and lineage-committed properties **simultaneously or separately** prior to reaching complete maturation. This capability spells out a remarkable property that speaks to the Emulation potency of these particular stem cells. We wish to highlight likewise their broad precursory abilities to supply virtually all of the main systems of the vertebrate organism, that is to say the skeletal system, the muscular system and the nervous system, the circulatory system which includes not only the hematic liquid tissue but also the blood vessel system. In regards to the contribution of HSCs/Progenitor Cells to Regeneration of Non-Hematopoietic Cells, Ji Yoon Lee and Seok-Ho Hong note:

"Hematopoietic Stem Cell Transplantation [HSCT] is feasible in myeloablated and non-myeloablated hosts, leading to a chimeric condition when allogeneic HSCT is carried out. In addition to using HSCT to induce hematopoietic cell reconstruction, cell therapy using [Bone-Marrow] BM-derived stem/progenitor cells has been continuously applied in regenerative medicine. Additionally, tissue-specific stem/progenitor cells in hematopoietic and non-hematopoietic tissues are also regarded as important therapeutic cell sources. *BM-derived stem/progenitor cells are able to differentiate into cells of other lineages in a process called transdifferentiation, as well as differentiation (32). Many reports have shown that HSCs can transdifferentiate into cells of other lineages, such as endothelial cells, cardiomyocytes, neural cells, and hepatocytes (33, 34).*"

Table 22.1 by the same authors presents the transdifferentiation capabilities of HSC's as chronologically discovered.

The prominent physiological impact of HSC's is revealed in the disease called Pearson Syndrome and other related maladies. Pearson Marrow-Pancreas Syndrome is another nomer for the same. The NIH describes its nosology in the following terms on Medline[29]:

"Pearson syndrome is a **severe** disorder that usually begins in infancy. It causes problems with the development of blood-forming (hematopoietic) cells in the bone marrow that have the potential to develop into different types of blood cells. For this reason, Pearson syndrome is considered a bone marrow failure disorder. Function of the pancreas and other organs can also be affected.

Item	Source of stem cells	Differentiation of HSCs/progenitors	Discovery Year
1	BM-hematopoietic cells	Microglia/Macroglia	1997
2	BM-myogenic progenitors	Muscle cells	1998
3	BM-Stem cells	Endothelial precursors	1999
4	BM-Stem cells	Hepatic oval cells	1999
5	BM-Stem cells	Neuronal cells	2000
6	BM-Stem cells	Hepatocytes	2000
7	Hematopoietic stem cells	Hepatocytes	2000
8	BM-Stem cells	Cardiac muscle	2001
9	BM-side population cells	Cardiac muscle, endothelium	2001
10	PB-stem cells	Hepatocytes/epithelial cells	2002
11	BM-Stem cells	Cardiac muscle	2005
12	BM-Stem cells	Lymphatic endothelial progenitors	2010
13	Corticalbone-derived stemcell	Myocytes/vascular cells	2013

Table 22.1 Transdifferentiation of HSCs or BM-derived progenitor cells to regenerative organs

Most affected individuals have a shortage of red blood cells (anemia), which can cause pale skin (pallor), weakness, and fatigue. Some of these individuals also have low numbers of white blood cells (neutropenia) and platelets (thrombocytopenia). Neutropenia can lead to frequent infections; thrombocytopenia sometimes causes easy bruising and bleeding. When visualized under the microscope, bone marrow cells from affected individuals may appear abnormal. Often, early blood cells (hematopoietic precursors) have multiple fluid-filled pockets called vacuoles. In addition, red blood cells in the bone

[29] https://medlineplus.gov/genetics/condition/pearson-syndrome/#synonyms

marrow can have an abnormal buildup of iron that appears as a ring of blue staining in the cell after treatment with certain dyes. These abnormal cells are called ring sideroblasts.

In people with Pearson syndrome, the pancreas does not work as well as usual. The pancreas produces and releases enzymes that aid in the digestion of fats and proteins. Reduced function of this organ can lead to high levels of fats in the liver (liver steatosis). The pancreas also releases insulin, which helps maintain correct blood sugar levels. A small number of individuals with Pearson syndrome develop diabetes, a condition characterized by abnormally high blood sugar levels that can be caused by a shortage of insulin. In addition, affected individuals may have scarring (fibrosis) in the pancreas.

People with Pearson syndrome have a reduced ability to absorb nutrients from the diet (malabsorption), and most affected infants have an inability to grow and gain weight at the expected rate (failure to thrive). Another common occurrence in people with this condition is buildup in the body of a chemical called lactic acid (lactic acidosis), which can be life-threatening. In addition, liver and kidney problems can develop in people with this condition. Some people with Pearson syndrome have droopy eyelids (ptosis), vision problems, hearing loss, seizures, or movement disorders.

About half of children with this severe disorder die in infancy or early childhood due to severe lactic acidosis or liver failure. Many of those who survive develop signs and symptoms later in life of a related disorder called Kearns-Sayre syndrome. This condition causes weakness of muscles around the eyes and other problems." …

"Kearns-Sayre syndrome[30] is a condition that affects many parts of the body, especially the eyes. The features of Kearns-Sayre syndrome usually appear before age 20, and the condition is diagnosed by a few characteristic signs and symptoms. People with Kearns-Sayre syndrome have progressive external ophthalmoplegia, which is weakness or paralysis of the eye muscles that impairs eye movement and causes drooping eyelids (ptosis). Affected individuals also have an eye condition called pigmentary retinopathy, which results from breakdown (degeneration) of the light-sensing tissue at the back of the eye (the retina) that gives it a speckled and streaked appearance. The retinopathy may cause loss of vision. In addition, people with Kearns-Sayre syndrome have at least one of the following signs or symptoms: abnormalities of the electrical signals that control the heartbeat (cardiac conduction defects), problems with coordination and balance that cause unsteadiness while walking (ataxia), or abnormally high levels of protein in the fluid that surrounds and protects the brain and spinal cord (the cerebrospinal fluid or CSF).

People with Kearns-Sayre syndrome may also experience muscle weakness in their limbs, deafness, kidney problems, or a deterioration of cognitive functions (dementia).

---

[30] https://medlineplus.gov/genetics/condition/kearns-sayre-syndrome/

Affected individuals often have short stature. In addition, diabetes mellitus is occasionally seen in people with Kearns-Sayre syndrome.

When the muscle cells of affected individuals are stained and viewed under a microscope, these cells usually appear abnormal. The abnormal muscle cells contain an excess of structures called mitochondria and are known as ragged-red fibers. A related condition called ophthalmoplegia-plus may be diagnosed if an individual has many of the signs and symptoms of Kearns-Sayre syndrome but not all the criteria are met."

The above paints a very broad dysfunctional picture stemming from hematopoiesis deficiency. Given the stature of HSC's in the configuration of the vertebrate organism, it is not difficult to understand the severity and lethality of diseases resulting from a dysregulation in their ontogenetic onset.

22.3. Why a 9-Month Course for Human Gestation?

In the life sciences, it is probably true that the question of reason for a 9-month course of human gestation under the stem cells directives has seldom if ever been posed. It is a question that we can finally pose and reasonably explicate along the exegetic line of the 9-point metric inherent to the quanto-geometric spectrum canon. As discussed at the beginning of this study, the construct of time duration, whether in a month period or else, is a vacuous notion that must yield to the more accurate construct of a rotational phase of the planet habitat. A 9-month event is nothing but an event occurring over 9 rotation cycles of planet earth. What the ontogenetic program in a human stem cell appears to undertake is a connection to phylogenesis as realized over the astrophysical background or habitat substrate. We have well seen that phylogenetic development obeys quanto-geometric spectrum canon, thus the 9-point metric (or supermetric as dubbed) underlying the mathematical framework. Ontogenetic emulation appears to embrace this metric in the case of human gestation, to the extent of a guarantee for tighter surrogation of phylogenesis.

Effectively, a detailed analysis of embryogenesis from blastomere to blastula to embryo to fetus, all of the quanto-geometric vectors are there to mediate the entire process of biogeometric development of the offspring. We have not contemplated the inclusion of such studies in this writing and only intend to focus on the physical life constant which is an essential figure for this Shell Symmetry. Make no mistake however that the 9-month course that subtends human gestation is directly connected to the quanto-geometric 9-point super metric.

22.4. Framing Life with Numeric Discipline

In Biophysics, the fundamental variables of Life deriving from ontogeny which would allow a towering description of life have inarguably remained hitherto unknown. However, the calculus that has been utilized to frame the developmental evolution of stem cells, in particular HSC's, does offer a path to an exacting mathematical-physics description by making contact with the fundamental physical variables at play through the QG analytic prism. In tracking single hematopoietic stem cells in vivo using high-throughput sequencing in conjunction with viral genetic barcoding, Rong Lu et al make the following interesting observations:

"Interesting clonal correlations were observed between granulocytic/monocytic progenitor (GMP) and other hematopoietic populations. When compared to its upstream progenitors, GMP correlates the best with its immediate upstream progenitor, Flk2+ multipotent progenitor (MPP/Flk2+). When compared to intermediate progenitor cells from similar differentiation stages in the hematopoietic hierarchy, GMP is correlated more closely with the megakaryotic/erythroid progenitor (MEP) than with the common lymphocyte progenitor (CLP). This is consistent with previous studies that suggest GMP and MEP both belong to myeloid lineages and share a common myeloid progenitor (CMP). When compared to blood cells, GMP correlates most closely with its progeny granulocyte (Gr). In general, the clonal correlations of the hematopoietic populations reflect the progression and divergence of the hematopoietic hierarchy as characterized by previous studies. This clonal correlation parameter can be used as a novel indicator to determine the developmental relationships of different cell populations. For instance, it can be applied to studies of cancer metastasis to identify the primary and secondary metastatic sites. As we have shown, the relationship between various cell populations can be directly inferred from comparisons between their barcode distributions."

The *Institute for Frontier Life and Medical Sciences* at Kyoto University offers a notable set of data correlating HSC's and viral genomics[31] and found for them a Pearson Correlation Coefficient of ***0.298841***, an exceptional number, when it comes to stamping phylogenesis or emulative ontogenesis. First a brief reminder about Pearson correlation.

22.5. Generic Bivariate Statistical Analysis

In Bivariate Statistical Analysis, the Pearson Correlation coefficient, often referred to as the Pearson *R* test, takes the shape of a statistical formula which measures the strength between variables and relationships. To determine how strong the relationship is between two sampled variables under examination, one needs to find the coefficient value, which can range between *-1.00* and *1.00* and is usually signified by *r*. Where the coefficient takes the value of *-1.0,* it indicates a perfectly negative correlation between the sampled variables, *0.0* is for just no correlation, and *1.0* is for a very strong or perfectly positive linear correlation. A positive correlation translates in that as the values of *x* variables increase, the values of the *y* variables narrowly increase as well. A negative correlation translates in that as one variable increases, the other decreases. The formula to compute a Pearson Correlation coefficient is:

[31] https://genomics.virus.kyoto-u.ac.jp/immuno-navigator/index.php/bin/tfbs_9606/bin/download2.cgi?cell=1&probe2=218238_at&probe=201377_at&entrez=9898&o=9606

$$r = \frac{\sum_{i=1}^{n} x_i y_i - \frac{1}{n}\sum_{i=1}^{n} x_i \sum_{i=1}^{n} y_i}{\sqrt{\sum_{i=1}^{n} x_i^2 - \frac{1}{n}\left(\sum_{i=1}^{n} x_i\right)^2}\sqrt{\sum_{i=1}^{n} y_i^2 - \frac{1}{n}\left(\sum_{i=1}^{n} y_i\right)^2}}$$

where x and y are the two variables under examination.

The Spearman rank correlation can also be used for an analysis of the association between such data. A Spearman rank coefficient is a Pearson correlation coefficient calculated with the *ranks* of the values of each of the 2 variables instead of their actual values. This rank correlation test can only be used in the event of there being at least 10 pairs of data (ideally at least 2 sets of 15). The formula to compute the Spearman's Rank Correlation coefficient is:

$$r_s = 1 - 6\sum D^2/n(n^2-1)$$

where r_s is the Spearman's Rank correlation coefficient, D are the differences between ranks, and n the number of pairs of measurements. It is important to note that the Pearson correlation coefficient is typically used for jointly normally distributed data or data that follow a bivariate normal distribution. For nonnormally distributed continuous data, for ordinal data, or for data with relevant outliers, a Spearman rank correlation is typically used as a measure of a simple monotonic association. Relevant to our study will be the Pearson Correlation coefficient which will be the one figure we retain, in accordance with the higher significance acquired by distribution normals and the profound typological meaning of the percentages (fractional parts of 1) associated with these normals in our Framework.

22.6. Theoretical Derivation of the Fundamental Life Constant

As we well know, the $M_3(s)$ Creation function realizes the suite of all foundational Shell constants in U_3. The last hitherto studied constant of the kind was the Van Der Vaals constant, the pivot for the nano Materials Shell, which is the value taken by the function for degree of freedom $n = 5$. When the U_3 spatial spread explodes from 5 to 6 degrees of freedom, that is to say at $n = 6$, the value taken by the function at that point s creates the foundational physical life constant for Shell 6, the Life Shell. Let's compute:

Value of s at that symmetry-breaking point is: $s = \sigma\frac{\sqrt{10}}{2}$

By replacement in the $M_3(s)$ function expressed below:

$$M_3(s) = -\frac{\sqrt{7}}{\sigma^4 4\sqrt{2\pi}} \cdot e^{-\frac{s^2}{2\sigma^2}} \cdot s^3$$

$$M_3(s_6) = -\frac{\sqrt{7}}{\sigma^4 4\sqrt{2\pi}} \cdot e^{\frac{-\left(\frac{\sigma\sqrt{10}}{2}\right)^2}{2\sigma^2}} \cdot \left(\frac{\sigma\sqrt{10}}{2}\right)^3$$

$$M_3(s_6) = -\frac{\sqrt{7}}{\sigma^4 4\sqrt{2\pi}} \cdot e^{\frac{-(\sigma^2)\left(\frac{10}{4}\right)}{2\sigma^2}} \cdot \frac{\sigma^3\left(\sqrt{10}\right)^3}{8}$$

$$M_3(s_6) = -\frac{\sqrt{7}}{\sigma^4 4\sqrt{2\pi}} \cdot e^{-\frac{10}{8}} \cdot \frac{\sigma^3\left(\sqrt{10}\right)^3}{8}$$

$$M_3(s_6) = -\frac{\sqrt{7}}{\sigma 4\sqrt{2\pi}} \cdot e^{-\frac{10}{8}} \cdot \frac{\left(\sqrt{10}\right)^3}{8}$$

$$\boxed{M_3(s_6) = 0.2988415670/\sigma}$$

To 7-digit approximation, this number is the same that has been measured by the "Frontier Life" group for the correlation between hematopoietic stem cell genomics and viral genomics. That a correlation exists between top-level precursory pluripotent cells of higher vertebrates and viral entities is a direct proof that animal stem cells are on the very life line or spectrum initiating with viral entities, which is a line of phylogenetic essence. That the correlation is a Pearson **normally distributed** correlation tells us that the distribution is of Quanto-Geometric nature. Finally, that the value of this experimental correlation matches the predicted QG value for the Physical Life constant grants us a solid proof that what the "Frontier Life" group innocuously measured was indeed the Physical Life constant, while their measurement in turn authenticates our numeric finding (Fig. 22.2).

To comprehend the scope and depth of the above, the biologist and the biophysicist ought to represent for themselves that viruses are at the top of the phylogenetic hierarchy, whilst hematopoietic stem cells are similarly at the top of the animal ontogenetic hierarchy. This parallel puts us directly on track of their common denominator. Further, the numeric value of the correlation of their genomic expressions must say something quintessential about their inner biogeometric architecture, since genes are in the very business of building that architecture. This quintessential pivot is nothing but the fundamental Physical Life constant in supreme governance of this Shell. Without this constant, the whole biogeometric construction endeavor of the living karyotic body, otherwise known as organogenesis at its top tier, cannot be understood. It is also clear that the QG biogeometric primitives, as discussed in previous Chapters, are equally key factors in the buildout of this architecture.

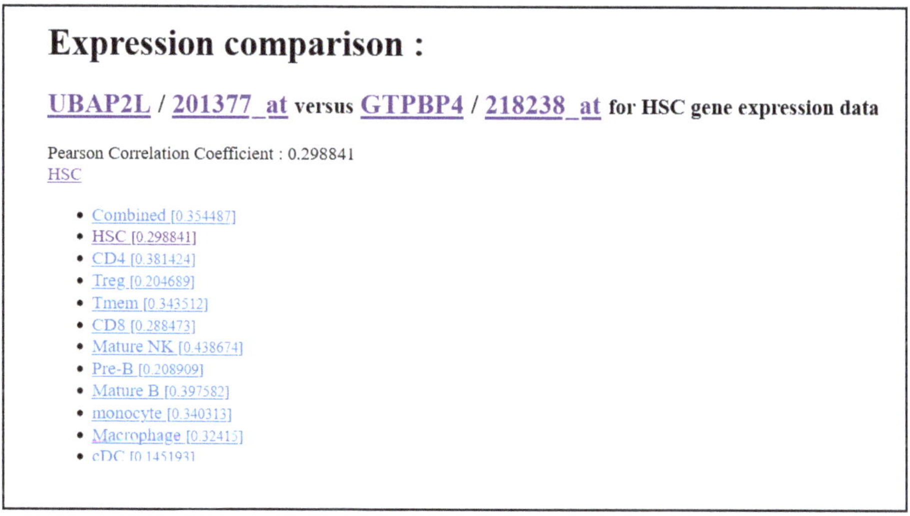

Expression comparison :

UBAP2L / 201377_at versus GTPBP4 / 218238_at for HSC gene expression data

Pearson Correlation Coefficient : 0.298841
HSC

- Combined [0.354487]
- HSC [0.298841]
- CD4 [0.381424]
- Treg [0.204689]
- Tmem [0.343512]
- CD8 [0.288473]
- Mature NK [0.438674]
- Pre-B [0.208909]
- Mature B [0.397582]
- monocyte [0.340313]
- Macrophage [0.32415]
- cDC [0.145193]

Fig. 22.2 Dataset of the viral/HSC genomic correlation by the Japanese Institute of Frontier Life and Medical Science as published on their website

In the final analysis, a Pearson Correlation coefficient helps us quantify the relationship between two sampled variables; thus, the quantity that correlates virus and hematopoietic stem cells, both as heads of the two spectral hierarchies of the living, understandably incarnates that scalar constant. The ultimate identity of said scalar must be the numeric pivot for genesis and development of life, in one word, the very physical life constant. There is no question that more may be uncovered along this investigative line.

22.7. How Combinatorics of Genic Pairing Underscore QG Phylogenetic Emulation

Many may or may not be surprised that the quantum causal regime described all throughout this publication in many different ways are at the center of experimentally known gene combinatorics in biota. The following is a brief recount of what is known, along with analytic interpretation in the light of QG covariant prescriptions.

Through cell division and fertilization, human animals acquire 46 chromosomes containing their unique DNA, namely 23 from the mother, and 23 from the father. Since human cells carry two copies of each chromosome, they have two versions of each gene.

These different versions of a gene are called alleles in genetic science nomenclature. However, before we move to consider genes inside of a chromosome and their matching, it is of import to heed at the first place the basic parity of chromosomic entities in the proteome. This pairing is realized at fertilization, as each chromosome of the spermatozoid alleles or pairs with the corresponding egg's chromosome. The pairing of chromosomes creates altogether the covariant ontic unit that is the basis of the biogeometric development to follow through, in essence the QG basis duplet.

Biogeometric development of the ontic duplet will be realized expectedly through covariance between its two constituent elements. The allele genes in a pair of chromosomes interact in such a way that they produce what is called phenotypic traits or gene expression. In other words, the ontic interaction results in extrinsic phenotypic expression. Genic alleles are described as either dominant or recessive depending on their associated traits, the science dictates. In fact, one should be keen to understand that it is not the extrinsic trait that determines whether the gene is dominant or recessive. The ontic trait and its covariant class pertain to the gene. It is within the covariant interaction that mutual influence of allele genes on one another negotiates the phenotypic outcome.

That in genetic science the attributes of dominant and recessive are descriptors of ontic status of alleles is an interesting syntactic choice which exactly meets the QG ontic descriptors of the basis duplet. If the offspring is going to be a male, then the allele coming from the father is dominant. But if the child is going to be a female, then the allele coming from the mother is the dominant one. Perhaps, the term recessive to mean shadowed could be best substituted by masked or eclipsed, since recessive tends to evoke a trend toward disappearance or extinction. One shall therefore intently keep in mind that recessive in this context only means masked and does not imply any hierarchical relation.

Consequently, the event that determines the covariant linkage attribute in terms of dominant or eclipsed (recessive) is the outcome of the sex-determining set of chromosomes. Sex-linked genes pertain to the 23rd pair of chromosomes, which are two XX chromosomes in females and the egg, and an XY pair instead in males and their spermatozoids. It is capital to retain that each has certain genetic traits that are absolutely specific to each one of them. Inasmuch as sex-related characteristics are of paramount significance in ontogeny, development beyond birth and into adult life, one can see why the attribution of dominant/recessive to alleles in embryonic chromosomic inventory at large depends on the signal coming from the 23rd pair. This event alone is the biochemical trigger that determines whether the set of alleles coming from the mother or instead the set of alleles coming from the father is going to be the dominant one in the covariant interaction between paired alleles. The set of alleles coming from the mother play the instantiator role of the scalar or massive tenet in the physical ontic duplet of the canon, while the set of alleles coming from the father plays the instantiator role of the wavefunction-propagator. While females will always pass down their X chromosome since it is the only kind they carry, males will pass down either an X or Y chromosome at random, and that is only the reason why males determine the offspring's sex. In that sense, and only that sense, the father's contribution is said to have a bigger impact on embryogenesis.

Consistent with QG covariant prescriptions, in the chromosomic cross-linkage system, there are cases where the two basis tenets, the two alleles that is, are covariantly equated in their mutual influence. In genetic science, this case is known as one where both alleles are considered to be dominant, being also called *codominance*. The phenotypic result of this class of covariance is that of both alleles being expressed equally. An example is the blood group AB which is a case of codominance of the A and B dominant alleles. It is clear that we are here facing an instance of QG quadratic covariance where the two tenets are of equated influence on one another, none gaining dominance in the relationship. Co-dominance is arguably a bit of a misnomer since one cannot logically have two dominant at the same time in a pair, if dominance means supremacy over the other. Equated covariance as understood quanto-geometrically would be a better qualifier for gene combinatorics resulting in hybrid phenotypic expression. Equated genic covariance (producing co-dominant expression) is for sure an important exegetic line when it comes to interpret hybrid physiological and anatomic traits. It is probably because of equated covariant expression, among other possible reasons, that geneticists came to the belief that there is no real indication that boys necessarily take after their father in physiognomic looks any more than they do their mothers.

Finally, the case does exist of orphan alleles meeting no other corresponding alleles. One can logically understand why in the case of a dominant orphan allele, the gene is expressed anyway, and likewise why an orphan recessive allele has or leads to no phenotypic expression at all (because it is *eclipsed*). It remains beyond the scope of this study to address other secondary or adjunct cases of genic covariance, this undertaking requiring the full breadth of the QG spectral cadre. What is clear is that the tripartite Norm of QG Quintessence is at the heart of the proteome's combinatorics. Being able to so closely marry the proteome's combinatorics, it is no surprise that the QG analytic canon has been able to theoretically compute the value of the fundamental Life constant with such a notable precision.

22.8. Conclusion

Posing the problem of a formal description of life in terms of information, as customarily posited, equates to raising an exegetic frame that is far too insufficient to thoroughly embrace this elusive and quite complex problem. The so-called mystery of the origin of Life is not exactly a question of genetic information per se as frequently posed, important though genomic protein coding might be. The customary biophysical description of a host of biological and biochemical material as efficient and diverse formal causes, even when wrapped in a mathematical coding coat of sorts, will not assertively materialize the connection to the first cause. In order to cast an authentic, that is to say connected phylogenetic and ontogenetic, description of life in adequate numeric terms, one ought to uncover, as we have done, the Symmetry Hierarchy that is the orchestrator of the configuration and processes of Life, along with the resulting Physical Life constant, hereby straightforwardly computed from Theory, as a central element of this orchestration. That is how elastoplasticity explodes from what has been considered flat physics of inanimate matter. Going forward, the means to comprehend in an exacting manner the core orchestration of Life has been decisively and irrevocably laid down with this exposé.

EPILOGUE

The Quanto-Geometry Theory comes at a period when effectual knowledge is more than ever held captive by the institutionalization of physical science and is further intently expressed in a mystifying language, for the most part, to the common mind, obviously at such effects.

This seclusion trend in the form of knowledge institutionalized in the West is at least as old as the episode among Pythagorean mathematicians who threatened to kill at high seas one of their peers, named Hippasus, because he had resolved to reveal to the rest of the world the secrecy of the irrational nature of $\sqrt{2}$, as legend has it. The remnants of secret societies and fraternities still active in most higher learning institutions and the dependence of other prestigious ones to royal dominion of another era are a reflection and a reminder of the mindset that had instituted these centers of knowledge.

As much as institutionalization of knowledge is ineluctable in any culture, the formation of a resulting sanctimonious knowledge cast, more often than not, ends up disparaging and blocking the dissemination of effectual knowledge to the common man. The criticism that we have addressed throughout this exposé to actors in the sphere of physical science could very well be turned to mathematicians alike, albeit for a different reason. They have taken the greatest care to obscure this important science to the average mind, to the point that most human live in the rebuke and fear of mathematical science, to the greatest detriment of our culture.

The intellectual evolution of the common man however, let alone his spiritual evolution, goes thru complete mastery of mathematical-physics conceptualization. There is no doubt that the discovery of those dimensions of abstraction and physical reality is a very delectable exercise to the intellect, just because natural canon resonates at all Shells of matter and life. The task ahead for mathematicians is a thorough reformulation of all of Mathematics on the basis of Abstract Number Theory from the perspective of the Higher Number Stack with its scalar and geometric signatures.

It is a rather remarkable fact, well demonstrated by the detail and fullness of this study, that physical matter does not directly develops in the mold of differential algebra, not even integrative processes or methodologies, given their importance in academic curriculae. The complexities that arise from the application of these methods, a taste of which is given in the widespread Lagrangian formulation, understandably result from a position of laying down uncertainties and walking backward from these uncertainties to higher certainty or certainties, by differential or integrative imputation. As natural as this process may appear, having become habit, it is only a choice in the ignorance of the fundamental canon in physical law, the Quanto-Geometric Grand Eigenfunction with its direct functional compositions. The Grand Eigenfunction crystallizes this one simple single rule behind Nature's phylogenesis, of which A. Einstein had a distinct and unrelenting intuition, not to mention the luminary R. Feynman. This rule represents a higher order of certainty and its discovery affords us the means for a direct, integral and all-transparent formulation of physical knowledge in a *positive* methodology.

It is another very remarkable fact, and to my mind completely unacceptable at that, that at more than a century from A. Einstein insights in the dynamics of the intangible and the time-free infinitesimal foreground uncovered by E. Schrödinger's quantum theory, the common man lives in aberrant intellectual prostration with regards to transcendental knowledge of the behavior of the Void. He lives in an obstructive illusion of time ubiquitous, completely severed from Nature's true drivers, and unable to relate matter to life for higher or transcendental ontology, both on an individual and collective basis. This comes to be the result of elitist mystification in the historic conduct of the knowledge cast in the West. We must defeat this trend that had in fact pervaded all cultures on the globe at different epochs, even those predating the onset of western civilization.

The positive order of Certainties, the very prolific order of things in the Void, the mathematical-physics Abstract Number Stack, the consequential time-free foreground and background that prelude the pivotal physical constants must all become subject matters for formal learning starting from *kindergarten* globally. This positive vision must be taught to human minds at their youngest stage for it to solidify at maturity through reiterative amplification across developmental and educational stages. There is no space civilization for mankind outside of this program.

Should this challenge find Western academia in paralysis or otherwise unwilling, there are clear signs that the citizenry of the world will resolutely move forward on own accord in the direction of promoting and disseminating grander and more open effectual and *positive* knowledge with all the information tools already available. This renaissance in mathematical and fundamental physical science is at any rate far more important than any pursuit in the resolution of other known open questions in the sciences, both because of its implications in the evolution of the global culture and because those sought-after answers will naturally follow. Let's do it!

POSTFACE

History of Quanto-Geometric Theory

The Quanto-Geometric Theory has been in the published literature since 1995 when the author released his French publication **La Quanto-Géométrie, Tome I, (ISBN 0-9647466-1-1)**. Written perhaps in an overly sophisticated language and in an extremely dense style for certain, frankly difficult at times to the intended audience, that first cast of the theory presented it on limited mathematical ground and more so on philosophical grounds. The author justly considers nowadays that first publication to simply represent the protogenesis of the Quanto-Geometric Theory.

A few years later in May of 1999, the author released the publication **The Human Mind Inside (ISBN 0-9647466-0-2)**, an English title addressing human mind's complex processes in the light of the Quanto-Geometric Model. This text, largely revised in August 2014, is available in eBook format **(ISBN 9781311207111)** and print format as well **(ISBN 978-0-9647466-3-3)**. A summary of what to become Chapter 3 Book I of the current publication series was already presented in the original 1999 text, amplifying the mathematical foundation of the Quanto-Geometric Theory. It is noteworthy that in said publication the notion of irrevocable dismissal of the time variable in the physical description of nature, one of the key signatures of the Theory and a notion that seems to be gaining traction of late in some corners of theoretical physics, was already in full display.

The current Treatise comes at a time when a yearning for renewal of perspectives in theoretical physics, amidst the consummated failure of other prominent frameworks, seems to pervade the academy of physics. If anything, this Treatise brings to the fore the heightened light of that anticipated candle of conceptual renaissance and the physics of problem resolution for certain. At the very least, this content will help spark creative and fresh debates whilst setting a solid launch pad for the new Physics.

Equally of note, the author released in October 2022 the monograph **Functional Architecture of the Human Mental:** *A Reference Psychophysics Treatise of Human Mentation and its Disorders*, co-authored with Caroline Stamu-O'Brien, MD. This 800-page tome brings an innovative mathematical-physics dimension to the understanding of human emotions, cognition, behavior and psychiatric diagnostic entities. The author may be reached at **quantogeometry@gmail.com (www.quantogeometry.org)**

Table of Physical Constants
NIST CODATA 2010

QUANTITY	VALUE	UNCERTAINTY	UNITS
{220} lattice spacing of silicon	192.015 5714 e-12	0.000 0032 e-12	m
alpha particle mass	6.644 656 75 e-27	0.000 000 29 e-27	kg
alpha particle mass energy equivalent	5.971 919 67 e-10	0.000 000 26 e-10	J
alpha particle mass energy equivalent in MeV	3727.379 240	0.000 082	MeV
alpha particle mass in u	4.001 506 179 125	0.000 000 000 062	u
alpha particle molar mass	4.001 506 179 125 e-3 0.000 000 000 062 e-3 kg mol^-1		
alpha particle-electron mass ratio	7294.299 5361	0.000 0029	
alpha particle-proton mass ratio	3.972 599 689 33	0.000 000 000 36	
Angstrom star	1.000 014 95 e-10	0.000 000 90 e-10	m
atomic mass constant	1.660 538 921 e-27	0.000 000 073 e-27	kg
atomic mass constant energy equivalent	1.492 417 954 e-10	0.000 000 066 e-10	J
atomic mass constant energy equivalent in MeV	931.494 061	0.000 021	MeV
atomic mass unit-electron volt relationship	931.494 061 e6	0.000 021 e6	eV
atomic mass unit-hartree relationship	3.423 177 6845 e7	0.000 000 0024 e7	E_h
atomic mass unit-hertz relationship	2.252 342 7168 e23	0.000 000 0016 e23	Hz
atomic mass unit-inverse meter relationship	7.513 006 6042 e14	0.000 000 0053 e14	m^-1
atomic mass unit-joule relationship	1.492 417 954 e-10	0.000 000 066 e-10	J
atomic mass unit-kelvin relationship	1.080 954 08 e13	0.000 000 98 e13	K
atomic mass unit-kilogram relationship	1.660 538 921 e-27	0.000 000 073 e-27	kg
atomic unit of 1st hyperpolarizability	3.206 361 449 e-53	0.000 000 071 e-53	C^3 m^3 J^-2
atomic unit of 2nd hyperpolarizability	6.235 380 54 e-65	0.000 000 28 e-65	C^4 m^4 J^-3
atomic unit of action	1.054 571 726 e-34	0.000 000 047 e-34	J s
atomic unit of charge	1.602 176 565 e-19	0.000 000 035 e-19	C
atomic unit of charge density	1.081 202 338 e12	0.000 000 024 e12	C m^-3
atomic unit of current	6.623 617 95 e-3	0.000 000 15 e-3	A
atomic unit of electric dipole mom.	8.478 353 26 e-30	0.000 000 19 e-30	C m
atomic unit of electric field	5.142 206 52 e11	0.000 000 11 e11	V m^-1
atomic unit of electric field gradient	9.717 362 00 e21	0.000 000 21 e21	V m^-2
atomic unit of electric polarizability	1.648 777 2754 e-41	0.000 000 0016 e-41	C^2 m^2 J^-1
atomic unit of electric potential	27.211 385 05	0.000 000 60	V
atomic unit of electric quadrupole mom.	4.486 551 331 e-40	0.000 000 099 e-40	C m^2
atomic unit of energy	4.359 744 34 e-18	0.000 000 19 e-18	J
atomic unit of force	8.238 722 78 e-8	0.000 000 36 e-8	N
atomic unit of length	000 000 17 e-10 m		
atomic unit of mag. dipole mom.	1.854 801 936 e-23	0.000 000 041 e-23	J T^-1
atomic unit of mag. flux density	2.350 517 464 e5	0.000 000 052 e5	T
atomic unit of magnetizability	7.891 036 607 e-29	0.000 000 013 e-29	J T^-2

atomic unit of mass	9.109 382 91 e-31	0.000 000 40 e-31	kg
atomic unit of mom.um	1.992 851 740 e-24	0.000 000 088 e-24	kg m s^-1
atomic unit of permittivity	1.112 650 056... e-10 (exact)	F m^-1	
atomic unit of time	000 000 012 e-17 s		
atomic unit of velocity	2.187 691 263 79 e6	0.000 000 000 71 e6	m s^-1
Avogadro constant	6.022 141 29 e23	0.000 000 27 e23	mol^-1
Bohr magneton	927.400 968 e-26	0.000 020 e-26	J T^-1
Bohr magneton in eV/T	5.788 381 8066 e-5	0.000 000 0038 e-5	eV T^-1
Bohr magneton in Hz/T	13.996 245 55 e9	0.000 000 31 e9	Hz T^-1
Bohr magneton in inverse meters per tesla	46.686 4498	0.000 0010	m^-1 T^-1
Bohr magneton in K/T	0.671 713 88	0.000 000 61	K T^-1
Bohr radius	0.529 177 210 92 e-10 0.000 000 000 17 e-10 m		
Boltzmann constant	1.380 6488 e-23	0.000 0013 e-23	J K^-1
Boltzmann constant in eV/K	8.617 3324 e-5	0.000 0078 e-5	eV K^-1
Boltzmann constant in Hz/K	2.083 6618 e10	0.000 0019 e10	Hz K^-1
Boltzmann constant in inverse meters per kelvin	69.503 476	0.000 063	m^-1 K^-1
characteristic impedance of vacuum	376.730 313 461...	(exact)	ohm
classical electron radius	2.817 940 3267 e-15	0.000 000 0027 e-15	m
Compton wavelength	2.426 310 2389 e-12	0.000 000 0016 e-12	m
Compton wavelength over 2 pi	386.159 268 00 e-15	0.000 000 25 e-15	m
conductance quantum	7.748 091 7346 e-5	0.000 000 0025 e-5	S
conventional value of Josephson constant	483 597.9 e9	(exact)	Hz V^-1
conventional value of von Klitzing constant	25 812.807	(exact)	ohm
Cu x unit	1.002 076 97 e-13	0.000 000 28 e-13	m
deuteron g factor	0.857 438 2308	0.000 000 0072	
deuteron mag. mom.	0.433 073 489 e-26	0.000 000 010 e-26	J T^-1
deuteron mag. mom. to Bohr magneton ratio	0.466 975 4556 e-3	0.000 000 0039 e-3	
deuteron mag. mom. to nuclear magneton ratio	0.857 438 2308	0.000 000 0072	
deuteron mass	3.343 583 48 e-27	0.000 000 15 e-27	kg
deuteron mass energy equivalent	3.005 062 97 e-10	0.000 000 13 e-10	J
deuteron mass energy equivalent in MeV	1875.612 859	0.000 041	MeV
deuteron mass in u	2.013 553 212 712	0.000 000 000 077	u
deuteron molar mass	2.013 553 212 712 e-3 0.000 000 000 077 e-3 kg mol^-1		
deuteron rms charge radius	2.1424 e-15	0.0021 e-15	m
deuteron-electron mag. mom. ratio	-4.664 345 537 e-4	0.000 000 039 e-4	
deuteron-electron mass ratio	3670.482 9652	0.000 0015	
deuteron-neutron mag. mom. ratio	-0.448 206 52	0.000 000 11	
deuteron-proton mag. mom. ratio	0.307 012 2070	0.000 000 0024	
deuteron-proton mass ratio	1.999 007 500 97	0.000 000 000 18	
electric constant	8.854 187 817... e-12 (exact)	F m^-1	
electron charge to mass quotient	-1.758 820 088 e11	0.000 000 039 e11	C kg^-1
electron g factor	-2.002 319 304 361 53 0.000 000 000 000 53		
electron gyromag. ratio	1.760 859 708 e11	0.000 000 039 e11	s^-1 T^-1
electron gyromag. ratio over 2 pi	28 024.952 66	0.000 62	MHz T^-1
electron mag. mom.	-928.476 430 e-26	0.000 021 e-26	J T^-1
electron mag. mom. anomaly	1.159 652 180 76 e-3	0.000 000 000 27 e-3	
electron mag. mom. to Bohr magneton ratio	-1.001 159 652 180 76 0.000 000 000 000 27		
electron mag. mom. to nuclear magneton ratio	-1838.281 970 90	0.000 000 75	
electron mass	9.109 382 91 e-31	0.000 000 40 e-31	kg
electron mass energy equivalent	8.187 105 06 e-14	0.000 000 36 e-14	J
electron mass energy equivalent in MeV	0.510 998 928	0.000 000 011	MeV
electron mass in u	5.485 799 0946 e-4	0.000 000 0022 e-4	u

electron molar mass	5.485 799 0946 e-7	0.000 000 0022 e-7	kg mol^-1
electron to alpha particle mass ratio	1.370 933 555 78 e-4	0.000 000 000 55 e-4	
electron to shielded helion mag. mom. ratio	864.058 257	0.000 010	
electron to shielded proton mag. mom. ratio	-658.227 5971	0.000 0072	
electron volt	1.602 176 565 e-19	0.000 000 035 e-19	J
electron volt-atomic mass unit relationship	1.073 544 150 e-9	0.000 000 024 e-9	u
electron volt-hartree relationship	3.674 932 379 e-2	0.000 000 081 e-2	E_h
electron volt-hertz relationship	2.417 989 348 e14	0.000 000 053 e14	Hz
electron volt-inverse meter relationship	8.065 544 29 e5	0.000 000 18 e5	m^-1
electron volt-joule relationship	1.602 176 565 e-19	0.000 000 035 e-19	J
electron volt-kelvin relationship	1.160 4519 e4	0.000 0011 e4	K
electron volt-kilogram relationship	1.782 661 845 e-36	0.000 000 039 e-36	kg
electron-deuteron mag. mom. ratio	-2143.923 498	0.000 018	
electron-deuteron mass ratio	2.724 437 1095 e-4	0.000 000 0011 e-4	
electron-helion mass ratio	1.819 543 0761 e-4	0.000 000 0017 e-4	
electron-muon mag. mom. ratio	206.766 9896	0.000 0052	
electron-muon mass ratio	4.836 331 66 e-3	0.000 000 12 e-3	
electron-neutron mag. mom. ratio	960.920 50	0.000 23	
electron-neutron mass ratio	5.438 673 4461 e-4	0.000 000 0032 e-4	
electron-proton mag. mom. ratio	-658.210 6848	0.000 0054	
electron-proton mass ratio	5.446 170 2178 e-4	0.000 000 0022 e-4	
electron-tau mass ratio	2.875 92 e-4	0.000 26 e-4	
electron-triton mass ratio	1.819 200 0653 e-4	0.000 000 0017 e-4	
elementary charge	1.602 176 565 e-19	0.000 000 035 e-19	C
elementary charge over h	2.417 989 348 e14	0.000 000 053 e14	A J^-1
Faraday constant	96 485.3365	0.0021	C mol^-1
Faraday constant for conventional electric current	96 485.3321	0.0043	C_90 mol^-1
Fermi coupling constant	1.166 364 e-5	0.000 005 e-5	GeV^-2
fine-structure constant	7.297 352 5698 e-3	0.000 000 0024 e-3	
first radiation constant	3.741 771 53 e-16	0.000 000 17 e-16	W m^2
first radiation constant for spectral radiance	1.191 042 869 e-16	0.000 000 053 e-16	W m^2 sr^-1
Hartree energy	4.359 744 34 e-18	0.000 000 19 e-18	J
Hartree energy in eV	27.211 385 05	0.000 000 60	eV
hartree-atomic mass unit relationship	2.921 262 3246 e-8	0.000 000 0021 e-8	u
hartree-electron volt relationship	27.211 385 05	0.000 000 60	eV
hartree-hertz relationship	6.579 683 920 729 e15 0.000 000 000 033 e15 Hz		
hartree-inverse meter relationship	2.194 746 313 708 e7	0.000 000 000 011 e7	m^-1
hartree-joule relationship	4.359 744 34 e-18	0.000 000 19 e-18	J
hartree-kelvin relationship	3.157 7504 e5	0.000 0029 e5	K
hartree-kilogram relationship	4.850 869 79 e-35	0.000 000 21 e-35	kg
helion g factor	-4.255 250 613	0.000 000 050	
helion mag. mom.	-1.074 617 486 e-26	0.000 000 027 e-26	J T^-1
helion mag. mom. to Bohr magneton ratio	-1.158 740 958 e-3	0.000 000 014 e-3	
helion mag. mom. to nuclear magneton ratio	-2.127 625 306	0.000 000 025	
helion mass	5.006 412 34 e-27	0.000 000 22 e-27	kg
helion mass energy equivalent	4.499 539 02 e-10	0.000 000 20 e-10	J
helion mass energy equivalent in MeV	2808.391 482	0.000 062	MeV
helion mass in u	3.014 932 2468	0.000 000 0025	u
helion molar mass	3.014 932 2468 e-3	0.000 000 0025 e-3	kg mol^-1
helion-electron mass ratio	5495.885 2754	0.000 0050	
helion-proton mass ratio	2.993 152 6707	0.000 000 0025	

hertz-atomic mass unit relationship	4.439 821 6689 e-24	0.000 000 0031 e-24	u
hertz-electron volt relationship	4.135 667 516 e-15	0.000 000 091 e-15	eV
hertz-hartree relationship	1.519 829 846 0045 e-16 0.000 000 000 0076 e-16 E_h		
hertz-inverse meter relationship	3.335 640 951... e-9	(exact)	m^-1
hertz-joule relationship	6.626 069 57 e-34	0.000 000 29 e-34	J
hertz-kelvin relationship	4.799 2434 e-11	0.000 0044 e-11	K
hertz-kilogram relationship	7.372 496 68 e-51	0.000 000 33 e-51	kg
inverse fine-structure constant	137.035 999 074	0.000 000 044	
inverse meter-atomic mass unit relationship	1.331 025 051 20 e-15 0.000 000 000 94 e-15 u		
inverse meter-electron volt relationship	1.239 841 930 e-6	0.000 000 027 e-6	eV
inverse meter-hartree relationship	4.556 335 252 755 e-8 0.000 000 000 023 e-8 E_h		
inverse meter-hertz relationship	299 792 458	(exact)	Hz
inverse meter-joule relationship	1.986 445 684 e-25	0.000 000 088 e-25	J
inverse meter-kelvin relationship	1.438 7770 e-2	0.000 0013 e-2	K
inverse meter-kilogram relationship	2.210 218 902 e-42	0.000 000 098 e-42	kg
inverse of conductance quantum	12 906.403 7217	0.000 0042	ohm
Josephson constant	483 597.870 e9	0.011 e9	Hz V^-1
joule-atomic mass unit relationship	6.700 535 85 e9	0.000 000 30 e9	u
joule-electron volt relationship	6.241 509 34 e18	0.000 000 14 e18	eV
joule-hartree relationship	2.293 712 48 e17	0.000 000 10 e17	E_h
joule-hertz relationship	1.509 190 311 e33	0.000 000 067 e33	Hz
joule-inverse meter relationship	5.034 117 01 e24	0.000 000 22 e24	m^-1
joule-kelvin relationship	7.242 9716 e22	0.000 0066 e22	K
joule-kilogram relationship	1.112 650 056... e-17 (exact)	kg	
kelvin-atomic mass unit relationship	9.251 0868 e-14	0.000 0084 e-14	u
kelvin-electron volt relationship	8.617 3324 e-5	0.000 0078 e-5	eV
kelvin-hartree relationship	3.166 8114 e-6	0.000 0029 e-6	E_h
kelvin-hertz relationship	2.083 6618 e10	0.000 0019 e10	Hz
kelvin-inverse meter relationship	69.503 476	0.000 063	m^-1
kelvin-joule relationship	1.380 6488 e-23	0.000 0013 e-23	J
kelvin-kilogram relationship	1.536 1790 e-40	0.000 0014 e-40	kg
kilogram-atomic mass unit relationship	6.022 141 29 e26	0.000 000 27 e26	u
kilogram-electron volt relationship	5.609 588 85 e35	0.000 000 12 e35	eV
kilogram-hartree relationship	2.061 485 968 e34	0.000 000 091 e34	E_h
kilogram-hertz relationship	1.356 392 608 e50	0.000 000 060 e50	Hz
kilogram-inverse meter relationship	4.524 438 73 e41	0.000 000 20 e41	m^-1
kilogram-joule relationship	8.987 551 787... e16	(exact)	J
kilogram-kelvin relationship	6.509 6582 e39	0.000 0059 e39	K
lattice parameter of silicon	543.102 0504 e-12	0.000 0089 e-12	m
Loschmidt constant (273.15 K, 100 kPa)	2.651 6462 e25	0.000 0024 e25	m^-3
Loschmidt constant (273.15 K, 101.325 kPa)	2.686 7805 e25	0.000 0024 e25	m^-3
mag. constant	12.566 370 614 e-7 (exact)	N A^-2	
mag. flux quantum	2.067 833 758 e-15	0.000 000 046 e-15	Wb
Mo x unit	1.002 099 52 e-13	0.000 000 53 e-13	m
molar gas constant	8.314 4621	0.000 0075	J mol^-1 K^-1
molar mass constant	1 e-3	(exact)	kg mol^-1
molar mass of carbon-12	12 e-3	(exact)	kg mol^-1
molar Planck constant	3.990 312 7176 e-10	0.000 000 0028 e-10	J s mol^-1
molar Planck constant times c	0.119 626 565 779	0.000 000 000 084	J m mol^-1
molar volume of ideal gas (273.15 K, 100 kPa)	22.710 953 e-3	0.000 021 e-3	m^3 mol^-1

molar volume of ideal gas (273.15 K, 101.325 kPa)	22.413 968 e-3	0.000 020 e-3	m^3 mol^-1
molar volume of silicon	12.058 833 01 e-6	0.000 000 80 e-6	m^3 mol^-1
muon Compton wavelength	11.734 441 03 e-15	0.000 000 30 e-15	m
muon Compton wavelength over 2 pi	1.867 594 294 e-15	0.000 000 047 e-15	m
muon g factor	-2.002 331 8418	0.000 000 0013	
muon mag. mom.	-4.490 448 07 e-26	0.000 000 15 e-26	J T^-1
muon mag. mom. anomaly	1.165 920 91 e-3	0.000 000 63 e-3	
muon mag. mom. to Bohr magneton ratio	-4.841 970 44 e-3	0.000 000 12 e-3	
muon mag. mom. to nuclear magneton ratio	-8.890 596 97	0.000 000 22	
muon mass	1.883 531 475 e-28	0.000 000 096 e-28	kg
muon mass energy equivalent	1.692 833 667 e-11	0.000 000 086 e-11	J
muon mass energy equivalent in MeV	105.658 3715	0.000 0035	MeV
muon mass in u	0.113 428 9267	0.000 000 0029	u
muon molar mass	0.113 428 9267 e-3	0.000 000 0029 e-3	kg mol^-1
muon-electron mass ratio	206.768 2843	0.000 0052	
muon-neutron mass ratio	0.112 454 5177	0.000 000 0028	
muon-proton mag. mom. ratio	-3.183 345 107	0.000 000 084	
muon-proton mass ratio	0.112 609 5272	0.000 000 0028	
muon-tau mass ratio	5.946 49 e-2	0.000 54 e-2	
natural unit of action	1.054 571 726 e-34	0.000 000 047 e-34	J s
natural unit of action in eV s	6.582 119 28 e-16	0.000 000 15 e-16	eV s
natural unit of energy	8.187 105 06 e-14	0.000 000 36 e-14	J
natural unit of energy in MeV	0.510 998 928	0.000 000 011	MeV
natural unit of length	386.159 268 00 e-15	0.000 000 25 e-15	m
natural unit of mass	9.109 382 91 e-31	0.000 000 40 e-31	kg
natural unit of mom.um	2.730 924 29 e-22	0.000 000 12 e-22	kg m s^-1
natural unit of mom.um in MeV/c	0.510 998 928	0.000 000 011	MeV/c
natural unit of time	1.288 088 668 33 e-21 0.000 000 000 83 e-21 s		
natural unit of velocity	299 792 458	(exact)	m s^-1
neutron Compton wavelength	1.319 590 9068 e-15	0.000 000 0011 e-15	m
neutron Compton wavelength over 2 pi	0.210 019 415 68 e-15 0.000 000 00 17 e-15 m		
neutron g factor	-3.826 085 45	0.000 000 90	
neutron gyromag. ratio	1.832 471 79 e8	0.000 000 43 e8	s^-1 T^-1
neutron gyromag. ratio over 2 pi	29.164 6943	0.000 0069	MHz T^-1
neutron mag. mom.	-0.966 236 47 e-26	0.000 000 23 e-26	J T^-1
neutron mag. mom. to Bohr magneton ratio	-1.041 875 63 e-3	0.000 000 25 e-3	
neutron mag. mom. to nuclear magneton ratio	-1.913 042 72	0.000 000 45	
neutron mass	1.674 927 351 e-27	0.000 000 074 e-27	kg
neutron mass energy equivalent	1.505 349 631 e-10	0.000 000 066 e-10	J
neutron mass energy equivalent in MeV	939.565 379	0.000 021	MeV
neutron mass in u	1.008 664 916 00	0.000 000 000 43	u
neutron molar mass	1.008 664 916 00 e-3	0.000 000 000 43 e-3	kg mol^-1
neutron to shielded proton mag. mom. ratio	-0.684 996 94	0.000 000 16	
neutron-electron mag. mom. ratio	1.040 668 82 e-3	0.000 000 25 e-3	
neutron-electron mass ratio	1838.683 6605	0.000 0011	
neutron-muon mass ratio	8.892 484 00	0.000 000 22	
neutron-proton mag. mom. ratio	-0.684 979 34	0.000 000 16	
neutron-proton mass difference	2.305 573 92 e-30	0.000 000 76 e-30	
neutron-proton mass difference energy equivalent	2.072 146 50 e-13	0.000 000 68 e-13	
neutron-proton mass difference energy equivalent in MeV	1.293 332 17	0.000 000 42	

neutron-proton mass difference in u	0.001 388 449 19	0.000 000 000 45	
neutron-proton mass ratio	1.001 378 419 17	0.000 000 000 45	
neutron-tau mass ratio	0.528 790	0.000 048	
Newtonian constant of gravitation	6.673 84 e-11	0.000 80 e-11	m^3 kg^-1 s^-2
Newtonian constant of gravitation over h-bar c	6.708 37 e-39	0.000 80 e-39	(GeV/c^2)^-2
nuclear magneton	5.050 783 53 e-27	0.000 000 11 e-27	J T^-1
nuclear magneton in eV/T	3.152 451 2605 e-8	0.000 000 0022 e-8	eV T^-1
nuclear magneton in inverse meters per tesla	2.542 623 527 e-2	0.000 000 056 e-2	m^-1 T^-1
nuclear magneton in K/T	3.658 2682 e-4	0.000 0033 e-4	K T^-1
nuclear magneton in MHz/T	7.622 593 57	0.000 000 17	MHz T^-1
Planck constant	6.626 069 57 e-34	0.000 000 29 e-34	J s
Planck constant in eV s	4.135 667 516 e-15	0.000 000 091 e-15	eV s
Planck constant over 2 pi	1.054 571 726 e-34	0.000 000 047 e-34	J s
Planck constant over 2 pi in eV s	6.582 119 28 e-16	0.000 000 15 e-16	eV s
Planck constant over 2 pi times c in MeV fm	197.326 9718	0.000 0044	MeV fm
Planck length	1.616 199 e-35	0.000 097 e-35	m
Planck mass	2.176 51 e-8	0.000 13 e-8	kg
Planck mass energy equivalent in GeV	1.220 932 e19	0.000 073 e19	GeV
Planck temperature	1.416 833 e32	0.000 085 e32	K
Planck time	5.391 06 e-44	0.000 32 e-44	s
proton charge to mass quotient	9.578 833 58 e7	0.000 000 21 e7	C kg^-1
proton Compton wavelength	1.321 409 856 23 e-15 0.000 000 000 94 e-15 m		
proton Compton wavelength over 2 pi	0.210 308 910 47 e-15 0.000 000 000 15 e-15 m		
proton g factor	5.585 694 713	0.000 000 046	
proton gyromag. ratio	2.675 222 005 e8	0.000 000 063 e8	s^-1 T^-1
proton gyromag. ratio over 2 pi	42.577 4806	0.000 0010	MHz T^-1
proton mag. mom.	1.410 606 743 e-26	0.000 000 033 e-26	J T^-1
proton mag. mom. to Bohr magneton ratio	1.521 032 210 e-3	0.000 000 012 e-3	
proton mag. mom. to nuclear magneton ratio	2.792 847 356	0.000 000 023	
proton mag. shielding correction	25.694 e-6	0.014 e-6	
proton mass	1.672 621 777 e-27	0.000 000 074 e-27	kg
proton mass energy equivalent	1.503 277 484 e-10	0.000 000 066 e-10	J
proton mass energy equivalent in MeV	938.272 046	0.000 021	MeV
proton mass in u	1.007 276 466 812	0.000 000 000 090	u
proton molar mass	1.007 276 466 812 e-3 0.000 000 000 090 e-3 kg mol^-1		
proton rms charge radius	0.8775 e-15	0.0051 e-15	m
proton-electron mass ratio	1836.152 672 45	0.000 000 75	
proton-muon mass ratio	8.880 243 31	0.000 000 22	
proton-neutron mag. mom. ratio	-1.459 898 06	0.000 000 34	
proton-neutron mass ratio	0.998 623 478 26	0.000 000 000 45	
proton-tau mass ratio	0.528 063	0.000 048	
quantum of circulation	3.636 947 5520 e-4	0.000 000 0024 e-4	m^2 s^-1
quantum of circulation times 2	7.273 895 1040 e-4	0.000 000 0047 e-4	m^2 s^-1
Rydberg constant	10 973 731.568 539	0.000 055	m^-1
Rydberg constant times c in Hz	3.289 841 960 364 e15 0.000 000 000 017 e15 Hz		
Rydberg constant times hc in eV	13.605 692 53	0.000 000 30	eV
Rydberg constant times hc in J	2.179 872 171 e-18	0.000 000 096 e-18	J
Sackur-Tetrode constant (1 K, 100 kPa)	-1.151 7078	0.000 0023	
Sackur-Tetrode constant (1 K, 101.325 kPa)	-1.164 8708	0.000 0023	
second radiation constant	1.438 7770 e-2	0.000 0013 e-2	m K

shielded helion gyromag. ratio	2.037 894 659 e8	0.000 000 051 e8	s^-1 T^-1
shielded helion gyromag. ratio over 2 pi	32.434 100 84	0.000 000 81	MHz T^-1
shielded helion mag. mom.	-1.074 553 044 e-26	0.000 000 027 e-26	J T^-1
shielded helion mag. mom. to Bohr magneton ratio	-1.158 671 471 e-3	0.000 000 014 e-3	
shielded helion mag. mom. to nuclear magneton ratio	-2.127 497 718	0.000 000 025	
shielded helion to proton mag. mom. ratio	-0.761 766 558	0.000 000 011	
shielded helion to shielded proton mag. mom. ratio	-0.761 786 1313	0.000 000 0033	
shielded proton gyromag. ratio	2.675 153 268 e8	0.000 000 066 e8	s^-1 T^-1
shielded proton gyromag. ratio over 2 pi	42.576 3866	0.000 0010	MHz T^-1
shielded proton mag. mom.	1.410 570 499 e-26	0.000 000 035 e-26	J T^-1
shielded proton mag. mom. to Bohr magneton ratio	1.520 993 128 e-3	0.000 000 017 e-3	
shielded proton mag. mom. to nuclear magneton ratio	2.792 775 598	0.000 000 030	
speed of light in vacuum	299 792 458	(exact)	m s^-1
standard acceleration of gravity	9.806 65	(exact)	m s^-2
standard atmosphere	101 325	(exact)	Pa
standard-state pressure	100 000	(exact)	Pa
Stefan-Boltzmann constant	5.670 373 e-8	0.000 021 e-8	W m^-2 K^-4
tau Compton wavelength	0.697 787 e-15	0.000 063 e-15	m
tau Compton wavelength over 2 pi	0.111 056 e-15	0.000 010 e-15	m
tau mass	3.167 47 e-27	0.000 29 e-27	kg
tau mass energy equivalent	2.846 78 e-10	0.000 26 e-10	J
tau mass energy equivalent in MeV	1776.82	0.16	MeV
tau mass in u	1.907 49	0.000 17	u
tau molar mass	1.907 49 e-3	0.000 17 e-3	kg mol^-1
tau-electron mass ratio	3477.15	0.31	
tau-muon mass ratio	16.8167	0.0015	
tau-neutron mass ratio	1.891 11	0.000 17	
tau-proton mass ratio	1.893 72	0.000 17	
Thomson cross section	0.665 245 8734 e-28	0.000 000 0013 e-28	m^2
triton g factor	5.957 924 896	0.000 000 076	
triton mag. mom.	1.504 609 447 e-26	0.000 000 038 e-26	J T^-1
triton mag. mom. to Bohr magneton ratio	1.622 393 657 e-3	0.000 000 021 e-3	
triton mag. mom. to nuclear magneton ratio	2.978 962 448	0.000 000 038	
triton mass	5.007 356 30 e-27	0.000 000 22 e-27	kg
triton mass energy equivalent	4.500 387 41 e-10	0.000 000 20 e-10	J
triton mass energy equivalent in MeV	2808.921 005	0.000 062	MeV
triton mass in u	3.015 500 7134	0.000 000 0025	u
triton molar mass	3.015 500 7134 e-3	0.000 000 0025 e-3	kg mol^-1
triton-electron mass ratio	5496.921 5267	0.000 0050	
triton-proton mass ratio	2.993 717 0308	0.000 000 0025	
unified atomic mass unit	1.660 538 921 e-27	0.000 000 073 e-27	kg
von Klitzing constant	25 812.807 4434	0.000 0084	ohm
weak mixing angle	0.2223	0.0021	
Wien frequency displacement law constant	5.878 9254 e10	0.000 0053 e10	Hz K^-1
Wien wavelength displacement law constant	2.897 7721 e-3	0.000 0026 e-3	m K

CODATA RECOMMENDED VALUES OF THE FUNDAMENTAL PHYSICAL CONSTANTS: 2018 - *NIST SP 961 (May 2019)*

An extensive list of constants is available on the NIST Physics Laboratory Web site physics.nist.gov/constants. For numerical values a number in parentheses, if present, is the one-standard-deviation uncertainty in the last two digits. For units with square brackets the full descriptions of m−1 and m are cycles per meter and meter per cycle, respectively. For the first radiation constant the full description of m2 is m−2 (m/cycle).

Quantity	Symbol	Numerical value	Unit
speed of light in vacuum	c	299 792 458 (exact)	m s^{-1}
Newtonian constant of gravitation	G	$6.674\,30(15) \times 10^{-11}$	m^3 kg^{-1} s^{-2}
Planck constant	h	$6.626\,070\,15 \times 10^{-34}$ (exact)	J Hz^{-1}
in eV s		$4.135\,667\,696\ldots \times 10^{-15}$	eV Hz^{-1}
	$\hbar$	$1.054\,571\,817\ldots \times 10^{-34}$	J s
in eV s		$6.582\,119\,569\ldots \times 10^{-16}$	eV s
elementary charge	e	$1.602\,176\,634 \times 10^{-19}$ (exact)	C
vacuum magnetic permeability $4\pi\alpha\hbar/e^2c$	μ_0	$1.256\,637\,062\,12(19) \times 10^{-6}$	N A^{-2}
$\mu_0/(4\pi \times 10^{-7})$		$1.000\,000\,000\,55(15)$	N A^{-2}
vacuum electric permittivity $1/\mu_0 c^2$	ϵ_0	$8.854\,187\,8128(13) \times 10^{-12}$	F m^{-1}
Josephson constant $2e/h$	K_J	$483\,597.848\,4\ldots \times 10^9$	Hz V^{-1}
von Klitzing constant $\mu_0 c/2\alpha = 2\pi\hbar/e^2$	R_K	$25\,812.807\,45\ldots$	Ω
magnetic flux quantum $2\pi\hbar/(2e)$	Φ_0	$2.067\,833\,848\ldots \times 10^{-15}$	Wb
Bohr magneton $e\hbar/2m_e$	μ_B	$9.274\,010\,0783(28) \times 10^{-24}$	J T^{-1}
in eV T^{-1}		$5.788\,381\,8060(17) \times 10^{-5}$	eV T^{-1}
nuclear magneton $e\hbar/2m_p$	μ_N	$5.050\,783\,7461(15) \times 10^{-27}$	J T^{-1}
in eV T^{-1}		$3.152\,451\,258\,44(96) \times 10^{-8}$	eV T^{-1}
fine-structure constant $e^2/4\pi\epsilon_0\hbar c$	α	$7.297\,352\,5693(11) \times 10^{-3}$	
inverse fine-structure constant	α^{-1}	$137.035\,999\,084(21)$	
Rydberg frequency $\alpha^2 m_e c^2/2h = E_h/2h$	cR_∞	$3.289\,841\,960\,2508(64) \times 10^{15}$	Hz
energy equivalent in eV		$13.605\,693\,122\,994(26)$	eV
Rydberg constant	R_∞	$10\,973\,731.568\,160(21)$	[m^{-1}]
Bohr radius $\hbar/\alpha m_e c = 4\pi\epsilon_0\hbar^2/m_e e^2$	a_0	$5.291\,772\,109\,03(80) \times 10^{-11}$	m
Hartree energy $\alpha^2 m_e c^2 = e^2/4\pi\epsilon_0 a_0 = 2(cR_\infty)h$	E_h	$4.359\,744\,722\,2071(85) \times 10^{-18}$	J
in eV		$27.211\,386\,245\,988(53)$	eV
electron mass	m_e	$9.109\,383\,7015(28) \times 10^{-31}$	kg
in u		$5.485\,799\,090\,65(16) \times 10^{-4}$	u
energy equivalent in MeV	$m_e c^2$	$0.510\,998\,950\,00(15)$	MeV
electron-muon mass ratio	m_e/m_μ	$4.836\,331\,69(11) \times 10^{-3}$	
electron-proton mass ratio	m_e/m_p	$5.446\,170\,214\,87(33) \times 10^{-4}$	
electron charge to mass quotient	$-e/m_e$	$-1.758\,820\,010\,76(53) \times 10^{11}$	C kg^{-1}
reduced Compton wavelength $\hbar/m_e c = \alpha a_0$	$\lambda\!\!\!^{-}_C$	$3.861\,592\,6796(12) \times 10^{-13}$	m
Compton wavelength	λ_C	$2.426\,310\,238\,67(73) \times 10^{-12}$	[m]
classical electron radius $\alpha^2 a_0$	r_e	$2.817\,940\,3262(13) \times 10^{-15}$	m
Thomson cross section $(8\pi/3)r_e^2$	σ_e	$6.652\,458\,7321(60) \times 10^{-29}$	m^2
electron magnetic moment	μ_e	$-9.284\,764\,7043(28) \times 10^{-24}$	J T^{-1}
to Bohr magneton ratio	μ_e/μ_B	$-1.001\,159\,652\,181\,28(18)$	
to nuclear magneton ratio	μ_e/μ_N	$-1838.281\,971\,88(11)$	
electron magnetic moment anomaly $\|\mu_e\|/\mu_B - 1$	a_e	$1.159\,652\,181\,28(18) \times 10^{-3}$	
electron g-factor $-2(1 + a_e)$	g_e	$-2.002\,319\,304\,362\,56(35)$	
electron-proton magnetic moment ratio	μ_e/μ_p	$-658.210\,687\,89(20)$	
muon mass in u	m_μ	$0.113\,428\,9259(25)$	u
energy equivalent in MeV	$m_\mu c^2$	$105.658\,3755(23)$	MeV
muon-electron mass ratio	m_μ/m_e	$206.768\,2830(46)$	
muon magnetic moment	μ_μ	$-4.490\,448\,30(10) \times 10^{-26}$	J T^{-1}
to Bohr magneton ratio	μ_μ/μ_B	$-4.841\,970\,47(11) \times 10^{-3}$	
to nuclear magneton ratio	μ_μ/μ_N	$-8.890\,597\,03(20)$	
muon magnetic moment anomaly			
$\|\mu_\mu\|/(e\hbar/2m_\mu) - 1$	a_μ	$1.165\,920\,89(63) \times 10^{-3}$	

Quantity	Symbol	Numerical value	Unit
muon g-factor $-2(1 + a_\mu)$	g_μ	$-2.002\,331\,8418(13)$	
muon-proton magnetic moment ratio	μ_μ/μ_p	$-3.183\,345\,142(71)$	
proton mass	m_p	$1.672\,621\,923\,69(51) \times 10^{-27}$	kg
in u		$1.007\,276\,466\,621(53)$	u
energy equivalent in MeV	$m_p c^2$	$938.272\,088\,16(29)$	MeV
proton-electron mass ratio	m_p/m_e	$1836.152\,673\,43(11)$	
proton magnetic moment	μ_p	$1.410\,606\,797\,36(60) \times 10^{-26}$	J T^{-1}
to nuclear magneton ratio	μ_p/μ_N	$2.792\,847\,344\,63(82)$	
proton magnetic shielding correction $1 - \mu'_p/\mu_p$ (H_2O, sphere, 25 °C)	σ'_p	$2.5689(11) \times 10^{-5}$	
proton gyromagnetic ratio $2\mu_p/\hbar$	γ_p	$2.675\,221\,8744(11) \times 10^{8}$	s^{-1} T^{-1}
		$42.577\,478\,518(18)$	MHz T^{-1}
shielded proton gyromagnetic ratio $2\mu'_p/\hbar$ (H_2O, sphere, 25 °C)	γ'_p	$2.675\,153\,151(29) \times 10^{8}$	s^{-1} T^{-1}
		$42.576\,384\,74(46)$	MHz T^{-1}
neutron mass in u	m_n	$1.008\,664\,915\,95(49)$	u
energy equivalent in MeV	$m_n c^2$	$939.565\,420\,52(54)$	MeV
neutron-proton mass ratio	m_n/m_p	$1.001\,378\,419\,31(49)$	
neutron magnetic moment	μ_n	$-9.662\,3651(23) \times 10^{-27}$	J T^{-1}
to nuclear magneton ratio	μ_n/μ_N	$-1.913\,042\,73(45)$	
deuteron mass in u	m_d	$2.013\,553\,212\,745(40)$	u
energy equivalent in MeV	$m_d c^2$	$1875.612\,942\,57(57)$	MeV
deuteron-proton mass ratio	m_d/m_p	$1.999\,007\,501\,39(11)$	
deuteron magnetic moment	μ_d	$4.330\,735\,094(11) \times 10^{-27}$	J T^{-1}
to nuclear magneton ratio	μ_d/μ_N	$0.857\,438\,2338(22)$	
helion (^3He nucleus) mass in u	m_h	$3.014\,932\,247\,175(97)$	u
energy equivalent in MeV	$m_h c^2$	$2808.391\,607\,43(85)$	MeV
shielded helion magnetic moment (gas, sphere, 25 °C)	μ'_h	$-1.074\,553\,090(13) \times 10^{-26}$	J T^{-1}
to Bohr magneton ratio	μ'_h/μ_B	$-1.158\,671\,471(14) \times 10^{-3}$	
to nuclear magneton ratio	μ'_h/μ_N	$-2.127\,497\,719(25)$	
alpha particle mass in u	m_α	$4.001\,506\,179\,127(63)$	u
energy equivalent in MeV	$m_\alpha c^2$	$3727.379\,4066(11)$	MeV
Boltzmann constant	k	$1.380\,649 \times 10^{-23}$ (exact)	J K^{-1}
Avogadro constant	N_A	$6.022\,140\,76 \times 10^{23}$ (exact)	mol^{-1}
atomic mass constant $\frac{1}{12}m(^{12}C) = 1$ u	m_u	$1.660\,539\,066\,60(50) \times 10^{-27}$	kg
energy equivalent in MeV	$m_u c^2$	$931.494\,102\,42(28)$	MeV
Faraday constant $N_A e$	F	$96\,485.332\,12\ldots$	C mol^{-1}
molar gas constant $N_A k$	R	$8.314\,462\,618\ldots$	J mol^{-1} K^{-1}
in eV K^{-1}		$8.617\,333\,262\ldots \times 10^{-5}$	eV K^{-1}
molar volume of ideal gas RT/p ($T = 273.15$ K, $p = 101.325$ kPa)	V_m	$22.413\,969\,54\ldots \times 10^{-3}$	m^3 mol^{-1}
Stefan-Boltzmann constant $\pi^2 k^4/60\hbar^3 c^2$	σ	$5.670\,374\,419\ldots \times 10^{-8}$	W m^{-2} K^{-4}
first radiation constant $2\pi hc^2$	c_1	$3.741\,771\,852\ldots \times 10^{-16}$	[W m^2]
second radiation constant hc/k	c_2	$1.438\,776\,877\ldots \times 10^{-2}$	[m K]
Wien displacement law constant $b = \lambda_{max} T = c_2/4.965\,114\,231\ldots$	b	$2.897\,771\,955\ldots \times 10^{-3}$	[m K]
Cu x unit: $\lambda(\mathrm{Cu\,K}\alpha_1)/1\,537.400$	xu(Cu Kα_1)	$1.002\,076\,97(28) \times 10^{-13}$	m
Mo x unit: $\lambda(\mathrm{Mo\,K}\alpha_1)/707.831$	xu(Mo Kα_1)	$1.002\,099\,52(53) \times 10^{-13}$	m

BIBLIOGRAPHY

Acosta, Virgilio, Cowan, Clyde L., and Graham, B.J., **Essentials of Modern Physics**, Harper & Row Publishers, 1973

Alexander Medvinsky, Stanislav Rybtsov, Samir Taoudi, **Embryonic origin of the adult hematopoietic system:advances and questions**, The Company of Biologists, doi.org/10.1242/dev.040998, 2011

Asher Peres, **Critique of the Wheeler-Dewitt Equation**, arXiv:gr-qc/9704061, 1997

Baez, John, **Hypercharge and Weak Isospin, - Elementary Particles**, University of California, Lecture Notes, 2003

Barbour, J., **The End of Time: The Next Revolution in Physics**, Oxford University Press, 1999, ISBN 0-297-81985-2

Basini, G. and Capozzielo, S., **Questioning the Quark Model – Strong Interaction, Gravitation and Time Arrows – An Approach to Asymptotic Freedom,** Europhysics Letters, pp. 635-641 (2003)

Bogdan, P, Scholz, C. and Zetsche, F., Particles and Nuclei: An Introduction to Physical Concepts, Springer, 2008

Born, Max, **Deciphering the Cosmic Number: The Strange Friendship of Wolfgang Pauli and Carl Jung**, W. W. Norton & Co, 2009

Braibant, Sylvie, Gracomelli, Giorgio and Spurio, Fauricio, **Particles and Fundamental Interactions: An Introduction to Particle Physics**, Springer, 2012

Brown, Kevin, **Reflections on Relativity**, 2016

Caroll, S., **From Eternity to Here: The Quest for the Ultimate Theory of Time**, October 26, 2010 Penguin Group

Caroll, S., **The Cosmological Constant**, Encyclopedia of Astronomy and Astrophysics, IOP Publishing Ltd, 2006

Charles, Mortimer, E., **Chemistry**, Wadsworth Publishing, 1983

Clay, Ellen, Clark, Brad and Schneider Vic, **Semicontinuous Groups and Separation Properties**, University of Southwestern Mississipi Lafayette, 1990

Connes, A., Rovelli, C., **Von Neumann Algebra, Automorphisms and Time-thermodynamics Relation in General Covariant Quantum Theories**, arXiv:gr-qc/9406019v1 June 1994
Courant, R.; Hilbert, D. **Methods of Mathematical Physics**

Courant, R.; Hilbert, D. **Methods of Mathematical Physics**

Cruz, Andoni, Chamizo, Jose, Garritz, Diana, **Estructura Atómica: Un Enfoque Químico**, Fondo Educativo Interamericano, 1986

D. Halliday, R. Resnick, J. Walker, **Fundamentals of Physics**, 5th Edition, John Wiley & Sons, 1997

D. Page, W. Wooters, **Evolution without Evolution: Dynamics Described by Stationary Observables,** Phys. Rev. D27, 2885, (1983).

Daniels, Farrington, and Albert, Robert A., **Physical Chemistry**, John Wiley and Sons, 1963

Davis, Tamara, and Griffen, Brendan, **Cosmological Constant**, Scholarpedia, 2010

Dirac, Paul, M., **The Evolution of the Physicist's Picture of Nature**, Scientific American, May 1963

Drake, Kyle, Feinberg, Michael, Guild, David and Turetsky, Emma, **Representations of the Symmetry Group of Spacetime**, 2009

Edwards, B., Larson, R., and Hostetler, R. **Calculus of a Single Variable**, D.C. Heath and Company, 1994.

Efimov, Sergey P., **Formulas for Magnetic Moments of the Proton and Neutron**, Bauman Moscow State Technical University, 2012

Einstein, Albert, **The Foundation of the General Theory of Relativity**, Annalen der Physik 354

Ekaterina Moreva, et al, **Time from Quantum Entanglement: An Experimental Illustration**, arXiv: 1310.4691v1 [quant-ph] 17 Oct 2013

Farwaz T. Ulabi, Michel M. Maharbiz, **Circuits**, National Technology and Science Press, 2009

Fernflores Francisco, **The Equivalence of Mass and Energy, Stanford Encyclopedia of Philosophy**, 2012

Feynman, Richard, P., **Theory of Fundamental Processes**, W.A. Benjamin, 1962

Feynman, Richard, Pl, **QED: The Strange Theory of Light and Matter**, Princeton University Press, 1985

Friedlander, G., and Macias, E., and Kennedy, J., **Nuclear and Radiochemistry**, John Riley & Sons, 1981

George B. Arfken & Hans J. Weber, **Mathematical Methods for Physicists**, 6th edition, Academic Press, 2005

Giovanni Sansone (translated by Ainsley H. Diamond), **Orthogonal Functions**, Interscience Publishers, 1959

Giovanni Sansone (translated by Ainsley H. Diamond), **Orthogonal Functions**, Interscience Publishers, 1959

Gross, J. David, **The Role of Symmetry in Fundamental Physics**, National Academy of Science, vol 93, No 25, 1996

H. McKean and V. Moll, **Elliptic Curves,** Cambridge University Press, 1999

Halliday, David, Resnick, Robert and Walker, Jearl, **Fundamentals of Physics**, John Wiley and Sons, 1997

Hawking, S., **Une Brève Histoire du Temps**, Editions J'ai lu, 1989

Hedrick, Philip W., **Population Biology: The Evolution and Ecology of Populations**, Jones and Bartlett Publishers, 1984

Hervik, Sigbjorn, **Einstein's General Theory of Relativity: With Modern Applications in Cosmology**, Springer Science and Business Media, 2007

Hobson, Art, **Physics: Concepts and Connections**, Pearson Addison-Wesley, 2010

Hooft, Gerard 't, **The Conceptual Basis of Quantum Field Theory**, Handbook of the Philosophy of Science, Elsevier, 2010

James A Coker, **Recent Advances in Understanding Extremophiles**, F1000Research 2019

Jay C. Brown, William W. Newcomb, Gail W. Wertz, **Helical Virus Structure: The Case of the Rhabdovirus,** Virus ISSN 1999-4915, University of Virginia School of Medicine

Jean Rostand, Andrée Tetry, **La Vie**, Librairie Larousse, 1962

Ji Yoon Lee1, Seok-Ho Hong, **Hematopoietic Stem Cells and Their Roles in Tissue Regeneration,** International Journal of Stem Cells Vol. 13, No. 1, 2020

John C. Baez, Emory F. Bunn, **The Meaning of Einstein's Equation**, E.F.B., 2006

John M. DeSesso, **The arrogance of teratology: A brief chronology of attitudes throughout history**, Birth Defects Research Volume 111, Issue 3 p. 123-141, 2018

Katznelson, Y., **An Introduction to Harmonic Analysis**, 3rd Edition. Cambridge University Press, 2004

Lide, D. R., **A Century of Excellence in Measurements, Standards and Technology**, CRC Press, 2001

Markowitz, W., Hall, R., G. and Essen, L., **Frequency of Cesium in Terms of Ephemeris Time**, Physical Review Letters, 1958

Marshak, R. E. and Sudarhan E. C. G., **Remarks on the Possible Existence of a Neutral Muon, University of Rochester**, 1957

McTaggart J. M. E., **The Unreality of Time**, University Press, 1921

Michiel JL de Hoon, Patrick Eichenberger, Dennis Vitkup, **Hierarchical evolution of the bacterial sporulation network**, Center for Computational Biology and Bioinformatics, Columbia University, 2010

Milonni, Peter, W., **The Quantum Vacuum: An Introduction to Quantum Electrodynamics**, Academic Press, Inc., 1994

Organisation Intergouvernementale de la Convention du Mètre, **The International System of Units (SI),** 8th Edition, 2006

P.W. Atkins, **Quanta: A handbook of concepts**, Oxford University Press, 1974

Peleg, Y., Pnini, R.; Zaarur, E.; Hecht, E., **Quantum Mechanics** (Schaum's Outline Series), 2nd Edition, McGraw Hill, 2010

Peskin, Michael, E., and Schroeder, Daniel, V., **Introduction to Quantum Field Theory, Sarat Book House**, 2005

Peskin, Michael, E., and Schroeder, Daniel, V., **Introduction to Quantum Field Theory,** Sarat Book House, 2005

Rabia Khan, Fernanda Cristina Petersen and Sudhanshu Shekhar, **Commensal Bacteria: An Emerging Player in Defense Against Respiratory Pathogens**, Frontiers in Immunology, doi: 10.3389/fimmu.2019.01203, 2019

Ragheb, M., **Nuclear Processes: The Strong Force**, University of Illinois, 2013

Rees, Martin, Just Six Numbers: **The Deep Forces That Shape the Universe**, Weidenfeld & Nicolson, 1999

Richmond, A., **Calculus For Electronics**, McGraw-Hill Book Company, 1972.

Robert L. Oldershaw, **Towards A Resolution of The Vacuum Energy Density Crisis**, arXiv:0901.3381, 2009

Rong Lu, Norma F. Neff, Stephen R. Quake, Irving L. Weissman, **Tracking single hematopoietic stem cells in vivo using high-throughput sequencing in conjunction with viral genetic barcoding**, Nat Biotechnol. doi:10.1038/nbt.1977.

Schrödinger, E., **Structure of Spacetime**, Cambridge University Press, London, 1950.

Sharma, VK. **Adaptive Significance of Circadian Clocks,** Chronobiology International. **20** (6): 901–19. doi:10.1081/CBI-120026099. PMID 14680135. November 2003

Shouguo Gao, Zhijie Wu, Jeerthi Kannan, Liza Mathews, Xingmin Feng, Sachiko Kajigaya, Neal S. Young, **Comparative Transcriptomic Analysis of the Hematopoietic System between Humanand Mouse by Single Cell RNA Sequencing**, Cells, doi:10.3390/cells10050973, 2021

Smolin L. **Time Reborn: From the Crisis in Physics to the Future of the Universe**, 2013

Smolin, Lee, **The Trouble with Physics: The Rise of String Theory, the Fall of a Science, What Comes Next**, Mariner Books, 2007

Sunder, V.S. **Functional Analysis: Spectral Theory**, Birkhäuser Verlag, 1997

The Regents of the University of Colorado, **Biological Science: Molecules to Men**, Houghton Mifflin Co, 1968

Thomas, R., and Rosa, A., **The Analysis and Design of Linear Circuits**, Prentice-Hall, Inc. 1994.

Tortora, Gerard J., Funke, Berdell R. and Case, Christine L., **Microbiology An Introduction**, Benjamin Cummings Publishing, 1982

Vvedensky, G. G., **Group Theory**, 2001

Weber L. Robert, Manning V. Kenneth, White Marsh W, **Physique Générale**, McGraw-Hill Canada, 1967

Weber, Robert L et al, **Physique Générale**, Mc Graw Hill, 1967

Wilson, J. and Hawkers, F. F. B., **Laser Principles and Applications**, Prentice Hall, 1987

Woit, Peter, **Not Even Wrong: The Failure of String Theory and the Search for Unity in Physical Law**, Basic Books, 2006

Yang, C. N. and Mills R. L., **Conservation of Isotopic Spin and Isotopic Gauge Invariance**, Physical Review Letters, 1954

Yitzhak Katznelson, **An Introduction to Harmonic Analysis**, 3rd Edition. Cambridge University Press, 2004

Young, Nicholas, **An Introduction to Hilbert Space**, Cambridge University Press, 1988

Zee Anthony, **Quantum Field Theory in a Nutshell**, Princeton University Press, 2010

APPENDIX

I.- Dirac Lagrangian For the Hydrogen Atom

Dirac's Lagrangian energy density function for the Hydrogen atom:

$$\psi_{nlm}(r,\vartheta,\varphi)=\sqrt{\left(\frac{2}{na_0}\right)^3\frac{(n-l-1)!}{2n[(n+l)!]}}\bullet e^{-\rho/2}\rho^l L_{n-l-1}^{2l+1}(\rho)\cdot Y_{lm}(\vartheta,\varphi)$$

where r, θ, φ are the radial and angular variables, ρ the sum of the square of 4 sub-functions of these variables or spinor, n the principal quantum number, l the angular momentum quantum number and m the magnetic quantum number and ψ the main wave density function itself.

II.- QG Atomic p-Orbital Function

$$\mathrm{M}_{[s,x]_2}=[(s^2+x^2-1)]^2\cdot\left(-(s+x)\cdot e^{\frac{-(s^2+x^2)}{2}}\cdot\frac{1}{\sqrt{2\pi}}\right)^2$$

where s and x are the independent variable of the 3-space function M.

III.- Computation of Theoretically Derived Van der Vaals a Constant

Compute the value of the $M_3(s)$ Creation function at s_3.

$$s=\sigma\sqrt{8/5}\,,\ y_P(dq/ds)=M_3(s)=-\frac{\sqrt{7}}{\sigma^4 4\sqrt{2\pi}}\cdot e^{-\frac{s^2}{2\sigma^2}}\cdot s^3$$

$$M_3(s_3) = -\frac{\sqrt{7}}{\sigma^4 4\sqrt{2\pi}} \cdot e^{-\frac{\left(\sigma\sqrt{8/5}\right)^2}{2\sigma^2}} \cdot \left(\sigma\sqrt{8/5}\right)^3$$

$$= -\frac{\sqrt{7}}{\sigma^4 4\sqrt{2\pi}} \cdot e^{-\frac{\sigma^2 8/5}{2\sigma^2}} \cdot \sigma^3 \left(\sqrt{8/5}\right)^3$$

$$= -\frac{\sqrt{7}}{\sigma 4\sqrt{2\pi}} \cdot e^{-\frac{8}{10}} \cdot 2.0238577025077$$

$$= -\frac{\sqrt{7}}{\sigma 4\sqrt{2\pi}} \cdot 0.44932896411 \cdot 2.0238577025077$$

$$= -\frac{\sqrt{7}}{\sigma 4\sqrt{2\pi}} \cdot 0.909377884988444$$

$$= -\frac{\sqrt{7}}{\sigma 10.026513098524} \cdot 0.909377884988444$$

$$= -\frac{0.263875515352797}{\sigma} \cdot 0.909377884988444$$

$$= -\frac{0.23996255805}{\sigma}$$

INDEX

2

3

4

A

B

C

D

E

F

G

H

I

J

K

L

M

N

O

P

Q

R

S

W

X

Z

The author may be reached via email sent to:

jjc@qgmindarchitects.com

FROM THE SAME AUTHOR

The Human Mind Inside: *ISBN 978-0-9647466-3-3*
Quanto-Geometry – Vol I: *ISBN 978-0-9647466-4-0*
Quanto-Geometry – Vol II: *ISBN 978-0-9647466-6-4*
Quanto-Geometry – Vol III: *ISBN 978-0-9647466-9-5*
Religion is Cosmic Culture: *ISBN 978-0-4630-8695-7*
La Quanto-Géométrie: *ISBN 0-9647466-1-1*
Functional Architecture of the Human Mental: *ISBN 978-0-9647466-8-9 (co-author Caroline Stamu-O'Brien, MD)*
AI Threat to the Human Knowledge Base: *ISBN 979-8-201-16016-6*

www.ingramcontent.com/pod-product-compliance
Ingram Content Group UK Ltd.
Pitfield, Milton Keynes, MK11 3LW, UK
UKHW022205051026
15624UKWH00001B/9

9 780964 746626